S0-CFN-173

Introduction to Optics

Third Edition

Frank L. Pedrotti, S. J.

St. Louis University

Leno S. Pedrotti

Center for Occupational
Research and Development (CORD)

Emeritus Professor of Physics
Air Force Institute of Technology

Leno M. Pedrotti

University of Dayton

PEARSON
Prentice
Hall

Upper Saddle River, New Jersey 07458

Library of Congress Cataloging-in-Publication Data

Pedrotti, Frank L., 1932-
   Introduction to optics.-- 3rd ed. / Frank L. Pedrotti, Leno M. Pedrotti, Leno S. Pedrotti.
     p. cm.
   Includes bibliographical references and index.
   ISBN 0-13-149933-5
   1. Optics. I. Pedrotti, Leno Matthew. II. Pedrotti, Leno S., 1927-III. Title.
   QC355.3.P43 2007
   535--dc22

                       2006041697

Senior Editor: *Erik Fahlgren*
Editor-in-Chief, Science: *Daniel Kaveney*
Assistant Managing Editor: *Beth Sweeten*
Associate Editor: *Christian Botting*
Senior Managing Editor, Art Production and Management: *Patricia Burns*
Manager, Production Technologies: *Matthew Haas*
Managing Editor, Art Management: *Abigail Bass*
Editorial Assistant: *Jessica Berta*
Production Editor: *Cindy Miller/GGS Book Services*
AV Editor: *Greg Dulles*
Art Studio: *Laserwords Private Limited*
Manufacturing Buyer: *Alan Fischer*
Manufacturing Manager: *Alexis Heydt-Long*
Art Director: *Jayne Conte*
Cover Design: *Bruce Kenselaar*
Cover Image: *Separated images of two incoherent point sources*
Cover Photo: *Springer-Verlag GmbH & Co KG*

 © 2007, 1993, 1987 Pearson Education, Inc.
Pearson Prentice Hall
Pearson Education, Inc.
Upper Saddle River, NJ 07458

All rights reserved. No part of this book may be reproduced, in any form
or by any means, without permission in writing from the publisher.

Pearson Prentice Hall™ is a trademark of Pearson Education, Inc.

Printed in the United States of America
10 9 8 7 6 5 4 3 2 1

ISBN 0-13-149933-5

Pearson Education LTD., *London*
Pearson Education Australia PTY, Limited, *Sydney*
Pearson Education Singapore, Pte. Ltd.
Pearson Education North Asia Ltd, *Hong Kong*
Pearson Education Canada, Ltd., *Toronto*
Pearson Educación de Mexico, S.A. de C.V.
Pearson Education—Japan, *Tokyo*
Pearson Education Malaysia, Pte. Ltd.
Pearson Education, *Upper Saddle River,* New Jersey

*To our families and friends
for their timely support during the
preparation of this textbook.*

Brief Contents

Contents

Part II

Physical Constants

Speed of light	$c = 2.998 \times 10^8$ m/s
Electron charge	$e = 1.602 \times 10^{-19}$ C
Electron rest-mass	$m = 9.109 \times 10^{-31}$ kg
Planck constant	$h = 6.626 \times 10^{-34}$ J-s
Boltzmann constant	$k = 1.3805 \times 10^{-23}$ J/K
Permittivity of vacuum	$\varepsilon_0 = 8.854 \times 10^{-12}$ C^2/N-m^2
Permeability of vacuum	$\mu_0 = 4\pi \times 10^{-7}$ T-m/A

Preface

INTRODUCTION

The field of optics impacts an ever-expanding range of applications in physics, engineering, and technology. The parallel emergence of lasers, fiber optics, nonlinear devices, and a variety of semiconductor sources and detectors in the 1960s initiated a continuing period of rapid development in applied and theoretical optics. The need for a variety of updated optics texts with different approaches and emphases is apparent, both for students of optics and for practitioners who need an occasional review of the basics.

With *Introduction to Optics* we propose to teach introductory modern optics at an intermediate level. In order to use this text, students should have a preparatory background that includes a calculus-based introductory physics sequence and at least two semesters of calculus. The material in this text encompasses the traditional areas of classical optics and many topics in modern optics. The organization of the material in this text is intended to facilitate its use in a variety of one-semester courses, with different emphases. In addition, the text includes more than enough material to be used in a full-year optics course for physics or engineering students at the sophomore, junior, or senior undergraduate level.

Specific features of the text, in terms of coverage beyond traditional areas, include extensive use of 2×2 matrices in dealing with ray tracing, polarization, and multiple thin-film interference; three chapters devoted to lasers; a separate chapter on the optics of the eye; and individual chapters on holography, coherence, fiber optics, interferometry, Fourier optics, nonlinear optics, and Fresnel equations. We have organized the text into two parts. Part I includes chapters that form the basis of a typical one-semester optics course designed for sophomore or junior physics majors. Chapters in Part I are designed to be covered in sequence, although subsets of these chapters can be omitted to suit specific curricular needs. Part II contains chapters that are largely self-contained and, upon completion of a subset of chapters in Part I, can be covered in a variety of sequences. We hope that this organization will allow teachers to select specialized content to suit individual curricular needs. Possible orderings of chapter coverage appropriate for several possible optics courses with different emphases are provided later in this Preface.

IMPROVEMENTS IN THE THIRD EDITION

The Third Edition of this text differs from the Second Edition in that several topics have been added or expanded and the order of presentation of the material has been changed to provide teachers with greater freedom to tailor the text to meet specific curricular requirements. New features of the Third Edition include the following:

- The treatment of wave equations (Chapter 4) has been expanded with the addition of qualitative descriptions of cylindrical and Gaussian beams and a brief introduction to light polarization.
- The coverage of optical interferometry (Chapter 8) has been expanded by providing a more detailed description of the operation and applications of the Fabry-Perot interferometer and by adding a description of gravitational wave detectors.
- New descriptions of other modern optics applications appear throughout the text. For example, we have added a section on wavelength-division-multiplexing in fiber communication systems in Chapter 10 and we have added sections on liquid-crystal displays and plasma displays in Chapter 17.
- A new chapter (Chapter 26, "Laser Operation") provides both a *quantitative*, rate-equation treatment of laser operation and a qualitative introduction to the semiconductor laser. The four chapters—Chapter 6, "Properties of Lasers," Chapter 26, "Laser Operation," Chapter 27, "Characteristics of Laser Beams," and Chapter 28, "Selected Modern Applications,"—taken together constitute a rather comprehensive introduction to laser optics for undergraduates.
- Based on extensive reviewer feedback we have carefully revised each chapter, sharpening and clarifying discussions and mathematical developments throughout.
- Chapter 28, an expansion and revision of the Second Edition chapter "Laser Applications," includes new sections describing ultrashort pulse production, laser cooling and trapping, optical parametric oscillators, and near-field microscopy.
- More than 100 problems have been added.
- For Instructors, we once again offer an Instructor's Solutions Manual (0-13-173918-2) containing fully worked-out solutions to all end-of-chapter problems. Instructors can also visit the *Introduction to Optics, 3rd Edition* information page on Prentice Hall's online catalog (www.prenhall.com <www.prenhall.com>) to download additional instructor resources, including electronic versions of the figures from *Introduction to Optics, 3rd Edition*.

Nearly all of the material from the Second Edition remains in the Third Edition. We have sharpened and clarified discussions and mathematical discussions throughout.

TEXT ORGANIZATION

The text is divided into two parts. The individual chapters in Part II primarily rely only on material covered in Part I and so can be covered in a variety of different sequences. As an indication of the flexibility of the text, we offer four possible one-semester course organizations.

Traditional "Balanced" Optics Course

Prerequisites: Two or three semesters of introductory physics and two semesters of calculus

Chapter 1   Nature of Light
Chapter 2   Geometrical Optics
Chapter 3   Optical Instrumentation
Chapter 4   Wave Equations
Chapter 5   Superposition of Waves
Chapter 6   Properties of Lasers
Chapter 7   Interference of Light
Chapter 8   Optical Interferometry
Chapter 9   Coherence
Chapter 10 Fiber Optics
Chapter 11 Fraunhofer Diffraction
Chapter 13 Fresnel Diffraction
Chapter 15 Production of Polarized Light

Traditional Course: Geometrical Optics Emphasis

Prerequisites: Two or three semesters of introductory physics and two semesters of calculus

Chapter 1   Nature of Light
Chapter 2   Geometrical Optics
Chapter 3   Optical Instrumentation
Chapter 18 Matrix Methods in Paraxial Optics
Chapter 19 Optics of the Eye
Chapter 20 Aberration Theory
Chapter 4   Wave Equations
Chapter 5   Superposition of Waves
Chapter 6   Properties of Lasers
Chapter 7   Interference of Light
Chapter 8   Optical Interferometry
Chapter 10 Fiber Optics

Physical Optics

Prerequisites: Two or three semesters of introductory physics, two semesters of calculus, and one semester of intermediate electromagnetics

Chapter 1   Nature of Light
Chapter 4   Wave Equations
Chapter 5   Superposition of Waves
Chapter 6   Properties of Lasers
Chapter 7   Interference of Light
Chapter 8   Optical Interferometry
Chapter 9   Coherence
Chapter 11 Fraunhofer Diffraction
Chapter 12 The Diffraction Grating
Chapter 13 Fresnel Diffraction
Chapter 14 Matrix Treatment of Polarization
Chapter 15 Production of Polarized Light
Chapter 16 Holography
Chapter 21 Fourier Optics

Laser Optics

Prerequisites: Two or three semesters of introductory physics, two semesters of calculus, and one semester of intermediate electromagnetics

Chapter 4   Wave Equations
Chapter 5   Superposition of Waves
Chapter 6   Properties of Lasers
Chapter 26 Laser Operation
Chapter 7   Interference of Light
Chapter 8   Optical Interferometry
Chapter 9   Coherence
Chapter 10 Fiber Optics
Chapter 11 Fraunhofer Diffraction
Chapter 14 Matrix Treatment of Polarization
Chapter 15 Production of Polarized Light
Chapter 18 Matrix Methods in Paraxial Optics
Chapter 27 Characteristics of Laser Beams
Chapter 28 Selected Modern Applications

Other chapter sequences are, of course, possible. For example, for advanced undergraduates with two semesters of electromagnetic theory, the Laser Optics course could be altered by replacing Chapters 4, 5, and 7 with Chapters 16 ("Holography"), 22 ("Theory of Multilayer Films"), and 24 ("Nonlinear Optics and the Modulation of Light"). A variety of two-semester, two-quarter, or three-quarter sequences are also possible.

ACKNOWLEDGMENTS

We wish to thank the many teachers who have inspired us with an interest in optics and teaching, and the many students who have motivated us to teach with clarity and efficiency.

First and Second Edition

For their very helpful reading of portions of the manuscript for the First Edition, we are indebted to Hugo Weichel, James Tucci, Hajime Sakai, Arthur H. Guenther, and Thomas B. Greenslade. For their suggestions leading to the Second Edition we wish to thank the team of reviewers selected by Prentice Hall: Joel Blatt, Florida Institute of Technology; James Boger, Oregon Institute of Technology; Harry Daw, New Mexico State University; Edward Eyler, University of Delaware; and Daniel Wilkins, University of Nebraska. For his review and suggestions in the chapter on the eye, we are also pleased to acknowledge and thank Dr. Michael Pedrotti, O.D. We also wish to thank Judy Lawson, CORD, for her sketch of Einstein that graces page 1.

Third Edition

Many of the changes made in the Third Edition of the book grew in part from the thoughtful suggestions made by the reviewers: Ralph Alexander, University of Missouri at Rolla; Ramendra Bahuguna, San Jose State University; Robert M. Bunch, Rose-Hulman Institute of Technology; Henry J. Leckenby, Texas A&M, Kingsville; Peter G. LoPresti, University of Tulsa; Sergey Mirov, University of Alabama, Birmingham; B. Paul Padley, Rice University; Jie Shan, Case Western University; John E. Sohl, Weber State University; Rick Trebino, Georgia Tech; and Paul Voytas, Wittenberg University. We are deeply indebted to this group of reviewers for their valuable suggestions. From this group we wish to extend special thanks to John Sohl and Robert Bunch for their careful reading of the entire Third Edition. We would also like to thank Robert Brecha, Bruce Craver, and Peter Powers from the University of Dayton for suggestions and advice leading to some of the new material in the Third Edition and Mark Whitney and Kathy Kral of CORD for artistic and logistical help.

Finally, we express our gratitude to the editorial and production staff of Prentice Hall for their enthusiastic and professional support. In particular, we are indebted to our acquisitions editors, Holly Hodder for the First Edition, Ray Henderson for the Second Edition, and Erik Fahlgren for the Third Edition, and to our production editors, Kathleen Lafferty for the first two editions and Kevin Bradley and Cindy Miller (of GGS Book Services) for the Third Edition. All efforts were made to ensure accuracy, but responsibility for all errors lie with authors.

Frank L. Pedrotti, S.J.
Leno S. Pedrotti
Leno M. Pedrotti

1 Nature of Light

"They could but make the best of it and went around with woebegone faces, sadly complaining that on Mondays, Wednesdays, and Fridays, they must look on light as a wave; on Tuesdays, Thursdays, and Saturdays, as a particle. On Sundays, they simply prayed."

The Strange Story of the Quantum
Banesh Hoffmann, 1947

INTRODUCTION

The words cited above—taken from a 1947 popular primer on the quantum world—delighted many readers who were just then coming into contact with ideas related to the nature of light and quanta. Hoffmann's amusing and informative account—involving in part the wave-particle twins "tweedledum" and "tweedledee"—captured nicely the level of frustration felt in those days about the true nature of light. And today, some 60 years later, the puzzle of tweedledum and tweedledee lingers. What is light? What is a photon? Indeed, in October of 2003, The Optical Society of America devoted a special issue of *Optics and Photonic News* to the topic "The Nature of Light: What is a Photon?" In this issue,[1] a number of renowned scientists, through five penetrating essays, accepted the challenge of describing the photon. Said Arthur Zajonc, in his lead article titled "Light Reconsidered":

> Light is an obvious feature of everyday life, and yet light's true nature has eluded us for centuries. Near the end of his life, Albert Einstein wrote, "All the 50 years of conscious brooding have brought me no closer to the answer to the question: What are light quanta?" We are today in the same state of "learned ignorance" with respect to light as was Einstein.

---

[1] "The Nature of Light: What is a Photon?" *OPN Trends*, Vol 3., No. 1, October 2003.

The evolution in our understanding of the physical nature of light forms one of the most fascinating accounts in the history of science. Since the dawn of modern science in the sixteenth and seventeenth centuries, light has been pictured either as particles or waves—seemingly incompatible models—each of which enjoyed a period of prominence among the scientific community. In the twentieth century it became clear that somehow light was both wave and particle, yet it was precisely neither. For some time this perplexing state of affairs, referred to as the *wave-particle duality*, motivated the greatest scientific minds of our age to find a resolution to these apparently contradictory models of light. In a formal sense, the solution was achieved through the creation of *quantum electrodynamics*, one of the most successful theoretical structures in the annals of physics. However, many scientists would agree, a comfortable *understanding* of the true nature of light is somewhat more elusive.

In our account of the developing understanding of light and photons, we will be content to sketch briefly a few of the high points. Certain areas of physics once considered to be disciplines apart from optics—electricity and magnetism, and atomic physics—are very much involved in this account. This alone suggests that the resolution achieved also constitutes one of the great unifications in our understanding of the physical world. The final result is that light and subatomic particles, like electrons, are both considered to be manifestations of energy and are governed by the same set of formal principles. In this introductory chapter, we begin with a brief history of light, addressing it alternately as particle and wave. Along the way we meet the great minds that championed one viewpoint or the other. We follow this account with several basic relationships—borrowed from quantum physics and the special theory of relativity—that describe the properties of subatomic particles, like electrons, and the *photon*. We close this chapter with an introductory glance at the electromagnetic spectrum and a survey of the *radiometric* units we use to describe the properties of electromagnetic radiation.

1-1 A BRIEF HISTORY[2]

In the seventeenth century the most prominent advocate of a *particle theory* of light was Isaac Newton, the same creative giant who had erected a complete science of mechanics and gravity. In his treatise *Optics*, Newton clearly regarded rays of light as streams of very small particles emitted from a source of light and traveling in straight lines. Although Newton often argued forcefully for positing hypotheses that were derived only from observation and experiment, here he himself adopted a particle hypothesis, believing it to be adequately justified by his experience. Important in his considerations was the observation that light seemed to cast sharp *shadows* of objects, in contrast to water and sound waves, which bend around obstacles in their paths. At the same time, Newton was aware of the phenomenon now referred to as *Newton's rings*. Such light patterns are not easily explained by viewing light as a stream of particles traveling in straight lines. Newton maintained his basic particle hypothesis, however, and explained the phenomenon by endowing the particles themselves with what he called "fits of easy reflection and easy transmission," a kind of periodic motion due to the attractive and repulsive forces imposed by material obstacles. Newton's eminence as a scientist was such that his point of view dominated the century that followed his work.

Christian Huygens, a Dutch scientist contemporary with Newton, championed a view (in his *Treatise on Light*) that considered light as a wave, spreading out from a light source in all directions and propagating through an all-pervasive elastic medium called the *ether*. He was impressed, for example,

---

[2]A more in-depth historical account may be found, for example, in Vasco Ronchi, *The Nature of Light* (Cambridge: Harvard University Press, 1970).

by the experimental fact that when two beams of light intersected, they emerged unmodified, just as in the case of two water or sound waves. Adopting a wave theory, Huygens was able to derive the *laws of reflection and refraction* and to explain *double refraction* in calcite as well.

Within two years of the centenary of the publication of Newton's *Optics*, the Englishman Thomas Young performed a decisive experiment that seemed to demand a wave interpretation, turning the tide of support to the wave theory of light. It was the *double-slit experiment*, in which an opaque screen with two small, closely spaced openings was illuminated by monochromatic light from a small source. The "shadows" observed formed a complex interference pattern like those produced with water waves.

Victories for the wave theory continued up to the twentieth century. In the mood of scientific confidence that characterized the latter part of the nineteenth century, there was little doubt that light, like most other classical areas of physics, was well understood. In 1821, Augustin Fresnel published results of his experiments and analysis, which required that light be a transverse wave. On this basis, double refraction in calcite could be understood as a phenomenon involving *polarized light*. It had been assumed that light waves in an ether were necessarily longitudinal, like sound waves in a fluid, which cannot support transverse vibrations. For each of the two components of polarized light, Fresnel developed the *Fresnel equations*, which give the amplitude of light reflected and transmitted at a plane interface separating two optical media.

Working in the field of electricity and magnetism, James Clerk Maxwell synthesized known principles in his set of four *Maxwell equations*. The equations yielded a prediction for the speed of an *electromagnetic wave* in the ether that turned out to be the *measured speed of light*, suggesting its electromagnetic character. From then on, light was viewed as a particular region of the electromagnetic spectrum of radiation. The experiment (1887) of Albert Michelson and Edward Morley, which attempted to detect optically the earth's motion through the ether, and the special theory of relativity (1905) of Albert Einstein were of monumental importance. Together they led inevitably to the conclusion that the assumption of an ether was superfluous. The problems associated with transverse vibrations of a *wave in a fluid* thus vanished.

If the nineteenth century served to place the wave theory of light on a firm foundation, that foundation was to crumble as the century came to an end. The wave-particle controversy was resumed with vigor. Again, we mention only briefly some of the key events along the way. Difficulties in the wave theory seemed to show up in situations that involved the *interaction of light with matter*. In 1900, at the very dawn of the twentieth century, Max Planck announced at a meeting of the German Physical Society that he was able to derive the correct blackbody radiation spectrum only by making the curious assumption that atoms emitted light in discrete energy chunks rather than in a continuous manner. Thus *quanta* and *quantum mechanics* were born. According to Planck, the energy E of a quantum of electromagnetic radiation is proportional to the frequency ν of the radiation:

$$E = h\nu \tag{1-1}$$

where the constant of proportionality h, *Planck's constant*, has the very small value of 6.63×10^{-34} J-s. Five years later, in the same year that he published his theory of special relativity, Albert Einstein offered an explanation of the *photoelectric effect*, the emission of electrons from a metal surface when irradiated with light. Central to his explanation was the conception of light as a stream of light quanta whose energy is related to frequency by Planck's equation (1-1). Then in 1913, the Danish physicist Niels Bohr once more incorporated the *quantum of radiation* in his explanation of the *emission and absorption* processes of the hydrogen atom, providing a physical basis for understanding the hydrogen spectrum. Again in 1922, the model of light quanta came to the rescue for Arthur Compton, who explained the scattering of X-rays from electrons as particle-like collisions between light quanta and electrons in which both energy and momentum were conserved. In 1926, the chemist Gilbert Lewis suggested the name "photon" for the "quantum of light" and it has been so identified ever since.

All such victories for the photon or particle model of light indicated that light could be treated as a kind of particle, possessing both energy and momentum. It was Louis de Broglie who saw the other side of the picture. In 1924, he published his speculations that subatomic *particles* are endowed with *wave properties*. He suggested, in fact, that a particle with momentum p had an associated wavelength of

$$\lambda = \frac{h}{p} \tag{1-2}$$

where h was, again, Planck's constant. Experimental confirmation of de Broglie's hypothesis appeared during the years 1927–1928, when Clinton Davisson and Lester Germer in the United States and Sir George Thomson in England performed experiments that could only be interpreted as *the diffraction of a beam of electrons.*

Thus, the wave-particle duality came full circle. *Light behaves like waves in its propagation and in the phenomena of interference and diffraction; however, it exhibits particle-like behavior when exchanging energy with matter, as in the Compton and photoelectric effects.* Similarly, electrons often behaved like particles, as observed in the pointlike scintillations of a phosphor exposed to a beam of electrons; in other situations they were found to behave like waves, as in the diffraction produced by an electron microscope.

1-2 PARTICLES AND PHOTONS

Photons and electrons that behaved both as particles and as waves seemed at first an impossible contradiction, since particles and waves are very different entities indeed. Gradually it became clear, to a large extent through the reflections of Niels Bohr and especially in his *principle of complementarity*, that photons and electrons were neither waves nor particles, but something more complex than either.

In attempting to explain physical phenomena, it is natural that we appeal to well-known physical models like waves and particles. As it turns out, however, the complete nature of a photon or an electron is not exhausted by either model. In certain situations, wavelike attributes may predominate; in other situations, particle-like attributes stand out. We know of no simpler physical model that is adequate to handle all cases.

Quantum mechanics describes both light and matter and, together with special relativity, predicts that the momentum, p, wavelength, λ, and speed, v, for both material particles and photons are given by the same general equations:

$$p = \frac{\sqrt{E^2 - m^2c^4}}{c} \tag{1-3}$$

$$\lambda = \frac{h}{p} = \frac{hc}{\sqrt{E^2 - m^2c^4}} \tag{1-4}$$

$$v = \frac{pc^2}{E} = c\sqrt{1 - \frac{m^2c^4}{E^2}} \tag{1-5}$$

In these equations, m is the *rest mass* and E is the *total energy*, the sum of the rest-mass energy mc^2 and kinetic energy E_K, that is, the work done to accelerate the particle from rest to its measured speed. The proper expression for kinetic energy is no longer simply $E_K = \frac{1}{2}mv^2$, but rather is $E_K = mc^2(\gamma - 1)$, where $\gamma = 1/\sqrt{1 - (v^2/c^2)}$. This relativistic expression[3] for kinetic energy E_K approaches $\frac{1}{2}mv^2$ for $v \ll c$.

A crucial difference between particles like electrons and neutrons and particles like photons is that the latter have *zero rest mass*. Equations (1-3) to (1-5) then take the simpler forms for photons:

$$p = \frac{E}{c} \qquad (1\text{-}6)$$

$$\lambda = \frac{h}{p} = \frac{hc}{E} \qquad (1\text{-}7)$$

$$v = \frac{pc^2}{E} = c \qquad (1\text{-}8)$$

Thus, while nonzero rest-mass particles like electrons have a *limiting* speed of c, Eq. (1-8) shows that zero rest-mass particles like photons must travel with the constant speed c. The energy of a photon is not a function of its speed but rather of its frequency, as expressed in Eq. (1-1) or in Eqs. (1-6) and (1-7), taken together. Notice that for a photon, because of its zero rest mass, there is no distinction between its total energy and its kinetic energy. The following example helps clarify the differences in the momentum, wavelength, and speed of electrons and photons of the same total energy.

Example 1-1

An electron is accelerated to a kinetic energy E_K of 2.5 MeV. (a) Determine its relativistic momentum, de Broglie wavelength, and speed. (b) Determine the same properties for a photon having the same total energy as the electron.

Solution

The electron's total energy E must be the sum of its rest mass energy mc^2 and its kinetic energy E_K. The rest mass energy is
$mc^2 = (9.11 \times 10^{-31}\ \text{kg})(3 \times 10^8\ \text{m/s})^2 = 8.19 \times 10^{-14}\ \text{J}$.
Since $1\ \text{eV} = 1.6 \times 10^{-19}\ \text{J}$, we have $mc^2 = 5.11 \times 10^5\ \text{eV} = 0.511\ \text{MeV}$. Thus,

$$E = mc^2 + E_K = 0.511\ \text{MeV} + 2.5\ \text{MeV} = 3.011\ \text{MeV}$$

or

$$E = 3.011 \times 10^6\ \text{eV} \times (1.602 \times 10^{-19}\ \text{J/eV}) = 4.82 \times 10^{-13}\ \text{J}$$

The other quantities are then calculated in order. Working with SI units we obtain, from Eq. (1-3):

$$p = \frac{\sqrt{E^2 - (mc^2)}}{c} = \frac{\sqrt{(4.82 \times 10^{-13}\ \text{J})^2 - (8.19 \times 10^{-14}\ \text{J})^2}}{3 \times 10^8\ \text{m/s}}$$
$$= 1.58 \times 10^{-21}\ \text{kg·m/s}$$

from Eq. (1-4):

---

[3]This discussion is not meant to be a condensed tutorial on relativistic mechanics, but, with the help of Eqs. (1-3) to (1-8), a summary of some basic relations that unify particles of matter and light.

$$\lambda = \frac{h}{p} = \frac{6.626 \times 10^{-34} \text{ J·s}}{1.58 \times 10^{-21} \text{ kg·m/s}} = 4.19 \times 10^{-13} \text{ m} = 0.419 \text{ pm}$$

and from Eq. (1-5):

$$v = \frac{pc^2}{E} = \frac{(1.58 \times 10^{-21} \text{ kg·m/s})(3 \times 10^8 \text{ m/s})^2}{4.82 \times 10^{-13} \text{ J}}$$

$$= 2.95 \times 10^8 \text{ m/s}$$

For the photon, with $m = 0$, we get instead,
from Eq. (1-6):

$$p = \frac{E}{c} = \frac{4.82 \times 10^{-13} \text{ J}}{3 \times 10^8 \text{ m/s}} = 1.61 \times 10^{-21} \text{ kg·m/s}$$

from Eq. (1-7):

$$\lambda = \frac{h}{p} = \frac{6.626 \times 10^{-34} \text{ J·s}}{1.61 \times 10^{-21} \text{ kg·m/s}} = 0.412 \text{ pm}$$

and from Eq. (1-8):

$$v = c = 3.00 \times 10^8 \text{ m/s}$$

There is another important distinction between electrons and photons. Electrons obey *Fermi-Dirac statistics*, whereas photons obey *Bose-Einstein statistics*. A consequence of Fermi-Dirac statistics is that no two electrons in the same interacting system can be in the same *state*, that is, have precisely the same physical properties. Bose-Einstein statistics impose no such prohibition, so that identical photons with the same energy, momentum, and polarization can occur together in large numbers, as they do, for example, in a laser cavity.

In the theory called *quantum electrodynamics*, which combines the principles of quantum mechanics with those of special relativity, photons interact only with charges. An electron, for example, is capable of both absorbing and emitting a photon. There is no conservation law for photons as there is for the charge associated with particles. As indicated in the preceding example, in this theory the wave-particle duality becomes reconciled in the sense that both classical waves (i.e., light) and classical particles (i.e., electrons) are seen to have the same basic nature, which is neither wholly wave nor wholly particle. Essential distinctions between photons and electrons are removed and both are subject to the same general principles. Nevertheless, the complementary aspects of particle and wave descriptions of light remain, justifying our use of one or the other when appropriate. The wave description of light will be found adequate to describe most of the optical phenomena treated in this text.

In this brief comparison we have remarked on some of the differences and similarities between classical particles and light and have provided several fundamental relations that apply to both. The notion that light interacts with matter by exchanging photons of definite energy, momentum, and polarization will serve us well several chapters hence when we consider laser operation.

1-3 THE ELECTROMAGNETIC SPECTRUM

In this text we are concerned with the properties and applications of light. Following the pivotal work of James Clerk Maxwell, for whom the equations that govern electricity and magnetism are named, "light" is identified as an *electromagnetic wave* having a frequency in the range that human eyes can detect and interpret. All electromagnetic waves are made up of time-varying electric and magnetic fields. Electromagnetic (EM) waves are produced by accelerating charge distributions, carry energy, and exert forces on charged particles upon which they impinge. The properties of electromagnetic waves are discussed in more detail in Chapter 4 and elsewhere throughout this text. In the last two sections of this introductory chapter, we wish to introduce only the most basic characteristics of electromagnetic waves.

In free (that is, empty) space all electromagnetic waves travel with the same speed, commonly given the symbol c. This speed emerges naturally from Maxwell's equations and is given, approximately, as $c = 3 \times 10^8$ m/s. As with all waves, the frequency of an electromagnetic wave is determined by the frequency of the source of the wave. An electromagnetic disturbance that propagates through space as a wave may be *monochromatic*, that is, characterized for practical purposes by a single frequency, or *polychromatic*, in which case it is represented by many frequencies, either discrete or in a continuum. The distribution of energy among the various constituent waves is called the *spectrum* of the radiation, and the adjective *spectral* implies a dependence on wavelength. Various regions of the *electromagnetic spectrum* are referred to by particular names, such as *radio waves, cosmic rays, light*, and *ultraviolet radiation*, because of differences in the way they are produced or detected. Most of the common descriptions of the various frequency ranges are given in Figure 1-1, in which the electromagnetic spectrum is

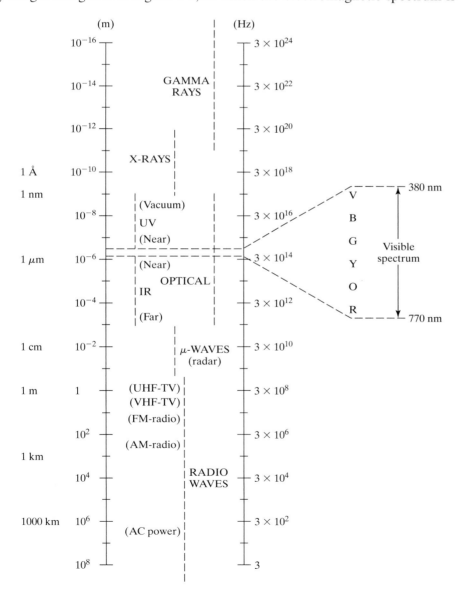

Figure 1-1 Electromagnetic spectrum, arranged by wavelength in meters and frequency in hertz. The narrow portion occupied by the visible spectrum is highlighted.

displayed in terms of both frequency ν and wavelength, λ. Recall that these two quantities are related, as with all types of wave motion, through the velocity c:

$$c = \lambda\nu \tag{1-9}$$

As indicated in Figure 1-1, common units for wavelength are the *angstrom* ($1 \text{ Å} = 10^{-10}$ m), the *nanometer* (1 nm $= 10^{-9}$ m), and the *micrometer* (1μm$= 10^{-6}$ m). The regions ascribed to various types of waves, as shown, are not precisely bounded. Regions may overlap, as in the case of the continuum from X-rays to gamma rays. The choice of label will depend on the manner in which the radiation is either produced or used. The narrow range of electromagnetic waves from approximately 380 to 770 nm is capable of producing a visual sensation in the human eye and is properly referred to as "light." It is not surprising that this *visible region* of the spectrum corresponds to the frequencies of electromagnetic radiation that predominate in the output of the sun. Humans "see" different wavelengths of light as different colors. The visible spectrum of colors ranges from red (long-wavelength end) to violet (short-wavelength end) and is bounded by the invisible *ultraviolet* and *infrared* regions, as shown. The three regions taken together comprise the *optical spectrum*, that region of the electromagnetic spectrum of special interest in a textbook on optics. In addition, atoms and molecules have resonant frequencies in this optical spectrum and so EM waves in this frequency range interact most strongly with atoms and molecules. We now provide brief sketches of the various invisible regions of the electromagnetic spectrum.

Ultraviolet

On the short-wavelength side of visible light, this electromagnetic region spans wavelengths ranging from 380 nm down to 10 nm. Ultraviolet light is sometimes subdivided into three categories: UV-A refers to the wavelength range 380–315 nm, UV-B to the range 315–280 nm and UV-C to the range 280–10 nm. The sun emits significant amounts of electromagnetic radiation in all three UV bands but, due to absorption in the *ozone layer* of the earth's atmosphere, roughly 99% of the UV radiation that reaches the earth's surface is in the UV-A band. UV radiation from the sun is linked to a variety of health risks. UV-A radiation, generally regarded as the least harmful of the three UV bands, can contribute to skin aging and is possibly linked to some forms of skin cancer. UV-A radiation does not contribute to sunburns. UV-B radiation has been linked to a variety of skin cancers and contributes to the sunburning process. The link between UV-B radiation and skin cancer is a primary reason for the concern related to ozone depletion, which is believed to be in part caused by human use of so-called chlorofluorocarbon (CFC) compounds. Ozone (O_3) is formed when UV-C radiation reacts with oxygen in the stratosphere and, as mentioned, plays an important role in the filtering of UV-B and UV-C from the electromagnetic radiation that reaches the earth's surface. CFC compounds can participate in chemical processes that lead to the conversion of ozone into "ordinary" oxygen (O_2). The concern that CFCs and other similar chemicals may contribute to the depletion of the ozone layer and thus increase the risk of skin cancer led to protocols calling for the reduction of the use of refrigerants, aerosol sprays, and other products that release these chemicals into the atmosphere. Sunblock and sunscreen lotions are intended in part to block harmful UV-B radiation. On the other hand, UV radiation has the beneficial effect of inducing vitamin D production in the skin.

X-rays

X-rays are EM waves with wavelengths in the 10 nm to 10^{-4} nm range. These can be produced when high-energy electrons strike a metal target and are used as diagnostic tools in medicine to see bone structure and as treatments for certain cancers. X-ray diffraction serves as a probe of the lattice structure of crystalline solids and X-ray telescopes provide important information from astronomical objects.

Gamma Rays

This type of EM radiation has its origin in nuclear radioactive decay and certain other nuclear reactions. Gamma rays have very short wavelengths in the range from 0.1 nm to 10^{-14} nm. Like X-rays, penetrating gamma rays find use in the medical area, often in the treatment of localized cancers.

Infrared Radiation

On the long-wavelength side of the visible spectrum, infrared (IR) radiation has wavelengths spanning the region from 770 nm to 1 mm. Objects in thermal equilibrium at terrestrial temperatures emit radiation that has its energy output peak in the IR range. Consequently, infrared radiation is sometimes termed "heat radiation" and finds application in nightvision scopes that detect the IR emitted from objects in absolute "darkness" and in infrared photography wherein objects at different temperatures (and so with different peak wavelengths of emitted radiation) are imaged as areas of contrasting brightness. Images such as these can be used to map the temperature variation across the surface of the earth, for example. Infrared radiation is used as a treatment for sore muscles and joints, and, more recently, lasers that emit IR radiation have been used to treat the eye for vision abnormalities. Infrared radiation is also used in optical fiber communication systems and in a variety of remote control devices.

Microwaves

Beyond infrared radiation we find microwaves, with wavelengths from 1 mm to 30 cm or so. Microwave ovens, which have become a common kitchen appliance, use microwaves to heat food. In addition, microwaves play an important role in radar systems both on the ground and in the air, in telecommunications, and in spectroscopy.

Radio Waves

Radio waves are long-wavelength EM radiations produced, for example, by electrons oscillating in conductors that form antennas of various shapes. Radio waves have wavelengths ranging from meters to thousands of meters. Used commonly in radio and television broadcasts, they include the AM radio band (540–1600 kHz) with wavelengths ranging from 188 to 556 m as well as the FM radio band (88–108 MHz) with wavelengths from 2.78 to 3.41 m.

We have indicated that EM waves may lose and gain energy only in discrete amounts that are multiples of the energy associated with the energy quanta that have come to be called photons. Equation (1-1) gives the energy of a photon as $h\nu$. When EM wave energy is detected, the detector can record only energies that are multiples of a photon's energy. As the following example indicates, for macroscopic light sources, the energy of a photon is typically far less than the total detected energy, and so, in such a case, the restriction that the detected energy must be only a multiple of a photon's energy goes unnoticed. Since the energy per photon decreases with increased wavelength, for a given total energy, the energy "graininess" is less for long-wavelength radiation than for short-wavelength radiation. To understand the interaction of light with individual atoms and molecules, it is important to keep in mind that EM waves gain and lose energy in discrete amounts proportional to the frequency of the radiation. Consider the following example.

Example 1-2

A certain sensitive radar receiver detects an electromagnetic signal of frequency 100 MHz and power (energy/time) 6.63×10^{-16} J/s.

  a. What is the wavelength of a photon with this frequency?
  b. What is the energy of a photon in this signal? Express this energy in J and in eV.
  c. How many photons/s would arrive at the receiver in this signal?
  d. What is the energy (in J and in eV) of a visible photon of wavelength 555 nm?
  e. How many visible ($\lambda = 555$ nm) photons/s would correspond to a detected power of 6.63×10^{-16} J/s?
  f. What is the energy (in J and in eV) of an X-ray of wavelength 0.1 nm?
  g. How many X-ray ($\lambda = 0.1$ nm) photons/s would correspond to a detected power of 6.63×10^{-16} J/s?

Solution

  a. $\lambda = c/\nu = \dfrac{3 \times 10^8 \text{ m/s}}{100 \times 10^6 \text{ Hz}} = 3 \text{ m}$

  b. $E = h\nu = (6.63 \times 10^{-34} \text{ J} \cdot \text{s})(100 \times 10^6 \text{ Hz}) = 6.63 \times 10^{-26} \text{ J}$

  $E = 6.63 \times 10^{-26} \text{ J}\left(\dfrac{1 \text{ eV}}{1.6 \times 10^{-19} \text{ J}}\right) = 4.14 \times 10^{-7} \text{ eV}$

  c. The number of detected photons per second N would be

  $$N = \frac{\text{Power}}{\text{Energy/photon}} = \frac{6.63 \times 10^{-16} \text{ J/s}}{6.3 \times 10^{-26} \text{ J}} = 10^{10}/\text{s}$$

  So each photon contributes but one part in 10 billion of the total power in the radar wave even for this very weak signal. In such a case, the "graininess" of the power in the signal is likely to go undetected.

  d. $E_{555} = h\nu = \dfrac{hc}{\lambda} = \dfrac{(6.63 \times 10^{-34} \text{ J} \cdot \text{s})(3 \times 10^8 \text{ m/s})}{555 \times 10^{-9} \text{ m}}$

  $= 3.58 \times 10^{-19} \text{ J} = 2.2 \text{ eV}$

  e. $N_{555} = \dfrac{\text{Power}}{\text{Energy/photon}} = \dfrac{6.63 \times 10^{-16} \text{ J/s}}{3.58 \times 10^{-19} \text{ J}} = 1850/\text{s}$. The effect of addition or removal of a single photon would perhaps be noticeable.

  f. $E_{\text{X-ray}} = h\nu = \dfrac{hc}{\lambda} = \dfrac{(6.63 \times 10^{-34} \text{ J} \cdot \text{s})(3 \times 10^8 \text{ m/s})}{0.1 \times 10^{-9} \text{ m}}$

  $= 1.99 \times 10^{-15} \text{ J} = 12{,}400 \text{ eV}$

  g. $N_{\text{X-ray}} = \dfrac{\text{Power}}{\text{Energy/photon}} = \dfrac{6.63 \times 10^{-16} \text{ J/s}}{1.99 \times 10^{-15} \text{ J}} = 0.33/\text{s}.$

  One X-ray would be detected every 3 seconds or so. The discreteness of the energy of light quanta would be very evident in this case.

In this example we have introduced the notion of a detector and the energy and power carried by an electromagnetic wave. The energy carried by an EM wave can be specified in many related ways: the power, power per unit area, and power per unit solid angle, for example. To quantify these characteristics of EM waves, we turn to the topic of radiometry.

1-4 RADIOMETRY

Radiometry is the science of measurement of electromagnetic radiation. In this discussion we are content to present briefly the *radiometric quantities* or physical terms used to characterize the energy content of radiation.

Many radiometric terms have been introduced and used in the optics literature; however, we include here only those approved International System (SI) units. These terms and their units are summarized in Table 1-1.[4]

Radiometric quantities appear either without subscripts or with the subscript e (*electromagnetic*). The terms *radiant energy*, Q_e(J = joules), *radiant energy density*, w_e(J/m^3), and *radiant flux* or *radiant power*, Φ_e(W = watts = J/s), need no further explanation. *Radiant flux density* at a surface, measured in watts per square meter, may be either emitted (scattered, reflected) from a surface, in which case it is called *radiant exitance*, M_e, or incident onto a surface, in which case it is called *irradiance*, E_e. The radiant flux (Φ_e) emitted per unit of solid angle (ω) by a point source in a given direction (Figure 1-2) is called the *radiant intensity*, I_e. This quantity, often confused with *irradiance*, is given by

$$I_e = \frac{d\Phi}{d\omega} \tag{1-10}$$

Differential solid angle $d\omega$ measured in steradians (sr) is defined in Figure 1-2. The radiant intensity I_e from a sphere radiating Φ_e watts (W) of power uniformly in all directions, for example, is $\Phi_e/4\pi$ (W/sr), since the total surrounding solid angle is 4π sr.

TABLE 1-1 RADIOMETRIC TERMS

Term	Symbol (units)	Defining equation
Radiant energy	Q_e(J = W · s)	—
Radiant energy density	w_e(J/m^3)	$w_e = dQ_e/dV$
Radiant flux, Radiant power	Φ_e(W)	$\Phi_e = dQ_e/dt$
Radiant exitance	M_e(W/m^2)	$M_e = d\Phi_e/dA$
Irradiance	E_e(W/m^2)	$E_e = d\Phi_e/dA$
Radiant intensity	I_e(W/sr)	$I_e = d\Phi_e/d\omega$
Radiance	$L_e\left(\dfrac{W}{sr \cdot m^2}\right)$	$L_e = dI_e/dA \cos\theta$

Abbreviations: J, joule; W, watt; m, meter; sr, steradian.

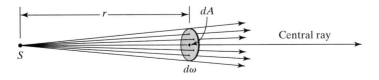

Figure 1-2 The radiant intensity is the flux through the cross section dA per unit of solid angle. Here the solid angle $d\omega = dA/r^2$.

---

[4]The introduction "in the abstract" of so many new units, some rarely used and others misused, is not very palatable pedagogically. Table 1-1 is meant to serve as a convenient summary that can be referred to when needed.

The familiar inverse-square law of radiation from a point source, illustrated in Figure 1-3, is now apparent by calculating the irradiance of a point source on a spherical surface surrounding the point, of solid angle 4π sr and surface area $4\pi r^2$. Thus,

$$E_e = \frac{d\Phi_e}{dA} = \frac{\Phi_e}{A} = \frac{4\pi I_e}{4\pi r^2} = \frac{I_e}{r^2}, \quad \text{point source} \tag{1-11}$$

The *radiance*, L_e, describes the radiant intensity per unit of projected area, perpendicular to the specified direction, and is given by

$$L_e = \frac{dI_e}{dA\cos\theta} = \frac{d^2\Phi_e}{d\omega(dA\cos\theta)} \tag{1-12}$$

The importance of the radiance is suggested in the following considerations. Suppose a plane radiator or reflector is perfectly *diffuse*, by which we mean that it radiates uniformly in all directions. The radiant intensity is measured for a fixed solid angle defined by the fixed aperture A_p at some distance r from the radiating surface, shown in Figure 1-4. The aperture might be the input aperture of a detecting instrument measuring all the flux that so enters. When viewed at $\theta = 0°$, along the normal to the surface, a certain maximum

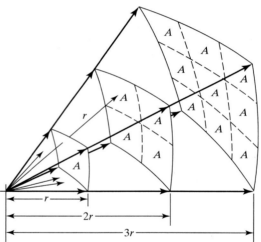

Figure 1-3 Illustration of the inverse-square law. The flux leaving a point source within any solid angle is distributed over increasingly larger areas, producing an irradiance that decreases inversely with the square of the distance.

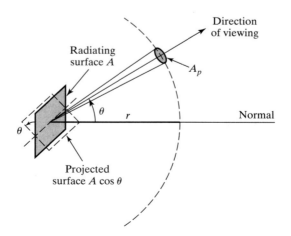

Figure 1-4 Radiant flux collected along a direction making an angle θ with the normal to the radiating surface. The projected area of the surface ($A\cos\theta$) is shown by the dashed rectangle.

intensity $I(0)$ is observed. As the aperture is moved along the circle of radius r, thereby increasing the angle θ, the cross section of radiation presented by the surface decreases in such a way that

$$I(\theta) = I(0) \cos \theta \qquad (1\text{-}13)$$

a relation called *Lambert's cosine law*. If the radiance is determined at each angle θ, it is found to be constant, because the intensity must be divided by the projected area $A \cos \theta$ such that the cosine dependence cancels:

$$L_e = \frac{I(\theta)}{A \cos \theta} = \frac{I(0) \cos \theta}{A \cos \theta} = \frac{I(0)}{A} = \text{constant} \qquad (1\text{-}14)$$

Thus, when a radiating (or reflecting) surface has a radiance that is independent of the viewing angle, the surface is said to be perfectly diffuse, or a *Lambertian surface*.

We show next that the radiance has the same value at any point along a ray propagating in a uniform, nonabsorbing medium. Figure 1-5 pictures a narrow beam of radiation in such a medium, including a central ray and a small bundle of surrounding rays (not shown) that pass through the elemental areas dA_1 and dA_2 situated at different points along the beam. The central ray makes angles of θ_1 and θ_2, respectively, relative to the area normals, as shown. The solid angle $d\omega_1 = dA_2 \cos \theta_2/r^2$, where $dA_2 \cos \theta_2$ represents the projection of area dA_2 normal to the central ray. According to Eq. (1-12), the radiance L_1 at dA_1 is given by

$$L_1 = \frac{d^2\Phi_1}{d\omega_1(dA_1 \cos \theta_1)} = \frac{d^2\Phi_1}{(dA_2 \cos \theta_2/r^2)(dA_1 \cos \theta_1)} \qquad (1\text{-}15)$$

By a similar argument, in which we reverse the roles of dA_1 and dA_2 in the figure,

$$L_2 = \frac{d^2\Phi_2}{d\omega_2(dA_2 \cos \theta_2)} = \frac{d^2\Phi_2}{(dA_1 \cos \theta_1/r^2)(dA_2 \cos \theta_2)} \qquad (1\text{-}16)$$

For a nonabsorbing medium, the power associated with the radiation passing through the continuous bundle of rays remains constant, that is, $d\Phi_1 = d\Phi_2$, so that we can conclude from Eqs. (1-15) and (1-16) that $L_1 = L_2$. It follows that the radiance of the beam is also the radiance of the source, at the initial point of the beam, or $L_1 = L_2 = L_0$.

Suppose, referring to Figure 1-6, that we wish to know the quantity of radiant power reaching an element of area dA_2 on surface S_2 due to the source element dA_1 on surface S_1. The line joining the elemental areas, of

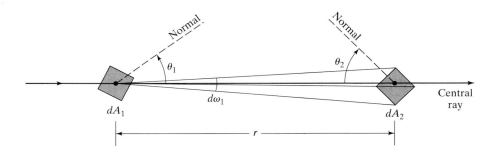

Figure 1-5 Geometry used to show the invariance of the radiance in a uniform, lossless medium.

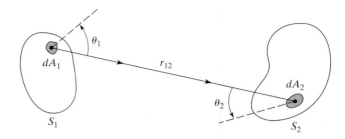

Figure 1-6 General case of the illumination of one surface by another radiating surface. Each elemental *radiating area* dA_1 contributes to each elemental *irradiated area* dA_2.

length r_{12}, makes angles of θ_1 and θ_2 with the respective normals to the surfaces, as shown. The radiant power is $d^2\Phi_{12}$, a second-order differential because both the source and receptor are elemental areas. By Eq. (1-15) or Eq. (1-16),

$$d^2\Phi_{12} = \frac{L\,dA_1\,dA_2 \cos\theta_1 \cos\theta_2}{r_{12}^2}$$

and the total radiant power at the entire second surface due to the entire first surface is, by integration,

$$\Phi_{12} = \int\limits_{A_1} \int\limits_{A_2} \frac{L \cos\theta_1 \cos\theta_2\,dA_1\,dA_2}{r_{12}^2} \qquad (1\text{-}17)$$

By adding powers rather than amplitudes in this integration, we have tacitly assumed that the radiation source emits incoherent radiation. We shall say more about coherent and incoherent radiation later. Now, let's try an example.

Example

Consider a $\Phi_e = 5$ milliwatt Helium-Neon laser emitting a "pencil-like" beam with a divergence angle α of 1.3 milliradians. The laser cavity is designed so that the beam emerges from a surface area $\Delta A_S = 2.5 \times 10^{-3}$ cm^2 at the output mirror. See the sketch shown in Figure 1-7.

  a. Determine the solid angle $\Delta\omega$ in terms of R and α.

  b. Determine the radiance of this laser light source in units of $\dfrac{\text{W}}{\text{cm}^2 \cdot \text{sr}}$.

Solution

  a. The solid angle $\Delta\omega$ is equal to $\Delta A_T/R^2$, where
      $$\Delta A_T = \pi r_T^2 = \pi(R\tan(\alpha/2))^2 \simeq \pi R^2(\alpha/2)^2,$$

      since $\tan(\alpha/2) \simeq \alpha/2$ for small angles.

      Thus, $\Delta\omega = \dfrac{(\pi R^2 \alpha^2)/4}{R^2} = \dfrac{\pi\alpha^2}{4}$, independent of the value of R. That is

      $$\Delta\omega = \frac{\pi\alpha^2}{4} = \frac{\pi(0.0013)^2}{4}\text{ sr} = 1.33 \times 10^{-6}\text{ sr}$$

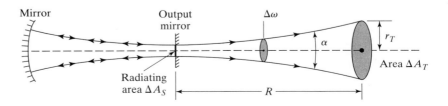

Figure 1-7 Example 1-3.

b. Making use of the defining equation for radiance L_e in Table 1-1, we obtain $L_e = \dfrac{\Delta I_e}{\Delta A_S \cos\theta}$, where we have replaced the differentials dI_e and dA_s by the small quantities ΔI_e and ΔA_s. Here $\Delta I_e = \dfrac{\Phi_e}{\Delta\omega}$ and $\theta = 0$, since the laser beam direction is normal to ΔA_S. So,

$$\Delta L_e = \frac{\Phi_e}{\Delta A_S\,\Delta\omega} = \frac{5\times10^{-3}\,\text{W}}{(2.5\times10^{-3}\,\text{cm}^2)(1.33\times10^{-6}\,\text{sr})}$$

$$= 1.5\times10^6 \frac{\text{W}}{\text{cm}^2\cdot\text{sr}}$$

a robust value indeed!

PROBLEMS

1-1 Calculate the de Broglie wavelength of (a) a golf ball of mass 50 g moving at 20 m/s and (b) an electron with kinetic energy of 10 eV.

1-2 The threshold of sensitivity of the human eye is about 100 photons per second. The eye is most sensitive at a wavelength of around 550 nm. For this wavelength, determine the threshold in watts of power.

1-3 What is the energy, in electron volts, of light photons at the ends of the visible spectrum, that is, at wavelengths of 380 and 770 nm?

1-4 Determine the wavelength and momentum of a photon whose energy equals the rest-mass energy of an electron.

1-5 Show that the rest-mass energy of an electron is 0.511 MeV.

1-6 Show that the relativistic momentum of an electron, accelerated through a potential difference of 1 million volts, can be conveniently expressed as 1.422 MeV/c, where c is the speed of light.

1-7 Show that the wavelength of a photon, measured in angstroms, can be found from its energy, measured in electron volts, by the convenient relation

$$\lambda(\mathring{A}) = \frac{12,400}{E(\text{eV})}$$

1-8 Show that the relativistic kinetic energy,

$$E_K = mc^2(\gamma - 1)$$

reduces to the classical expression $\frac{1}{2}mv^2$, when $v \ll c$.

1-9 A proton is accelerated to a kinetic energy of 2 billion electron volts (2 GeV). Find (a) its momentum, (b) its de Broglie wavelength, and (c) the wavelength of a photon with the same total energy.

1-10 Solar radiation is incident at the earth's surface at an average of 1000 W/m^2 on a surface normal to the rays. For a mean wavelength of 550 nm, calculate the number of photons falling on 1 cm^2 of the surface each second.

1-11 Two parallel beams of electromagnetic radiation with different wavelengths deliver the same power to equivalent surface areas normal to the beams. Show that the numbers of photons striking the surfaces per second for the two beams are in the same ratio as their wavelengths.

1-12 Calculate the band of frequencies of electromagnetic radiation capable of producing a visual sensation in the normal eye.

1-13 What is the length of a half-wave dipole antenna designed to broadcast FM radio waves at 100 MHz?

1-14 A so-called "rabbit-ears" TV antenna is made of a pair of adjustable rods that can spread apart at different angles. If the rods are each adjusted to a quarter wavelength for a TV channel that has a middle frequency of 90 MHz, how long are the rods?

1-15 A soprano's voice is sent by radio waves to a listener in a city 90 km away.

   a. How long does it take for the soprano's voice to reach the listener?
   b. In the same time interval, how far from the soprano has the sound wave in the auditorium traveled? Take the speed of sound to be 340 m/s.

1-16 A small, monochromatic light source, radiating at 500 nm, is rated at 500 W.

   a. If the source radiates uniformly in all directions, determine its radiant intensity.
   b. If the surface area of the source is 5 cm^2, determine the radiant excitance.
   c. What is the irradiance on a screen situated 2 m from the source, with its surface normal to the radiant flux?
   d. If the receiving screen contains a hole with diameter 5 cm, how much radiant flux gets through?

1-17 A 1.5-mW helium-neon laser beam delivers a spot of light 5 mm in diameter across a room 15 m wide. The beam radiates from a small circular area of diameter 0.5 mm at the output mirror of the laser. Assume that the beam irradiance is constant across the diverging beam.

   a. What is the beam divergence angle of this laser?
   b. Into what solid angle is the laser sending its beam?
   c. What is the irradiance at the spot on the wall 15 m from the laser?
   d. What is the radiance of the laser?

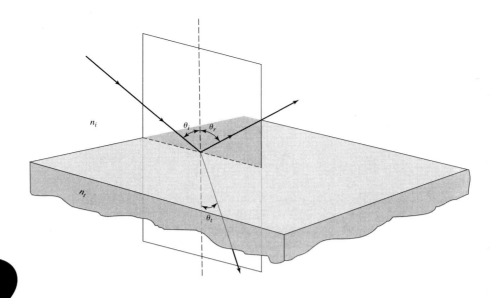

2 *Geometrical Optics*

INTRODUCTION

The treatment of light as wave motion allows for a region of approximation in which the wavelength is considered to be negligible compared with the dimensions of the relevant components of the optical system. This region of approximation is called *geometrical optics*. When the wave character of the light may not be so ignored, the field is known as *physical optics*. Thus, geometrical optics forms a special case of physical optics in a way that may be summarized as follows:

$$\lim_{\lambda \to 0} \{\text{physical optics}\} = \{\text{geometrical optics}\}$$

Since the wavelength of light—around 500 nm—is very small compared to ordinary objects, early unrefined observations of the behavior of a light beam passing through apertures or around obstacles in its path could be handled by geometrical optics. Recall that the appearance of *distinct shadows* influenced Newton to assert that the apparent rectilinear propagation of light was due to a stream of light corpuscles rather than a wave motion. Wave motion characterized by longer wavelengths, such as those in water waves and sound waves, was known to give distinct bending around obstacles. Newton's model of light propagation, therefore, seemed not to allow for the existence of a wave motion with very small wavelengths. There was in fact already evidence of some degree of bending, even for light waves, in the time of Isaac Newton. The Jesuit Francesco Grimaldi had noticed the fine structure in the edge of a shadow, a structure not explainable in terms of the rectilinear propagation of light. This bending of light waves around the edges of an obstruction came to be called *diffraction*, a topic we explore in detail in later chapters.

Within the approximation represented by geometrical optics, light is understood to travel out from its source along straight lines, or *rays*. The ray is then simply the path along which light energy is transmitted from one point to another in an optical system. The ray is a useful construct, although abstract in the sense that a light beam, in practice, cannot be narrowed down indefinitely to approach a straight line. A pencil-like laser beam is perhaps the best actual approximation to a ray of light. (When an aperture through which the beam is passed is made small enough, however, even a laser beam begins to spread out in a characteristic diffraction pattern.) When a light ray traverses an optical system consisting of several homogeneous media in sequence, the optical path is a sequence of straight-line segments. Discontinuities in the line segments occur each time the light is reflected or refracted. The laws of geometrical optics that describe the subsequent direction of the rays are the Law of Reflection and the Law of Refraction.

Law of Reflection
When a ray of light is reflected at an *interface* dividing two optical media, the reflected ray remains within the *plane of incidence*, and the angle of reflection θ_r equals the angle of incidence θ_i. The *plane of incidence* is the plane containing the incident ray and the surface normal at the point of incidence.

Law of Refraction (Snell's Law)
When a ray of light is refracted at an interface dividing two transparent media, the transmitted ray remains within the plane of incidence and the sine of the angle of refraction θ_t is directly proportional to the sine of the angle of incidence θ_i. These two laws are summarized in Figure 2-1, which depicts the general case in which an incident ray is partially reflected and partially transmitted at a plane interface separating two transparent media.

2-1 HUYGENS' PRINCIPLE

The Dutch physicist Christian Huygens envisioned light as a series of pulses emitted from each point of a luminous body and propagated in relay fashion

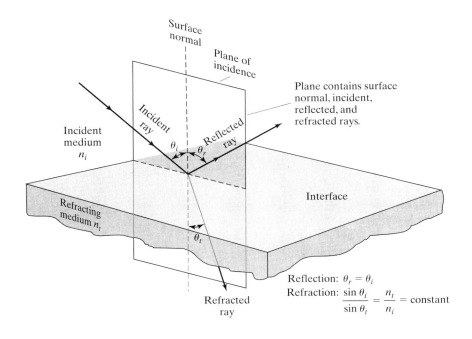

Reflection: $\theta_r = \theta_i$

Refraction: $\dfrac{\sin \theta_i}{\sin \theta_t} = \dfrac{n_t}{n_i} = $ constant

Figure 2-1   Reflection and refraction at an *interface* between two optical media. Incident, reflected, and refracted rays are shown in the *plane of incidence*.

by the particles of the ether, an elastic medium filling all space. Consistent with his conception, Huygens imagined each point of a propagating disturbance as capable of originating new pulses that contributed to the disturbance an instant later. To show how his model of light propagation implied the laws of geometrical optics, he enunciated a fruitful principle that can be stated as follows: Each point on the leading surface of a wave disturbance—the wavefront—may be regarded as a secondary source of spherical waves (or *wavelets*), which themselves progress with the speed of light in the medium and whose envelope at a later time constitutes the new wavefront. Simple applications of the principle are shown in Figure 2-2 for a plane and spherical wave. In each case, AB forms the initial wave disturbance or wavefront, and $A'B'$ is the new wavefront at a time t later. The radius of each wavelet is, accordingly, vt, where v is the speed of light in the medium. Notice that the new wavefront is tangent to each wavelet at a single point. According to Huygens, the remainder of each wavelet is to be disregarded in the application of the principle. Indeed, were the remainder of the wavelet considered to be effective in propagating the light disturbance, Huygens could not have derived the law of rectilinear propagation from his principle. To see this more clearly, refer to Figure 2-3, which shows a spherical wave disturbance originating at O and incident upon an aperture with an opening SS'. According to the notion

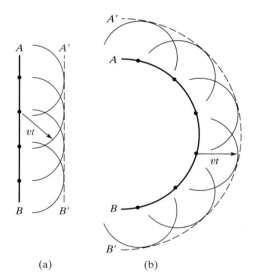

(a)   (b)

Figure 2-2 Illustration of Huygens' principle for (a) plane and (b) spherical waves.

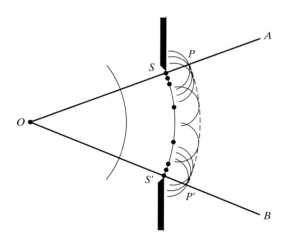

Figure 2-3 Huygens' construction for an obstructed wavefront.

of rectilinear propagation, the lines OA and OB form the sharp edges of the shadow to the right of the aperture. Some of the wavelets that originate from points of the wavefront (arc SS'), however, overlap into the region of shadow. According to Huygens, however, these are ignored and the new wavefront ends abruptly at points P and P', precisely where the extreme wavelets originating at points S and S' are tangent to the new wavefront. In so disregarding the effectiveness of the overlapping wavelets, Huygens avoided the possibility of diffraction of the light into the region of geometric shadow. Huygens also ignored the wavefront formed by the back half of the wavelets, since these wavefronts implied a light disturbance traveling in the opposite direction. Despite weaknesses in this model, remedied later by Fresnel and others, Huygens was able to apply his principle to prove the laws of both reflection and refraction, as we show in what follows.

Figure 2-4a illustrates the Huygens construction for a narrow, parallel beam of light to prove the law of reflection. Huygens' principle must be modified slightly to accommodate the case in which a wavefront, such as AC, encounters a plane interface, such as XY, at an angle. Here the angle of incidence of the rays AD, BE, and CF relative to the perpendicular PD is θ_i. Since points along the plane wavefront do not arrive at the interface simultaneously, allowance is made for these differences in constructing the wavelets that determine the reflected wavefront. If the interface XY were not present, the Huygens construction would produce the wavefront GI at the instant ray CF reached the interface at I. The intrusion of the reflecting surface, however, means that during the same time interval required for ray CF to progress from F to I, ray BE has progressed from E to J and then a distance equivalent to JH after reflection. Thus, a wavelet of radius $JN = JH$ centered at J is drawn above the reflecting surface. Similarly, a wavelet of radius DG is drawn centered at D to represent the propagation after reflection of the lower part of the beam. The new wavefront, which must now be tangent to these wavelets at points M and N, and include the point I, is shown as KI in the figure. A representative reflected ray is DL, shown perpendicular to the reflected wavefront. The normal PD drawn for this ray is used to define angles of incidence and reflection for the beam. The construction makes clear the equivalence between the angles of incidence and reflection, as outlined in Figure 2-4a.

Similarly, in Figure 2-4b, a Huygens construction is shown that illustrates the law of refraction. Here we must take into account a different speed of light in the upper and lower media. If the speed of light in vacuum is c, we express the speed in the upper medium by the ratio c/n_i, where n_i is a constant that characterizes the medium and is referred to as the *refractive index*. Similarly, the speed of light in the lower medium is c/n_t. The points D, E, and F on the incident wavefront arrive at points D, J, and I of the plane interface XY at different times. In the absence of the refracting surface, the wavefront GI is formed at the instant ray CF reaches I. During the progress of ray CF from F to I in time t, however, the ray AD has entered the lower medium, where its speed is, let us say, slower. Thus, if the distance DG is $v_i t$, a wavelet of radius $v_t t$ is constructed with center at D. The radius DM can also be expressed as

$$DM = v_t t = v_t \left(\frac{DG}{v_i} \right) = \left(\frac{n_i}{n_t} \right) DG$$

Similarly, a wavelet of radius $(n_i/n_t)\, JH$ is drawn centered at J. The new wavefront KI includes point I on the interface and is tangent to the two wavelets at points M and N, as shown. The geometric relationship between the angles θ_i and θ_t, formed by the representative incident ray AD and refracted ray DL, is *Snell's law*, as outlined in Figure 2-4b. Snell's law of refraction may be expressed as

$$n_i \sin \theta_i = n_t \sin \theta_t \tag{2-1}$$

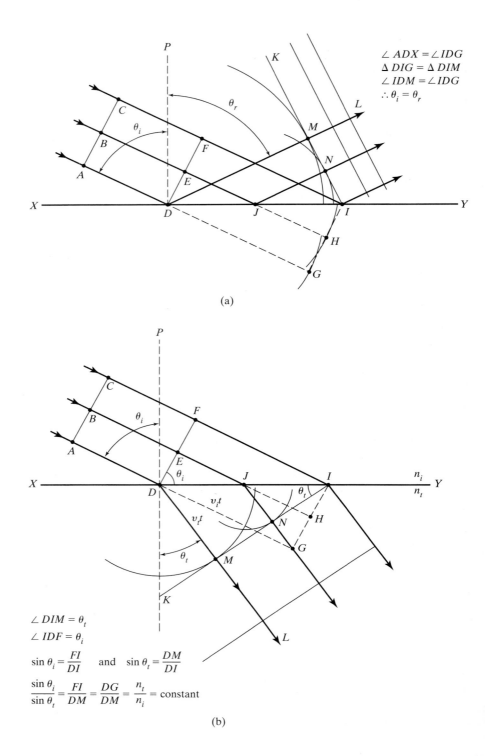

(a)

(b)

Figure 2-4 (a) Huygens' construction to prove the law of reflection. (b) Huygens' construction to prove the law of refraction.

2-2 FERMAT'S PRINCIPLE

The laws of geometrical optics can also be derived, perhaps more elegantly, from a different fundamental hypothesis. The root idea had been introduced by Hero of Alexandria, who lived in the second century B.C. According to Hero, when light is propagated between two points, it takes the shortest path. For propagation between two points in the same uniform medium, the path is clearly the straight line joining the two points. When light from the first point A, Figure 2-5, reaches the second point B after reflection from a plane surface, however, the same principle predicts the law of reflection, as follows. Figure 2-5

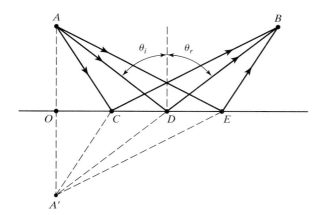

Figure 2-5 Construction to prove the law of reflection from Hero's principle.

shows three possible paths from A to B, including the correct one, ADB. Consider, however, the arbitrary path ACB. If point A' is constructed on the perpendicular AO such that $AO = OA'$, the right triangles AOC and $A'OC$ are equal. Thus, $AC = A'C$ and the distance traveled by the ray of light from A to B via C is the same as the distance from A' to B via C. The shortest distance from A' to B is obviously the straight line $A'DB$, so the path ADB is the correct choice taken by the actual light ray. Elementary geometry shows that for this path, $\theta_i = \theta_r$. Note also that to maintain $A'DB$ as a single straight line, the reflected ray must remain within the plane of incidence, that is, the plane of the page.

The French mathematician Pierre de Fermat generalized Hero's principle to prove the law of refraction. If the terminal point B lies below the surface of a second medium, as in Figure 2-6, the correct path is definitely not the shortest path or straight line AB, for that would make the angle of refraction equal to the angle of incidence, in violation of the empirically established law of refraction. Appealing to the "economy of nature," Fermat supposed instead that the ray of light traveled the path of least *time* from A to B, a generalization that included Hero's principle as a special case. If light travels more slowly in the second medium, as assumed in Figure 2-6, light bends at the interface so as to take a path that favors a shorter time in the second medium, thereby minimizing the overall transit time from A to B. Mathematically, we are required to minimize the total time,

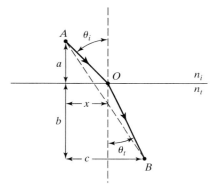

Figure 2-6 Construction to prove the law of refraction from Fermat's principle.

$$t = \frac{AO}{v_i} + \frac{OB}{v_t}$$

where v_i and v_t are the velocities of light in the incident and transmitting media, respectively. Employing the Pythagorean theorem and the distances defined in Figure 2-6, we have $AO = \sqrt{a^2 + x^2}$ and $OB = \sqrt{b^2 + (c - x)^2}$, so that

$$t = \frac{\sqrt{a^2 + x^2}}{v_i} + \frac{\sqrt{b^2 + (c - x)^2}}{v_t}$$

Since other choices of path change the position of point O and therefore the distance x, we can minimize the time by setting $dt/dx = 0$:

$$\frac{dt}{dx} = \frac{x}{v_i\sqrt{a^2 + x^2}} - \frac{c - x}{v_t\sqrt{b^2 + (c - x)^2}} = 0$$

Again from Figure 2-6, in the two right triangles containing AO and OB, respectively, the angles of incidence and refraction can be conveniently introduced into the preceding condition, since $\sin \theta_i = \dfrac{x}{\sqrt{a^2 + x^2}}$ and $\sin \theta_t = \dfrac{c - x}{\sqrt{b^2 + (c - x)^2}}$, giving

$$\frac{dt}{dx} = \frac{\sin \theta_i}{v_i} - \frac{\sin \theta_t}{v_t} = 0$$

Simplifying the equation set equal to zero, we obtain at once $v_t \sin \theta_i = v_i \sin \theta_t$. Introducing the refractive indices of the media through the relation $v = c/n$, we arrive at Snell's law:

$$n_i \sin \theta_i = n_t \sin \theta_t$$

Fermat's principle, like that of Huygens, required refinement to achieve more general applicability. Situations exist where the actual path taken by a light ray may represent a maximum time or even one of many possible paths, all requiring equal time. As an example of the latter case, consider light propagating from one focus to the other inside an ellipsoidal mirror, along any of an infinite number of possible paths. Since the ellipse is the locus of all points whose combined distances from the two foci is a constant, all paths are indeed of equal time. A more precise statement of Fermat's principle, which requires merely an extremum relative to neighboring paths, may be given as follows: The actual path taken by a light ray in its propagation between two given points in an optical system is such as to make its optical path equal, in the first approximation, to other paths closely adjacent to the actual one.

With this formulation, Fermat's principle falls in the class of problems called *variational calculus*, a technique that determines the form of a function that minimizes a definite integral. In optics, the definite integral is the integral of the time required for the transit of a light ray from starting to finishing points.[1]

2-3 PRINCIPLE OF REVERSIBILITY

Refer again to the cases of reflection and refraction pictured in Figures 2-5 and 2-6. If the roles of points A and B are interchanged, so that B is the source of light rays, Fermat's principle of least time must predict the same path as determined for the original direction of light propagation. In general, then, any actual ray of light in an optical system, if reversed in direction, will retrace the same path backward. This principle of *reversibility* will be found very useful in various applications to be dealt with later.

2-4 REFLECTION IN PLANE MIRRORS

Before discussing the formation of images in a general way, we discuss the simplest—and experientially, the most accessible—case of images formed by plane mirrors. In this context it is important to distinguish between *specular reflection* from a perfectly smooth surface and *diffuse reflection* from a granular or rough surface. In the former case, all rays of a parallel beam incident on the surface obey the law of reflection from a plane surface and therefore reflect as a parallel beam; in the latter case, though the law of reflection is obeyed locally for each ray, the microscopically granular surface results in

---

[1] It is of interest to note here that a similar principle, called *Hamilton's principle of least action* in mechanics, that calls for a minimum of the definite integral of the Lagrangian function (the kinetic energy minus the potential energy), represents an alternative formulation of the laws of mechanics and indeed implies Newton's laws of mechanics themselves.

rays reflected in various directions and thus a diffuse scattering of the originally parallel rays of light. Every plane surface will produce some such scattering, since a perfectly smooth surface can only be approximated in practice. The treatment that follows assumes the case of specular reflection.

Consider the specular reflection of a single light ray OP from the xy-plane in Figure 2-7a. By the law of reflection, the reflected ray PQ remains within the plane of incidence, making equal angles with the normal at P. If the path OPQ is resolved into its x-, y-, and z-components, it is clear that the direction of ray OP is altered by the reflection only along the z-direction, and then in such a way that its z-component is simply reversed. If the direction of the incident ray is described by its unit vector, $\hat{\mathbf{r}}_1 = (x, y, z)$, then the reflection causes

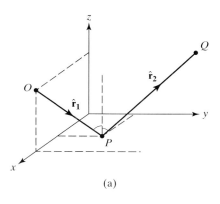

$$\hat{\mathbf{r}}_1 = (x, y, z) \longrightarrow \hat{\mathbf{r}}_2 = (x, y, -z)$$

It follows that if a ray is incident from such a direction as to reflect sequentially from all three rectangular coordinate planes, as in the "corner reflector" of Figure 2-7b,

$$\hat{\mathbf{r}}_1 = (x, y, z) \longrightarrow \hat{\mathbf{r}}_2 = (-x, -y, -z)$$

and the ray returns precisely parallel to the line of its original approach. A network of such corner reflectors ensures the exact return of a beam of light—a headlight beam from highway reflectors, for example, or a laser beam from a mirror on the moon.

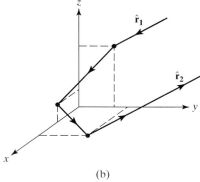

Figure 2-7 Geometry of a ray reflected from a plane.

Image formation in a plane mirror is illustrated in Figure 2-8a. A point object S sends rays toward a plane mirror, which reflect as shown. The law of reflection ensures that pairs of triangles like SNP and $S'NP$ are equal, so all reflected rays appear to originate at the *image point* S', which lies along the normal line SN, and at such a depth that the *image distance* $S'N$ equals the *object distance* SN. The eye sees a point image at S' in exactly the same way it would see a real point object placed there. Since none of the actual rays of light lies below the mirror surface, the image is said to be a *virtual image*. The image S' *cannot* be projected on a screen as in the case of a *real image*. All points of an extended object, such as the arrow in Figure 2-8b, are imaged by a plane mirror in similar fashion: Each object point has its image point along its normal to the mirror surface and as far below the reflecting surface as the object point lies above the surface. Note that the image position does not depend on the position of the eye. Further, the construction of Figure 2-8b makes clear that the image size is identical with the object size, giving a *magnification* of unity. In addition, the transverse orientation of object and image are the same. A right-handed object, however, appears left-handed in its image. In Figure 2-8c, where the mirror does not lie directly below the object, the mirror plane may be extended to determine the position of the image as seen by an eye positioned to receive reflected rays originating at the object. Figure 2-8d illustrates multiple images of a point object O formed by two perpendicular mirrors. Images I_1 and I_2 result from single reflections in the two mirrors, but a third image I_3 results from sequential reflections from both mirrors.

2-5 REFRACTION THROUGH PLANE SURFACES

Consider light ray (1) in Figure 2-9a, incident at angle θ_1 at a plane interface separating two transparent media characterized, in order, by refractive indices n_1 and n_2. Let the angle of refraction be the angle θ_2. Snell's law, which now takes the form

$$n_1 \sin \theta_1 = n_2 \sin \theta_2 \qquad (2\text{-}2)$$

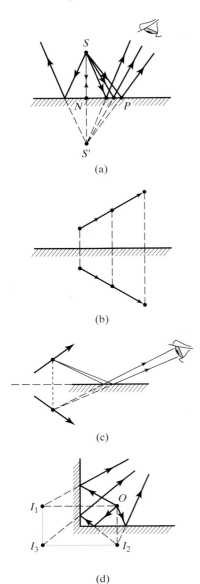

Figure 2-8   Image formation in a plane mirror.

Figure 2-9   Geometry of rays refracted by a plane interface.

requires an angle of refraction such that refracted rays bend away from the normal, as shown in Figure 2-9a, for rays 1 and 2, when $n_2 < n_1$. For $n_2 > n_1$, on the other hand, the refracted ray bends toward the normal. The law also requires that ray 3, incident normal to the surface ($\theta_1 = 0$), be transmitted without change of direction ($\theta_2 = 0$), regardless of the ratio of refractive indices.

In Figure 2-9a, the three rays shown originate at a source point S below an interface and emerge into an upper medium of lower refractive index, as in the case of light emerging from water ($n_1 = 1.33$) into air ($n_2 = 1.00$). A unique image point is not determined by these rays because they have no common intersection or virtual image point below the surface from which they appear to originate after refraction, as shown by the dashed line extensions of the refracted rays. For rays making a small angle with the normal to the surface, however, a reasonably good image can be located. In this approximation, where we allow only such *paraxial rays*[2] to form the image, the angles of incidence and refraction are both small, and the approximation

$$\sin\theta \cong \tan\theta \cong \theta \text{ (in radians)}$$

is valid. From Eq. (2-2), Snell's law can be approximated by

$$n_1 \tan\theta_1 \cong n_2 \tan\theta_2 \qquad (2\text{-}3)$$

and taking the appropriate tangents from Figure 2-9b, we have

$$n_1\left(\frac{x}{s}\right) = n_2\left(\frac{x}{s'}\right)$$

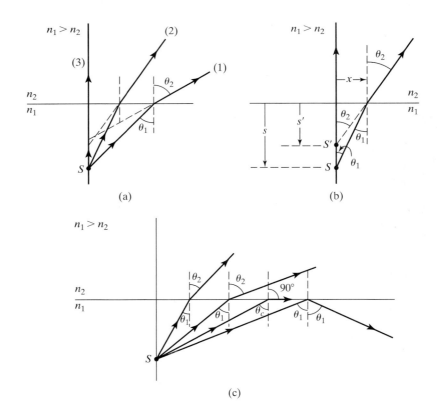

[2]In general, a paraxial ray is one that remains near the central axis of the image-forming optical system, thus making small angles with the optical axis.

The image point occurs at the vertical distance s' below the surface given by

$$s' = \left(\frac{n_2}{n_1}\right)s \tag{2-4}$$

where s is the corresponding depth of the object. Thus, objects underwater, viewed from directly overhead, appear to be nearer the surface than they actually are, since in this case $s' = (1/1.33)\, s = (3/4)\, s$. Even when the viewing angle θ_2 is not small, a reasonably good retinal image of an underwater object is formed because the aperture or pupil of the eye admits only a small bundle of rays while forming the image. Since these rays differ very little in direction, they will appear to originate from approximately the same image point. However, the depth of this image will not be 3/4 the object depth, as for paraxial rays, and in general will vary with the angle of viewing.

Rays from the object that make increasingly larger angles of incidence with the interface must, by Snell's law, refract at increasingly larger angles, as shown in Figure 2-9c. A critical angle of incidence θ_c is reached when the angle of refraction reaches 90°. Thus, from Snell's law,

$$\sin \theta_c = \left(\frac{n_2}{n_1}\right)\sin 90 = \frac{n_2}{n_1}$$

or

$$\theta_c = \sin^{-1}\left(\frac{n_2}{n_1}\right) \tag{2-5}$$

For angles of incidence $\theta_1 > \theta_c$, the incident ray experiences *total internal reflection*, as shown. For angle of incidence $\theta_1 < \theta_c$ both refraction and reflection occur. The reflected rays for this case are not shown in Figure 2.9c. This phenomenon is essential in the transmission of light along glass fibers by a series of total internal reflections, as discussed in Chapter 10. Note that the phenomenon does not occur unless $n_1 > n_2$, so that θ_c can be determined from Eq. (2-5).

We return to the nature of images formed by refraction at a plane surface when we deal with such refraction as a special case of refraction from a spherical surface.

2-6 IMAGING BY AN OPTICAL SYSTEM

We discuss now what is meant by an image in general and indicate the practical and theoretical factors that render an image less than perfect. In Figure 2-10, let the region labeled "optical system" include any number of reflecting and/or refracting surfaces, of any curvature, that may alter the direction of rays leaving an *object point O*. This region may include any number of intervening media, but we shall assume that each individual medium is homogeneous and isotropic, and so characterized by its own refractive index. Thus rays spread out radially in all directions from object point O, as shown, in real *object space*, which precedes the first reflecting or refracting surface of the optical system. The family of spherical surfaces normal to the rays are the *wavefronts*, the locus of

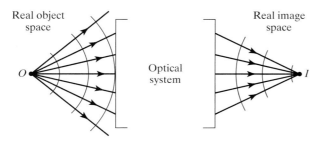

Figure 2-10 Image formation by an optical system.

points such that each ray contacting a wavefront represents the same transit time of light from the source. In real object space the rays are diverging and the spherical wavefronts are expanding. Suppose now that the optical system redirects these rays in such a way that on leaving the optical system and entering real *image space*, the wavefronts are contracting and the rays are converging to a common point that we define to be the *image point, I*. In the spirit of Fermat's principle, we can say that since every such ray starts at O and ends at I, every such ray requires the same transit time. These rays are said to be *isochronous*. Further, by the *principle of reversibility*, if I is the object point, each ray will reverse its direction but maintain its path through the optical system, and O will be the corresponding image point. The points O and I are said to be *conjugate* points for the optical system. In an ideal optical system, every ray from O intercepted by the system—and only these rays—also passes through I. To image an actual object, this requirement must hold for every object point and its conjugate image point.

Nonideal images are formed in practice because of (1) light scattering, (2) aberrations, and (3) diffraction. Some rays leaving O do not reach I due to reflection losses at refracting surfaces, diffuse reflections from reflecting surfaces, and scattering by inhomogeneities in transparent media. Loss of rays by such means merely diminishes the brightness of the image; however, some of these rays are scattered through I from nonconjugate object points, degrading the image. When the optical system itself cannot produce the one-to-one relationship between object and image rays required for perfect imaging of all object points, we speak of *system aberrations*. Such aberrations are treated in some detail in Chapter 20. Finally, since every optical system intercepts only a portion of the wavefront emerging from the object, the image cannot be perfectly sharp. Even if the image were otherwise perfect, the effect of using a limited portion of the wavefront leads to diffraction and a blurred image, which is said to be *diffraction limited*. This source of imperfect image formation, discussed further in the sections under diffraction, represents a fundamental limit to the sharpness of an image that cannot be entirely overcome. This difficulty rises from the wave nature of light. Only in the unattainable limit of geometrical optics, where $\lambda \rightarrow 0$, would diffraction effects disappear entirely.

Reflecting or refracting surfaces that form perfect images are called *Cartesian surfaces*. In the case of reflection, such surfaces are the conic sections, as shown in Figure 2-11. In each of these figures, the roles of object and image points may be reversed by the principle of reversibility. Notice that in Figure 2-11b, the image is virtual. In Figure 2-11c, the parallel reflected rays are said to form an image "at infinity." In each case, one can show that Fermat's principle, requiring isochronous rays between object and image points, leads to a condition that is equivalent to the geometric definition of the corresponding conic section.

Cartesian surfaces that produce perfect imaging by refraction may be more complicated. Let us ask for the equation of the appropriate refracting surface that images object point O at image point I, as illustrated in Figure 2-12. There an arbitrary point P with coordinates (x, y) is on the required surface Σ. The requirement is that every ray from O, like OPI, refracts and passes through the image I. Another such ray is evidently OVI, normal to the surface at its vertex point V. By Fermat's principle, these are isochronous rays. Since the media on either side of the refracting surface are characterized by different refractive indices, however, the isochronous rays are not equal in length. The transit time of a ray through a medium of thickness x with refractive index n is

$$t = \frac{x}{v} = \frac{nx}{c}$$

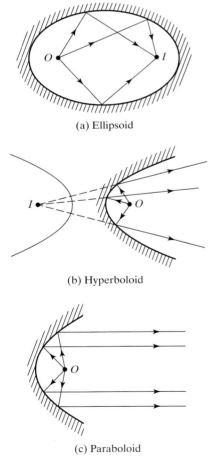

(a) Ellipsoid

(b) Hyperboloid

(c) Paraboloid

Figure 2-11 Cartesian reflecting surfaces showing conjugate object and image points.

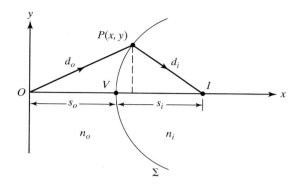

Figure 2-12 Cartesian refracting surface which images object point O at image point I.

Therefore, equal times imply equal values of the product nx, called the *optical path length*. In the problem at hand, then, Fermat's principle requires that

$$n_o d_o + n_i d_i = n_o s_o + n_i s_i = \text{constant} \qquad (2\text{-}6)$$

where the distances are defined in Figure 2-12. In terms of the (x, y)-coordinates of P, the first sum of Eq. (2-6) becomes

$$n_0(x^2 + y^2)^{1/2} + n_i[y^2 + (s_o + s_i - x)^2]^{1/2} = \text{constant} \qquad (2\text{-}7)$$

The constant in the equation is determined by the middle member of Eq. (2-6), $n_o s_o + n_i s_i$, which can be calculated once the specific problem is defined. Equation (2-7) describes the *Cartesian ovoid* of revolution shown in Figure 2-13a.

In most cases, however, the image is desired in the same optical medium as the object. This goal is achieved by a lens that refracts light rays twice, once at each surface, producing a real image outside the lens. Thus it is of particular interest to determine the Cartesian surfaces that render every object ray parallel after the first refraction. Such rays incident on the second surface can then be refracted again to form an image. The solutions to this problem are illustrated in Figure 2-13b and c. Depending on the relative magnitudes of the refractive indices, the appropriate refracting surface is either a hyperboloid ($n_i > n_o$) or an ellipsoid ($n_o > n_i$), as shown.

The first of these corresponds to the usual case of an object in air. A double hyperbolic lens then functions as shown in Figure 2-14. Note, however, that the aberration-free imaging so achieved applies only to object point O at the correct distance from the lens and on axis. For nearby points, imaging is not perfect. The larger the actual object, the less precise is its image. Because images of actual objects are not free from aberrations and because hyperboloid surfaces are *difficult to grind* exactly, most optical surfaces are spherical.[3] The *spherical aberrations* so introduced are accepted as a compromise when weighed against the relative ease of fabricating spherical surfaces. In the remainder of this chapter, we examine, in detail, spherical reflecting and refracting surfaces and, more briefly, cylindrical reflecting and refracting surfaces. Note that a plane surface can be treated as a special case of a cylindrical or a spherical surface in the limit that the *radius of curvature R* of either type of surface tends to infinity.

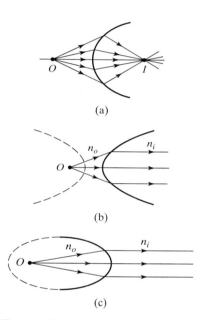

Figure 2-13 Cartesian refracting surfaces. (a) Cartesian ovoid images O at I by refraction. (b) Hyperbolic surface images object point O at infinity when O is at one focus and $n_i > n_o$. (c) Ellipsoid surface images object point O at infinity when O is at one focus and $n_o > n_i$.

2-7 REFLECTION AT A SPHERICAL SURFACE

Spherical mirrors may be either concave or convex relative to an object point O, depending on whether the center of curvature C is on the same or opposite side of the reflecting surface. In Figure 2-15 the mirror shown is convex, and two

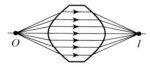

Figure 2-14 Aberration-free imaging of point object O by a double hyperbolic lens.

---

[3]The refinement of lens construction using injection molding technology has eased the production of lenses with aspherical surfaces.

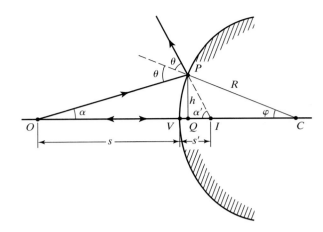

Figure 2-15 Reflection at a spherical surface.

rays of light originating at O are drawn, one normal to the spherical surface at its vertex V and the other an arbitrary ray incident at P. The first ray reflects back along itself; the second reflects at P as if from a plane tangent at P, satisfying the law of reflection. The two reflected rays diverge as they leave the mirror. The intersection of the two rays (extended backward) determines the image point I conjugate to O. The image is *virtual*, located behind the mirror surface.

Object and image distances from the vertex are shown as s and s', respectively. A perpendicular of height h is drawn from P to the axis at Q. We seek a relationship between s and s' that depends only on the radius of curvature R of the mirror. As we shall see, such a relation is possible only to first-order approximation of the sines and cosines of the angles made by the object and image rays to the spherical surface. This means that in place of the expansions

$$\sin \varphi = \varphi - \frac{\varphi^3}{3!} + \frac{\varphi^5}{5!} - \cdots$$

and

$$\cos \varphi = 1 - \frac{\varphi^2}{2!} + \frac{\varphi^4}{4!} + \cdots \tag{2-8}$$

we consider the first terms only and write

$$\sin \varphi \cong \varphi \quad \text{and} \quad \cos \varphi \cong 1 \tag{2-9}$$

relations that can be accurate enough if the angle φ is small enough.[4] This approximation leads to *first-order*, or *Gaussian*, optics, after Karl Friedrich Gauss, who in 1841 developed the foundations of the subject. Returning now to the problem at hand, notice that two angular relationships may be obtained from Figure 2-15, because the exterior angle of a triangle equals the sum of its interior angles. These are

$$\theta = \alpha + \varphi \quad \text{and} \quad 2\theta = \alpha + \alpha'$$

which combine to give

$$\alpha - \alpha' = -2\varphi \tag{2-10}$$

Using the small-angle approximation, the angles of Eq. (2-10) can be replaced by their tangents, yielding

$$\frac{h}{s} - \frac{h}{s'} = -2\frac{h}{R}$$

[4]For example, for angles φ around $10°$, the approximation leads to errors around 1.5%.

where we have also neglected the axial distance VQ, small when angle φ is small. Cancellation of h produces the desired relationship,

$$\frac{1}{s} - \frac{1}{s'} = -\frac{2}{R} \tag{2-11}$$

If the spherical surface is chosen to be concave instead, the center of curvature would be to the left. For certain positions of the object point O, it is then possible to find a real image point also to the left of the mirror. In these cases, the resulting geometric relationship analogous to Eq. (2-11) consists of terms that are all positive. It is possible, by employing an appropriate sign convention, to represent all cases by the single equation

$$\frac{1}{s} + \frac{1}{s'} = -\frac{2}{R} \tag{2-12}$$

The sign convention to be used in conjunction with Eq. (2-12) is as follows. Assume the light propagates from left to right:

1. The *object distance s* is positive when O is to the left of V, corresponding to a real object. When O is to the right, corresponding to a virtual object, s is negative.
2. The *image distance s'* is positive when I is to the left of V, corresponding to a real image, and negative when I is to the right of V, corresponding to a virtual image.
3. The *radius of curvature R* is positive when C is to the right of V, corresponding to a convex mirror, and negative when C is to the left of V, corresponding to a concave mirror.

These rules[5] can be quickly summarized by noticing that positive object and image distances correspond to real objects and real images and that convex mirrors have positive radii of curvature. Applying Rule 2 to Figure 2-15, we see that the general Eq. (2-12) becomes identical with Eq. (2-11), a special case derived in conjunction with Figure 2-15. Virtual objects occur only with a sequence of two or more reflecting or refracting elements and are considered later.

The spherical mirror described by Eq. (2-12) yields, for a plane mirror with $R \to \infty$, $s' = -s$, as determined previously. The negative sign implies a virtual image for a real object. Notice also in Eq. (2-12) that object distance and image distance appear symmetrically, implying their interchangeability as conjugate points. For an object at infinity, incident rays are parallel and $s' = -R/2$, as illustrated in Figure 2-16a and b for both concave ($R < 0$) and convex ($R > 0$) mirrors. The image distance in each case is defined as the *focal length f* of the mirrors. Thus,

$$f = -\frac{R}{2} \begin{cases} >0, & \text{concave mirror} \\ <0, & \text{convex mirror} \end{cases} \tag{2-13}$$

and the mirror equation can be written, more compactly, as

$$\frac{1}{s} + \frac{1}{s'} = \frac{1}{f} \tag{2-14}$$

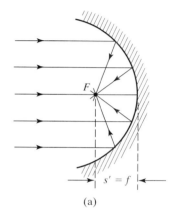

(a)

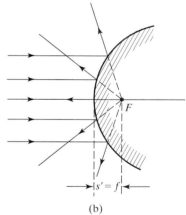

(b)

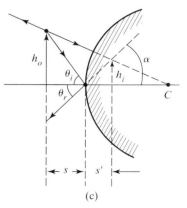

(c)

Figure 2-16 Location of focal points (a) and (b) and construction to determine magnification (c) of a spherical mirror.

---

[5]Although this set of sign conventions is widely used, the student is cautioned that other schemes exist. No one with a continuing involvement in optics can hope to escape confronting other conventions, nor should the matter be beyond the mental flexibility of the serious student to accommodate.

The focal point F, located a focal length f from the vertex of the mirror, and shown in Figure 2-16a and b, serves as an important construction point in graphical ray-tracing techniques, which we discuss following Example 2-1.

In Figure 2-16c, a construction is shown that allows the determination of the transverse magnification. The object is an extended object of transverse dimension h_o. The image of the top of the object arrow is located by two rays whose behavior on reflection is known. The ray incident at the vertex must reflect to make equal angles with the axis. The other ray is directed toward the center of curvature along a normal and so must reflect back along itself. The intersection of the two reflected rays occurs behind the mirror and locates a virtual image of dimension h_i there. Because of the equality of the three angles shown, it follows that

$$\frac{h_o}{s} = \frac{h_i}{s'}$$

The lateral magnification m is defined by the ratio of lateral image size to corresponding lateral object size, so that

$$|m| = \frac{h_i}{h_o} = \frac{s'}{s} \qquad (2\text{-}15)$$

Extending the sign convention to include magnification, we assign a $(+)$ magnification to the case where the image has the same orientation as the object and a $(-)$ magnification where the image is inverted relative to the object. To produce a $(+)$ magnification in the construction of Figure 2-16c, where s' must itself be negative, we modify Eq. (2-15) to give the general form

$$m = -\frac{s'}{s} \qquad (2\text{-}16)$$

The following example illustrates the correct use of the sign convention.

Example 2-1

An object 3 cm high is placed 20 cm from (a) a convex and (b) a concave spherical mirror, each of 10-cm focal length. Determine the position and nature of the image in each case.

Solution

a. Convex mirror: $f = -10$ cm and $s = +20$ cm.

$$\frac{1}{s} + \frac{1}{s'} = \frac{1}{f} \quad \text{or} \quad s' = \frac{fs}{s - f} = \frac{(-10)(20)}{(20) - (-10)} = -6.67 \text{ cm}$$

$$m = -\frac{s'}{s} = -\frac{-6.67}{20} = +0.333 = \frac{1}{3}$$

The image is virtual (because s' is negative), 6.67 cm to the right of the mirror vertex, and is erect (because m is positive) and $\frac{1}{3}$ the size of the object, or 1 cm high.

b. Concave mirror: $f = +10$ cm and $s = +20$ cm.

$$s' = \frac{fs}{s - f} = \frac{(10)(20)}{20 - 10} = +20 \text{ cm}$$

$$m = -\frac{s'}{s} = -\frac{20}{20} = -1$$

The image is real (because s' is positive), 20 cm to the left of the mirror vertex, and is inverted (because m is negative) and the same size as the object, or 3 cm high. Image and object happen to be at $2f = 20$ cm, the center of curvature of the mirror.

The location and nature of the image formed by a mirror can be determined by graphical ray-trace techniques. Figure 2-17 illustrates how three key rays—labeled 1, 2, and 3—each leaving a point P at the tip of an object, can be drawn to locate the conjugate image point P'. In fact, under the conditions for which Eqs. (2-12) through (2-16) are valid, the paths of *any* two rays leaving P are sufficient to locate the conjugate image point P'. A third ray serves as a convenient check on the accuracy of the first two chosen rays. The three key rays discussed in connection with Figure 2-17 are chosen as the basis of the graphical ray-trace technique because, once the mirror center of curvature C, the focal point F, and vertex V are located along the optical axis of a spherical mirror, these three rays can be drawn using only a straightedge device. The conjugate image point P' marks the tip of the image—the entire image then lies between P' and the point on the optical axis directly above or below P'.

Refer to Figure 2-17a, b, and c in connection with the following description of how the three key rays can be drawn. Note the difference in each ray

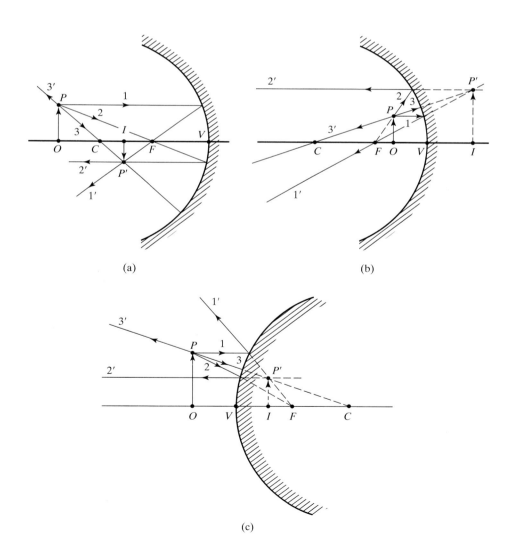

(a)

(b)

(c)

Figure 2-17 Ray diagrams for spherical mirrors. (a) Real image, concave mirror. The object distance is greater than the focal length. (b) Virtual image, concave mirror. The object distance is less than the focal length. (c) Virtual image, convex mirror.

trace, depending on the object location before or after points C and F, and on the geometry of the mirror surface, concave or convex.

- *Ray 1.* This ray leaves point P as a ray parallel to the optical axis, strikes the mirror, reflects and passes through the focal point F of a *concave* mirror—as in Figure 2-17a and b. Or, as in Figure 2-17c, it strikes a *convex* mirror and reflects as if it came from the focal point F behind the mirror. In each case, after reflection this ray is labeled $1'$.
- *Ray 2.* This ray leaves point P, passes through F, strikes a *concave* mirror, and is reflected as a ray parallel to the optical axis, as in Figure 2-17a. Or, as in Figure 2-17b, it leaves point P as if it is coming from the point F to its left (dotted line), strikes the *concave* mirror, and reflects as a parallel ray. Or, as in Figure 2-17c, for a *convex* mirror, the ray leaves point P heading toward focal point F behind the mirror, strikes the mirror, and reflects as a parallel ray. In each case, after reflection, this ray is labeled $2'$.
- *Ray 3.* This ray leaves point P in Figure 2-17a, passes through point C for the *concave* mirror, strikes the mirror, and reflects back along itself. Or, as in Figure 2-17b—still for a *concave* mirror—ray 3 appears to come from the point C to its left, strikes the mirror, and reflects back along itself. Or, as in Figure 2-17c, for a *convex* mirror, it heads toward point C behind the mirror, strikes the mirror, and reflects back along itself. In each case, after reflection, this ray is labeled $3'$.

To understand how these rays locate the conjugate image point P' that marks the tip of the image, it is useful to imagine that these three rays arrive at the eye of one viewing the image. For the case shown in Figure 2-17a, the three rays $1'$, $2'$, and $3'$ intersect at a real image point as they progress away from the mirror and toward the viewer. For the arrangements shown in Figure 2-17b and 2-17c, the rays $1'$, $2'$, and $3'$ appear to originate from a point of intersection (a virtual image point) located behind the mirror. The real or apparent point of intersection is interpreted as the emanation point of these rays. That is, the viewer "sees" the tip of an image at point P'.

2-8 REFRACTION AT A SPHERICAL SURFACE

We turn now to a similar treatment of refraction at a spherical surface, choosing in this case the concave surface of Figure 2-18. Two rays are shown emanating from object point O. One is an axial ray, normal to the surface at its vertex and so refracted without change in direction. The other ray is an arbitrary ray incident at P and refracting there according to Snell's law,

$$n_1 \sin \theta_1 = n_2 \sin \theta_2 \qquad (2\text{-}17)$$

The two refracted rays appear to emerge from their common intersection, the image point I. In triangle CPO, the exterior angle $\alpha = \theta_1 + \varphi$. In triangle CPI, the exterior angle $\alpha' = \theta_2 + \varphi$. Approximating for paraxial rays and substituting for θ_1 and θ_2 in Eq. (2-17), we have

$$n_1(\alpha - \varphi) = n_2(\alpha' - \varphi) \qquad (2\text{-}18)$$

Next, writing the tangents for the angles by inspection of Figure 2-18, where again we may neglect the distance QV in the small angle approximation,

$$n_1\left(\frac{h}{s} - \frac{h}{R}\right) = n_2\left(\frac{h}{s'} - \frac{h}{R}\right)$$

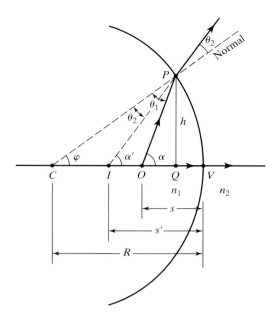

Figure 2-18 Refraction at a spherical surface for which $n_2 > n_1$.

or

$$\frac{n_1}{s} - \frac{n_2}{s'} = \frac{n_1 - n_2}{R} \qquad (2\text{-}19)$$

Employing the *same sign convention* as introduced for mirrors (i.e., positive distances for real objects and images and negative distances for virtual objects and images), the virtual image distance $s' < 0$ and the radius of curvature $R < 0$. If these negative signs are understood to apply to these quantities for the case of Figure 2-18, a general form of the refraction equation may be written as

$$\frac{n_1}{s} + \frac{n_2}{s'} = \frac{n_2 - n_1}{R} \qquad (2\text{-}20)$$

which holds equally well for convex surfaces. When $R \to \infty$, the spherical surface becomes a plane refracting surface, and

$$s' = -\left(\frac{n_2}{n_1}\right)s \qquad (2\text{-}21)$$

where s' is the apparent depth determined previously. For a real object ($s > 0$), the negative sign in Eq. (2-21) indicates that the image is virtual. The lateral magnification of an extended object is simply determined by inspection of Figure 2-19. Snell's law requires, for the ray incident at the vertex V and in the small-angle approximation, $n_1\theta_1 = n_2\theta_2$ or, using tangents for angles,

$$n_1\left(\frac{h_o}{s}\right) = n_2\left(\frac{h_i}{s'}\right)$$

The lateral magnification is, then,

$$m = \frac{h_i}{h_o} = -\frac{n_1 s'}{n_2 s} \qquad (2\text{-}22)$$

where the negative sign is attached to give a negative value corresponding to an inverted image. For the case of a plane refracting surface, Eq. (2-21) may

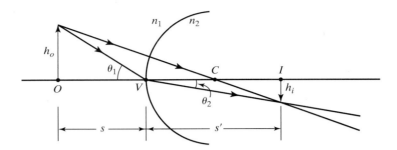

Figure 2-19 Construction to determine lateral magnification at a spherical refracting surface.

be incorporated into Eq. (2-22), giving $m = +1$. Thus, the images formed by plane refracting surfaces have the same lateral dimensions and orientation as the object.

Example 2-2

As an extended example of refraction by spherical surfaces, refer to Figure 2-20. In (a), a real object is positioned in air, 30 cm from a convex spherical surface of radius 5 cm. To the right of the interface, the refractive index is that of water. Before constructing representative rays, we first find the image distance and lateral magnification of the image, using Eqs. (2-20) and (2-22). Equation (2-20) becomes

$$\frac{1}{30} + \frac{1.33}{s'_1} = \frac{1.33 - 1}{5}$$

giving $s'_1 = +40$ cm. The positive sign indicates that the image is real and so is located to the right of the surface, where real rays of light are refracted. Equation (2-22) becomes

$$m = -\frac{(1)(+40)}{(1.33)(+30)} = -1$$

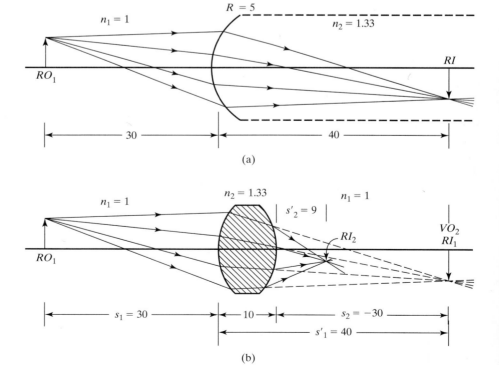

Figure 2-20 Example of refraction by spherical surfaces. (All distances are in cm.) (a) Refraction by a single spherical surface. (b) Refraction by a thick lens. Subscripts 1 and 2 refer to refractions at the first and second surfaces, respectively.

indicating an inverted image, equal in size to that of the object. Figure 2-20a shows the image, as well as several rays, which are now determined. In this example we have assumed that the medium to the right of the spherical surface extends far enough so that the image is formed inside it, without further refraction. Let us suppose now (Figure 2-20b) that the second medium is only 10 cm thick, forming a *thick lens*, with a second, concave spherical surface, also of radius 5 cm. The refraction by the first surface is, of course, unaffected by this change. Inside the lens, therefore, rays are directed as before to form an image 40 cm to the right of the first surface. However, these rays are intercepted and refracted by the second surface to produce a different image, as shown. Since the convergence of the rays striking the second surface is determined by the position of the first image, its location now specifies the appropriate object distance to be used for the second refraction. We call the real image formed by surface (1) a *virtual object* for surface (2). Then, by the *sign convention established previously*, we make the virtual object distance, relative to the second surface, a negative quantity when using Eqs. (2-20) and (2-22). For the second refraction, then, Eq. (2-20) becomes

$$\frac{1.33}{-30} + \frac{1}{s'_2} = \frac{1 - 1.33}{-5}$$

or $s' = +9$ cm. The magnification, according to Eq. (2-22), is

$$m = \frac{(-1.33)(+9)}{(1)(-30)} = +\frac{2}{5}$$

The final image is, then, 2/5 the lateral size of its (virtual) object and apears with the same orientation. Relative to the original object, the final image is 2/5 as large and inverted.

In general, whenever a train of reflecting or refracting surfaces is involved in the processing of a final image, the individual reflections and/or refractions are considered in the order in which light is actually incident upon them. The object distance of the nth step is determined by the image distance for the $(n-1)$th step. If the image of the $(n-1)$ step is not actually formed, it serves as a *virtual object* for the nth step.

2-9 THIN LENSES

We now apply the preceding method to discover the thin-lens equation. As in the example of Figure 2-20, two refractions at spherical surfaces are involved. The simplification we make is to neglect the thickness of the lens in comparison with the object and image distances, an approximation that is justified in most practical situations. At the first refracting surface, of radius R_1,

$$\frac{n_1}{s_1} + \frac{n_2}{s'_1} = \frac{n_2 - n_1}{R_1} \tag{2-23}$$

and at the second surface, of radius R_2,

$$\frac{n_2}{s_2} + \frac{n_1}{s'_2} = \frac{n_1 - n_2}{R_2} \tag{2-24}$$

We have assumed that the lens faces the same medium of refractive index n_1 on both sides. Now the second object distance, in general, is given by

$$s_2 = t - s'_1 \tag{2-25}$$

where t is the thickness of the lens. Notice that this relationship produces the correct sign of s_2, as in Figure 2-20, and also when the intermediate image falls inside or to the left of the lens. In the thin-lens approximation, neglecting t,

$$s_2 = -s_1'$$ (2-26)

When this value of s_2 is substituted into Eq. (2-24) and Eqs. (2-23) and (2-24) are added, the terms n_2/s_1' cancel and there results

$$\frac{n_1}{s_1} + \frac{n_1}{s_2'} = (n_2 - n_1)\left(\frac{1}{R_1} - \frac{1}{R_2}\right)$$

Now s_1 is the original object distance and s_2' is the final image distance, so we may drop their subscripts and write simply

$$\frac{1}{s} + \frac{1}{s'} = \frac{n_2 - n_1}{n_1}\left(\frac{1}{R_1} - \frac{1}{R_2}\right)$$ (2-27)

The *focal length* of the thin lens is defined as the image distance for an object at infinity, or the object distance for an image at infinity, giving

$$\frac{1}{f} = \frac{n_2 - n_1}{n_1}\left(\frac{1}{R_1} - \frac{1}{R_2}\right)$$ (2-28)

Equation (2-28) is called the *lensmaker's equation* because it predicts the focal length of a lens fabricated with a given refractive index and radii of curvature and used in a medium of refractive index n_1. In most cases, the ambient medium is air, and $n_1 = 1$. The thin-lens equation, in terms of the focal length, is then

$$\frac{1}{s} + \frac{1}{s'} = \frac{1}{f}$$ (2-29)

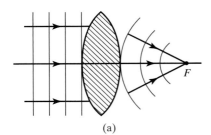

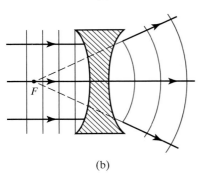

Figure 2-21 Lens action on plane wavefronts of light. (a) Converging lens (positive focal length). (b) Diverging lens (negative focal length).

Wavefront analysis for plane wavefronts, as shown in Figure 2-21, indicates that a lens thicker in the middle causes convergence, and one thinner in the middle causes divergence of the incident parallel rays. The portion of the wavefront that must pass through the thicker region is delayed relative to the other portions. Converging lenses are characterized by positive focal lengths and diverging lenses by negative focal lengths, as is evident from the figure, where the images are real and virtual, respectively.

Sample ray diagrams for converging (or *convex*) and diverging (or *concave*) lenses are shown in Figure 2-22. The thin lenses are best represented, for purposes of ray construction, by a vertical line with edges suggesting the general shape of the lens—ordinary arrowheads for converging lenses, inverted arrowheads for diverging lenses. Graphical methods of locating images, as with spherical mirrors in Figure 2-17, make use of *three key rays*. This procedure is outlined next and illustrated in Figures 2-22 and 2-23. The three rays leaving the tip of the object change direction due to refraction at the thin-lens interfaces. The redirected rays can be used to locate the image.

- *Ray 1.* A ray leaving the tip of the object, parallel to the optical axis, undergoing refraction at the lens surfaces and passing through the *right* focal point F of a *converging* lens, as in Figure 2-22a. Or, as in Figure 2-22b, a parallel ray which refracts at the lens surfaces as if coming directly from the *left* focal point F of a *diverging* lens.
- *Ray 2.* A ray leaving the tip of the object and passing through the *left* focal point F of a *converging* lens, undergoing refraction at the lens surfaces,

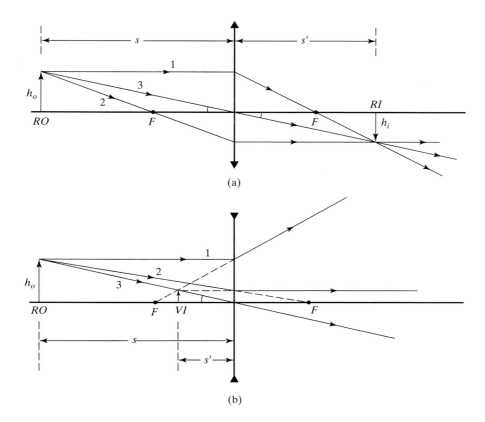

Figure 2-22 Ray diagrams for image formation by a convex lens (a) and a concave lens (b).

and emerging parallel to the axis as in Figure 2-22a. Or, as in Figure 2-22b, a ray leaving the tip of the object, directed toward the *right* focal point F of a *diverging* lens, undergoing refraction at the lens and emerging parallel to the axis.

- *Ray 3.* A ray leaving the tip of the object and passing directly through the center of a converging or diverging lens, emerging unaltered, as in Figure 2-22a or 2-22b.

The viewer, located at the far right in Figure 2-22a and 2-22b, receives these rays as if they have come directly from an object and so "sees" the tip of the image at the point where the backwards extensions of these rays either intersect or appear to intersect. Any two rays are sufficient to locate the image; the third ray may be drawn as a check on the accuracy of the graphical trace.

In constructing ray diagrams, as in Figure 2-22, observe that, except for the central ray (ray 3), each ray refracted by a convex lens bends toward the axis and each ray refracted by a concave lens bends away from the axis. From either diagram, the angles subtended by object and image at the center of the lens are seen to be equal. For either the real image *RI* in (a) or the virtual image *VI* in (b), it follows that

$$\frac{h_o}{s} = \frac{h_i}{s'}$$

and lateral magnification

$$|m| = \left| \frac{h_i}{h_o} \right| = \left| \frac{s'}{s} \right|$$

In accordance with the sign convention adopted here, the magnification should be the negative of the ratio of the image and object distances since, in case (a), $s > 0$, $s' > 0$, and $m < 0$ because the image is inverted; in case (b),

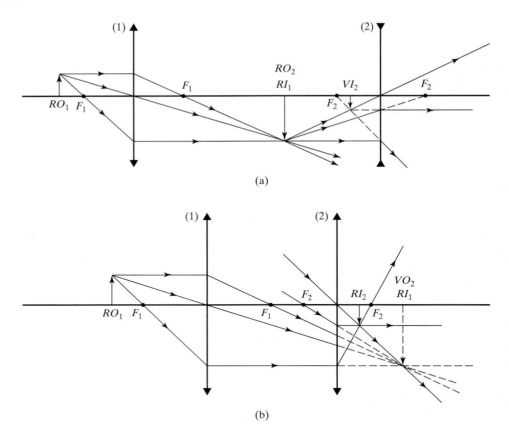

Figure 2-23 (a) Formation of a virtual image VI_2 by a two-element train of a convex lens (1) and concave lens (2). (b) Formation of a real image RI_2 by a train of two convex lenses. The intermediate image RI_1 serves as a virtual object VO_2 for the second lens.

$s > 0$, $s' < 0$, and $m > 0$. In either case, then,

$$m = -\frac{s'}{s} \qquad (2\text{-}30)$$

Further ray-diagram examples for *a train of two lenses* are illustrated in Figure 2-23 and a calculation involving image formation in two lenses is given in Example 2-3.

Example 2-3

Find and describe the intermediate and final images produced by a two-lens system such as the one sketched in Figure 2-23a. Let $f_1 = 15$ cm, $f_2 = 15$ cm, and their separation be 60 cm. Let the object be 25 cm from the first lens, as shown.

Solution

The first lens is convex: $f_1 = +15$ cm, $s_1 = 25$ cm.

$$\frac{1}{s_1} + \frac{1}{s_1'} = \frac{1}{f} \quad \text{or} \quad s_1' = \frac{s_1 f}{s_1 - f} = \frac{(25)(15)}{25 - 15} = +37.5 \text{ cm}$$

$$m_1 = -\frac{s_1'}{s_1} = -\frac{37.5}{25} = -1.5$$

Thus, the first image is real (because s_1' is positive), 37.5 cm to the right of the first lens, inverted (because m is negative), and 1.5 times the size of the object.

The second lens is concave: $f_2 = -15$ cm. Since real rays of light diverge from the first real image, it serves as a real object for the second lens, with $s_2 = 60 - 37.5 = +22.5$ cm to the left of the lens. Then,

$$s_2' = \frac{s_2 f}{s_2 - f} = \frac{(22.5)(-15)}{(22.5) - (-15)} = -9 \text{ cm}$$

$$m_2 = -\frac{s_2'}{s_2} = -\frac{-9}{22.5} = +0.4$$

Thus, the final image is virtual (because s_2' is negative), 9 cm to the *left* of the seconds lens, erect *with respect to its own object* (because m is positive), and 0.4 times its size. The *overall* magnification is given by $m = m_1 m_2 = (-1.5)(0.4) = -0.6$. Thus, the final image is inverted relative to the *original* object and 6/10 its lateral size. All these features are exhibited qualitatively in the ray diagram of Figure 2-23a.

Table 2-1 and Figure 2-24 provide a convenient summary of image formation in lenses and mirrors.

2-10 VERGENCE AND REFRACTIVE POWER

Another way of interpreting the thin-lens equation is useful in certain applications, including optometry. The interpretation is based on two considerations. In the thin-lens equation,

$$\frac{1}{s} + \frac{1}{s'} = \frac{1}{f} \tag{2-31}$$

TABLE 2-1   SUMMARY OF GAUSSIAN MIRROR AND LENS FORMULAS

	Spherical surface	Plane surface
Reflection	$\dfrac{1}{s} + \dfrac{1}{s'} = \dfrac{1}{f}, f = -\dfrac{R}{2}$ $m = -\dfrac{s^i}{s}$ Concave: $f > 0, R < 0$ Convex : $f < 0, R > 0$	$s' = -s$ $m = +1$
Refraction Single surface	$\dfrac{n_1}{s} + \dfrac{n_2}{s'} = \dfrac{n_2 - n_1}{R}$ $m = -\dfrac{n_1 s'}{n_2 s}$ Concave: $R < 0$ Convex : $R > 0$	$s' = -\dfrac{n_2}{n_1} s$ $m = +1$
Refraction Thin lens	$\dfrac{1}{s} + \dfrac{1}{s'} = \dfrac{1}{f}$ $\dfrac{1}{f} = \dfrac{n_2 - n_1}{n_1} \left(\dfrac{1}{R_1} - \dfrac{1}{R_2} \right)$ $m = -\dfrac{s'}{s}$ Concave: $f < 0$ Convex : $f > 0$	

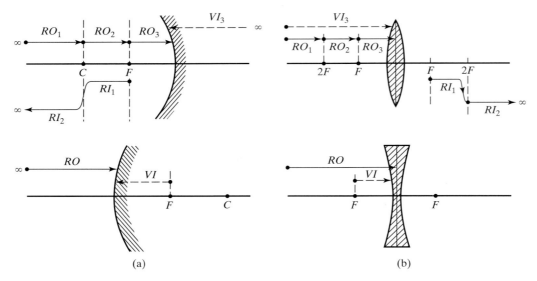

Figure 2-24   Summary of image formation by (a) spherical mirrors and (b) thin lenses. The location, nature, magnification, and orientation of the image are indicated or suggested. The letters R and V refer to *real* and *virtual, O* and *I* to *object* and *image*. Changes in elevation of the horizontal lines suggest the magnification in the various regions.

notice that (1) the reciprocals of distances in the left member add to give the reciprocal of the focal length and (2) the reciprocals of the object and image distances describe the curvature of the wavefronts incident at the lens and centered at the object and image positions O and I, respectively. A plane wavefront, for example, has a curvature of zero. In Figure 2-25 spherical waves expand from the object point O and attain a curvature, or *vergence, V,* given by $1/s$, when they intercept the thin lens. On the other hand, once refracted by the lens, the wavefronts contract, in Figure 2-25a, and expand further, in Figure 2-25b, to locate the real and virtual image points shown. The curvature, or vergence, V', of the wavefronts as they emerge from the lens is $1/s'$. The change in curvature from object space to image space is due to the *refracting power P* of the lens, given by $1/f$. With these definitions, Eq. (2-31) may be written

$$V + V' = P \tag{2-32}$$

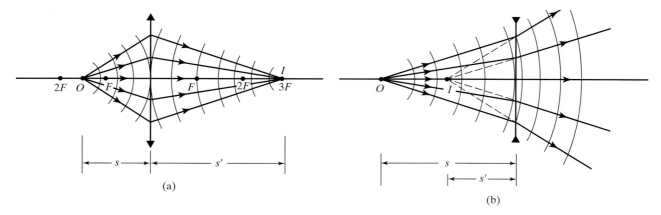

Figure 2-25   Change in curvature of wavefronts on refraction by a thin lens. (a) Convex lens. (b) Concave lens.

The units of the terms in Eq. (2-32) are reciprocal lengths. When the lengths are measured in *meters*, their *reciprocals* are said to have units of *diopters* (D). Thus, the refracting power of a lens of focal length 20 cm is said to be $\frac{1}{0.2 \text{ m}} = 5$ diopters. This alternative point of view emphasizes the degree of wave curvature or ray convergence rather than object and image distances. Accordingly, the degree of convergence V' of the image rays is determined by the original degree of convergence V of the object rays and the refracting power P of the lens, that is, the power to change incident wave curvature. Eq. (2-32) can also be applied to the case of refraction at a single surface, Eq. (2-20), in which case the refractive indices in object and image space need not be 1. In this event, the *power of the refracting surface* is $(n_2 - n_1)/R$.

This approach is useful for another reason. When thin lenses are placed together, *in contact*, the focal length f of the combination, treated as a single thin lens, can be found in terms of the focal lengths $f_1, f_2, \ldots$ of the individual lenses. For example, with two such lenses back-to-back, we write the lens equations

$$\frac{1}{s_1} + \frac{1}{s_1'} = \frac{1}{f_1} \quad \text{and} \quad \frac{1}{s_2} + \frac{1}{s_2'} = \frac{1}{f_2}$$

Since the image distance for the first lens plays the role of the object distance for the second lens, we may write

$$s_2 = -s_1'$$

and, adding the two equations,

$$\frac{1}{s_1} + \frac{1}{s_2'} = \frac{1}{f_1} + \frac{1}{f_2} = \frac{1}{f}$$

The reciprocals of the individual focal lengths, therefore, add to give the reciprocal of the overall focal length f of the pair. In general, for several thin lenses, *in direct contact*,

$$\frac{1}{f} = \frac{1}{f_1} + \frac{1}{f_2} + \frac{1}{f_3} + \cdots \qquad (2\text{-}33)$$

Expressed in diopters, the refractive powers simply add:

$$P = P_1 + P_2 + P_3 + \cdots \qquad (2\text{-}34)$$

In a nearsighted eye, the refracted (converging) power of the eye is too great, so that a real image is formed in front of the retina. By reducing the convergence with a number of diverging lenses placed in front of the eye, until an object is clearly focused, an optometrist can determine the net diopter specification of the single corrective lens needed by simply adding the diopters of these test lenses. In a farsighted eye, the natural converging power of the eye is not strong enough, and additional converging power must be added in the form of spectacles with a converging lens.

2-11 NEWTONIAN EQUATION
FOR THE THIN LENS

When object and image distances are measured relative to the focal points F of a lens, as by the distances x and x' in Figure 2-26, an alternative form of the thin-lens equation results, called the *Newtonian form*. In the figure, the two rays shown determine two right triangles, joined by the focal point, on each side of the lens. Since each pair constitutes similar triangles, we may set up proportions between sides that represent the lateral magnification:

$$\frac{h_i}{h_o} = \frac{f}{x} \quad \text{and} \quad \frac{h_i}{h_o} = \frac{x'}{f}$$

Introducing a negative sign for the magnification, due to the inverted image,

$$m = -\frac{f}{x} = -\frac{x'}{f} \tag{2-35}$$

The two parts of Eq. (2-35) also constitute the Newtonian form of the thin-lens equation,

$$xx' = f^2 \tag{2-36}$$

This equation is somewhat simpler than Eq. (2-29) and is found to be more convenient in certain applications.

2-12 CYLINDRICAL LENSES

Spherical lenses and mirrors with circular cross sections are far more common in optical systems than are *cylindrical lenses*. Nevertheless, cylindrical lenses are important, for example, in the field of optometry for correcting the visual defect known as astigmatism, as well as in novel visual displays where it is useful to image points as lines. We close this chapter on geometrical optics with a brief look at this special type of lens.

The *optical axis* for a spherical lens is an *axis of symmetry* since rotation of the lens through an arbitrary angle about the optical axis leaves the lens looking just as it did before the rotation. Because the orientation of the surface curvature does not change in such a rotation, its optical behavior remains unchanged. This rotational symmetry simplifies the analysis of the imaging properties of such a spherical lens. On the other hand, a cylindrical lens—shaped like a section of a soft drink can, sliced down the side from top to bottom—lacks rotational symmetry about its optical axis. As a consequence, a cylindrical lens has asymmetric focusing properties, as will be seen later in greater detail. Whereas a spherical lens produces a point image of a point

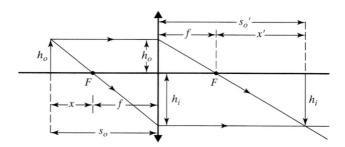

Figure 2-26 Construction used to derive Newton's equations for the thin lens.

object, a cylindrical lens produces a line image of a point object. Because of these properties, a spherical lens is said to be *stigmatic*, and the cylindrical lens *astigmatic*.

Consider first a spherical lens, as shown in Figure 2-27a and b. Orthogonal vertical and horizontal axes are shown as solid diametrical lines through the geometric lens center. Parallel rays of light passing through the vertical axis (see Figure 2-27a) and through the horizontal axis (see Figure 2-27b) are handled identically by the lens, converging them to a common focus at F.

The lens can be rotated through an arbitrary angle about its optical axis with the same result. Thus, the focusing properties of a spherical lens are invariant to rotation about its central (optical) axis.

Next, consider the *convex* and *concave* cylindrical lenses shown in Figure 2-28. One surface of the lens is cylindrical while the opposite is plane.[6] Thus, the curved surface has a definite, finite radius of curvature, whereas the plane surface has an infinite radius of curvature. In Figure 2-29, two vertical slices or sections are shown perpendicular to the axis of a convex cylindrical lens. Through each section, three representative rays are drawn. The operation of this lens is clearly asymmetric. Focusing occurs for rays along a vertical section but not for rays along a horizontal section, where the lens presents no curvature. Thus, rays 1, 2 and 3 focus to point A, and rays 4, 5 and 6 focus to point B. However, there is no focusing of rays in a horizontal section, such as the pairs of rays 1 and 4, 2 and 5, or 3 and 6. Other vertical sections would produce other points along the focused line image AB in the same way. Notice that the line image AB so formed is always parallel to the cylinder axis. This important feature is also shown in Figure 2-30, where the line image is real for a cylindrical convex lens and virtual for a cylindrical concave lens. From these figures, it is evident that the length of the line image is equal to the axial length of the lens, assuming that rays of light parallel to the optical axis enter along the entire extent of the lens. If an aperture is placed in front of the lens to limit the bundle of light rays through the lens, the height of the line image is just the aperture dimension along the cylinder axis, or the *effective height* of the lens.

In Figure 2-29, the line image formed is the result of an object point "at infinity," which produces parallel rays at the lens. In Figure 2-31, the object point O is near the lens, producing diverging rays of light incident on the lens. Still, if the lens is thin, focusing occurs along the vertical sections, as shown. Rays 1 and 3, in the left vertical section of Figure 2-31, focus at A; rays 2 and 4 in the right vertical section focus at B. However, no focusing occurs for rays 1, 5, and 2 along the horizontal section. Because of the divergence of the rays entering the lens, however, the length of the focused line image AB is no longer equal to the effective length CL of the lens. The divergence of the extreme rays at each end of the lens now determines an image that is *longer* than the length of the lens. The image length AB can be found from a simple, geometrical argument that is apparent in Figure 2-32a, a view of the central horizontal section in Figure 2-31 as seen from above. If the effective length of the cylindrical lens is CL, then by similar triangles it follows that

$$\frac{AB}{CL} = \frac{s + s'}{s}$$

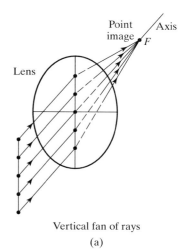

Vertical fan of rays

(a)

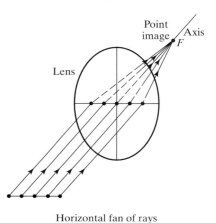

Horizontal fan of rays

(b)

Figure 2-27 Parallel rays of light focused by a spherical lens. Because of its axis of symmetry relative to rotation about an axis through its center, the lens treats (a) vertical and (b) horizontal fans of rays similarly, producing in each case a point image at the same location. Each ray refracts twice through the lens, once at each surface. For simplicity, only one refraction is shown.

---

[6]To be more precise, we are speaking of a *plano-convex* or *plano-concave* cylindrical lens as shown in Figure 2-28. Generally speaking, both surfaces of the lens might be cylindrical. In such a case, the behavior of the lens as a whole, due to the addition of the powers of the two surfaces, may not reduce to that of the simple *plano-convex* or *plano-concave* lens described here.

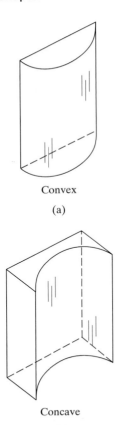

Convex

(a)

Concave

(b)

Figure 2-28 Cylindrical lenses shown as sections of a solid and hollow cylindrical rod.

or

$$AB = \left(\frac{s + s'}{s}\right)CL \qquad (2\text{-}37)$$

Subject to our sign convention, this relation is a general form for the plano-cylindrical lens that handles all cases, with s and s' object and image distances, respectively, and AB always positive.

Example 2-4

A thin plano-cylindrical lens in air has a radius of curvature of 10 cm, a refractive index of 1.50, and an axial length of 5 cm. Light from a point object is incident on the convex cylindrical surface from a distance of 25 cm to the left of the lens. Find the position and length of the line image formed by the lens.

Solution

As given, $s = 25$ cm, $R = 10$ cm, n(lens) $= 1.50$ and $CL = 5$ cm. Using the spherical surface refraction equation (see Table 2-1),

$$\frac{n_1}{s} + \frac{n_2}{s'} = \frac{n_2 - n_1}{R}$$

and

$$AB = \left(\frac{s + s'}{s}\right)CL$$

together with the sign convention—positive for real objects and images, negative for virtual objects and images, positive R for convex surface.

Entering values, we have for the first convex surface at entry, $\frac{1}{25} +$ $\frac{1.50}{s'} = \frac{1.50 - 1.00}{10}$, which gives $s' = 150$ cm, real. And for the second

Figure 2-29 Focusing property of a convex cylindrical lens. Rays through a vertical section, such as rays 1, 2, and 3, come to a common focus, but rays through a horizontal section, such as rays 1 and 4, do not. Parallel rays form a line image that is parallel to the cylinder axis. For simplicity, refraction is shown only at the front surface and spherical aberration for non-paraxial rays is ignored.

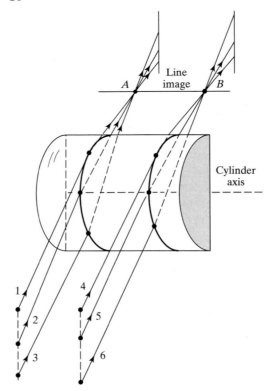

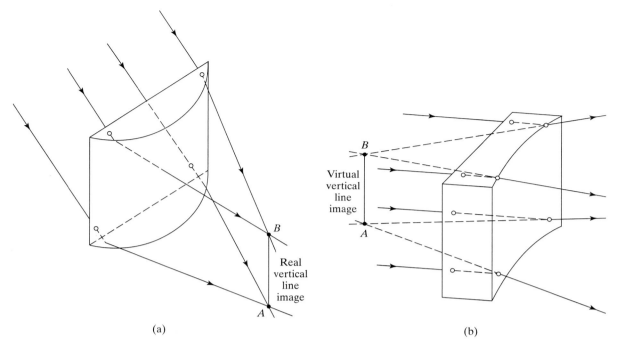

Figure 2-30   Formation of line images by cylindrical lenses for light incident from a distant object. In (a) the convex lens forms a real image. In (b), the concave lens forms a virtual image. In either case, the line image is parallel to the cylinder axis.

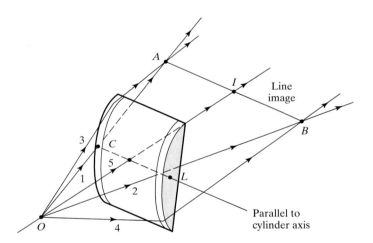

Figure 2-31   Formation of a line image AB by a convex cylindrical lens when the object is a point O at a finite distance from the lens. In this case, the line image AB is longer than the axial length of the lens, CL.

(plane) surface at exit, we obtain $\dfrac{1.50}{-150} + \dfrac{1.0}{s'} = 0$, which gives $s' = 100$ cm.

Then with Eq. (2-37), $AB = \left(\dfrac{25 + 100}{25}\right) 5 \text{ cm} = 25 \text{ cm}.$

Thus, the line image is parallel to the cylindrical axis, enlarged to 25 cm, and located 1 m from the lens. If the lens is rotated about its optical axis, the line image also rotates, remaining always parallel to the cylindrical axis.

Looking again at Figure 2-31, imagine a screen placed on the exit side of the lens so as to capture the light from the lens. We have argued that when the screen is at the distance s' from the lens, one sees a focused line image AB on the screen, in this case with a horizontal orientation. As the screen is moved

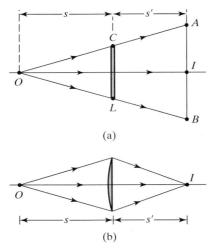

Figure 2-32 (a) Light rays in a top view of the horizontal (nonfocusing) section of the lens in Figure 2-31. (b) Light rays in a side view of the vertical (focusing) section of the lens in Figure 2-31.

further than s' from the lens, one sees an unfocused blur that has the general shape of the aperture—either of the rectangular cross section of the lens alone or of the lens with an aperture placed against it. Further, it should be evident from Figure 2-31 that, as a screen, initially positioned just behind the lens, moves toward the line image *AB,* the horizontal dimension (width) of the blur increases and its vertical dimension (height) decreases. As the screen moves beyond the line image, its width continues to increase but its height now also increases due to the divergence of the rays after focusing. If an aperture placed in front of the lens is circular, these blur images are elliptical in shape, with changing major and minor axes formed by the width and height of the blur. If the aperture is square, the blurs are rectangular in shape. Widths and heights of the blur pattern can be found at any position of the screen using the geometry apparent in Figure 2-32a and b, respectively. This behavior can be observed easily in the laboratory.

Up to this point we have been dealing with a cylindrical lens whose axis is either horizontal or vertical. Of course, the cylinder axis can be oriented at any angle. An astigmatic eye, for example, while it possesses predominantly spherical optics, might have a cylindrical axis component whose axis could be horizontal, vertical, or some angle in between. To deal with cylindrical lenses and astigmatism in a general way, then, we must be able to determine the effect of combining cylindrical lenses having arbitrary orientations with each other and with spherical lenses. It turns out that two cylindrical lenses can produce the same effect as a sphero-cylindrical lens. Lens prescriptions for vision correction are, in fact, expressed in terms of combinations of spherical and cylindrical lenses. This subject is treated further elsewhere.[7]

PROBLEMS

2-1 Derive an expression for the transit time of a ray of light that travels a distance x_1 through a medium of index n_1, a distance x_2 through a medium of index $n_2,\ldots,$ and a distance x_m through a medium of index n_m. Use a summation to express your result.

2-2 Deduce the Cartesian oval for perfect imaging by a refracting surface when the object point is on the optical x-axis 20 cm from the surface vertex and its conjugate image point lies 10 cm inside the second medium. Assume the refracting medium to have an index of 1.50 and the outer medium to be air. Find the equation of the intersection of the oval with the xy-plane, where the origin of the coordinates is at the object point. Generate a table of (x, y)-coordinates for the surface and plot, together with sample rays.

2-3 A double convex lens has a diameter of 5 cm and zero thickness at its edges. A point object on an axis through the center of the lens produces a real image on the opposite side. Both object and image distances are 30 cm, measured from a plane bisecting the lens. The lens has a refractive index of 1.52. Using the equivalence of optical paths through the center and edge of the lens, determine the thickness of the lens at its center.

2-4 Determine the minimum height of a wall mirror that will permit a 6-ft person to view his or her entire height. Sketch rays from the top and bottom of the person, and determine the proper placement of the mirror such that the full image is seen, regardless of the person's distance from the mirror.

2-5 A ray of light makes an angle of incidence of 45° at the center of the top surface of a transparent cube of index 1.414. Trace the ray through the cube.

2-6 To determine the refractive index of a transparent plate of glass, a microscope is first focused on a tiny scratch in the upper surface, and the barrel position is recorded. Upon further lowering the microscope barrel by 1.87 mm, a focused image of the scratch is seen again. The plate thickness is 1.50 mm. What is the reason for the second image, and what is the refractive index of the glass?

2-7 A small source of light at the bottom face of a rectangular glass slab 2.25 cm thick is viewed from above. Rays of light totally internally reflected at the top surface outline a circle of 7.60 cm in diameter on the bottom surface. Determine the refractive index of the glass.

---

[7]See F. L. Pedrotti and L. S. Pedrotti, *Optics and Vision* (Upper Saddle River, N. J.: Prentice Hall, Inc., 1998).

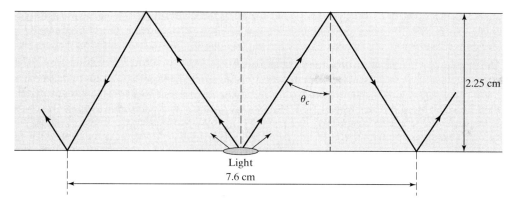

Figure 2-33 Problem 2-7

2-8 Show that the lateral displacement s of a ray of light penetrating a rectangular plate of thickness t is given by

$$s = \frac{t \sin(\theta_1 - \theta_2)}{\cos \theta_2}$$

where θ_1 and θ_2 are the angles of incidence and refraction, respectively. Find the displacement when $t = 3$ cm, $n = 1.50$, and $\theta_1 = 50°$.

2-9 A meter stick lies along the optical axis of a convex mirror of focal length 40 cm, with its nearer end 60 cm from the mirror surface. How long is the image of the meter stick?

2-10 A glass hemisphere is silvered over its curved surface. A small air bubble in the glass is located on the central axis through the hemisphere 5 cm from the plane surface. The radius of curvature of the spherical surface is 7.5 cm, and the glass has an index of 1.50. Looking along the axis into the plane surface, one sees two images of the bubble. How do they arise and where do they appear?

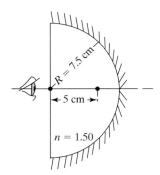

Figure 2-34 Problem 2-10

2-11 A concave mirror forms an image on a screen twice as large as the object. Both object and screen are then moved to produce an image on the screen that is three times the size of the object. If the screen is moved 75 cm in the process, how far is the object moved? What is the focal length of the mirror?

2-12 A sphere 5 cm in diameter has a small scratch on its surface. When the scratch is viewed through the glass from a position directly opposite, where does the scratch appear and what is its magnification? Assume $n = 1.50$ for the glass.

2-13 a. At what position in front of a spherical refracting surface must an object be placed so that the refraction produces parallel rays of light? In other words, what is the focal length of a single refracting surface?
 b. Since real object distances are positive, what does your result imply for the cases $n_2 > n_1$ and $n_2 < n_1$?

2-14 A small goldfish is viewed through a spherical glass fishbowl 30 cm in diameter. Determine the apparent position and magnification of the fish's eye when its actual position is (a) at the center of the bowl and (b) nearer to the oberver, halfway from center to glass, along the line of sight. Assume that the glass is thin enough so that its effect on the refraction may be neglected.

2-15 A small object faces the convex spherical glass window of a small water tank. The radius of curvature of the window is 5 cm. The inner back side of the tank is a plane mirror, 25 cm from the window. If the object is 30 cm outside the window, determine the nature of its final image, neglecting any refraction due to the thin glass window itself.

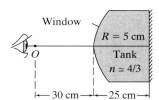

Figure 2-35 Problem 2-15

2-16 A plano-convex lens having a focal length of 25.0 cm is to be made with glass of refractive index 1.520. Calculate the radius of curvature of the grinding and polishing tools to be used in making this lens.

2-17 Calculate the focal length of a thin meniscus lens whose spherical surfaces have radii of curvature of magnitude 5 and 10 cm. The glass is of index 1.50. Sketch both positive and negative versions of the lens.

2-18 One side of a fish tank is built using a large-aperture thin lens made of glass ($n = 1.50$). The lens is equiconvex, with radii of curvature 30 cm. A small fish in the tank is 20 cm from the lens. Where does the fish appear when viewed through the lens? What is its magnification?

2-19 Two thin lenses have focal lengths of −5 and +20 cm. Determine their equivalent focal lengths when (a) cemented together and (b) separated by 10 cm.

2-20 Two identical, thin, plano-convex lenses with radii of curvature of 15 cm are situated with their curved surfaces in contact at their centers. The intervening space is filled with oil of refractive index 1.65. The index of the glass is 1.50. Determine the focal length of the combination. (*Hint*: Think of the oil layer as an intermediate thin lens.)

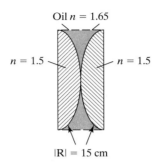

Oil $n = 1.65$

$n = 1.5$ $n = 1.5$

$|R| = 15$ cm

Figure 2-36   Problem 2-20

2-21 An eyepiece is made of two thin lenses each of +20-mm focal length, separated by a distance of 16 mm.

  a. Where must a small object be positioned so that light from the object is rendered parallel by the combination?

  b. Does the eye see an image erect relative to the object? Is it magnified? Use a ray diagram to answer these questions by inspection.

2-22 A diverging thin lens and a concave mirror have focal lengths of equal magnitude. An object is placed $(3/2)f$ from

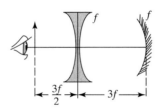

Figure 2-37   Problem 2-22

the diverging lens, and the mirror is placed a distance $3f$ on the other side of the lens. Using Gaussian optics, determine the *final* image of the system, after two refractions (a) by a three-ray diagram and (b) by calculation.

2-23 A small object is placed 20 cm from the first of a train of three lenses with focal lengths, in order, of 10, 15, and 20 cm. The first two lenses are separated by 30 cm and the last two by 20 cm. Calculate the final image position relative to the last lens and its linear magnification relative to the original object when (a) all three lenses are positive, (b) the middle lens is negative, (c) the first and last lenses are negative. Provide ray diagrams for each case.

2-24 A convex thin lens with refractive index of 1.50 has a focal length of 30 cm in air. When immersed in a certain transparent liquid, it becomes a negative lens with a focal length of 188 cm. Determine the refractive index of the liquid.

2-25 It is desired to project onto a screen an image that is four times the size of a brightly illuminated object. A plano-convex lens with $n = 1.50$ and $R = 60$ cm is to be used. Employing the Newtonian form of the lens equations, determine the appropriate distance of the object and screen from the lens. Is the image erect or inverted? Check your results using the ordinary lens equations.

2-26 Three thin lenses of focal lengths 10 cm, 20 cm, and −40 cm are placed in contact to form a single compound lens.

  a. Determine the powers of the individual lenses and that of the unit, in diopters.

  b. Determine the vergence of an object point 12 cm from the unit and that of the resulting image. Convert the result to an image distance in centimeters.

2-27 A lens is moved along the optical axis between a fixed object and a fixed image screen. The object and image positions are separated by a distance L that is more than four times the focal length of the lens. Two positions of the lens are found for which an image is in focus on the screen, magnified in one case and reduced in the other. If the two lens positions differ by distance D, show that the focal length of the lens is given by $f = (L^2 − D^2)/4L$. This is *Bessel's method* for finding the focal length of a lens.

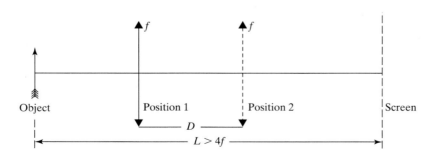

Figure 2-38   Problem 2-27

2-28 An image of an object is formed on a screen by a lens. Leaving the lens fixed, the object is moved to a new position and the image screen moved until it again receives a focused image. If the two object positions are S_1 and S_2 and if the transverse magnifications of the image are M_1 and M_2, respectively, show that the focal length of the lens is given by

$$f = \frac{(S_2 - S_1)}{\left(\dfrac{1}{M_1} - \dfrac{1}{M_2}\right)}$$

This is *Abbe's method* for finding the focal length of a lens.

2-29 Derive the law of reflection from Fermat's principle by minimizing the distance of an arbitrary (hypothetical) ray from a given source point to a given receiving point.

2-30 Determine the ratio of focal lengths for two identical, thin, plano-convex lenses when one is silvered on its flat side and the other on its curved side. Light is incident on the unsilvered side.

2-31 Show that the minimum distance between an object and its image, formed by a thin lens, is 4f. When does this occur?

2-32 A ray of light traverses successively a series of plane interfaces, all parallel to one another and separating regions of differing thickness and refractive index.

   a. Show that Snell's law holds between the first and last regions, as if the intervening regions did not exist.

   b. Calculate the net lateral displacement of the ray from point of incidence to point of emergence.

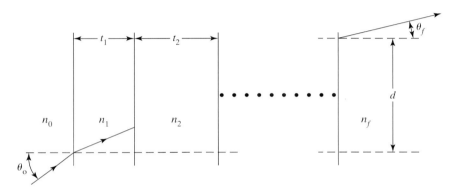

Figure 2-39   Problem 2-32

2-33 A parallel beam of light is incident on a plano-convex lens that is 4 cm thick. The radius of curvature of the spherical side is also 4 cm. The lens has a refractive index of 1.50 and is used in air. Determine where the light is focused for light incident on each side.

2-34 A spherical interface, with radius of curvature 10 cm, separates media of refractive index 1 and $\frac{4}{3}$. The center of curvature is located on the side of the higher index. Find the focal lengths for light incident from each side. How do the results differ when the two refractive indices are interchanged?

2-35 An airplane is used in aerial surveying to make a map of ground detail. If the scale of the map is to be 1:50,000 and the camera used has a focal length of 6 in., determine the proper altitude for the photograph.

2-36 Light rays emanating in air from a point object on axis strike a plano-cylindrical lens with its convex surface facing the object. Describe the line image by length and location if the lens has a radius of curvature of 5 cm, a refractive index of 1.60, and an axial length of 7 cm. The point object is 15 cm from the lens.

2-37 A plano-cylindrical lens in air has a curvature of 15 cm and an axial length of 2.5 cm. The refractive index of the lens is 1.52. Find the position and length of the line image formed by the lens for a point object 20 cm from the lens. Light from the object is incident on the convex cylindrical surface of the lens.

2-38 A plano-cylindrical lens in air has a radius of curvature of 10 cm, a refractive index of 1.50, and an axial length of 5 cm. Light from a point object is incident on the concave, cylindrical surface from a distance of 25 cm to the left of the lens. Find the position and length of the image formed by the lens.

2-39 A plano-concave cylindrical lens is used to form an image of a point object 20 cm from the lens. The lens has a refractive index of 1.50, a radius of curvature of 20 cm, and an axial length of 2 cm. Describe as completely as possible the line image of the point.

2-40 Consider the plano-convex cylindrical lens in problem 2-36. If the point object is only 6 cm from the lens, describe the line image.

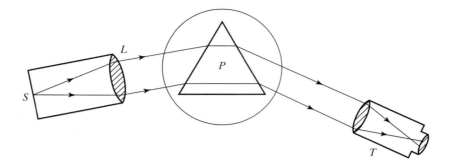

3

Optical Instrumentation

INTRODUCTION

The principles of geometrical optics, developed earlier, are applied in this chapter in order to discuss several practical optical instruments. The discussion begins with an introduction to the operation of *stops*, *pupils*, and *windows*, of great practical importance to *light control* in optical instrumentation. We follow this with a brief overview of *aberrations* and then examine in turn, the optics and operation of prisms, cameras, eyepieces, microscopes, and telescopes.

3-1 STOPS, PUPILS, AND WINDOWS

We have been studying ways to trace rays through an optical system using the step-by-step application of Gaussian formulas and ray tracing. However, not every light ray from an object point, directed toward or into an optical system, reaches the final image. Depending on the location of the object point and the ray angle, many of these rays are blocked by the limiting apertures of lenses and mirrors or by physical apertures intentionally inserted into the optical system. An *aperture*, in its broadest sense, is an opening defined by a geometrical boundary. In this section, we wish to concentrate on the effects of such spatial limitations of light beams in an optical system.

The apertures dealt with are often purposely inserted into an optical system to achieve various practical purposes. We will see in Chapter 20 how apertures can be used to modify the effects of spherical aberration, astigmatism, and distortion. In other applications, apertures may be introduced to produce a sharp border to the image, like the sharp outline we see looking into the eyepiece of an optical instrument. Apertures may also be used to shield the image from undesirable light scattered from optical components. In any case, apertures are inevitably present because every lens or mirror has a finite diameter that effectively introduces an aperture into the system.

The presence of apertures in an optical system influences its image-forming properties in two important ways—by limiting the *field of view* and by controlling the *image brightness*. The first limitation determines how much of the surface of a broad object can be seen by looking back through the optical system; the second determines how bright the image can be, that is, how much irradiance (W/m^2) reaches the image. Both of these limitations depend directly on the optical behavior of bundles of rays that leave points on an object and thread their way through an optical system—and its aperture—to a conjugate image point.

There exists in the optical industry today many different practical optical systems, each with many choices available for the placement of apertures. In this introduction we limit ourselves to a basic description of how apertures can affect both image brightness and field of view. We examine first the role of *aperture stops* and their related *pupils* in fixing image brightness, then the role of *field stops* and their related *windows* in determining a field of view.

Although the actions of stops, pupils, and windows depend quite clearly on the principles of geometrical optics, the details can at times be confusing. In the descriptive text and correlated figures that follow, it will benefit the reader to coordinate the reading of the text and examination of associated figures closely. Thus, for example, in the text and correlated Figures 3-1a, b, and c, we define carefully the meaning of aperture stops (AS), entrance pupils (E_nP), and exit pupils (E_xP) and illustrate their effects in several simple optical systems.

Image Brightness: Aperture Stops and Pupils

Aperture Stop (AS)   The *aperture stop* of an optical system is the actual physical component that limits the size of the *maximum cone* of rays—from an axial object point to a conjugate image point—that can be processed by the entire system. Thus it controls the brightness of the image. The diaphragm of a camera and the iris of the human eye are examples of aperture stops. Another example is found in the telescope, in which the first, or objective, lens determines how much light is accepted by the telescope to form a final image on the retina of the eye. In the telescope, then, the objective lens is the aperture stop of the optical system. However, the aperture stop is not always identical with the first component of an optical system. For example, refer to Figure 3-1a. As shown, the aperture stop (AS) in front of the lens determines the extreme (or *marginal*) rays that can be accepted by the lens. But if the object OO' is moved to a position nearer to AS, at some point the lens rim becomes the limiting aperture. This position is just the point of intersection between the optical axis and a line drawn from the lens rim through the edge of the aperture stop AS. In this case, the angle ($L'OL$) subtended by the lens rim at O becomes smaller than the angle ($M'OM$) subtended by the edge of the aperture, and so we designate the lens as the aperture stop.

Entrance Pupil (E_nP)   The *entrance pupil* is the *limiting aperture* (opening) that the light rays "see," looking into the optical system from any object point. In Figure 3-1a, this is simply the aperture stop itself, so in this case, AS and E_nP are identical. To see that this is not always the case, look at Figure 3-1b, where the aperture stop is behind the lens (a *rear stop*), as in most cameras. Which component now limits the cone of light rays? You can see that it is that component whose aperture edges limit rays from O to their smallest angle relative to the axis. Looking into the optical system from object space, one

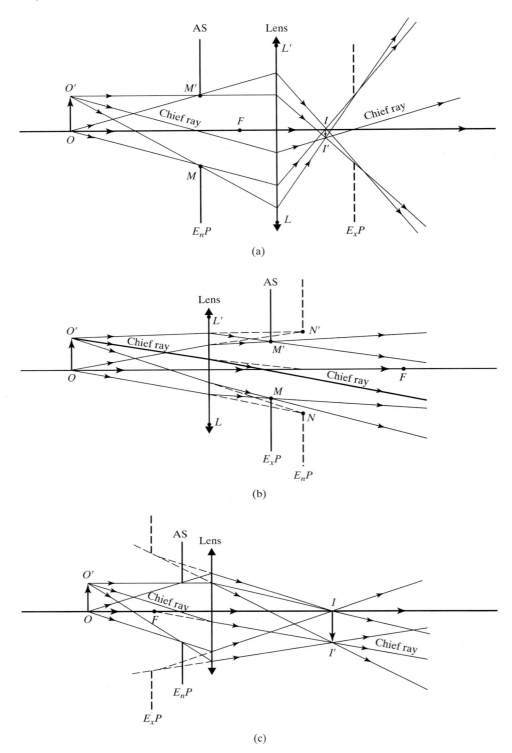

Figure 3-1 Limitation of light rays by various combinations of a positive lens and aperture.

sees the lens directly but sees the AS through the lens. In other words, the *effective aperture* due to the AS is *its image* formed by the lens, the dashed line marked E_nP. Since rays from O, directed toward this virtual aperture, make a smaller angle ($N'ON$) than rays directed toward the lens edge ($L'OL$), this virtual aperture serves as the effective entrance pupil for the system. Notice that rays from O, directed toward the edges of E_nP, are in fact first refracted by the lens so as to pass through the edges of the real aperture stop. This must be the case, since AS and E_nP are, by definition, conjugate planes: The edges of E_nP are the images of the edges of AS. This example illustrates the general rule: *The entrance pupil E_nP is the image of the controlling aperture stop formed by the imaging elements preceding it.*[1] When the controlling aperture stop is the first such element (a *front stop*), it serves itself as the entrance pupil.

Another example, in which an aperture placed in front of the lens functions as the AS for the system, is shown in Figure 3-1c. It is different from Figure 3-1a in that the aperture is placed *inside* the focal length of the lens. Nevertheless, the aperture is the AS for the system because it, not the lens, limits the system rays to their smallest angle with the axis. Furthermore, it is the E_nP of the system because it is the first element encountered by the light from the object.

Exit Pupil (E_xP)   We have described the E_nP of an optical system as the image of the AS one sees by looking into the optical system from the object. If one looks into the optical system from the image, another image of the AS can be seen that appears to limit the output beam size. This image is called the *exit pupil* of the optical system. Thus, *the exit pupil is the image of the controlling aperture stop formed by the imaging elements following it* (or *to the right* of it in our figures). The rear stop in Figure 3-1b is automatically the E_xP for the system because it is the last optical component. According to our definition of the E_xP, the exit pupil is the optical conjugate of the AS; the E_xP and AS are conjugate planes. It follows that the E_xP is also conjugate with the E_nP. In Figure 3-1a, the E_xP is the real image of the E_nP; in Figure 3-1c, it is the virtual image. Notice that in each case, rays intersecting the edges of the entrance pupil also (actually, or when extended) intersect the edges of the exit pupil.

In a system like that of Figure 3-1a, a screen held at the position of the exit pupil receives a sharp image of the circular opening of the aperture stop. If the system represents the eyepiece of some optical instrument, the exit pupil is matched in position and diameter to the pupil of the eye. Notice further that if the screen is moved closer to the lens, it intercepts a sharp image II' of the object OO'. The exit pupil is seen to limit the solid angle of rays forming each point of the image and therefore determines the image brightness, point by point.

Chief Ray   The *chief*, or *principal, ray* is a ray from an object point that passes through the *axial point*, in the *plane of the entrance pupil*. Given the conjugate nature of the entrance pupil with both the aperture stop and the exit pupil, this ray must also pass (actually or when extended) through their axial points. The chief ray in the cone of rays leaving object point O' is shown in all three systems of Figure 3-1. The chief ray in the cone of rays leaving the *axial point O* always coincides with the optical axis.

Before adding to our collection of new concepts that arise from a consideration of apertures in optical systems, we consider a system slightly more complex than those of Figure 3-1. In Figure 3-2, we specify a particular optical system consisting of two lenses, $L1$ and $L2$, with an aperture A placed between them, as shown. The first question to be answered is: Which element serves as the effective AS for the whole system? The answer to this question is not always obvious. It can always be answered, however, by determining which of the actual elements in the given system—in this case, A, $L1$, or $L2$— has an *entrance pupil* that confines rays to their *smallest angle with the axis*, as seen from the object point. To decide which candidate presents the limiting aperture, it is necessary to find the entrance pupil for each by imaging each one through that part of the optical system *lying to its left*:

---

[1]"Preceding" is used in the sense that light must pass through those imaging elements first. If we always use light rays directed from left to right, we can simply say, "by all imaging elements to its left."

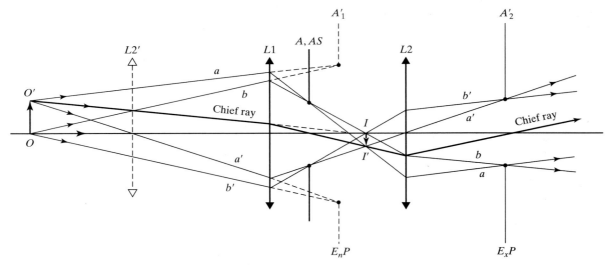

Figure 3-2 Limitation of light rays in an optical system consisting of two positive lenses and an aperture A. The labels a, a', b and b' assist in tracking the various rays.

L2: By ray diagram or by calculation, the image of lens $L2$, formed by $L1$ (as if light went from right to left), is $L2'$, and is real. Both its location and size (magnification) are shown.

A: The image of aperture A backward through $L1$ is virtual and is shown as A_1'.

L1: Since lens $L1$ is the first element, it acts as its own entrance pupil.

The three candidates for entrance pupils, $L2'$, $L1$, and A_1', are next viewed from the axial point O. Since the rim of A_1' subtends the smallest angle at O, as is clear from the figure, we conclude that its conjugate aperture A is the aperture stop (AS) of the system.

Once the AS is identified, *it is imaged through the optical elements to its right* to find the *exit pupil*. In this case, aperture A is imaged through $L2$ to form $A2'$. The *chief ray*, together with its two marginal rays, a and a' is drawn from the tip O' of the object. Notice that the chief ray passes (actually or when extended) through the centers of AS and its conjugate planes, E_nP and E_xP. The chief ray from O' intersects the optical axis at A, at A_1' (which is virtual), and at A_2'. The two cones–defined by the pairs of rays a, a' and b, b'–emerging from the points O and O' are limited by the size of the entrance pupil A_1' and just make it through the exit pupil A_2'. The image of OO' formed by $L1$ is shown as II'; the final image (not shown) is virtual, since the rays from either O or O' diverge on leaving $L2$.

Field of View: Field Stops and Windows

In describing the limitations of a cone of rays from an axial object point, we have seen that entrance and exit pupils are related to the aperture stop and so govern the brightness of the image. As mentioned earlier, apertures also determine the *field of view* handled by the system. The controlling element in this connection is called the *field stop*, and it is related to an *entrance window* and an *exit window* in the same way that the aperture stop is related to entrance and exit pupils.

A simple experience of a limitation in the field of view is that of looking through an ordinary window. The edges of the window determine how much of the outdoors we can see. This field of view can be described in terms of the lateral dimensions of the object viewed, or in terms of the angular extent of the window, relative to the line of sight. One can talk about the field of view in terms of the object being viewed or in terms of the image formed at the viewer (on the retina, in this example).

To see how an aperture restricts the field of view—using diagrams that could well be applied to the case of window and eye lens, just discussed—look at Figure 3-3. In part (a), the optical system is a single aperture A placed in front of a single lens. Object and image planes are also shown. Rays from an axial point O are limited in angle by the aperture and focused by the lens at point O'. The same is true for the off-axis point T and its image, T'. In both cases, the lens is large enough to intercept the entire cone of rays admitted by A. If the object plane is uniformly bright and the aperture is a circular hole, then a circle of radius $O'T'$ is uniformly illuminated in the image plane.

However, if one considers object points below T, the upper rays from such points, passing through the aperture, miss the lens. Such a point, U, is shown in part (b) of the figure, the same optical system as (a), but redrawn for clarity. It is chosen such that the chief or central ray of the bundle from U just misses the top of the lens. About half the beam is lost so that image point U' receives only about half as much light as points O' and T'. Thus, the circular image begins to dim as its radius increases. This partial shielding of the outer portion of the image by the aperture for off-axis object points is called *vignetting*. Excessive vignetting may make the image of a point appear astigmatic. Finally, object point V is chosen such that all its rays through the aperture A miss the lens entirely. The lateral field of view processed by this optical system is at most a circle of radius OV. It is often defined as the smaller circle of radius OU if one considers the usable field of view as that which consists of all object points that produce image points having at least half the maximum

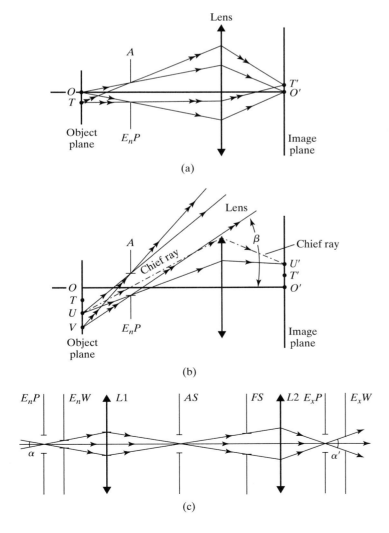

Figure 3-3 Referring to the same optical system, diagrams (a) and (b) illustrate both the way in which an aperture limits the field of view and the process by *vignetting*. Diagram (c) is an example of a more complex optical system, showing the angular field of view in object and image space, α and α', respectively.

irradiance near the center. One can then also define the *angular field of view* as twice the angle β made by the chief ray with the axis at the center of the opening represented by the entrance pupil. The lens itself, in this case, acts both as the field stop and the entrance window. In other, more complex, systems, the relevant quantities may be described as follows.

Field Stop (FS)   The field stop is the aperture that controls the field of view by limiting the solid angle formed by chief rays. *As seen from the center of the entrance pupil, the field stop (or its image) subtends the smallest angle.* When the edge of the field of view is to be sharply delineated, the field stop should be placed in an image plane so that it is sharply focused along with the final image. A simple example of such a field stop is the opening directly in front of the film that outlines the final image in a camera. Intentional limitation of the field of view using an aperture is desirable when either far-off axis imaging is of unacceptable quality due to aberrations or when vignetting severely reduces the illumination in the outer portions of the image.

Entrance Window (E_nW)   *The entrance window is the image of the field stop formed by all optical elements preceding it.* The entrance window delineates the lateral dimensions of the object to be viewed, as in the viewfinder of a camera, and its angular diameter determines the angular field of view. When the field stop is located in an image plane, the entrance window lies in the conjugate object plane, where it outlines directly the lateral dimensions of the object field imaged by the optical system.

Exit Window (E_xW)   The *exit window is the image of the field stop formed by all optical elements following it.* To an observer in image space, the exit window seems to limit the area of the image in the same way as an outdoor scene appears limited by the window of a room.

In Figure 3-3c, the locations of the stops, pupils, and windows are shown in a more complex optical system consisting of two lenses and two apertures. The first aperture is the AS of the system and, as we have seen, is related to an entrance pupil, its image in $L1$, and an exit pupil, its image in $L2$. The second aperture is the field stop, FS, with its corresponding images through the lenses: the entrance window to the left and the exit window to the right. The field of view in object space can then be described by α, the angle subtended by the entrance window at the center of the entrance pupil. Similarly, the field of view in image space can be described by α', the angle subtended by the exit window at the center of the exit pupil. We see that the size of the field imaged by the optical system is effectively determined by the entrance window and, actually, by the size of the field stop. Notice that since E_nW and E_xW are both images of the FS, they are conjugate planes. Thus, the same bundle of rays that fills the entrance window also fills the field stop and the exit window.

The Summary of Terms that follows is provided as a convenient reference for a subject that requires patience, and experience with many examples, to master.

SUMMARY OF TERMS

Brightness

Aperture stop AS:	The *real* element in an optical system that limits the size of the cone of rays accepted by the system from an axial object point.
Entrance pupil E_nP:	The image of the aperture stop formed by the optical elements (if any) that precede it.
Exit pupil E_xP:	The image of the aperture stop formed by the optical elements (if any) that follow it.

Field of view

Field stop FS:	The *real* element that limits the angular field of view formed by an optical system.

Entrance window E_nW: The image of the field stop formed by the optical elements (if any) that precede it.

Exit window E_xW: The image of the field stop formed by the optical elements (if any) that follow it.

The following example will provide practice with the thin-lens formula $1/s + 1/s' = 1/f$ and procedures for determining stops and pupils for a specific optical system. As you follow through the steps in the solution, be sure to verify the correct use of the sign convention related to the object distance s, image distance s', focal length f, and image magnification $m = -s'/s$ for each calculation.

Example 3-1

An optical system (see sketch in Figure 3-4 below) is made up of a positive thin lens L_1 of diameter 6 cm and focal length $f_1 = 6$ cm, a negative thin lens L_2 of diameter 6 cm and focal length $f_2 = -10$ cm, and an aperture A of diameter 3 cm. The aperture A is located 3 cm in front of lens L_1, which is located 4 cm in front of lens L_2. An object OP, 3 cm high is located 18 cm to the left of L_1.

Problem

a. Determine which element (A, L_1, or L_2) serves as the aperture stop AS.
b. Determine size and location of the entrance and exit pupils.
c. Determine the location and size of the intermediate image of OP formed by L_1 and the final image formed by the system.
d. Using a scale of 1 cm $= \frac{1}{4}$ in., draw a diagram of the optical system and locate to scale, on the drawing, the two pupils, intermediate image $O'P'$ and final image $O''P''$.
e. Draw the chief ray from object point P to its conjugate in the final image, P''.

Solution

a. Determine first which element (A, L_1, or the image of L_2 in L_1) subtends the smallest half-angle from rim to point O.

Elements A and L_1 have no "optics" to their left, so each subtends a half-angle directly:

$$\text{For element } A: \theta_A \simeq 1.5/15 = 0.1 \text{ rad}$$

$$\text{For element } L_1: \theta_{L_1} \simeq 3/18 = 0.17 \text{ rad}$$

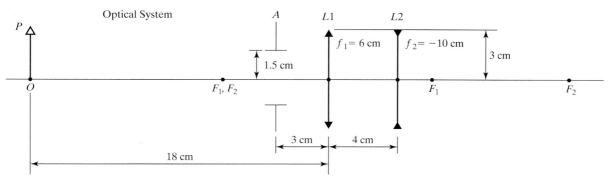

Figure 3-4   Sketch of optical system of Example 3-1.

Find the image of L_2 in L_1 with light traveling right to left so that $s = 4$ cm, $f = 6$ cm. Thus, $\frac{1}{4} + \frac{1}{s'} = \frac{1}{6}$ gives $s' = -12$ cm. So the image is virtual, 12 cm to the right of L_1, of half-size given by $(3 \text{ cm}) \times \left(\frac{-s'}{s}\right) = (3 \text{ cm})\left(-\frac{-12}{4}\right) = 9$ cm. (This virtual image is shown as L_2' in the final drawing.) So, for image L_2':

$$\theta_{L_2'} \simeq 9/30 = 0.3 \text{ rad}$$

Comparing half-angles, element A subtends the smallest half-angle and so serves as the aperture stop AS.

b. Location and size of entrance and exit pupils.

Entrance Pupil E_nP: There are no optics to the *left* of the aperture stop A, so it serves also as the entrance pupil E_nP.

Exit Pupil E_xP: L_1 and L_2 are to the *right* of AS, so we must image AS through both to locate the position and size of the exit pupil E_xP.

Through lens L_1: $s_1 = 3$ cm, $f_1 = 6$ cm so $\frac{1}{3} + \frac{1}{s_1'} = \frac{1}{6}$ gives $s_1' = -6$ cm. This is a virtual image that serves as the object for lens L_2, with $s_2 = 6$ cm $+ 4$ cm $= 10$ cm and $f_2 = -10$ cm. Thus, $\frac{1}{10} + \frac{1}{s_2'} = \frac{1}{-10}$ yields $s_2' = -5$ cm. So the exit pupil E_xP is located 5 cm to the left of L_2 or 1 cm to the left of L_1. Its size is $(3 \text{ cm})\left(-\frac{-6}{3}\right)\left(-\frac{-5}{10}\right) = 3$ cm, as shown in the final drawing.

c. Locating the intermediate and final image.

OP imaged through L_1: $s_1 = 18$ cm, $f_1 = 6$ cm; $\frac{1}{18} + \frac{1}{s_1'} = \frac{1}{6}$ gives $s_1' = 9$ cm right of L_1 or 5 cm to the right of L_2. The size of $O'P'$, the intermediate image, is thus $(3 \text{ cm})\left(-\frac{9}{18}\right) = -1.5$ cm. So, $O'P'$ is inverted, 1.5 cm long, and 5 cm to the right of L_2. (But it never forms there due to the presence of L_2.)

O'P' imaged through L_2: s_2, a virtual object for L_2, equals -5 cm. So, $\frac{1}{-5} + \frac{1}{s_2'} = \frac{1}{-10}$ gives $s_2' = 10$ cm. The final image $O''P''$ is then 10 cm to the right of L_2. Its size, based on $O'P'$, is $(-1.5 \text{ cm})\left(-\frac{10}{-5}\right) = -3$ cm. So, $O''P''$, as shown in the final drawing, is real, inverted, 3 cm long (same as the object), and 10 cm from L_2.

d. The final drawing, based on an original scale of 1 cm $= \frac{1}{4}$in., with all items of interest, is shown in Figure 3-5.

e. The chief ray, from point P to conjugate point P'', is shown in the final drawing. Note that it leaves P, passes through M, the center of AS and E_nP, undergoes refraction at L_1, heads for $O'P'$, refracts again at L_2 before reaching $O'P'$, and heads for $O''P''$, the final image. Note also that the segment of the chief ray from L_2 to P'', if traced backward, will appear to be coming from point N, the center of the exit pupil E_xP. Thus, the chief ray involves the centers of AS, E_nP, and E_xP, as defined.

3-2 A BRIEF LOOK AT ABERRATIONS

All aberrations lead to a blurring of an image formed by an optical system, thereby frustrating the optical designer intent on producing an ideal image—that is, a faithful point-by-point re-creation of corresponding object points. *Chromatic* aberration and the five *monochromatic* aberrations (spherical aberration, coma, astigmatism, curvature of field, and distortion) and their corrections are considered in detail in Chapter 20. There we shall see that aberrations occur largely as a result of the form and shape of lenses and

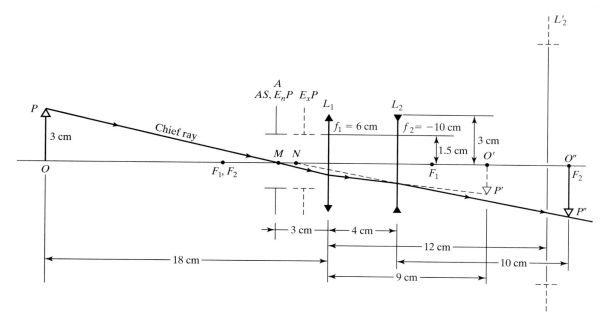

Figure 3-5 Solution to Example 3-1 (d).

mirrors and as a result of rays from object points striking spherical surfaces at angles that exceed those set by the paraxial approximation.

A brief description of spherical and chromatic aberration will serve us here as a useful background for the treatment of optical elements and instruments that conclude this chapter. In Chapter 20 we provide a more complete description of these two aberrations and their remedies, as well as a discussion of the other four monochromatic aberrations.

Spherical Aberration

In the paraxial ray approximation, all rays emanating from an object point, after reflecting from a spherical mirror or passing through a lens with spherical surfaces, either intersect at, or to the viewer appear to intersect at, a common image point. In fact, rays emanating from an object point that are incident on an optical element (spherical mirror or lens) at different distances from the optical axis, after reflection or refraction, either intersect, or to the viewer appear to intersect, at different positions. The result is that point objects are not imaged as points but rather as small blurred lines. The effect of spherical aberration for a concave spherical mirror that images an axial object point is shown in Figure 3-6a. Note that rays incident on the mirror at symmetrically placed points Q converge at axial point M, whereas rays incident on the mirror at points P converge at axial point N. Consequently, light rays from a single point O form a blurred image along the line segment containing M and N. Figure 3-6b illustrates the effect of spherical aberration in a thin-lens system that forms a blurred line image of an axial point object. In the case shown, rays from axial point object O that encounter the lens at symmetrically placed points P converge at axial point N, while rays from O that encounter the lens at points Q converge at axial point M. As in the spherical mirror case, the result is a blurred line image along the line segment containing M and N. The constructions shown in Figure 3-6a and 3-6b suggest that "stopping down" the optical system, so that only nearly paraxial rays get through the entrance pupil, will limit the effect of spherical aberration. This remedy for spherical aberration is, of course, accompanied by a reduction in image brightness. Another remedy for spherical aberration is achieved by combining positive and negative lenses in an arrangement such that the spherical aberration from one tends to cancel that from the other.

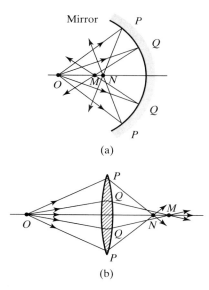

Figure 3-6 Spherical aberration (exaggerated) in (a) a spherical mirror and (b) a thin lens. For both arrangements, the rays from object point O fail to converge at a single image point.

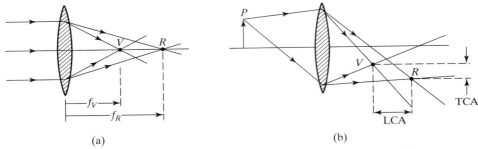

Figure 3-7   Chromatic aberration (exaggerated) for (a) parallel rays of white light incident on a thin lens and (b) white light incident on a thin lens from an off-axis point P.

Chromatic Aberration

Chromatic aberration results because the index of refraction of a material differs for different wavelengths. Since the focal length of a lens is dependent on the index of refraction of the lens material (see the "lensmaker's" equation, Eq. (2-28)), focal lengths and image positions differ for different wavelength components of the light used in the optical system. Thus, polychromatic light from a point object images not as a point but as a series of points, one for each distinct wavelength.

Figure 3-7 illustrates two related aspects of chromatic aberration. In Figure 3-7a, parallel rays of light focus nearer the lens, at point V, for violet light and further from the lens, at point R, for red light. Thus, *white* light coming from a single distant object point fails to image as a single point. Rather, the different wavelength components refract to form image points between the focal lengths f_V and f_R, as indicated in Figure 3-7a. Figure 3-7b shows a slightly different view of chromatic aberration evident in the behavior of white light incident on a lens from an off-axis point P. The violet and red components of the white light leaving point P refract differently at the lens and so converge at different image points, labeled V and R, respectively. The amount of chromatic aberration, in such a case, can be described by two distances, shown in Figure 3-7b, one called the *longitudinal* chromatic aberration (LCA) and the other the *lateral* or *transverse* chromatic aberration (TCA).

Chromatic aberration in lenses can be effectively reduced by using multiple refractive elements of opposite powers, as we show in detail in Chapter 20. Of course, images formed in mirrors do not suffer from chromatic aberration since the focal length of a mirror is independent of wavelength.

3-3 PRISMS

Angular Deviation of a Prism

The top half of a double-convex, spherical lens can form an image of an axial object point within the paraxial approximation, as shown in Figure 3-8. If the lens surfaces are flat, a prism is formed, and paraxial rays can no longer produce a unique image point. It is nevertheless helpful in some cases to think of a prism as functioning approximately like one-half of a convex lens.

In the following we derive the relationships that describe exactly the progress of a single ray of light through a prism. The bending that occurs at each face is determined by Snell's law. The degree of bending is a function of the refractive index of the prism and is, therefore, a function of the wavelength of the incident light. The variation of refractive index and light speed with wavelength is called *dispersion* and is discussed later. For the present, we assume monochromatic light, which has its own characteristic refractive index in the prism medium. The relevant angles describing the progress of the ray through the prism are defined in Figure 3-9. Angles of incidence and refraction at each prism face are shown relative to the normals constructed at the point of intersection with the light ray. The total angular deviation δ of the ray

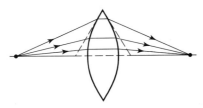

Figure 3-8   Focusing due to half of a convex lens approximates the action of a prism.

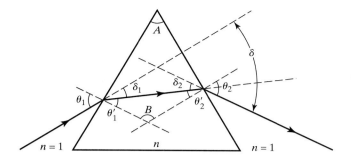

Figure 3-9 Progress of an arbitrary ray through a prism.

due to the action of the prism as a whole is the sum of the angular deviations δ_1 and δ_2 at the first and second faces, respectively. Snell's law at each prism face requires that

$$\sin \theta_1 = n \sin \theta_1' \qquad (3\text{-}1)$$

$$n \sin \theta_2' = \sin \theta_2 \qquad (3\text{-}2)$$

Inspection will show that the following geometrical relations must hold between the angles:

$$\delta_1 = \theta_1 - \theta_1' \qquad (3\text{-}3)$$

$$\delta_2 = \theta_2 - \theta_2' \qquad (3\text{-}4)$$

$$B = 180 - \theta_1' - \theta_2' = 180 - A \qquad (3\text{-}5)$$

$$A = \theta_1' + \theta_2' \qquad (3\text{-}6)$$

The two members of Eq. (3-5) follow because the sum of the angles of a triangle is 180° and because the sum of the angles of a quadrilateral must be 360°. Notice that the angles A and B and the two right angles formed by the normals with the prism sides constitute such a quadrilateral.

Using Eqs. (3-1) through (3-6), a programmable calculator or computer may easily be programmed to perform the sequential operations that finally determine the *angle of deviation*, δ. Assuming that the *prism angle A* and refractive index n are given, then the stepwise calculation for a ray incident at an angle θ_1 is as follows:

$$\theta_1' = \sin^{-1}\left(\frac{\sin \theta_1}{n}\right) \qquad (3\text{-}7)$$

$$\delta_1 = \theta_1 - \theta_1' \qquad (3\text{-}8)$$

$$\theta_2' = A - \theta_1' \qquad (3\text{-}9)$$

$$\theta_2 = \sin^{-1}(n \sin \theta_2') \qquad (3\text{-}10)$$

$$\delta = \theta_1 + \theta_2 - \theta_1' - \theta_2' \qquad (3\text{-}11)$$

The variation of deviation with angle of incidence for $A = 30°$ and $n = 1.50$ is shown in Figure 3-10. Notice that a minimum deviation occurs for $\theta_1 = 23°$. Refraction by a prism under the condition of minimum deviation is most often utilized in practice. We may argue rather neatly that when minimum deviation occurs, the ray of light passes symmetrically through the prism, making it unnecessary to subscript angles, as shown in Figure 3-11. Suppose this were not the case, and minimum deviation occurred for a nonsymmetrical case, as in Figure 3-9. Then if the ray were reversed, following the same path backward, it would have the same total deviation as the forward ray, which we

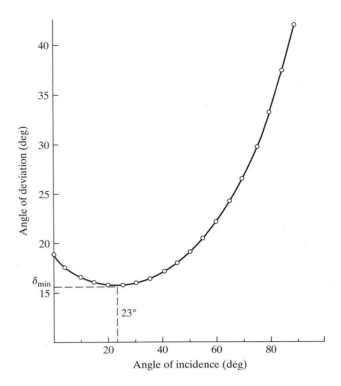

Figure 3-10 Graph of total deviation versus angle of incidence for a light ray through a prism with $A = 30°$ and $n = 1.50$. Minimum deviation occurs for an angle of 23°.

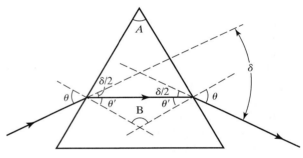

Figure 3-11 Progress of a ray through a prism under the condition of minimum deviation.

have supposed to be a minimum. Hence there would be two angles of incidence, θ_1 and θ_2, producing minimum deviation, contrary to experience. The geometric relations simplify in this case. From Eq. (3-11),

$$\delta = 2\theta - 2\theta' \tag{3-12}$$

and from Eq. (3-6),

$$A = 2\theta' \tag{3-13}$$

Together these allow us to write

$$\theta' = \frac{A}{2} \quad \text{and} \quad \theta = \frac{\delta + A}{2} \tag{3-14}$$

so that Eq. (3-1) becomes

$$\sin\left(\frac{A + \delta}{2}\right) = n\sin\left(\frac{A}{2}\right)$$

or

$$n = \frac{\sin[(A + \delta)/2]}{\sin(A/2)} \tag{3-15}$$

Eq. (3-15) provides a method of determining the refractive index of a material that can be produced in the form of a prism. Measurement of both prism angle and minimum deviation of the sample determines n. An approximate form of Eq. (3-15) follows for the case of small prism angles and, consequently, small deviations. Approximating the sine of the angles by the angles in radians, we may write

$$ n \cong \frac{(A + \delta)/2}{A/2} $$

or

$$ \delta \cong A(n - 1), \text{ minimum deviation, small } A, \text{ prism in air} \qquad (3\text{-}16) $$

For $A = 15°$, the deviation given by Eq. (3-16) is correct to within about 1%. For $A = 30°$, the error is about 5%.

Dispersion

The minimum deviation of a monochromatic beam through a prism is given implicitly by Eq. (3-15) in terms of the refractive index. The refractive index, however, depends on the wavelength, so that it would be better to write n_λ for this quantity. As a result, the total deviation is itself wavelength dependent, which means that various wavelength components of the incident light are separated on refraction from the prism. A typical *normal dispersion* curve and the nature of the resulting color separation are shown in Figure 3-12. Notice that shorter wavelengths have larger refractive indices and, therefore, smaller speeds in the prism. Consequently, violet light is deviated most in refraction through the prism. The dispersion indicated in the graph n versus λ of Figure 3-12 is called "normal" dispersion. When the refracting medium has characteristic excitations that absorb light of wavelengths within the range of the dispersion curve, the curve is monotonically decreasing, as shown, but has a positive slope in the wavelength region of the absorption. When this occurs, the term *anomalous dispersion* is used, although there is nothing anomalous about it. (This subject is treated further in Chapter 25.) The normal dispersion curve shown is typical but varies somewhat for different materials. An empirical relation that approximates the curve, introduced by Augustin Cauchy, is

$$ n_\lambda = A + \frac{B}{\lambda^2} + \frac{C}{\lambda^4} + \cdots \qquad (3\text{-}17) $$

where $A, B, C, \ldots$ are empirical constants to be fitted to the dispersion data of a particular material. Often the first two terms are sufficient to provide a reasonable fit, in which case experimental knowledge of n at two distinct wavelengths is sufficient to determine values of A and B that represent the dispersion. The *dispersion*, defined as $dn/d\lambda$, is then approximately, using Cauchy's formula, $dn/d\lambda = -2B/\lambda^3$.

It is important to distinguish dispersion from deviation. Although prism materials of large n produce a large deviation at a given wavelength, the dispersion or separation of neighboring wavelengths need not be correspondingly large. Figure 3-13 depicts extreme cases illustrating the distinction. Historically, dispersion has been characterized by using three wavelengths of light near the middle and ends of the visible spectrum. They are called *Fraunhofer lines*. These lines were among those that appeared in the solar spectrum studied by J. von Fraunhofer. Their wavelengths, together with refractive indices, are given in Table 3-1. The F and C *dark lines* are due to absorption by hydrogen atoms, and the D dark line is due to absorption by the

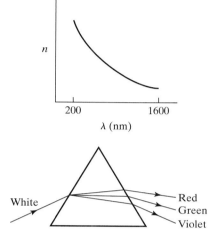

Figure 3-12 Typical normal dispersion curve and consequent color separation for white light refracted through a prism.

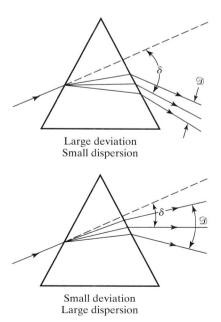

Figure 3-13 Extreme cases showing the dispersion 𝒟 for three wavelengths and the deviation δ for the intermediate wavelength.

TABLE 3-1   FRAUNHOFER LINES

λ (nm)	Characterization	n Crown glass	Flint glass
486.1	*F*, blue	1.5286	1.7328
589.2	*D*, yellow	1.5230	1.7205
656.3	*C*, red	1.5205	1.7076

sodium atoms in the sun's outer atmosphere.[2] Using the thin prism at minimum deviation for the *D* line, for example, the ratio of angular spread of the *F* and *C* wavelengths to the deviation of the *D* wavelength, as suggested in Figure 3-13, is

$$\frac{\mathcal{D}}{\delta} = \frac{n_F - n_C}{n_D - 1}$$

This measure of the ratio of dispersion $\mathcal{D}$ to deviation δ is defined as the *dispersive power* Δ, so that

$$\Delta = \frac{n_F - n_C}{n_D - 1} \tag{3-18}$$

Using Table 3-1, we may calculate the dispersive power of crown glass to be 1/65, while that of flint glass is 1/29, more than twice as great. The reciprocal of the dispersive power is known as the *Abbe number*.

Prism Spectrometers

An analytical instrument employing a prism as a dispersive element, together with the means of measuring the prism angle and the angles of deviation of various wavelength components in the incident light, is called a *prism spectrometer*. Its essential components are shown in Figure 3-14. Light to be analyzed is focused onto a narrow slit *S* and then collimated by lens *L* and refracted by the prism *P*, which typically rests on a rotatable indexed platform. Rays of light corresponding to each wavelength component emerge mutually parallel after refraction by the prism and are viewed by a telescope focused for infinity. As the telescope is rotated around the indexed prism table, a focused image of the slit is seen for each wavelength component at its corresponding angular deviation. The deviation δ is measured relative to the telescope position when viewing the slit without the

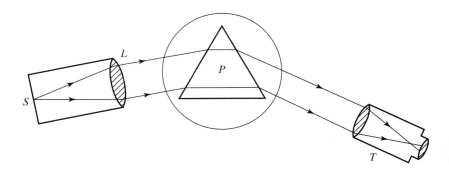

Figure 3-14   Essentials of a spectrometer.

[2]Because the yellow sodium *D* line is a doublet (589.0 and 589.6 nm), the more monochromatic *d* line of helium at 587.56 nm is often preferred to characterize the center of the visible spectrum. The green line of mercury at 546.07, which lies nearer to the center of the luminosity curve (see Chapter 19), is also used.

prism in place. When the instrument is used for visual observations without the capability of measuring the angular displacement of the spectral "lines," it is called a *spectroscope*. If means are provided for recording the spectrum, for example, with a photographic film in the focal plane of the telescope objective, the instrument is called a *spectrograph*. When the prism is made of some type of glass, its wavelength range is limited by the absorption of glass outside the visible region. To extend the usefulness of the spectrograph farther into the ultraviolet, for example, prisms made from quartz (SiO_2) and fluorite (CaF_2) have been used. Wavelengths extending further into the infrared can be handled by prisms made of salt (NaCl, KCl) and sapphire (Al_2O_3).

Chromatic Resolving Power

If the wavelength difference between two components of the light incident on a prism is allowed to decrease, the ability of the prism to resolve them will ultimately fail. The resolving power of a prism spectrograph thus represents an important performance parameter, which we shall evaluate in this section. Imagine two spectral lines formed on a photographic film in a prism spectrograph. The lines are images of the slit, so that for precise wavelength measurements the entrance slit should be kept as narrow as possible consistent with the requirement of adequate illumination of the film. Even with the narrowest of slit widths, however, the spectral line image is found to possess a width, directly traceable to the limitation that the edges of the collimating lens or prism face impose on the light beam. The phenomenon is therefore due to the *diffraction* of light, treated later. Since the line images have an irreducible width due to diffraction, as $\Delta\lambda$ decreases and the lines approach one another, a point is reached where the two lines appear as one, and the *limit of resolution* of the instrument is realized. No amount of magnification of the images can produce a higher resolution or enhancement of the ability to discriminate between two such closely spaced spectral lines.

Consider Figure 3-15a, in which a monochromatic parallel beam of light is incident on a prism, such that it fills the prism face. Employing Fermat's principle, the ray *FTW* is isochronous with ray *GX*, since they begin and end on the same plane wavefronts, *GF* and *XW*, respectively. Their optical paths can be equated to give

$$FT + TW = nb$$

where b is the base of the prism and n is the refractive index of the prism, corresponding to the wavelength λ. If a second neighboring wavelength component λ' is now also present in the incident beam, such that $\lambda' - \lambda = \Delta\lambda$, the component λ' will be associated with a different refractive index, $n' = n - \Delta n$. For normal dispersion, Δn will be a small, positive quantity. The

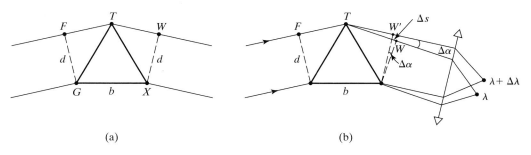

Figure 3-15 Constructions used to determine chromatic resolving power of a prism. (a) Refraction of monochromatic light. (b) Refraction of two wavelength components separated by $\Delta\lambda$.

emerging wavefronts for the two components, shown in Figure 3-15b, are thus separated by a small angular difference, $\Delta\alpha$, and are accordingly focused at different points in the focal plane of the telescope objective. Fermat's principle, applied to the second component λ', gives

$$FT + TW - \Delta s = (n - \Delta n)b$$

Subtracting the last two equations, we conclude

$$\Delta s = b\,\Delta n \tag{3-19}$$

or, introducing the dispersion,

$$\Delta s = b\left(\frac{dn}{d\lambda}\right)\Delta\lambda \tag{3-20}$$

Equation (3-20) now relates the path difference Δs to the wavelength difference $\Delta\lambda$. The angular difference $\Delta\alpha$ can also be introduced, using

$$\Delta\alpha = \frac{\Delta s}{d} = \left(\frac{b}{d}\right)\left(\frac{dn}{d\lambda}\right)\Delta\lambda \tag{3-21}$$

where d is the beam width. We appeal now to *Rayleigh's criterion*, which determines the limit of resolution of the diffraction-limited line images. This criterion is explained and used in the later treatment of diffraction, where it is shown that the minimum separation $\Delta\alpha$ of the two wavefronts, such that the images formed are just barely resolvable, is given by

$$\Delta\alpha = \frac{\lambda}{d} \tag{3-22}$$

Combining Eqs. (3-21) and (3-22), therefore,

$$\frac{\lambda}{d} = \left(\frac{b}{d}\right)\left(\frac{dn}{d\lambda}\right)\Delta\lambda$$

or the minimum wavelength separation permissible for resolvable images is

$$(\Delta\lambda)_{min} = \frac{\lambda}{b(dn/d\lambda)} \tag{3-23}$$

The *resolving power* provides an alternate way of describing the resolution limit of the instrument. By definition, the resolving power $\Re$

$$\Re = \frac{\lambda}{(\Delta\lambda)_{min}} = b\frac{dn}{d\lambda} \tag{3-24}$$

where we have incorporated Eq. (3-23). Since dispersion is limited by the glass, prism resolving power might be improved by increasing the base b. However, this technique soon requires impractically large and heavy prisms. The dispersion $dn/d\lambda$ may be calculated, for example, from the Cauchy formula for the prism material, using Eq. (3-17).

Example 3-2

Determine the resolving power and minimum resolvable wavelength difference for a prism made from flint glass with a base of 5 cm.

Solution

With the help of Table 3-1 we can calculate an approximate average value of the dispersion $\dfrac{\Delta n}{\Delta \lambda}$ for $\lambda = 550$ nm as

$$\frac{\Delta n}{\Delta \lambda} = \frac{n_F - n_D}{\lambda_F - \lambda_D} = \frac{1.7328 - 1.7205}{486 - 589} = -1.19 \times 10^{-4}\, \text{nm}^{-1}$$

Thus, the resolving power is

$$\Re = b\left(\frac{dn}{d\lambda}\right) = (0.05 \times 10^9\, \text{nm})(1.19 \times 10^{-4}\, \text{nm}^{-1}) = 5971$$

The minimum resolvable wavelength difference in the region around 550 nm is, then,

$$(\Delta \lambda)_{\min} = \frac{\lambda}{\Re} = \frac{5550\, \text{Å}}{5971} \cong 1\, \text{Å}$$

Although grating spectrographs achieve higher resolving powers, they are generally more wasteful of light. Furthermore, they produce higher-order images of the same wavelength component, which can be confusing when interpreting spectral records. These instruments are discussed later.

Prisms with Special Applications

Prisms may be combined to produce *achromatic* overall behavior, that is, the net dispersion for two given wavelengths may be made zero, even though the deviation is not zero. On the other hand, a *direct vision prism*, Figure 3-16b, accomplishes zero deviation for a particular wavelength while at the same time providing dispersion. Schematics involving combinations of these two prism types are shown in Figure 3-16. The arrangement of prisms in Figure 3-16a, combined so that one prism cancels the dispersion of the other, can also be reversed so that the dispersion is additive, providing double dispersion.

A prism design useful in spectrometers is one that produces a constant deviation for all wavelengths as they are observed or detected. One example is the *Pellin-Broca prism*, illustrated in Figure 3-17. A collimated beam of light

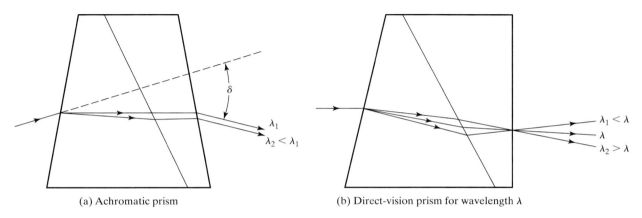

(a) Achromatic prism   (b) Direct-vision prism for wavelength λ

Figure 3-16   Nondispersive and nondeviating prisms.

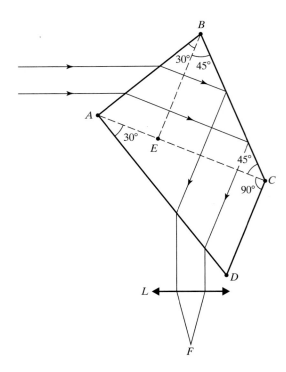

Figure 3-17 Pellin-Broca prism of constant deviation.

enters the prism at face AB and departs at face AD, making an angle of 90° with the incident direction. The dashed lines are merely added to assist in analyzing the operation of the prism, a single structure. Of the incident wavelengths, only one will refract at the precise angle that conforms to the case of minimum deviation, as shown, with the light rays parallel to the prism base AE. At face BC, total internal reflection occurs to direct the light beam into the prism section ACD, where it again traverses under the condition of minimum deviation. Since the prism section BEC serves only as a mirror, the beam passes effectively with minimum deviation through sections ABE and ACD, which together constitute a prism of 60° apex angle. In use, the spectral line is observed or recorded at F, the focal point of lens L. Thus, an observing telescope may be rigidly mounted. Instead, the prism is rotated on its prism table (or about an axis normal to the page), and as it rotates, various wavelengths in the incident beam successively meet the condition of incidence angle for minimum deviation, producing the path indicated, with focus at F. The prism rotation may be calibrated in terms of angle, or better, in terms of wavelength.

Reflecting Prisms

Total internally-reflecting prisms are frequently used in optical systems, both to alter the direction of the optical axis and to change the orientation of an image. Of course, prisms alone cannot produce images. When used in conjunction with image-forming elements, the light incident on the prism is first collimated and rendered normal to the prism face to avoid prismatic aberrations in the image. Plane mirrors may substitute for the reflecting prisms, but the prism's reflecting faces are easier to keep free of contamination, and the process of total internal reflection is capable of higher reflectivity. The stability in the angular relationship of prism faces may also be an important advantage in some applications. Some examples of reflecting prisms in use are illustrated in Figure 3-18. The *Porro prism*, Figure 3-18d, consists of two right-angle prisms, oriented in such a way that the face of one prism is partially revealed to receive the incident light and the face of the second prism is partially revealed to output the refracted light. The prism halves are separated in the figure to clarify its action. Images are inverted in both vertical and horizontal directions by the pair, so that the Porro prism is commonly used in binoculars to produce erect images.

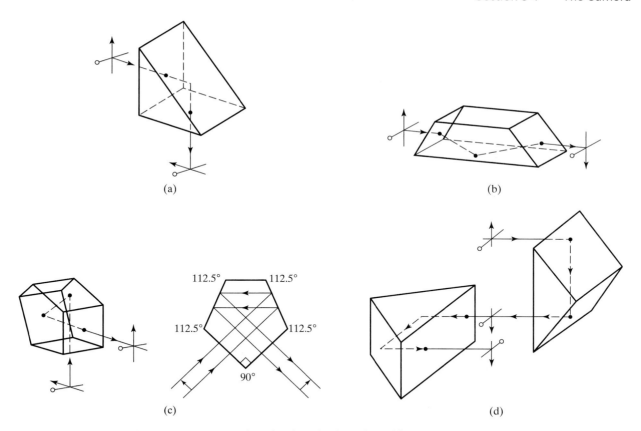

(a)

(b)

112.5°   112.5°

112.5°   112.5°

90°

(c)

(d)

Figure 3-18 Image manipulation by reflecting prisms. (a) Right-angle prism. (b) Dove prism. (c) Penta prism; pentagonal cross section. (d) Porro prism.

3-4 THE CAMERA

The simplest type of camera is the pinhole camera, illustrated in Figure 3-19a. Light rays from an object are admitted into a light-tight box and onto a photographic film through a tiny pinhole, which may be provided with any simple means of shuttering, such as a piece of black tape. An image of the object is projected on the back wall of the box, which is lined with a piece of film.

As stated earlier, an image point is determined ideally when every ray from a given object point, each processed by the optical system, intersects at the corresponding image point. A pinhole does no focusing and actually blocks out most of the rays from each object point. Because of the smallness of the pinhole, however, every point in the image is reached only by rays that originate at *approximately* the same point of the object, as in Figure 3-19b. Alternatively, every object point sends a bundle of rays to the screen, which are limited by the small pinhole and so form a small circle of light on the screen, as in Figure 3-19a. The overlapping of these circles of light due to each object point maps out an image whose sharpness depends on the diameter of the individual circles. If they are too large, the image is blurred. Thus, as the pinhole is reduced in size, the image improves in clarity, until a certain pinhole size is reached. As the pinhole is reduced further, the images of each object point actually grow larger again due to diffraction, with consequent degradation of the image. Experimentally, one finds that the optimum pinhole size is around 0.5 mm when the pinhole-to-film distance is around 25 cm. The pinhole itself must be accurately formed in as thin an aperture as possible. A pinhole in aluminum foil, supported by a larger aperture, works well. The primary advantage of a pinhole camera (other than its elegant simplicity!) is that, since there is no focusing involved, all objects near and far are in focus

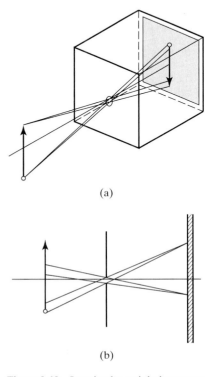

(a)

(b)

Figure 3-19 Imaging by a pinhole camera.

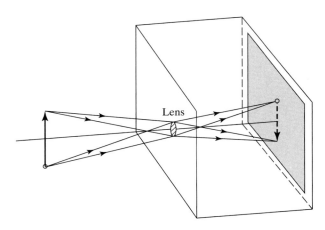

Figure 3-20   Simple camera.

on the screen. In other words, the *depth of field* of the camera is unlimited. The primary disadvantage is that, since the pinhole admits so little of the available light, exposure times must be long. The pinhole camera is not useful in freezing the action of moving objects. The pinhole-to-film distance, while not critical, does affect the sharpness of the image and the field of view. As this distance is reduced, the angular aperture seen by the film is larger, so that more of the scene is recorded, with corresponding decrease in size of any feature of the scene. Also, the image circles decrease in size, producing a clearer image.

If the pinhole aperture is opened sufficiently to accommodate a converging lens, we have the basic elements of the ordinary camera (Figure 3-20). The most immediate benefits of this modification are (1) an increase in the brightness of the image due to the focusing of *all* the rays of light from each object point onto its conjugate image point and (2) an increase in sharpness of the image, also due to the focusing power of the lens. The lens-to-film distance is now critical and depends on the object distance and lens focal length. For distant objects, the film must be situated in the focal plane of the lens. For closer objects, the focus falls beyond the film. Since the film plane is fixed, a focused image is procured by allowing the lens to be moved farther from the film, that is, by "focusing" the camera. The extreme possible position of the lens determines the nearest distance of objects that can be handled by the camera. "Close-ups" can be managed by changing to a lens with shorter focal length. Thus, the focal length of the lens determines the subject area received by the film and the corresponding image size. In general, image size is proportional to focal length. A *wide-angle lens* is a short focal-length lens with a large field of view. A *telephoto lens* is a long focal-length lens, providing magnification at the expense of subject area. The telephoto lens avoids a correspondingly "long" camera by using a positive lens, separated from a second negative lens of shorter focal length, such that the combination remains positive.

Also important to the operation of the camera is the size of its aperture, which admits light to the film. In most cameras, the aperture is variable and is coordinated with the exposure time (shutter speed) to determine the total exposure of the film to light from the scene. The light power incident at the image plane (irradiance E_e in watts per square meter) depends directly on (1) the area of the aperture and inversely on (2) the size of the image. If, as in Figure 3-21, the aperture is circular with diameter D and the energy of the light is assumed to be distributed uniformly over a corresponding image circle of diameter d, then

$$E_e \propto \frac{\text{area of aperture}}{\text{area of image}} = \frac{D^2}{d^2} \qquad (3\text{-}25)$$

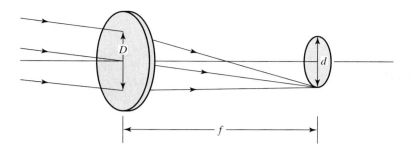

Figure 3-21 Illumination of image. The aperture (not shown) determines the useful diameter D of the lens.

The image size, as in Figure 3-21, is proportional to the focal length of the lens, so we can write

$$E_e \propto \left(\frac{D}{f}\right)^2 \tag{3-26}$$

The quantity f/D is the *relative aperture* of the lens (also called *f*-number or *f*/stop), which we symbolize by the letter A,

$$A \equiv \frac{f}{D} \tag{3-27}$$

but is, unfortunately, usually identified by the symbol f/A. For example, a lens of 4-cm focal length that is stopped down to an aperture of 0.5 cm has a relative aperture of $A = 4/0.5 = 8$. This aperture is usually referred to by photographers as $f/8$. The irradiance is now

$$E_e \propto \frac{1}{A^2} \tag{3-28}$$

Most cameras provide selectable apertures that sequentially change the irradiance at each step by a factor of 2. The corresponding *f*-numbers, then, form a geometric series with ratio $\sqrt{2}$, as in Table 3-2. Larger aperture numbers correspond to smaller exposures. Since the total *exposure* (J/m^2) of the film depends on the product of irradiance $\left(\frac{\text{J}}{\text{m}^2 \cdot \text{s}}\right)$ and time (s), a desirable total exposure may be met in a variety of ways. Accordingly, if a particular film (whose *speed* is described by an ISO number) is perfectly exposed by light from a particular scene with a shutter speed of $\frac{1}{50}$ s and a relative aperture of $f/8$, it will also be perfectly exposed by any other combination that gives the same total exposure, for example, by choosing a shutter speed of $\frac{1}{100}$ s and an aperture of $f/5.6$. The change in shutter speed cuts the total exposure in half, but opening the aperture to the next *f*/stop doubles the exposure, leaving no change in net exposure.

TABLE 3-2 STANDARD RELATIVE APERTURES AND IRRADIANCE AVAILABLE ON CAMERAS

$A = f$-number	$(A = f\text{-number})^2$	E_e
1	1	E_0
1.4	2	$E_0/2$
2	4	$E_0/4$
2.8	8	$E_0/8$
4	16	$E_0/16$
5.6	32	$E_0/32$
8	64	$E_0/64$
11	128	$E_0/128$
16	256	$E_0/256$
22	512	$E_0/512$

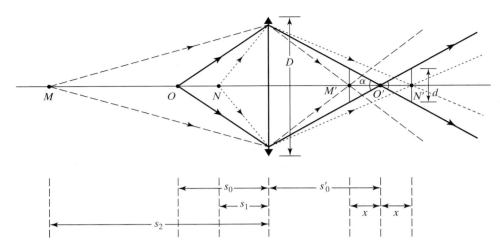

Figure 3-22 Construction illustrating depth of field *MN*. Object and image spaces are not shown to the same scale.

The particular combination of shutter speed and relative aperture chosen for an optimum total exposure is not always arbitrary. The shutter speed must be fast, of course, to capture an action shot without blurring the image. The choice of relative aperture also affects another property of the image, the *depth of field*. To define this quantity precisely, we utilize Figure 3-22, which shows an axial object point O at distance s_0 from a lens being imaged at O', a distance s_0' on the other side. All objects in the object plane are precisely focused in the image plane, disregarding the usual lens aberrations. Objects closer to (s_1), and farther from (s_2), the lens, however, send bundles of rays that focus farther $(s_0' + x)$ from and closer $(s_0' - x)$ to the image plane, respectively. Thus, a flat film, situated at distance s_0' from the lens, intercepts *circles of confusion* corresponding to such object points. If the diameters of these circles are small enough, the resultant image is still acceptable. Suppose the largest acceptable diameter is d, as shown, such that all images within a distance x of the precise image are suitably "in focus." The *depth of field* is then said to be the interval MN in object space conjugate to the interval $M'N'$, as shown. Notice that although the interval $M'N'$ is symmetric about s_0' in image space, the depth of field interval (MN) is not symmetric about s_0 in object space.

The near-point and far-point distances, s_1 and s_2, of the depth of field (MN) can be determined once the allowable blurring parameter d is chosen and the lens is specified by focal length and relative aperture. The angle α in Figure 3-22 may be specified in two ways,

$$\tan \alpha \cong \frac{D}{s_0'} \quad \text{and} \quad \tan \alpha \cong \frac{d}{x}$$

so that

$$x \cong \frac{ds_0'}{D} \tag{3-29}$$

It is then required to find, from the lens equation, the object distance s_1 corresponding to image distance $s_0' + x$ and the object distance s_2 corresponding to image distance $s_0' - x$. After a moderate amount of algebra, one finds

$$s_1 = \frac{s_0 f(f + Ad)}{f^2 + Ads_0} \tag{3-30}$$

$$s_2 = \frac{s_0 f(f - Ad)}{f^2 - Ads_0} \tag{3-31}$$

where the aperture is $A = f/D$. The depth of field, $MN = s_2 - s_1$, can be expressed as

$$\text{depth of field} = \frac{2Ads_0(s_0 - f)f^2}{f^4 - A^2d^2s_0^2} \tag{3-32}$$

Acceptable values of the circle diameter d depend on the quality of the photograph desired. A slide that will be projected or a negative that will be enlarged requires better original detail and hence a smaller value for d. For most photographic work, d is of the order of thousandths of an inch.

Example 3-3

A 5 cm focal length lens with an $f/16$ aperture is used to image an object 9 ft away. The blurring diameter in the image is chosen to be $d = 0.04$ mm.

Problem

Determine the location of the near point (s_1), far point (s_2), and the depth of field.

Solution

Based on Figure 3-22 and the given data, we have:

$$s_0 = 9 \text{ ft} \simeq 275 \text{ cm} \qquad d = 0.004 \text{ cm}$$
$$f = 5 \text{ cm} \qquad\qquad A = 16$$

For the near point, using Eq. (3-30),

$$s_1 = \frac{s_0 f(f + Ad)}{f^2 + Ads_0} = \frac{(275)(5)[5 + 16(0.004)]}{25 + 16(0.004)(275)}\text{cm} = 163.5 \text{ cm} \simeq 5.4 \text{ ft}$$

For the far point, using Eq. (3-31),

$$s_2 = \frac{s_0 f(f - Ad)}{f^2 - Ads_0} = \frac{(275)(5)[5 - 16(0.004)]}{25 - 16(0.004)(275)}\text{cm} = 1103 \text{ cm} \simeq 30 \text{ ft}$$

Thus, the depth of field, MN in Figure 3-22, is about 25 ft for a 5-cm focal length lens imaging an object 9 ft away. In effect this lens will image all objects from 5 ft to 30 ft with an acceptable sharpness.

Most cameras are equipped with a depth-of-field scale from which values of s_1 and s_2 can be read, once the object distance and aperture are selected. According to Eq. (3-32), depth of field is greater for smaller apertures (larger f-numbers), shorter focal lengths, and longer shooting distances.

The camera lens is called upon to perform a prodigious task. It must provide a large field of view, in the range of 35° to 65° for normal lenses and as large as 120° or more for wide-angle lenses. The image must be in focus and reasonably free from aberrations over the entire area of the film in the focal plane. The aberrations that must be reduced to an acceptable degree are, in addition to chromatic aberration, the five monochromatic aberrations: spherical aberration, coma, astigmatism, curvature of field, and distortion. Since a corrective measure for one type of aberration often causes greater degradation in the image due to another type of aberration, the optical solution represents one of many possible compromise lens designs. The labor involved in the design of a suitable lens that meets particular specifications within acceptable limits has been reduced considerably with the help of computer programming. Human ingenuity is nevertheless an essential component in the design task, since there is more than one optical solution to a given set of specifications. The demands made upon a photographic lens cannot all be met using a single element. Various stages in solving the lens design problem are illustrated in Figure 3-23a, from the single-element meniscus lens, which may still be found in simple cameras, to the four-element Tessar lens. The use of a symmetrical placement of lenses, or groups of lenses, with respect to the aperture is often a distinctive feature of such lens designs. In such placements, one group may reverse the aberrations introduced by the other, reducing overall image degradation due to factors

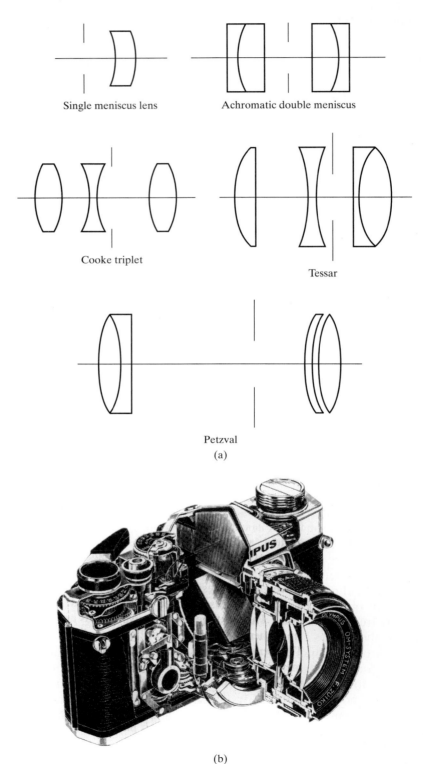

Single meniscus lens

Achromatic double meniscus

Cooke triplet

Tessar

Petzval

(a)

(b)

Figure 3-23   (a) Camera lens design. (b) Cutaway view of a 35-mm camera, revealing the multiple element lens. (Courtesy Olympus Corp., Woodbury, N.Y.)

such as coma, distortion, and lateral chromatic aberration. The multiple-element lens in a 35-mm camera is shown in the cutaway photo (Figure 3-23b).

3-5 SIMPLE MAGNIFIERS AND EYEPIECES

The simple magnifier is essentially a positive lens used to read small print, in which case it is often called a *reading glass*, or to assist the eye in examining small detail in a real object. It is often a simple convex lens but may be a doublet or a triplet, thereby providing for higher-quality images.

Figure 3-24 illustrates the working principle of the *simple magnifier*. A small object of dimension h, when examined by the *unaided* eye, is assumed to be held at the *near point* of the normal eye—nearest position of distinct vision—at position (a), 25 cm from the eye. At this position the object subtends an angle α_0 at the eye. To project a larger image on the retina, the simple magnifier is inserted and the object is moved physically closer to position (b), where it is at or just inside the focal point of the lens. In this position, the lens forms a virtual image subtending a larger angle α_M at the eye. The *angular magnification*[3] of the simple magnifier is defined to be the ratio α_M/α_0. In the paraxial approximation, the angles may be represented by their tangents, giving

$$\frac{\alpha_M}{\alpha_0} = \frac{h/s}{h/25} = \frac{25}{s}$$

If the image is viewed at infinity, $s = f$ and

$$M = \frac{25}{f} \qquad \text{image at infinity} \qquad (3\text{-}33)$$

At the other extreme, if the virtual image is viewed at the nearpoint of the eye, then $s' = -25$ cm, and from the thin-lens equation,

$$s = \frac{25f}{25 + f}$$

giving a magnification of

$$M = \frac{25}{f} + 1 \quad \text{image at normal near point} \qquad (3\text{-}34)$$

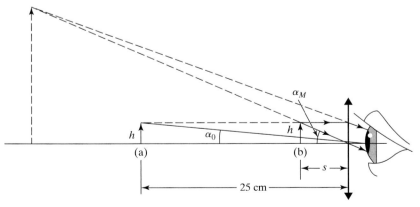

Figure 3-24 Operation of a simple magnifier.

---

[3]When viewing virtual images with optical instruments, the images may be at great distances, even "at infinity," when rays entering the eye are parallel. In such cases, lateral magnifications also approach infinity and are not very useful. The more convenient *angular magnification* is clearly a measure of the image size formed on the retina and is used to describe magnification when eyepieces are involved, as in microscopes and telescopes.

The actual angular magnification depends, then, on the particular viewer, who will move the simple magnifier until the virtual image is seen comfortably. For small focal lengths, Eqs. (3-33) and (3-34) do not differ greatly, and in citing magnifications, Eq. (3-33) is most often used. Simple magnifiers may have magnifications in the range of 2× to 10×, although the achievement of higher magnifications usually requires a lens corrected for aberrations.

In general, when magnifiers are used to aid the eye in viewing images formed by prior components of an optical system, they are called *oculars*, or *eyepieces*. The real image formed by the objective lens of a microscope, for example, serves as the object that is viewed by the eyepiece, whose angular magnification contributes to the overall magnification of the instrument. To provide quality images, the ocular is corrected to some extent for aberrations and, in particular, to reduce transverse chromatic aberration. To accomplish this improvement, two lenses are most often used. The effective focal length f of two thin lenses, separated by a distance L, is given, as shown in Chapter 18, by

$$\frac{1}{f} = \frac{1}{f_1} + \frac{1}{f_2} - \frac{L}{f_1 f_2} \tag{3-35}$$

where f_1 and f_2 represent the individual focal lengths of the pair. By the lensmaker's formula, for lenses made of the same glass,

$$\frac{1}{f_1} = (n - 1)\left(\frac{1}{R_{11}} - \frac{1}{R_{12}}\right) = (n - 1)K_1 \tag{3-36}$$

and

$$\frac{1}{f_2} = (n - 1)\left(\frac{1}{R_{21}} - \frac{1}{R_{22}}\right) = (n - 1)K_2 \tag{3-37}$$

where the expressions in parentheses involving the radii of curvature of the lens surfaces are symbolized by constants K_1 and K_2, respectively. Incorporating Eqs. (3-36) and (3-37) into Eq. (3-35),

$$\frac{1}{f} = (n - 1)K_1 + (n - 1)K_2 - L(n - 1)^2 K_1 K_2 \tag{3-38}$$

To correct for transverse chromatic aberration, we require that the effective focal length of the pair remain independent of refractive index,[4] or

$$\frac{d(1/f)}{dn} = 0$$

From Eq. (3-38),

$$\frac{d(1/f)}{dn} = K_1 + K_2 - 2LK_1 K_2 (n - 1) = 0$$

---

[4]Some longitudinal chromatic aberration remains because the principal planes of the system do not coincide. Refer to Chapter 20 and the related discussion on chromatic aberration.

This condition is met, therefore, when the lenses are separated by the distance

$$L = \frac{1}{2}\left[\frac{1}{K_1(n-1)} + \frac{1}{K_2(n-1)}\right]$$

or, more simply, when

$$L = \tfrac{1}{2}(f_1 + f_2) \qquad (3\text{-}39)$$

This condition is valid independent of the lens shapes, leaving the choice of shapes as latitude for compensating other aberrations.

Both the *Huygens* and *Ramsden eyepieces*, Figures 3-25 and 3-26, incorporate the design feature required by Eq. (3-39); that is, plano-convex lenses are separated by half the sum of their focal lengths. In the diagram of Figure 3-25, the focal length of the field lens, FL, is approximately 1.7 times the focal length of the eye lens, or ocular, EL. The primary image "observed" by the eyepiece is in this case a virtual object (VO) for the field lens, since its virtual position falls between the lenses. The field lens then forms a real image (RI) that is viewed by the eye lens. When the real image falls in the focal plane of the eye lens, the magnified image is viewed at infinity by the eye located at the exit pupil. Note that the Huygens eyepiece cannot be used as an ordinary magnifier. If crosshairs or a reticle with a scale is used with the eyepiece to make possible quantitative measurements, then to be in focus with the image formed by the ocular EL, the crosshairs must be placed in the focal plane of RI, conveniently attached to the field or aperture stop placed there (Figure 3-27). The image of the crosshairs does not share in the image quality provided by the eyepiece as a whole, however, because the eye lens alone is involved in forming the image. This is not a problem in the Ramsden eyepiece, Figure 3-26, in

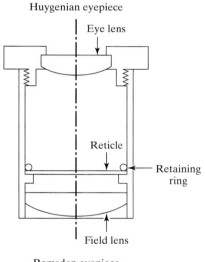

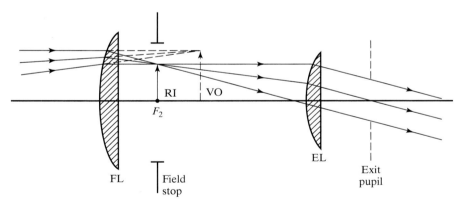

Figure 3-25 Huygens eyepiece.

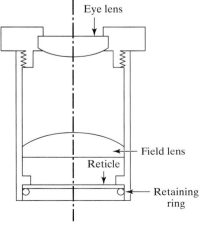

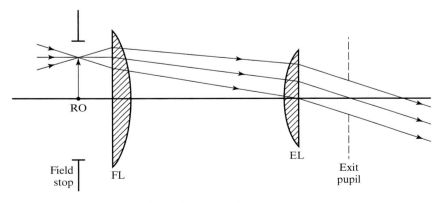

Figure 3-26 Ramsden eyepiece.

Figure 3-27 Construction of Huygens and Ramsden eyepieces.

which both the primary and intermediate images are located just in front of the field lens. In this eyepiece, the lenses have the same focal length f and, according to Eq. (3-39), are separated by f. Ideally, when the real object, RO, falls at the position of the first lens, rays emerge from the eyepiece parallel to one another, giving a virtual magnified image at infinity. Thus a reticle, the field stop, and field lens are all essentially in the "same plane." A disadvantage of this arrangement is that the surface of the lens is then also in focus, including dust and smudges. By using a lens separation slightly smaller than f, the reticle is in focus at a position slightly in front of the lens, as shown in the ray diagram of Figure 3-26 and in Figure 3-27. With a lens separation somewhat less than f, however, the requirement on L that corrects for transverse chromatic aberration is somewhat violated. A modification of the Ramsden eyepiece that almost eliminates chromatic defects is the *Kellner eyepiece*, which replaces the Ramsden eye lens with an achromatic doublet. Other eyepieces have also been designed to achieve higher magnifications and wider fields.

Example 3-4

A Huygens eyepiece uses two lenses having focal lengths of 6.25 cm and 2.50 cm, respectively. Determine their optimum separation in reducing chromatic aberration, their equivalent focal length, and their angular magnification when viewing an image at infinity.

Solution

The optimum separation is given by

$$L = \tfrac{1}{2}(f_1 + f_2) = \tfrac{1}{2}(6.25 + 2.50) = 4.375 \text{ cm}$$

The equivalent focal length is found from

$$\frac{1}{f} = \frac{1}{f_1} + \frac{1}{f_2} - \frac{L}{f_1 f_2} = \frac{1}{6.25} + \frac{1}{2.50} - \frac{4.375}{(6.25)(2.50)}$$

which gives $f = 3.57$ cm. The angular magnification is

$$M = \frac{25}{f} = \frac{25}{3.57} = 7$$

In designing eyepieces, one usually desires an exit pupil that is not much greater than the size of the pupil of the eye, so that radiance is not lost. Recall that, in this instance, the exit pupil is an image of the entrance pupil as formed by the ocular and that the ratio of entrance to exit pupil diameters equals the magnification. Since the entrance pupil is determined by preceding optical elements in the optical system (the diameter of the objective lens, in a simple telescope), this requirement places a limit on the magnifying power of the eyepiece and, thus, a lower limit on its focal length.

The important specifications of an eyepiece, assuming its aberrations are within acceptable limits for a particular application, include the following:

1. Angular magnification, given by $25/f$, where f is the focal length in centimeters. Available values are 4× to 25×, corresponding to focal lengths of 6.25 to 1 cm or less.
2. Eye relief, that is, the distance from eye lens to exit pupil. Available eyepieces have eye reliefs in the range of 6 to 26 mm.
3. Field-of-view, or size of the primary image that the eyepiece can cover, in the range of 6 to 30 mm.

3-6 MICROSCOPES

The magnification of small objects accomplished by the simple magnifier is increased further by the compound microscope. In its simplest form, the instrument consists of two positive lenses, an objective lens of small focal length that faces the object and a magnifier functioning as an eyepiece. The eyepiece "looks" at the real image formed by the objective. Referring to Figure 3-28, where the object lies outside the focal length f_o of the objective, a real image I is formed within the microscope tube. After coming to a focus at I, the light rays continue to the eyepiece, or ocular lens. For visual observations, the intermediate image is made to occur at or just inside the first focal point f_e of the eyepiece. The eye positioned near the eyepiece—at the E_xP—then sees a virtual image, inverted and magnified, as shown. The objective lens functions as the aperture stop and entrance pupil of the optical system. The image of the objective formed by the eyepiece is then the exit pupil, which locates the position of maximum radiant energy density and thus the optimum position for the entrance pupil of the eye. A special aperture, functioning as a field stop, is placed at the position of the intermediate image I. The eye then sees both in focus together, giving the field of view a sharply defined boundary. If a camera is attached to the microscope, a real final image is required. In this case, the intermediate image I must be located outside the ocular focal length f_e.

Total Magnification

When the final image is viewed by the eye, the magnification of the microscope may be defined as in the case of the simple magnifier. Thus, the angular magnification for an image viewed at infinity is

$$M = \frac{25}{f_{\text{eff}}} \tag{3-40}$$

where f_{eff} (in cm) is the effective focal length of the two lenses, separated by a distance d, and given by Eq. (3-35).

$$\frac{1}{f_{\text{eff}}} = \frac{1}{f_o} + \frac{1}{f_e} - \frac{d}{f_o f_e} \tag{3-41}$$

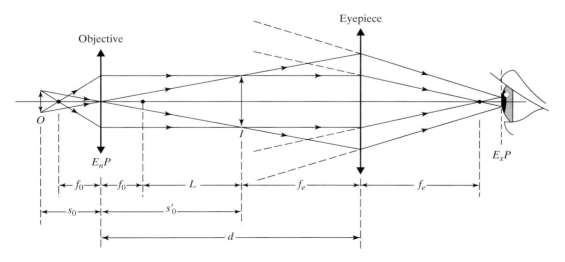

Figure 3-28   Image formation in a compound microscope.

Substituting Eq. (3-41) into Eq. (3-40),

$$M = \frac{25(f_e + f_o - d)}{f_o f_e} \tag{3-42}$$

Based on an algebraic manipulation of the thin-lens equation, however, we can show that the ratio of image to object distance, s'_o/s_o, for the objective lens is

$$\frac{s'_o}{s_o} = \frac{d - f_e - f_o}{f_o} \tag{3-43}$$

where we have used the fact that $s'_o = d - f_e$, evident in the diagram. Incorporating Eq. (3-43) into Eq. (3-42),

$$M = -\left(\frac{s'_o}{s_o}\right)\left(\frac{25}{f_e}\right) \tag{3-44}$$

showing that the total magnification is just the product of the linear magnification of the objective (s'_o/s_o) multiplied by the angular magnification of the eyepiece ($25/f_e$) when viewing the final image at infinity. The negative sign indicates an inverted image. Comparing Figure 3-28 with Figure 2-26—the geometry associated with Newton's equation for a thin lens—we see that the magnitude of the lateral magnification is given by

$$|m| = \left|\frac{h_i}{h_o}\right| = \left|\frac{s'_o}{s_o}\right| = \frac{x'}{f_o} = \frac{L}{f_o} \tag{3-45}$$

since $x' = L$ is the distance between the objective image and its second focal point, as shown. The magnification of the microscope may then be expressed, perhaps more conveniently, as

$$M = -\left(\frac{25}{f_e}\right)\left(\frac{L}{f_o}\right) \tag{3-46}$$

In many microscopes, the length L is standardized at 16 cm. The focal lengths f_e and f_o are themselves effective focal lengths of multielement lenses, appropriately corrected for aberrations.

Example 3-5

A microscope has an objective of 3.8-cm focal length and an eyepiece of 5-cm focal length. If the distance between the lenses is 16.4 cm, find the magnification of the microscope.

Solution

$$L = d - f_o - f_e = 16.4 - 3.8 - 5 = 7.6 \text{ cm}$$

and

$$M = -\left(\frac{25}{f_e}\right)\left(\frac{L}{f_o}\right) = -\left(\frac{25}{5}\right)\left(\frac{7.6}{3.8}\right) = -10, \qquad \text{a magnification of } 10\times$$

Numerical Aperture

To collect more light and produce brighter images, cones of rays from the object, intercepted by the objective lens (usually the aperture stop), should

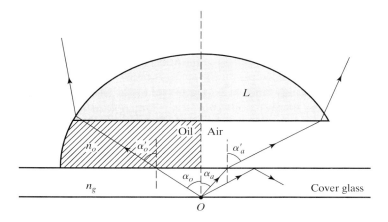

Figure 3-29   Microscope objective, illustrating the increased light-gathering power of an oil-immersion lens.

be as large as possible. As magnifications increase and the focal lengths and diameters of the objective lenses decrease corespondingly, the solid angle of useful rays from the object also decreases. In Figure 3-29, the useful light rays originating at the object point O, passing through a thin cover glass and then air to the first element of the objective lens L, make an intitial half-angle of α_a on the right of the optical axis. Due to refraction at the glass-air interface, rays making a larger angle than α_a do not reach the lens. This limitation is somewhat relieved by using a transparent fluid "coupler" whose index matches as closely as possible that of the glass. On the left of the optical axis in the diagram, a layer of oil is shown, and a larger half-angle α_o is possible. Typically, the cover glass index is 1.522 and the oil index is 1.516, providing an excellent match. The light-gathering capability of the objective lens is thus increased by increasing the refractive index in object space.

A measure of this capability is the product of half-angle and refractive index, called the *numerical aperture* N. A. where

$$\text{N. A.} = n \sin \alpha \qquad (3\text{-}47)$$

The numerical aperture is an invariant in object space, due to Snell's law. That is, in the case of air,

$$\text{N. A.} = n_g \sin \alpha_a = \sin \alpha_a'$$

and when an *oil-immersion objective* is used,

$$\text{N. A.} = n_g \sin \alpha_o = n_o \sin \alpha_o'$$

The maximum value of the numerical aperture when air is used is unity, but when object space is filled with a fluid of index n, the maximum numerical aperture may be increased up to the value of n. In practice, the limit is around 1.6. The numerical aperture is an alternative means of defining a relative aperture or of describing how "fast" a lens is. As shown previously, image brightness is inversely proportional to the square of the f-number. Here also, image brightness is proportional to the square of the numerical aperture. The numerical aperture is an important design parameter also because it limits the resolving power and the depth of focus of the lens. The resolving power is proportional to the numerical aperture, whereas the depth of focus is inversely proportional to the square of the numerical aperture. Most microscopes use objectives with numerical apertures in the approximate range of 0.08 to 1.30.

Biological specimens are covered with a cover glass of 0.17- or 0.18-mm thickness. For objectives with numerical apertures over 0.30, the cover glass has increasing influence on the image quality, since it introduces a large degree of spherical aberration when oil immersion is not involved. Thus, a *biological objective* compensates for the aberration introduced by a cover glass. In contrast, a *metallurgical objective* is designed without such compensation. Objectives may be classified broadly in relation to the corrections introduced into their design. For low magnifications, with focal lengths in the range of 8 to 64 mm, *achromatic objectives* are generally used. Such objectives are chromatically corrected, usually for the Fraunhofer *C* (red) and *F* (blue) wavelengths, and spherically corrected, at the Fraunhofer *D* (sodium yellow) wavelength. For higher magnifications, objective lenses with focal lengths in the range of 4 to 16 mm incorporate some fluorite elements, which together with the glass elements provide better correction over the visual spectrum. When the correction is nearly perfect throughout the visual spectrum, the objectives are said to be *apochromatic*. Since correction is more crucial at high magnifications, apochromats are usually objectives with focal lengths in the range of 1.5 to 4 mm. For even higher magnifications, the objective is usually designed as an *immersion objective*. Modern techniques and materials have also made possible *flat-field objectives* that essentially eliminate field curvature over the useful portion of the field. With ultraviolet immersion microscopes, it is customary to replace the oil with glycerine and the optical glass elements with fluorite and quartz elements because of their higher transmissivity at short wavelengths.

This discussion should make it clear that high-quality microscopes today are designed as a whole and usually for a specific use. The design of an objective or an eyepiece is directly related to the performance of other optical elements in the instrument, often including a relay lens within the body tube of the microscope as well. Thus it is generally not possible to interchange objectives and eyepieces between different model microscopes without loss or deterioration of the image.

Figure 3-30 illustrates the optical components in a standard microscope and the detailed processing of light rays through the instrument.

3-7 TELESCOPES

Telescopes may be broadly classified as *refracting* or *reflecting*, according to whether lenses or mirrors are used to produce the image. There are, in addition, *catadioptric* systems that combine refracting and reflecting surfaces. Telescopes may also be distinguished by the erectness or inversion of the final image and by either a visual or photographic means of observation.

Refracting Telescopes
Figures 3-31 and 3-32 show two refracting telescope types, producing, respectively, inverted and erect images. The *Keplerian telescope* in Figure 3-31 is often referred to as an *astronomical telescope* since inversion of astronomical objects in the images produced creates no difficulties. The *Galilean telescope*, illustrated in Figure 3-32, produces an erect image by means of an eyepiece of negative focal length. In either case, nearly parallel rays of light from a distant object are collected by a positive objective lens, which forms a real image in its focal plane. The objective lens, being larger in diameter than the pupil of the eye, permits the collection of more light and makes visible those point sources such as stars that might otherwise not be detected. The objective lens is usually a doublet, corrected for chromatic aberration. The real image formed by the objective is observed with an eyepiece, represented in the figures as a single lens. This intermediate image, located at or near the focal point of the ocular, serves as a real object (RO) for the ocular in the astronomical telescope and a virtual object (VO) in the case of the Galilean telescope. In either case, the light is refracted by the eyepiece in order to produce parallel, or nearly parallel, light rays. An eye placed near the ocular views an image at infinity but with an angular magnification given by the ratio of the

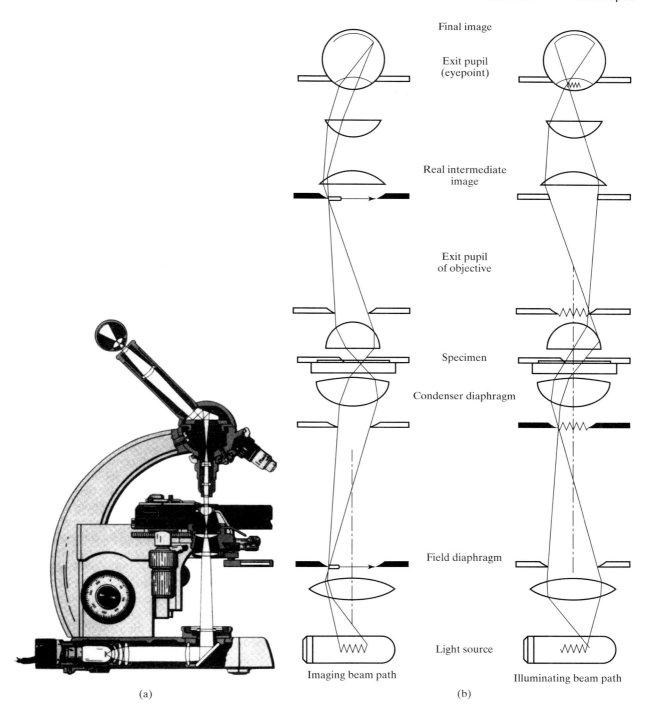

Final image

Exit pupil
(eyepoint)

Real intermediate
image

Exit pupil
of objective

Specimen

Condenser diaphragm

Field diaphragm

Light source

Imaging beam path

Illuminating beam path

(a)

(b)

Figure 3-30 (a) Standard microscope illustrating Koehler illumination. (b) Schematic showing detailed ray traces through the instrument both for object illumination and image formation. (Courtesy Carl Zeiss, Inc., Thornwood, N.Y.)

angles α'/α, as shown. The object subtends the angle α at the unaided eye and the angle α' at the eyepiece.

From the two right triangles formed by the intermediate image and the optical axis, it is evident that the angular magnification is

$$M = \frac{\alpha'}{\alpha} = -\frac{f_o}{f_e} \qquad (3\text{-}48)$$

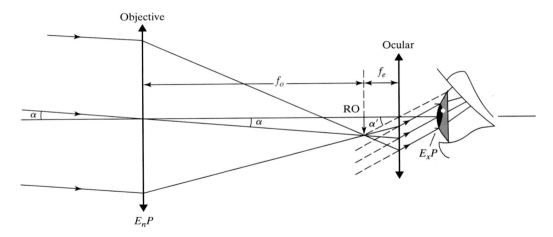

Figure 3-31   Astronomical telescope.

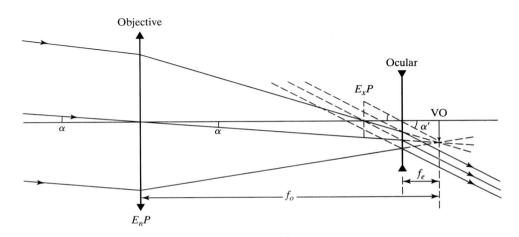

Figure 3-32   Galilean telescope.

The negative sign is introduced, as usual, to indicate that the image is inverted in Figure 3-31, where $f_e > 0$, and is erect in Figure 3-32, where $f_e < 0$. In either case, the length L of the telescope is given by

$$L = f_o + f_e \qquad (3\text{-}49)$$

permitting a short Galilean telescope, a circumstance that makes this design convenient in the *opera glass*. The astronomical telescope may be modified to produce an erect image by the insertion of a third positive lens whose function is simply to invert the intermediate image, but this lengthens the telescope by at least four times the focal length of the additional lens. Image inversion may also be achieved without additional length, as in binoculars, through the use of inverting Porro prisms, discussed previously.

The objective lens of either telescope functions as the aperture stop and entrance pupil, whose image in the ocular is then the exit pupil, as shown. In the astronomical telescope, the exit pupil is situated just outside the eyepiece and is designed to match the size of the pupil of the eye. A telescope should produce an exit pupil at sufficient distance from the eyepiece to produce a comfortable *eye relief*. Greater ease of observation is also achieved if the exit pupil is a little larger in diameter than the eye pupil, allowing for some relative motion between eye and eyepiece. Notice that in the Galilean telescope the exit pupil falls inside the eyepiece, where it is inaccessible to the eye. This represents a disadvantage of the Galilean telescope, leading to a restriction in the field of view. Notice also that a field stop with reticle can be employed at the location of the intermediate image in the astronomical telescope, whereas no such arrangement is possible in conjunction with the Galilean telescope. The diameter of the exit pupil D_{ex} is simply related to the diameter of the objective lens D_{obj} through the angular magnification, as follows. Since the exit pupil is the image of the entrance pupil formed by the eyepiece, we may write for the linear, transverse magnification either

$$m_e = \frac{D_{ex}}{D_{obj}} \qquad (3\text{-}50)$$

or, employing the Newtonian form of the magnification,

$$m_e = -\frac{f}{x} = -\frac{f_e}{f_0} \qquad (3\text{-}51)$$

where x is the distance of the object (objective lens) from the focal point of the eyepiece, or f_0. Combining Eqs. (3-48), (3-50), and (3-51),

$$m_e = \frac{1}{M} = \frac{D_{ex}}{D_{obj}}$$

so that

$$D_{ex} = \frac{D_{obj}}{M} \qquad (3\text{-}52)$$

Thus, the diameter of the bundle of parallel rays filling the objective lens is greater by a factor of M than the diameter of the bundle of rays that pass through the exit pupil. It should be pointed out that the image is not, therefore, brighter by the same proportion, however, because the apparent size of the image increases by the same factor M. The brightness of the image cannot be greater than the brightness of the object; in fact, it is less bright due to inevitable light losses due to reflections from lens surfaces.

Binoculars (Figure 3-33) afford more comfortable telescopic viewing, allowing both eyes to remain active. In addition, the use of Porro or other prisms to produce erect final images also permits the distance between objective lenses to be greater than the interpupillary distance, enhancing the stereoscopic effect produced by ordinary binocular vision. The designation "6 × 30" for binoculars means that the angular magnification M produced is 6× and the diameter of the objective lens is 30 mm. Using Eq. (3-52), we conclude that the exit pupil for this pair of binoculars is 5 mm, a good match for the normal pupil diameter. For night viewing, when the pupils are somewhat larger, a rating of 7 × 50, producing an exit pupil diameter of 7 mm, would be preferable.

Figure 3-33 Cutaway view of binoculars revealing compound objective and ocular lenses and image-inverting prism. (Courtesy Carl Zeiss, Inc., Thornwood, N.Y.)

Example 3-6

Determine the eye relief and field of view for the 6 × 30 binoculars just described. Assume an objective focal length of 15 cm and a field lens (eyepiece) diameter of 1.50 cm.

Solution

The focal length of the ocular is found from

$$f_e = -\frac{f_o}{M} = -\frac{15}{-6} = 2.5 \text{ cm}$$

The eye relief is the distance of the exit pupil from the eyepiece. Since the exit pupil is the image of the objective formed by the eyepiece, the eye relief is the image distance s', given by

$$s' = \frac{sf}{s-f} = \frac{Lf_e}{L-f_e} = \frac{(f_o + f_e)f_e}{(f_o + f_e) - f_e} = \frac{(15 + 2.5)(2.5)}{15} = 2.92 \text{ cm}$$

The angular field of view from the objective subtends both the object on one side and the field lens of the eyepiece on the other. Thus, for objects at a standard distance of 1000 yd,

$$\theta = \frac{h}{s} = \frac{D_f}{L}$$

or

$$h = s\theta = \frac{sD_f}{L} = \frac{(3000 \text{ ft})(1.50)}{15 + 2.5} = 257 \text{ ft at 1000 yd}$$

Reflection Telescopes

Larger-aperture objective lenses provide greater light-gathering power and resolution. Large homogeneous lenses are difficult to produce without optical defects, and their weight is difficult to support. These problems, as well as the elimination of chromatic aberrations, are solved by using curved, reflecting surfaces in place of lenses. The largest telescopes, like the Hale 200-in. reflector on Mount Palomar, use such mirrors. Such large reflecting telescopes are usually employed to examine very faint astronomical objects and use the integrating power of photographic plates, exposed over long time intervals, in observations.

Several basic designs for reflecting telescopes are shown in Figure 3-34. In the *Newtonian* design (a), a *parabolic* mirror is used to focus accurately all parallel rays to the same primary focal point, f_p. Before focusing, a plane mirror is used to divert the converging rays to a secondary focal point, f_s, near the body of the telescope, where an eyepiece is located to view the image. The use of a parabolic mirror avoids both chromatic and spherical aberration, but coma is present for off-axis points, severely limiting the useful field of view. In the 200-in. Hale telescope, the flat mirror can be dispensed with and the rays allowed to converge at their primary focus. This telescope is large enough so that the observer can be mounted on a specially built platform situated just behind the primary focus (Figure 3-35). Of course, any obstruction placed inside the telescope reduces the cross section of the incident light waves contributing to the image. In the *Cassegrain* design (Figure 3-34b), the secondary mirror is hyperboloidal convex in shape, reflecting light from the primary mirror through an aperture in the primary mirror to a secondary focus, where it is conveniently viewed or recorded. The hyperboloidal surface permits perfect imaging between the primary and secondary focal points, which function as the foci of the hyperboloid. Such accurate imaging is also possible when the secondary mirror is concave ellipsoidal, as in the *Gregorian* telescope (Figure 3-34c). The primary and secondary focal points of this telescope are now the foci of the ellipsoid.

The Schmidt Telescope

Perhaps the most celebrated catadioptric telescope is due to a design of Bernhardt Schmidt. He sought to remove the spherical aberration of a primary

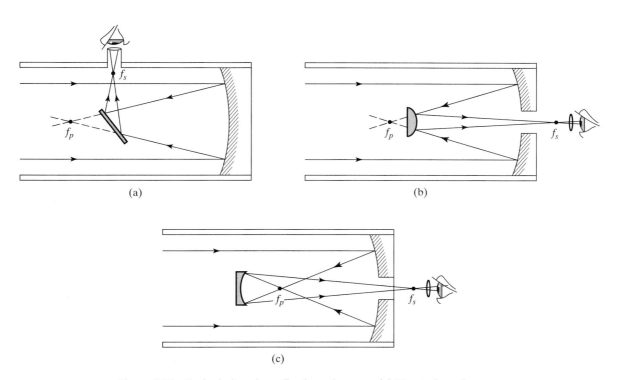

Figure 3-34 Basic designs for reflecting telescopes. (a) Newtonian telescope. (b) Cassegrain telescope. (c) Gregorian telescope.

Figure 3-35 Hale telescope (200-in.) showing observer in prime-focus cage and reflecting surface of 200-in. mirror. (California Institute of Technology.)

spherical mirror by using a thin refracting correcting plate at the aperture of the telescope. To understand his design, refer to Figure 3-36. A concave primary reflector in (a) receives small bundles of parallel rays from various directions. Each bundle enters at the aperture, which is located at the center of curvature of the primary mirror. Since the axis of any bundle may be considered an optical axis, there are no off-axis points and thus coma and astigmatism do not enter into the aberrations of the system. When the bundles are small, each bundle consists of paraxial rays that focus at the same distance from the mirror, a distance equal to its focal length, or half the radius of curvature of the mirror. The locus of such image points is then the spherical surface indicated by the dashed line. However, when the bundles are large, as shown in (b), spherical aberration occurs, which produces a shorter focus for rays reflecting from the outer zones of the mirror relative to the optical axis of the bundle. Schmidt designed a transparent correcting plate, to be placed at the aperture, whose function was to bring the focus of all zones to the same point on the spherical focal surface, as indicated in (c). The shape suggested in the figure is designed to make the focal point of all zones agree with the focal point of a zone whose radius is 0.707 of the aperture radius, the usual choice. The resulting *Schmidt optical system* is therefore highly corrected for coma, astigmatism, and spherical aberration. Because the correcting plate is situated at the center of curvature of the mirror, it presents approximately the same optics to parallel beams arriving from different directions and so permits a wide field of view. Residual aberrations are due to errors in the actual fabrication of the correcting plate and because the plate does not present precisely the same cross section, and therefore the same correction, to beams entering from different directions. One disadvantage is that the focal plane is spherical, requiring a careful shaping of photographic films and plates. Notice also that with the correcting plate attached at twice the focal length of the mirror, the telescope is twice as long as the telescopes described previously in Figure 3-34. Nevertheless, the *Schmidt camera*, as it is often called, has been highly successful and has spawned a large number of variants, including designs to flatten the field near the focal plane.

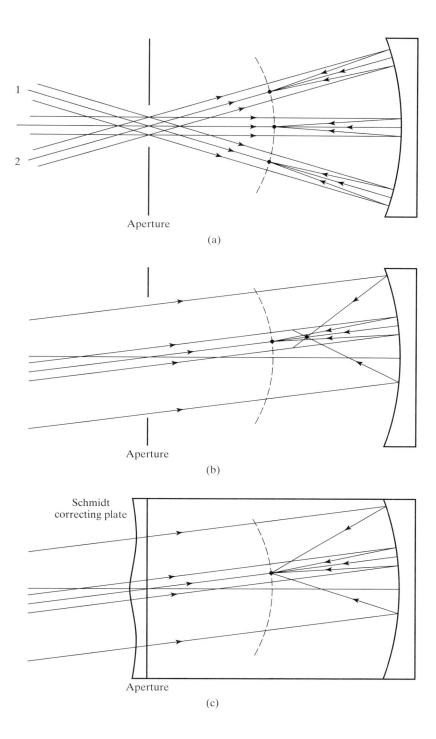

(a)

(b)

Schmidt
correcting plate

Aperture

(c)

Figure 3-36 The Schmidt optical system.

PROBLEMS

3-1 An object measures 2 cm high above the axis of an optical system consisting of a 2-cm aperture stop and a thin convex lens of 5-cm focal length and 5-cm aperture. The object is 10 cm in front of the lens and the stop is 2 cm in front of the lens. Determine the position and size of the entrance and exit pupils, as well as the image. Sketch the chief ray and the two extreme rays through the optical system, from the top of the object to its conjugate image point.

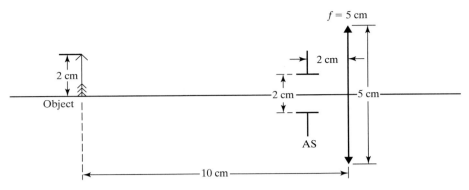

Figure 3-37   Problem 3-1.

3-2 Repeat problem 3-1 for an object 4 cm high, with a 2-cm aperture stop and a thin convex lens of 6-cm focal length and 5-cm aperture. The object is 14 cm in front of the lens and the stop is 2.50 cm behind the lens.

3-3 Repeat problem 3-1 for an object 2 cm high, with a 2-cm aperture stop and a thin convex lens of 6-cm focal length and 5-cm aperture. The object is 14 cm in front of the lens and the stop is 4 cm in front of the lens.

3-4 An optical system, centered on an optical axis, consists of (left to right)

1. Source plane
2. Thin lens L_1 at 40 cm from the source plane
3. Aperture A at 20 cm farther from L_1
4. Thin lens L_2 at 10 cm farther from A
5. Image plane

Lens L_1 has a focal length of 40/3 cm and a diameter of 2 cm; lens L_2 has a focal length of 20/3 cm and a diameter of 2 cm; aperture A has a centered circular opening of 0.5-cm diameter.

  a. Sketch the system.
  b. Find the location of the image plane.
  c. Locate the aperture stop and entrance pupil.
  d. Locate the exit pupil.
  e. Locate the field stop, the entrance window, and the exit window.
  f. Determine the angular field of view.

3-5 Refer back to the extended example in the text, involving both a positive and a negative lens, of focal lengths 6 cm and −10 cm, respectively. For the identical optical system, already partially analyzed,

  a. Determine the location and size of the field stop, FS.
  b. Determine the location and size of the entrance and exit windows.
  c. Using the chief ray from object point P to image point P'' as shown in the example, draw the two marginal rays from P to P'', which, with the chief ray, define the cone of light that successfully gets through the optical system.

3-6 Plot a curve of total deviation angle versus entrance angle for a prism of apex angle 60° and refractive index 1.52.

3-7 A parallel beam of white light is refracted by a 60° glass prism in a position of minimum deviation. What is the angular separation of emerging red ($n = 1.525$) and blue (1.535) light?

3-8 a. Approximate the Cauchy constants A and B for crown and flint glasses, using data for the C and F Fraunhofer lines from Table 3-1. Using these constants and the Cauchy relation approximated by two terms, calculate the refractive index of the D Fraunhofer line for each case. Compare your answers with the values given in the table.

  b. Calculate the dispersion in the vicinity of the Fraunhofer D line for each glass, using the Cauchy relation.

  c. Calculate the chromatic resolving power of crown and flint prisms in the vicinity of the Fraunhofer D line, if each prism base is 75 mm in length. Also calculate the minimum resolvable wavelength interval in this region.

3-9 An equilateral prism of dense barium crown glass is used in a spectroscope. Its refractive index varies with wavelength, as given in the table:

nm	n
656.3	1.63461
587.6	1.63810
486.1	1.64611

  a. Determine the minimum angle of deviation for sodium light of 589.3 nm.
  b. Determine the dispersive power of the prism.
  c. Determine the Cauchy constants A and B in the long wavelength region; from the Cauchy relation, find the dispersion of the prism at 656.3 nm.
  d. Determine the minimum base length of the prism if it is to resolve the hydrogen doublet at 656.2716- and 656.2852-nm wavelengths. Is the project practical?

3-10 A prism of 60° refracting angle gives the following angles of minimum deviation when measured on a spectrometer: *C* line, 38°20′; *D* line, 38°33′; *F* line, 39°12′. Determine the dispersive power of the prism.

3-11 The refractive indices for certain crown and flint glasses are

Crown: $n_C = 1.527$, $n_D = 1.530$, $n_F = 1.536$
Flint: $n_C = 1.630$, $n_D = 1.635$, $n_F = 1.648$

The two glasses are to be combined in a double prism that is a direct-vision prism for the *D* wavelength. The refracting angle of the flint prism is 5°. Determine the required angle of the crown prism and the resulting angle of dispersion between the *C* and the *F* rays. Assume that the prisms are thin and the condition of minimum deviation is satisfied.

3-12 An achromatic thin prism for the *C* and *F* Fraunhofer lines is to be made using the crown and flint glasses described in Table 3-1. If the crown glass prism has a prism angle of 15°, determine (a) the required prism angle for the flint glass and (b) the resulting "mean" deviation for the *D* line.

3-13 A perfectly diffuse, or *Lambertian*, surface has the form of a square, 5 cm on a side. This object radiates a total power of 25 W into the forward directions that constitute half the total solid angle of 4π. A camera with a 4-cm focal length lens and stopped down to *f*/8 is used to photograph the object when it is placed 1 m from the lens.

   a. Determine the radiant exitance, radiant intensity, and radiance of the object. (See Table 1-1.)
   b. Determine the radiant flux delivered to the film.
   c. Determine the irradiance at the film.

3-14 Investigate the behavior of Eq. (3-32), giving the dependence of the depth of field on aperture, focal length, and object distance. With the help of a calculator or computer program, generate curves showing each dependence.

3-15 A camera is used to photograph three rows of students at a distance 6 m away, focusing on the middle row. Suppose that the image defocusing or blur circles due to object points in the first and third rows is to be kept smaller than a typical silver grain of the emulsion, say 1 μm. At what object distance nearer and farther than the middle row does an unacceptable blur occur if the camera has a focal length of 50 mm and is stopped down to an *f*/4 setting?

3-16 A telephoto lens consists of a combination of two thin lenses having focal lengths of +20 cm and −8 cm, respectively. The lenses are separated by a distance of 15 cm. Determine the focal length of the combination, distance from negative lens to film plane, and image size of a distant object subtending an angle of 2° at the camera.

3-17 A 5-cm focal length camera lens with *f*/4 aperture is focused on an object 6 ft away. If the maximum diameter of the circle of confusion is taken to be 0.05 mm, determine the depth of field of the photograph.

3-18 The sun subtends an angle of 0.5° at the earth's surface, where the irradiance is about 1000 W/m² at normal incidence. What is the irradiance of an image of the sun formed by a lens with diameter 5 cm and focal length 50 cm?

3-19 **a.** A camera uses a convex lens of focal length 15 cm. How large an image is formed on the film of a 6-ft-tall person 100 ft away?
   b. The convex lens is replaced by a telephoto combination consisting of a 12-cm focal length convex lens and a concave lens. The concave lens is situated in the position of the original lens, and the convex lens is 8 cm in front of it. What is the required focal length of the concave lens such that distant objects form focused images on the same film plane? How much larger is the image of the person using this telephoto lens?

3-20 The lens on a 35-mm camera is marked "50 mm, 1:1.8."
   a. What is the maximum aperture diameter?
   b. Starting with the maximum aperture setting, supply the next three *f*-numbers that would allow the irradiance to be reduced to $\frac{1}{3}$ the preceding at each successive stop.

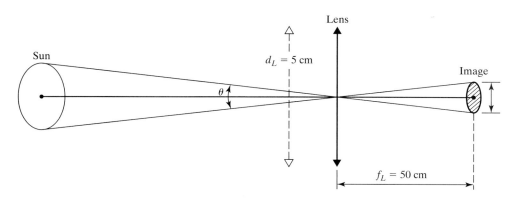

Figure 3-38   Problem 3-18.

c. What aperture diameters correspond to these *f*-numbers?

d. If a picture is taken at maximum aperture and at $\frac{1}{100}$ s, what exposure time at each of the other openings provides equivalent total exposures?

3-21 The magnification given by Eq. (3-33) is also valid for a double-lens eyepiece if the equivalent focal length given by Eq. (3-35) is used. Show that the magnification of a double-lens eyepiece, designed to satisfy the condition for the elimination of chromatic aberration, is, for an image at infinity,

$$M = 12.5\left(\frac{1}{f_1} + \frac{1}{f_2}\right)$$

3-22 A magnifier is made of two thin plano-convex lenses, each of 3-cm focal length and spaced 2.8 cm apart. Find (a) the equivalent focal length and (b) the magnifying power for an image formed at the near point of the eye.

3-23 The objective of a microscope has a focal length of 0.5 cm and forms the intermediate image 16 cm from its second focal point.

a. What is the overall magnification of the microscope when an eyepiece rated at 10× is used?

b. At what distance from the objective is a point object viewed by the microscope?

3-24 A homemade compound microscope has, as objective and eyepiece, thin lenses of focal lengths 1 cm and 3 cm, respectively. An object is situated at a distance of 1.20 cm from the objective. If the virtual image produced by the eyepiece is 25 cm from the eye, compute (a) the magnifying power of the microscope and (b) the separation of the lenses.

3-25 Two thin convex lenses, when placed 25 cm apart, form a compound microscope whose apparent magnification is 20. If the focal length of the lens representing the eyepiece is 4 cm, determine the focal length of the other.

3-26 A level telescope contains a *graticule*—a circular glass on which a scale has been etched—in the common focal plane of objective and eyepiece so that it is seen in focus with a distant object. If the telescope is focused on a telephone pole 30 m away, how much of the post falls between millimeter marks on the graticule? The focal length of the objective is 20 cm.

3-27 A pair of binoculars is marked "7 × 35." The focal length of the objective is 14 cm, and the diameter of the field lens of the eyepiece is 1.8 cm. Determine (a) the angular magnification of a distant object, (b) the focal length of the ocular, (c) the diameter of the exit pupil, (d) the eye relief, and (e) the field of view in terms of feet at 1000 yd.

3-28 a. Show that when the final image is not viewed at infinity, the angular magnification of an astronomical telescope may be expressed by

$$M = -\frac{m_{oc}f_{obj}}{s''}$$

where m_{oc} is the linear magnification of the ocular and s'' is the distance from the ocular to the final image.

b. For such a telescope using two converging lenses with focal lengths of 30 cm and 4 cm, find the angular magnification when the image is viewed at infinity and when the image is viewed at a near point of 25 cm.

3-29 The moon subtends an angle of 0.5° at the objective lens of an astronomical telescope. The focal lengths of the objective and ocular lenses are 20 cm and 5 cm, respectively. Find the diameter of the image of the moon viewed through the telescope at near point of 25 cm.

3-30 An opera glass uses an objective and eyepiece with focal lengths of +12 cm and −4.0 cm, respectively. Determine the length (lens separation) of the instrument and its magnifying power for a viewer whose eyes are focused (a) for infinity and (b) for a near point of 30 cm.

3-31 An astronomical telescope is used to project a real image of the moon onto a screen 25 cm from an ocular of 5-cm focal length. How far must the ocular be moved from its normal position?

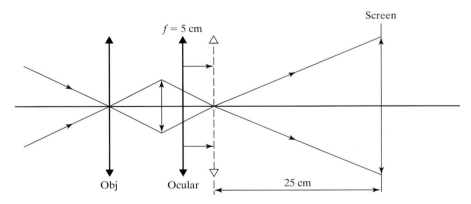

Figure 3-39   Problem 3-31.

3-32 a. The Ramsden eyepiece of a telescope is made of two positive lenses of focal length 2 cm each and also separated by 2 cm. Calculate its magnifying power when viewing an image at infinity.
b. The objective of the telescope is a 30-cm positive lens, with a diameter of 4.50 cm. Calculate the overall magnification of the telescope.
c. What is the position and diameter of the exit pupil?
d. The diameter of the eyepiece field lens is 2 cm. Determine the angle defining the field of view of the telescope.

3-33 Show that the angular magnification of a Newtonian reflecting telescope is given by the ratio of objective to ocular

focal lengths, as it is for a refracting telescope when the image is formed at infinity.

3-34 The primary mirror of a Cassegrain reflecting telescope has a focal length of 12 ft. The secondary mirror, which is convex, is 10 ft from the primary mirror along the principal axis and forms an image of a distant object at the vertex of the primary mirror. A hole in the primary mirror permits viewing the image with an eyepiece of 4-in. focal length, placed just behind this mirror. Calculate the focal length of the secondary convex mirror and the angular magnification of the instrument.

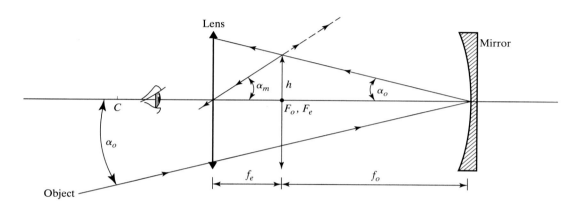

Figure 3-40   Problem 3-33.

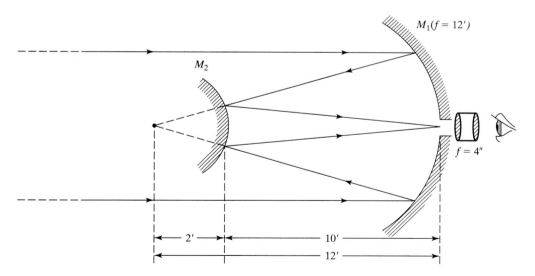

Figure 3-41   Problem 3-34.

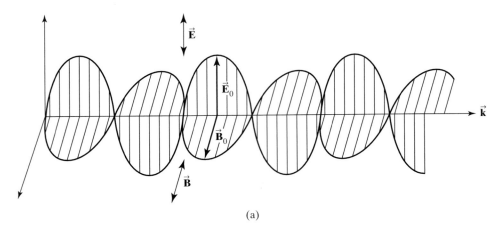

(a)

4

Wave Equations

INTRODUCTION

In this chapter we develop mathematical expressions for wave motion in general but concentrate on the most useful special case, the harmonic wave. Harmonic wave functions are then adapted to represent electromagnetic waves, which include light waves. Results from electromagnetism describing the physics of electromagnetic waves are borrowed to enable a determination of the energy delivered by such waves.

4-1 ONE-DIMENSIONAL WAVE EQUATION

The most general form of a one-dimensional traveling wave, and the differential equation it satisfies, can be determined in the following way. Consider first a one-dimensional wave pulse of arbitrary (but time-independent) shape, described by $y' = f(x')$, fixed to a (moving) coordinate system $O'(x', y')$, as in Figure 4-1a. Consider next that the O' system, together with the pulse, moves to the right along the x-axis at uniform speed v relative to a fixed coordinate system, $O(x, y)$, as in Figure 4-1b. Here the coordinate y could, for example, represent the transverse displacement from equilibrium of a string stretched out along the x-direction. As it moves, the pulse maintains its shape. Any point on the pulse, such as P, can be described by either of two coordinates, x or x', where $x' = x - vt$. The y-coordinate is identical in either system. From the point of view of the stationary coordinate system, then, the moving pulse has the mathematical form

$$y = y' = f(x') = f(x - vt)$$

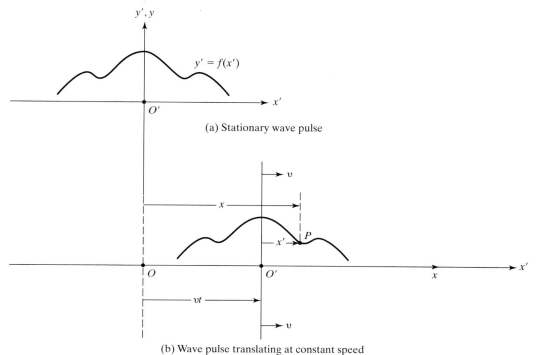

(a) Stationary wave pulse

(b) Wave pulse translating at constant speed

Figure 4-1   Translating wave pulses.

If the pulse moves to the left, the sign of v must be reversed, so that we may write

$$y = f(x \pm vt) \tag{4-1}$$

as the general form of a traveling wave. Notice that we have assumed $x = x'$ at $t = 0$. The original shape of the pulse, $y' = f(x')$, does not vary but is simply translated along the x-direction by the amount vt at time t. The function f is any function whatsoever, so that, for example,

$$y = A \sin(k[x - vt])$$
$$y = A(x + vt)^2$$
$$y = e^{k(x-vt)}$$

all represent traveling waves. Only the first, however, represents the important case of a periodic wave.

We wish to find next the partial differential equation that is satisfied by all such waves, regardless of the particular function f. Since y is a function of two variables, x and t, we use the chain rule of partial differentiation and write

$$y = f(x')$$

where

$$x' = x \pm vt$$

so that

$$\partial x'/\partial x = 1 \quad \text{and} \quad \partial x'/\partial t = \pm v$$

Employing the chain rule, the spatial derivative is

$$\frac{\partial y}{\partial x} = \frac{\partial f}{\partial x'}\frac{\partial x'}{\partial x} = \frac{\partial f}{\partial x'}$$

Repeating the procedure to find the second derivative,

$$\frac{\partial^2 y}{\partial x^2} = \frac{\partial}{\partial x}\left(\frac{\partial y}{\partial x}\right) = \frac{\partial(\partial y/\partial x)}{\partial x'}\frac{\partial x'}{\partial x} = \frac{\partial}{\partial x'}\left(\frac{\partial f}{\partial x'}\right) = \frac{\partial^2 f}{\partial x'^2}$$

Similarly, the temporal derivatives are found:

$$\frac{\partial y}{\partial t} = \frac{\partial f}{\partial x'}\frac{\partial x'}{\partial t} = \pm v\frac{\partial f}{\partial x'}$$

$$\frac{\partial^2 y}{\partial t^2} = \frac{\partial}{\partial t}\left(\frac{\partial y}{\partial t}\right) = \frac{\partial(\partial y/\partial t)}{\partial x'}\frac{\partial x'}{\partial t} = \frac{\partial}{\partial x'}\left(\pm v\frac{\partial f}{\partial x'}\right)(\pm v) = v^2\frac{\partial^2 f}{\partial x'^2}$$

Combining the results for the two second derivatives, we arrive at the one-dimensional differential wave equation,

$$\frac{\partial^2 y}{\partial x^2} = \frac{1}{v^2}\frac{\partial^2 y}{\partial t^2} \tag{4-2}$$

Any wave of the form of Eq. (4-1) must satisfy Eq. (4-2), regardless of the physical nature of the wave itself. Thus, to determine whether a given function of x and t represents a traveling wave, it is sufficient to show either that it is of the general form of Eq. (4-1) or that it satisfies Eq. (4-2).

4-2 HARMONIC WAVES

Of special importance are *harmonic* waves that involve the sine or cosine functions,

$$y = A_{\cos}^{\sin}[k(x \pm vt)] \tag{4-3}$$

where A and k are constants that can be varied without changing the harmonic character of the wave. These are periodic waves, representing smooth patterns that repeat themselves endlessly. Such waves are generated by undamped oscillators undergoing simple harmonic motion. In addition, the sine and cosine functions together form a *complete set* of functions; that is, a linear combination of terms like those in Eq. (4-3) can be found to represent any periodic waveform. Such a series of terms is called a *Fourier series* and is treated further in Section 9-1. Thus combinations of harmonic waves are capable of representing more complicated waveforms, even a series of rectangular pulses or square waves.

Since $\sin x = \cos(x - \pi/2)$, the only difference between the sine and cosine functions is a relative translation of $\pi/2$ radians. It is sufficient in what follows, therefore, to treat only one of these functions. Accordingly, a section of a sine wave is pictured in Figure 4-2. In Figure 4-2a, a section of a wave with *amplitude A* is shown at a fixed time, as in a snapshot; in Figure 4-2b, the time variations of the wave are pictured at a fixed point x along the wave. In Figure 4-2a, the repetitive spatial unit of the wave is shown as the *wavelength* λ. Because of this periodicity, increasing all x by λ should reproduce the same wave. Mathematically, the wave is reproduced because the argument of the sine function is advanced by 2π. Symbolically,

$$A \sin k[(x + \lambda) + vt] = A \sin[k(x + vt) + 2\pi]$$

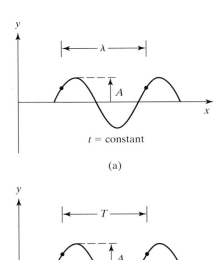

Figure 4-2 Extension of a sine wave in space and time. (a) Section of a sine wave at a fixed time. (b) Section of a sine wave at a fixed point.

or

$$A \sin(kx + k\lambda + kvt) = A \sin(kx + kvt + 2\pi)$$

It follows that $k\lambda = 2\pi$, so that the *propagation constant k* contains information regarding the wavelength.

$$k = \frac{2\pi}{\lambda} \qquad (4\text{-}4)$$

Alternatively, if the wave is viewed from a fixed position, as in Figure 4-2b, it is periodic in time with a repetitive temporal unit called the *period T*. Increasing all t by T, the waveform is exactly reproduced, so that

$$A \sin k[x + v(t + T)] = A \sin[k(x + vt) + 2\pi]$$

or

$$A \sin(kx + kvt + kvT) = A \sin(kx + kvt + 2\pi)$$

Clearly, $kvT = 2\pi$, and we have an expression that relates the period T to the propagation constant k and wave velocity v. The same information is included in the relation

$$v = \nu\lambda \qquad (4\text{-}5)$$

where we have used Eq. (4-4) together with the reciprocal relation between period T and frequency ν

$$\nu = \frac{1}{T} \qquad (4\text{-}6)$$

Related descriptions of wave parameters are often used. The combination $\omega = 2\pi\nu$ is called the *angular frequency*, and the reciprocal of the wavelength $\kappa = 1/\lambda$ is called the *wave number*. Note that the propagation constant k is related to the *spatial period* (i.e., the wavelength) of the wave in the same way that the angular frequency ω is related to the temporal period T. Therefore, the propagation constant k is the *spatial frequency* of the wave. With these relationships it is easy to show the equivalence of the following common forms for harmonic waves:

$$y = A_{\cos}^{\sin}[k(x \pm vt)] \qquad (4\text{-}7)$$

$$y = A_{\cos}^{\sin}\left[2\pi\left(\frac{x}{\lambda} \pm \frac{t}{T}\right)\right] \qquad (4\text{-}8)$$

$$y = A_{\cos}^{\sin}[(kx \pm \omega t)] \qquad (4\text{-}9)$$

In any case, the argument of the sine or cosine, which depends on space and time, is called the *phase*, φ. For example, in Eq. (4-7),

$$\varphi = k(x \pm vt) \qquad (4\text{-}10)$$

When x and t change together in such a way that φ is constant, the displacement $y = A \sin \varphi$ is also constant. The condition of constant phase evidently describes the motion of a fixed point on the waveform, which moves with the velocity of the wave. Thus if φ is constant,

$$d\varphi = 0 = k(dx \pm vdt)$$

and

$$\frac{dx}{dt} = \mp v$$

confirming that v represents the wave velocity, which is in the negative x-direction when $\varphi = k(x + vt)$ and in the positive x-direction when $\varphi = k(x - vt)$.

Notice that the waveforms of Eqs. (4-7) through (4-9) that use the sine function all represent waves for which $y = 0$ at position $x = 0$ and time $t = 0$. All of the cosine waveforms in these equations represent waves for which $y = A$ at position $x = 0$ and time $t = 0$. As pointed out previously, both situations could be handled by either the sine or cosine function if an angle of $\pi/2$ is added to the phase. In general, to accommodate any arbitrary initial displacement, some angle φ_0 must be added to the phase. For example, Eq. (4-7) with the sine function becomes

$$y = A \sin[k(x \pm vt) + \varphi_0]$$

Now suppose our initial boundary conditions are such that $y = y_0$ when $x = 0$ and $t = 0$. Then

$$y = A \sin \varphi_0 = y_0$$

from which the required *initial phase angle* φ_0 can be calculated as

$$\varphi_0 = \sin^{-1}\left(\frac{y_0}{A}\right)$$

The waveforms in Eqs. (4-7) to (4-9) can be generalized further to yield any initial displacement, therefore, by the addition of an initial phase angle φ_0 to the phase. In many cases, the precise phase of the wave is not of interest. Then φ_0 can be set equal to zero for simplicity.

Example 4-1

A traveling wave in a string has a displacement from equilibrium given as a function of distance along the string x and time t as

$$y(x, t) = (0.35 \text{ m}) \sin[(3\pi/\text{m}) \times - (10\pi/\text{s})t + \pi/4)]$$

Determine the wavelength, frequency, velocity, and initial phase angle. Also find the displacement at $x = 10$ cm and $t = 0$.

Solution

By comparison with Eq. (4-9), $k = 3\pi/\text{m}$ and $\omega = 10\pi/\text{s}$. Thus,

$$\lambda = \frac{2\pi}{k} = \frac{2}{3} \text{ m} \quad \text{and} \quad \nu = \frac{\omega}{2\pi} = 5 \text{ Hz}$$

The initial phase $(x = 0, t = 0)$ is $\pi/4$. The velocity of the wave may be found from $v = \lambda\nu = (2/3)5$ m/s $= 3.33$ m/s in the positive x-direction (due to the negative sign in the phase). One can also set the phase $\varphi = (3\pi/\text{m})x - (10\pi/\text{s})t + \pi/4$ equal to a constant so that

$$d\varphi = (3\pi/\text{m})dx - (10\pi/\text{s})dt = 0$$

or $v = dx/dt = 10\pi/3\pi$ m/s $= +3.33$ m/s. Furthermore, the displacement at $x = 10$ cm, t $= 0$ is

$$y = (0.1 \text{ m}, 0) = (0.35 \text{ m}) \sin\left(0.3\pi + \frac{\pi}{4}\right) = +0.346 \text{ m}$$

4-3 COMPLEX NUMBERS

In many situations it is useful to represent harmonic waves in complex-number notation. To this end, we first review briefly some important relations involving complex numbers.

A complex number $\tilde{z}$ is expressed as the sum of its *real* and *imaginary* parts,

$$\tilde{z} = a + ib \qquad (4\text{-}11)$$

where

$$a = \operatorname{Re}(\tilde{z}) \quad \text{and} \quad b = \operatorname{Im}(\tilde{z})$$

are real numbers and $i = \sqrt{-1}$. The form of the complex number given by Eq. (4-11) can also be cast into polar form. Referring to Figure 4-3, the complex number $\tilde{z}$ is represented in terms of its real and imaginary parts along the corresponding axes. The magnitude of $\tilde{z}$, symbolized by $|\tilde{z}|$, also called its *absolute value* or *modulus*, is given by the Pythagorean theorem as

$$|\tilde{z}|^2 = a^2 + b^2 \qquad (4\text{-}12)$$

Since from Figure 4-3, $a = |\tilde{z}| \cos \theta$ and $b = |\tilde{z}| \sin \theta$, it is also possible to express $\tilde{z}$ by

$$\tilde{z} = |\tilde{z}|(\cos \theta + i \sin \theta)$$

The expression in parentheses is, by Euler's formula,

$$e^{i\theta} = \cos \theta + i \sin \theta \qquad (4\text{-}13)$$

so that

$$\tilde{z} = |\tilde{z}|e^{i\theta} \qquad (4\text{-}14)$$

where

$$\theta = \tan^{-1}\left(\frac{b}{a}\right) \qquad (4\text{-}15)$$

The *complex conjugate* $\tilde{z}^*$ is simply the complex number $\tilde{z}$ with i replaced by $-i$. Thus if $\tilde{z} = a + ib$,

$$\tilde{z}^* = a - ib \quad \text{or} \quad \tilde{z}^* = |\tilde{z}|e^{-i\theta} \qquad (4\text{-}16)$$

where the asterisk is used to denote the complex conjugate. A very useful minitheorem is that the product of a complex number with its complex conjugate equals the square of its absolute value. Using the polar form,

$$\tilde{z}\tilde{z}^* = (|\tilde{z}|e^{i\theta})(|\tilde{z}|e^{-i\theta}) = |\tilde{z}|^2 \qquad (4\text{-}17)$$

Finally, it will be helpful to list the values of $e^{i\theta}$, using Euler's formula, Eq. (4-13), for frequently occurring special cases. These are given in Figure 4-4, together with a mnemonic device to assist in recalling them quickly.

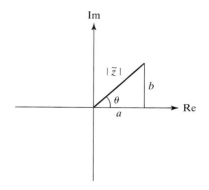

Figure 4-3 Graphical representation of a complex number along real (Re) and imaginary (Im) axes.

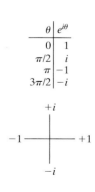

Figure 4-4 Frequently used values of $e^{i\theta}$.

4-4 HARMONIC WAVES AS COMPLEX FUNCTIONS

Using Euler's formula, it is possible to express a harmonic wave as the real (or imaginary) part of the complex function

$$\tilde{y} = Ae^{i(kx-\omega t)} \tag{4-18}$$

so that

$$y = \text{Re}\,(\tilde{y}) = A\cos(kx - \omega t) \quad \text{or} \quad y = \text{Im}\,(\tilde{y}) = A\sin(kx - \omega t) \tag{4-19}$$

Note that any equation that involves only terms that are linear in $\tilde{y}$ and its derivatives will also hold for $y = \text{Re}\,(\tilde{y})$ or $y = \text{Im}\,(\tilde{y})$. Many mathematical manipulations can be carried out more simply with exponential functions than with trigonometric functions. As a result, it is common practice to use the complex waveform Eq. (4-18) to represent a harmonic wave when doing calculations and then to take the real or imaginary part of this complex function to recover the physical wave represented by one of the forms in Eq. (4-19).

4-5 PLANE WAVES

We wish now to generalize the harmonic wave equation further so that it can represent a waveform propagating along any direction in space. Since an arbitrary direction involves the three spatial coordinates x, y, and z, we represent the wave "displacement" or disturbance by ψ rather than y; for example,

$$\psi = A\sin(kx - \omega t) \tag{4-20}$$

It is important to note that ψ need not represent only physical displacements but could represent any quantity that varies in space and time such as the difference of air pressure from its equilibrium value (as in a sound wave) or the strength of an electric or magnetic field (as in a light wave). Equation (4-20) represents a traveling wave moving along the $+x$-direction. At fixed time (for simplicity we take $t = 0$), the wave is described by

$$\psi = A\sin kx \tag{4-21}$$

When $x = $ constant, the phase $\varphi = kx = $ constant. Thus, the surfaces of constant phase are a *family of planes* perpendicular to the x-axis. These surfaces of constant phase are often called the *wavefronts* of the disturbance. For concreteness, consider a plane sound wave propagating along the x-direction through a sample of air. The propagation of this sound wave alters the air pressure P as it passes. Let

$$\psi = P - P_0 = (10\,\text{N/m}^2)\sin[(2\pi/\text{m})x - (680\pi/\text{s})t]$$

represent the difference in the air pressure from its equilibrium value ($P_0 \approx 10^5\,\text{N/m}^2$). In Figure 4-5, ψ is plotted as a function of x at a fixed time $t = 0$ and several wavefronts associated with the plane sound wave are depicted at this same time. Note that at all points on a given plane wavefront ψ have the same value. In addition, ψ has the same value on all wavefronts separated by a wavelength $\lambda = 1$ m. A motion picture depiction of the sound wave of Figure 4-5 would show the wavefronts moving in the positive x-direction at the wave speed 340 m/s. Clearly, plane waves, which have planar wavefronts that are infinite in extent, are approximations to real waves, which have limited extent in directions that are transverse to the propagation direction. Treating

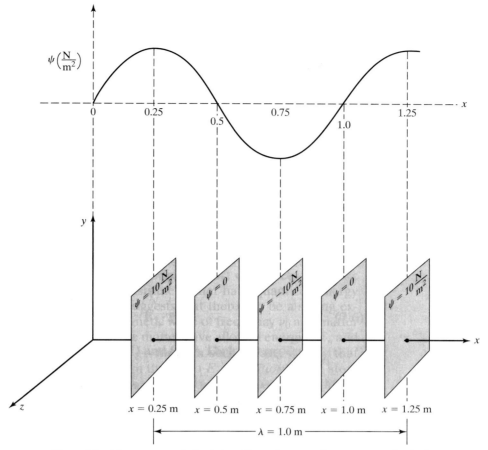

Figure 4-5 Air pressure variation induced by a plane sound wave propagating in the x-direction. The plot shows the difference in air pressure from its equilibrium value as a function of x at t = 0, and several planar wavefronts associated with the sound wave are depicted at the same time. Note that the wavefronts associated with adjacent maxima are separated by one wavelength.

a wave as a simple plane wave is a useful approximation if a portion of the wave has nearly planar wavefronts over the region of interest.

Consider again the wave represented in Eq. (4-20). Since, for this waveform, the wave disturbance at an arbitrary point in space, defined by the vector $\vec{r}$ in Figure 4-6a, is the same as for the point x along the x-axis, where $x = r \cos \theta$. Eq. (4-21) may then be written as

$$\psi = A \sin(kr \cos \theta)$$

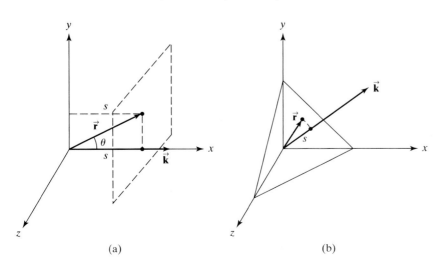

(a)   (b)

Figure 4-6 Generalization of the plane wave to an arbitrary direction. The wave direction is given by the vector $\bar{\mathbf{k}}$ along the x-axis in (a) and an arbitrary direction in (b).

Equation 4-20 can therefore be generalized if the propagation constant, whose magnitude $2\pi/\lambda$ has already been determined in Eq. (4-4), is now considered to be a vector quantity, pointing in the direction of propagation. Then $kr \cos \theta = \vec{\mathbf{k}} \cdot \vec{\mathbf{r}}$, and the harmonic wave of Eq. (4-20) becomes

$$\psi = A \sin(\vec{\mathbf{k}} \cdot \vec{\mathbf{r}} - \omega t) \tag{4-22}$$

In this form, Eq. (4-22) can represent plane waves propagating in any arbitrary direction given by $\vec{\mathbf{k}}$, as shown in Figure 4-6b. In the general case,

$$\vec{\mathbf{k}} \cdot \vec{\mathbf{r}} = xk_x + yk_y + zk_z = kr \cos \theta \equiv ks$$

where (k_x, k_y, k_z) are the components of the propagation direction and (x, y, z) are the components of the point in space where the displacement ψ is evaluated and s is the component of the position vector along the direction of the propagation of the wave. Note that, in general, s represents the distance along a waveform measured along a direction that is perpendicular to the wavefronts associated with the wave.

A general harmonic wave in three-dimensions can be expressed in complex form as

$$\psi = A e^{i(\vec{\mathbf{k}} \cdot \vec{\mathbf{r}} - \omega t)} \tag{4-23}$$

(Recall that the physical waveform is described by the real or imaginary part of the complex form.) The partial differential equation satisfied by such three-dimensional waves is a generalization of Eq. (4-2) in the form

$$\frac{\partial^2 \psi}{\partial x^2} + \frac{\partial^2 \psi}{\partial y^2} + \frac{\partial^2 \psi}{\partial z^2} = \frac{1}{v^2} \frac{\partial^2 \psi}{\partial t^2} \tag{4-24}$$

as can easily be verified by computing the second partial derivatives of ψ from Eq. (4-23). The wave Eq. (4-24) is often written more compactly by separating the spatial second derivatives from the wave function ψ by treating them as operators:

$$\left(\frac{\partial^2}{\partial x^2} + \frac{\partial^2}{\partial y^2} + \frac{\partial^2}{\partial z^2} \right)\psi = \frac{1}{v^2} \frac{\partial^2 \psi}{\partial t^2}$$

The entire operator in parentheses is known as the *Laplacian operator*,

$$\nabla^2 \equiv \frac{\partial^2}{\partial x^2} + \frac{\partial^2}{\partial y^2} + \frac{\partial^2}{\partial z^2}$$

and Eq. (4-24) becomes simply

$$\nabla^2 \psi = \frac{1}{v^2} \frac{\partial^2 \psi}{\partial t^2} \tag{4-25}$$

4-6 SPHERICAL WAVES

Harmonic wave disturbances emanating from a point source in a homogeneous medium travel at equal rates in all directions. As shown in Figure 4-7, surfaces of constant phase, that is, wavefronts, are then spherical surfaces centered at the source. Such waves, which are of course also solutions to

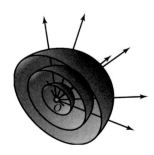

Figure 4-7 Portions of three spherical wavefronts emanating from a point source O. The rays indicate that the direction of energy propagation is radially outward from O.

Eq. (4-25), can be represented by the complex waveform

$$\psi = \left(\frac{A}{r}\right)e^{i(kr-\omega t)} \qquad (4\text{-}26)$$

Here, r is the radial distance from the point source to a given point on the waveform and, as for plane waves, $k = 2\pi/\lambda$ and $\omega = 2\pi/T$.

Note that the constant A appearing in Eq. (4-26) is not the overall amplitude of the wave, which is instead given by A/r. The spherical wave, as it propagates further from the source, decreases in amplitude, in contrast to a plane wave for which the amplitude is constant. If the amplitude at distance r from the point source is A/r, then the irradiance (W/m^2) of the wave there is proportional to $(A/r)^2$. That is, Eq. (4-26) encodes the familiar *inverse square law* of propagation for spherical wave disturbances. Clearly, Eq. (4-26) is not valid as r approaches zero but rather describes the disturbance at a finite distance from a small physical source. Over a small enough region (or sufficiently far from the source), the spherical wavefronts associated with a spherical wave are approximately planar. For this reason, waves emanating from point sources can be adequately described by plane waveforms when the region of interest is small compared to the distance from the point source.

4-7 OTHER HARMONIC WAVEFORMS

Cylindrical Waves

Another useful complex waveform represents a cylindrical wave in which the wavefronts are outward-moving cylindrical surfaces surrounding a line of symmetry, as shown in Figure 4-8. Such a wave takes the form,

$$\psi = \frac{A}{\sqrt{\rho}}e^{i(k\rho \pm \omega t)} \qquad (4\text{-}27)$$

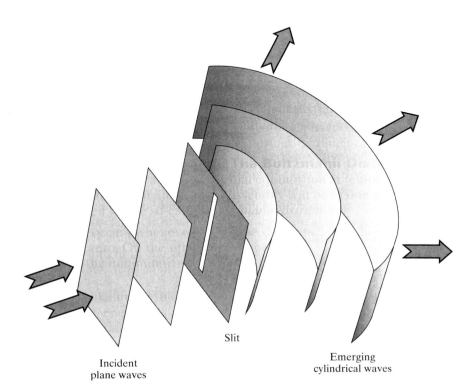

Slit

Incident
plane waves

Emerging
cylindrical waves

Figure 4-8 Plane waves incident on a slit generate cylindrical waves.

Here, ρ represents the perpendicular distance from the line of symmetry to a point on the waveform. That is, if the z-axis is the line of symmetry, then $\rho = \sqrt{x^2 + y^2}$. Waves of this form are not exact solutions to the wave equation given in Eq. (4-25) and so do not exactly represent physical waves but rather are approximately valid for large ρ. Still, they are useful forms that approximate the wave that emerges from a slit illuminated by a plane wave.

Gaussian Beams

Another important family of (single-frequency) approximate solutions to the differential wave Eq. (4-25) consists of the rather complicated but important *Hermite-Gaussians*. Hermite-Gaussian waveforms are, to an excellent approximation, produced by laser systems that use spherical mirrors to form the laser cavity and are beamlike, in the sense that the beam irradiance is strongly confined in the transverse direction. We defer a detailed discussion of these beams until Chapter 27 but sketch and indicate some of the most important features of the simplest Hermite-Gaussian waveform in Figure 4-9. The parameter $w(z)$, shown in Figure 4-9, is often called the *spot size* and marks the transverse distance from the axis of the beam to the point at which the irradiance falls to $e^{-2} \approx 0.135$ of the maximum irradiance that occurs on the symmetry axis of the beam. Note that the beam spreads while maintaining nearly spherical wavefronts that change radius of curvature as the beam propagates. The minimum spot size w_0 occurs at the so-called *beam waist*, where the wavefronts are planar. The location and size of the beam waist is determined by the nature of the laser cavity (and subsequent focusing elements) that form the beam. Note that the half-angle beam divergence is larger for beams with smaller beam waists. This is an important general feature of the propagation of electromagnetic waves that we shall return to and elucidate in Chapters 11 and 27. In regions close to the symmetry axis, the Gaussian beam can be adequately described by planar waveforms.

4-8 ELECTROMAGNETIC WAVES

The harmonic waveforms discussed so far can represent any type of wave disturbance that varies in a sinusoidal manner. Some familiar examples of such disturbances are waves on a string, water waves, and sound waves. The disturbance ψ may refer to transverse displacements of a string or longitudinal

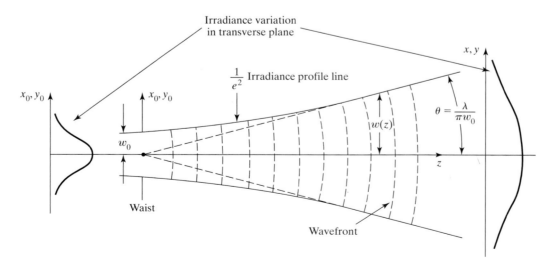

Figure 4-9 Gaussian beam propagating in the z-direction. The spot size at the beam waist (planar wavefront) is defined as w_0. The half-angle beam divergence $\theta = \lambda/(\pi w_0)$ is valid only in the far field. Note the change in transverse irradiance as the beam propagates to the right.

pressure variations due to a sound wave propagating in a gas—as mentioned earlier. In general, harmonic waves are produced by sources that oscillate in a periodic fashion. Charged particles oscillating with a regular frequency emit harmonic electromagnetic waves. For electromagnetic waves (including light), ψ can stand for either of the varying electric or magnetic fields that together constitute the wave. Figure 4-10a depicts a plane electromagnetic wave traveling in some arbitrary direction. From Maxwell's equations, which describe such waves, we know that the harmonic variations of the electric and magnetic fields are always perpendicular to one another and to the direction of propagation given by $\vec{\mathbf{k}}$, as suggested by the orthogonal set of axes in Figure 4-10a. These variations may be described by the harmonic waveforms

$$\vec{\mathbf{E}} = \vec{\mathbf{E}}_0 \sin(\vec{\mathbf{k}} \cdot \vec{\mathbf{r}} - \omega t) \qquad (4\text{-}28)$$

$$\vec{\mathbf{B}} = \vec{\mathbf{B}}_0 \sin(\vec{\mathbf{k}} \cdot \vec{\mathbf{r}} - \omega t) \qquad (4\text{-}29)$$

where $\vec{\mathbf{E}}$ and $\vec{\mathbf{B}}$ represent the electric and magnetic fields, respectively, and $\vec{\mathbf{E}}_0$ and $\vec{\mathbf{B}}_0$ are their amplitudes. Each component of the wave travels with the same propagation vector $\vec{\mathbf{k}}$ and frequency ω and thus with the same wavelength and speed. Furthermore, electromagnetic theory tells us that the field amplitudes are related by $E_0 = cB_0$, where c is the speed of the wave. Figure 4-10b shows a plane wave propagating along the positive z-direction with the electric field varying along the x-direction and the magnetic field varying along the y-direction.

At any specified time and place,

$$E = cB \qquad (4\text{-}30)$$

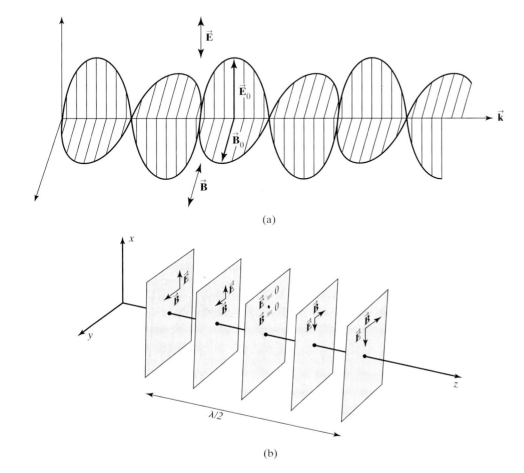

(a)

(b)

Figure 4-10 Plane electromagnetic wave described by Eqs. (4-28) and (4-29). (a) The electric field $\vec{\mathbf{E}}$, magnetic field $\vec{\mathbf{B}}$, and propagation vector $\vec{\mathbf{k}}$ are everywhere mutually perpendicular. (b) Wavefronts for a (linearly polarized) plane electromagnetic wave.

In free space, the velocity c is given by

$$c = \frac{1}{\sqrt{\varepsilon_0 \mu_0}} \tag{4-31}$$

where the constants ε_0 and μ_0 are, respectively, the permittivity and permeability of vacuum. Measured values for these constants, $\varepsilon_0 = 8.8542 \times 10^{-12}\,(\text{C}\cdot\text{s})^2/\text{kg}\cdot\text{m}^3$ and $\mu_0 = 4\pi \times 10^{-7}\,\text{kg}\cdot\text{m}/(\text{A}\cdot\text{s})^2$, provide an indirect method of determining the speed of electromagnetic waves in free space and yield a value of $c = 2.998 \times 10^8$ m/s. Recall, from Chapter 1, that *light* is the term for electromagnetic radiation that human eyes can "see." Humans see different wavelengths of light as different colors. In Chapter 1 we noted that light wavelengths range from 380 nm (violet) to 770 nm (red).

An electromagnetic wave, of course, represents the transmission of energy. The energy density, u_E in J/m^3, associated with the electric field in free space is

$$u_E = \frac{1}{2}\varepsilon_0 E^2 \tag{4-32}$$

and the energy density associated with the magnetic field in free space is

$$u_B = \frac{1}{2}\frac{1}{\mu_0}B^2 \tag{4-33}$$

These expressions, easily derived for the static electric field of an ideal capacitor and the static magnetic field of an ideal solenoid, are generally valid. Incorporating Eqs. (4-30) and (4-31) into either of the Eqs. (4-32) or (4-33), u_E and u_B are shown to be equal. For example, starting with Eq. (4-33),

$$u_B = \frac{1}{2}\frac{1}{\mu_0}B^2 = \frac{1}{2}\frac{1}{\mu_0}\left(\frac{E}{c}\right)^2 = \left(\frac{1}{2}\frac{\varepsilon_0 \mu_0}{\mu_0}\right)E^2 = \frac{1}{2}\varepsilon_0 E^2 = u_E \tag{4-34}$$

The energy of an electromagnetic wave is therefore divided equally between its constituent electric and magnetic fields. The total energy density is the sum

$$u = u_E + u_B = 2u_E = 2u_B$$

or

$$u = \varepsilon_0 E^2 = \left(\frac{1}{\mu_0}\right)B^2 \tag{4-35}$$

Consider next the rate at which energy is transported by the electromagnetic wave, or its *power*. In a time Δt, the energy transported through a cross section of area A (Figure 4-11) is the energy associated with the volume ΔV of a rectangular volume of length $c\,\Delta t$. Thus,

$$\text{power} = \frac{\text{energy}}{\Delta t} = \frac{u\,\Delta V}{\Delta t} = \frac{u(Ac\,\Delta t)}{\Delta t} = ucA \tag{4-36}$$

or the power transferred per unit area, S, is

$$S = uc \tag{4-37}$$

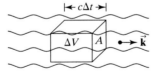

Figure 4-11 Energy flow of an electromagnetic wave. In time Δt, the energy enclosed in the rectangular volume ΔV flows across the surface A.

We now express the energy density u in terms of E and B, as follows, making use of Eqs. (4-31) and (4-35):

$$u = \sqrt{u}\sqrt{u} = (\sqrt{\varepsilon_0}E)\left(\frac{B}{\sqrt{\mu_0}}\right) = \frac{\varepsilon_0}{\sqrt{\varepsilon_0\mu_0}}EB = \varepsilon_0 cEB \qquad (4\text{-}38)$$

Inserting this result into Eq. (4-37),

$$S = \varepsilon_0 c^2 EB \qquad (4\text{-}39)$$

The power per unit area, S, when assigned the direction of propagation, is called the *Poynting vector*. Since this direction is the same as that of the cross product of the orthogonal vectors, $\vec{E}$ and $\vec{B}$, we can write, finally,

$$\vec{S} = \varepsilon_0 c^2 \vec{E} \times \vec{B} \qquad (4\text{-}40)$$

Note that since this relation involves the product of two waveforms, it does not hold for waveforms written in complex form. Because of the rapid variation of the electric and magnetic fields, whose frequencies are 10^{14} to 10^{15} Hz in the visible spectrum, the magnitude of the Poynting vector in Eq. (4-39) is also a rapidly varying function of time. In most cases, a time average of the power delivered per unit area is all that is required. This quantity is called the *irradiance*, E_e.[1]

$$E_e = \langle|\vec{S}|\rangle = \varepsilon_0 c^2 \langle E_0 B_0 \sin^2(\vec{k}\cdot\vec{r} \pm \omega t)\rangle \qquad (4\text{-}41)$$

where the angle brackets denote a time average and we have expressed the fields as sine functions of the phase. The average of the functions $\sin^2\theta$ or $\cos^2\theta$ over a period is easily shown to be $1/2$, so that

$$E_e = \tfrac{1}{2}\varepsilon_0 c^2 E_0 B_0$$
$$E_e = \tfrac{1}{2}\varepsilon_0 c E_0^2$$
$$E_e = \tfrac{1}{2}\left(\frac{c}{\mu_0}\right)B_0^2 \qquad (4\text{-}42)$$

The alternative forms of Eq. (4-42) are expressed for the case of free space. They apply also to a medium of refractive index n if ε_0 is replaced by $n^2\varepsilon_0$ and c is replaced by the velocity c/n. Notice that these changes leave the first of the alternative forms invariant.

Example 4-2

A laser beam of radius 1 mm carries a power of 6 kW. Determine its average irradiance and the amplitude of its E and B fields.

Solution

The average irradiance

$$E_e = \frac{\text{power}}{\text{area}} = \frac{6000}{\pi(10^{-3})^2} = 1.91 \times 10^9 \text{ W/m}^2$$

From Eq. (4-42),

$$E_0 = \left(\frac{2E_e}{\varepsilon_0 c}\right)^{1/2} = \left[\frac{2(1.91 \times 10^9)}{\varepsilon_0 c}\right]^{1/2} = 1.20 \times 10^6 \text{ V/m}$$

and, from Eq. (4-30),

$$B_0 = \frac{E_0}{c} = \frac{1.20 \times 10^6}{c} = 4.00 \times 10^{-3} \text{ T}$$

---

[1]To avoid confusion of electric field with irradiance, throughout most of the remainder of this text we will use the symbol I, rather than E_e, to denote irradiance.

4-9 LIGHT POLARIZATION

As we have noted, the fields associated with electromagnetic waves are vector quantities such that, at every point in the wave, the electric field, the magnetic field, and the direction of energy propagation are mutually perpendicular, with the direction of energy propagation being the direction of $\vec{E} \times \vec{B}$. In order to completely specify the electromagnetic wave, it is sufficient to specify the electric field since the magnetic field and Poynting vector can be determined once $\vec{E}$ is known. The direction of the electric field is known as the *polarization* of the wave. For example, consider an electric field propagating in the positive z-direction and polarized in the x-direction,

$$\vec{E} = E_0 \sin(kz - \omega t)\hat{x} \qquad (4\text{-}43)$$

According to Maxwell's equations, the magnetic field associated with this electric field would be

$$\vec{B} = \left(\frac{1}{c}\right)E_0 \sin(kz - \omega t)\hat{y}$$

and Eq. (4-40) would give the Poynting vector as

$$\vec{S} = \varepsilon_0 c E_0^2 \sin^2(kz - \omega t)\hat{z}$$

The polarization of an electromagnetic wave determines the direction of the force that the electromagnetic wave exerts on charged particles in the path of the wave through application of the *Lorentz force law*. This law states that the electromagnetic force on a particle of charge Q moving with velocity $\vec{V}$ in an electromagnetic field is

$$\vec{F} = Q(\vec{E} + \vec{V} \times \vec{B})$$

Unless the speed of the charged particle is a significant fraction of the speed of light, the magnitude of the electric force on the particle will be much larger than that of the magnetic force. The electric force on the charged particle is along the direction of the polarization of the wave and so must be perpendicular to the direction of propagation of the wave. Many optical applications depend critically on the nature and manipulation of the polarization of electromagnetic waves. In summary, electromagnetic waves are produced by oscillating (in general, simply accelerating) charge distributions and carry energy (and momentum) as they travel. These waves exert forces on charged particles in the wave path.

Linear and Elliptical Polarizations

The nature, production, and some applications of polarized light are treated in detail in Chapters 14 and 15 and so we conclude this section with but a brief discussion of the basic nature of the polarization of harmonic electromagnetic fields. The electric field of Eq. (4-43) is said to be linearly polarized along the x-direction since the direction of the electric field is always along the x-direction. As illustrated in Figure 4-12a, light may be linearly polarized along any line that is perpendicular to the direction of wave propagation. In the figure, the electric field $\vec{E}$, made up of equal parts along the x- and y-directions, oscillates along a line making an angle of 45° with the x-axis. The electric field vector in the $z = 0$ plane is shown every eighth of a period over one period, $T = 2\pi/\omega$.

In general, electric fields may be elliptically polarized in the sense that, over time, the electric field vector traces out an ellipse as the wave propagates. The special case of an electromagnetic wave that is circularly polarized is illustrated in Figure 4-12b. In this graph, the component of the electric field

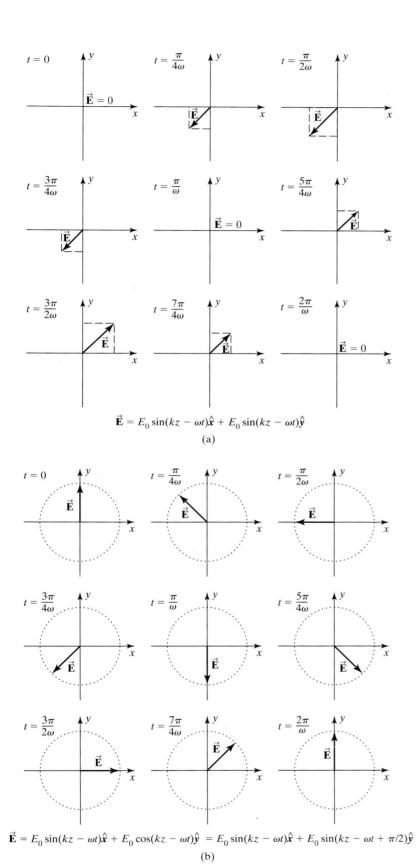

$$\vec{\mathbf{E}} = E_0 \sin(kz - \omega t)\hat{\mathbf{x}} + E_0 \sin(kz - \omega t)\hat{\mathbf{y}}$$

(a)

$$\vec{\mathbf{E}} = E_0 \sin(kz - \omega t)\hat{\mathbf{x}} + E_0 \cos(kz - \omega t)\hat{\mathbf{y}} = E_0 \sin(kz - \omega t)\hat{\mathbf{x}} + E_0 \sin(kz - \omega t + \pi/2)\hat{\mathbf{y}}$$

(b)

Figure 4-12 Electric field polarization. Evolution of the electric field vector over one period at a fixed plane $z = 0$ is shown for a wave with (a) linear polarization and (b) circular polarization.

along the y-direction is always $\pi/2$ out of phase with the x-component of the electric field, leading to circular polarization. Note that both circular and linear polarizations represent limiting cases of elliptical polarizations. An investigation of the general case of arbitrary elliptical polarization is left as an exercise (see problem 4-24).

Unpolarized Light

Often the individual atoms in a source, at a given instant, emit light with differing random polarizations. The light coming from such a source is then a superposition of electromagnetic fields with differing and randomly distributed polarizations. Such light is said to be *randomly polarized* or, commonly, *unpolarized*. If a certain electromagnetic field consists of the superposition of fields with many different polarizations, of which one or more predominates, we say the field is *partially polarized*. Polarized light can be produced by passing unpolarized light through one of a variety of optical systems that transmit only a particular polarization of light. Such polarizing agents are discussed in some detail in Chapter 15. In addition, laser sources, as we discuss in the Chapter 6, can produce polarized light.

4-10 DOPPLER EFFECT

The familiar *Doppler effect* for sound waves has its counterpart in light waves, but with an important difference. Recall that when dealing with sound waves, the apparent frequency of a source increases or decreases depending on the motion of both source and observer along the line joining them. The frequency shift due to a moving *source* is based physically on a change in transmitted wavelength. The frequency shift due to a moving *observer* is based physically on the change in speed of the sound waves relative to the observer. The two effects are physically distinct and described by different equations. They are also essentially different from the case of light waves. The difference between the Doppler effect in sound and light waves is more than the difference in wave speeds. Whereas sound waves propagate through a material medium, light waves propagate in vacuum. As soon as the medium of propagation is removed, there is no longer a physical basis for the distinction between moving observer and moving source. There is *one* relative motion between them that determines the frequency shift in the Doppler effect for light. The derivation of the Doppler effect for light requires the theory of special relativity and so is not carried out here. The result[2] is expressed by

$$\frac{\lambda'}{\lambda} = \sqrt{\frac{1 - \dfrac{v}{c}}{1 + \dfrac{v}{c}}} \tag{4-44}$$

where λ' is the Doppler-shifted wavelength and v is the relative velocity between source and observer. The sign of v is positive when they are approaching one another and negative when they are separating from one another. When $v \ll c$, Eq. (4-44) is approximated by

$$\frac{\lambda'}{\lambda} = 1 - \frac{v}{c} \tag{4-45}$$

---

[2]Robert Resnick, *Basic Concepts in Relativity and Early Quantum Mechanics* (New York: John Wiley and Sons, 1972), Ch. 2.

The Doppler effect is especially important when used to determine the speed of astronomical sources emitting electromagnetic radiation. The *redshift* is the shift in wavelength of such radiation toward longer wavelengths, due to a relative speed of the source *away* from us.

Example 4-3

Light from a distant galaxy shows the characteristic lines of the oxygen spectrum, except that the wavelengths are shifted from their values as measured using laboratory sources. In particular, the line expected at 513 nm shows up at 525 nm. What is the speed of the galaxy relative to the earth?

Solution

Here, $\lambda = 513$ nm and $\lambda' = 525$ nm. Thus, using Eq. (4-45),

$$\frac{525}{513} = 1 - \frac{v}{c}$$

$$v = -0.0234c = -7020 \text{ km/s}$$

Since the apparent λ is larger (the frequency less), the galaxy is moving away from the earth with a speed of approximately 7020 km/s.

Another situation in which the Doppler effect is of pivotal importance is the *Doppler broadening* of the spectral lines associated with the light emitted by the fast-moving atoms of a gas. Since such atoms have a range of velocities relative to a laboratory detector, a range of detected frequencies, corresponding to the range of atomic velocities, will result even if the atoms emit nearly single-frequency electromagnetic radiation. Finally, we mention *Doppler weather radar*, in which a source emits an electromagnetic radio wave towards moving raindrops and other particulates. In turn, the particulates reflect the radio wave back towards the source. The difference in wavelength of the emitted waves and that of those detected after reflection is directly related to the velocity of the particulates relative to the source of the radar.

PROBLEMS

4-1 A pulse of the form $y = ae^{-bx^2}$ is formed in a rope, where a and b are constants and x is in centimeters. Sketch this pulse. Then write an equation that represents the pulse moving in the negative direction at 10 cm/s.

4-2 A transverse wave pulse, described by

$$y = \frac{4\text{m}^3}{x^2 + 2\text{m}^2}$$

is initiated at $t = 0$ in a stretched string.

   a. Write an equation describing the displacement $y(x,t)$ of the traveling pulse as a function of time t and position x if it moves with a speed of 2.5 m/s in the negative x-direction.

   b. Plot the pulse at $t = 0$, $t = 2$ s, and $t = 5$ s.

4-3 Consider the following mathematical expressions, where distances are in meters:

   1. $y(z,t) = A \sin^2[4\pi(t/\text{s} + z/\text{m})]$
   2. $y(x,t) = A(x/\text{m} - t/\text{s})^2$
   3. $y(x,t) = A/(Bx^2 - t)$

   a. Which qualify as traveling waves? Prove your conclusion.
   b. If they qualify, give the magnitude and direction of the wave velocity.

4-4 If the following represents a traveling wave, determine its velocity (magnitude and direction), where distances are in meters.

$$y = \frac{(100 \text{ m})e^{[x^2/\text{m}^2 - 20(x/\text{m})(t/\text{s}) + 100 t^2/\text{s}^2]}}{x/\text{m} - 10t/\text{s}}$$

4-5 A harmonic traveling wave is moving in the negative z-direction with an amplitude (arbitrary units) of 2, a wavelength of 5 m, and a period of 3 s. Its displacement at the origin is zero at time zero. Write a wave equation for this wave (a) that exhibits directly both wavelength and period; (b) that exhibits directly both propagation constant and velocity; (c) in complex form.

4-6 **a.** Write the equation of a harmonic wave traveling along the x-direction at $t = 0$ if it is known to have an amplitude of 5 m and a wavelength of 50 m.

   b. Write an expression for the disturbance at $t = 4$ s if it is moving in the negative x-direction at 2 m/s.

4-7 For a harmonic wave given by

$$y = (10 \text{ cm}) \sin[(628.3/\text{cm})x - (6283/\text{s})t]$$

determine (a) wavelength; (b) frequency; (c) propagation constant; (d) angular frequency; (e) period; (f) velocity; (g) amplitude.

4-8 Use the constant phase condition to determine the velocity of each of the following waves in terms of the constants A, B, C, and D. Distances are in meters and time in seconds. Verify your results dimensionally.

a. $f(y,t) = A(y - Bt)$
b. $f(x,t) = A(Bx + Ct + D)^2$
c. $f(z,t) = A \exp(Bz^2 + BC^2t^2 - 2BCzt)$

4-9 A harmonic wave traveling in the $+x$-direction has, at $t = 0$, a displacement of 13 units at $x = 0$ and a displacement of -7.5 units at $x = 3\lambda/4$. Write the equation for the wave at $t = 0$.

4-10 a. Show that if the maximum positive displacement of a sinusoidal wave occurs at distance x_0 cm from the origin when $t = 0$, its initial phase angle φ_0 is given by

$$\varphi_0 = \frac{\pi}{2} - \left(\frac{2\pi}{\lambda}\right)x_0$$

where the wavelength λ is in centimeters.
b. Determine the initial phase and sketch the wave when $\lambda = 10$ cm and $x_0 = 0, \frac{5}{6}, \frac{5}{2}, 5,$ and $-\frac{1}{2}$ cm.
c. What are the appropriate initial phase angles for (b) when a cosine function is used instead?

4-11 By finding appropriate expressions for $\vec{k} \cdot \vec{r}$, write equations describing a sinusoidal plane wave in three dimensions, displaying wavelength and velocity, if propagation is

a. along the $+z$-axis
b. along the line $x = y, z = 0$
c. perpendicular to the planes $x + y + z = $ constant

4-12 Show that if $\tilde{z}$ is a complex number, (a) Re $(\tilde{z}) = (\tilde{z} + \tilde{z}^*)/2$; (b) Im$(\tilde{z}) = (\tilde{z} - \tilde{z}^*)/2i$; (c) $\cos\theta = (e^{i\theta} + e^{-i\theta})/2$; (d) $\sin\theta = (e^{i\theta} - e^{-i\theta})/2i$.

4-13 Show that a wave function, expressed in complex form, is shifted in phase (a) by $\pi/2$ when multiplied by i and (b) by π when multiplied by -1.

4-14 Two waves of the same amplitude, speed, and frequency travel together in the same region of space. The resultant wave may be written as a sum of the individual waves.

$$\psi(y,t) = A \sin(ky + \omega t) + A \sin(ky - \omega t + \pi)$$

With the help of complex exponentials, show that

$$\psi(y,t) = 2A \cos(ky)\sin(\omega t)$$

4-15 The energy flow to the earth's surface associated with sunlight is about 1.0 kW/m². Find the maximum values of E and B for a wave of this power density.

4-16 A light wave is traveling in glass of index 1.50. If the electric field amplitude of the wave is known to be 100 V/m, find

(a) the amplitude of the magnetic field and (b) the average magnitude of the Poynting vector.

4-17 The solar constant is the radiant flux density (irradiance) from the sun at the surface of the earth's atmosphere and is about 0.135 W/cm². Assume an average wavelength of 700 nm for the sun's radiation that reaches the earth. Find (a) the amplitude of the $\vec{E}$- and $\vec{B}$-fields; (b) the number of photons that arrive each second on each square meter of a solar panel; (c) a harmonic wave equation for the $\vec{E}$-field of the solar radiation, inserting all constants numerically.

4-18 a. The light from a 220-W lamp spreads uniformly in all directions. Find the irradiance of these optical electromagnetic waves and the amplitude of their $\vec{E}$-field at a distance of 10 m from the lamp. Assume that 5% of the lamp energy is converted to light.
b. Suppose a 2000-W laser beam is concentrated by a lens into a cross-sectional area of about 1×10^{-6} cm². Find the corresponding irradiance and amplitudes of the $\vec{E}$- and $\vec{B}$-fields there.

4-19 Show that, in order to conserve flux, the amplitude of a cylindrical wave must vary inversely with $\sqrt{r}$.

4-20 Show that Eq. (4-45) for the Doppler effect follows from Eq. (4-44) when $v \ll c$.

4-21 How fast does one have to approach a red traffic light to see a green signal? So that we all get the same answer, say that a good red is 640 nm and a good green is 540 nm.

4-22 A quasar near the limits of the observed universe to date shows a wavelength that is 4.80 times the wavelength emitted by the same molecules on the earth. If the Doppler effect is responsible for this shift, what velocity does it determine for the quasar?

4-23 Estimate the Doppler broadening of the 706.52-nm line of helium when the gas is at 1000 K. Use the root-mean-square velocity of a gas molecule given by

$$v_{rms} = \sqrt{\frac{3RT}{M}}$$

where R is the gas constant, T the Kelvin temperature, and M the molecular weight.

4-24 Consider the electric field,

$$\vec{E} = E_x \sin(kz - \omega t + \varphi_{0x})\hat{x}$$
$$+ E_y \sin(kz - \omega t + \varphi_{0y})\hat{y}$$

Produce plots like those in Figure 4-12 that show the evolution of the electric field vector, at the plane $z = 0$, as a function of time over one complete temporal cycle for the following cases.

a. $E_x = 2E_y, \varphi_{0x} = \varphi_{0y} = 0$
b. $E_x = 2E_y, \varphi_{0x} = 0, \varphi_{0y} = \pi/2$
c. $E_x = 2E_y, \varphi_{0x} = 0, \varphi_{0y} = -\pi/2$
d. $E_x = 2E_y, \varphi_{0x} = \pi/4, \varphi_{0y} = -\pi/4$
e. $E_x = 2E_y, \varphi_{0x} = 0, \varphi_{0y} = -\pi/4$

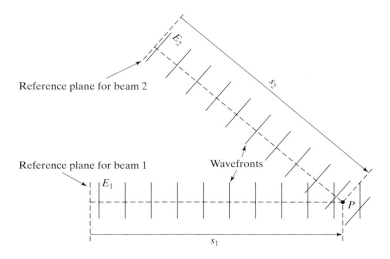

Reference plane for beam 2

Reference plane for beam 1

Wavefronts

5

Superposition of Waves

INTRODUCTION

In Chapter 4, equations describing waves of given amplitude, wavelength, and frequency were developed. Quite commonly, it is necessary to deal with situations in which two or more such waves arrive at the same point in space or exist together along the same direction. Several important cases of the combined effects of two or more harmonic waves are treated in this chapter. The first case deals with the superposition of harmonic waves of differing amplitudes and phases but with the same frequency. The analysis shows that the resultant is just another harmonic wave having the same frequency. This leads to an important difference between the irradiance attainable from randomly phased and coherent harmonic waves. The chapter next treats standing waves that result from the superposition of a harmonic wave with its reflected counterpart. We end by considering the superposition of waves of slightly different frequencies and relate this analysis to the phenomenon of beats and to the distinction between the phase and group velocities of an electromagnetic waveform.

5-1 SUPERPOSITION PRINCIPLE

To explain the combined effects of waves successfully one must ask specifically: What is the net displacement ψ at a point in space where waves with independent displacements ψ_1 and ψ_2 exist together? In most cases of interest, the correct answer is given by the *superposition principle*: The resultant displacement is the sum of the separate displacements of the constituent waves:

$$\psi = \psi_1 + \psi_2 \qquad (5\text{-}1)$$

Using this principle, the resultant wave amplitude and irradiance (W/m^2) can be calculated and verified by measurement.

The same principle can be stated more formally as follows. If ψ_1 and ψ_2 are independently solutions of the wave equation,

$$\nabla^2 \psi = \frac{1}{v^2} \frac{\partial^2 \psi}{\partial t^2}$$

then the *linear combination*,

$$\psi = a\psi_1 + b\psi_2$$

where a and b are constants, is also a solution.

The superposition of electromagnetic (EM) waves may be expressed in terms of their electric or magnetic fields by the vector equations,

$$\vec{E} = \vec{E}_1 + \vec{E}_2 \quad \text{and} \quad \vec{B} = \vec{B}_1 + \vec{B}_2$$

In general, the orientation of the electric or magnetic fields must be taken into account. The superposition of waves at a point where their electric fields are orthogonal, for example, does not yield the same result as the case where they are parallel. A more formal accounting of the vector nature of $\vec{E}$ in the superposition of two EM waves will be taken up in Chapter 7. For the present, we treat electric fields as scalar quantities. This treatment is strictly valid for cases where the individual $\vec{E}$ vectors are parallel; it is often applied in cases where they are nearly parallel. The treatment is valid also for cases of unpolarized light, in which the $\vec{E}$ field can be represented by two (randomly phased) orthogonal components. In that case, the scalar theory applies to each component and its parallel counterpart in the superposing waves, and thus to the entire wave.

When two or more light fields of very large amplitude mix together in a material, the resultant light field may not be simply the superposition of the individual fields. In such a case, the interaction with the material can produce nonlinear effects. The possibility of producing high-energy densities, using laser light, has facilitated the study and use of such effects, making *nonlinear optics* an important branch of modern optics. An introduction to this subject is taken up in Chapter 24.

5-2 SUPERPOSITION OF WAVES OF THE SAME FREQUENCY

The first case of superposition to be considered is the situation in which two harmonic plane waves of the same frequency combine, at a particular point in space P, to form a resultant wave disturbance as shown in Figure 5-1. We permit the two waves to differ in amplitude and phase. We take the waves, at the superposition point P, to have the forms

$$E_1 = E_{01} \cos(ks_1 - \omega t + \varphi_1) \qquad (5\text{-}1)$$

$$E_2 = E_{02} \cos(ks_2 - \omega t + \varphi_2) \qquad (5\text{-}2)$$

Here, s_1 and s_2 represent the directed distances measured along the propagation directions of each light wave from the reference planes at which the phases of the individual waves are φ_1 and φ_2 at time $t = 0$.

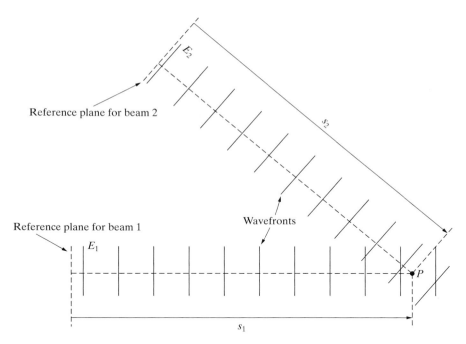

Reference plane for beam 2

Reference plane for beam 1

E_1

Wavefronts

s_2

P

s_1

Figure 5-1 Superposition of two plane waves at point P. The directed distances s_1 and s_2 are the perpendicular distances from the reference planes to the point of superposition.

To simplify notation we introduce the constant phases,

$$\alpha_1 = ks_1 + \varphi_1 \tag{5-3}$$
$$\alpha_2 = ks_2 + \varphi_2 \tag{5-4}$$

The time variations of the EM waves at the given point can thus be expressed by

$$E_1 = E_{01} \cos(\alpha_1 - \omega t) \tag{5-5}$$
$$E_2 = E_{02} \cos(\alpha_2 - \omega t) \tag{5-6}$$

Two such waves, intersecting at the fixed point P, differ in phase by

$$\alpha_2 - \alpha_1 = k(s_2 - s_1) + (\varphi_2 - \varphi_1)$$

due to a path difference, given by the first term, and an initial phase difference, given by the second term. By the superposition principle, the resultant electric field E_R at the point P is

$$E_R = E_1 + E_2 = E_{01} \cos(\alpha_1 - \omega t) + E_{02} \cos(\alpha_2 - \omega t) \tag{5-7}$$

Three cases illustrating the nature of the superposition, at a fixed point in space, of two harmonic waves of the same frequency are illustrated in Figure 5-2 and discussed next.

Constructive Interference

Figure 5-2a illustrates the case of *constructive interference* in which the individual waves being superposed are "in step" in the sense that the peaks of the waves occur at the same times. In this case, the resultant wave is in step with the individual waves being superposed. And the amplitude of the resultant wave is simply the sum $(E_{01} + E_{02})$ of the amplitudes of the individual waves.

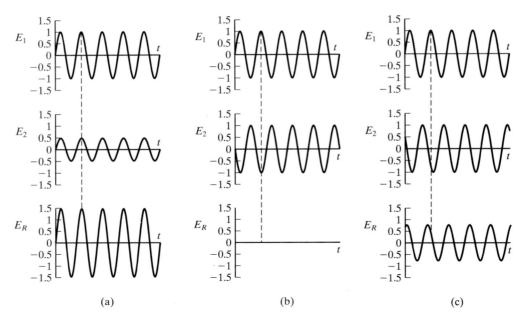

Figure 5-2 Three cases of the superposition of waves of the same frequency at a fixed point in space. In each case, $E_R = E_1 + E_2$. The vertical dotted lines mark the values of E_1, E_2, and E_R at a specific time. (a) Constructive interference; E_1 and E_2 are in phase and the amplitude of the resultant wave is simply the sum of the amplitudes of E_1 and E_2. (b) Destructive interference; E_1 and E_2 differ in phase by π and the amplitude of the resultant wave is the difference in the amplitudes of E_1 and E_2. For the case shown, the amplitudes of E_1 and E_2 are equal, so, at the point of interference, the resultant disturbance is zero. (c) General superposition; the amplitude of the resultant wave is neither the sum nor the difference of the amplitudes of E_1 and E_2.

Two waves of the forms given in Eqs. (5-5) and (5-6) will constructively interfere if the difference $(\alpha_2 - \alpha_1)$ in their phase constants is $m(2\pi)$, where m is an integer. In this case, the resultant wave can be formed as

$$E_R = E_1 + E_2 = E_{01} \cos(\alpha_1 - \omega t) + E_{02} \cos(\alpha_1 + 2m\pi - \omega t)$$
$$= (E_{01} + E_{02}) \cos(\alpha_1 - \omega t)$$

Destructive Interference

If two harmonic waves of the same frequency are "out of step" in the sense that the peaks of one always coincide with the troughs of the other, the waves are said to *destructively interfere*. In this case, the waves add to form a resultant wave that has an amplitude that is (in magnitude) the difference of the amplitudes of the waves being superposed and which is in step with the individual wave of largest amplitude. Two waves of the forms given in Eqs. (5-5) and (5-6) will destructively interfere if the difference in their phase constants is an odd multiple of π, that is, if $\alpha_2 - \alpha_1 = (2m + 1)\pi$, where m is an integer. In this case,

$$E_R = E_1 + E_2 = E_{01} \cos(\alpha_1 - \omega t) + E_{02} \cos(\alpha_1 + (2m + 1)\pi - \omega t)$$
$$= (E_{01} - E_{02}) \cos(\alpha_1 - \omega t)$$

Note that when waves of equal amplitude destructively interfere, the resultant disturbance, at the point of interference, is zero, as shown in Figure 5-2b.

General Superposition

The general case of the superposition of two waves that are neither in step nor out of step is depicted in Figure 5-2c. In this case, the amplitude of the resultant wave is intermediate in magnitude between the magnitude of the sum and that of the difference of the amplitudes of the waves being superposed. As shown in Figure 5-2c, the resultant wave is not, in general, in step with either of the waves being superposed. For this general case, the resultant field of Eq. (5-7) can be simplified by writing the fields in complex form and using a phasor diagram to aid in the addition of the individual fields. Proceeding thus we write,

$$E_R = \text{Re}(E_{01}e^{i(\alpha_1 - \omega t)} + E_{02}e^{i(\alpha_2 - \omega t)}) = \text{Re}(e^{-i\omega t}(E_{01}e^{i\alpha_1} + E_{02}e^{i\alpha_2}))$$

Defining

$$E_0 e^{i\alpha} = E_{01}e^{i\alpha_1} + E_{02}e^{i\alpha_2} \qquad (5\text{-}8)$$

permits the resultant wave to be expressed in the standard form,

$$E_R = \text{Re}(E_0 e^{i(\alpha - \omega t)}) = E_0 \cos(\alpha - \omega t)$$

The amplitude and phase of the resultant field can be found conveniently by treating the complex numbers in Eq. (5-8) as vectors, as indicated in the phasor diagrams of Figure 5-3. Recall (see Section 4-3) that a complex quantity can be represented as a vector in a phasor diagram in which the real and imaginary parts of the complex quantity are, respectively, the horizontal and vertical components of its vector representation. See Figure 5-3a. In such a representation, the length of the vector is the magnitude of the complex quantity and the angle that the vector makes with the horizontal axis is the phase of the complex quantity. From Figure 5-3b, the components of the resultant phasor $E_0 e^{i\alpha}$ are

$$E_0 \cos \alpha = E_{01} \cos \alpha_1 + E_{02} \cos \alpha_2$$

and

$$E_0 \sin \alpha = E_{01} \sin \alpha_1 + E_{02} \sin \alpha_2$$

The cosine law may be applied to Figure 5-3a, yielding an expression for the amplitude of the resultant field E_0,

$$E_0^2 = E_{01}^2 + E_{02}^2 + 2E_{01}E_{02} \cos(\alpha_2 - \alpha_1) \qquad (5\text{-}9)$$

and from Figure 5-3b, the phase angle of the resultant field is given clearly by

$$\tan \alpha = \frac{E_{01} \sin \alpha_1 + E_{02} \sin \alpha_2}{E_{01} \cos \alpha_1 + E_{02} \cos \alpha_2} \qquad (5\text{-}10)$$

In summary, we have shown that,

$$E_R = E_{01} \cos(\alpha_1 - \omega t) + E_{02} \cos(\alpha_2 - \omega t) = E_0 \cos(\alpha - \omega t) \qquad (5\text{-}11)$$

where the amplitude E_0 of the resultant field is given by Eq. (5-9) and the phase α of the resultant field is given by Eq. (5-10).

This graphical procedure could be extended to accommodate any number of component waves of the same frequency, as shown in Figure 5-4 for four such waves. The diagram makes apparent the proper generalization of Eqs. (5-10) and (5-9) for N such harmonic waves:

$$\tan \alpha = \frac{\displaystyle\sum_{i=1}^{N} E_{0i} \sin \alpha_i}{\displaystyle\sum_{i=1}^{N} E_{0i} \cos \alpha_i} \qquad (5\text{-}12)$$

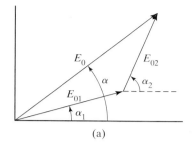

(a)

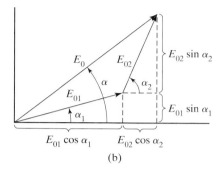

(b)

Figure 5-3 Phasor diagrams useful for determining the sum of two harmonic waves. (a) Vector addition used to determine the sum phasor. (b) Phasor components.

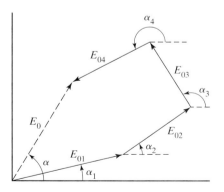

Figure 5-4 Phasor diagram for four harmonic waves of the same frequency. Superposition produces a resultant wave of the same frequency, with amplitude E_0 and phase α.

and by the Pythagorean theorem,

$$E_0^2 = \left(\sum_{i=1}^{N} E_{0i} \sin \alpha_i \right)^2 + \left(\sum_{i=1}^{N} E_{0i} \cos \alpha_i \right)^2 \tag{5-13}$$

Eq. (5-13) may be cast into a form that looks more like a generalization of the cosine law in Eq. (5-9). Expanding each term,

$$\left(\sum_{i=1}^{N} E_{0i} \sin \alpha_i \right)^2 = \sum_{i=1}^{N} E_{0i}^2 \sin^2 \alpha_i + 2\sum_{j>i}^{N}\sum_{i=1}^{N} E_{0i}E_{0j} \sin \alpha_i \sin \alpha_j \tag{5-14}$$

$$\left(\sum_{i=1}^{N} E_{0i} \cos \alpha_i \right)^2 = \sum_{i=1}^{N} E_{0i}^2 \cos^2 \alpha_i + 2\sum_{j>i}^{N}\sum_{i=1}^{N} E_{0i}E_{0j} \cos \alpha_i \cos \alpha_j \tag{5-15}$$

The first term of the right members is the sum of the squares of the individual terms of the series in the left members. The double sums represent all the cross products, excluding—by the use of notation $j > i$—the self-products already accounted for in the first term and also avoiding a duplication of products already tallied by the factor 2. Adding Eqs. (5-14) and (5-15),

$$E_0^2 = \sum_{i=1}^{N} E_{0i}^2(\sin^2 \alpha_i + \cos^2 \alpha_i) + 2\sum_{j>i}^{N}\sum_{i=1}^{N} E_{0i}E_{0j}(\cos \alpha_i \cos \alpha_j + \sin \alpha_i \sin \alpha_j)$$

The expressions in parentheses are equivalent to unity in the first term and are equivalent to $\cos(\alpha_j - \alpha_i)$ in the second, so that, finally,

$$E_0^2 = \sum_{i=1}^{N} E_{0i}^2 + 2\sum_{j>i}^{N}\sum_{i=1}^{N} E_{0i}E_{0j} \cos(\alpha_j - \alpha_i) \tag{5-16}$$

Summarizing, the sum of N harmonic waves of identical frequency is again a harmonic wave of the same frequency, with amplitude given by Eq. (5-13) or (5-16) and phase given by Eq. (5-12).

Example 5-1

Determine the result of the superposition of the following harmonic waves:

$E_1 = 7 \cos(\pi/3 - \omega t)$, $E_2 = 12 \sin(\pi/4 - \omega t)$, and $E_3 = 20 \cos(\pi/5 - \omega t)$

Solution

To make all phase angles consistent, first change the sine wave to a cosine wave:

$E_2 = 12 \cos(\pi/4 - \pi/2 - \omega t) = 12 \cos(-\pi/4 - \omega t)$. Then, using Eq. (5-13),

$$E_0^2 = \left[7 \sin\left(\frac{\pi}{3}\right) + 12 \sin\left(-\frac{\pi}{4}\right) + 20 \sin\left(\frac{\pi}{5}\right) \right]^2$$
$$+ \left[7 \cos\left(\frac{\pi}{3}\right) + 12 \cos\left(-\frac{\pi}{4}\right) + 20 \cos\left(\frac{\pi}{5}\right) \right]^2$$

or $E_0^2 = 9.333^2 + 28.166^2$ and $E_0 = 29.67$. The same result can be found using Eq. (5-16), which would take the form

$$E_0^2 = 7^2 + 12^2 + 20^2$$
$$+ 2\left[(7 \times 12) \cos\left(-\frac{\pi}{4} - \frac{\pi}{3}\right) + (7 \times 20) \cos\left(\frac{\pi}{5} - \frac{\pi}{3}\right) \right.$$
$$\left. + (12 \times 20) \cos\left(\frac{\pi}{5} - \left(-\frac{\pi}{4}\right)\right) \right]$$

The phase angle of the resulting harmonic wave is found using Eq. (5-12). Since the sums forming the numerator and denominator have already been calculated in the first part, we have

$$\tan \alpha = \frac{9.333}{28.166} \quad \text{and} \quad \alpha = 0.32 \text{ rad}$$

Thus, the resulting harmonic wave $E_R = E_0 \cos(\alpha - \omega t)$ can be written as

$$E_R = 29.67 \cos(0.32 - \omega t)$$

5-3 RANDOM AND COHERENT SOURCES

The effort expended in achieving the form of Eq. (5-16) pays immediate dividends in enabling us to distinguish rather neatly two important cases of superposition: (1) the case of N randomly phased sources of equal amplitude and frequency, where N is a large number, and (2) the case of N coherent sources of the same type. In the first instance, if phases are random, the phase differences $(\alpha_i - \alpha_j)$ are also random. The sum of cosine terms in Eq. (5-16) then approaches zero as N increases, because terms are equally divided between positive and negative fractions ranging from -1 to $+1$. This leaves

$$E_0^2 = \sum_{i=1}^{N} E_{0i}^2 = N E_{01}^2 \tag{5-17}$$

because there are N sources of *equal amplitude*. Thus for N randomly phased sources, the squares of the individual amplitudes add to produce the square of the resultant amplitude. Recalling that the irradiance (W/m^2) is proportional to the square of the amplitude of the electric field, we can say that *the resultant irradiance of N identical but randomly phased sources is the sum of the individual irradiances.* On the other hand, if the N sources are *coherent, and in phase,* so that all α_i are equal, then Eq. (5-16) becomes

$$E_0^2 = \sum_{i=1}^{N} E_{0i}^2 + 2\sum_{j>i}^{N}\sum_{i=1}^{N} E_{0i} E_{0j}$$

since all of the cosine factors are unity in this case. The right side should be recognizable as the square of the sum of the individual amplitudes. For the case of equal amplitudes,

$$E_0^2 = \left(\sum_{i=1}^{N} E_{0i}\right)^2 = (N E_{01})^2 = N^2 E_{01}^2 \tag{5-18}$$

Here the individual amplitudes (*rather than the irradiances*) simply add to produce a resultant $E_0 = N E_{01}$. Thus, *the resultant irradiance of N identical coherent sources, radiating in phase with each other, is N^2 times the irradiance of the individual sources.* Notice that in this case the result does not require that N be a large number. We conclude that the irradiance of 100 coherent in-phase sources, for example, is 100 times greater than the more usual case of

100 random sources. If E is interpreted as the amplitude of a compressional wave, the result holds for sound intensities as well.

5-4 STANDING WAVES

Another important case of superposition arises when a given wave exists in both forward and reverse directions along the same medium. This condition occurs most frequently when one or the other of the oppositely directed waves experiences a reflection at some point along its path, as in Figure 5-5a. For concreteness, consider the situation shown in Figure 5-5a in which a wave traveling in the negative x-direction (solid line) encounters a barrier and is reflected (dashed line). Let us assume for the moment an ideal situation in which none of the energy is lost on reflection nor absorbed by the transmitting medium. This permits us to write both waves with the same amplitude. The oppositely directed waves can then be written as

$$E_1 = E_0 \sin(\omega t + kx)\ldots \quad \text{to the left} \tag{5-19}$$

$$E_2 = E_0 \sin(\omega t - kx - \varphi_R)\ldots \quad \text{to the right} \tag{5-20}$$

Here, φ_R is included to account for possible phase shifts upon reflection. Note that, for convenience, we use sine functions to represent the individual waves. Additionally, we have chosen to write the temporal part before the spatial part in arguments of the sine functions of Eqs. (5-19) and (5-20). This is useful if one wishes to associate φ_R with the phase shift caused by reflection.

The resultant wave in the medium, by the principle of superposition, is

$$E_R = E_1 + E_2 = E_0[\sin(\omega t + kx) + \sin(\omega t - kx - \varphi_R)] \tag{5-21}$$

It is expedient in this case to define

$$\beta_+ = \omega t + kx \quad \text{and} \quad \beta_- = \omega t - kx - \varphi_R$$

and employ the trigonometric identity

$$\sin \beta_+ + \sin \beta_- \equiv 2 \sin\tfrac{1}{2}(\beta_+ + \beta_-) \cos\tfrac{1}{2}(\beta_+ - \beta_-)$$

Applied to Eq. (5-21), this leads to the result

$$E_R = 2E_0 \cos\left(kx + \frac{\varphi_R}{2}\right) \sin\left(\omega t - \frac{\varphi_R}{2}\right) \tag{5-22}$$

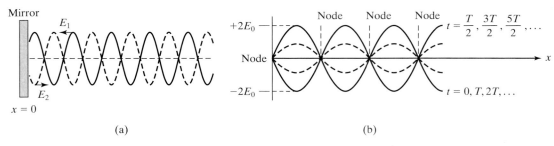

Figure 5-5 Standing waves. (a) A typical standing wave situation occurs when a wave E_1 and its reflection E_2 exist along the same medium. For the case shown, a π phase shift has occurred upon reflection so that a node (zero displacement) will exist at the mirror. (b) Resultant displacement of a standing wave, shown at various instants. The solid lines represent the maximum displacement of the wave. The displacement at the nodes is always zero.

The case shown in Figure 5-5a, in which there is a π phase shift upon reflection, corresponds to the physically interesting case of the reflection of an electromagnetic field from a plane conducting mirror (and the analogous case of the reflection of a transverse wave in a string from a rigid boundary). In these cases, $\varphi_R/2 = \pi/2$ and the resultant field takes the form

$$E_R = (2E_0 \sin kx)\cos \omega t \qquad (5\text{-}23)$$

As shown in 5-5b, Eq. (5-23) represents a *standing wave*. Interpretation is facilitated by regarding the quantity in parentheses as a space-dependent amplitude. At any point x along the medium, the oscillations are given by

$$E_R = A(x)\cos \omega t$$

where $A(x) = 2E_0 \sin kx$. There exist values of x for which $A(x) = 0$, and thus $E_R = 0$ for all t. These values occur whenever

$$\sin kx = 0, \quad \text{or} \quad kx = \frac{2\pi x}{\lambda} = m\pi, \qquad m = 0, \pm 1, \pm 2, \ldots$$

or

$$x = m\left(\frac{\lambda}{2}\right) = 0, \frac{\lambda}{2}, \lambda, \frac{3\lambda}{2}, \ldots \qquad (5\text{-}24)$$

Such points are called the *nodes* of the standing wave and are separated by half a wavelength. At various times, the standing wave will appear as sine waves of various amplitudes, like those shown in Figure 5-5b. Although their amplitudes vary with time, all pass through zero at the fixed nodal points. E_R has its maximum value at all points when $\cos \omega t = \pm 1$, or when

$$\omega t = 2\pi\nu t = \left(\frac{2\pi}{T}\right)t = m\pi$$

Thus, the outer envelope of the standing wave occurs at times

$$t = m\left(\frac{T}{2}\right) = 0, \frac{T}{2}, T, \frac{3T}{2}, \ldots$$

where T is the period. There are also periodic times when the standing wave is everywhere zero, since $\cos \omega t = 0$ for $t = T/4, 3T/4, \ldots$.

We have concentrated on the important case in which the field suffers a π phase shift upon reflection. In general, as can be seen from an examination of Eq. (5-22), as long as the amplitude of the reflected wave is the same as that of the incident wave, a standing wave will result. In the more general case of complete reflection with an arbitrary phase shift, the positions of the nodes will be shifted from the case depicted in Figure 5-5b but the nodes will still be separated by $\lambda/2$. The times at which the form is everywhere zero or everywhere at its maximum displacement also change. The principal features of the standing wave, however, remain unaffected.

In the treatment of lasers in Chapter 6, we will find that laser light is generated in *laser cavities*, which often take the form of two highly reflecting mirrors surrounding a gain medium. The light in such a cavity then consists of counterpropagating electromagnetic waves that form standing waves. It is typically the case that the electromagnetic boundary conditions at the mirror surfaces require that the electric field be zero at the mirrors, which then implies that standing wave nodes occur at the mirrors. This situation is illustrated in Figure 5-6. As can be seen from the figure and by examination of Eq. (5-24), the requirement that nodes exist at the mirror positions restricts the wavelengths

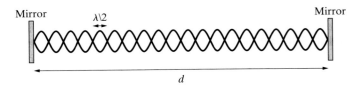

Figure 5-6 Standing wave mode of a laser cavity with mirror spacing d. Each loop of the standing wave envelope is of length $\lambda/2$. In a typical laser cavity, about 1 million half-waves fit into the length of the cavity.

that can be supported by the cavity to discrete values. If the distance between the cavity mirrors is d, the cavity will support standing waves with wavelengths λ_m that satisfy the relation

$$d = m\left(\frac{\lambda_m}{2}\right) \qquad (5\text{-}25)$$

where m is a nonzero integer. That is, the standing wave *normal modes* of the cavity have wavelengths such that an integer number of half-wavelengths "fit" into the cavity length.

Equation (5-25) can be used to determine the frequencies of the standing wave modes of a laser cavity. Solving Eq. (5-25) for the wavelengths of the standing wave modes gives

$$\lambda_m = \frac{2d}{m}$$

The frequencies ν_m of the standing wave modes can be found from the fundamental relationship between the frequency, wavelength, and speed c of the wave:

$$\nu_m = \frac{c}{\lambda_m} = m\frac{c}{2d} \qquad (5\text{-}26)$$

In passing, we note that the analysis just given is strictly valid only for cavities with plane mirrors but is indicative of the behavior of cavities with spherical mirrors (see Chapter 27) as well. In general, the output of a laser will consist of those frequencies given in Eq. (5-26) that the laser gain medium is capable of supporting. The following example elaborates on this point.

Example 5-2

A certain He-Ne laser cavity of the type shown in Figure 5-6 has a mirror separation of 30 cm. The helium-neon laser gain medium is capable of supporting laser light of wavelengths in the range from $\lambda_1 = 632.800$ nm to $\lambda_2 = 632.802$ nm. Find:

a. The approximate number m of half-wavelengths that fit into the cavity
b. The range of frequencies supported by the helium-neon gain medium
c. The difference in the frequencies of adjacent standing wave modes of the cavity
d. The number of standing wave modes that will likely be present in the laser output

Solution

a. From Eq. (5-25): $m = \dfrac{2d}{\lambda_1} \approx \dfrac{(2 \times 0.3\ \text{m})}{632.8 \times 10^{-9}\ \text{m}} = 948{,}166.$ (Here, we have rounded down to an integer.)

b. The frequency range $\Delta \nu_{\text{gain}}$ supported by the gain medium is

$$\Delta \nu_{\text{gain}} = \frac{c}{\lambda_1} - \frac{c}{\lambda_2}$$

$$= (3 \times 10^8 \text{ m/s})\left(\frac{1}{632.800 \times 10^{-9} \text{ m}} - \frac{1}{632.802 \times 10^{-9} \text{ m}}\right)$$

$$= 1.50 \times 10^9 \text{ Hz}$$

c. Using Eq. (5-26): $\nu_{m+1} - \nu_m = (m + 1)\dfrac{c}{2d} - m\dfrac{c}{2d} = \dfrac{c}{2d} =$

$$\frac{3 \times 10^8 \text{ m/s}}{2(0.3 \text{ m})} = 5 \times 10^8 \text{ Hz}$$

d. The number $\mathcal{N}$ of standing wave modes that will likely be present in the laser output is the ratio of the frequency range supported by the gain medium to the separation between standing wave modes:

$$\mathcal{N} = \frac{\Delta \nu_{\text{gain}}}{\nu_{m+1} - \nu_m} = \frac{1.5 \times 10^9 \text{ Hz}}{5 \times 10^8 \text{ Hz}} = 3$$

Unlike traveling waves, standing waves transmit no energy. All the energy in the wave goes into sustaining the oscillations between nodes, at which points forward and reverse waves cancel. If not all of the field incident on a boundary is reflected, the incident and reflected waves will not add to form a perfect standing wave. In such a case, the superposed incident and reflected wave may be written profitably as the sum of a standing wave (which transmits no energy) and a traveling wave that carries the energy that is absorbed by or transmitted through the boundary. In this way, we may model the fields in laser cavities with partially transmitting mirrors.

5-5 THE BEAT PHENOMENON

Yet another case of superposition, with important applications in optics, is that of waves of the same or comparable amplitude but differing in frequency. Differences in frequency imply differences in wavelength and, if the medium is dispersive, differences in velocity. In this section we will consider the case of a nondispersive medium in which (by definition) waves of different frequency travel with the same speed. Consider the superposition of two such waves of different frequency and wavelength but with the same speed through the medium:

$$E_1 = E_0 \cos(k_1 x - \omega_1 t) \tag{5-27}$$

$$E_2 = E_0 \cos(k_2 x - \omega_2 t) \tag{5-28}$$

The superposition of these waves, which are traveling together in a given medium, is

$$E_R = E_1 + E_2 = E_0[\cos(k_1 x - \omega_1 t) + \cos(k_2 x - \omega_2 t)]$$

Making use of the trigonometric identity,

$$\cos \alpha + \cos \beta \equiv 2 \cos\tfrac{1}{2}(\alpha + \beta)\cos\tfrac{1}{2}(\alpha - \beta) \tag{5-29}$$

and identifying

$$\alpha = k_1 x - \omega_1 t$$
$$\beta = k_2 x - \omega_2 t$$

we have

$$E_R = 2E_0 \cos\left[\frac{(k_1 + k_2)}{2}x - \frac{(\omega_1 + \omega_2)}{2}t\right]\cos\left[\frac{(k_1 - k_2)}{2}x - \frac{(\omega_1 - \omega_2)}{2}t\right]$$

(5-30)

Now let

$$\omega_p = \frac{\omega_1 + \omega_2}{2}, \qquad k_p = \frac{k_1 + k_2}{2}$$

(5-31)

and

$$\omega_g = \frac{\omega_1 - \omega_2}{2}, \qquad k_g = \frac{k_1 - k_2}{2}$$

(5-32)

Then,

$$E_R = 2E_0 \cos(k_p x - \omega_p t)\cos(k_g x - \omega_g t)$$

(5-33)

Equation (5-33) represents a product of two cosine waves. The first possesses a frequency ω_p and propagation constant k_p that are, respectively, the averages of the frequencies and propagation constants of the component waves. The second cosine factor represents a wave with frequency ω_g and propagation constant k_g that are much smaller by comparison, since differences of the original values are taken in Eq. (5-32). With $\omega_p \gg \omega_g$, plots of the cosine functions versus time may appear like those of Figure 5-7a, calculated at the same point x_0. The slowly varying cosine function is a factor that ranges between $+1$ and -1 for various t. The product of this cosine factor and the overall fixed amplitude $2E_0$ may be regarded as the *slowly varying amplitude* of the resultant wave. That is, the overall effect is that the low-frequency wave serves as an envelope modulating the high-frequency wave, as shown in Figure 5-7b. The higher-frequency cosine factor, in the resultant waveform, is sometimes referred to as the *carrier wave* to distinguish it from the lower-frequency *envelope wave*. The dashed lines depict the envelope of the resulting wave disturbance. Such a wave disturbance exhibits the phenomenon of *beats*. Because the square of the displacement of the wave at any time is a measure of its irradiance, the energy delivered by the traveling sequence of pulses in Figure 5-7b is itself pulsating at a *beat frequency*, ω_b. The figure shows that the beat frequency is twice the frequency of the modulating envelope, or

$$\omega_b = 2\omega_g = 2\left(\frac{\omega_1 - \omega_2}{2}\right) = \omega_1 - \omega_2$$

(5-34)

From Eq. (5-34) we see that the beat frequency is simply the difference frequency for the two waves. In the case of sound, this is the usual beat frequency heard when two tuning forks are made to vibrate simultaneously, equal to the difference in fork frequencies. The phenomenon of beats provides a sensitive method of measuring the difference in frequencies of two signals of nearly the same frequency. Two guitarists may ensure that their guitars are "in tune" with each other by striking a note and listening for the beat note. One or the other of the guitarists may then adjust the tension in the guitar string until the beat note disappears, indicating that the guitars are in tune. In the optical arena, the beat phenomenon can be used to measure the difference between the emitted radar wave and the Doppler-shifted return signal in a Doppler weather radar system (see Section 4-10) or as part of a feedback loop designed to ensure that two sources have the same frequency.

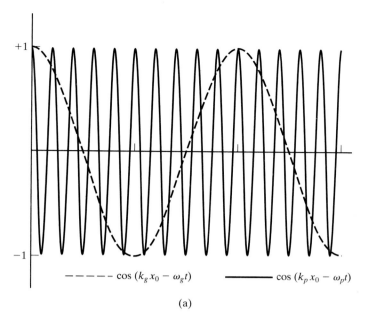

$$-------\cos\left(k_g x_0 - \omega_g t\right) \qquad ———— \cos\left(k_p x_0 - \omega_p t\right)$$

(a)

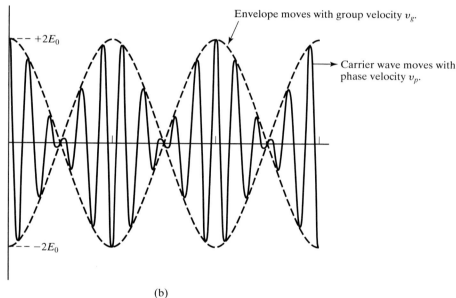

Envelope moves with group velocity v_g.

Carrier wave moves with phase velocity v_p.

$+2E_0$

$-2E_0$

(b)

Figure 5-7 (a) Separate plots of the cosine factors of Eq. (5-33) at $x = x_0$, where $\omega_p \gg \omega_g$. (b) Modulated wave representing Eq. (5-33) at $x = x_0$. The terms *group velocity* and *phase velocity* are introduced in the next section.

5-6 PHASE AND GROUP VELOCITIES

In general, any pulse of light can be viewed as a superposition of harmonic waves of different frequencies.[1] Generally, the duration of a pulse is inversely proportional to the range of frequencies of the harmonic waves that superpose to form the pulse. That is, narrower pulses are composed of harmonic waves with a wider range of frequencies. Even so-called monochromatic light, unless it has infinite spatial extent and has been in existence for all time, possesses a spread of wavelengths, however narrow, about its average wavelength. Electromagnetic waves of different frequencies travel with slightly different speeds through a given medium. As we have noted previously, this is known as *dispersion*. Even "transparent" media are at least weakly absorptive and so must also be dispersive.

---

[1]*Fourier analysis* provides techniques for dealing with the superposition of many harmonic components. This subject is introduced in Chapter 9.

Thus, the interaction with the medium that results in absorption generally also affects the speed of the wave. In a train of wave pulses, the high-amplitude pulses occur at times and positions at which the crests of the component harmonic waves are coincident and so constructively interfere. The regions of low field amplitude between the pulses result from the juxtaposition of constituent waves more or less out of phase. If all of the harmonic waves in the pulse train move with the same speed, then the positions of constructive interference (i.e., the pulses) also move at this speed. However, if the harmonic components move with different speeds, the positions of net constructive interference have a more complicated relationship to the speeds of the constituent harmonic waves. The *phase velocity* of an electromagnetic signal is a measure of the velocity of the harmonic waves that constitute the signal. The *group velocity* of the signal is the velocity at which the positions of maximal constructive interference propagate.

The analysis of the preceding section on the phenomenon of beats serves as a useful starting point for a quantitative discussion of the phase and group velocities of a wave pulse. Any two wavelength components of such a light beam, moving through a dispersive medium, can be represented by Eqs. (5-27) and (5-28) and thus produce a resultant like the one pictured in Figure 5-7b. The velocity of the higher-frequency carrier wave in the resultant waveform of Eq. (5-33), as well as that of the lower-frequency envelope wave, can be found from the general relation for velocity,

$$v = \nu\lambda = \frac{\omega}{k} \tag{5-35}$$

The velocity of the higher-frequency carrier wave in the wave form of Eq. (5-33), is known is the phase velocity,

$$v_p = \frac{\omega_p}{k_p} = \frac{\omega_1 + \omega_2}{k_1 + k_2} \cong \frac{\omega}{k} \tag{5-36}$$

where the final member is an approximation in the case $\omega_1 \cong \omega_2 = \omega$ and $k_1 \cong k_2 = k$ for neighboring frequency and wavelength components in a continuum. On the other hand, the velocity of the envelope, called the *group velocity*, is

$$v_g = \frac{\omega_g}{k_g} = \frac{\omega_1 - \omega_2}{k_1 - k_2} \cong \frac{d\omega}{dk} \tag{5-37}$$

assuming again that the differences between frequencies and propagation constants are small. Now group velocity $v_g = d\omega/dk$ and phase velocity $v_p = \omega/k$ need not be the same. Referring back to Figure 5-7b of the previous section of this chapter, if $v_p > v_g$, the high-frequency waves would appear to have a velocity to the right relative to the envelope, also in motion. These waves would seem to disappear at the right node and be generated at the left node of each pulse. If $v_p < v_g$, their relative motion would, of course, be reversed. When $v_p = v_g$, the high-frequency waves and envelope would move together at the same rate, without relative motion. The relation between group and phase velocities can be found as follows. Substituting Eq. (5-36) into Eq. (5-37) and performing the differentiation of a product,

$$v_g = \frac{d\omega}{dk} = \frac{d}{dk}(kv_p)$$

$$v_g = v_p + k\left(\frac{dv_p}{dk}\right) \tag{5-38}$$

When the velocity of a wave does not depend on wavelength, that is, in a nondispersive medium, $dv_p/dk = 0$, and phase and group velocities are equal. This is the case of light propagating in a vacuum, where $v_p = v_g = c$. In dispersive media, however, $v_p = c/n$, where the refractive index n is a function of λ or k. Then $n = n(k)$, and

$$\frac{dv_p}{dk} = \frac{d}{dk}\left(\frac{c}{n}\right) = \frac{-c}{n^2}\left(\frac{dn}{dk}\right)$$

When incorporated into Eq. (5-38), we have an alternate relation between phase and group velocities,

$$v_g = v_p\left[1 - \frac{k}{n}\left(\frac{dn}{dk}\right)\right] \qquad (5\text{-}39)$$

Again, using $k = 2\pi/\lambda$ and $dk = -(2\pi/\lambda^2)\,d\lambda$, Eq. (5-39) can be reformulated as

$$v_g = v_p\left[1 + \frac{\lambda}{n}\left(\frac{dn}{d\lambda}\right)\right] \qquad (5\text{-}40)$$

In regions of normal dispersion, $dn/d\lambda < 0$ and $v_g < v_p$.

These results, derived here for the case of two wave components, hold in general for a number of waves with a narrow range of frequencies. As mentioned, if harmonic waves with a wide range of frequencies constitute the waveform, the result will be narrow pulses separated by relatively longer intervals of low field amplitude, as shown in Figure 5-8. Still, the sum of a larger number of closely grouped harmonic components can still be characterized both by a phase velocity, the average velocity of the individual waves, and by the group velocity, the velocity of the modulating waveform itself. Since the latter determines the speed with which energy is transmitted, it is the directly measurable speed of the waves. When carrier waves are modulated to contain information, as in amplitude modulation (AM) of radio waves, we may speak of the group velocity as the signal velocity, which is usually less than the phase velocity of the carrier waves. When light pulses, consisting of a number of harmonic waves extending over a range of frequencies, are transmitted through a dispersive medium, the velocity

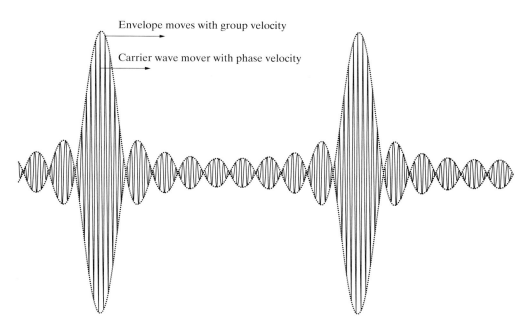

Figure 5-8   Snapshot of a waveform that is the sum of 10 equal-amplitude harmonic waves with frequency spacing about 1/50 of the average frequency of the constituent harmonic waves. The dotted-line envelope moves with the group velocity and the high-frequency carrier wave moves with the phase velocity.

of the group is the velocity of the pulses and is different from the velocity of the individual harmonic waves. In the formalism of quantum mechanics, the electron itself is represented by a localized wave packet that can be decomposed into a number of harmonic waves with a range of wavelengths. The measured velocity of the electron is the velocity of the wave packet, that is, the group velocity of the constituent waves. Recently, the relation between the group and phase velocities of light signals has gained prominence as researchers have succeeded in preparing propagation media with dispersive properties such that the group velocity of the light pulses traveling in the nearly transparent media approaches zero.[2]

The dependence of index of refraction on wavelength can take a variety of forms, depending upon the nature of the medium through which the signal is propagating. In Example 5-3 we consider one such dependence.

Example 5-3

For wavelengths in the visible spectrum, the index of refraction of a certain type of crown glass can be approximated by the relation $n(\lambda) = 1.5255 + (4825 \text{ nm}^2)/\lambda^2$.

a. Find the index of refraction of this glass for 400 nm light, 500 nm light, and 700 nm light.
b. Find the phase velocity, in this glass, for a pulse with frequency components centered around 500 nm.
c. Find the group velocity, in this glass, for a pulse with frequency components centered around 500 nm.

Solution

a. The indices are

$$n_{400} = 1.5255 + (4825 \text{ nm}^2)/(400 \text{ nm})^2 = 1.5557$$

$$n_{500} = 1.5255 + (4825 \text{ nm}^2)/(500 \text{ nm})^2 = 1.5448$$

$$n_{700} = 1.5255 + (4825 \text{ nm}^2)/(700 \text{ nm})^2 = 1.5353$$

b. The phase velocity is $v_p = \dfrac{c}{n_{500}} = \dfrac{3 \times 10^8 \text{ m/s}}{1.5448} = 1.942 \times 10^8 \text{ m/s}$.

c. The group velocity for this pulse would be

$$v_g = v_p \left[1 + \frac{\lambda}{n}\left(\frac{dn}{d\lambda}\right)\right]\Bigg|_{\lambda=500 \text{ nm}} = v_p \left[1 + \frac{\lambda}{n}\left(\frac{4825 \text{ nm}^2}{\lambda^3}\right)(-2)\right]\Bigg|_{\lambda=500 \text{ nm}}$$

$$= v_p \left[1 - \frac{9650 \text{ nm}^2}{n\lambda^2}\right]\Bigg|_{\lambda=500 \text{ nm}}$$

$$v_g = (1.942 \times 10^8 \text{ m/s})\left[1 - \frac{9650 \text{ nm}^2}{(1.5448)(500 \text{ nm})^2}\right] = 1.893 \times 10^8 \text{ m/s}$$

The group and phase velocities do not differ greatly for visible pulses propagating through glass because the index of refraction of glass varies only slightly across the visible spectrum.

The notion of a group velocity is sensible only so long as the pulses "hold together" as the constituent harmonic waves propagate. In the general case, pulses spread and distort as they propagate due to the differing speeds of the

---

[2]For reviews of this phenomenon, see Kirk T. McDonald, *Am. J. Phys.*, Vol. 68, No. 293 (2000) and Barbara Gross Levi, *Phys. Today*, Vol. 54, No. 17 (2001).

different harmonic components. In cases where the pulses break apart, as will happen if the phase velocities of the component harmonic waves differ by a sufficient amount, no single group velocity can be assigned to the signal.

PROBLEMS

5-1 Two plane waves are given by

$$E_1 = \frac{5E_0}{[(3/\text{m})x - (4/\text{s})t]^2 + 2} \quad \text{and}$$

$$E_2 = \frac{-5E_0}{[(3/\text{m})x + (4/\text{s})t - 6]^2 + 2}$$

a. Describe the motion of the two waves.
b. At what instant is their superposition everywhere zero?
c. At what point is their superposition always zero?

5-2 a. Show in a phasor diagram the following two harmonic waves:

$$E_1 = 2 \cos \omega t \quad \text{and} \quad E_2 = 7 \cos\left(\frac{\pi}{4} - \omega t\right)$$

b. Determine the mathematical expression for the resultant wave.

5-3 Find the resultant of the superposition of two harmonic waves in the form

$$E = E_0 \cos(\alpha - \omega t)$$

with amplitudes of 3 and 4 and phases of $\pi/6$ and $\pi/2$, respectively. Both waves have a period of 1 s.

5-4 Two waves traveling together along the same line are given by

$$y_1 = 5 \sin\left[\omega t + \frac{\pi}{2}\right]$$

$$y_2 = 7 \sin\left[\omega t + \frac{\pi}{3}\right]$$

Write the form of the resultant wave.

5-5 Plot and write the equation of the superposition of the following harmonic waves:

$$E_1 = \sin\left(\frac{\pi}{18} - \omega t\right), E_2 = 3 \cos\left(\frac{5\pi}{9} - \omega t\right), \text{ and}$$

$$E_3 = 2 \sin\left(\frac{\pi}{6} - \omega t\right), \text{ where the period of each is 2 s.}$$

5-6 One hundred antennas are putting out identical waves, given by

$$E = 0.02 \cos(\varepsilon - \omega t) \text{ V/m}$$

The waves are brought together at a point. What is the amplitude of the resultant when (a) all waves are in phase (coherent sources) and (b) the waves have random phase differences?

5-7 Two plane waves of the same frequency and with vibrations in the z-direction are given by

$$\psi(x, t) = (4 \text{ cm}) \cos\left(\frac{\pi}{3 \text{ cm}}x - \frac{20}{\text{s}}t + \pi\right)$$

$$\psi(y, t) = (2 \text{ cm}) \cos\left(\frac{\pi}{4 \text{ cm}}y - \frac{20}{\text{s}}t + \pi\right)$$

Write the resultant wave form expressing their superposition at the point $x = 5$ cm and $y = 2$ cm.

5-8 Beginning with the relation between group velocity and phase velocity in the form

$$v_g = v_p - \lambda(dv_p/d\lambda)$$

(a) express the relation in terms of n and ω and (b) determine whether the group velocity is greater or less than the phase velocity in a medium having a normal dispersion.

5-9 The *dispersive power* of glass is defined as the ratio $(n_F - n_C)/(n_D - 1)$, where C, D, and F refer to the Fraunhofer wavelengths, $\lambda_c = 6563$ Å, $\lambda_D = 5890$ Å, and $\lambda_F = 4861$ Å. Find the approximate group velocity in glass whose dispersive power is $1/30$ and for which $n_D = 1.50$.

5-10 The dispersion curve of glass can be represented approximately by Cauchy's empirical equation, $n = A + B/\lambda^2$. Find the phase and group velocities for light of 500-nm wavelength in a particular glass for which $A = 1.40$ and $B = 2.5 \times 10^6$ Å^2.

5-11 The dielectric constant K of a gas is related to its index of refraction by the relation $K = n^2$.

a. Show that the group velocity for waves traveling in the gas may be expressed in terms of the dielectric constant by

$$v_g = \frac{c}{\sqrt{K}}\left[1 - \frac{\omega}{2K}\frac{dK}{d\omega}\right]$$

where c is the speed of light in vacuum.
b. An empirical relation giving the variation of K with frequency is

$$K = 1 + [A/(\omega_0^2 - \omega^2)]$$

where A and ω_0 are constants for the gas. If the second term is very small compared to the first, show that

$$v_g \cong c\left[\frac{\omega^2 A}{(\omega_0^2 - \omega^2)^2}\right]$$

5-12 a. Show that group velocity can be expressed as

$$v_g = v_p - \lambda\left(\frac{dv_p}{d\lambda}\right)$$

b. Find the group velocity for plane waves in a dispersive medium, for which $v_p = A + B\lambda$, where A and B are constants. Interpret the result.

5-13 Waves on the ocean have different velocities, depending on their depth. Long-wavelength waves, traveling deep in the ocean, have a speed given approximately by

$$v_p = \left(\frac{g\lambda}{2\pi}\right)^{1/2}$$

where g is the acceleration of gravity. Short-wavelength waves, corresponding to surface ripples, have a velocity given approximately by

$$v_p = \left(\frac{2\pi T}{\lambda \rho}\right)^{1/2}$$

where ρ is the density and T is the surface tension. Show that the group velocity for long-wavelength waves is 1/2 their phase velocity and the group velocity for short-wavelength waves is 3/2 their phase velocity.

5-14 A laser emits a monochromatic beam of wavelength λ, which is reflected normally from a plane mirror, receding at a speed v. What is the beat frequency between the incident and reflected light?

5-15 Standing waves are produced by the superposition of the wave

$$y = (7 \text{ cm}) \sin\left[2\pi\left(\frac{t}{T} - \frac{2x}{\pi \text{ cm}}\right)\right]$$

and its reflection in a medium whose absorption is negligible. For the resultant wave, find the amplitude, wavelength, length of one loop, velocity, and period.

5-16 A medium is disturbed by an oscillation described by

$$y = (3 \text{ cm}) \sin\left(\frac{\pi x}{10 \text{ cm}}\right) \cos\left(\frac{50\pi}{\text{s}} t\right)$$

a. Determine the amplitude, frequency, wavelength, speed, and direction of the component waves whose superposition produces this result.
b. What is the internodal distance?
c. What are the displacement, velocity, and acceleration of a particle in the medium at $x = 5$ cm and $t = 0.22$ s?

5-17 Express the plane waves of Eqs. (5-19) and (5-20) in the complex representation. In this form, show that the superposition of the waves is the standing wave given by Eq. (5-22).

5-18 A certain argon-ion laser can support lasing over a frequency range of 6 GHz. Estimate the number of standing wave modes that might be in the laser output if the laser cavity is 1 m long.

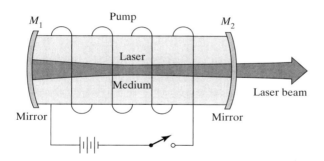

6

Properties of Lasers

INTRODUCTION

The laser is perhaps the most important optical device to be developed in the past 50 years. Since its arrival in the 1960s, rather quietly and unheralded outside the scientific community, it has provided the stimulus to make optics one of the most rapidly growing fields in science and technology today.

The word *laser* is an acronym that stands for **l**ight **a**mplification by the **s**timulated **e**mission of **r**adiation. The key words here are *amplification* and *stimulated emission*. A laser system converts *pump* energy (which may be electrical or optical in nature) into intense, highly directional, nearly mono-chromatic, electromagnetic wave energy. Albert Einstein laid the theoretical foundation of laser action as early as 1916, when he was the first to predict the existence of a radiative process called *stimulated emission*. His theoretical work, however, remained largely unexploited until 1954, when C. H. Townes and co-workers developed a **m**icrowave **a**mplifier based on **s**timulated **e**mission of **r**adiation. It was called a *maser*. Shortly thereafter, in 1958, A. Schawlow and C. H. Townes adapted the principle of masers to light in the visible region, and in 1960, T. H. Maiman built the first laser device. Maiman's laser used a ruby crystal as the laser amplifying medium and a two-mirror cavity as the optical resonator. Within months of the arrival of Maiman's ruby laser, which emitted deep red light at a wavelength of 694.3 nm, A. Javan and associates developed the first gas laser, the helium-neon laser, which emitted light in both the infrared (at 1.15 μm) and visible (at 632.8 nm) spectral regions.

Following the discovery of the ruby and helium-neon (He-Ne) lasers, many other laser devices, using different amplifying media and producing output at different wavelengths, were developed in rapid succession. Still, for the greater part of the 1960s, the laser was viewed by the world of industry and technology as a scientific curiosity. It was referred to in jest as "a solution in search of a problem." Since that time, however, the number of industrial and research applications of the laser has increased rapidly. Currently, new laser applications are discovered almost weekly. Together with the fiber-optic and semiconductor optoelectronic devices, the laser has revolutionized optics and the optics industry.

In this chapter, we first describe the basic processes involved in the interaction of light with matter and discuss the nature of light emitted by non-laser light sources. We then examine the essential elements of a laser system, describe the basic operation of the laser, and list the unique characteristics of laser light. Finally, by way of a summary, a table is provided that lists many of the more common lasers, along with their important operating parameters and characteristics. In Chapters 26, we present a more in-depth treatment of laser operation, systems, and applications. In Chapter 27, we examine, in detail, the propagation characteristics of the electromagnetic waves emitted by laser sources.

6-1 ENERGY QUANTIZATION IN LIGHT AND MATTER

Electromagnetic fields result when charged particles are accelerated. Charged particles, in turn, are accelerated by electromagnetic fields. Laser light is a manifestation of a particular interaction between charged particles and electromagnetic fields. An understanding of this interaction necessitates a discussion of the *quantization* of the energies of electromagnetic fields and atoms.

Energy Quantization of Electromagnetic Fields

As we discussed in Chapter 1, the energy of electromagnetic radiation of frequency ν is quantized in units of $h\nu$, where $h = 6.63 \times 10^{-34}$ J·s is Planck's constant. These units of electromagnetic energy—as we have mentioned—are called *photons*. Generally, the total energy E_n^{EM} stored in an electromagnetic field of frequency ν is given by,

$$E_n^{EM} = \frac{1}{2}h\nu + nh\nu \qquad n = 0, 1, 2, \ldots \qquad (6\text{-}1)$$

where n is the number of photons in the field. A photon should be thought of as a quantum of energy associated with the entire electromagnetic field. Note that the lowest possible energy, which occurs when there are no photons in the field (i.e., when $n = 0$), is not zero but rather is $h\nu/2$. Such a field is the *ground state* of the electromagnetic field and corresponds to total darkness. This state of total darkness is referred to as the *electromagnetic vacuum*. The fact that the electromagnetic vacuum has energy (even if this energy cannot be transferred to another system) is relevant to the discussion of spontaneous emission given later. Since the electromagnetic field can only take on energies given by Eq. (6-1), an electromagnetic wave gives up and receives energy in multiples of the photon energy $h\nu$. As Example 6-1 illustrates, the energy of a photon is much less than the total energy stored in a typical macroscopic electromagnetic field and so the graininess of the electromagnetic field often goes unnoticed.

Example 6-1

Find the approximate number of photons emitted in 1 s from a source that emits 1 W of light power at a wavelength of 500 nm.

Solution

The output power P is the rate of energy emission. The total energy, E_{TOT}, emitted in a time interval t of 1 s can then be written as

$$E_{TOT} = Pt = (1\ \text{W})(1\ \text{s}) = 1\ \text{J}$$

This emitted energy is a multiple n of the photon energy $h\nu = hc/\lambda$. That is,

$$E_{TOT} = nh\nu = nhc/\lambda$$

The number of photons n emitted in 1 s is therefore

$$n = \frac{E_{TOT}}{hc}\lambda = \frac{1\ \text{J}}{(6.63 \times 10^{-34}\ \text{J}\cdot\text{s})(3 \times 10^8\ \text{m/s})}(5 \times 10^{-7}\ \text{m}) = 2.5 \times 10^{18}$$

For this light source, one photon more or less is unlikely to be noticed.

Energy Quantization in Matter

Atoms are composed of charged particles and so interact with electromagnetic fields. The atoms and molecules that constitute matter have quantized energy levels. The law of energy conservation suggests that there can be a strong exchange of energy between an electromagnetic wave of frequency ν_0 and matter only when some of the constituents of the matter have allowed energies E_n and E_m such that $E_n - E_m = E_{n+1}^{EM} - E_n^{EM} = h\nu_0$.[1] In such a case, we say that the electromagnetic field is *resonant* with the E_n to E_m *transition* of the atom or molecule. The different energy levels of atoms are associated with different configurations of the electrons that surround the nucleus of the atom. Molecules have energy levels associated with the energy of electronic configuration, rotational kinetic energy of the molecule, and energy stored in vibration of the molecule. In addition, solids and liquids can have energy levels associated not with individual atoms and molecules, but with the entire solid. Solids and liquids sometimes have continuous bands of energy.

The allowed energies of electrons in the wide variety of types of atoms and molecules cannot be summarized in a simple formula. To illustrate the basic nature of the quantization of allowed energies of the electrons in atoms,[2] consider the allowed energies of the electron in the simplest atom, hydrogen. A bound electron in hydrogen can only have energies given by the relation

$$E_n = -\frac{13.6\ \text{eV}}{n^2} \qquad n = 1, 2, 3 \ldots \qquad (6\text{-}2)$$

These allowed energies are often displayed in an energy level diagram like the one shown in Figure 6-1. The ground state, with the lowest possible energy $E_1 = -13.6\ \text{eV}$, corresponds to the electron in its most stable state in which it is closest to the nucleus. *Excited states* of higher energy correspond to the electron in "orbits" further from the nucleus. An electron with energy greater than zero is no longer bound to the proton in the nucleus of the hydrogen atom. That is, an energy of 13.6 eV must be given to an electron in the ground state of hydrogen in order to *ionize* the hydrogen atom. It is important to note that the energy of a free electron is *not quantized*. As a result, *free electrons* can interact with photons of any frequency. In Example 6-2 we explore the quantization conditions for the electromagnetic field and the hydrogen atom.

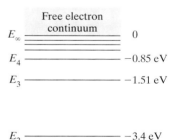

Figure 6-1 Allowed energies of the electron in the hydrogen atom. Notice that a free electron that has been stripped from the hydrogen atom can have any energy.

---

[1]Actually, the situation is considerably more complicated than this. For example, nonlinear optical processes involve the exchange of two or more photons with matter at a given time.

[2]It is awkward to refer to the energy levels of atoms or molecules or liquids or solids, so, at this point, we begin to refer to the energy levels of atoms with the understanding that the discussion at hand may also be applicable for molecules, liquids, and solids.

Example 6-2

a. Calculate the energy difference, in eV and in J, between the ground state energy E_1 and the first excited state energy E_2 of the hydrogen atom.

b. What is the frequency and wavelength of the photon with the same energy as the energy difference $E_2 - E_1$ of part (a)?

Solution

a. Using Eq. (6-2),

$$E_2 - E_1 = -\frac{13.6 \text{ eV}}{2^2} - \left(-\frac{13.6 \text{ eV}}{1^2}\right)$$

$$= -3.4 \text{ eV} + 13.6 \text{ eV} = 10.2 \text{ eV}$$

$$E_2 - E_1 = (10.2 \text{ eV})\left(\frac{1.6 \times 10^{-19} \text{ J}}{\text{eV}}\right) = 1.63 \times 10^{-18} \text{ J}$$

b. The frequency of a photon with this energy is

$$E_2 - E_1 = h\nu_0 \Rightarrow \nu_0 = \frac{E_2 - E_1}{h} = \frac{1.63 \times 10^{-18} \text{ J}}{6.63 \times 10^{-34} \text{ J} \cdot \text{s}} = 2.46 \times 10^{15} \text{ Hz}$$

The wavelength of this photon is

$$\lambda_0 = \frac{c}{\nu_0} = \frac{3 \times 10^8 \text{ m/s}}{2.46 \times 10^{15} \text{ Hz}} = 1.22 \times 10^{-7} \text{ m} = 122 \text{ nm}$$

Lineshape Function

We have, so far, treated the energy levels of atoms as if they were truly discrete. In fact, these energy levels typically have a narrow but finite width ΔE that arises from the inevitable interaction of the atom with its environment. Widths of energy levels in atoms vary greatly but a typical value is on the order of 10^{-7} eV. This width in the allowed energy levels in turn relaxes the restriction that a photon must have a precise energy that just matches the difference in discrete allowed energy levels. Rather, the photon energy must be within a very small range of energies, near the nominal energy difference of two levels, in order to interact with the atom. Roughly, if states with nominal energies E_n and E_m have energy widths ΔE_n and ΔE_m, respectively, a photon should have energy in the range

$$h\nu = E_n - E_m - \frac{(\Delta E_n + \Delta E_m)}{2} \text{ to } h\nu = E_n - E_m + \frac{(\Delta E_n + \Delta E_m)}{2}$$

in order to significantly interact with the atom. The center frequency of the transition is $\nu_0 = (E_n - E_m)/h$. Thus, we can say that only fields of frequencies given by

$$\nu = \frac{(E_n - E_m)}{h} \pm \frac{(\Delta E_n + \Delta E_m)}{2h} \equiv \nu_0 \pm \frac{\Delta\nu}{2}$$

are likely to have a significant interaction with the pair of atomic levels with energies E_n and E_m. Notice that we have defined the frequency *linewidth* $\Delta\nu$ associated with transitions between a pair of atomic levels as

$$\Delta\nu = \frac{\Delta E_n + \Delta E_m}{h} \tag{6-3}$$

It is common to refer to the frequency ν_0 as the frequency at *line center*. The linewidths of different atomic transitions vary over a wide range of values but a linewidth in the range $10^6 - 10^9$ Hz for a transition associated with a photon of nominal frequency $10^{14} - 10^{15}$ Hz is typical. The likelihood of the atom with energy levels E_n and E_m interacting with a photon of frequency ν is proportional to a *lineshape function* $g(\nu)$, which is, typically, a nearly symmetric function of width $\Delta\nu$ that peaks at $\nu = \nu_0$. A typical lineshape function is shown in Figure 6-2. By convention, the lineshape function $g(\nu)$ is normalized so that

$$\int_{\text{all }\nu} g(\nu)\, d\nu = 1$$

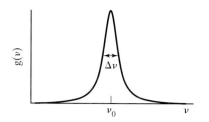

Figure 6-2 Lineshape function $g(\nu)$ for an atomic transition between energy levels of nominal energy difference $h\nu_0$. The linewidth $\Delta\nu$ of the transition is the full width at half maximum of the lineshape function.

The various physical processes that affect the form of the lineshape function are discussed in some detail in Chapter 26. As we build towards a discussion of laser light, we first review the nature of nonlaser light sources. Such a review naturally begins with a discussion of the interaction of light and matter in thermal equilibrium.

6-2 THERMAL EQUILIBRIUM AND BLACKBODY RADIATION

When a system is in *thermal equilibrium* with its surroundings, there is no *net* energy flow between the system and its surroundings. In such a case, the system and its surroundings can be characterized by a common temperature, T. When two systems at different temperatures are brought together, there is a net energy flow from the system at higher temperature to the system at lower temperature. When the two systems come to a common temperature, the rates of energy flow between the systems become equal. Since electrons and protons are charged particles and constituents of atoms, all atoms can emit and absorb light energy. Absorption and emission of electromagnetic waves is therefore an important mode of energy exchange between systems. In this section we will borrow relations from thermodynamics that describe atoms and electromagnetic fields in thermal equilibrium at a given temperature T. This discussion of thermal equilibrium will form a useful background for a discussion of the laser, in which neither the gain medium nor the electromagnetic field is in thermal equilibrium with its surroundings.

Atoms in Thermal Equilibrium: The Boltzmann Distribution

Consider an assembly of atoms in thermal equilibrium at temperature T. The likelihood P_i that a given atom in this assembly will be in one of the states of energy E_i is given by the so-called *Boltzmann distribution*

$$P_i = P_1 e^{-(E_i - E_1)/k_B T} \tag{6-4}$$

Here, P_1 is the likelihood that the atom will be in the ground state of the system which has energy E_1, $k_B = 1.38 \times 10^{-23}$ J/K $= 8.62 \times 10^{-5}$ eV/K is Boltzmann's constant, and the temperature is measured in Kelvins. Recall that the lowest possible temperature (so-called *absolute zero*) has the value 0 on the Kelvin temperature scale and that the temperature in Kelvins is related to the temperature in Celsius degrees via the relation $T_{\text{Kelvins}} = T_{\text{Celsius}} + 273$. For example, the freezing temperature of water is $T = 0°C = 273$ K. The

Boltzmann distribution[3] indicates that *in thermal equilibrium, atoms are more likely to be in states with lower energies.* In Example 6-3 we show that in thermal equilibrium at room temperature essentially all hydrogen atoms will be in their electronic ground states.

Example 6-3

   a. Find the ratio of the likelihood P_2 that a hydrogen atom will be in one of its excited states with energy E_2 (see Figure 6-1) to the likelihood P_1 that a hydrogen atom will be in its ground state of energy E_1 if the atoms are in thermal equilibrium at room temperature (293 K).
   b. Find the temperature at which the ratio of the likelihoods P_2/P_1 is 1/1000.

Solution

   a. Using Eq. (6-4) and the values of the ground state and first excited state energies of hydrogen found from Eq. (6-2) gives

$$\frac{P_2}{P_1} = e^{-(E_2 - E_1)/k_B T} = e^{-(-3.4 + 13.6)/(8.62 \times 10^{-5} \cdot 293)} = 4.1 \times 10^{-176}!$$

That is, it is very likely that all of the hydrogen atoms will be in the ground electronic state at room temperature. In problem 6-3 you will be asked to show that a significant number of hydrogen *molecules* will be in the first excited *rotational* energy state at room temperature. In a hydrogen atom there are eight distinct ways of combining the orbital and spin angular momentum of the electron to yield the same energy E_2. Therefore, the ratio of the likelihood that a hydrogen atom will have energy E_2 to the likelihood it will have energy E_1 is a factor of 8 larger than the fraction given here. (See footnote 3.)

   b. Using Eq. (6-4) again,

$$\frac{P_2}{P_1} = e^{-(10.2 \text{ eV})/[(8.62 \times 10^{-5} \text{ eV/K})T]} = 0.001$$

so that

$$-(10.2 \text{ eV})/[(8.62 \times 10^{-5} \text{ eV/K})T] = \ln(0.001)$$

and,

$$T = 17{,}100 \text{ K}$$

This temperature is, roughly, a factor of 3 more than the surface temperature of the sun.

EM Waves in Thermal Equilibrium

One of the early triumphs of quantum mechanics was the prediction of the experimentally verified relation giving the wavelength distribution (i.e., the *spectrum*) associated with electromagnetic radiation in thermal equilibrium with a *blackbody* at temperature T. A *blackbody* is an ideal absorber: All radiation falling on a blackbody, irrespective of wavelength or angle of incidence, is completely absorbed. It follows that a blackbody is also a perfect emitter:

---

[3]If several distinct physical states have the same energy, the energy level is said to be degenerate. If the number of distinct states that have energy E_i is g_i and the ground state of the system is not degenerate, then the likelihood that the system will be in any of the states with energy E_i is $P_i = g_i P_1 e^{-(E_i - E_1)/k_B T}$.

No body at the same temperature can emit more radiation at any wavelength or into any direction than a blackbody. Blackbodies are approached in practice by blackened surfaces and by tiny apertures in radiating cavities. An excellent example of a blackbody is the surface formed by the series of sharp edges of a stack of razor blades. The array of blade edges effectively traps the incident light, resulting in almost perfect absorption.

The correct form for the spectral exitance M_λ of a blackbody was first derived in 1900 by Max Planck, who found it necessary to postulate quantization in the process of radiation and absorption by the blackbody. Planck found the spectral exitance associated with a blackbody at temperature T to be

$$M_\lambda = \frac{2\pi hc^2}{\lambda^5}\left(\frac{1}{e^{hc/\lambda k_B T} - 1}\right) \qquad (6\text{-}5)$$

The spectral exitance M_λ is the power per unit area per unit wavelength interval emitted by a source. This quantity is plotted in Figure 6-3 for different temperatures. The spectral radiant exitance is seen to increase with absolute temperature at each wavelength. The peak exitance also shifts toward shorter wavelengths with increasing temperature, falling into the visible spectrum (between dashed vertical lines) at $T = 5000$ and 6000 K. The variation of $\lambda_{\max}$, the wavelength at which M_λ peaks, with the temperature can be found by differentiating M_λ with respect to λ and setting this equal to zero. The result is the *Wien displacement law*, given by

$$\lambda_{\max}T = \frac{hc}{5k_B} = 2.898 \times 10^3 \, (\mu\text{m} \cdot \text{K}) \qquad (6\text{-}6)$$

and is indicated in Figure 6-3 by the dashed curve. If the spectral exitance of Eq. (6-5) is integrated over all wavelengths, the total radiant exitance or area

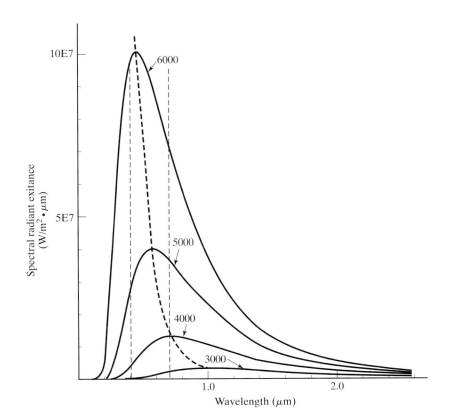

Figure 6-3 Blackbody radiation spectral distribution at four Kelvin temperatures. The vertical dashed lines mark the visible spectrum, and the dashed curve connecting the peaks of the four curves illustrates the Wien displacement law. (Note that 5E7 = 5×10^7.)

under the blackbody radiation curve at temperature T is found to be

$$M = \int_0^\infty M_\lambda \, d\lambda = \sigma T^4 \tag{6-7}$$

This relation is known as the *Stefan-Boltzmann law*, and σ, the Stefan-Boltzmann constant, is equal to 5.67×10^{-8} W/(m$^2 \cdot$ K^4).

The radiation from real surfaces is always less than that of the blackbody, or *Planckian source*, and is accounted for quantitatively by the *emissivity*, ε. Distinguishing now between the radiant exitance M of a measured specimen and that of a blackbody M_{bb} at the same temperature, we define

$$\varepsilon(T) = \frac{M}{M_{bb}} \tag{6-8}$$

If the radiant exitance of the blackbody and the specimen are compared in various narrow wavelength intervals, a spectral emissivity is calculated, which is not, in general, a constant. In those special cases where the emissivity is independent of wavelength, the specimen is said to be a *graybody*. In this instance, the spectral exitance of the specimen is proportional to that of the blackbody and their curves are the same except for a constant factor. The spectral radiation from a heated tungsten wire, for example, is close to that of a graybody with $\varepsilon = 0.4 - 0.5$.

Blackbody radiation is used to establish a color scale in terms of absolute temperature alone. The *color temperature* of a specimen of light is then the temperature of the blackbody with the closest spectral energy distribution. In this way, a candle flame ($\lambda_{max} \sim 1500$ nm) can be said to have a color temperature of 1900 K, whereas the sun ($\lambda_{max} = 500$ nm) has a typical color temperature of 5800 K. Sources at room temperature (293 K) emit electromagnetic radiation with a peak wavelength in the infrared region, $\lambda_{max} = 9890$ nm.

6-3 NONLASER SOURCES OF ELECTROMAGNETIC RADIATION

Before we examine the production and properties of laser light, we make a brief survey of some important nonlaser light sources. In order for an assembly of atoms or particles to be treated as a blackbody or a graybody, the assembly must be able to emit and absorb a continuous range of frequencies. This requires that there be available energy states which differ by a continuous range of energies. Such a situation can occur in solids and liquids in which the myriad modes of interaction between neighboring atoms lead to a nearly continuous range of possible energies of the atoms or molecules in the material. Hot gasses, in which some of the atoms are ionized, also constitute a system that can have a continuous emission spectrum. Since an ionized (free) electron can have any energy, the range of energies associated with a hot gas are continuous. Colder dilute gasses, on the other hand, have allowed energies that correspond only to the allowed transitions of the electronic, rotational, or vibrational energy states of the constituents of the gas. Such a gas only emits and absorbs light of particular, nearly discrete, energies. In many typical cases the spectrum of light emitted by a gas has a continuous background overlaid with a *line spectrum* corresponding to particular transitions between discrete energy levels. The way in which energy is distributed in the radiation determines the color of the light and, consequently, the color of surfaces seen under the light. Anyone who has used a camera is aware that the actual color response of film depends on the type of light used to illuminate the subject.

The following brief survey of nonlaser sources of light is not intended to be comprehensive; rather it is intended to serve as a backdrop for the discussion of laser light that follows in subsequent sections.

Sunlight and Skylight

Daylight is a combination of sunlight and skylight. Direct light from the sun has a spectral distribution that is clearly different from that of skylight, which has a predominantly blue hue. A plot of spectral solar irradiance is given in Figure 6-4. The nature of extraterrestrial solar radiation indicates that the sun behaves approximately as a blackbody with a temperature of 6000 K at its center and 5000 K at its edge. However, the radiation received at the earth's surface is modified by absorption in the earth's atmosphere. The annual average of total irradiance just outside the earth's atmosphere is the *solar constant*, 1350 W/m². The average total irradiance reaching the earth's surface on a clear day is about 1000 W/m². Note that the spectrums of the extraterrestrial sunlight and the sunlight reaching the earth both consist of a continuous blackbody background "interrupted" by dips. These dips are the result of selective absorption by specific elements in the sun's outer layers and the earth's atmosphere.

Cosmic Background Radiation

In 1965, Arno Penzias and Robert Wilson at the Bell Labs discovered that the earth is bathed in isotropic blackbody radiation with a spectral irradiance consistent with a color temperature of 2.7 K. This radiation is believed to have originated early in the development of the universe, when the universe was hot and dense. The subsequent expansion of the universe lowered the temperature of the radiation while maintaining the blackbody character of the spectral irradiance of this background radiation. This cosmic background radiation is important evidence supporting the *Big Bang* theory of the origin of the universe.

Incandescent Sources

Optical sources that use light produced by a material heated to incandescence by an electric current are called *incandescent lamps*. Radiation arises

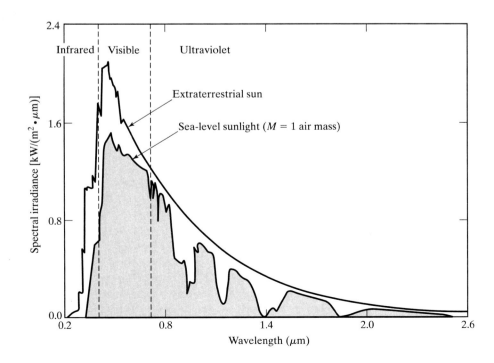

Figure 6-4 Solar spectral irradiance above the atmosphere and on a horizontal surface at sea level: clear day, sun at zenith.

from the de-excitation of the atoms or molecules of the material after they have been thermally excited. The energy is emitted over a broad continuum of wavelengths. Commercially available *blackbody sources* consist of cavities equipped with a small hole. Radiation from the small hole has an emissivity that is essentially constant and equal to unity. Such sources are available at operating temperatures from that of liquid nitrogen ($-196°$C) to 3000°C. Incandescent sources particularly useful in the infrared include the *Nernst glower*. This source is a cylindrical tube or rod of refractory material (zirconia, yttria, thoria) heated by an electric current and useful from the visible to around 30 μm. The Nernst glower behaves like a graybody with an emissivity greater than 0.75. When the material is a rod of bonded silicon carbide, the source is called a *globar*, approximating a graybody with an average emissivity of 0.88 (see Figure 6-5).

The *tungsten filament lamp* is a popular source for optical instrumentation designed to use continuous radiation in the visible and into the infrared region. The lamp is available in a wide variety of filament configurations and of bulb and base shapes. The filament is in coil or ribbon form, the ribbon providing a more uniform radiating surface. The bulb is usually a glass envelope, although quartz is used for higher-temperature operation. Radiation over the visible spectrum approximates that of a graybody, with emissivities approaching unity for tightly coiled filaments. Light output (typically given in lumens) depends both on the filament temperature and the electrical input power (wattage). During operation, tungsten gradually evaporates from the filament and deposits on the inner bulb surface, leaving a dark film that can decrease the light output by as much as 18% during the life of the lamp. This process also weakens the filament and increases its electrical resistance. The presence of an inert gas, usually nitrogen or argon, introduced at about 8/10 of the atmospheric pressure, helps to slow down the evaporation. More recently this problem has been minimized by the addition of a halogen vapor (iodine, bromine) to the gas in the *quartz-halogen* or *tungsten-halogen* lamp. The halogen vapor functions in a regenerative cycle to keep the bulb free of tungsten. Iodine reacts with the deposited tungsten to form the gas tungsten iodide, which then dissociates at the hot filament to redeposit the tungsten and free the iodine for repeated operation. A typical spectral irradiance curve for a 100-W quartz-halogen filament source is given in Figure 6-6. The lamp approximates a 3000-K graybody source, providing a useful continuum from 0.3 to 2.5 μm. In *tungsten arc lamps*, an arc discharge between two tungsten electrodes heats the electrodes to incandescence in an atmosphere of argon, providing a spectral distribution of radiation like that of tungsten lamps at 3100 K.

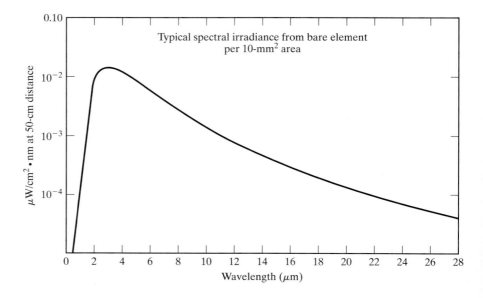

Figure 6-5 Globar infrared source, providing continuous usable emission from 1 to over 25 μm at a temperature variable up to 1000 K. The source is a 6.2-mm diameter silicon carbide resistor. (Oriel Corp., General Catalogue, Stratford, Conn.)

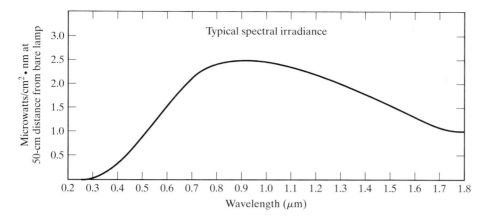

Figure 6-6 Spectral irradiance from a 100-W quartz halogen lamp, providing continuous radiation from 0.3 to 2.5 μm. (Oriel Corp., General Catalogue, Stratford, Conn.)

Discharge Lamps

The *discharge lamp* depends for its radiation output on the dynamics of an electrical discharge in a gas. A current is passed through the ionized gas between two electrodes sealed in a glass or quartz tube. (Glass envelopes absorb ultraviolet radiation below about 300 nm, whereas quartz transmits down to about 180 nm.) An electric field accelerates electrons sufficiently to ionize the vapor atoms. The source of the electrons may be a heated cathode (thermionic emission), a strong field applied at the cathode (field emission), or the impact of positive ions on the cathode (secondary emission). De-excitation of the excited vapor atoms provides a release of energy in the form of photons of radiation. High-pressure and high-current operation generally results in a continuous spectral output, in addition to *spectral lines* characteristic of the vapor. The width of the spectral lines is a direct indication of the linewidth $\Delta \nu$ of the particular transition leading to that spectral line. At lower pressure and current, sharper spectral lines appear, and the background continuum is minimal. When sharp spectral lines are desired, the lamp is designed to operate at low temperature, pressure, and current. The *sodium arc lamp*, for example, provides radiation almost completely confined to a narrow "yellow" band due to the spectral lines at 589.0 and 589.6 nm. The low-pressure *mercury discharge tube* is often used to provide, with the help of isolating filters, strong monochromatic radiation at wavelengths of 404.7 and 435.8 nm (violet), 546.1 nm (green), and 577.0 and 579.1 nm (yellow). Other gases or vapors may be used to provide spectral lines of other desired wavelengths. For the highest spectral purity, particular isotopes of the gas are used.

When high intensity rather than spectral purity is desired, other designs become available. Perhaps the oldest source of this kind is the *carbon arc*, still widely used in searchlights and motion picture projectors. The high-current arc is formed between two carbon rods in air. The source has a spectral distribution close to that of a graybody at 6000 K. A wide range of spectral outputs is possible by using different materials in the core of the carbon rod. When the arc is enclosed in an atmosphere of vapor at high pressure, the lamp is a *compact short-arc source* and the radiation is divided between line and continuous spectra. See Figure 6-7 for a sketch of this type of lamp and its housing. The most useful of these lamps, designed to operate from 50 W to 25 kW, are the high-pressure *mercury arc lamp*, with comparatively weak background radiation but strong spectral lines and a good source of ultraviolet; the *xenon arc lamp*, with practically continuous radiation from the near-ultraviolet through the visible and into the near-infrared; and the *mercury-xenon arc lamp*, providing essentially the mercury spectrum but with xenon's contribution to the continuum and its own strong spectral emission in the 0.8- to 1-μm range. Spectral emission curves for Xe and Hg-Xe lamps are shown in Figures 6-8 and 6-9.

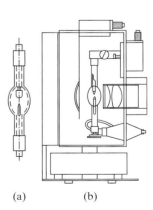

(a)   (b)

Figure 6-7 High-intensity, compact short-arc light source. (a) Compact arc lamp. (b) Lamp installed in housing, showing back reflector and focusing system. (The Ealing Corp.)

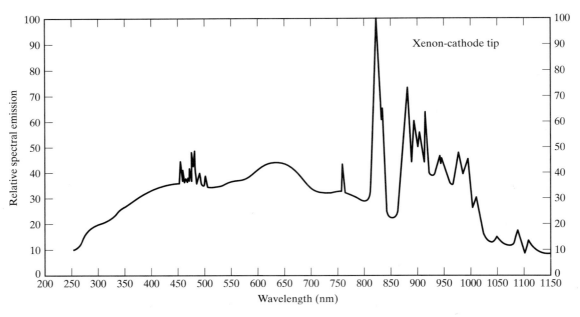

Figure 6-8   Spectral emission for xenon compact arc lamp. (Canrad-Hanovia, Inc.)

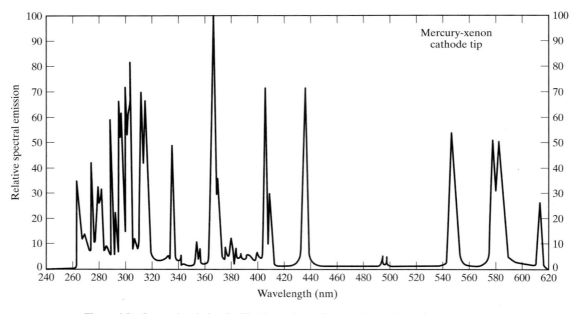

Figure 6-9   Spectral emission for Hg-Xe arc lamp. (Canrad-Hanovia, Inc.)

Flash tubes represent a high output source of visible and near-infrared radiation, produced by a rapid discharge of stored electrical energy through a gas-filled tube. The gas is most often xenon. The *photoflash* tube, in contrast, provides high-intensity, short-duration illumination by the rapid combustion of metallic (aluminum or zirconium) foil or wire in a pure oxygen atmosphere. Flash lamps and arc lamps are often used as optical pumps for laser systems (like Neodynium:YAG) using solid-state gain media.

The familiar *fluorescent lamps* use low-pressure, low-current electrical discharges in mercury vapor. The ultraviolet radiation from excited mercury atoms is converted to visible light by stimulating fluorescence in a phosphor coating on the inside of the glass-envelope surface. Spectral outputs depend on the particular phosphor used. "Daylight" lamps, for example, use a mixture of zinc beryllium silicate and magnesium tungstate.

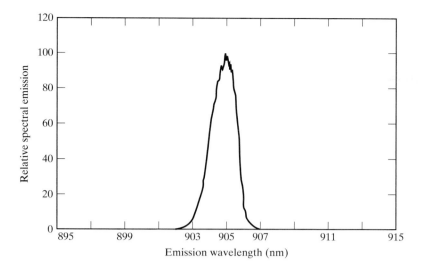

Figure 6-10 Spectral output from a GaAs light-emitting diode.

Light-Emitting Diodes

A light-emitting diode (LED) is an important light source quite different from those just described. A LED is a solid-state device employing a *p-n* junction in a semiconducting crystal. The device is hermetically sealed in an optically centered package. When a small bias voltage is applied in the forward direction, optical energy is produced by the recombination of electrons and holes in the vicinity of the junction. Popular LEDs include the infrared GaAs device, with a peak output wavelength near 900 nm, and the visible SiC device, with peak output at 580 nm. LEDs provide narrow spectral emission bands, as is evident in Figure 6-10. Solid solutions of similar compound semiconductor materials produce output in a variety of spectral regions when the composition of the alloy is varied.

6-4 EINSTEIN'S THEORY OF LIGHT-MATTER INTERACTION

In 1916, Einstein showed that the existence of thermal equilibrium between light and matter could be explained by positing but three basic interaction processes. These processes are known as *stimulated absorption*, *stimulated emission*, and *spontaneous emission*. In this section we discuss these processes and introduce the so-called *Einstein A and B coefficients* that govern their rates of occurrence. We now turn to brief descriptions of the three basic processes, illustrated in Figure 6-11, that describe the interaction of light with matter. It is useful to keep in mind the situation depicted in Figure 6-12a, in which a nearly monochromatic field of frequency ν' and irradiance I is incident on a sample of atoms that have two levels of energy E_1 and E_2 that are nearly *resonant* with the incident light so that $E_2 - E_1 \approx h\nu'$. The lineshape function $g(\nu)$ of the 2 to 1 transition in relation to the frequency ν' of the incident

Figure 6-11 Three basic processes that affect the passage of radiation through matter.

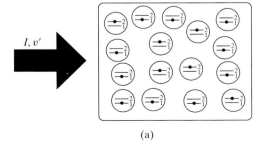

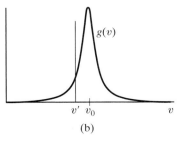

Figure 6-12 (a) Monochromatic light of irradiance I and frequency ν' incident on an assembly of atoms with energy levels 2 and 1 nearly resonant with the energy of a photon in the incident light. (b) Lineshape function for the 2 to 1 transition of the atoms in (a).

light is shown in Figure 6-12b. Further, let the *population densities* of the two atomic energy levels be N_2 and N_1, respectively. The population density of an energy level is the number per unit volume of atoms that are in that energy level.

Stimulated Absorption

Stimulated absorption, or simply absorption, is the process by which electromagnetic waves transfer energy to matter. An atom *can* be raised from an initial state with energy E_m to a final state with energy E_n by absorption of a photon of frequency ν' provided that the photon energy satisfies the approximate relation $h\nu' \approx E_n - E_m = h\nu_0$. The stimulated absorption process is illustrated in Figure 6-11a. In this figure, an atom originally in its lowest possible energy state E_1, called its *ground state*, is raised to an *excited state* of energy E_2 by absorption of a photon of energy, $E_2 - E_1 = h\nu'$. The rate of occurrence per unit volume of stimulated absorption, $R_{\text{St. Abs.}}$, when a monochromatic field of frequency ν' and irradiance I is incident on an assembly of atoms like the one depicted in Figure 6-11a is

$$R_{\text{St. Abs.}} = B_{12}g(\nu')(I/c)N_1 \qquad (6\text{-}9)$$

Here, B_{12} is the Einstein B coefficient for stimulated absorption. Note that the rate of stimulated absorption is proportional to B_{12}, the irradiance I of the incident field, the lineshape function evaluated at the frequency of the input field $g(\nu')$, and the population density N_1 of the lower of the two levels involved in the transition.

Stimulated Emission

As shown in Figure 6-11b, when a photon of energy $h\nu' \approx E_2 - E_1 = h\nu_0$ encounters an atom initially in an excited state E_2, it can "stimulate" the atom to drop to the lower state, E_1. In the process, the atom releases a photon of the *same energy, direction, phase, and polarization* as that of the incident photon. As a result, the energy of the electromagnetic wave passing by the atom is increased by one quantum $h\nu'$ of energy. It is stimulated emission that makes possible the amplification of light within a laser system. That stimulated emission produces a "twin" photon, which accounts for the unique degree of monochromaticity, directionality, and coherence associated with laser light. The rate of occurrence per unit volume of stimulated emission, $R_{\text{St. Em.}}$, when a monochromatic field of frequency ν' and irradiance I is incident on an assembly of atoms like the one depicted in Figure 6-12a is

$$R_{\text{St. Em.}} = B_{21}g(\nu')(I/c)N_2 \qquad (6\text{-}10)$$

The parameter B_{21} is the Einstein B coefficient for stimulated emission.

Spontaneous Emission

Spontaneous emission, illustrated in Figure 6-11c, can take place if an atom is in an excited state even when there are no photons incident on the atom. In the process shown, an atom in an excited state with energy E_2 "spontaneously" gives up its energy and falls to the state with energy E_1 and a photon of energy $h\nu \approx E_2 - E_1 = h\nu_0$ is released. The photon is emitted in a random direction. The likelihood that the spontaneously emitted photon will have a given frequency ν is proportional to the lineshape function $g(\nu)$. That is, the spontaneous emission from a sample of atoms all in the same excited state will occur with a range of frequencies characterized by the linewidth $\Delta\nu$. This behavior is to be contrasted with the stimulated emission process, which produces only photons that have the same frequency as that of the incident field. The spectrum of spontaneous emission has the same frequency dependence as the lineshape function $g(\nu)$. The line-spectra components of the spectra of Figures (6-8) and (6-9) are due primarily to spontaneous emission, and so the width of these line features is also the width of the lineshape function for the pair of levels associated with that particular spectral line. The line spectrum of a weakly excited dilute gas is sometimes called the *fluorescence spectrum* of the gas. Spontaneous emission occurs even in the presence of an incident electromagnetic wave that causes stimulated emission. In such a case, the field radiated by the atom is composed of some photons originating from the spontaneous emission process and some photons originating from the stimulated emission process. All of the stimulated emission photons have the same frequency and the same direction as the incident electromagnetic wave, whereas the spontaneous photons are emitted in—more or less—any direction with a range of frequencies described by the lineshape function.

Fundamentally, spontaneous emission is a result of an interaction with the electromagnetic vacuum described in Section 6-1. That is, even a field containing no photons still has an effect on an atom in an excited state. Spontaneous emission is sometimes aptly referred to as *vacuum-stimulated emission*. The electromagnetic vacuum has $\frac{1}{2}h\nu$ of energy at all frequencies, in all directions, and is randomly phased and so can induce the atom to emit photons in any direction and with any frequency that the atom can emit. Of course, there is no absorption stimulated by the electromagnetic vacuum because the vacuum, being the ground state of the electromagnetic field, can provide no energy quanta to the atom. The spontaneous emission rate per unit volume is

$$R_{\text{Sp. Em.}} = A_{21}N_2 \tag{6-11}$$

Here, A_{21} is the Einstein A coefficient for the 2 to 1 transition.

Relations Between the Einstein *A* and *B* Coefficients

Einstein was able to establish a relation between the Einstein A and B coefficients by showing that thermal equilibrium will exist between a radiation field and an assembly of atoms if the following relations hold:

$$\frac{A_{21}}{B_{21}} = 8\pi h\nu^3/c^3 \tag{6-12}$$

and

$$B_{12} = B_{21} \tag{6-13}$$

These relations are necessary for thermal equilibrium between an assembly of atoms and a radiation field to exist but hold generally as well, since the Einstein A and B coefficients are characteristics of the atomic levels. We review Einstein's derivation of these relations in Chapter 26 and note here that they also follow from a direct quantum mechanical treatment of the interaction of an electromagnetic field with an atom. For the purpose of the present discussion, Eq. (6-13) is most relevant. Note that the rate coefficients for stimulated emission and stimulated absorption are equal. This equality of the rate coefficients, and Eqs. (6-9) and (6-10), implies that the ratio of the overall rate of stimulated emission to that of stimulated absorption is the ratio of the population densities of the upper and lower energy levels. That is,

$$\frac{R_{\text{St. Em.}}}{R_{\text{St. Abs}}} = \frac{B_{21}g(\nu')(I/c)N_2}{B_{12}g(\nu')(I/c)N_1} = \frac{N_2}{N_1} \tag{6-14}$$

Now the population densities in an assembly of atoms in thermal equilibrium are proportional to the likelihoods that a given atom will be in a particular energy state. Therefore, the Boltzmann distribution of Eq. (6-4) can be used to write

$$P_2/P_1 = N_2/N_1 = e^{-(E_2-E_1)/k_BT} < 1$$

Therefore, stimulated absorption will occur more often than stimulated emission in an assembly of atoms in thermal equilibrium with its environment. As a result, an assembly of atoms in thermal equilibrium will always be a net absorber of incident radiation. In order for an assembly of atoms to *amplify* an incident electromagnetic field, pump energy must be supplied to the atoms in order to drive the atom out of thermal equilibrium and preferentially populate the upper energy level so that $N_2 > N_1$ and $R_{\text{St. Em.}} > R_{\text{St. Abs.}}$ Such an amplifying or *gain* medium plays a central role in laser action. When a level of higher energy has a greater population density than that of a level of lower energy, we say that a *population inversion* exists. Of course, spontaneous emission occurs in addition to the stimulated processes and so a significant amount of light may be emitted by an absorbing medium, but this light is emitted in any direction and with any frequency within the linewidth of the transition.

6-5 ESSENTIAL ELEMENTS OF A LASER

The laser device is an optical oscillator that emits an intense, highly collimated beam of coherent radiation. The device consists of three essential elements: an external energy source or *pump*, a *gain medium*, and an optical cavity, or *resonator*. These three elements are shown schematically in Figure 6-13: as a unit in Figure 6-13a and separately in Figure 6-13b, c, and d. Laser systems with moderate or high power outputs also typically require a cooling system.

The Pump

The pump is an external energy source that produces a population inversion in the laser gain medium. As explained in the previous section, amplification of a light wave or photon radiation field will occur only in a medium that exhibits a population inversion between two energy levels. In this case, the rate of stimulated emission will exceed that of stimulated absorption and the irradiance of the light will increase during each pass through the medium. Without the pump energy, the light wave would be attenuated during each pass through the medium.

Pumps can be optical, electrical, chemical, or thermal in nature, so long as they provide energy that can be coupled into the laser medium to excite the atoms and create the required population inversion. For gas lasers, such

Figure 6-13 Essential elements of a laser. (a) Integral laser device with output laser beam. (b) External energy source, or *pump*. The pump creates a population inversion in the laser medium. The pump can be an optical, electrical, chemical, or thermal energy source. The battery and helix pictured are only symbolic. (c) Empty optical cavity, or *resonator*, bounded by two mirrors. (d) Active cavity containing a *gain medium*. Population inversion and stimulated emission work together in the laser medium to produce amplification of light.

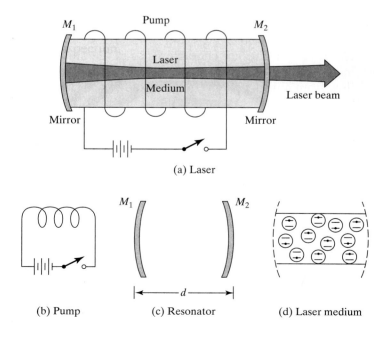

as the He-Ne laser, the most commonly used pump mechanism is an *electrical discharge*. The important parameters governing this type of pumping are the electron excitation cross sections and the lifetimes of the various energy levels. In some gas lasers, the free electrons generated in the discharge process collide with and excite the laser atoms, ions, or molecules directly. In others, excitation occurs by means of inelastic atom-atom (or molecule-molecule) collisions. In this latter approach, a mixture of two gases is used such that the two different species of atoms, say A and B, have excited states A^* and B^* that coincide. Energy may be transferred from one excited species to the other species in a process whose net effect can be symbolized by the relation $A^* + B \rightarrow A + B^*$. Atom A originally receives its excitation energy from a free electron or by some other excitation process. A notable example is the He-Ne laser, where the laser-active neon (Ne) atoms are excited by resonant transfer of energy from helium (He) atoms in a metastable state. The helium atoms receive their energy from free electrons via collisions.

Although there are numerous other pumps or excitation processes, we cite one more process which has some historical significance. The first laser, developed by T. Maiman at the Hughes Research Laboratories in 1960, was a pulsed ruby laser, which operated at the visible red wavelength of 694.3 nm. Figure 6-14 shows a drawing of the ruby laser device. To excite the Cr^{+3} impurity ions in the ruby rod, Maiman used a helical flashlamp filled with xenon gas. This particular method of exciting the laser medium is known as *optical pumping*. Solid and liquid gain media are typically optically pumped either by a flashlamp or another laser.

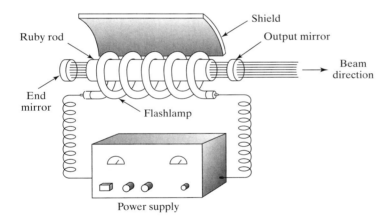

Figure 6-14 Components of a ruby laser system. The shield helps to reflect light from the flashlamp back into the ruby rod.

The Gain Medium

Laser systems are typically named by the makeup of the gain medium used in the device. The participating energy levels in the gain medium, which may be a gas, liquid, or solid, determine the wavelength of the laser radiation. Because of the large selection of laser media, the range of available laser wavelengths extends from the ultraviolet well into the infrared region, sometimes to wavelengths that are a sizable fraction of a millimeter. Laser action has been observed in over half of the known elements, with more than a thousand laser transitions in gases alone. Two of the most widely used transitions in gases are the 632.8-nm visible radiation from neon and the 10.6-μm infrared radiation from the CO_2 molecule. Other commonly used laser media and their operating wavelengths are listed in Table 6-1 at the end of this chapter.

In some lasers, the amplifying medium consists of two parts, the laser host medium and the laser atoms. For example, the host of the Nd:YAG laser is a crystal of yttrium aluminum garnet (commonly called YAG), whereas the laser atoms are the trivalent neodymium ions. In gas lasers consisting of mixtures of gases, the distinction between host and laser atoms is generally not made.

The most important requirement of the amplifying medium is its ability to support a population inversion between two energy levels of the laser atoms. This is accomplished by exciting (or pumping) more atoms into the higher energy level than exist in the lower level. As mentioned earlier, in the absence of pumping, there will be no population inversion between any two energy levels of a laser medium. Pumping, sometimes vigorous pumping, is required to produce the "unnatural" condition of a population inversion. As it turns out, though, due to the widely different lifetimes of available atomic energy levels, only certain pairs of energy levels with appropriate spontaneous lifetimes can be "inverted," even with vigorous pumping.

The Resonator

Given a suitable pump and a laser medium that can be inverted, the third basic element is a resonator, an optical "feedback device" that directs photons back and forth through the laser (amplifying) medium. The resonator, or optical cavity, in its most basic form consists of a pair of carefully aligned plane or curved mirrors centered along the optical axis of the laser system, as shown in Figure 6-13. One of the mirrors is chosen with a reflectivity as close to 100% as possible. The other is selected with a reflectivity somewhat less than 100% to allow part of the internally reflecting beam to escape and become the useful laser output beam.

As we discussed in Section 5-4, a laser cavity consisting of two flat mirrors separated by an optical distance d will only support standing wave modes of wavelengths λ_m and frequencies ν_m that satisfy the condition $d = m\lambda/2 = mc/2\nu$, where m is a (typically large) positive integer. Therefore, the frequencies of the modes of such a cavity are

$$\nu_m = m\frac{c}{2d} \qquad (6\text{-}15)$$

As we have noted, one of the mirrors in a laser cavity must be partially transmitting in order to allow for laser output. As a result, the cavity will support fields with a narrow range of frequencies near the standing wave frequencies given in Eq. (6-15). The laser resonator, then, in addition to acting as a feedback device, also acts as a frequency filter. Only electromagnetic fields that have frequencies near the resonant frequency of the lasing transition (and so can experience significant gain) and very near a standing wave frequency of the cavity (and so experience low loss) will be present in the laser output.

Typically, laser mirrors have spherical surfaces, and so the stable repetitive field patterns (i.e., the modes of the cavity) are more complicated than the plane standing wave modes produced by flat mirror cavities discussed earlier. In general, the geometry of the mirrors and their separation determine the mode structure of the electromagnetic field within the laser cavity. The exact distribution of the electric field pattern across the wavefront of the emerging laser beam, and thus the transverse irradiance of the beam, depends on the construction of the resonator cavity and mirror surfaces. Many different transverse irradiance patterns, called *TEM modes*, can be present in the output laser beam. By suppressing the gain of the higher-order modes—those with intense electric fields near the edges of the beam—the laser can be made to operate in a single fundamental mode, the TEM_{00} mode. The transverse variation of the irradiance of this TEM_{00} mode is *Gaussian* in shape, with a peak irradiance at the center and an exponentially decreasing irradiance toward the edges. This Gaussian mode was described briefly in Section 4-7. Properties of the fundamental TEM_{00} mode and the "higher-order" TEM modes are discussed in detail in Chapter 27.

The Cooling System

Overall *efficiency* is an important operating characteristic of a laser system. The overall efficiency of a laser system is the ratio of the total power required to pump the laser—sometimes called the *wall-plug power*—to the optical output power of the laser. Typical efficiencies (see Table 6-1 at the end of this chapter) range from fractions of a percent to 25% or so. Many important high-power laser systems have an efficiency of less than a percent. Pump energy that does not result in laser output inevitably degrades into thermal energy. As an example, consider an argon ion laser with a 10-W output. If the overall efficiency of this laser system is 0.05%, the total power used is $P_{used} = (10\text{ W})/(5 \times 10^{-4}) = 2 \times 10^4$ W. Of this total power, 99.95% is wasted as heat energy. If this heat energy is not removed from the system, the components of the system will be damaged or degraded.

Laser systems with solid gain media are typically cooled by surrounding the gain medium (and sometimes the optical pump) in a cooling jacket. Water, or a cooling oil, flows through the jacket, removing heat from the laser system. Laser systems with gas or liquid gain media can be cooled by this same mechanism, or by flowing the lasing medium itself through the cavity and cooling it before returning it to the cavity where it is again pumped. This method of cooling is sometimes used in carbon dioxide and dye lasers. Lasers with lower heat losses can sometimes be sufficiently cooled by forced air. Low-power lasers such as the He-Ne laser often need no external cooling system. In a high-power laser, the cooling system is an essential part of the system.

6-6 SIMPLIFIED DESCRIPTION OF LASER OPERATION

We have described briefly the basic elements that comprise the laser device. How do these elements—pump, medium, and resonator—work together to produce the laser output? Photons of a certain resonant energy must be created in the laser cavity, must interact with atoms, and must be amplified via stimulated emission, all while bouncing back and forth between the mirrors of the resonator. We can gain a reasonably accurate, though qualitative, understanding of laser operation by studying Figures 6-15 and 6-16. Figure 6-15a shows, in four steps, what happens to a typical atom in the laser medium during the creation of a laser photon. Figure 6-15b shows the actual energy level diagram for a helium-neon laser, with the four steps described in Figure 6-15a clearly identified. Figure 6-16 shows the same four-step process while focusing on the behavior of the atoms in the laser medium and the photon population in the laser cavity. Let us now examine these figures in turn.

In step 1 of Figure 6-15a, energy from an appropriate pump is coupled into the laser medium. The pump energy and rate is sufficiently high to excite a large number of atoms from the ground state E_0 to several excited states, collectively labeled E_3. Once at these levels, the atoms spontaneously decay, through various chains, back to the ground state E_0. Many, however, preferentially start the trip back by a very fast (usually radiationless) decay from pump levels E_3 to a special level, E_2. This decay process is shown in step 2. Level E_2 is labeled as the "upper laser level." It is special in the sense that it has a long lifetime. Whereas most excited levels in an atom might decay in times of the order of 10^{-8} s, level E_2 is *metastable*, with a typical lifetime of the order of 10^{-3} s, hundreds of thousands of times longer than other levels. Thus, as atoms funnel rapidly from pump levels E_3 to E_2, they begin to pile up at the metastable level, which functions as a bottleneck. In the process, N_2 grows to a large value. When level E_2 does decay, say by spontaneous emission, it does so to level E_1, labeled the "lower laser level." Level E_1 is an ordinary level that decays to the ground state quite rapidly, so the population N_1 cannot build to a large value. The net effect is the production of a population inversion ($N_2 > N_1$) required for light amplification via stimulated emission.

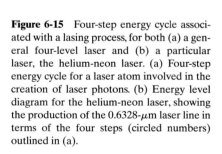

Figure 6-15 Four-step energy cycle associated with a lasing process, for both (a) a general four-level laser and (b) a particular laser, the helium-neon laser. (a) Four-step energy cycle for a laser atom involved in the creation of laser photons. (b) Energy level diagram for the helium-neon laser, showing the production of the 0.6328-μm laser line in terms of the four steps (circled numbers) outlined in (a).

Once the population inversion has been established and a photon of nearly resonant energy $h\nu' \cong E_2 - E_1$ passes by any one of the N_2 atoms in the upper laser level (step 3), stimulated emission can occur. When it does, laser amplification begins. Note carefully that a photon of resonant energy $E_2 - E_1$ can *also* stimulate absorption from level E_1 to level E_2, thereby losing itself in the process. Since N_2 is greater than N_1, however, and $B_{21} = B_{12}$, the rate for stimulated emission, $B_{21}g(\nu')(I/c)N_2$, exceeds that for stimulated absorption, $B_{12}g(\nu')(I/c)N_1$. Thus, light amplification occurs during each pass through the gain medium. In that event there is a steady increase in the incident resonant photon population and lasing continues. This is shown schematically in step 3, where the incident resonant photon approaching from the "left" leaves the vicinity of an N_2 atom in duplicate. In step 4, one of the inverted N_2 atoms, which dropped to level E_1 during the stimulated emission process, now decays rapidly to the ground state E_0. If the pump is still operating, this atom is ready to repeat the cycle, thereby ensuring a steady population inversion and a constant laser beam output. It is important to note that although the irradiance increases with each pass through the inverted gain medium, it decreases each time it encounters the output mirror of the resonator. So long as the gain per round-trip exceeds the loss per round-trip, the irradiance in the cavity continues to grow. As the irradiance grows the population inversion $N_2 - N_1$ necessarily decreases since an excess of stimulated emission creates photons at the expense of the population of the upper lasing level. *Therefore as the irradiance increases, the population inversion in the gain medium decreases.* This process is known as *gain saturation*. Eventually, the irradiance grows sufficiently to reduce the population inversion to the point that the gain per round-trip becomes equal to the loss per round-trip. When this occurs, the irradiance no longer grows and so the population inversion maintains a steady value. This is the *steady-state operating condition* for the laser.

In Figure 6-15b, the pump energy (step 1) is supplied by an electrical discharge in the low-pressure gas mixture, thereby elevating ground state helium atoms to higher energy states, one of which is represented by the 2^1S level. Then by resonant collisional energy transfer—made possible because the 2^1S level of helium is nearly equal to the $3s_2$ level of neon—step 2 is achieved as excited helium atoms transfer their energy over to ground state neon atoms, raising them to the neon $3s_2$ level. This process produces the population inversion required for effective amplification via stimulated emission of radiation.

The stimulated emission process (step 3) occurs between the neon levels $3s_2$ and $2p_4$, the transition with the highest probability[4] from $3s_2$ to any of the ten $2p$ states. This transition gives rise to photons of wavelength 0.6328 μm, photons that are amplified via stimulated emission and form the common red beam characteristic of helium-neon lasers. Finally, in step 4, the neon atom in energy state $2p_4$ decays by spontaneous emission to the $1s$ ground level. Once back in the ground state, it is again available to undergo collision with an excited helium atom and to repeat the cycle. Figure 6-15b relates the four steps to the emission of the He-Ne 0.6328 μm laser line, but other transitions from the $3s$ to the $2s$ and $2p$ levels have also been made to lase. One such transition, leading to the 1.1523 μm line, is indicated in the figure.

We now repeat the description of the buildup towards steady-state laser action with emphasis shifted to the evolution of the light field within the optical cavity. To aid a discussion of this evolution, consider Figure 6-16. In 6-16a, the laser medium is shown situated between the mirrors of the optical resonator. Mirror 1 is essentially 100% reflecting, and Mirror 2 is partially reflecting and partially transmitting. Most of the atoms in the laser medium are in the ground state. This is shown by the black dots. In Figure 6-16b, external energy (for example, light from a flashlamp or from an electrical discharge) is pumped into the medium, leading to a population inversion. Atoms occupying the upper laser level (state E_2 of Figure 6-15a) are shown by empty circles. The light amplification process is initiated in Figure 6-16c when excited atoms in the upper laser level E_2 spontaneously decay to level E_1.

---

[4]A readable, comprehensive discussion of the helium-neon laser, with energy level diagrams and transition probabilities, is given in G. H. B. Thompson, *Physics of Semiconductor Laser Devices* (New York: Wiley-Interscience, 1980).

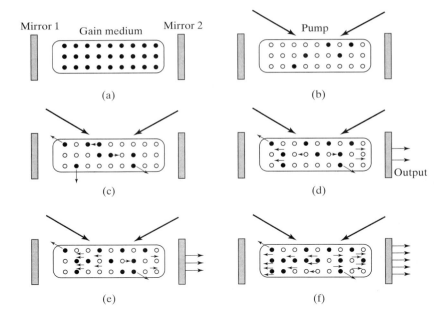

Figure 6-16 Time development of the start-up of laser oscillation in a typical laser cavity. (a) Quiescent laser. (b) Pumping of the gain medium creates a population inversion. (c) Spontaneous emission initiates stimulated emission. (d) Light amplification and gain saturation begin. (e) Gain saturation continues. (f) Established steady-state laser operation.

Since this is spontaneous emission, the photons given off in the process radiate out randomly in all directions. Many, therefore, leave through the sides of the laser cavity and are lost. Nevertheless, there will generally be several photons—let us call them "seed" photons—directed *along* the optical axis of the laser. These are the horizontal arrows shown in Figure 6-16c. With the seed photons of correct (resonant) energy accurately directed between the mirrors and many atoms still in the upper laser level E_2, the stage for stimulated emission is set. As the seed photons pass by the atoms in the upper laser level, stimulated emission adds identical photons in the same direction, providing an increasing population of coherent photons that bounce back and forth between the mirrors. In this process, of course, the number of atoms in the upper laser level is reduced. That is, gain saturation occurs. This buildup of the intracavity light and reduction of the population density N_2 of atoms in the upper lasing level is shown in Figures 6-16d and 6-16e. Since output Mirror 2 is partially transparent, a fraction of the photons incident on the mirror pass out through the mirror. These photons constitute the external laser beam. Those that do not leave through the output mirror are reflected, recycling back and forth through the cavity gain medium. In Figure 6-16f the steady-state operation of the laser is illustrated. In steady-state *continuous wave* (cw) operation, the population inversion is just sufficient to maintain a gain per cavity round-trip that offsets the loss per round-trip.

In summary, then, the laser process depends on the following:

1. A *population inversion* between two appropriate energy levels in the laser medium. This is achieved by the pumping process and the existence of a metastable upper laser state.
2. *Seed photons* of proper energy and direction, coming from the ever-present spontaneous emission process between the two laser energy levels. These initiate the *stimulated emission* process.
3. An *optical cavity* that confines and directs the growing number of resonant energy photons back and forth through the laser medium, continually exploiting the population inversion to create more and more stimulated emission, thereby creating more and more photons directed back and forth between the mirrors.
4. *Gain saturation* that follows from the fact that as the number of photons in the cavity grows, the rate of stimulated emission increases and so the

population inversion in the gain medium decreases. When the population inversion decreases to the level at which the gain per round-trip through the cavity is equal to the loss per round-trip through the cavity, the laser settles into steady-state continuous wave operation.

5. Coupling a certain fraction of the laser light out of the cavity through the *output coupler mirror* to form the external laser beam.

6-7 CHARACTERISTICS OF LASER LIGHT

Monochromaticity

Although no light can be truly monochromatic, laser light comes far closer than any other light source to reaching this ideal limit. The degree of monochromaticity of a light source can be specified by giving the linewidth of the radiation. The laser linewidth $\Delta \nu_L$ is the full width at half maximum (FWHM) of the spectral irradiance associated with the radiation. We have noted that the fluorescence lines from a weakly excited gas originate from spontaneous emission and so have linewidths that are the same as the width of the lineshape function $g(\nu)$ of the atomic transition involved in the fluorescence. The output from a laser is primarily *stimulated emission*, which produces photons of nominally identical frequencies. The fundamental limit to the narrowness of a laser line results from the fact that some of the randomly-phased *spontaneous emission* from the gain medium also exits the laser output mirror and, when mixed with the stimulated emission output, leads to a finite linewidth. This fundamental linewidth is sometimes called the Schawlow-Townes linewidth.[5] In practice, the linewidth of a laser is significantly larger than the limit set by the mixing of spontaneous emission into the laser output. The linewidth of a single mode in the output of a laser is typically governed by environmental noise such as mechanical vibrations, which change the cavity length, or index of refraction variations in the gain medium. Both of these mechanisms change the frequencies corresponding to the standing wave modes of the cavity. An example involving the He-Ne laser is instructive. The fluorescence linewidth of the neon 0.6328 μm lasing transition in the He-Ne laser is about 1.5 GHz. The operating linewidth of a typical single-mode helium-neon laser ranges from about 1 kHz to 1 MHz. The Schawlow-Townes linewidth of the He-Ne laser line is on the order of 10^{-3} Hz and so makes a negligible contribution to the operating linewidth. The operating linewidth of a single-mode He-Ne laser is 1000 to 1 million times narrower than the fluorescence linewidth associated with the neon transition.

Coherence

The optical property of light that most distinguishes the laser from other light sources is *coherence*. Coherence is a measure of the degree of phase correlation that exists in the radiation field of a light source at different locations and different times. This property of electromagnetic radiation is discussed in detail in Chapter 9. Here we give a brief qualitative description of this important feature of laser light. It is often described in terms of a *temporal coherence*, which is a measure of the degree of monochromaticity of the light, and a *spatial coherence*, which is a measure of the uniformity of phase across the optical wavefront. To obtain a qualitative understanding of temporal and spatial coherence, consider an ideal monochromatic point source of light. A portion of the wavefronts produced by such an ideal point source is indicated in Figure 6-17. The electromagnetic field produced by this ideal monochromatic point source has perfect temporal and spatial coherence. The temporal coherence is perfect

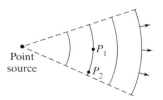

Figure 6-17 Portion of the wavefronts associated with a perfectly coherent light field produced by an ideal monochromatic point source.

[5]See, for example, M. O. Scully and M. S. Zubairy, *Quantum Optics* (Cambridge, UK: Cambridge University Press, 1997).

because, since the wave has a single frequency, knowledge of the phase at a given point (say, P_1) at time t_1 allows one to predict with complete confidence the phase of the field at point P_1 at some later time t_2. The spatial coherence of the wavefield is also perfect since along each wavefront the variation of the relative phase of the field is zero. Thus, knowledge of the phase at point P_1 at time t_1 allows one to predict with perfect confidence the phase of the field at this same time t_1 at a spatially distinct point P_2 along the same wavefront. If the frequency of the point source varied in a random fashion, the temporal coherence of the wavefield would be reduced. If there were several nearby point sources emitting light of the same frequency but with random relative phases, the spatial coherence of the resulting wavefield would be reduced. To produce a field that is both temporally and spatially coherent, neighboring point sources of light must produce light of the same frequency and correlated phase. This is precisely what occurs in a laser gain medium due to the stimulated emission process caused by the recycling of light within the laser cavity.

A high degree of light coherence is necessary in interferometry and holography, which are both discussed later in this text. Nonlaser light sources emit light primarily via the uncorrelated spontaneous emission action of many atoms. The result is the generation of incoherent light. To achieve some measure of coherence with a nonlaser source, two modifications to the emitted light can be made. First, a pinhole can be used with the light source to limit the spatial extent of the source. Second, a narrow-band filter can be used to decrease significantly the linewidth of the light. Each modification improves the coherence of the light given off by the source, but at the expense of a drastic loss of light energy.

In contrast, as mentioned, a laser source, by the very nature of its production of light via stimulated emission, ensures both a narrow-band output and a high degree of phase correlation. Recall that in the process of stimulated emission, each photon added to the stimulating radiation has a phase, polarization, energy, and direction *identical* to that of the amplified light wave in the laser cavity. The laser light thus created and emitted is both temporally and spatially coherent. Figure 6-18 summarizes the basic ideas of coherence for nonlaser and laser sources.

The mixing of spontaneous emission into the laser output and environmental noise fluctuations prevent laser light sources from emitting perfectly coherent light. Still, typical lasers have spatial and temporal coherences far superior to that for light from other sources. The transverse spatial coherence of a single-mode laser beam extends across the full width of the beam, whatever that might be. The temporal coherence, also called "longitudinal spatial coherence," is many orders of magnitude above that of any ordinary light source. The *coherence time* t_c of a laser is a measure of the average time interval over which one can continue to predict the correct phase of the laser beam at a given point in space. The *coherence length* L_c is related to the coherence time by the equation $L_c = ct_c$, where c is the speed of light. Thus the coherence length is the average length of light beam along which the phase of the wave remains unchanged. For a single-mode He-Ne laser, the coherence time is of the order of milliseconds (compared with about 10^{-11} s for light from a sodium discharge lamp), and the coherence length for the same laser is thousands of kilometers (compared with fractions of a centimeter for the sodium lamp).

Directionality

When one sees the thin, pencil-like beam of a laser for the first time, one is struck immediately by the high degree of beam directionality. No other light source, with or without the help of lenses or mirrors, generates a beam of such precise definition and minimum angular spread. The high degree of directionality of a single-mode laser beam is due to the geometrical design of the laser cavity and to the fact that the stimulated emission process produces twin

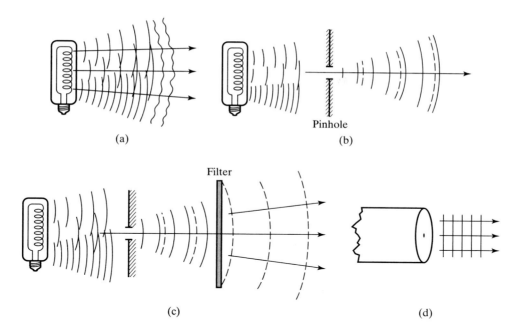

(a)   (b)

(c)   (d)

Figure 6-18 A tungsten lamp requires a pinhole and filter to produce partially coherent light. The light from a laser is naturally coherent. (a) Tungsten lamp. The
tungsten lamp is an extended source that emits many wavelengths. The emission
lacks both temporal and spatial coherence. The wavefronts are irregular and
change shape in a haphazard manner. (b) Tungsten lamp with pinhole. An ideal
pinhole limits the extent of the tungsten source and improves the spatial coherence
of the light. However, the light still lacks temporal coherence since all wavelengths
are present. Power in the beam has been decreased. (c) Tungsten lamp with pinhole
and filter. Adding a good narrow-band filter further reduces the power but improves the temporal coherence. Now the light is "coherent," but the available
power is far below that initially radiated by the lamp. (d) Laser. Light coming from
the laser has a high degree of spatial and temporal coherence. In addition, the output power can be very high.

photons. Figure 6-19 shows a specific cavity design and an external laser
beam with a (far-field) angular spread signified by the angle θ. The cavity mirrors shown are shaped with surfaces concave toward the cavity, thereby "focusing" the reflecting light back into the cavity and forming a *beam waist* of
radius w_0 at one position in the cavity. The nature of the beam inside the
laser cavity and its characteristics outside the cavity are discussed in some detail in Chapter 27. Here we simply note that if a laser output consists of the
fundamental TEM_{00} mode, the divergence angle will be

$$\theta = \frac{\lambda}{\pi w_0} \qquad (6\text{-}16)$$

where θ designates the *half-angle* beam spread.

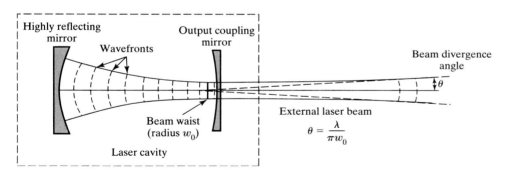

Figure 6-19 External and internal laser
beam for a given cavity. The divergence
angle θ associated with a field of beam-
waist radius w_0 is indicated.

It is important to note that the angular spread increases as the beam waist is made smaller. This general behavior is very similar to the behavior of a beam that passes through a circular aperture. We will see in Chapter 11 that the *far-field* divergence angle due to *diffraction* of a beam passing through a circular aperture of radius r is $\theta_{\text{diff}} = 0.61\lambda/r$. The beam-waist radius w_0 is determined by the design of the laser cavity and depends on the radii of curvature of the two mirrors and the distance between the mirrors. Therefore, one can build lasers with a given beam-waist radius and, consequently, a given beam divergence. We explore this relationship in Example 6-4.

Example 6-4

a. A He-Ne laser ($0.6328\ \mu$m) has an internal beam waist of radius near 0.25 mm. Determine the beam divergence angle θ.

b. Since we can control the beam-waist radius w_0 by laser cavity design and "select" the wavelength by choosing different laser media, what lower limit might we expect for the beam divergence? Suppose we design a laser with a beam waist of 0.25-cm radius and a wavelength of 200 nm. By what factor is the beam divergence decreased?

Solution

a.

$$\theta = \frac{\lambda}{\pi w_0} = \frac{0.6328\ \mu\text{m} \times 10^{-9}\ \text{m}}{\pi(2.5 \times 10^{-4}\ \text{m})} = 8 \times 10^{-4}\ \text{rad}$$

This is a typical laser beam divergence, indicating that the beam radius increases about 8 cm every 100 m.

b.

$$\theta = \frac{\lambda}{\pi w_0} = \frac{200 \times 10^{-9}\ \text{m}}{\pi(2.5 \times 10^{-3}\ \text{m})} = 2.55 \times 10^{-5}\ \text{rad}$$

This represents, roughly, a 30-fold decrease in beam spread over the He-Ne laser described in part (a). This beam radius would increase about 8 cm every 3130 m.

Note from Figure 6-19 that, near the beam waist, a Gaussian TEM_{00} mode laser field acts much like a plane wave of truncated transverse dimension. That is, the phase fronts are nearly planar and parallel in this region.

Laser Source Irradiance

The irradiance (power per unit area) of a typical laser is far greater than other sources of electromagnetic radiation largely due to the directionality and compactness of the laser beam. For example, lightbulbs spread their output uniformly in all directions so that the irradiance 1 m from a lightbulb with a light power output of 10 W would be

$$I = P/A = \frac{P}{4\pi r^2} = \frac{10\ \text{W}}{4\pi(1\ \text{m}^2)}$$

$$= 0.796\ \text{W/m}^2 \qquad \text{Irradiance 1 m from a 10-W lightbulb}$$

The output from a He-Ne laser, on the other hand, is concentrated in the thin beam of light emerging from the laser. One meter from the output of a He-Ne laser the beam radius might be about 2 mm. For such a situation, the irradiance

1 m from a He-Ne laser with a much smaller output power of 1 mW would be

$$I = P/A = \frac{0.001 \text{ W}}{\pi(0.002 \text{ m})^2}$$

$$= 79.6 \text{ W/m}^2 \qquad \text{Irradiance 1 m from a 1-mW He-Ne laser}$$

High-power lasers may have a continuous output of 10^5 W with a beam radius of 1 cm. The irradiance of such a laser would be 3.18×10^8 W/m^2. Recall that a laser system converts pump energy into laser output and so the average power output of a laser is always less than the average pump power. However, the laser power is concentrated in a monochromatic directional beam of small cross-sectional area and so laser irradiances can be very high.

Focusability

Nonlaser sources must have a significant transverse extent in order to produce a significant amount of light. Therefore, the images of these sources formed by lenses and mirrors have finite sizes governed by the laws of geometrical optics. The amount of light at the image position is determined by the amount of light from the source intercepted by the lens. As a result, a significant amount of light from these sources cannot be "focused" to a small spot. By contrast, the small transverse extent of laser beams allows a lens or mirror to intercept essentially all of the power in the beam. In addition, since the laser beam has a high degree of directionality, it behaves (near the beam waist at any rate) like a bundle of parallel rays coming from a point object at infinity. As a result, nearly all of the laser power is concentrated at the focal point of the lens or mirror. The diameter of the focused spot is limited by lens aberrations and diffraction but can be roughly as small as the wavelength of the laser light. We will discuss the limits to the "focusability" of laser beams in Chapter 27. The ways in which laser and nonlaser light is focused are illustrated in Figure 6-20.

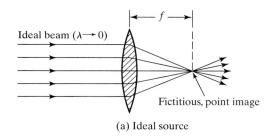

(a) Ideal source

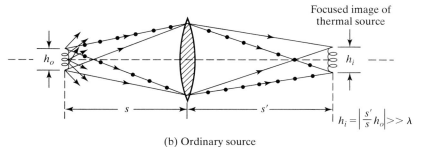

(b) Ordinary source

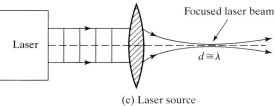

(c) Laser source

Figure 6-20 Focused beams from various sources. (a) Ideal, collimated beam is focused to a fictitious "point" in accordance with geometrical optics. (b) Incoherent radiation from a thermal source is "focused" to a demagnified image of size $h_i \gg \lambda$. (c) Coherent laser beam is focused to a diffraction-limited spot of diameter $d \cong \lambda$.

Laser energy focused onto small target areas makes it possible to drill tiny holes in hard, dense material, make tiny cuts or welds, make high-density recordings, and generally carry out industrial or medical procedures in target areas only a wavelength or two in size. In ophthalmology, for example, where Nd:YAG lasers are used in ocular surgery, target irradiances of 10^9 to 10^{12} W/cm^2 are required. Such irradiance levels are readily developed with the help of beam expanders and suitable focusing optics.

Pulsed Operation

We have thus far described only continuous wave (cw) lasers in which the laser system delivers a laser beam of constant irradiance. Many important applications of lasers require that the laser output be pulsed, in the sense that the laser output turns on and off in very short time periods. A pulsed laser delivers bursts of radiation with durations (*pulse widths*) as small as a few femtoseconds. In Chapter 26 we will discuss Q-switching and mode locking, which are the two primary methods used to pulse the output of a laser. Pulsed laser output is useful for controlling the delivery of laser energy in materials-processing applications, in time-of-flight distance measurements, in tracking rapid changes in the properties of systems, and in many other applications.

6-8 LASER TYPES AND PARAMETERS

To this point we have examined the basic processes involved in the interaction of light with matter, identified the essential parts that make up a laser, described in a general way how a laser operates, and studied the characteristics that make lasers such a unique source of light. Now, by way of summary, we turn our attention to the identification of some of the common lasers in existence today and to the parameters that distinguish them from one another.

A careful examination of Table 6-1 serves as an introduction to the state of laser technology. For each laser listed, the entries include data on pump mechanism, emission wavelength, output power (or in some cases, energy per pulse), nature of output, beam diameter, beam divergence, and operating efficiency. Both pulsed and continuously operating (cw) lasers are represented. Taken as a whole, Table 6-1 includes lasers whose wavelengths vary from 193 nm (deep ultraviolet) to 10.6 μm (far infrared); whose cw power outputs vary from 0.1 mW to 20 kW; whose beam divergences vary from 0.2 mrad (circular cross section) to 200 $\times$ 600 mrad (oval cross section); and whose overall efficiencies (laser energy out divided by pump energy in) vary from less than 0.1% to 20%. We shall revisit this table in later chapters when we discuss applications of some of the laser systems listed here.

We note in closing that the rather qualitative description of laser properties given in this chapter is supplemented by much more quantitative treatments of laser operation and laser beam properties given in Chapters 26 and 27. Important applications of lasers are discussed in Chapters 26 and 28 and in many other places throughout this text. In addition, the nature and operation of diode lasers, which differ in many respects from the lasers discussed in this chapter, are treated briefly in Chapter 26.

TABLE 6-1 LASER PARAMETERS FOR SEVERAL COMMON LASERS

Gain medium	Pump type	Wavelength	Power/Energy	Output type	Beam diameter	Beam divergence	Efficiency	Cooling
Gas, atomic								
Helium Neon	electric discharge	0.6328 μm, others	0.1–50 mW	cw	0.5–2.5 mm	0.5–3 mrad	<0.1%	air
Helium Cadmium	electric discharge	325 nm, 441.6 nm, others	5–150 mW	cw	0.2–2 mm	1–3 mrad	<0.1%	air
Gas, ion								
Argon	electric discharge	several from 350–530 nm, main lines: 488 nm, 514.5 nm	2 mW–20 W	cw (or mode-locked)	0.6–2 mm	0.4–1.5 mrad	<0.1%	water or forced air
Krypton	electric discharge	several from 350–800 nm, main line: 647.1 nm	5 mW–6 W	cw (or mode-locked)	0.6–2 mm	0.4–1.5 mrad	<0.05%	water or forced air
Gas, molecular								
Carbon Dioxide	electric discharge	10.6 μm	3 W–20 kW	cw or long pulse	3–50 mm	1–3 mrad	5–15%	flowing gas
Nitrogen	electric discharge	337.1 nm	1–300 mW (average)	pulsed	2 × 3–6 × 30 mm (rectangular)	1–3 × 7 mrad	<0.1%	flowing gas
Gas, excimer								
Argon Fluoride	short-pulse electric discharge	193 nm	up to 50 W (average)	pulsed	2 × 4–25 × 30 mm (rectangular)	2–6 mrad	<1%	air or water
Krypton Fluoride	short-pulse electric discharge	248 nm	up to 100 W (average)	pulsed	2 × 4–25 × 30 mm (rectangular)	2–6 mrad	<2%	air or water
Xenon Chloride	short-pulse electric discharge	308 nm	up to 150 W (average)	pulsed	2 × 4–25 × 30 mm (rectangular)	2–6 mrad	<2.5%	air or water
Xenon Fluoride	short-pulse electric discharge	351 nm	up to 30 W (average)	pulsed	2 × 4–25 × 30 mm (rectangular)	2–6 mrad	<2%	air or water

TABLE 6-1 Continued

Gain medium	Pump type	Wavelength	Power/Energy	Output type	Beam diameter	Beam divergence	Efficiency	Cooling
Liquid								
Various Dyes	other lasers, flashlamp	tunable 300–1000 nm	20 mW–1W (average)	cw or (ultrashort) pulsed	1–20 mm	0.3–2 mrad	1–20%	dye flow or water
Solid-State								
Nd:YAG	flashlamp, arc lamp, diode laser	1.064 μm	up to 10 kW (average)	cw or pulsed	0.7–10 mm	0.3–25 mrad	0.1–2% (5–8%, diode pumped)	air or water
Nd:glass	flashlamp	1.06 μm	0.1–100 J per pulse	pulsed	3–25 mm	3–10 mrad	1–5%	water
Alexandrite	flashlamp	tunable, 700–818 nm	<100 W average power	cw or pulsed	a few mm	a few mrad	0.5%	air or water
Ti-sapphire	flashlamp, diode laser, doubled Nd:YAG	tunable, 660–1000 nm	~2 W average power	cw or (ultrashort) pulsed	a few mm	a few mrad	comparable to Nd:YAG	air or water
Erbium:Fiber	flashlamp, diode laser	1.55 μm	1–100 W	cw or pulsed	a few mm	a few mrad	comparable to Nd:YAG	air
Semiconductor Lasers								
GaAs, GaAlAs	electric current, optical pumping	780–900 nm, composition dependent	1 mW to several watts, diode arrays up to 100 kW	cw or pulsed	N/A (diverges too rapidly)	200 × 600 mrad (oval in shape)	1–50%	air, heat sink
InGaAsP	electric current, optical pumping	1100–1600 nm, composition dependent	1 mW to ~1 W	cw or pulsed	N/A (diverges too rapidly)	200 × 600 mrad (oval in shape)	1–20%	air, heat sink

PROBLEMS

6-1 The Lyman series in the line spectra of atomic hydrogen is the name for the light emitted from transitions from excited states to the $n = 1$ hydrogen ground state. The Balmer series refers to the light emitted from transitions from excited states with $n \geq 3$ to the $n = 2$ energy state.

   a. Find the wavelengths of the three shortest-wavelength photons in the Lyman series. In what range of the electromagnetic spectrum are the spectral lines of the Lyman series?

   b. Find the wavelengths of the three shortest-wavelength photons in the Balmer series. In what range of the electromagnetic spectrum are the spectral lines of the Balmer series?

6-2 a. Will a photon of energy 5 eV likely be absorbed by a hydrogen atom originally in its ground state?

   b. What is the range of photon wavelengths that could ionize a hydrogen atom that is originally in its ground state?

   c. What is the range of photon wavelengths that could ionize a hydrogen atom that is originally in its $n = 2$ energy state?

6-3 The allowed rotational energies E_l^{rot} of a diatomic molecule are given by

$$E_l = \frac{l(l + 1)\hbar^2}{2I}$$

In this expression l is the rotational quantum number and can take the values $l = 0, 1, 2\ldots$; I is the rotational inertia of the molecule about an axis through its center of mass; and $\hbar = h/2\pi$. The equilibrium separation of the two atoms in a diatomic hydrogen molecule H_2 is about 0.074 nm. The mass of each hydrogen atom is about 1.67×10^{-27} kg.

   a. Show that the rotational inertia of the hydrogen molecule about an axis through its center of mass is about $I = 4.6 \times 10^{-48}$ kg·m².

   b. Find the difference in energy between the first excited rotational energy state and the ground rotational state. That is, find $E_1^{\text{rot}} - E_0^{\text{rot}}$. Express the answer in both J and eV.

   c. Find the relative likelihood $P_{l=1}/P_{l=0}$ that a hydrogen molecule will be in its first excited rotational state in thermal equilibrium at room temperature, $T = 293$ K. (Ignore possible state degeneracies.)

6-4 The allowed energies E_k^{vib} associated with the vibration of a diatomic molecule are given by

$$E_k^{\text{vib}} = (k + 1/2)hf$$

Here, k is the vibrational quantum number and can take the values $k = 0, 1, 2\ldots$ and f is the resonant frequency of the vibration. In a simple model of diatomic hydrogen H_2, the resonant vibration frequency can be taken as $f = 1.3 \times 10^{14}$ Hz.

   a. Find the difference in energy between the first excited vibrational energy state and the ground vibrational state of diatomic hydrogen. That is, find $E_1^{\text{vib}} - E_0^{\text{vib}}$. Express the answer in both J and eV.

   b. Find the relative likelihood $P_{k=1}/P_{k=0}$ that a hydrogen molecule will be in its first excited vibrational state in thermal equilibrium at room temperature, $T = 293$ K.

6-5 Referring to problems 6-3 and 6-4 and Eq. (6-2), construct an energy level diagram for the H_2 molecule that shows the first vibrational and rotational states associated with the ground electronic state of the molecule. (*Hint:* The molecule can be vibrating and rotating at the same time.)

6-6 Consider an assembly of atoms that have two energy levels separated by an energy corresponding to a wavelength of 0.6328 μm, as in the He-Ne laser. What is the ratio of the population densities of these two energy levels if the assembly of atoms is in thermal equilibrium as a temperature of $T = 300$ K?

6-7 The rate of decay of an assembly of atoms with population density N_2 at excited energy level E_2 when spontaneous emission is the only important process is

$$\left(\frac{dN_2}{dt}\right)_{\text{spont}} = -A_{21}N_2$$

Show that an initial population density N_{20} decreases to a value N_{20}/e in a time τ equal to $1/A_{21}$. That is, show that A_{21} is the inverse of the *lifetime* of the atomic level.

6-8 Derive the Wien displacement law from the Planck blackbody spectral radiance formula.

6-9 Derive the Stefan-Boltzmann law from the Planck blackbody spectral radiance formula. (*Hint:* Use a substitution of $x = hc/\lambda k_B T$ to facilitate the integration.)

6-10 a. At what wavelength does a blackbody at 6000 K radiate the most per unit wavelength?

   b. If the blackbody is a 1-mm diameter hole in a cavity radiator at this temperature, find the power radiated through the hole in the narrow wavelength region 5500–5510 Å.

6-11 At a given temperature, $\lambda_{\max} = 550$ nm for a blackbody cavity. The cavity temperature is then increased until its total radiant exitance is doubled. What is the new temperature and the new $\lambda_{\max}$?

6-12 What must be the temperature of a graybody with emissivity of 0.45 if it is to have the same total radiant exitance as a blackbody at 5000 K?

6-13 Plot the spectral exitance M_λ for a graybody of emissivity 0.4 in thermal equilibrium at 451°F, the temperature at which paper burns.

6-14 Why should one expect lasing at ultraviolet wavelengths to be more difficult to attain than lasing at infrared wavelengths? Develop your answer based on the ratio A_{21}/B_{21} and the meaning of the A_{21} and B_{21} coefficients.

6-15 The gain bandwidth of the lasing transition (that is, the width of the atomic lineshape $g(\nu)$ associated with the transition) in a Nd:YAG gain medium is about $\Delta\nu = 1.2 \times 10^{11}$ Hz. Express this bandwidth as a wavelength range $\Delta\lambda$. Use Table 6-1 to find the center wavelength of the Nd:YAG lasing transition.

6-16 The output of an argon ion laser can consist of a number of modes of frequency that match the cavity resonance condition and are within the gain bandwidth of the lasing transition. The gain bandwidth of the lasing transition is roughly equal to the width of the atomic lineshape function $g(\nu)$ associated with the lasing transition. Take the gain bandwidth of an argon ion laser to be 2 GHz and the linewidth of an individual mode from the argon ion laser to be 100 kHz. The coherence time of a light beam is roughly equal to the reciprocal of the spread of frequencies present in light. Find the coherence time and the coherence length of the argon ion laser if

a. The laser output consists of a single mode.
b. The laser output consists of a number of modes with frequencies spread across the gain bandwidth of the lasing transition.

6-17 Find the number of standing wave cavity modes within the gain bandwidth of the argon ion laser of problem 6-16 if the laser system uses a resonator with flat mirrors separated by a distance $d = 0.5$ m.

6-18 A He-Ne laser has a beam waist (diameter) equal to about 1 mm.

a. What is its beam-spread angle in the far field?
b. Estimate the diameter of this beam after it has propagated over a distance of 1 km.

6-19 To what diameter spot should a He-Ne laser of power 10 mW be focused if the irradiance in the spot is to be the same as the sun's irradiance at the surface of the earth? (The irradiance of the sun at the earth's surface is about 1000 W/m^2.)

6-20 For a Nd:YAG laser, there are four pump levels located at 1.53 eV, 1.653 eV, 2.119 eV, and 2.361 eV above the ground state energy level.

a. What is the wavelength associated with the photon energy required to populate each of the pump levels?

b. Knowing that a Nd:YAG laser emits photons of wavelength 1.064 μm, determine the *quantum efficiency* associated with each of the four pump levels. (The quantum efficiency is the ratio of the energy of a single pump event to that of an output photon.)

6-21 To operate a Nd:YAG laser, 2500 W of "wall-plug" power are required for a power supply that drives the arc lamps. The arc lamps provide pump energy to create the population inversion. The overall laser system, from power in (to the power supply) to power out (laser output beam), is characterized by the following component efficiencies:

80%—power supply operation
30%—arc lamps for pump light energy
70%—optical reflectors for concentrating pump light on laser rod
15%—for spectral match of pump light to Nd:YAG pump levels
50%—due to internal cavity/rod losses

a. Taking the efficiencies into account sequentially as they "occur," how much of the initial 2500 W is available for power in the output beam?
b. What is the overall operational efficiency (wall-plug efficiency) for this laser?

6-22 Table 6-1 indicates that diode lasers have a large divergence angle. Why is this reasonable?

6-23 As the irradiance within a laser cavity increases in the build up to steady state, does the population inversion in the gain medium increase or decrease? Explain.

6-24 Why is a Nd:YAG laser system that uses a diode laser as a pump more efficient that a Nd:YAG laser system that uses an arc lamp as a pump? See Table 6-1.

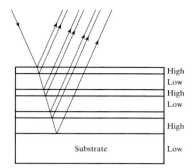

7 *Interference of Light*

INTRODUCTION

Like standing waves and beats, treated in Chapter 5, the phenomenon of interference depends on the superposition of two or more individual waves under
rather strict conditions that will soon be clarified. When interest lies primarily
in the effects of enhancement or diminution of light waves, due precisely to
their superposition, these effects are usually said to be due to the interference
of light. When conditions of enhancement, or *constructive interference*, and
diminution, or *destructive interference*, alternate in a spatial display, the interference is said to produce a pattern of *fringes*, as in the double-slit interference
pattern. The same conditions may lead to the enhancement of one visible wavelength interval or color at the expense of the others, in which case interference
colors are produced, as in oil slicks and soap films. The simplest explanation of
these phenomena can be undertaken successfully by treating light as a wave
motion. In this and following chapters, several such applications, considered
under the general heading of interference, are presented.

7-1 TWO-BEAM INTERFERENCE

We consider first the interference of two *plane waves* of the same frequency,
represented by $\vec{\mathbf{E}}_1$ and $\vec{\mathbf{E}}_2$. We may express the two electric fields at a point P
where the fields are combined as

$$\vec{\mathbf{E}}_1 = \vec{\mathbf{E}}_{01} \cos(ks_1 - \omega t + \phi_1) \qquad (7\text{-}1)$$

$$\vec{\mathbf{E}}_2 = \vec{\mathbf{E}}_{02} \cos(ks_2 - \omega t + \phi_2) \qquad (7\text{-}2)$$

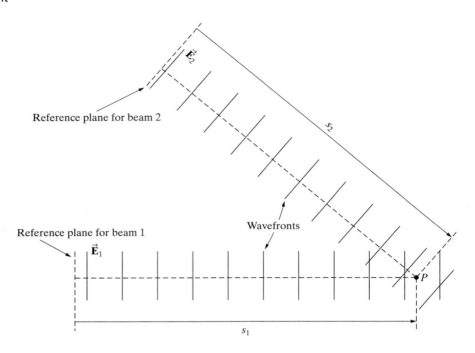

Figure 7-1 Two-beam interference.

In these relations $k = 2\pi/\lambda$, and s_1 and s_2 can be taken to be the distances traveled by each beam along its respective path from its source to the observation point P. (See Figure 7-1.) Then ϕ_1 and ϕ_2 represent the phases of these waves at their respective sources at time $t = 0$. These waves combine to produce a disturbance at point P, whose electric field $\vec{E}_p$ is given by the principle of superposition,

$$\vec{E}_p = \vec{E}_1 + \vec{E}_2$$

It should be noted that $\vec{E}_1$ and $\vec{E}_2$ are rapidly varying functions with optical frequencies of order 10^{14} to 10^{15} Hz for visible light. Thus both $\vec{E}_1$ and $\vec{E}_2$ average to zero over very short time intervals. Measurement of the waves by their effect on the eye or some other light detector depends on the energy of the light beam. The radiant power density, or *irradiance*, E_e (W/m^2), measures the time average of the square of the wave amplitude. In practice, the time average is carried out by a detector. The averaging time for the eye is on the order of 1/30 of a second; other detectors have averaging times as short as a nanosecond. In general, the averaging time of physical detectors greatly exceeds an optical period ($10^{-14} - 10^{-15}$ s).

Unfortunately, the standard symbol for irradiance, except for the subscript, is the same as that for the electric field. To avoid confusion, we use here the symbol I for irradiance, so that

$$I = \varepsilon_0 c \langle \vec{E} \cdot \vec{E} \rangle \tag{7-3}$$

Thus, the resulting irradiance at P is given by

$$I = \varepsilon_0 c \langle \vec{E}_p^2 \rangle = \varepsilon_0 c \langle \vec{E}_p \cdot \vec{E}_p \rangle$$
$$= \varepsilon_0 c \langle (\vec{E}_1 + \vec{E}_2) \cdot (\vec{E}_1 + \vec{E}_2) \rangle$$

or

$$I = \varepsilon_0 c \langle \vec{E}_1 \cdot \vec{E}_1 + \vec{E}_2 \cdot \vec{E}_2 + 2\vec{E}_1 \cdot \vec{E}_2 \rangle \tag{7-4}$$

In Eq. (7-4), the first two terms correspond to the irradiances of the individual waves, I_1 and I_2. The last term depends on an interaction of the waves and is called the *interference term*, I_{12}. We may then write

$$I = I_1 + I_2 + I_{12} \qquad (7\text{-}5)$$

If light behaved without interference, like classical particles, we would then expect $I = I_1 + I_2$. The presence of the third term I_{12} is indicative of the wave nature of light, which can produce enhancement or diminution of the irradiance through interference. Notice that when $\vec{\mathbf{E}}_1$ and $\vec{\mathbf{E}}_2$ are orthogonal, so that their dot product vanishes, no interference results. When the electric fields are parallel, on the other hand, the interference term makes its maximum contribution. Two beams of unpolarized light produce interference because each can be resolved into orthogonal components of $\vec{\mathbf{E}}$ that can then be paired off with similar components of the other beam. Each component produces an interference term with $\vec{\mathbf{E}}_1 \| \vec{\mathbf{E}}_2$ ($\vec{\mathbf{E}}_1$ parallel to $\vec{\mathbf{E}}_2$).

Consider the interference term,

$$I_{12} = 2\varepsilon_0 c \langle \vec{\mathbf{E}}_1 \cdot \vec{\mathbf{E}}_2 \rangle \qquad (7\text{-}6)$$

where $\vec{\mathbf{E}}_1$ and $\vec{\mathbf{E}}_2$ are given by Eqs. (7-1) and (7-2). Their dot product,

$$\vec{\mathbf{E}}_1 \cdot \vec{\mathbf{E}}_2 = \vec{\mathbf{E}}_{01} \cdot \vec{\mathbf{E}}_{02} \cos(ks_1 - \omega t + \phi_1)\cos(ks_2 - \omega t + \phi_2)$$

can be simplified in an instructive manner using a trigonometric identity. To this end, let us define

$$\alpha \equiv ks_1 + \phi_1 \quad \text{and} \quad \beta \equiv ks_2 + \phi_2$$

so that

$$2\vec{\mathbf{E}}_1 \cdot \vec{\mathbf{E}}_2 = 2\vec{\mathbf{E}}_{01} \cdot \vec{\mathbf{E}}_{02} \cos(\alpha - \omega t)\cos(\beta - \omega t)$$

The identity $2\cos(A)\cos(B) = \cos(A + B) + \cos(B - A)$ helps us cast the time average of $2\vec{\mathbf{E}}_1 \cdot \vec{\mathbf{E}}_2$ as

$$2\langle \vec{\mathbf{E}}_1 \cdot \vec{\mathbf{E}}_2 \rangle = \vec{\mathbf{E}}_{01} \cdot \vec{\mathbf{E}}_{02}[\langle \cos(\alpha + \beta - 2\omega t) \rangle + \langle \cos(\beta - \alpha) \rangle]$$

The first time average in this relation is taken over a rapidly oscillating cosine function and so is zero. Thus,

$$2\langle \vec{\mathbf{E}}_1 \cdot \vec{\mathbf{E}}_2 \rangle = \vec{\mathbf{E}}_{01} \cdot \vec{\mathbf{E}}_{02}\langle \cos(\beta - \alpha) \rangle = \vec{\mathbf{E}}_{01} \cdot \vec{\mathbf{E}}_{02}\langle \cos(k(s_2 - s_1) + \phi_2 - \phi_1) \rangle$$
$$\equiv \vec{\mathbf{E}}_{01} \cdot \vec{\mathbf{E}}_{02}\langle \cos \delta \rangle$$
$$(7\text{-}7)$$

where we have defined the phase difference between $\vec{\mathbf{E}}_2$ and $\vec{\mathbf{E}}_1$ as

$$\delta = k(s_2 - s_1) + \phi_2 - \phi_1 \qquad (7\text{-}8)$$

For purely monochromatic fields, δ is time-independent, in which case $\langle \cos \delta \rangle = \cos \delta$. However, as we will discuss, for real fields, which are not perfectly monochromatic, care must be taken in treating this time average. Combining Eqs. (7-6) and (7-7),

$$I_{12} = \varepsilon_0 c \vec{\mathbf{E}}_{01} \cdot \vec{\mathbf{E}}_{02}\langle \cos \delta \rangle \qquad (7\text{-}9)$$

The irradiance terms I_1 and I_2 of Eq. (7-5) can be shown to yield

$$I_1 = \varepsilon_0 c \langle \vec{\mathbf{E}}_1 \cdot \vec{\mathbf{E}}_1 \rangle = \varepsilon_0 c E_{01}^2 \langle \cos^2(\alpha - \omega t) \rangle = \frac{1}{2} \varepsilon_0 c E_{01}^2 \qquad (7\text{-}10)$$

and

$$I_2 = \varepsilon_0 c \langle \vec{\mathbf{E}}_2 \cdot \vec{\mathbf{E}}_2 \rangle = \varepsilon_0 c E_{02}^2 \langle \cos^2(\beta - \omega t) \rangle = \frac{1}{2} \varepsilon_0 c E_{02}^2 \qquad (7\text{-}11)$$

In Eqs. (7-10) and (7-11) we used the fact that the time average of the square of a rapidly oscillating sinusoidal function is 1/2. In Eq. (7-9) when $E_{01} \| E_{02}$, their dot product is identical with the product of their magnitudes E_{01} and E_{02}. These may be expressed in terms of I_1 and I_2 by the use of Eqs. (7-10) and (7-11), and when combined with Eq. (7-9) results in

$$I_{12} = 2\sqrt{I_1 I_2} \langle \cos \delta \rangle \qquad (7\text{-}12)$$

so that we may write, finally,

$$I = I_1 + I_2 + 2\sqrt{I_1 I_2} \langle \cos \delta \rangle \qquad (7\text{-}13)$$

Notice that once we have made the assumption that the $\vec{\mathbf{E}}$ fields are parallel, the treatment becomes much the same as the scalar theory of Chapter 5.

Interference of Mutually Incoherent Fields

In practice, for electric fields $\vec{\mathbf{E}}_1$ and $\vec{\mathbf{E}}_2$ originating from different sources, the time average in Eq. (7-13) is zero. This occurs because no source is perfectly monochromatic. To model real sources, Eqs. (7-1) and (7-2) must be modified to account for departures from monochromaticity. One way to do this is to allow the phases ϕ_1 and ϕ_2 to be functions of time. For laser sources, these phases would typically be random functions of time that vary on a time scale much longer than an optical period but still shorter than typical detector averaging times. The interference term I_{12}, in this case, takes the form,

$$2\sqrt{I_1 I_2} \langle \cos(k(s_2 - s_1) + \phi_2(t) - \phi_1(t)) \rangle$$

As stated, for real detectors and for all but those laser sources with state-of-the-art frequency stability, the time average in the preceding relation will be zero. In such a case we say that the sources are *mutually incoherent* and the detected irradiance will be

$$I = I_1 + I_2 \qquad \text{Mutually incoherent beams}$$

It is often said, therefore, that light beams from independent sources, even if both sources are the same kind of laser, do not interfere with each other. In fact, these fields do interfere but the interference term averages to zero over the averaging times of most real detectors.

Interference of Mutually Coherent Beams

If light from the same laser source is split and then recombined at a detector, the time average in Eq. (7-13) need not be zero. This occurs because the departures from monochromaticity of each beam, while still present, will be correlated since both beams come from the same source. In this case, the phase difference $\phi_2(t) - \phi_1(t)$ will be strictly zero if the beams travel paths of *equal duration* before being recombined at the detector. In such a case, δ is a constant and the interference term takes the form,

$$2\sqrt{I_1 I_2}\langle\cos(k(s_2 - s_1) + \phi_1(t) - \phi_1(t))\rangle = 2\sqrt{I_1 I_2}\cos(k(s_2 - s_1))$$
$$= 2\sqrt{I_1 I_2}\cos\delta$$

Even if the electric fields travel paths that differ in duration by a time δt, the phase difference resulting from the departure from monochromaticity, $\phi_1(t) - \phi_1(t + \delta t)$, will still be nearly zero so long as δt is less than the so-called *coherence* time, τ_0, of the source. Qualitatively, the coherence time of the source is the time interval over which departures from monochromaticity are small. In Chapter 9 we show that the coherence time of a source is inversely proportional to the range of frequencies, $\Delta\nu$, of the components that make up the electric field. That is,

$$\tau_0 = \frac{1}{\Delta\nu}$$

Associated with the coherence time of a source is a coherence length, $l_t = c\tau_0$, which is the distance that the electric field travels in a coherence time. For a white light source the coherence length is about $1\ \mu$m; laser sources have coherence lengths that range from tens of centimeters to tens of kilometers. Throughout the rest of this chapter, we will presume that the difference in the lengths of paths traveled by beams originating from the same source is considerably less than the coherence length of the source. In such a case, the electric fields are said to be *mutually* coherent and the irradiance of the combined fields will have the form

$$I = I_1 + I_2 + 2\sqrt{I_1 I_2}\cos\delta \qquad \text{Mutually coherent beams} \qquad (7\text{-}14)$$

where δ is the total phase difference at the point of recombination of the beam. As we have noted, if the beams originate from the same source, this phase difference accumulates as a result of a difference in path lengths traveled by the respective beams. In many cases of interest, other factors can lead to a phase difference between the beams as well. Important mechanisms of this sort include differing phase shifts due to reflection from beam splitters and differing indices of refraction in the separate paths taken by the two beams. Depending on whether $\cos\delta > 0$ or $\cos\delta < 0$ in Eq. (7-14), the interference term either augments or diminishes the sum of the individual irradiances I_1 and I_2, leading to constructive or destructive interference, respectively. Since the relative distances traveled by the two beams will, in general, differ for different observation points in the region of overlap, the phase difference δ will also differ for different observation points. Typically, $\cos\delta$ will take on alternating maximum and minimum values, and *interference fringes*, spatially separated, will occur in the observation plane.

To be more specific, when $\cos\delta = +1$, constructive interference yields the maximum irradiance

$$I_{\max} = I_1 + I_2 + 2\sqrt{I_1 I_2} \qquad (7\text{-}15)$$

This condition occurs whenever the phase difference $\delta = 2m\pi$, where m is any integer or zero. On the other hand, when $\cos\delta = -1$, destructive interference yields the minimum, or background, irradiance

$$I_{\min} = I_1 + I_2 - 2\sqrt{I_1 I_2} \qquad (7\text{-}16)$$

a condition that occurs whenever $\delta = (2m + 1)\pi$. A plot of irradiance I versus phase δ, in Figure 7-2a, exhibits periodic fringes. Destructive interference is complete, that is, cancellation is complete, when $I_1 = I_2 = I_0$. Then, Eqs. (7-15) and (7-16) give

$$I_{\max} = 4I_0 \quad \text{and} \quad I_{\min} = 0$$

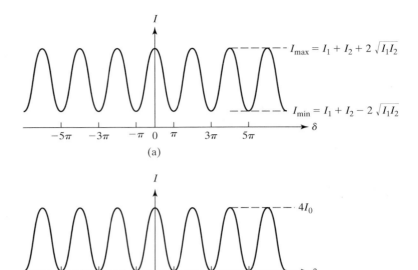

Figure 7-2 Irradiance of interference fringes as a function of phase difference δ. Visibility is enhanced in (b), where the background irradiance $I_{min} = 0$ when $I_1 = I_2$.

Resulting fringes, shown in Figure 7-2b, now exhibit better contrast. A measure of *fringe contrast*, called *visibility*, with values between 0 and 1, is given by the quantity

$$\text{visibility} = \frac{I_{max} - I_{min}}{I_{max} + I_{min}} \tag{7-17}$$

In the experimental utilization of fringe patterns, it is therefore usually desirable to ensure that the interfering beams have the same amplitudes.

Another useful form of Eq. (7-14), for the case of interfering beams of equal amplitude so that $I_1 = I_2 = I_0$, is found by writing

$$I = I_0 + I_0 + 2\sqrt{I_0^2}\cos \delta = 2I_0(1 + \cos \delta)$$

and then making use of the trigonometric identity

$$1 + \cos \delta \equiv 2 \cos^2\left(\frac{\delta}{2}\right)$$

The irradiance for two equal interfering beams is then

$$I = 4I_0 \cos^2\left(\frac{\delta}{2}\right) \tag{7-18}$$

Notice that energy is not conserved at each point of the superposition, that is, $I \neq 2I_0$, but that over at least one spatial period of the fringe pattern $I_{av} = 2I_0$. This situation is typical of interference and diffraction phenomena: If the power density falls below the average at some points, it rises above the average at other points in such a way that the total pattern satisfies the principle of energy conservation.

Example 7-1

Consider two interfering beams with parallel electric fields that are superposed. Take the electric fields of the individual beams to be

$$E_1 = 2 \cos(ks_1 - \omega t) \quad (\text{kV/m})$$

$$E_2 = 5 \cos(ks_2 - \omega t) \quad (\text{kV/m})$$

Let us determine the irradiance contributed by each beam acting alone and that due to their mutual interference at a point where their path difference is such that $k(s_2 - s_1) = \pi/12$. We have

$$I_1 = \tfrac{1}{2}\varepsilon_0 c E_{01}^2 = \tfrac{1}{2}\varepsilon_0 c (2000)^2 = 5309 \text{ W/m}^2$$

$$I_2 = \tfrac{1}{2}\varepsilon_0 c E_{02}^2 = \tfrac{1}{2}\varepsilon_0 c (5000)^2 = 33{,}180 \text{ W/m}^2$$

$$I_{12} = 2\sqrt{I_1 I_2}\cos\delta = 2\sqrt{(5309 \times 33180)}\cos(\pi/12) = 25{,}640 \text{ W/m}^2$$

To find the visibility near this point of recombination, we must calculate

$$I_{\max} = I_1 + I_2 + 2\sqrt{I_1 I_2} = 5309 + 33180 + 2\sqrt{(5309 \times 33180)}$$

$$= 65{,}034 \text{ W/m}^2$$

$$I_{\min} = I_1 + I_2 - 2\sqrt{I_1 I_2} = 5309 + 33180 - 2\sqrt{(5309 \times 33180)}$$

$$= 11{,}945 \text{ W/m}^2$$

The visibility is then given by Eq. (7-17), or

$$\text{visibility} = \frac{65{,}034 - 11{,}945}{65{,}034 + 11{,}945} = 0.690$$

If the amplitudes of the two waves were equal, then $I_{\max} = 4I_0$, $I_{\min} = 0$, and the visibility would be 1.

In the analysis leading to the irradiance that results from the superposition of two mutually coherent beams, Eq. (7-14), we assumed that the individual beams were plane waves described by Eqs. (7-1) and (7-2). In fact, the analysis holds for any sort of harmonic wave (e.g., spherical, cylindrical, or Gaussian). However, for these types of waves, the amplitudes E_{01} and E_{02} (and so the irradiances I_1 and I_2) depend on the distance from the source to the observation point.

7-2 YOUNG'S DOUBLE-SLIT EXPERIMENT

The decisive experiment performed by Thomas Young in 1802 is shown schematically in Figure 7-3. Monochromatic light is first allowed to pass through a single small hole in order to approximate a single point source S. The light spreads out in spherical waves from the source S according to Huygens' principle and is allowed to fall on a plane with two closely spaced holes, S_1 and S_2. In a modern version of this experiment, a laser is typically used to illuminate the two holes. In either case, the holes become two coherent sources of light, whose interference can be observed on a screen some distance away. If the two holes are equal in size, light waves emanating from the holes have comparable amplitudes, and the irradiance at any point of superposition is given by Eq. (7-18). Referring to Figure 7-3, we will now develop an expression for the irradiance at observation points such as P on a screen that is a distance L from the plane containing the two holes S_1 and S_2. The phase difference δ between the two waves arriving at the observation point P must be determined to calculate the resultant irradiance there. Clearly, if $S_2 P - S_1 P = s_2 - s_1 = m\lambda$, the waves will arrive in phase, and maximum irradiance or brightness results. If $s_2 - s_1 = (m + \tfrac{1}{2})\lambda$, the requisite condition for destructive interference or darkness is met. Practically speaking, the hole separation a is much smaller than the screen distance L, allowing a simple expression for the path distance, $s_2 - s_1$. Using P as a center, let an arc $S_1 Q$ be drawn of radius s_1 so that it intersects the line $S_2 P$ at Q. Then $s_2 - s_1$ is equal to the segment Δ, as shown. The first approximation is to regard arc $S_1 Q$ as a straight-line segment that

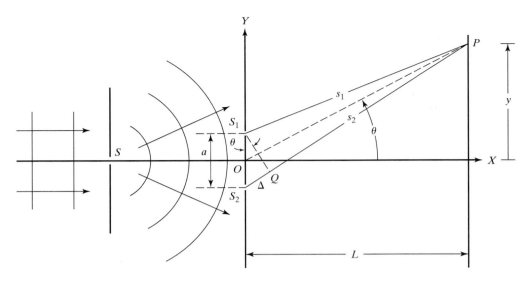

Figure 7-3 Schematic for Young's double-slit experiment. The holes S_1 and S_2 are usually slits, with the long dimensions extending into the page. The hole at S is not necessary if the source is a spatially coherent laser.

forms one leg of the right triangle $S_1 S_2 Q$. If θ is the angle between the line segments $S_1 S_2$ and $S_1 Q$, then $\Delta = a \sin \theta$. The second approximation identifies the angle θ with the angle between the optical axis OX and the line drawn from the midpoint O between holes to the point P at the screen. Observe that the corresponding sides of the two angles θ are related such that $OX \perp S_1 S_2$, and OP is almost exactly perpendicular to $S_1 Q$.

The condition for *constructive interference* at a point P on the screen is, then, to a very good approximation

$$s_2 - s_1 = \Delta = m\lambda \cong a \sin \theta \tag{7-19}$$

whereas for *destructive interference*,

$$\Delta = \left(m + \tfrac{1}{2}\right)\lambda \cong a \sin \theta \tag{7-20}$$

where m is zero or of integral value. Typically, at observation points of interest, the electric field *amplitudes* of the beams originating from the two slits are nearly the same so that the irradiance on the screen, at a point determined by the angle θ, is found using Eq. (7-18) and the relationship between path difference Δ and phase difference δ,

$$\delta = k(s_2 - s_1) = \frac{2\pi}{\lambda}\Delta$$

The result is

$$I = 4I_0 \cos^2\left(\frac{\pi\Delta}{\lambda}\right) = 4I_0 \cos^2\left(\frac{\pi a \sin \theta}{\lambda}\right)$$

For points P near the optical axis, where $y \ll L$, we may approximate further: $\sin \theta \cong \tan \theta \cong y/L$, so that

$$I = 4I_0 \cos^2\left(\frac{\pi a y}{\lambda L}\right) \tag{7-21}$$

By allowing the cosine function in Eq. (7-21) to become alternately ± 1 and 0, the conditions expressed by Eqs. (7-19) and (7-20) for constructive and destructive interference are reproduced.

Arguing now from Eq. (7-19) and the small angle relation $\sin \theta \cong \tan \theta \cong y/L$, we find the bright fringe positions to be given by

$$y_m = \frac{m\lambda L}{a}, \qquad m = 0, \pm 1, \pm 2, \ldots \qquad (7\text{-}22)$$

Consequently, there is a constant separation between irradiance maxima, corresponding to successive values of m, given by

$$\Delta y = y_{m+1} - y_m = \frac{\lambda L}{a} \qquad (7\text{-}23)$$

with minima situated midway between the maxima. Thus, fringe separation is proportional both to wavelength and screen distance and inversely proportional to the hole spacing. Reducing the hole spacing expands the fringe pattern formed by each color. Measurement of the fringe separation provides a means of determining the wavelength of the light. The single hole, used to secure a degree of spatial coherence, may be eliminated if laser light, both highly monochromatic and spatially coherent, is used to illuminate the double slit. In the observational arrangement just described, fringes are observed on a screen placed perpendicular to the optical axis at some distance from the aperture, as indicated in Figure 7-4. Fringe maxima coincide with integral orders of m, and fringe minima fall halfway between adjacent maxima.

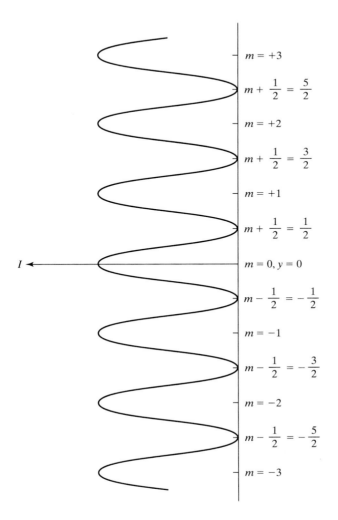

Figure 7-4 Irradiance versus distance from the optical axis for a double-slit fringe pattern. The *order* of the interference pattern is indicated by m, with integral values of m determining positions of fringe maxima.

Example 7-2

Laser light passes through two identical and parallel slits, 0.2 mm apart. Interference fringes are seen on a screen 1 m away. Interference maxima are separated by 3.29 mm. What is the wavelength of the light? How does the irradiance at the screen vary, if the contribution of one slit alone is I_0?

Solution

From Eq. (7-23),

$$\lambda = a\Delta y/L = (0.0002 \text{ m})(3.29 \times 10^{-3} \text{ m})/(1 \text{ m})$$

$$= 6.58 \times 10^{-7} \text{ m} = 658 \text{ nm}$$

According to Eq. (7-21), $I = 4I_0 \cos^2[\pi a y/\lambda L]$. In this case,

$$I = 4I_0 \cos^2[\pi(0.0002)y/(658 \times 10^{-9})(1\text{m})] = 4I_0 \cos^2[(955/\text{m})y]$$

An alternative way to view the formation of bright (B) positions of constructive interference and dark (D) positions of destructive interference is shown in Figure 7-5. The crests and valleys of spherical waves from S_1 and S_2 are shown approaching the screen. Along directions marked B, wave crests (or wave valleys) from both slits coincide, producing maximum irradiance. Along directions marked D, on the other hand, the waves are seen to be out of step by half a wavelength, and destructive interference results.

Obviously, fringes should be present in all the space surrounding the holes, where light from the holes is allowed to interfere, though the irradiance is greatest in the forward direction. If we imagine two coherent point sources of light radiating in all directions, then the condition given by Eq. (7-19) for bright fringes,

$$s_2 - s_1 = m\lambda \tag{7-24}$$

defines a family of bright fringe surfaces in the space surrounding the holes. To visualize this set of surfaces, we may take advantage of the inherent symmetry in the arrangement. In Figure 7-6, the intersection of several bright fringe surfaces with a plane that includes the two sources is shown, each surface corresponding

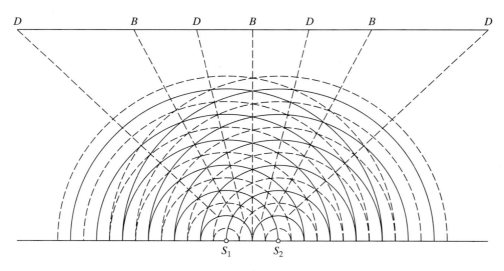

Figure 7-5   Alternating bright and dark interference fringes are produced by light from two coherent sources. Along directions where crests (solid circles) from S_1 intersect crests from S_2, brightness (B) results. Along directions where crests meet valleys (dashed circles), darkness (D) results.

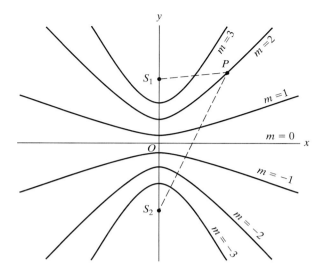

Figure 7-6 Bright fringe surfaces for two coherent point sources. The distances from S_1 and S_2 to any point P on a bright fringe surface differ by an integral number of wavelengths. The surfaces are generated by rotating the pattern about the y-axis.

to an integral value of order m. The surfaces are hyperbolic, since Eq. (7-24) is precisely the condition for a family of hyperbolic curves with parameter m. Inasmuch as the y-axis is an axis of symmetry, the corresponding bright fringe surfaces are generated by rotating the entire pattern about the y-axis. One should then be able to visualize the intercept of these surfaces with the plane of an observational screen placed anywhere in the vicinity. In particular, a screen placed perpendicular to the OX axis, as in Figure 7-3, intercepts hyperbolic arcs that appear as straight-line fringes near the axis, whereas a screen placed perpendicular to the OY axis shows concentric circular fringes centered on the axis. Because the fringe system extends throughout the space surrounding the two sources, the fringes are said to be *nonlocalized*.

The holes S, S_1, and S_2 of Figure 7-3 are usually replaced by parallel, narrow slits (oriented with their long sides perpendicular to the page in Figure 7-3) to illuminate more fully the interference pattern. The effect of the array of point sources along the slits, each set producing its own fringe system as just described, is simply to elongate the pattern parallel to the fringes, without changing their geometrical relationships. This is true even when two points along a source slit are not mutually coherent.

7-3 DOUBLE-SLIT INTERFERENCE WITH VIRTUAL SOURCES

Interference fringes may sometimes appear in arrangements when only one light source is present. It is possible, through reflection or refraction, to produce virtual images that, acting together or with the actual source, behave as two coherent sources that can produce an interference pattern. Figures 7-7 to 7-9 illustrate three such examples. These examples are not only of some historic importance; they also serve to impress us with the variety of ways unexpected fringe patterns may appear in optical experiments, especially when the extremely coherent light of a laser is being used.

Lloyd's Mirror
In Figure 7-7, interference fringes are produced due to the superposition of light at the screen that originates at the actual source S and, by reflection, also originates effectively from its virtual source S' below the surface of the plane mirror MM'. Where the direct and reflected beams strike the screen, fringes will appear. The position of bright fringes is given by Eq. (7-22), where a is

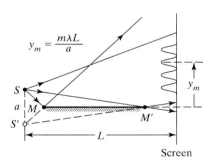

Figure 7-7 Interference with Lloyd's mirror. Coherent sources are the point source S and its virtual image, S'.

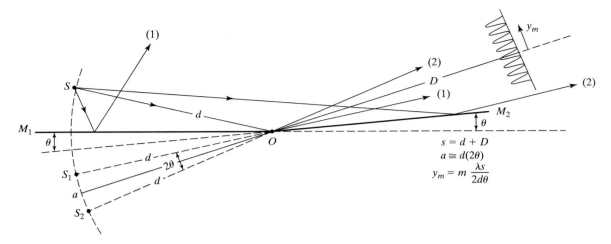

Figure 7-8 Interference with Fresnel's mirrors. Coherent sources S_1 and S_2 are the two virtual images of point source S, formed in the two plane mirrors M_1 and M_2. Direct light from S is not allowed to reach the screen.

In the figure:

$$s = d + D$$
$$a \cong d(2\theta)$$
$$y_m = m\frac{\lambda s}{2d\theta}$$

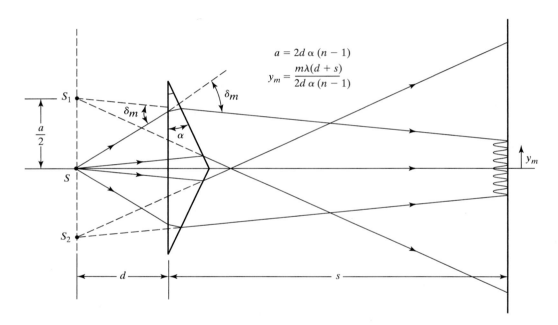

In the figure:

$$a = 2d\,\alpha\,(n-1)$$
$$y_m = \frac{m\lambda(d+s)}{2d\,\alpha\,(n-1)}$$

Figure 7-9 Interference with Fresnel's biprism. Coherent sources are the virtual images S_1 and S_2 of source S, formed by refraction in the two halves of the prism.

twice the distance of source S above the mirror plane. The arrangement is known as *Lloyd's mirror*. If the screen were to contact the mirror at M', the fringe at M' would be found to be dark. Since at this point the optical-path difference between the two interfering beams vanishes, one might expect a bright fringe. The contrary experimental result—a dark fringe—is explained by requiring a phase shift of π for the air-glass reflection.[1]

Fresnel's Mirrors
Another closely related arrangement is *Fresnel's mirrors*, Figure 7-8. Interference occurs between the light reflected from each of two mirrors, M_1 and M_2,

---

[1]A theoretical explanation for phase changes on reflection results from an analysis based on Maxwell's equations and requires identification of the state of polarization of the light. The subject is treated in Chapter 23.

inclined at a small relative angle θ. Two rays reflected from each are shown labeled as (1) from M_1 and (2) from M_2. Interference fringes appear in the region of overlap. Interference effectively occurs between the two coherent virtual images S_1 and S_2, acting as sources. Once the virtual image separation a is related to the tilt angle θ and to the distance d from actual source to the intersection of the mirrors at O, the fringe pattern may again be described by Eq. (7-22). The screen is shown at distance D from point O.

Fresnel's Biprism

Figure 7-9 shows *Fresnel's biprism*, which refracts light from a small source S in such a way that it appears to come from two coherent, virtual sources, S_1 and S_2. Extreme rays for refraction at the top and bottom halves are shown. Interference fringes are seen in the overlap region on the screen. In practice, the prism angle α is very small, of the order of a degree. One of the rays (shown) passes through the wedge in a symmetrical fashion, making equal entrance and exit angles with the two sides and satisfying the condition for minimum deviation. For this ray the deviation angle δ_m is given by $\delta_m = \alpha(n - 1)$. The geometry of this particular ray provides a means of approximately determining the virtual source separation a in terms of prism index n and angle α:

$$a = 2d\delta_m = 2d\alpha(n - 1) \tag{7-25}$$

Interference fringes are then described by Eq. (7-22), as usual.

7-4 INTERFERENCE IN DIELECTRIC FILMS

The familiar appearance of colors on the surface of oily water and soap films and the beautiful iridescence often seen in mother-of-pearl, peacock feathers, and butterfly wings are associated with the interference of light in single or multiple thin surface layers of transparent material. There exists a variety of situations in which such interference can take place, affecting the nature of the interference pattern and the conditions under which it can be observed. Variables in the situation include the size and spectral width of the source and the shape and reflectance of the film.

Consider the case of a film of transparent material bounded by parallel planes, such as might be formed by an oil slick, a metal oxide layer, or an evaporated coating on a flat, glass substrate (Figure 7-10). A beam of light incident on the film surface at A divides into reflected and refracted portions. This separation of the original light into two parts, preliminary to recombination and interference, is usually referred to as *amplitude division*, in contrast

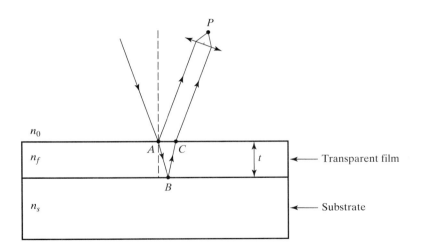

Figure 7-10 Double-beam interference from a film. Rays reflected from the top and bottom plane surfaces of the film are brought together at P by a lens.

to a situation like Young's double slit, in which separation is said to occur by *wavefront division*. The refracted beam reflects again at the film-substrate interface B and leaves the film at C, in the same direction as the beam reflected at A. Part of the beam may reflect internally again at C and continue to experience multiple reflections within the film layer until it has lost its irradiance. There will thus exist multiple parallel beams emerging from the top surface, although with rapidly diminishing amplitudes. Unless the reflectance of the film is large, a good approximation to the more complex situation of multiple reflection (Section 7-9) is to consider only the first two emerging beams. The two parallel beams leaving the film at A and C can be brought together by a converging lens, the eye, for example. The two beams intersecting at P superpose and interfere. Since the two beams travel different paths from point A onward, a relative phase difference develops that can produce constructive or destructive interference at P. The optical path difference Δ, *in the case of normal incidence*, is the additional path length ABC traveled by the refracted ray times the refractive index of the film. Thus,

$$\Delta = n(AB + BC) = n(2t) \qquad (7\text{-}26)$$

where t is the film thickness. For example, if $2nt = \lambda_0$, the wavelength of the light in vacuum, the two interfering beams, on the basis of optical-path difference alone, would be in phase and produce constructive interference. However, an additional phase difference, due to the phenomenon of phase changes on reflection, must be considered. Suppose that $n_f > n_0$ and $n_f > n_s$. In fact, often $n_0 = n_s$ because the media bounding the film are identical, as in the case of a water film (soap bubble) in air. Then the reflection at A occurs with light going from a lower index n_0 toward a higher index n_f, a condition usually called *external reflection*. The reflection at B, on the other hand, occurs for light going from a higher index n_f toward a lower index n_s, a condition called *internal reflection*. A relative phase shift of π occurs between the externally and internally reflected beams, so that, equivalently, an additional path difference of $\lambda/2$ is introduced between the two beams. The net optical-path difference between the beams is then $\lambda + \lambda/2$, which puts them precisely out of phase, and destructive interference results at P. If, instead, both reflections are external ($n_0 < n_f < n_s$) or if both reflections are internal ($n_0 > n_f > n_s$), no relative phase difference due to reflection needs to be taken into account. In that case, constructive interference occurs at P.

A frequent use of such single-layer films is in the production of *antireflecting coatings* on optical surfaces. In most cases, the light enters the film from air, so that $n_0 = 1$. Furthermore, if $n_s > n_f$, no relative phase shift between the two reflected beams occurs, and the optical-path difference alone determines the type of interference to be expected. If the film thickness is $\lambda_f/4$, where λ_f is the wavelength of the light in the film, then $2t = \lambda_f/2$ and the optical-path difference $2n_f t = \lambda_0/2$, since $\lambda_0 = n_f \lambda_f$. Destructive interference occurs at this wavelength and to some extent at neighboring wavelengths, which means that the light reflected from such a film is the incident spectrum minus the wavelength region around λ_0. If the incident light is white and λ_0 is in the visible region, the reflected light is colored. Extinction of a region of the spectrum by nonreflecting films of $\lambda/4$ thickness is, of course, more effective if the amplitudes of the two reflected beams are equal. In general, all one can say is that for constructive interference the two amplitudes add (being in phase), and for destructive interference the amplitudes subtract (being exactly out of phase). For the difference to be zero, that is, for destructive interference to be complete, the amplitudes must be equal. We show later (Chapter 23) that, in the case of normal incidence, the *reflection coefficient* (or ratio of reflected to incident electric field amplitudes) is given by

$$r = \frac{1 - n}{1 + n} \qquad (7\text{-}27)$$

where the *relative index* $n = n_2/n_1$. The amplitudes of the electric field reflected internally and externally from the film of Figure 7-10 are then equal, assuming a nonabsorbing film, if the relative indices are equivalent for these cases, that is, if

$$\frac{n_f}{n_0} = \frac{n_s}{n_f} \quad \text{or} \quad n_f = \sqrt{n_0 n_s} \tag{7-28}$$

Since usually $n_0 = 1$, the requirement that reflected beams be of equal amplitude is met by choosing a film whose refractive index is the square root of the substrate's refractive index. A suitable film material for the application may or may not exist, and some compromise is made. For example, to reduce the reflectance of lenses employed in optical instruments handling white light, the film thickness of $\lambda/4$ is determined with a λ in the center of the visible spectrum or wherever the detection system is most sensitive. In the case of the eye, this is the yellow-green portion near 550 nm. Assuming $n = 1.50$ for the glass lens, ideally $n_f = \sqrt{1.50} = 1.22$. The nearest practical film material with a matching index is MgF_2, with $n = 1.38$. For an antireflection coating of this type, the reduced reflected light near the middle of the spectrum results in a predominance of the blue and red ends of the spectrum, so that the coatings appear purple in reflected light.

As another example, consider a multilayer stack of alternating high-low index dielectric films (Figure 7-11). If each film has an optical thickness of $\lambda_f/4$, a little analysis shows that in this case all emerging beams are in phase. Multiple reflections in the region of λ_0 increase the total reflected intensity and the quarter-wave stack performs as an efficient mirror. Such multilayer stacks can be designed to satisfy extinction or enhancement of reflected light over a greater portion of the spectrum than would a single-layer film. They are treated in greater detail in Chapter 22.

Returning now to the single-layer film, we want first to generalize the conditions for constructive and destructive interference by calculating the optical-path difference in the case incident rays *are not normal*. Figure 7-12 illustrates a ray incident on a film at an angle θ_i. The phase difference at points C and D between emerging beams is due to the optical path difference between paths AD and ABC. After points C and D are reached, the respective beams are parallel and in the same medium, so that no further phase difference

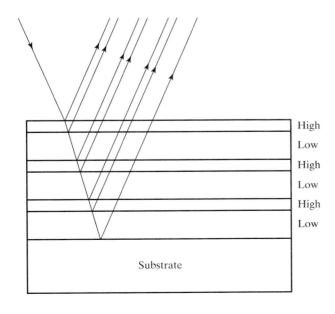

High
Low
High
Low
High
Low

Substrate

Figure 7-11 Multilayer dielectric mirror of alternating high and low index. Each film is $\lambda_f/4$ in optical thickness.

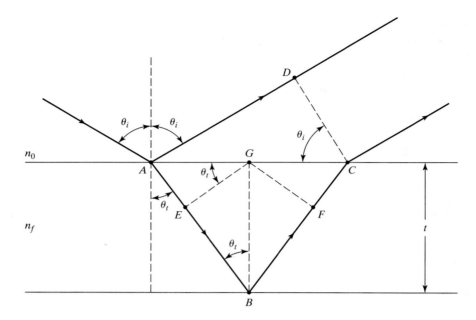

Figure 7-12 Single-film interference with light incident at arbitrary angle θ_i.

occurs. To assist in the calculation, point G is shown midway between A and C at the foot of the altitude BG in the isosceles triangle ABC. Points E and F are determined by constructing the perpendiculars GE and GF to the ray paths AB and BC, respectively. The optical-path difference between the emerging beams is, then,

$$\Delta = n_f(AB + BC) - n_0(AD)$$

where n_f and n_0 are the refractive indices of film and external medium, as shown.

It is helpful to break the distances AB and BC into parts and rearrange terms, resulting in

$$\Delta = [n_f(AE + FC) - n_0AD] + n_f(EB + BF) \tag{7-29}$$

The quantity in square brackets vanishes, as we now show. By Snell's law,

$$n_0 \sin \theta_i = n_f \sin \theta_t \tag{7-30}$$

In addition, by inspection,

$$AE = AG \sin \theta_t = \left(\frac{AC}{2}\right) \sin \theta_t \tag{7-31}$$

and

$$AD = AC \sin \theta_i \tag{7-32}$$

From Eq. (7-31) and incorporating, in turn, Eqs. (7-32) and (7-30),

$$2AE = AC \sin \theta_t = AD\left(\frac{\sin \theta_t}{\sin \theta_i}\right) = AD\left(\frac{n_0}{n_f}\right)$$

so that

$$n_0AD = 2n_fAE = n_f(AE + FC) \tag{7-33}$$

which was to be proved. There remains, then, from Eq. (7-29),

$$\Delta = n_f(EB + BF) = 2n_fEB \qquad (7\text{-}34)$$

The length EB is related to the film thickness t by $EB = t \cos \theta_t$, so we have, finally,

$$\Delta = 2n_ft \cos \theta_t \qquad (7\text{-}35)$$

The optical-path difference Δ is economically expressed by Eq. (7-35) in terms of the angle of refraction, not the angle of incidence, which of course can be recovered through Snell's law, Eq. (7-30). Notice that for normal incidence, $\theta_i = \theta_t = 0$ and $\Delta = 2n_ft$, as expected. The corresponding phase difference is $\delta = k\Delta = (2\pi/\lambda_0)\,\Delta$. The net phase difference must also take into account possible phase differences that arise on reflection, as discussed previously. Nevertheless, if we call Δ_p the optical-path difference given by Eq. (7-35) and Δ_r the equivalent path difference arising from phase change on reflection, we can state quite generally the conditions for

$$\text{constructive interference:} \quad \Delta_p + \Delta_r = m\lambda \qquad (7\text{-}36)$$

and

$$\text{destructive interference:} \quad \Delta_p + \Delta_r = \left(m + \tfrac{1}{2}\right)\lambda \qquad (7\text{-}37)$$

where $m = 0, 1, 2, \ldots$.

    If, for example, constructive interference results between the two parts of a single beam incident at angle θ_i, the same condition will hold for all beams incident at the same angle. This is possible if the source is an extended source, as in Figure 7-13. Independent point sources S_1, S_2, and S_3 are shown, all contributing to the intensity of the light at P. Since these sources are not coherent, interference is sustained only between pairs of reflected rays originating from the same source. If the lens aperture becomes too small to admit two such beams, such as (a) and (b) from S_1, no interference is detected. This may happen, for example, if the film thickness and, therefore, the spatial separation of two interfering beams—such as (a) and (b)—are increased, while the pupil of the eye viewing the reflected light is limited in size. Without a focusing device, these *virtual fringes* do not appear. They are called *localized fringes* because they are, so to speak, localized at infinity. Recall that *nonlocalized fringes* (Figure 7-6) are, in contrast, formed everywhere. Fringes formed as in Figure 7-13 are also

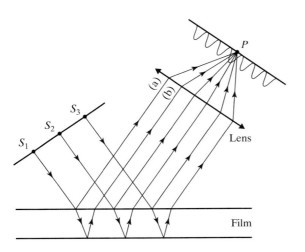

Figure 7-13 Interference by a dielectric film with an extended source. Fringes of equal inclination are focused by a lens.

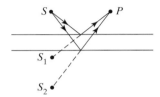

Figure 7-14 Interference by a dielectric film with a point source. Real, nonlocalized fringes appear as in the two-point source pattern of Figure 7-6. Refraction has been ignored.

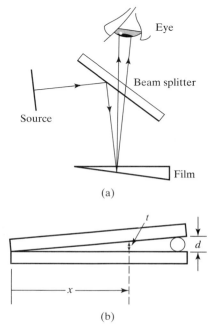

(a)

(b)

Figure 7-15 Interference from a wedge-shaped film, producing localized fringes of equal thickness. (a) Viewing assembly. (b) Air wedge formed with two microscope slides.

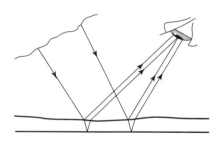

Figure 7-16 Interference by an irregular film illuminated by an extended source. Variations in film thickness, as well as angle of incidence, determine the wavelength region reinforced by interference.

referred to as *Haidinger fringes*, or *fringes of equal inclination*, since they are formed by parallel incident beams from an extended source. If a different inclination is chosen, parallel rays from the various source points are incident on the film at a different angle, reflect as parallel rays from the film at a different angle, and all focus at some other point where they interfere, according to the conditions expressed by Eqs. (7-36) and (7-37).

The fringes of equal inclination just described are not possible if the source is a point or is very small, since every ray of light from the source to the film must, in that case, arrive at a different angle of incidence (Figure 7-14). Fringes of a different kind are nonetheless formed. Since rays are reflected to any point P from the two film surfaces as if they originated at the virtual sources S_1 and S_2, this may be considered an instance of the two-point source pattern already discussed in connection with Figure 7-6. Real, nonlocalized fringes are formed in the space above the film. If the source of light is a laser, the fringe pattern is clearly visible on a screen placed anywhere in the vicinity of the film. The condition for interference is just that of the two-source interference pattern, where the slit separation is the distance between virtual sources S_1 and S_2. In Figure 7-14, S_1 and S_2 are located approximately by ignoring refraction in the film.

7-5 FRINGES OF EQUAL THICKNESS

If the film is of varying thickness t, the optical-path difference $\Delta = 2 n_f t \cos \theta_t$ varies even without variation in the angle of incidence. Thus, if the direction of the incident light is fixed, say at normal incidence, a bright or dark fringe will be associated with a particular thickness for which Δ satisfies the condition for constructive or destructive interference, respectively. For this reason, fringes produced by a variable-thickness film are called *fringes of equal thickness*. A typical arrangement for viewing these fringes is shown in Figure 7-15a. An extended source is used in conjunction with a beam splitter set at an angle of 45° to the incident light. The beam splitter in this position enables light to strike the film at normal incidence, while at the same time providing for the transmission of part of the reflected light into the detector (eye). Fringes, often called *Fizeau fringes*, are seen localized at the film, from which the interfering rays diverge. At normal incidence, $\cos \theta_t = 1$ and $\Delta = 2n_f t$. Thus the condition for bright and dark fringes, Eqs. (7-36) and (7-37), is

$$2n_f t + \Delta_r = \begin{cases} m\lambda, & \text{bright} \\ \left(m + \frac{1}{2}\right)\lambda, & \text{dark} \end{cases} \qquad (7\text{-}38)$$

where Δ_r is either $\lambda/2$ or 0, depending on whether there is or is not a relative phase shift of π between the rays reflected from the top and bottom surfaces of the film. One way of forming a suitable wedge for experimentation is to use two clean, glass microscope slides, wedged apart at one end by a thin spacer, perhaps a hair, as in Figure 7-15b. The resulting air layer between the slides shows Fizeau fringes when the slides are illuminated by monochromatic light. For this film, the two reflections are from glass to air (internal reflection) and from air to glass (external reflection), so that Δ_r in Eq. (7-38) is $\lambda/2$. As t increases in a linear fashion along the length of the slides from $t = 0$ to $t = d$, Eq. (7-38) is satisfied for consecutive orders of m, and a series of equally spaced, alternating bright and dark fringes will be seen by reflected light. These fringes are virtual, localized fringes and cannot be projected onto a screen.

If the extended source of Figure 7-15a is the sky and white light is incident at some angle on a film of variable thickness, as in Figure 7-16, the film may appear in a variety of colors, like an oil slick after a rain. Suppose that in a small region of the film the thickness is such as to produce constructive interference for wavelengths in the red portion of the spectrum at some order m. If the wavelengths at which constructive interference occurs again for orders $m + 1$ and $m - 1$ are outside the visible spectrum, the reflected light appears red. This can occur readily for low orders and therefore for thin films.

7-6 NEWTON'S RINGS

Since Fizeau fringes are fringes of equal thickness, their contours directly reveal any nonuniformities in the thickness of the film. Figure 7-17a shows how this circumstance can be put to practical use in determining the quality of the spherical surface of a lens, for example, in an arrangement in which the Fizeau fringes have come to be referred to as *Newton's rings*. An air wedge, formed between the spherical surface and an optically flat surface, is illuminated with normally incident monochromatic light, such as from a laser, or from a sodium or mercury lamp with a filter. Equal-thickness contours for a perfectly spherical surface, and therefore the fringes viewed, are concentric circles around the point of contact with the optical flat. At that point, $t = 0$ and the path difference between reflected rays is $\lambda/2$, as a result of reflection. The center of the fringe pattern thus appears dark, and Eq. (7-38) gives $m = 0$ for the order of the destructive interference. Irregularities in the surface of the lens show up as distortions in the concentric ring pattern. This arrangement can also be used as an optical means of measuring the radius of curvature of the lens surface. A geometrical relation exists between the radius r_m of the mth-order dark fringe, the corresponding air-film thickness t_m, and the radius of curvature R of the air film or the lens surface. Referring to Figure 7-17b and making use of the Pythagorean theorem, we have

$$R^2 = r_m^2 + (R - t_m)^2$$

or

$$R = \frac{r_m^2 + t_m^2}{2t_m} \tag{7-39}$$

The radius of the mth dark ring is measured and the corresponding thickness of the air wedge is determined from the interference condition of Eq. (7-38). Thus, R can be found. A little thought should convince one that light transmitted through the film and the optical flat will also show circular interference fringes. As shown in Figure 7-18, the pattern differs in two important respects from the

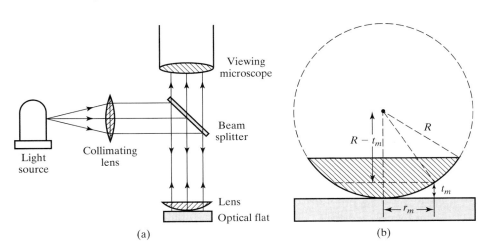

(a)   (b)

Figure 7-17 (a) Newton's rings apparatus. Interference fringes of equal thickness are produced by the air wedge between lens and optical flat. (b) Essential geometry for production of Newton's rings.

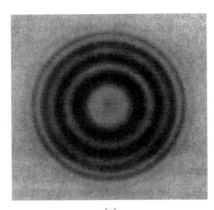

(a)

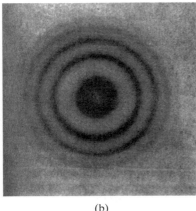
(b)

Figure 7-18 Newton's rings in (a) reflected light and (b) transmitted light are complementary. (From M. Cagnet, M. Francon, and J. C. Thrierr, *Atlas of Optical Phenomenon*, Plate 9, Berlin: Springer-Verlag, 1962.)

fringes by reflected light. First, the fringes show poor contrast, because the two transmitted beams with largest amplitudes have quite different values and result in incomplete cancellation. Second, the center of the fringe pattern is bright rather than dark, and the entire fringe system is complementary to the system by reflection.

Example 7-3

A plano-convex lens ($n = 1.523$) of $\frac{1}{8}$ diopter power is placed, convex surface down, on an optically flat surface as shown in Figure 7-17a. Using a traveling microscope and sodium light ($\lambda = 589.3$ nm), interference fringes are observed. Determine the radii of the first and tenth *dark rings*. ·

Solution

In this case, $\Delta_r = \lambda/2$, so that Eq. (7-38) leads to an air-film thickness at the mth dark ring given by $t_m = m\lambda/2n_f$. Since the film is air, $n_f = 1$ and $t_m = m\lambda/2$. The ring radii are given by Eq. (7-39). On neglecting the very small term in t_m^2, this is $r_m^2 = 2Rt_m$. The radius of curvature of the convex surface of the lens is found from the lensmaker's equation:

$$\frac{1}{f} = (n - 1)\left(\frac{1}{R_1} - \frac{1}{R_2}\right)$$

With $f = 8$ m, $n = 1.523$, and $R_2 \rightarrow \infty$, this gives $R = 4.184$ m. Then,

$$r_m^2 = 2Rt_m = 2R\left(\frac{m\lambda}{2}\right) = mR\lambda$$

$$r_1^2 = (1)(4.184)(589.3 \times 10^{-9})\ \text{m}^2 = 2.466 \times 10^{-6}\ \text{m}^2$$

$$r_{10}^2 = (10)(4.184)(589.3 \times 10^{-9})\ \text{m}^2 = 24.66 \times 10^{-6}\ \text{m}^2$$

or $r_1 = 1.57$ mm and $r_{10} = 4.97$ mm.

It is ironic that the phenomenon we have been describing, involving so intimately the wave nature of light, should be known as Newton's rings after one who championed the corpuscular theory of light. Probably the first measurement of the wavelength of light was made by Newton, using this technique. Consistent with his corpuscular theory, however, Newton interpreted this quantity as a measurement of the distance between the "easy fits of reflection" of light corpuscles.

7-7 FILM-THICKNESS MEASUREMENT BY INTERFERENCE

Fringes of equal thickness provide a sensitive optical means for measuring thin films. A sketch of one possible arrangement is shown in Figure 7-19. Suppose the film F to be measured has a thickness d. The film has been deposited on some substrate S. Monochromatic light is channeled from a light source LS through a fiber-optic light pipe LP to a right-angle beam-splitting prism BS, which transmits one beam to a flat mirror M and the other to the film surface. After reflection, each is transmitted by the beam splitter into a microscope MS, where they are allowed to interfere. Equivalently, the beam reflected from the mirror M can be considered to arise from its virtual image M'. The virtual mirror M' is constructed by imaging M through the beam-splitter reflecting plane. This construction makes it clear that the interference

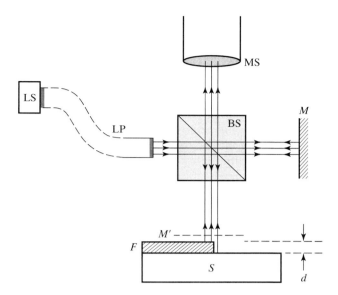

Figure 7-19 Film-thickness measurement. Interference fringes produced by light reflected from the film surface and substrate allow a determination of the film thickness d.

pattern results from interference due to the air film between the reflecting plane at M' and the film F. In practice, mirror M can be moved toward or away from the beam splitter to equalize optical-path lengths and can be tilted to make M' more or less parallel to the film surface. Furthermore, the beam splitter and mirror assembly form one unit that can be attached to the microscope in place of its objective lens. When M' and the film surface are not precisely parallel, the usual Fizeau fringes due to a wedge will be seen through the microscope, which has been prefocused on the film. The light beam striking the film is allowed to cover the edge of the film F, so that two fringe systems are seen side by side, corresponding to air films that differ by the required thickness at their juncture. Figure 7-20a shows a typical photograph of the

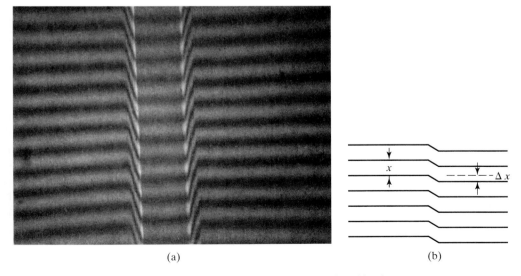

(a)   (b)

Figure 7-20   (a) Photograph of interference fringes produced by the arrangement shown in Figure 7-19. The trough-like depression evident in the interference pattern was made by evaporating the film over a thin, straight wire. (b) Sketch (not to scale) of the left side of the trough shown in the photo. The fringe pattern shifts by an amount Δx at the film edge. (Photo by J. Feldott.)

fringe systems, made through a microscope. The translation of one fringe system relative to the other provides a means of determining d, as follows.

For normal incidence, bright fringes satisfy Eq. (7-36),

$$\Delta_p + \Delta_r = 2nt + \Delta_r = m\lambda$$

where t represents the thickness of the air film at some point. If the air-film thickness now changes by an amount $\Delta t = d$, the order of interference m changes accordingly, and we have

$$2\Delta t = 2d = (\Delta m)\lambda$$

where we have set $n = 1$ for an air film. Increasing the thickness t by $\lambda/2$, for example, changes the order of any fringe by $\Delta m = 1$, that is, the fringe pattern translates by one whole fringe. For a shift of fringes of magnitude Δx (Figure 7-20b) the change in m is given by $\Delta m = \Delta x/x$, resulting in

$$d = (\Delta x/x)(\lambda/2) \tag{7-40}$$

Since both fringe spacing x and fringe shift Δx can be measured with a stable microscope—or from a photograph like that of Figure 7-20—the film thickness d is determined. When using monochromatic light, the net shift of fringe systems is ambiguous because a shift $\Delta x = 0.5x$, for example, will look exactly like a shift $\Delta x = 1.5x$. This ambiguity may be removed in one of two ways. If the shift is more than one fringe width, this situation is apparent when viewing white-light fringes, formed in the same way. The superposition of colors that form the white-light fringes creates a pattern whose center at $m = 0$ is unique, serving as an unambiguous index of fringe location. The integral shift of fringe patterns is then easily seen and can be combined with the monochromatic measurement of Δx described previously. A second method is to prepare the film so that its edge is not sharp but tails off gradually. In this case, each fringe of one set can be followed down the film edge into the corresponding fringe of the second set, as in Figure 7-20. If the film cannot be provided with a gradually tailing edge, a thin film of silver, for example, can be evaporated over both the film and substrate. The step in the metal film will usually be somewhat sloped, but the total step will be the same as the thickness of the film to be measured. A one-to-one correspondence between individual fringes of each set can then be made visually.

7-8 STOKES RELATIONS

In order to account for the multiple internal reflections in a thin film, we must develop some relations for the reflection and transmission coefficients for electric fields incident on an interface between two different media. We begin with an argument owing to Sir George Stokes, which yields information concerning the amplitudes of reflected and transmitted portions of a plane wavefront incident on a plane refracting surface, as in Figure 7-21a. Let E_i represent the amplitude of the incident light. We define reflection and transmission coefficients[2] by

$$r = \frac{E_r}{E_i}, \qquad t = \frac{E_t}{E_i} \tag{7-41}$$

---

[2]We will have occasion later to also use *reflectance* (R) and *transmittance* (T), defined as the ratio of the corresponding irradiances. Although $R = r^2$, $T \neq t^2$, as explained in Section 23-4.

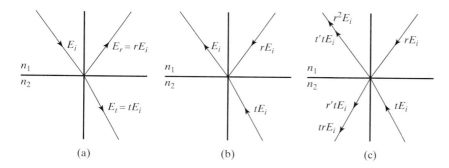

Figure 7-21   Figures used in deriving Stokes relations.

so that at the interface, E_i is divided into a reflected part, $E_r = rE_i$, and a transmitted part, $E_t = tE_i$ as shown. For a ray incident from the second medium, we define similar quantities, which we distinguish with prime notation, r' and t'. According to the principle of ray reversibility, the situation shown in Figure 7-21b must also be valid. In general, however, two rays incident at the interface, as in Figure 7-21b, each result in a reflected and a transmitted ray, all of which are shown, with appropriate amplitudes, in Figure 7-21c. We conclude that the situations depicted in Figure 7-21b and c must be physically equivalent, so that we can write

$$E_i = (r^2 + t't)E_i$$

and

$$0 = (r't + tr)E_i$$

or

$$tt' = 1 - r^2 \tag{7-42}$$
$$r = -r' \tag{7-43}$$

Equations (7-42) and (7-43) are the *Stokes relations* between amplitude coefficients for angles of incidence related through Snell's law. Equation (7-43) states that the amplitudes of reflected beams for rays incident from either direction are the same in magnitude but differ by a π phase shift. This becomes clearer if Eq. (7-43) is written in the equivalent form, $r = e^{i\pi}r'$. This result agrees with the predictions of the more complete *Fresnel equations*, treated in Chapter 23. Both the Fresnel theory and experiments, such as Lloyd's mirror, establish the fact that the phase shift occurs for the ray incident on the interface from the side of higher velocity or lower index. This wave phenomenon has its analogy in the reflection of waves from the fixed end of a rope. Both of the Stokes relations will be needed in the discussion that follows.

7-9 MULTIPLE-BEAM INTERFERENCE IN A PARALLEL PLATE

We return now to the problem of reflections from a thin film, already considered in a two-beam approximation in Section 7-4. For concreteness, we consider the case of a parallel plate of thickness t and index of refraction n_f surrounded by air on both sides. Consider the multiple reflections of the narrow beam of light of amplitude E_0 and angle of incidence θ_i, as shown in Figure 7-22. The reflection and transmission amplitude coefficients are r and t at an external reflection and r' and t' at an internal reflection. The amplitude of each segment of the beam can be assigned by multiplying the previous amplitude by the appropriate reflection or transmission coefficient, beginning with the incident wave of amplitude E_0 and working progressively through

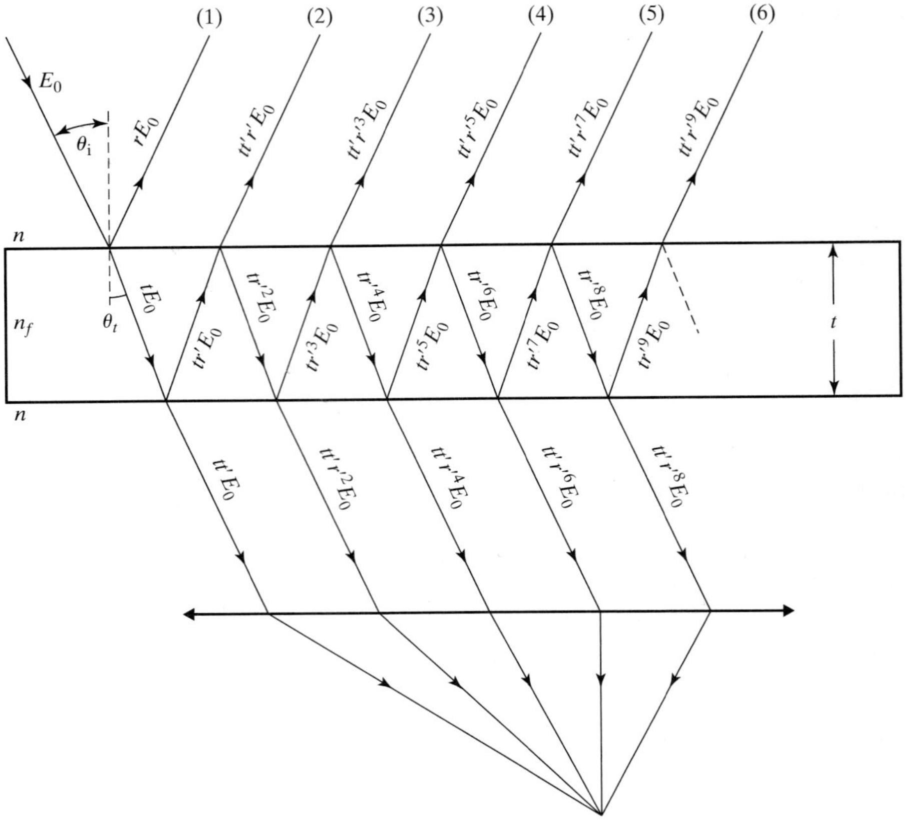

Figure 7-22 Multiply reflected and transmitted beams in a parallel plate.

the train of reflections. Multiple parallel beams emerge from the top and from the bottom of the plate. Multiple-beam interference takes place when either set is focused to a point by a converging lens, as shown for the transmitted beam. Having originated from a single beam, the multiple beams are coherent. Further, if the incident beam is near normal, the beams are brought together with their E vibrations nearly parallel.

We consider the superposition of the reflected beams from the top of the plate. According to Eq. (7-35), the phase difference between successive reflected beams is given by

$$\delta = k\Delta, \quad \text{where } \Delta = 2n_f t \cos \theta_t \qquad (7\text{-}44)$$

If the incident ray is expressed as $E_0 e^{i\omega t}$, the successive reflected rays can be expressed by appropriately modifying both the amplitude and phase of the initial wave. Referring to Figure 7-22, these are

$$E_1 = (rE_0)e^{i\omega t}$$
$$E_2 = (tt'r'E_0)e^{i(\omega t - \delta)}$$
$$E_3 = (tt'r'^3 E_0)e^{i(\omega t - 2\delta)}$$
$$E_4 = (tt'r'^5 E_0)e^{i(\omega t - 3\delta)}$$

and so on. A little inspection of these equations shows that the Nth such reflected wave can be written as

$$E_N = (tt'r'^{(2N-3)}E_0)e^{i[\omega t - (N-1)\delta]} \qquad (7\text{-}45)$$

a form that holds good for all but E_1, which never traverses the plate. When these waves are superposed, therefore, the resultant E_R may be written as

$$E_R = \sum_{N=1}^{\infty} E_N = rE_0e^{i\omega t} + \sum_{N=2}^{\infty} tt'E_0r'^{(2N-3)}e^{i[\omega t - (N-1)\delta]}$$

Factoring a bit, we have

$$E_R = E_0e^{i\omega t}\left[r + tt'r'e^{-i\delta}\sum_{N=2}^{\infty}r'^{(2N-4)}e^{-i(N-2)\delta}\right]$$

The summation is now in the form of a geometric series,

$$\sum_{N=2}^{\infty} x^{N-2} = 1 + x + x^2 + \cdots$$

where

$$x = r'^2e^{-i\delta}$$

Since $|x| < 1$, the series converges to the sum $S = 1/(1 - x)$. Thus,

$$E_R = E_0e^{i\omega t}\left(r + \frac{tt'r'e^{-i\delta}}{1 - r'^2e^{-i\delta}}\right)$$

Making use next of the Stokes relations, Eqs. (7-42) and (7-43),

$$E_R = E_0e^{i\omega t}\left[r - \frac{(1 - r^2)re^{-i\delta}}{1 - r^2e^{-i\delta}}\right]$$

After simplifying,

$$E_R = E_0e^{i\omega t}\left[\frac{r(1 - e^{-i\delta})}{1 - r^2e^{-i\delta}}\right]$$

The irradiance, I_R, of the resultant beam is proportional to the square of the amplitude, E_R, which is itself complex, so we calculate $|E_R|^2 = E_R E_R^*$, or

$$|E_R|^2 = E_0^2r^2\left[\frac{e^{i\omega t}(1 - e^{-i\delta})}{1 - r^2e^{-i\delta}}\right]\left[\frac{e^{-i\omega t}(1 - e^{i\delta})}{1 - r^2e^{i\delta}}\right]$$

After processing the product of the bracketed terms and making use of the identity,

$$2\cos\delta \equiv (e^{i\delta} + e^{-i\delta})$$

there results

$$|E_R|^2 = E_0^2(2r^2) \cdot \left(\frac{1 - \cos\delta}{1 + r^4 - 2r^2\cos\delta}\right) \qquad (7\text{-}46)$$

or, in terms of irradiance,

$$I_R = \left[\frac{2r^2(1 - \cos\delta)}{1 + r^4 - 2r^2\cos\delta}\right]I_i \qquad (7\text{-}47)$$

where I_i represents the irradiance of the incident beam, and we have used the proportionality

$$\frac{I_R}{I_i} = \frac{|E_R|^2}{|E_0|^2} \qquad (7\text{-}48)$$

A similar treatment of the transmitted beams leads to the resultant transmitted irradiance,

$$I_T = \left[\frac{(1 - r^2)^2}{1 + r^4 - 2r^2 \cos \delta}\right]I_i \tag{7-49}$$

Equation (7-49) also follows more directly by combining Eq. (7-47) with the relation $I_R + I_T = I_i$, required by the conservation of energy for nonabsorbing films.

A minimum in reflected irradiance occurs, according to Eq. (7-47), when $\cos \delta = 1$, or when

$$\delta = 2\pi m \quad \text{and} \quad \Delta = 2n_f t \cos \theta_t = m\lambda \tag{7-50}$$

Necessarily, this must also be the condition for a transmission maximum. Equation (7-49) gives $I_T = I_i$, as expected. A study of Figure 7-22, or the equations describing the set of reflected beams, shows that in the case of a reflection minimum, the second reflected beam and all subsequent beams are in phase with one another but exactly out of phase with the first reflected beam. Since the net reflected irradiance vanishes, there is a perfect cancellation of the first beam with the sum of all the remaining beams. The two-beam approximation works well, then, if the amplitude of the second beam is close to the amplitude of the first beam. Our equations show that their ratio is

$$\left|\frac{E_2}{E_1}\right| = \left|\frac{tt'r'E_0}{rE_0}\right| = 1 - r^2$$

which is close to unity when r^2 is small. For normal incidence on glass of index $n = 1.5$, $r^2 = 0.04$. Thus, 96% of the cancellation occurs between the first two reflected beams alone, and the two-beam treatment is well justified.

Reflection maxima occur, in the other extreme, when $\cos \delta = -1$, or when

$$\delta = \pi, 3\pi, \ldots = \left(m + \tfrac{1}{2}\right)2\pi$$

and

$$\Delta = 2n_f t \cos \theta_t = \left(m + \tfrac{1}{2}\right)\lambda \tag{7-51}$$

In this case, Eqs. (7-47) and (7-49) yield

$$I_R = \left[\frac{4r^2}{(1 + r^2)^2}\right]I_i \tag{7-52}$$

$$I_T = \left[\frac{(1 - r^2)}{(1 + r^2)}\right]^2 I_i \tag{7-53}$$

It is easily verified that $I_R + I_T = I_i$. Also, note that the denominator of the expression on the right-hand side of Eq. (7-49) is smallest when $\cos \delta = -1$, so that this is the condition for a transmission minimum. Therefore, Eq. (7-52) does indeed give the maximum reflected intensity. Parallel plates, such as the one studied here, can be used as Fabry-Perot interferometers. This application is treated in much more detail in Chapter 8.

PROBLEMS

7-1 Two mutually coherent beams having parallel electric fields are described by

$$E_1 = 3 \cos\left(ks_1 - \omega t + \frac{\pi}{5} \right)$$

$$E_2 = 4 \cos\left(ks_2 - \omega t + \frac{\pi}{6} \right)$$

with amplitudes in kV/m. The beams interfere at a point P where the phase difference due to path is $\pi/3$ (the first beam having the longer path). At the point of superposition, calculate (a) the irradiances I_1 and I_2 of the individual beams; (b) the irradiance I_{12} due to their interference; (c) the net irradiance; (d) the fringe visibility.

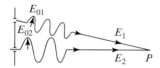

Figure 7-23   Problem 7-1.

7-2 Two harmonic light waves with amplitudes of 1.6 and 2.8 interfere at some point P on a screen. What visibility results there if (a) their electric field vectors are parallel and (b) if they are perpendicular?

7-3 The ratio of the amplitudes of two beams forming an interference fringe pattern is 2/1. What is the visibility? What ratio of amplitudes produces a visibility of 0.5?

7-4 a. Show that if one beam of a two-beam interference setup has an irradiance of N times that of the other beam, the fringe visibility is given by

$$V = \frac{2\sqrt{N}}{N + 1}$$

b. Determine the beam irradiance ratios for visibilities of 0.96, 0.9, 0.8, and 0.5.

7-5 A mercury source of light is positioned behind a glass filter, which allows transmission of the 546.1-nm green light from the source. The light is allowed to pass through a narrow, horizontal slit positioned 1 mm above a flat mirror surface. Describe both qualitatively and quantitatively what appears on a screen 1 m away from the slit.

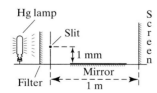

Figure 7-24   Problem 7-5.

7-6 Two slits are illuminated by light that consists of two wavelengths. One wavelength is known to be 436 nm. On a screen, the fourth minimum of the 436-nm light coincides with the third maximum of the other light. What is the wavelength of the other light?

7-7 In a Young's experiment, narrow double slits 0.2 mm apart diffract monochromatic light onto a screen 1.5 m away. The distance between the fifth minima on either side of the zeroth-order maximum is measured to be 34.73 mm. Determine the wavelength of the light.

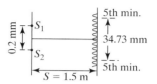

Figure 7-25   Problem 7-7.

7-8 A quasi-monochromatic beam of light illuminates Young's double-slit setup, generating a fringe pattern having a 5.6-mm separation between consecutive dark bands. The distance between the plane containing the apertures and the plane of observation is 7 m, and the two slits are separated by 1.0 mm. Sketch the experimental arrangement. Why is an initial single slit necessary? What is the wavelength of the light?

7-9 In an interference experiment of the Young type, the distance between slits is 0.5 mm, and the wavelength of the light is 600 nm.

a. If it is desired to have a fringe spacing of 1 mm at the screen, what is the proper screen distance?

b. If a thin plate of glass ($n = 1.50$) of thickness 100 microns is placed over one of the slits, what is the lateral fringe displacement at the screen?

c. What path difference corresponds to a shift in the fringe pattern from a peak maximum to the (same) peak half-maximum?

7-10 White light (400 to 700 nm) is used to illuminate a double slit with a spacing of 1.25 mm. An interference pattern falls on a screen 1.5 m away. A pinhole in the screen allows some light to enter a spectrograph of high resolution. If the pinhole in the screen is 3 mm from the central white fringe, where would one expect dark lines to show up in the spectrum of the pinhole source?

7-11 Sodium light (589.3 nm) from a narrow slit illuminates a Fresnel biprism made of glass of index 1.50. The biprism is twice as far from a screen on which fringes are observed as it is from the slit. The fringes are observed to be separated by 0.03 cm. What is the biprism angle α?

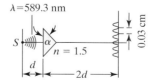

Figure 7-26   Problem 7-11.

7-12 The small angle θ between two plane, adjacent reflecting surfaces is determined by examining the interference fringes produced in a Fresnel mirror experiment. A source slit is parallel to the intersection between the mirrors and 50 cm away. The screen is 1 m from the same intersection, measured along the normal to the screen. When illuminated with sodium light (589.3 nm), fringes appear on the screen with a spacing of 0.5 mm. What is the angle θ?

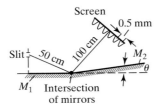

Figure 7-27   Problem 7-12.

7-13 The prism angle of a very thin prism is measured by observing interference fringes as in the Fresnel biprism technique. The distances from slit to prism and from prism to eye are in the ratio of 1:4. Twenty dark fringes are found to span a distance of 0.5 cm when green mercury light is used. If the refractive index of the prism is 1.50, determine the prism angle.

7-14 Light of continuously variable wavelength illuminates normally a thin oil (index of 1.30) film on a glass surface. Extinction of the reflected light is observed to occur at wavelengths of 525 and 675 nm in the visible spectrum. Determine the thickness of the oil film and the orders of the interference.

7-15 A thin film of MgF_2 ($n = 1.38$) is deposited on glass so that it is antireflecting at a wavelength of 580 nm under normal incidence. What wavelength is minimally reflected when the light is incident instead at 45°?

7-16 A nonreflecting, single layer of a lens coating is to be deposited on a lens of refractive index $n = 1.78$. Determine the refractive index of a coating material and the thickness required to produce zero reflection for light of wavelength 550 nm.

7-17 Remember that the energy of a light beam is proportional to the square of its amplitude.

a. Determine the percentage of light energy reflected in air from a single surface separating a material of index 1.40 for light of $\lambda = 500$ nm.

b. When deposited on glass of index 1.60, how thick should a film of this material be in order to reduce the reflected energy by destructive interference?

c. What is then the effective percent reflection from the film layer?

7-18 A soap film is formed using a rectangular wire frame and held in a vertical plane. When illuminated normally by laser light at 632.8 nm, one sees a series of localized interference fringes that measure 15 per cm. Explain their formation.

7-19 A beam of white light (a continuous spectrum from 400 to 700 nm, let us say) is incident at an angle of 45° on two parallel glass plates separated by an air film 0.001 cm thick. The reflected light is admitted into a prism spectroscope. How many dark "lines" are seen across the entire spectrum?

7-20 Two microscope slides are placed together but held apart at one end by a thin piece of tin foil. Under sodium light (589 nm) normally incident on the air film formed between the slides, one observes exactly 40 bright fringes from the edges in contact to the edge of the tin foil. Determine the thickness of the foil.

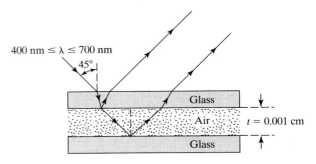

Figure 7-28   Problem 7-19.

7-21 Plane plates of glass are in contact along one side and held apart by a wire 0.05 mm in diameter, parallel to the edge in contact and 20 cm distant. Using filtered green mercury light ($\lambda = 546$ nm), directed normally on the air film between plates, interference fringes are seen. Calculate the separation of the dark fringes. How many dark fringes appear between the edge and the wire?

7-22 Show that the separation of the virtual sources I_1 and I_2 producing interference from a film of index n and uniform thickness t, when illuminated by a point source, is $2t/n$. Assume the film is in air and light is incident at near-normal incidence.

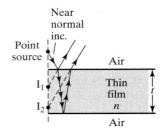

Figure 7-29   Problem 7-22.

7-23 Newton's rings are formed between a spherical lens surface and an optical flat. If the tenth bright ring of green light (546.1 nm) is 7.89 mm in diameter, what is the radius of curvature of the lens surface?

7-24 Newton's rings are viewed both with the space between lens and optical flat empty and filled with a liquid. Show that the ratio of the radii observed for a particular order fringe is very nearly the square root of the liquid's refractive index.

7-25 A Newton's ring apparatus is illuminated by light with two wavelength components. One of the wavelengths is 546 nm. If the eleventh bright ring of the 546-nm fringe system coincides with the tenth ring of the other, what is the second wavelength? What is the radius at which overlap takes place and the thickness of the air film there? The spherical surface has a radius of 1 m.

7-26 A fringe pattern, such as that in Figure 7-20, found using an interference microscope objective, is observed to have a regular spacing of 1 mm. At a certain point in the pattern, the fringes are observed to shift laterally by 3.4 mm. If the illumination is green light of 546.1 nm, what is the dimension of the "step" in the film that caused the shift?

7-27 A laser beam from a 1-mW He-Ne laser (632.8 nm) is directed onto a parallel film with an incident angle of 45°. Assume a beam diameter of 1 mm and a film index of 1.414. Determine (a) the amplitude of the E-vector of the incident beam; (b) the angle of refraction of the laser beam into the film; (c) the magnitudes of r' and tt', using the Stokes relations and a reflection coefficient, $r = 0.280$; (d) the independent amplitudes of the first three reflected beams and, by comparison with the incident beam, the percentage of radiant power density reflected in each; (e) the same information as in (d) for the first two transmitted beams; (f) the minimum thickness of film that would lead to total cancellation of the reflected beams when they are brought together at a point by a lens.

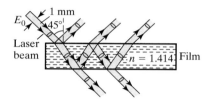

Figure 7-30   Problem 7-27.

7-28 a. Using Eq. (7-27) and the stokes relations, show that amplitudes of the first three reflected and first three transmitted beams from a parallel, nonabsorbing glass ($n = 1.52$) plate, when the incident beam is near normal and of unit amplitude, are given by

	(1)	(2)	(3)
reflected	0.206	0.198	0.0084
transmitted	0.957	0.041	0.0017

b. Show as a result that the first two reflected rays produce a visibility of 0.999, whereas the first two transmitted rays produce a visibility of only 0.085.

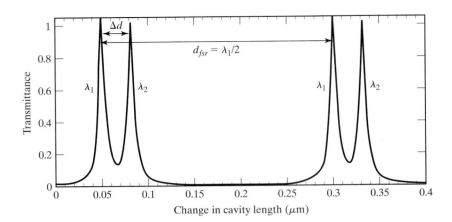

8

Optical Interferometry

INTRODUCTION

An instrument designed to exploit the interference of light and the fringe patterns that result from optical-path differences, in any of a variety of ways, is called an *optical interferometer*. This general description of the instrument reflects the wide variety of designs and uses of interferometers. Applications extend also to acoustic and radio waves, but here we are interested in the optical interferometer. In this chapter we discuss chiefly the Michelson and the Fabry-Perot interferometers and suggest only a few of their many applications.

To achieve interference between two coherent beams of light, an interferometer divides an initial beam into two or more parts that travel diverse optical paths and then reunite to produce an interference pattern. One criterion for broadly classifying interferometers distinguishes the manner in which the initial beam is separated. *Wavefront-division interferometers* sample different portions of the same wavefront of a coherent beam of light, as in the case of Young's double slit, or adaptations like those using Lloyd's mirror or Fresnel's biprism. *Amplitude-division interferometers* instead use some type of *beam splitter* that divides the initial beam into two parts. The Michelson interferometer is of this type. Usually the beam splitting is managed by a semireflecting metallic or dielectric film; it can also occur by frustrated total internal reflection at the interface of two prisms forming a cube, or by means of double refraction or diffraction. Another means of classification distinguishes between those interferometers that make use of the interference of two beams, as in the case of the Michelson interferometer, and those that operate with multiple beams, as in the Fabry-Perot interferometer.

8-1 THE MICHELSON INTERFEROMETER

The Michelson interferometer, first introduced by Albert Michelson in 1881, has played a vital role in the development of modern physics. This simple and versatile instrument was used, for example, to establish experimental evidence for the validity of the special theory of relativity, to detect and measure hyperfine structure in line spectra, to measure the tidal effect of the moon on the earth, and to provide a substitute standard for the meter in terms of wavelengths of light. Michelson himself pioneered much of this work.

A schematic of the Michelson interferometer is shown in Figure 8-1a. From an extended source of light S, beam 1 of light is split by a beam splitter (BS) by means of a thin, semitransparent front surface metallic or dielectric film, deposited on glass. The interferometer is therefore of the amplitude-splitting type. Reflected beam 2 and transmitted beam 3, of roughly equal amplitudes, continue to fully-reflecting mirrors $M2$ and $M1$, respectively, where their directions are reversed. On returning to the beam splitter, beam 2 is now transmitted and beam 3 is reflected by the semitransparent film so that they come together again and leave the interferometer as beam 4. The useful aperture of this double-beam interferometer is such that all rays striking $M1$ and $M2$ will be normal, or nearly so. Thus, beam 4 includes rays that have traveled different optical paths and will demonstrate interference. At least one of the mirrors is equipped with tilting adjustment screws that allow the surface of $M1$ to be made perpendicular to that of $M2$. One of the mirrors is also movable along the direction of the beam by means of an accurate track and micrometer screw. In this way, the difference between the optical paths of

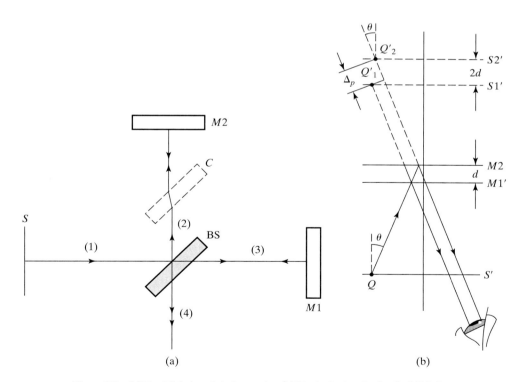

(a)   (b)

Figure 8-1   (a) The Michelson interferometer. (b) Equivalent optics for the Michelson interferometer.

beams 2 and 3 can be gradually varied. Notice that beam 3 traverses the beam splitter three times, whereas beam 2 traverses it only once. In some applications, where white light is used, it is essential that the optical paths of the two beams be made precisely equal. Although this can be accomplished at one wavelength by appropriately increasing the distance of $M2$ from BS, the correction would not suffice at another wavelength because of the dispersion of the glass. To compensate for all wavelengths at once, a compensator plate C made of the same material and dimensions as BS is inserted parallel to BS in the path of beam 2. Any small, remaining inequalities in optical paths can be removed by allowing the compensator to rotate, thus varying the optical path through the thickness of its glass plate.

The actual interferometer in Figure 8-1a possesses two optical axes at right angles to one another. A simpler but equivalent optical system, having a single optical axis, can be drawn by working with virtual images of source S and mirror $M1$ via reflection in the BS mirror. These positions are most simply found by regarding the assembly including S, $M1$, and beams 1 and 3 of Figure 8-1a as rotated counterclockwise by 90° about the point of intersection of the beams with the BS mirror. The resulting geometry is shown in Figure 8-1b. The new position of the source plane is S', and the new position of the mirror $M1$ is $M1'$. Light from a point Q on the source plane S' then effectively reflects from both mirrors $M2$ and $M1'$, shown parallel and with an optical path difference of d. The two reflected beams appear to come from the two virtual images, Q_1' and Q_2', of object point Q. Since the images $S1'$ and $S2'$ of the source plane in the mirrors must be separated by twice the mirror separation, the distance between Q_1' and Q_2' is $2d$, and the optical-path difference Δ_p between the two beams emerging from the interferometer is

$$\Delta_p = 2d \cos \theta \qquad (8\text{-}1)$$

where the angle θ measures the inclination of the beams relative to the optical axis. For a normal beam, $\theta = 0$ and $\Delta_p = 2d$. We expect this result, since, if one mirror is farther from BS than the other by a distance d, the extra distance traversed by the beam taking the longer route includes distance d twice, once before and once after reflection. If, in addition, $\Delta = m\lambda$, so that the two beams interfere constructively, it follows that they will do so repeatedly for every $\lambda/2$ translation of one of the mirrors so long as the separation Δ_p does not exceed the so-called *coherence length*, l_t, of the source. The coherence length is the length along a wave train over which the phase of the wave remains correlated. We show in Chapter 9 that the coherence length of the light emitted by a particular source is the ratio of the speed of light c to the spread of frequencies $\Delta \nu$ present in the source. The coherence length of typical laser sources ranges from tens of centimeters to tens of kilometers, as we mentioned in Chapter 7. Throughout this chapter—as we did in Chapter 7—we will assume that all effective path-length differences between interfering beams that originate from the same source are much less than the coherence length of the source.

The optical system of Figure 8-1b is now equivalent to the case of interference due to a plane, parallel air film, illuminated by an extended source. Virtual fringes of equal inclination may be seen by looking into the beam splitter along ray 4, with the eye or a telescope focused at infinity. Assuming that the two interfering beams are of equal amplitude, the irradiance of the fringe system of circles concentric with the optical axis is given, as in Eq. (7-18), by

$$I = 4I_0 \cos^2 \left(\frac{\delta}{2} \right) \qquad (8\text{-}2)$$

where the phase difference is

$$\delta = k\Delta = \left(\frac{2\pi}{\lambda}\right)\Delta \tag{8-3}$$

The net optical path difference is $\Delta = \Delta_p + \Delta_r$, as usual. A relative π phase shift between the two beams occurs because the reflection coefficients from opposite sides of a beam splitter differ by $-1 = e^{i\pi}$.[1] For dark fringes, then,

$$\Delta_p + \Delta_r = 2d\cos\theta + \frac{\lambda}{2} = \left(m + \frac{1}{2}\right)\lambda$$

or, more simply,

$$2d\cos\theta = m\lambda \qquad m = 0, 1, 2, \ldots \text{dark fringes} \tag{8-4}$$

If d is of such magnitude that the normal rays forming the center of the fringe system satisfy Eq. (8-4), that is, the center fringe is dark, then its order, given by

$$m_{\text{max}} = \frac{2d}{\lambda} \tag{8-5}$$

is a large integer. Neighboring dark fringes decrease in order outwards from the center of the pattern, as $\cos\theta$ decreases from its maximum value of 1. This ordering of fringes may be inverted for convenience by associating another integer p with each fringe of order m, where

$$p = m_{\text{max}} - m = \frac{2d}{\lambda} - m \tag{8-6}$$

Using Eq. (8-6) to replace m in Eq. (8-4), we arrive at

$$p\lambda = 2d(1 - \cos\theta) \qquad p = 0, 1, 2, \ldots \text{dark fringes} \tag{8-7}$$

where now the central fringe is of order zero and the neighboring fringes increase in order, outward from the center. Figure 8-2 illustrates the relationship between orders m and p for the arbitrary case where $m_{\text{max}} = 100$. Equation (8-4) or (8-7) indicates that, as d is varied, a particular point in the fringe pattern ($\theta = $ constant) will correspond to gradually changing values of order m or p.

Integral values occur whenever the point coincides with a dark fringe. Equivalently, this means that as d is varied, fringes of the pattern appear to move inward toward the center, where they disappear, or else move outward from the center, where they seem to originate, depending on whether the optical-path difference is decreasing or increasing. The motion of the fringe pattern thus reverses as one of the mirrors is moved continually through the point of zero path difference. Viewed in another way, Eq. (8-4) requires an increase in the angular separation $\Delta\theta$ of a given small fringe interval Δm as the mirror spacing d becomes smaller, since taking the differential of Eq. (8-4) leads to

$$|\Delta\theta| = \frac{\lambda\Delta m}{2d\sin\theta}$$

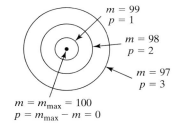

Figure 8-2   Alternate orderings of fringes.

---

[1]This conclusion assumes real beam-splitter transmission coefficients. The assumption is made to allow discussion of the fringe pattern in a concrete (and common) situation. It does not affect the validity of results like Eq. (8-8), for example, because measurements depend on the net *motion* of the fringe pattern, not on precisely where it is dark and where it is bright.

This means that the fringes are more widely separated when optical-path differences are small. In fact, if $d = \lambda/2$, then from Eq. (8-4), $m = \cos\theta$, and the entire field of view encompasses no more than one fringe! For a mirror translation Δd, the number Δm of fringes passing a point at or near the center of the pattern is, according to Eq. (8-4),

$$\Delta m = \frac{2\Delta d}{\lambda} \qquad (8\text{-}8)$$

Equation (8-8) suggests an experimental way of either measuring λ when Δd is known or calibrating the micrometer translation screw when λ is known.

Example 8-1

Fringes are observed due to monochromatic light in a Michelson interferometer. When the movable mirror is translated by 0.073 mm, a shift of 300 fringes is observed. What is the wavelength of the light? What displacement of the fringe system takes place when a flake of glass of index 1.51 and 0.005 mm thickness is placed in one arm of the interferometer? (Assume that the light beam is normal to the glass surface.)

Solution

Using Eq. (8-8),

$$\lambda = \frac{2\Delta d}{\Delta m} = \frac{(2)(0.073)}{300} = 4.87 \times 10^{-4} \text{ mm} = 487 \text{ nm}$$

With the glass inserted, one arm is effectively lengthened by a path difference of $\Delta d = n_g t - n_{air} t$, so that

$$\Delta m = \frac{2\Delta d}{\lambda} = \frac{2(n_g - 1)t}{\lambda} = \frac{2(0.51)(0.005 \times 10^{-3})}{487 \times 10^{-9}} = 10.5 \text{ fringes}$$

8-2 APPLICATIONS OF THE MICHELSON INTERFEROMETER

The Michelson interferometer is easily adaptable to the measurement of thin films, a technique essentially the same as that described in the preceding chapter. It can also be adapted to measure the index of refraction of a gas. An evacuable cell with plane, parallel windows is interposed in the path of beam 3 (Figure 8-1a) and is filled with a gas at a pressure and temperature for which its index of refraction is desired. The fringe system established under these conditions is monitored as the gas is gradually pumped out of the cell. A count Δm of the net fringe shift is related to the change in optical path during the decrease of the gas pressure. If the actual length of the cell is accurately known to be L, the change in optical path is given by

$$\Delta d = nL - L = L(n - 1) \qquad (8\text{-}9)$$

and using Eq. (8-8), it follows that the index can be determined from

$$n - 1 = \left(\frac{\lambda}{2L}\right)\Delta m \qquad (8\text{-}10)$$

Consider another direct application of the Michelson interferometer, the determination of wavelength difference between two closely spaced components of a spectral "line," λ and λ'. Each wavelength forms its own system of circular fringes according to Eq. (8-4). Suppose we view the circular systems near their center, so that $\cos \theta \cong 1$. Then for a given path difference d of the interferometer, the product $m\lambda$ is fixed, that is, $m\lambda = m'\lambda'$. When the fringe systems coincide, the pattern appears sharp, whereas when the fringes of one system in the region of observation lie midway between the fringes of the second system, the pattern appears rather uniform in brightness, or "washed out." The mirror movement Δd required between consecutive coincidences is related to the wavelength difference $\Delta \lambda$ as follows. At one coincidence, when fringes are "in step," the orders of the two systems corresponding to λ and λ' must be related by

$$m = m' + N$$

where N is an integer. If the optical-path difference at this time is d_1, then from Eq. (8-4),

$$\frac{2d_1}{\lambda} = \frac{2d_1}{\lambda'} + N \tag{8-11}$$

Let the optical-path difference be increased to d_2, when the next coincidence is found. Then,

$$m = m' + (N + 1)$$

or

$$\frac{2d_2}{\lambda} = \frac{2d_2}{\lambda'} + N + 1 \tag{8-12}$$

By subtracting Eq. (8-11) from Eq. (8-12) and by writing the mirror movement $\Delta d = d_2 - d_1$, we find

$$\lambda' - \lambda = \frac{\lambda \lambda'}{2 \Delta d} \tag{8-13}$$

Now since λ and λ' are very close, the wavelength difference of the two unresolved components can be approximated by

$$\Delta \lambda = \frac{\lambda^2}{2 \Delta d} \tag{8-14}$$

This technique is often employed in an optics laboratory course to measure the wavelength difference of 6 Å between the two components of the yellow "line" of sodium.

All the preceding discussion of the fringes from a Michelson interferometer has been in terms of virtual fringes of equal inclination. We have assumed that mirrors $M1$ and $M2$ are precisely perpendicular as shown in Figure 8-1a, or, what amounts to the same thing, precisely parallel in the equivalent optical system of Figure 8-1b. If the alignment is such that the air space between $M1'$ and $M2$ in Figure 8-1b is a wedge, fringes of equal thickness may be seen localized at the mirrors. These fringes will be straight, oriented parallel to the line that represents the intersection of $M1'$ and $M2$. If the wedge is of large angle, they will be curved in a way that can be shown to be hyperbolic arcs. Again, if the source is small, then real, nonlocalized fringes appear in the light emerging from the interferometer, as if formed by the two virtual images of the source in $M1'$ and $M2$. These fringes appear without effort when the intense, coherent light of a laser is used. These possibilities have already been discussed in the previous chapter, where we treated the various interference fringes that can be formed by illumination of a film. Figure 8-3 is a photograph showing the distortion of fringes of equal thickness produced by a candle flame when situated in one arm of a Michelson interferometer. Variations in temperature produce variations in optical-path length by changing the refractive index of the air.

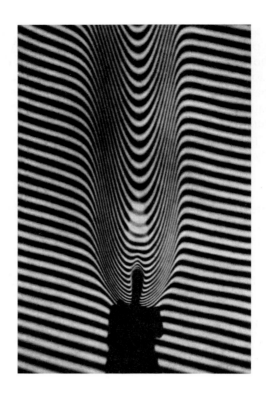

Figure 8-3 Deformation of fringes of equal thickness in the neighborhood of a candle flame. (From M. Cagnet, M. Francon, and J. C. Thrierr, *Atlas of Optical Phenomenon*, Plate 12, Berlin: Springer-Verlag, 1962.)

8-3 VARIATIONS OF THE MICHELSON INTERFEROMETER

Although there are many ways in which a beam of light may be split into two parts and reunited after traversing different paths, we examine briefly two variations that can be considered adaptations of the Michelson interferometer.

Twyman-Green Interferometer

A slight modification by Twyman and Green is shown in Figure 8-4a. Instead of using an extended source, this interferometer uses a point source together with a collimating lens $L1$, so that all rays enter the interferometer parallel to the optical axis, or $\cos \theta = 1$. The parallel rays emerging from the interferometer are brought to a focus by lens $L2$ at P, where the eye is placed. The circular fringes of equal inclination no longer appear; in their place are seen fringes of equal thickness. These fringes reveal imperfections in the optical system that cause variations in optical-path length. When no distortions appear in the plane wavefronts through the interferometer, uniform illumination is seen near P. If the interferometer components are of high quality, this system can

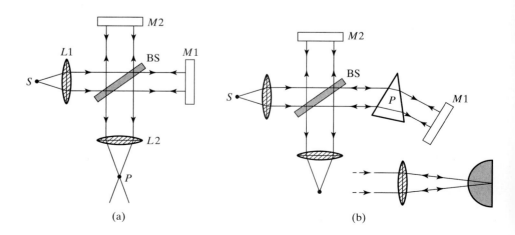

Figure 8-4 (a) Twyman-Green interferometer. (b) Twyman-Green interferometer used in the testing of a prism and a lens (inset).

(a)   (b)

be used to test the optical quality of another optical component, such as a prism, situated as shown in Figure 8-4b. Surface imperfections or internal variations in refractive index show up as a distortion of the fringe pattern. Lenses are tested for aberrations in the same way, once plane mirror $M1$ is replaced by a convex spherical surface that can reflect the refracted rays back along themselves, as suggested in the inset of Figure 8-4b.

Mach-Zehnder Interferometer

A more radical variation, sketched in Figure 8-5, is the Mach-Zehnder interferometer. The incident beam of roughly collimated light is divided into two beams at beam splitter BS. Each beam is again totally reflected by mirrors $M1$ and $M2$, and the beams are made coincident again by the semitransparent mirror $M3$. Path lengths of beams 1 and 2 around the rectangular system and through the glass of the beam splitters are identical. This interferometer has been used, for example, in aerodynamic research, where the geometry of air flow around an object in a wind tunnel is revealed through local variations of pressure and refractive index. A windowed test chamber, into which the model and a streamlined flow of air is introduced, is placed in path 1. An identical chamber is placed in path 2 to maintain equality of optical paths. The air-flow pattern is revealed by the fringe pattern. For such applications the interferometer must be constructed on a rather large scale. An advantage of the Mach-Zehnder over the Michelson interferometer is that, by appropriate small rotations of the mirrors, the fringes may be made to appear at the object being tested, so that both can be viewed or photographed together. In the Michelson interferometer, fringes appear localized on the mirror and so cannot be seen in sharp focus at the same time as a test object placed in one of its arms.

The Michelson, Twyman-Green, and Mach-Zehnder interferometers are all two-beam interference instruments that operate by division of amplitude. We turn now to an important case of a multiple-beam instrument, the Fabry-Perot interferometer.

8-4 THE FABRY-PEROT INTERFEROMETER

The Fabry-Perot interferometer makes use of an arrangement similar to the plane parallel plate, discussed in Section 7-9, to produce an interference pattern that results from the superposition of the multiple beams of the transmitted light. This instrument, probably the most adaptable of all interferometers, has been used, for example, in precision wavelength measurements, analysis of hyperfine spectral line structure, determination of refractive indices of gasses, and the calibration of the standard meter in terms of wavelengths. Although simple in structure, it is a high-resolution instrument that has proven to be a powerful tool in a wide variety of applications.

A possible arrangement is shown in Figure 8-6. Two thick glass or quartz plates are used to enclose a plane parallel "plate" of air between them, which forms the medium within which the beams are multiply reflected. The glass plates function as mirrors and the arrangement is often called

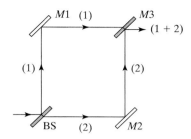

Figure 8-5   Mach-Zehnder interferometer.

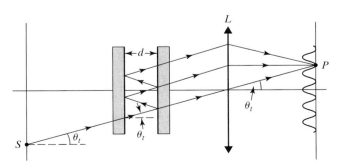

Figure 8-6   Fabry-Perot interferometer.

a *cavity*. The important surfaces of the glass plates are therefore the inner ones. Their surfaces are generally polished to a flatness of better than $\lambda/50$ and coated with a highly reflective layer of silver or aluminum. Silver films are most useful in the visible region of the spectrum, but their reflectivity drops off sharply around 400 nm, so that for applications below 400 nm, aluminum is usually used. Of course, the films must be thin enough to be partially transmitting. Optimum thicknesses for silver coatings are around 50 nm. The outer surfaces of the glass plate are purposely formed at a small angle relative to the inner faces (several minutes of arc are sufficient) to eliminate spurious fringe patterns that can arise from the glass itself acting as a parallel plate. The spacing, or thickness, d of the air layer, is an important performance parameter of the interferometer, as we shall see. When the spacing is fixed, the instrument is often referred to as an *etalon*.

Consider a narrow, monochromatic beam from an extended source point S making an angle (in air) of θ_t with respect to the optical axis of the system, as in Figure 8-6. The single beam produces multiple coherent beams in the interferometer, and the emerging set of parallel rays are brought together at a point P in the focal plane of the converging lens L. The nature of the superposition at P is determined by the path difference between successive parallel beams, $\Delta = 2n_f d \cos \theta_t$. Using $n_f = 1$ for air, the condition for brightness is

$$2d \cos \theta_t = m\lambda \tag{8-15}$$

Other beams from different points of the source but in the same plane and making the same angle θ_t with the axis satisfy the same path difference and also arrive at P. With d fixed, Eq. (8-15) is satisfied for certain angles θ_t, and the fringe system is the familiar concentric rings due to the focusing of fringes of equal inclination. When a collimating lens is used between source and interferometer, as shown in Figure 8-7a, every set of parallel beams entering the etalon must arise from the same source point. A one-to-one correspondence then exists between source and screen points. The screen may be the retina or a photographic plate. Figure 8-7b illustrates another arrangement, in which the source is small. Collimated light in this instance reaches the plates at a fixed angle θ_t ($\theta_t = 0$ is shown) and comes to a focus at a light detector. As the spacing d is varied, the detector records the interference pattern as a function of time in an *interferogram*. If, for example, the source light consists of two wavelength components, the output of the two systems is either a double set of circular fringes on a photographic plate or a plot of resultant irradiance

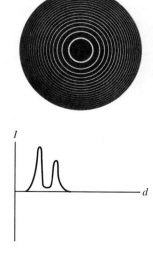

(a)

(b)

Figure 8-7 (a) Fabry-Perot interferometer, used with an extended source and a fixed plate spacing. A circular fringe pattern like the one shown may be photographed at the screen. (Photo from M. Cagnet, M. Francon, and J. C. Thrierr, *Atlas of Optical Phenomenon*, Plate 10, Berlin: Springer-Verlag, 1962.) (b) Fabry-Perot interferometer, used with a point source and a variable plate spacing. A detector at the focal point of the second lens records intensity as a function of plate spacing d. If a laser source is used, the lenses may not be needed.

I versus the plate spacing *d*, as suggested in Figure 8-7b. In many common applications the source is a laser, in which case the lenses shown in Figure 8-7b may not be needed. It is this last arrangement that we will discuss in the following sections.

8-5 FABRY-PEROT TRANSMISSION: THE AIRY FUNCTION

The irradiance transmitted through a Fabry-Perot interferometer can be calculated with the help of the analysis used to treat the parallel plate arrangement of Section 7-9. In this section we present an alternative method that can also be used to determine the loss rate of a laser cavity. Consider the arrangement of Figure 8-8. We will assume that the two mirrors that form the Fabry-Perot cavity are identical, are separated by a distance *d*, and have real (internal surface) electric-field reflection and transmission coefficients *r* and *t*. Further, we will assume that an electric field suffers no absorption upon encountering the cavity mirrors, so that

$$r^2 + t^2 = 1 \qquad \text{lossless mirrors} \qquad (8\text{-}16)$$

A useful parameter associated with the Fabry-Perot interferometer is the cavity round-trip time τ. The cavity round-trip time is the time needed for light to circulate once around the cavity and so is given by

$$\tau = 2d/v = 2nd/c$$

Here, $v = c/n$ is the speed of light in the medium filling the space between the mirrors, *n* is the index of refraction of this medium, and *c* is the speed of light in vacuum.

We wish to express the electric field E_T transmitted through the Fabry-Perot interferometer in terms of the field E_I incident on the interferometer, the reflection coefficient *r* of the cavity mirrors, and the length *d* of the cavity. In the analysis that follows, we will make use of the notion of a *propagation factor* $P_F(\Delta z, \Delta t)$. As we define it, the propagation factor is the ratio of an electric field $E(z, t)$ associated with a traveling monochromatic plane wave at position $z = z_0 + \Delta z$ and time $t = t_0 + \Delta t$ to the same electric field at position $z = z_0$ and time $t = t_0$. That is,

$$E(z_0 + \Delta z, t_0 + \Delta t) = P_F(\Delta z, \Delta t)E(z_0, t_0)$$

For example, for a plane monochromatic wave traveling in the $+z$ direction encountering no changes in optical media,

$$P_F(\Delta z, \Delta t) = \frac{E(z_0 + \Delta z, t_0 + \Delta t)}{E(z_0, t_0)} = \frac{E_0 e^{i[\omega(t_0 + \Delta t) - k(z_0 + \Delta z)]}}{E_0 e^{i(\omega t_0 - k z_0)}} = e^{i(\omega \Delta t - k \Delta z)}$$

$$(8\text{-}17)$$

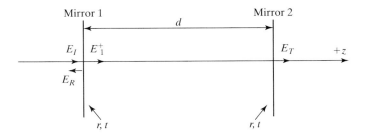

Figure 8-8 Schematic of a Fabry-Perot interferometer consisting of two mirrors with reflection and transmission coefficients *r* and *t*. The electric field incident on the interferometer from the left is E_I, the reflected field is E_R, the transmitted field is E_T, and E_1^+ is the right-going intracavity field at Mirror 1.

We choose not to include electric field changes caused by reflection from or transmission through mirrors in the definition of the propagation factor, but rather we will include these factors explicitly when we track changes to an electric field that encounters mirrors.

To determine the field transmitted through the Fabry-Perot cavity, it is convenient to first determine the amplitude of the intracavity right-going electric field shown as E_1^+ in Figure 8-8. Proceeding, we write the right-going (traveling in the $+z$ direction) electric field incident on the Fabry-Perot cavity from the left as

$$E_I = E_{0I}e^{i\omega t} \tag{8-18}$$

and the right-going electric field in the cavity, at the position of the first mirror as,

$$E_1^+ = E_{01}^+(t)e^{i\omega t} \tag{8-19}$$

Note that the amplitude of this field is, in general, time dependent to allow for the buildup or decay of the intracavity field as the incident field is turned on or off. At time $t + \tau$, the right-going intracavity field $E_1^+(t + \tau)$ can be formed as the sum of two parts. One part is the fraction of the incident field $tE_I(t + \tau)$ that is transmitted through Mirror 1 at this time. The other part is the fraction $r^2 P_F(\Delta z = 2d, \Delta t = \tau)E_1^+(t)$ of the entire right-going intracavity field that existed at Mirror 1 one cavity round-trip time τ earlier. This latter part has propagated around the cavity a distance $2d$ in a time τ, reflecting once from each mirror and returning back to Mirror 1 at time $t + \tau$. That is,

$$E_1^+(t + \tau) = tE_I(t + \tau) + r^2 P_F(\Delta z = 2d, \Delta t = \tau)E_1^+(t) \tag{8-20}$$

Using Eqs. (8-17) through (8-19) in Eq. (8-20) gives

$$E_{01}^+(t + \tau)e^{i\omega(t+\tau)} = tE_{0I}e^{i\omega(t+\tau)} + r^2 E_{01}^+(t)e^{i\omega t}e^{i(\omega\tau - 2kd)}. \tag{8-21}$$

Some time after the incident field is first directed onto the cavity, the intracavity electric field will settle down to a constant steady-state value. Once such a steady state has been reached, $E_{01}^+(t + \tau) = E_{01}^+(t) \equiv E_{01}^+$. In steady state, Eq. (8-21) can be solved for the intracavity right-going field amplitude E_{01}^+,

$$E_{01}^+ = \frac{t}{1 - r^2 e^{-i\delta}} E_{0I} \tag{8-22}$$

Here,

$$\delta = 2kd$$

is the round-trip phase shift.

The transmitted field E_T can be found by propagating the right-going cavity field E_1^+ at Mirror 1 through the cavity and out of Mirror 2,

$$E_T(t + \tau/2) = E_{0T}e^{i\omega(t+\tau/2)} = tP_F(\Delta z = d, \Delta t = \tau/2)E_1^+(t)$$

$$= tE_{01}^+ e^{i\omega t}e^{i(\omega\tau/2 - \delta/2)}$$

Using Eq. (8-22) in the preceding expression and performing some simplification leads to

$$E_{0T} = \frac{t^2 e^{-i\delta/2}}{1 - r^2 e^{-i\delta}} E_{0I} \tag{8-23}$$

Irradiance is proportional to the square of the magnitude of the field amplitude, $I_T \propto E_{0T}E_{0T}^*$, so the transmittance T of the Fabry-Perot cavity is

$$T \equiv \frac{I_T}{I_I} = \frac{E_{0T}E_{0T}^*}{E_{0I}E_{0I}^*} = \frac{t^4 e^{-i\delta/2}e^{i\delta/2}}{(1 - r^2 e^{-i\delta})(1 - r^2 e^{+i\delta})}$$

$$= \frac{t^4}{1 + r^4 - 2r^2 \cos\delta} = \frac{(1 - r^2)^2}{1 + r^4 - 2r^2 \cos\delta}$$

where we have used the lossless mirror condition $t^2 = 1 - r^2$ and one of the Euler identities. Note that this relation is in agreement with Eq. (7-49), which gives the transmittance of a parallel plate. Using Eq. (8-16), the trigonometric identity $\cos\delta = 1 - 2\sin^2(\delta/2)$, and simplifying a bit allows the transmittance to be put into the form of the Airy function,

$$T = \frac{1}{1 + [4r^2/(1 - r^2)^2]\sin^2(\delta/2)} \qquad (8\text{-}24)$$

Coefficient of Finesse

Fabry called the square-bracketed factor in Eq. (8-24), which is a function only of the reflection coefficient r of the mirrors, the *coefficient of finesse*, F:

$$F = \frac{4r^2}{(1 - r^2)^2} \qquad (8\text{-}25)$$

Equation (8-24) can then be expressed more compactly as

$$T = \frac{1}{1 + F \sin^2(\delta/2)} \qquad (8\text{-}26)$$

The coefficient of finesse is a sensitive function of the reflection coefficient r since, as r varies from 0 to 1, F varies from 0 to infinity. We show that F also represents a certain measure of *fringe contrast*, written as the ratio

$$\frac{(I_T)_{\max} - (I_T)_{\min}}{(I_T)_{\min}} = \frac{T_{\max} - T_{\min}}{T_{\min}} \qquad (8\text{-}27)$$

From the Airy formula, Eq. (8-26), T takes on its maximum value $T_{\max} = 1$, when $\sin(\delta/2) = 0$, and its minimum value $T_{\min} = 1/(1 + F)$, when $\sin(\delta/2) = \pm 1$. Thus,

$$\frac{(I_T)_{\max} - (I_T)_{\min}}{(I_T)_{\min}} = \frac{1 - 1/(1 + F)}{1/(1 + F)} = F \qquad (8\text{-}28)$$

Note that this measure of *fringe contrast*, the coefficient of finesse, differs from the related quantity, introduced in Chapter 7 as Eq. (7-17), called the *visibility*. The fringe profile may be plotted once a value of r is chosen. Such a plot, for several choices of r, is given in Figure 8-9. For each curve, we see that $T = T_{\max} = 1$ at $\delta = m(2\pi)$, and $T = T_{\min} = 1/(1 + F)$ at $\delta = (m + 1/2)2\pi$. Notice that $T_{\max} = 1$ regardless of r and that $T_{\min}$ is never zero but approaches this value as r approaches 1. For real mirrors with absorption losses, the maximum transmittance is somewhat less than unity. The transmittance peaks sharply at higher values of r as the phase difference approaches integral multiples of 2π, remaining near zero for most of the region between fringes. As r increases even more to an attainable value of 0.97, for example, F increases to 1078 and the fringe widths are less than a third of

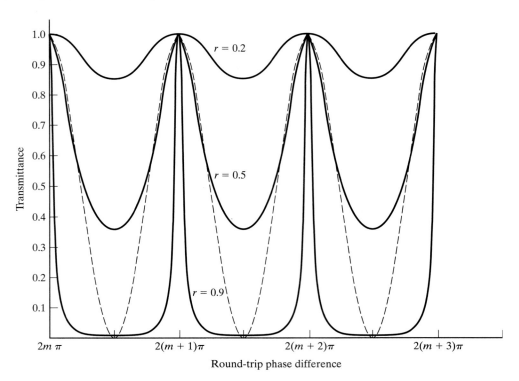

Figure 8-9   Fabry-Perot fringe profile. A plot of transmittance T versus round-trip phase difference δ for selected values of reflection coefficient r. Dashed lines represent comparable fringes from a Michelson interferometer.

their values at half-maximum for $r = 0.9$. The sharpness of these fringes is to be compared with the broader fringes from a Michelson interferometer, which have a simple $\cos^2(\delta/2)$ dependence on the phase (Eq. (8-2)). These are also shown in Figure 8-9 by the dashed lines, normalized to a maximum value of 1.

Finesse

The *coefficient of finesse F* is not to be confused with a second commonly used figure of merit $\mathcal{F}$, called simply the *finesse*:

$$\mathcal{F} = \frac{\pi\sqrt{F}}{2} = \frac{\pi r}{1 - r^2} \qquad (8\text{-}29)$$

We now show that the finesse $\mathcal{F}$ is the ratio of the separation between transmittance peaks to the full-width at half-maximum (FWHM) of the peaks. Equations (8-26) and (8-29) can be combined to write the transmittance as

$$T = \frac{1}{1 + (4\mathcal{F}^2/\pi^2)\sin^2(\delta/2)} \qquad (8\text{-}30)$$

The phase separation between adjacent transmittance peaks is sometimes called the *free spectral range* (FSR) of the cavity, δ_{fsr}. Thus,

$$\delta_{fsr} = \delta_{m+1} - \delta_m = (m + 1)2\pi - m2\pi = 2\pi$$

The half-width at half-maximum (HWHM) $\delta_{1/2}$ of the transmittance peaks (see Figure (8-10)) can be found from Eq. (8-30) by showing that when $T = 1/2$,

$$\sin^2(\delta/2) = \frac{\pi^2}{4\mathcal{F}^2} \qquad (8\text{-}31)$$

where

$$\delta = 2m\pi + \delta_{1/2}$$

Trigonometric identities and a small angle approximation can be used to verify that, at the half-maxima,

$$\sin^2(\delta/2) = \sin^2(m\pi + \delta_{1/2}/2) = \sin^2\left(\frac{\delta_{1/2}}{2}\right) \approx \left(\frac{\delta_{1/2}}{2}\right)^2 \qquad (8\text{-}32)$$

Combining Equations (8-31) and (8-32), we find that

$$\delta_{1/2} = \pi/\mathcal{F} \qquad (8\text{-}33)$$

Cavities with more highly reflecting mirrors have higher values for the finesse and so narrower transmittance peaks than do cavities with less highly reflecting mirrors. As suggested, the finesse of a cavity is the ratio of the free spectral range of the cavity to the FWHM of the cavity transmittance peaks:

$$\frac{\delta_{fsr}}{\text{FWHM}} = \frac{2\pi}{2\delta_{1/2}} = \mathcal{F} \qquad (8\text{-}34)$$

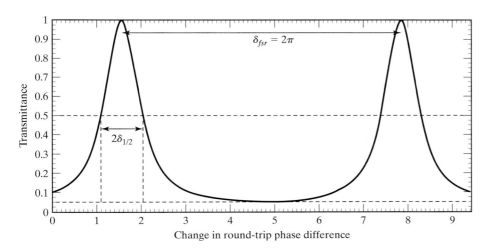

Figure 8-10 Transmittance T as a function of round-trip phase shift δ. The parameters used to produce this plot are discussed in Example 8-2.

The transmittance may be regarded as a function of the round-trip phase shift δ or any of the factors upon which δ depends, such as the mirror spacing (cavity length) d, the frequency ν (and so wavelength λ) of the input field, or the index of refraction n of the medium in the space between the mirrors. In different modes of operation, one of these quantities is typically varied while the others are held constant. Although the values of the free spectral range and the FWHM of the transmittance peaks depend, of course, on the chosen independent variable, the ratio of these quantities (i.e., the finesse) depends only on the reflectivities of the mirrors and so is a useful figure of merit for the Fabry-Perot cavity. We shall use the *term* free spectral range to refer to the separation between adjacent transmittance peaks regardless of the choice of independent variable but take care to *symbolically* differentiate between the free spectral ranges in the different modes of operation. For example, we shall give the free spectral range of a variable-length Fabry-Perot interferometer the symbol d_{fsr} and that of a variable-input-frequency Fabry-Perot interferometer the symbol ν_{fsr}.

Example 8-2

Estimate the coefficient of finesse F, the finesse $\mathcal{F}$, and the mirror reflectivity r for a Fabry-Perot cavity with the transmittance curve shown in Figure 8-10.

Solution

Using Eq. (8-27) and noting from Figure 8-10 that $T_{min} = 0.05$, the coefficient of finesse is found to be

$$F = \frac{T_{max} - T_{min}}{T_{min}} \approx \frac{1 - 0.05}{0.05} = 19$$

The finesse can be found either by extracting the FWHM from Figure 8-10 and using Eq. (8-34),

$$\mathcal{F} = \frac{\delta_{fsr}}{\text{FWHM}} \approx \frac{2\pi}{2.03 - 1.11} = 6.8$$

or by using Eq. (8-29),

$$\mathcal{F} = \frac{\pi\sqrt{F}}{2} \approx \frac{\pi\sqrt{19}}{2} = 6.8$$

The mirror reflection coefficient can be obtained from Eq. (8-29),

$$\mathcal{F} = \frac{\pi r}{(1 - r^2)} = 6.8$$

Rearranging gives

$$6.8r^2 + \pi r - 6.8 = 0$$

Taking the positive root of this quadratic reveals

$$r \approx 0.80$$

8-6 SCANNING FABRY-PEROT INTERFEROMETER

As noted earlier, a Fabry-Perot cavity is commonly used as a scanning interferometer. That is, the irradiance transmitted through a Fabry-Perot is measured as a function of the length of the cavity. An example of such a record that results from the use of a monochromatic incident field is shown in Figure 8-11a. There are many different methods used to change the length of the cavity in a controlled fashion. For example, if the Fabry-Perot interferometer consists of two mirrors separated by an air gap, the mirror separation can be controlled by means of a piezoelectric spacer, as shown in the Figure 8-11b. The transmittance is a maximum whenever

$$\delta = 2kd = 2\frac{2\pi}{\lambda}d = 2m\pi \qquad m = 0, \pm 1, \pm 2 \ldots$$

Rearrangement gives the condition for a maximum as

$$d_m = m\lambda/2 \tag{8-35}$$

Accordingly, the free spectral range in this mode of operation is

$$d_{fsr} = d_{m+1} - d_m = \lambda/2 \tag{8-36}$$

The cavity length change required to move from one transmittance peak to another is thus a measure of the wavelength of the source. In practice, however, this relation, by itself, is not used to experimentally determine the wavelength of the source because the length change cannot be measured with the desired accuracy. Instead, Eq. (8-36) can be used to calibrate the length change of the cavity in

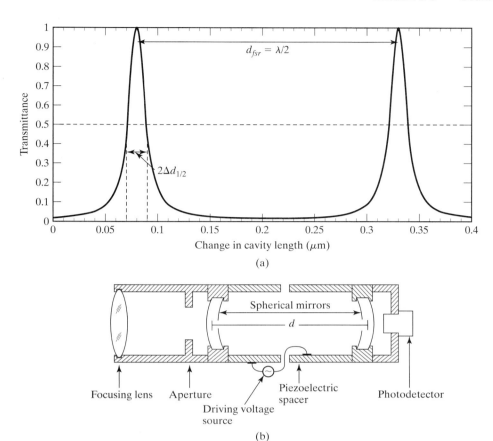

(a)

(b)

Figure 8-11 (a) Transmittance T as a function of the change in cavity length Δd, for a monochromatic input field. (b) Piezoelectric spacer used to control the mirror separation d.

order to, for example, determine the difference in wavelength of two closely spaced wavelength components in the input to the Fabry-Perot cavity.

An example of a record that would result when light of two different but closely spaced wavelengths λ_1 and λ_2 are simultaneously input into a Fabry-Perot cavity of nominal length $d = 5$ cm is shown in Figure 8-12.

If λ_1 and λ_2 are known to be, for example, very near a *nominal wavelength*, $\lambda = 500$ nm, this record can be used to accurately determine the difference in the two wavelengths. If it is known that the adjacent peaks in Figure 8-12 have the same mode number m, then the wavelengths must satisfy the relations

$$\lambda_1 = 2d_1/m$$

$$\lambda_2 = 2d_2/m$$

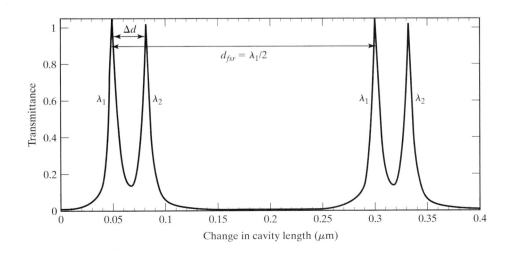

Figure 8-12 Fabry-Perot scan used to determine the difference in wavelength of two closely spaced wavelength components of the input field. The scan is for a nominal wavelength of 500 nm and a nominal mirror spacing of 5 cm.

The wavelength difference is, then,

$$\lambda_2 - \lambda_1 = \Delta\lambda = \frac{2}{m}(d_2 - d_1) = \frac{2}{(2d_1/\lambda_1)}\Delta d$$

Thus,

$$\frac{\Delta\lambda}{\lambda_1} = \frac{\Delta d}{d_1} \tag{8-37}$$

While it is true that the absolute length of the cavity is unlikely to be known to a high degree of accuracy, one can use the *nominal length*, $d \approx d_1$, of the cavity in this expression. Similarly, one typically replaces the wavelength λ_1 appearing in Eq. (8-37) by its nominal value λ. For the situation shown in Figure 8-12,

$$\Delta d \approx d_{fsr}/8 = \frac{\lambda}{16} = \frac{500 \times 10^{-9}\,\text{m}}{16} = 3.125 \times 10^{-8}\,\text{m}$$

so that for $d = 5$ cm,

$$\frac{\Delta\lambda}{\lambda} = \frac{\Delta d}{d} \approx \left(\frac{3.125 \times 10^{-8}\,\text{m}}{0.05\,\text{m}}\right) = 6.25 \cdot 10^{-7}$$

That is, this Fabry-Perot interferometer easily resolves a fractional difference in wavelength of less than one part in a million.

Resolving Power

The minimum wavelength difference, $\Delta\lambda_{\min}$, that can be determined in this manner is limited in part by the width of the transmittance peaks associated with the two wavelength components. A commonly used resolution criterion is that the minimum *resolvable* difference, $\Delta d_{\min}$, between the cavity lengths associated with the centers of the peaks of the transmittance functions of the two wavelength components is equal to the FWHM of these peaks. In this way, the crossover point of the two peaks will be not more than one-half of the maximum irradiance of either peak. This *resolution criterion*, $\Delta d \geq 2\,\Delta d_{1/2} \equiv \Delta d_{\min}$, is illustrated in Figure 8-13.

We now show that the minimum resolvable wavelength difference, $\Delta\lambda_{\min}$, can be compactly expressed in terms of the cavity finesse $\mathcal{F}$. As indicated by Eq. (8-29), the finesse of a Fabry-Perot cavity depends only on the reflection coefficient r of the cavity mirrors. As we mentioned, the finesse is a useful figure of merit because it is the ratio of the separation between adjacent transmittance peaks (that is, the cavity free spectral range) to the FWHM of a transmittance peak. Previously, as Eq. (8-34), we formed this ratio using the round-trip phase shift δ as the independent variable. Noting that $\delta = 2kd$, we now express the finesse using the cavity length d as the independent variable:

$$\mathcal{F} = \frac{\delta_{fsr}}{2\delta_{1/2}} = \frac{kd_{fsr}}{2k\,\Delta d_{1/2}} = \frac{d_{fsr}}{2\Delta d_{1/2}}$$

Therefore,

$$2\Delta d_{1/2} = \frac{d_{fsr}}{\mathcal{F}} = \frac{\lambda}{2\mathcal{F}}$$

Using the relation in Eq. (8-37) and imposing the resolution criterion illustrated in Figure 8-13, $\Delta d_{\min} = 2\Delta d_{1/2}$, leads to

$$\frac{\Delta\lambda_{\min}}{\lambda} = \frac{\Delta d_{\min}}{d} = \frac{2\Delta d_{1/2}}{d} = \frac{\lambda}{2d\mathcal{F}} \tag{8-38}$$

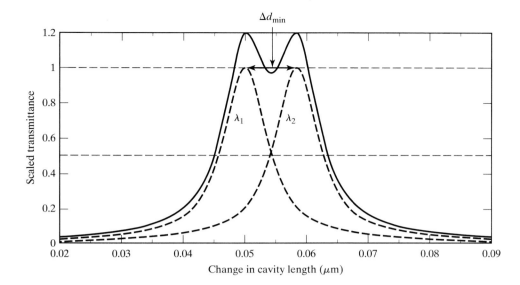

Figure 8-13 Scan of the (scaled) Fabry-Perot transmittance of two wavelength components of comparable strength. The dotted curves indicate the transmittance of the two wavelength components considered separately, and the solid curve is the scaled transmittance when both components are present in the input field. Note that these wavelength components are just barely resolved since the peaks are separated by a FWHM of either dotted curve.

The *resolving power*, $\mathcal{R}$, of a Fabry-Perot interferometer is the inverse of this ratio:

$$\mathcal{R} \equiv \frac{\lambda}{\Delta\lambda_{min}} = \frac{2d\mathcal{F}}{\lambda} = m\mathcal{F} \qquad (8\text{-}39)$$

Here, $m = 2d/\lambda$ is the mode number associated with the nominal wavelength λ and nominal cavity length d.

Large resolving powers are, of course, desirable. For the scanning Fabry-Perot interferometer, we see that large values occur when the mode number is large and for large values of the finesse, which occurs for reflection coefficients close to unity. Notice that to maximize the mode number m, Eq. (8-35) requires that the plate separation d be as large as possible.

Example 8-3

A Fabry-Perot interferometer has a 1-cm spacing between mirrors and a reflection coefficient of $r = 0.95$. For a wavelength around 500 nm, determine its mode number, its finesse, its minimum resolvable wavelength interval, and its resolving power.

Solution

Using Eqs. (8-35), (8-29), (8-38) and (8-39), we find

$$m = \frac{2d}{\lambda} = \frac{2(1 \times 10^{-2})}{500 \times 10^{-9}} = 40{,}000$$

$$\mathcal{F} = \frac{\pi r}{1 - r^2} = \frac{\pi(0.95)}{1 - 0.95^2} = 31$$

$$(\Delta\lambda)_{min} = \frac{\lambda}{m\mathcal{F}} = \frac{500 \text{ nm}}{(40{,}000)(31)} = 4 \times 10^{-4} \text{ nm}$$

$$\mathcal{R} = \frac{\lambda}{(\Delta\lambda)_{min}} = \frac{500}{4 \times 10^{-4}} = 1.2 \times 10^6$$

Good Fabry-Perot interferometers may be expected to have resolving powers of a million or more. This represents one to two orders of improvement over the performance of comparable prism and grating instruments. The photograph of the ring pattern of the mercury green line, revealing its fine structure, shown in Figure 8-14 illustrates the high-resolution performance of a Fabry-Perot instrument operated in the mode illustrated in Figure 8-7a.

We have determined the minimum wavelength separation that can be resolved with a Fabry-Perot interferometer. It is important to note that there is also a maximum wavelength separation, $\Delta\lambda_{\max}$, that can be resolved in an unambiguous manner. If the wavelength separation is too large, the transmittance peak associated with the mode number $m + 1$ of λ_1 will overlap the transmittance peak with mode number m associated with λ_2. The difference in cavity lengths associated with the transmittance peaks of the two wavelength components for the same mode number m is

$$\Delta d = m\lambda_2/2 - m\lambda_1/2 = m\Delta\lambda/2$$

The difference in cavity lengths associated with adjacent transmittance peaks for wavelength component λ_1 is the free spectral range of the variable-length Fabry-Perot interferometer,

$$d_{fsr} = (m + 1)\lambda_1/2 - m\lambda_1/2 = \lambda_1/2$$

The transmittance peak associated with the mode number $m + 1$ of λ_1 will overlap the transmittance peak with mode number m associated with λ_2 if $\Delta d = d_{fsr}$. That is, the overlap occurs if

$$m\Delta\lambda/2 = \lambda_1/2$$

Thus, the maximum wavelength separation that can be unambiguously resolved is

$$\Delta\lambda_{\max} = \lambda_1/m \approx \lambda/m$$

Here, λ is the nominal wavelength of the incident light composed of the two closely spaced wavelength components λ_1 and λ_2. We note that wavelength separations larger than λ/m can be measured with a Fabry-Perot cavity provided that one has additional knowledge of the wavelength separation so that the difference in mode number associated with adjacent transmission peaks can be unambiguously determined.

Figure 8-14 Fabry-Perot rings obtained with the mercury green line, revealing fine structure. (From M. Cagnet, M. Francon, and J. C. Thrierr, *Atlas of Optical Phenomenon*, Plate 10, Berlin: Springer-Verlag, 1962.)

It is interesting to note that the ratio of this maximum wavelength difference to the minimum resolvable wavelength difference is given by the finesse,

$$\frac{\Delta\lambda_{\max}}{\Delta\lambda_{\min}} = \frac{\lambda/m}{\lambda/(m\mathcal{F})} = \mathcal{F}$$

The fact that this ratio is the finesse is not surprising. The transmittance of a *fixed-length* Fabry-Perot interferometer considered as a function of a *variable-wavelength*-input field has transmittance peaks of FWHM equal to $\Delta\lambda_{\min} = \lambda/m\mathcal{F}$ and a peak separation (that is, a free spectral range) equal to $\Delta\lambda_{\max} = \lambda/m$. Thus, $\Delta\lambda_{\max}$ may be called the (wavelength) free spectral range λ_{fsr} of a Fabry-Perot interferometer. (See problem 8-23.)

Spherical, rather than flat, mirrors are often used in scanning Fabry-Perot interferometers. Spherical-mirror Fabry-Perot cavities are easier to align and fabricate and have greater light-gathering power than do flat-mirror cavities. However, spherical-mirror cavities also have a more complex transmittance spectrum than do the flat-mirror cavities just considered. Like cavities made from flat mirrors, spherical-mirror cavities have (so-called longitudinal) modes separated by the cavity free spectral range, but in addition they have (so-called transverse) modes associated with the relationship of the curvatures of the mirrors to the cavity length. The nature of the transverse modes of a cavity is discussed in more detail in Chapter 27. The more complicated mode structure associated with spherical-mirror Fabry-Perot cavities provides the possibility of additional markers that may be useful in the calibration of the Fabry-Perot interferometer.

8-7 VARIABLE-INPUT-FREQUENCY FABRY-PEROT INTERFEROMETERS

For the scanning Fabry-Perot cavity discussed in the previous section, the transmittance through the Fabry-Perot cavity is a function of the changing length of the cavity. A second variant of the Fabry-Perot interferometer uses a cavity of fixed length and a variable-frequency input field. In this mode of operation, the frequencies associated with the transmittance peaks provide frequency markers that can be used to monitor and calibrate the changing frequency of the input laser field. The free spectral range and FWHM of the transmittance T through a variable-input-frequency Fabry-Perot interferometer can be derived in a manner similar to that used in the discussion of the scanning Fabry-Perot interferometer of the last section. To do so, we should first relate the round-trip phase shift δ to the frequency of the input field ν. Making use of the fundamental relation $k = 2\pi/\lambda = 2\pi\nu$, the round-trip phase shift δ associated with an input field of frequency ν can be written as

$$\delta = 2kd = 4\pi(\nu/c)d$$

Thus, a record of the transmittance as a function of the variable input frequency will have maxima when the frequency of the input field has values that follow from the *resonance* condition,

$$\delta_m = 4\pi(\nu_m/c)d = 2m\pi \qquad m = 0, \pm 1, \pm 2 \ldots$$

That is, the resonant frequencies of the Fabry-Perot cavity are

$$\nu_m = mc/2d \qquad\qquad (8\text{-}40)$$

Note that we are assuming, here, that the index of refraction of the material in the space between the cavity mirrors is $n = 1$. In this mode of operation, the free spectral range of the interferometer is

$$\nu_{fsr} = \nu_{m+1} - \nu_m = c/2d \qquad (8\text{-}41)$$

In fact, the term *free spectral range* is most commonly applied for this case, that is, when the transmittance is considered as a function of input frequency. The FWHM $2\Delta\nu_{1/2}$ of the transmittance curves can be found from the basic expression for $\mathcal{F}$ and the relation between round-trip phase shift δ and frequency ν. That is,

$$\mathcal{F} = \frac{\delta_{fsr}}{2\delta_{1/2}} = \frac{4\pi(\nu_{fsr}/c)d}{2[4\pi(\Delta\nu_{1/2}/c)d]} = \frac{\nu_{fsr}}{2\Delta\nu_{1/2}}$$

so that

$$2\Delta\nu_{1/2} = \frac{\nu_{fsr}}{\mathcal{F}}$$

Using Eq. (8-41) and the expression for the finesse $\mathcal{F}$ given in Eq. (8-29) gives

$$2\Delta\nu_{1/2} = \frac{c}{2d}\frac{1 - r^2}{\pi r} \qquad (8\text{-}42)$$

The transmittance through a Fabry-Perot interferometer as a function of the frequency of the input field is shown in Figure 8-15. A Fabry-Perot cavity used in this manner is often characterized by a *quality factor, Q*, defined as the ratio of a nominal resonant frequency to the FWHM of the transmittance peaks,

$$Q = \frac{\nu}{2\Delta\nu_{1/2}} = \mathcal{F}\frac{\nu}{\nu_{fsr}} \qquad (8\text{-}43)$$

As noted, the transmittance of a Fabry-Perot interferometer, with an input laser field whose frequency is intentionally changed, can be used to calibrate the frequency change of the laser. The laser frequency could be changed, for example by changing the effective length of the laser cavity. Such a calibration procedure is useful, for example, in absorption spectroscopy.

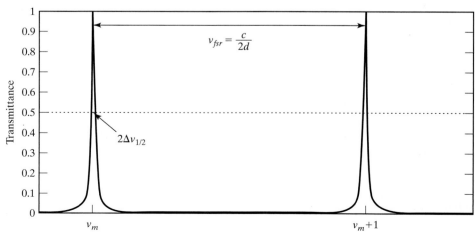

Figure 8-15 Transmittance T through a Fabry-Perot interferometer of fixed length d as a function of the variable frequency ν of the input field.

This application of a variable-input-frequency Fabry-Perot interferometer is explored in problem 8-22. Alternatively, as discussed at the end of the next section, the change in the transmittance through a fixed-length Fabry-Perot cavity induced by a change in the frequency of the laser input field can be used as a feedback signal to stabilize the frequency of the laser source.

In Example 8-4 we explore the relationships between various figures of merit for a variable-input-frequency Fabry-Perot interferometer.

Example 8-4

Consider the transmittance through a variable-input-frequency Fabry-Perot interferometer. Let the Fabry-Perot cavity have length $d = 5$ cm and finesse $\mathcal{F} = 30$. Take the nominal frequency of the laser to be $\nu = 5 \times 10^{14}$ Hz.

   a. Find the free spectral range, ν_{fsr}, of this Fabry-Perot cavity.
   b. Find the FWHM $2\Delta\nu_{1/2}$ of the transmittance peaks.
   c. Find the quality factor Q of this Fabry-Perot cavity.
   d. Estimate the smallest frequency change that could be easily monitored with this Fabry-Perot cavity.

Solution

   a. Using Eq. (8-41), $\nu_{fsr} = \dfrac{c}{2d} = \dfrac{3 \times 10^8 \text{ m/s}}{2(0.05 \text{ m})} = 3$ GHz.

   b. Using the expression for the finesse, $\mathcal{F} = \dfrac{\nu_{fsr}}{2\Delta\nu_{1/2}}$, we find

   $2\Delta\nu_{1/2} = \nu_{fsr}/\mathcal{F} = (3 \text{ GHz})/30 = 100$ MHz.

   c. Using Eq. (8-43), $Q = \dfrac{\nu}{2\Delta\nu_{1/2}} = \dfrac{5 \times 10^{14} \text{ Hz}}{10^8 \text{ Hz}} = 5 \times 10^6$

   d. If the frequency is originally adjusted to give maximum transmittance, a frequency change of $\Delta\nu = \Delta\nu_{1/2} = 50$ MHz would cause the transmittance to fall by a factor of 2. Thus, it would be easy to monitor a frequency change of 50 MHz with this Fabry-Perot.

8-8 LASERS AND THE FABRY-PEROT CAVITY

Laser cavities typically consist of two highly reflecting spherical mirrors and so have the same basic structure as spherical-mirror Fabry-Perot cavities. The frequencies for which a fixed-length Fabry-Perot cavity has maximum transmittance are also the frequencies for which the light generated in a laser medium, within the same cavity, would experience low loss. In addition, as we show later, the rate at which light energy stored in an optical cavity decreases over time due to transmission through and absorption by the cavity mirrors is directly related to the FWHM, $2\Delta\nu_{1/2}$, of the transmittance peaks of the same cavity used as a Fabry-Perot interferometer. This cavity loss rate, often called the *cavity decay rate* and given the symbol Γ, must be compensated for by the gain medium in order to maintain steady-state laser operation. The formalism introduced in Section 8-5 can be used to determine the rate at which the light energy stored in an optical cavity decreases over time. In particular, Eq. (8-21) can be adapted and used to develop an expression for the cavity loss rate. Let the field incident on a Fabry-Perot cavity be removed at time t_0. Further take the field in the cavity to be resonant with the cavity so that $\delta = 2m\pi$. Then for times $t > t_0$, Eq. (8-21) simplifies to

$$E_{01}^+(t + \tau) = r^2 E_{01}^+(t) \qquad (8\text{-}44)$$

If, during one round-trip time τ the change in the complex field amplitude E_{01}^+ is small compared to the amplitude itself, a Taylor series approximation can be used,

$$E_{01}^+(t + \tau) \approx E_{01}^+(t) + \tau \frac{d}{dt} E_{01}^+(t)$$

Using this in Eq. (8-44) and rearranging terms gives

$$\frac{d}{dt} E_{01}^+(t) = -\frac{1}{\tau}(1 - r^2)E_{01}^+(t)$$

One can verify by direct substitution that the solution to this differential equation is

$$E_{01}^+(t) = E_{01}^+(t_0)e^{-(1/\tau)(1 - r^2)(t - t_0)}$$

The right-going irradiance I^+ in the cavity is proportional to the square of the magnitude of the complex field amplitude of the right-going wave, so

$$I^+(t) = I^+(t_0)e^{-(2/\tau)(1 - r^2)(t - t_0)} \equiv I^+(t_0)e^{-\Gamma(t - t_0)}$$

That is, the cavity irradiance decays at the rate

$$\Gamma = \frac{2}{\tau}(1 - r^2) \qquad (8\text{-}45)$$

This sensible result indicates that, for lossless mirrors, the fractional irradiance loss $\Gamma\tau$ during each round-trip time τ is approximately $2(1 - r^2) = 2t^2$. The inverse of the cavity decay rate Γ is sometimes called the *photon lifetime*, τ_p, of the cavity. That is, the photon lifetime of a cavity is the time interval $(t - t_0)$ over which the energy stored in a cavity without gain or input decays to $1/e$ of its initial value. If the light in the cavity is sustained by an input as in a Fabry-Perot cavity, or by a pumped gain medium as in the case of a laser, τ_p is the approximate time that a given portion of the light field remains in the cavity. Note that the approximate number of round-trips, N_{rt}, that a portion of the light field makes before exiting the cavity is, then,

$$N_{rt} \approx \frac{\tau_p}{\tau} = \frac{1}{2(1 - r^2)} \qquad (8\text{-}46)$$

It is useful to note (see Eqs. (8-42) and (8-45)) that, for highly reflective mirrors (r close to 1), the cavity decay rate and the FWHM of the transmittance peaks $2\Delta\nu_{1/2}$ are simply related:

$$\Gamma = \frac{2}{\tau}(1 - r^2) = 2\pi r\left(\frac{c}{2d}\frac{1 - r^2}{\pi r}\right) = 2\pi r(2\Delta\nu_{1/2}) \cong 2\pi(2\Delta\nu_{1/2})$$

This leads us to a second definition of the cavity quality factor Q as the ratio of the operating resonant cavity frequency $\omega = 2\pi\nu$ to the cavity decay rate:

$$Q \approx \frac{2\pi\nu}{\Gamma} = \frac{\omega}{\Gamma}$$

In addition to the formal similarity between Fabry-Perot and laser cavities, Fabry-Perot interferometers can serve a variety of roles as diagnostic or control elements in optical systems. For example, an external scanning Fabry-Perot interferometer provides a means of investigating the mode structure of the output of a multimode laser. Two common uses of the Fabry-Perot as a control element are as a means of limiting a laser to single-mode operation and as a component in a laser frequency stabilization system. These are discussed next.

Mode Suppression with an Etalon

As noted, many laser systems permit so-called multimode operation. That is, the steady-state output of the laser includes electric fields with frequencies corresponding to many different cavity resonances. In some applications, it is preferable for the laser to have an output at only a single cavity resonant frequency. Such a single-mode laser has a longer coherence length than a multimode laser. A Fabry-Perot etalon of length d can be inserted into a laser cavity of length $l > d$ in order to suppress all but a single laser mode.

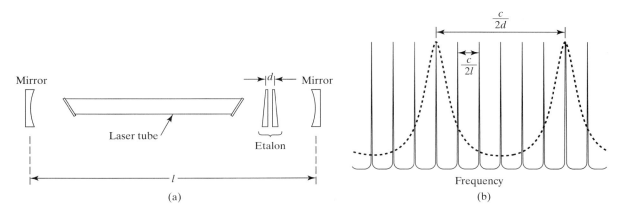

Figure 8-16   (a) Laser with intracavity etalon for single-mode operation. (b) Transmittance for laser cavity of length l (solid curve) and etalon of length d (dashed curve).

For a laser system like that shown in Figure 8-16a, a cavity mode of a given frequency will be present in the laser output only if it is amplified by the laser gain medium and satisfies also the low loss condition imposed by both the laser cavity and the etalon. The etalon, being much shorter than the laser cavity, has a free spectral range, $\nu_{fsr} = c/2d$, that is much larger than that of the laser cavity, $c/2l$. The length of the etalon can be chosen so that only a single etalon mode overlaps an existing cavity mode within the frequency range (the gain bandwidth) of laser operation. In addition, if the width of the etalon mode is less than the free spectral range of the cavity, only one cavity mode will be present in the laser output. Mode spacings in a typical laser system using an etalon for mode suppression are shown in Figure 8-16b. Tuning of the position of the etalon mode within the gain bandwidth can be accomplished by changing the effective etalon spacing d, for example, by piezoelectric control of the etalon spacing or by tilting the etalon. The use of an etalon to limit a laser to single mode operation is explored in Example 8-5.

Example 8-5

A certain argon-ion laser can support steady-state lasing over a range of frequencies of 6 GHz. That is, the gain bandwidth of the argon-ion laser is about 6 GHz. If the length of the laser cavity is $l = 1$ m, estimate the number of longitudinal cavity modes that might be present in the laser output. Also find the minimum length d of an etalon that could be used to limit this laser to single-mode operation.

Solution

The longitudinal cavity modes are separated by the free spectral range of the laser cavity,

$$\nu_{fsr}^{laser} = \frac{c}{2l} = \frac{3 \cdot 10^8 \text{ m/s}}{2(1 \text{ m})} = 0.15 \text{ GHz}$$

Therefore, the number of lasing modes would be given by

$$\# \text{ of lasing modes} \approx \frac{6 \text{ GHz}}{0.15 \text{ GHz}} = 40$$

To ensure single-mode operation, the free spectral range of the etalon must exceed the gain bandwidth. This requirement allows for a determination of the required etalon length d:

$$\nu_{fsr}^{etalon} = \frac{c}{2d} > 6 \text{ GHz}$$

$$d < \frac{c}{2(6 \cdot 10^9 \text{ Hz})} = \frac{3 \cdot 10^8 \text{ m/s}}{1.2 \cdot 10^{10} \text{ Hz}} = 2.5 \text{ cm}$$

Laser Frequency Stabilization

When embedded within a feedback loop, the Fabry-Perot cavity can be used to provide state-of-the-art frequency or length stabilization. For example, light output from a single-mode laser can be fed into a stabilized Fabry-Perot cavity adjusted to allow maximum transmission of this frequency of the laser light. When the laser frequency strays from the resonant frequency of the Fabry-Perot interferometer, the resultant dip in the transmittance of the Fabry-Perot can be used to initiate a feedback signal used to return the laser frequency to the resonant frequency of the Fabry-Perot cavity. Of course, such a system does not really stabilize the absolute frequency of the laser output but rather locks it to the resonant frequency of the Fabry-Perot. If the Fabry-Perot is in turn locked to a very stable frequency source of known frequency, absolute stabilization of the laser frequency is achieved.

8-9 FABRY-PEROT FIGURES OF MERIT

As we have seen, the Fabry-Perot interferometer is a flexible device that has many modes of operation. In Table 8-1 we list relations involving some figures of merit for Fabry-Perot cavities. In Table 8-2, representative values of these figures of merit, as well as some other quantities, are listed for different mirror reflection coefficients. Note that in Table 8-2 there are two rows each for the FSR and FWHM: The values in one set are pertinent when the transmittance varies as a result of changing the mirror spacing d, and the values in the other set apply when the transmittance varies as a result of changing the input frequency ν.

TABLE 8-1 FABRY-PEROT FIGURES OF MERIT.
Here r is the end mirror reflection coefficient, T is the Fabry-Perot transmittance, R is the resolving power of the Fabry-Perot with an input field of nominal wavelength λ whose mirror spacing d is varied, $2\,\Delta\nu_{1/2}$ is the FWHM of a transmittance peak when the frequency of the input is varied around frequency ν, Γ is the decay rate of the light within the Fabry-Perot cavity, τ_p is the photon lifetime of the cavity, and FSR stands for free spectral range.

Coefficient of Finesse	$F = \dfrac{4r^2}{(1-r^2)^2}$	$T = \dfrac{1}{1 + F\sin^2(\delta/2)}$	$F = \dfrac{T_{\max} - T_{\min}}{T_{\min}}$
Finesse	$\mathcal{F} = \dfrac{\pi\sqrt{F}}{2} = \dfrac{\pi r}{1-r^2}$	$\mathcal{F} = \dfrac{\text{FSR}}{\text{FWHM}}$	$\mathcal{R} \equiv \dfrac{\lambda}{\Delta\lambda_{\min}} = \dfrac{2d\mathcal{F}}{\lambda}$
Quality Factor	$Q = \dfrac{\nu}{\nu_{fsr}}\mathcal{F}$	$Q = \dfrac{\nu}{2\Delta\nu_{1/2}}$	$Q \approx \dfrac{\omega}{\Gamma} = \omega\tau_p$

TABLE 8-2 Fabry-Perot parameters for a cavity with a nominal spacing of $d = 5\,\text{cm}$, a nominal input wavelength of $\lambda = 500\,\text{nm}$, and a nominal frequency of $\nu = 6\cdot10^{14}\,\text{Hz}$. Photon lifetime and FWHM are quantities that are not applicable (NA) if the reflection coefficient is too low.

Mirror Reflection Coefficient	r	0.2	0.5	0.8	0.9	0.97	0.99
Coefficient of Finesse, F	$\dfrac{4r^2}{(1-r^2)^2}$	0.174	1.78	19.8	89.8	1080	9900
Finesse, $\mathcal{F}$	$\dfrac{\pi r}{1-r^2}$	0.655	2.09	6.98	14.9	51.6	156
Quality Factor, Q	$\dfrac{\nu}{(c/2d)}\mathcal{F}$	$1.31\cdot10^5$	$4.19\cdot10^5$	$1.40\cdot10^6$	$2.98\cdot10^6$	$1.03\cdot10^7$	$3.13\cdot10^7$
Photon Lifetime, τ_p (s)	$\dfrac{d}{c(1-r^2)}$	NA	NA	$4.63\cdot10^{-10}$	$8.77\cdot10^{-10}$	$2.82\cdot10^{-9}$	$8.38\cdot10^{-9}$
Resolving Power, $\mathcal{R}$	$\dfrac{2d\mathcal{F}}{\lambda}$	$1.31\cdot10^5$	$4.19\cdot10^5$	$1.40\cdot10^6$	$2.98\cdot10^6$	$1.03\cdot10^7$	$3.13\cdot10^7$
$\Delta\lambda_{\min}$ (nm)	$\dfrac{\lambda^2}{2d\mathcal{F}}$	$3.82\cdot10^{-3}$	$1.19\cdot10^{-3}$	$3.58\cdot10^{-4}$	$1.68\cdot10^{-4}$	$4.85\cdot10^{-5}$	$1.60\cdot10^{-5}$
FSR (Variable Spacing) (nm)	$\lambda/2$	250	250	250	250	250	250
FWHM (Variable Spacing) (nm)	$\dfrac{\lambda}{2\mathcal{F}}$	NA	NA	35.8	16.8	4.85	1.6
FSR (Variable Frequency) (GHz)	$\dfrac{c}{2d}$	3	3	3	3	3	3
FWHM (Variable Frequency) (GHz)	$\dfrac{c}{2d\mathcal{F}}$	NA	NA	0.43	0.202	0.0582	0.0192

8-10 GRAVITATIONAL WAVE DETECTORS

We conclude this chapter with a description of interferometers used for gravitational wave detection. At the time of this writing, members of the Laser Interferometer Gravitational Observatory (LIGO) project are building, at two different sites within the United States, interferometers designed to detect and study gravitational waves. Similar interferometers are being developed by scientists and engineers in Europe and Japan. Gravitational waves result from the acceleration of mass in a manner that is analogous to the generation of electromagnetic waves by the acceleration of charge. Gravitational waves exert time-varying forces on matter as they pass by. Because the gravitational force is so weak, gravitational waves coming from even the most dramatic astronomical events like the collision of black holes or the explosion of supernovae lead to extraordinarily small effects on earth. To date, gravitational waves have not been directly detected, but the interferometers currently being constructed are predicted to be sensitive enough to detect the gravitational waves from the dramatic events listed as well as from systems like rotating binary stars. Gravitational wave detection would open a new window to the universe in much the same way that the development of infrared, ultraviolet, and X-ray "telescopes" dramatically increased our store of knowledge regarding astronomical events. Information obtained from interferometers at widely separated locations will aid in distinguishing signals caused by gravitational waves from those caused by local environmental and instrument noise.

A schematic of the LIGO instruments being constructed and an aerial view of one of the installation sites are shown in Figure 8-17. Note that in order to achieve the desired sensitivity, the LIGO interferometer incorporates aspects of both the Michelson and Fabry-Perot interferometers in that it contains a Fabry-Perot cavity in each of the two arms of a Michelson interferometer. The distance between the hanging mirrors can vary in response to passing gravitational waves. Gravitational waves are predicted to be a form of transverse quadrupole radiation so that a wave propagating in a direction that is perpendicular to the plane of the interferometer will induce changes in length of opposite sign in the two arms of the interferometer. That is, if the length of one arm is being reduced, the length of the other arm will be increased due to the passage of the gravitational wave. Gravity waves propagating in other directions will also cause differing length changes in the two arms. The *gravitational strain h* induced in the lengths of the arms of the interferometer has the form

$$ h = \frac{\Delta L}{L} \tag{8-47} $$

where L is the nominal length of one arm of the interferometer and ΔL is the difference in the lengths of the arms caused by the passage of the gravitational wave.

In a Michelson interferometer, the differential length change ΔL in the arms of the interferometer leads to a change in the phase difference of the two beams coming from the interferometer arms and arriving at the detector. The size of the phase shift resulting from an astronomical event that produces a certain gravitational strain h can thus be increased by using interferometers with longer arms. It is predicted that in order to detect gravitational radiation from astronomical sources, sensitivities to strains of less than 10^{-21} (over a detection time of about 1 ms) are required. At first glance, this sensitivity would seem unachievable since it implies that length changes ΔL smaller than the size of an atomic nucleus ($\sim 10^{-15}$ m) would need to be measured in a device

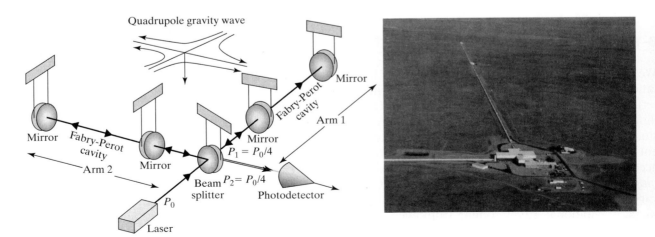

Figure 8-17   (a) Schematic of the LIGO interferometer. The mirrors are attached to hanging mounts, which approximate free masses. (b) Aerial view of the gravitational wave detector being built in Hanford, Washington. Courtesy of LIGO Laboratory.

with an arm length L of even 10 km. However, prototype devices with arm lengths of 40 m have been operated with noise levels corresponding to strains of about 2×10^{-19} for signals at about 450 Hz.[2]

In the setup illustrated in Figure 8-17a, the end mirrors of the two Fabry-Perot interferometers and the beam splitter are mounted on freely suspended masses. In one mode of operation, the lengths of the arms of the interferometer, in the absence of a signal, are adjusted so that destructive interference occurs at the detector. Detection of light, then, corresponds to the detection of a differential length change in the arms of the interferometer. One of the LIGO instruments uses interferometer arms that are 4 km long. The Fabry-Perot cavities effectively extend the length of the arms by causing the light from the laser to sample the cavity length many times before exiting to the detector. The use of the Fabry-Perot cavities in the interferometer arms increases the sensitivity of the device by a factor roughly equal to the number of round-trips in a photon lifetime of the Fabry-Perot cavities. For the mirror reflectivities of the LIGO device, this enhancement factor is about 50, making the effective length of an interferometer arm about 200 km. The generation of LIGO interferometers under construction in 2005 is predicted to be sensitive to gravitational strains of less than 10^{-21}, and the next generation of devices is predicted to have strain sensitivities of less than 10^{-22}, both for signals with frequencies in the range of 100–1000 Hz. To detect these tiny gravitational strains, environmental signals due to seismic activity and a variety of other sources must be either reduced in size or filtered. The filtering process is greatly aided by the use of interferometers at widely separated sites, which are unlikely to be subject to the same local environmental noise. In addition, the quadrupole nature of the gravitational waves leads to signals of unique signature. In Example 8-6 we show how one can estimate the signal power associated with a given gravitational strain.

Example 8-6

Assume that a gravitational wave causes a gravitational strain h of 10^{-21} in the arms of a gravitational wave detector like the one pictured in Figure 8-17a. Assume that the interferometer is set to a null (no detected power) in the absence of the gravitational strain. Calculate the phase difference of the light (of wavelength 488 nm) arriving at the detector from the two arms of the interferometer due to this gravitational strain, and use this result to estimate the detected power if the laser output power P_0 is 10 W. Assume that the nominal arm length is $L = 4$ km and that the light makes an average of 50 round-trips in each arm of the interferometer.

Solution

The total irradiance I_{det} at the photodetector is related to the irradiances I_1 and I_2 of the beams exiting the respective arms of the interferometer by the two-beam interference expression (Eq. (7-14)),

$$I_{det} = I_1 + I_2 + 2\sqrt{I_1 I_2} \cos \delta$$

Here, δ is the phase difference between the two beams arriving at the detector after traversing the interferometer arms. Since irradiance is power per unit area, the interference relation can be recast in terms of the detected power P_{det} and the powers of the beams exiting the two interferometer arms, P_1 and P_2. That is,

$$P_{det} = P_1 + P_2 + 2\sqrt{P_1 P_2} \cos \delta$$

---

[2]A. Abramovici et al., *Phys. Lett. A*, Vol. 218, 1996, 157–163.

Since the beams heading towards the detector encounter the 50-50 beam splitter as they enter and exit the respective interferometer arms,

$$P_1 = P_2 = P_0/4$$

and

$$P_\text{det} = \frac{P_0}{4} + \frac{P_0}{4} + 2\frac{P_0}{4}\cos\delta = \frac{P_0}{2}(1 + \cos\delta)$$

Note that the detected power varies between zero and the full laser power P_0 as $\cos\delta$ varies from -1 to 1. Since the detected power is to be zero in the absence of a strain caused by a gravitational wave, the phase shift can be written profitably as

$$\delta = \pi + \delta_g$$

where δ_g is the phase shift induced by the gravitational wave. Using this form for the phase difference between the two beams and using common trigonometric identities, the detected power can be written as

$$P_\text{det} = \frac{P_0}{2}(1 + \cos\delta) = \frac{P_0}{2}(1 + \cos(\pi + \delta_g)) = \frac{P_0}{2}(1 - \cos(\delta_g))$$
$$= P_0\sin^2(\delta_g/2)$$

For small arguments, the sine function can be approximated by its argument so that

$$P_\text{det} \approx P_0(\delta_g/2)^2$$

The phase difference induced by the gravitational wave is $\delta_g = k\Delta s$, where Δs is the difference in path lengths traveled by the beams passing through the two arms of the interferometer. This path difference Δs is

$$\Delta s \approx 2 \cdot 50\Delta L = 100hL$$

Here, $\Delta L = hL$ is the difference in the lengths of the interferometer arms (of nominal length L) induced by the gravitational wave. The factor of 50 accounts for the approximately 50 round-trips made by the light in the Fabry-Perot cavities in each interferometer arm, and the factor 2 accounts for the fact that the light traverses the length of an arm twice in one round-trip through the Fabry-Perot cavity in that arm. The phase shift δ_g induced by the gravitational strain is, therefore,

$$\delta_g \approx k\Delta s = \frac{2\pi}{\lambda}(100hL) = \frac{2\pi}{4.88\cdot10^{-7}\,\text{m}}(100)(10^{-21})(4000\,\text{m})$$
$$= 5.1510^{-9}\,\text{rad}$$

Using this in the final expression for the detected power,

$$P_\text{det} \approx P_0(\delta_g/2)^2 = (10\,\text{W})\left(\frac{5.15\cdot10^{-9}}{2}\right)^2 = 6.63\cdot10^{-17}\,\text{W}$$

This power corresponds to about 160 photons/s and, while small, is easily detected. However, even very low level environmental noise processes lead to power signals of this and greater levels. As noted, reliable detection of gravitational waves will require isolating the interferometer from environmental noise and separating the gravitational signal from the remaining environmental noise signals.

PROBLEMS

8-1 When one mirror of a Michelson interferometer is translated by 0.0114 cm, 523 fringes are observed to pass the crosshairs of the viewing telescope. Calculate the wavelength of the light.

8-2 When looking into a Michelson interferometer illuminated by the 546.1-nm light of mercury, one sees a series of straight-line fringes that number 12 per centimeter. Explain their occurrence.

8-3 A thin sheet of fluorite of index 1.434 is inserted normally into one beam of a Michelson interferometer. Using light of wavelength 589 nm, the fringe pattern is found to shift by 35 fringes. What is the thickness of the sheet?

8-4 Looking into a Michelson interferometer, one sees a dark central disk surrounded by concentric bright and dark rings. One arm of the device is 2 cm longer than the other, and the wavelength of the light is 500 nm. Determine (a) the order of the central disc and (b) the order of the sixth dark ring from the center.

8-5 A Michelson interferometer is used to measure the refractive index of a gas. The gas is allowed to flow into an evacuated glass cell of length L placed in one arm of the interferometer. The wavelength is λ.

    a. If N fringes are counted as the pressure in the cell changes from vacuum to atmospheric pressure, what is the index of refraction n in terms of N, λ, and L?

    b. How many fringes would be counted if the gas were carbon dioxide ($n = 1.00045$) for a 10-cm cell length, using sodium light at 589 nm?

8-6 A Michelson interferometer is used with red light of wavelength 632.8 nm and is adjusted for a path difference of 20 μm. Determine the angular radius of the (a) first (smallest-diameter) ring observed and (b) the tenth ring observed.

8-7 A polished surface is examined using a Michelson interferometer with the polished surface replacing one of the mirrors. A fringe pattern characterizing the surface contour is observed using He-Ne light of wavelength 632.8 nm. Fringe distortion over the surface is found to be less than one-fourth the fringe separation at any point. What is the maximum depth of polishing defects on the surface?

8-8 The plates of a Fabry-Perot interferometer have a reflection coefficient of $r = 0.99$. Calculate the minimum (a) resolving power and (b) plate separation that will accomplish the resolution of the two components of the H-alpha doublet of the hydrogen spectrum, whose separation is 1.360 nm at 656.3 nm.

8-9 A Fabry-Perot interferometer is to be used to resolve the mode structure of a He-Ne laser operating at 632.8 nm. The frequency separation between the modes is 150 MHz. The plates are separated by an air gap and have a reflectance (r^2) of 0.999.

    a. What is the coefficient of finesse of the instrument?

    b. What is the resolving power required?

    c. What plate spacing is required?

    d. What is the free spectral range of the instrument under these conditions?

    e. What is the minimum resolvable wavelength interval under these conditions?

8-10 A Fabry-Perot etalon is fashioned from a single slab of transparent material having a high refractive index ($n = 4.5$) and a thickness of 2 cm. The uncoated surfaces of the slab have a reflectance (r^2) of 0.90. If the etalon is used in the vicinity of wavelength 546 nm, determine (a) the highest-order fringe in the interference pattern, (b) the ratio T_{max}/T_{min}, and (c) the resolving power.

8-11 The separation of a certain doublet is 0.0055 nm at a wavelength of 490 nm. A variable-spaced Fabry-Perot interferometer is used to examine the doublet. At what spacing does the

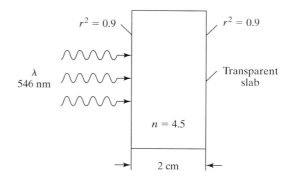

Figure 8-18   Problem 8-10.

mth order of one component coincide with the $(m + 1)$th order of the other?

8-12 White light is passed through a Fabry-Perot interferometer in the arrangement shown in Figure 8-19, where the detector is a spectroscope. A series of bright bands appear. When mercury light is simultaneously admitted into the spectroscope slit, 150 of the bright bands are seen to fall between the violet and green lines of mercury at 435.8 nm and 546.1 nm, respectively. What is the thickness of the etalon?

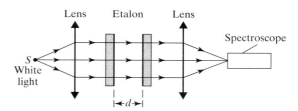

Figure 8-19   Problem 8-12.

8-13 Apply the reasoning used to calculate the finesse of a Fabry-Perot interferometer to the Michelson interferometer. Using the irradiance of Michelson fringes as a function of phase, calculate (a) the fringe separation; (b) the fringe width at half-maximum; (c) their ratio, the finesse.

8-14 Assume that in a Mach-Zehnder interferometer (Figure 8-5), the beam splitter and mirror $M3$ each transmit 80% and reflect 20% of the incident light. Compare the visibility (Eq. (7-17)) when observing the interference of the two emerging beams (shown) with the visibility that results from the two beams emerging from $M3$ along a direction at 90° relative to the first (not shown). For the second case, beam (1) is reflected and beam (2) is transmitted at $M3$.

8-15 Consider the Fabry-Perot cavity shown in Figure 8-8.

    a. With the method used in Section 8-5 to derive the Fabry-Perot transmittance, find the reflectance, $R = I_R/I_I$, of a Fabry-Perot cavity. (*Note*: The reflection coefficient for the external surface of the cavity mirror must be $-r$ if that from the internal surface is r and the transmission coefficients t are real.)

    b. Using the result from (a) and Eq. (8-24) (or an equivalent form), show that the sum of the irradiances reflected by and transmitted through the Fabry-Perot cavity is equal to the irradiance in the field incident on the Fabry-Perot. That is, show that $I_R + I_T = I_I$.

8-16 The reflectance R (see Problem 8-15) of a Fabry-Perot etalon is 0.6. Determine the ratio of transmittance of the etalon at maximum to the transmittance at halfway between maxima.

8-17 Find the transmittance, $T = I_T/I_I$, and the reflectance, $R = I_R/I_I$, of a Fabry-Perot cavity with mirrors of (internal) reflection coefficients r_1 and $r_2 \neq r_1$. Take the mirror separation to be d and see the note given in part (a) of Problem 8-15.

8-18 Consider the transmittance of the variable-input-frequency Fabry-Perot cavity shown in Figure 8-15. Assume that the Fabry-Perot cavity used has a length of 10 cm and that the nominal frequency of the laser input is 4.53×10^{14} Hz. Find

   a. The finesse, $\mathcal{F}$, of the cavity.
   b. The free spectral range, ν_{fsr}, of the transmittance.
   c. The FWHM, $2\Delta\nu_{1/2}$, of a transmittance peak.
   d. The quality factor, Q, of the cavity.
   e. The photon lifetime, τ_p, of the cavity.

8-19 Plot the transmittance, T, as a function of cavity length, d, for a scanning Fabry-Perot interferometer with a monochromatic input of wavelength 632.8 nm if the finesse, $\mathcal{F}$, of the cavity is 15. In the plot let d range from 5 cm to 5.000001 cm.

8-20 Find the values of all the quantities listed in the first column of Table 8-2 for a mirror reflection coefficient of 0.999.

8-21 Consider a light source consisting of two components with different wavelength λ_1 and λ_2. Let light from this source be incident on a scanning Fabry-Perot interferometer of nominal length $d = 5$ cm. Let the scaled transmittance through the Fabry-Perot as a function of the change in the cavity length be as shown in Figure 8-20a and 8-20b. Figure 8-20b

shows the first set of dual peaks of Figure 8-20a over a smaller length scale in order to allow a closer examination of the structure of the overlapping peaks.

   a. What is the nominal wavelength of the light source?
   b. Estimate the difference $\lambda_2 - \lambda_1$ in wavelength of the two components presuming that the overlapping transmittance peaks have the same mode number, $m_2 = m_1 = m$.
   c. Estimate the difference $\lambda_2 - \lambda_1$ in wavelength of the two components presuming that the overlapping transmittance peaks have mode numbers that differ by 1, so that $m_2 = m_1 + 1$.

8-22 In this problem we examine experimental absorption spectroscopy data. The output of a variable-frequency diode laser is divided at a beam splitter so that part of the laser beam is incident on a Fabry-Perot cavity of fixed length and part of the laser beam passes through a sample cell containing atmospheric oxygen, as shown in Figure 8-21a. An overlay of the scaled transmittance through the Fabry-Perot cavity (solid curve) and the scaled transmittance through the oxygen cell (curve made with + symbols) as functions of the laser frequency change is shown in Figure 8-21b. The dips in the transmittance through the oxygen cell indicate that the oxygen molecule strongly absorbs these frequencies. The free spectral range of the Fabry-Perot cavity used in the experiment was known to be 11.6 GHz. The free spectral range can be taken to be the distance between the tall transmittance peaks, indicated by the arrows in Figure 8-21b. (A spherical-mirror Fabry-Perot cavity was used in the experiment and so the transmittance includes peaks corresponding to both longitudinal and transverse modes.)

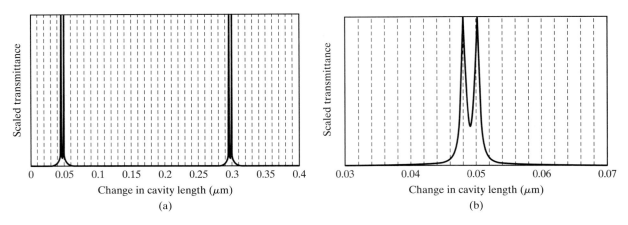

Figure 8-20   Problem 8-21.

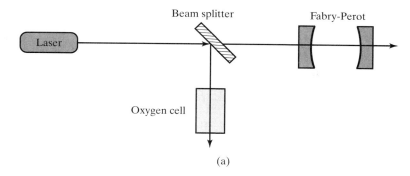

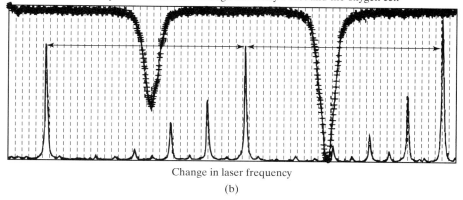

Figure 8-21   Problem 8-22. (a) Experimental arrangement. (b) Overlay of transmittance curves. (Courtesy of R. J. Brecha, Physics Department, University of Dayton.)

a. Estimate the difference in the frequencies of the two absorption dips shown in Figure 8-21b.

b. Estimate the "full-width-at-half-depth" of each absorption dip.

8-23 Consider the transmittance through a Fabry-Perot interferometer as a function of the variable wavelength λ of its input field. Show that the FWHM of the transmittance peaks is $2\Delta\lambda_{1/2} = \lambda/m\mathcal{F}$ and the separation between transmittance peaks is $\lambda_{fsr} = \lambda/m$. (Here $m = 2d/\lambda$, where d is the length of the Fabry-Perot interferometer.)

9

Coherence

INTRODUCTION

The term *coherence* is used to describe the correlation between phases of monochromatic radiations. Beams with random phase relationships are, generally speaking, incoherent beams, whereas beams with a constant phase relationship are coherent beams. The requirement of coherence between interfering beams of light, if they are to produce observable fringe patterns, was mentioned in connection with the interference mechanisms discussed in Chapter 7. We also discussed the relationship between coherence and the net irradiance of interfering beams in Chapter 5. There we concluded that in the superposition of in-phase coherent beams, individual amplitudes add together, whereas in the superposition of incoherent beams, individual irradiances add together. In this chapter, we examine the property of coherence in greater detail, distinguishing between *longitudinal coherence*, which is related to the spectral purity of the source, and *lateral* or *spatial coherence*, which is related to the size of the source. We also describe a quantitative measure of *partial coherence*, the condition under which most experimental measurements of interference take place. We begin our treatment with a brief description of Fourier analysis, which we will need in this chapter.

9-1 FOURIER ANALYSIS

When a number of harmonic waves of the same frequency are added together, even though they differ in amplitude and phase, the result is again a harmonic wave of the given frequency, as shown in Chapter 5. If the superposed waves differ in frequency as well, the result is periodic but anharmonic and

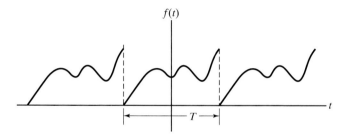

$f(t)$

Figure 9-1   Anharmonic function of time with period T.

may assume an arbitrary shape, such as that shown in Figure 9-1. An infinite variety of shapes may be synthesized in this way. The inverse process of decomposition of a given waveform into its harmonic components is called *Fourier analysis*.

The successful decomposition of a waveform into a series of harmonic waves is insured by the *theorem of Dirichlet*:

If $f(t)$ is a bounded function of period T with at most a finite number of maxima or minima or discontinuities in a period, then the *Fourier series*,

$$f(t) = \frac{a_0}{2} + \sum_{m=1}^{\infty} a_m \cos m\omega t + \sum_{m=1}^{\infty} b_m \sin m\omega t \qquad (9\text{-}1)$$

converges to $f(t)$ at all points where $f(t)$ is continuous and to the average of the right and left limits at each point where $f(t)$ is discontinuous.

In Eq. (9-1), m takes on integral values and $\omega = 2\pi\nu = 2\pi/T$, where T is the period of the arbitrary $f(t)$. The sine and cosine terms can be interpreted as harmonic waves with amplitudes of b_m and a_m, respectively, and frequencies of $m\omega$. The magnitudes of the coefficients or amplitudes determine the contribution each harmonic wave makes to the resultant anharmonic waveform. If Eq. (9-1) is multiplied by dt and integrated over one period T, the sine and cosine integrals vanish, and the result is

$$a_0 = \frac{2}{T} \int_{t_0}^{t_0+T} f(t)\, dt \qquad (9\text{-}2)$$

If Eq. (9-1) is multiplied throughout instead by $\cos n\omega t\, dt$, where n is any integer, and then integrated over a period, the only nonvanishing integral on the right side is the one including the coefficient a_n, and one finds

$$a_n = \frac{2}{T} \int_{t_0}^{t_0+T} f(t) \cos n\omega t\, dt \qquad (9\text{-}3)$$

Similarly, multiplying Eq. (9-1) by $\sin n\omega t\, dt$ and integrating gives

$$b_n = \frac{2}{T} \int_{t_0}^{t_0+T} f(t) \sin n\omega t\, dt \qquad (9\text{-}4)$$

Thus, once $f(t)$ is specified, each of the coefficients a_0, a_n, and b_n can be calculated, and the analysis is complete.

As an example, consider the Fourier analysis of the square wave shown in Figure 9-2 and represented over a period symmetric with the origin by

$$f(t) = \begin{cases} 0, & -T/2 < t < -T/4 \\ 1, & -T/4 < t < T/4 \\ 0, & T/4 < t < T/2 \end{cases}$$

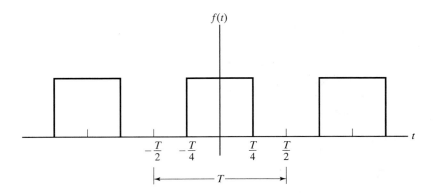

Figure 9-2 Square wave.

Since the function is even in t, the coefficients b_m are found to vanish, and only cosine terms (also even functions of t) remain. From Eqs. (9-2) and (9-3), we find

$$a_0 = 1$$

$$a_n = \left(\frac{2}{n\pi}\right)\sin\left(\frac{n\pi}{2}\right)$$

so that the Fourier series that converges to the square wave of Figure 9-2 as more terms are included in the summation is

$$f(t) = \frac{1}{2} + \sum_{m=1}^{\infty}\left[\left(\frac{2}{n\pi}\right)\sin\left(\frac{n\pi}{2}\right)\right]\cos m\omega t$$

Writing out the first few terms explicitly,

$$f(t) = \frac{1}{2} + \frac{2}{\pi}\left(\cos \omega t - \frac{1}{3}\cos 3\omega t + \frac{1}{5}\cos 5\omega t + \cdots\right)$$

Notice that the contribution of each successive term decreases because its amplitude decreases. Thus a finite number of terms may represent the function rather well. The more rapidly the series converges, the fewer are the terms needed for an adequate fit. Notice also that some amplitudes may be negative, that is, some harmonic waves must be subtracted from the sum to accomplish the convergence. The approximation to a square pulse obtained using a Fourier series representation with a finite number of terms is illustrated in Figure 9-3. Note that the approximation becomes increasingly better as the number of terms in the summation increases. Quite reasonably, fine features in the given $f(t)$, such as the corners of the square waves, require waves of smaller wavelength, or higher frequency components, to represent them. Accordingly, if the widths of the square waves were allowed to diminish,

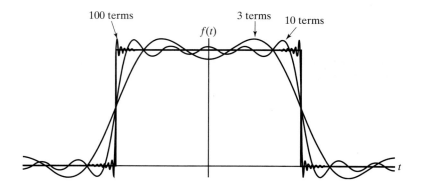

Figure 9-3 Fourier series approximations to a square wave. Approximations using the first 3, 10, and 100 terms of the summation are shown.

so that the individual squares approached spikes, one would expect a greater contribution from the high-frequency components for an adequate synthesis of the function.

   With the help of Euler's equation, the Fourier series given in general by Eq. (9-1), involving as it does both sine and cosine terms, can be expressed in complex notation using exponential functions. The result is

$$f(t) = \sum_{n=-\infty}^{+\infty} c_n e^{-in\omega t} \tag{9-5}$$

where now the coefficients are given by

$$c_n = \frac{1}{T} \int_{t_0}^{t_0+T} f(t) e^{in\omega t}\, dt \tag{9-6}$$

In cases where we wish instead to represent a nonperiodic function (cleverly interpreted mathematically as a periodic function whose period T approaches infinity), it is possible to generalize the Fourier series to a *Fourier integral*. For example, a single pulse is a nonperiodic function but can be interpreted as a periodic function whose period extends from $t = -\infty$ to $t = +\infty$. It can be shown that the discrete Fourier series now becomes an integral given by

$$f(t) = \int_{-\infty}^{+\infty} g(\omega) e^{-i\omega t}\, d\omega \tag{9-7}$$

where the coefficient

$$g(\omega) = \frac{1}{2\pi} \int_{-\infty}^{+\infty} f(t) e^{i\omega t}\, dt \tag{9-8}$$

The Fourier integral, Eq. (9-7), and the expression for its associated coefficient, Eq. (9-8), have a certain degree of mathematical symmetry and are together referred to as a *Fourier-transform pair*. Instead of a discrete spectrum of frequencies given by the Fourier series, Eq. (9-6), we are led to a continuous spectrum, as given by Eq. (9-8). In Figure 9-4, a sample discrete set of coefficients, as might be calculated from Eq. (9-6), is shown together with a continuous distribution approximated by the coefficients, such as might result from Eq. (9-8).

   It should be pointed out that if the function to be represented is a function of spatial position x with period L, say, rather than of time t with

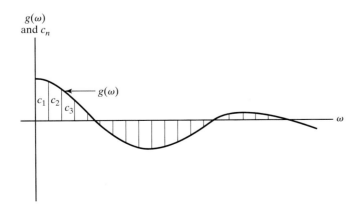

Figure 9-4   Fourier coefficients of a periodic function specify discrete harmonic components of amplitude c_n at frequency ω_n. The Fourier transform of a nonperiodic function requires instead a continuous frequency spectrum $g(\omega)$.

period T, then in Eqs. (9-1) through (9-8) T should be replaced by L and the temporal frequency $\omega = 2\pi/T$ should be replaced by the *spatial frequency*, $k = 2\pi/L$. For example, the Fourier transforms in Eqs. (9-7) and (9-8) become

$$f(x) = \int_{-\infty}^{+\infty} g(k)e^{-ikx}\,dk \tag{9-9}$$

$$g(k) = \frac{1}{2\pi} \int_{-\infty}^{+\infty} f(x)e^{ikx}\,dx \tag{9-10}$$

9-2 FOURIER ANALYSIS OF A FINITE HARMONIC WAVE TRAIN

The spectral resolution of an infinitely long sinusoidal wave is extremely simple: It is one term of the Fourier series, the term corresponding to the actual frequency of the wave. In this case, all other coefficients vanish. Sinusoidal waves without a beginning or an end are, however, mathematical idealizations. In practice, the wave is turned on and off at finite times. The result is a wave train of finite length, such as the one pictured in Figure 9-5. Fourier analysis of such a wave train must regard it as a nonperiodic function. Clearly, it cannot be represented by a single sine wave that has no beginning or end. Rather, the various harmonic waves that combine to produce the wave train must be numerous and so selected that they produce exactly the wave train during the time interval it exists and cancel exactly everywhere outside that interval. Evidently, the turning "on" and "off" of the wave adds many other spectral components to that of the temporary wave train itself. The use of the Fourier-transform integrals leads, in fact, to a *continuous* distribution of frequency components. What we have said here of a finite wave train is also true of any isolated pulse, regardless of its shape. We consider for simplicity the spectral resolution of a pulse that is, while it exists at some point, a harmonic wave. The problem must be handled, as suggested, by the Fourier integral transforms, Eqs. (9-7) and (9-8). We have placed the origin of the time frame, Figure 9-5, so that the wave train is symmetrical about it.

The wave train has a lifetime τ_0 and a frequency ω_0. Thus it may be represented by

$$f(t) = \begin{cases} e^{-i\omega_0 t}, & -\dfrac{\tau_0}{2} < t < \dfrac{\tau_0}{2} \\ 0, & \text{elsewhere} \end{cases} \tag{9-11}$$

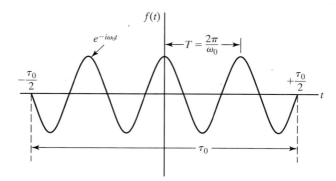

Figure 9-5 Finite harmonic wave train of lifetime τ_0 and period $2\pi/\omega_0$. The spatial extension of the pulse is $\ell_0 = c\tau_0$. The real part of $f(t)$ is plotted.

The frequency spectrum $g(\omega)$ is calculated from Eq. (9-8), with the specific function $f(t)$ of Eq. (9-11),

$$g(\omega) = \frac{1}{2\pi} \int_{-\infty}^{+\infty} f(t)e^{i\omega t}\, dt = \frac{1}{2\pi} \int_{-\tau_0/2}^{+\tau_0/2} e^{i(\omega-\omega_0)t}\, dt$$

Integrating, we have

$$g(\omega) = \left[\frac{e^{i(\omega-\omega_0)t}}{2\pi i(\omega-\omega_0)}\right]_{-\tau_0/2}^{+\tau_0/2}$$

$$g(\omega) = \frac{1}{\pi(\omega-\omega_0)}\left[\frac{e^{i(\omega-\omega_0)\tau_0/2} - e^{-i(\omega-\omega_0)\tau_0/2}}{2i}\right]$$

or, after using the identity,

$$e^{ix} - e^{-ix} \equiv 2i\sin x$$

$$g(\omega) = \frac{\sin[(\tau_0/2)(\omega-\omega_0)]}{\pi(\omega-\omega_0)} = \frac{\tau_0}{2\pi}\left\{\frac{\sin[(\tau_0/2)(\omega-\omega_0)]}{[(\tau_0/2)(\omega-\omega_0)]}\right\} \quad (9\text{-}12)$$

Calling $u = (\tau_0/2)(\omega-\omega_0)$, we then have $g(\omega) = (\tau_0/2\pi)[(\sin u)/u]$. The function $(\sin u)/u$, often called simply *sinc* (u), shows up frequently. It has the property that as u approaches 0, the function approaches a value of 1. Thus, from Eq. (9-12), we conclude that

$$\lim_{\omega\to\omega_0} g(\omega) = \frac{\tau_0}{2\pi} \quad (9\text{-}13)$$

Furthermore, the sinc function $(\sin u)/u$ vanishes whenever $\sin u = 0$, except at $u = 0$, the case already described by Eq. (9-13). In every other case, $\sin u = 0$ for $u = n\pi, n = \pm 1, \pm 2, \ldots$, and so

$$g(\omega) = 0 \quad \text{when} \quad \omega = \omega_0 \pm \frac{2n\pi}{\tau_0} \quad (9\text{-}14)$$

As ω increases (or decreases) from ω_0 then, $g(\omega)$ passes periodically through zero. The accompanying increase in the magnitude of u, or of the denominator of Eq. (9-12), gradually decreases the amplitude of an otherwise harmonic variation. These results are all displayed in Figure 9-6, where the origin of the frequency spectrum is chosen at its point of symmetry, $\omega = \omega_0$. When the

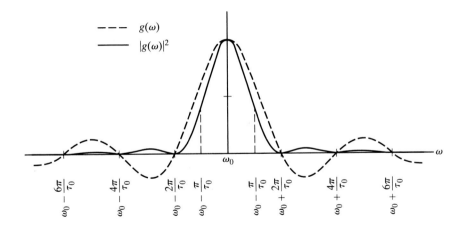

Figure 9-6 Fourier transform of the finite harmonic wave train of Figure 9-5. The dashed line gives the amplitude of the frequency spectrum and the solid line gives its square, the power spectrum. The curves have been normalized to the same maximum amplitude.

amplitude $g(\omega)$ is squared, the resulting curve is the *power spectrum*, shown as the solid curve in Figure 9-6. Although frequencies far from ω_0 contribute to the power spectrum, the bulk of the energy of the wave train is clearly carried by the frequencies present in the central maximum, of width $4\pi/\tau_0$. Notice that the shorter the wave train of Figure 9-5, that is, the smaller the lifetime τ_0, the wider is the central maximum of Figure 9-6. This means that the harmonic waves making important contributions to the actual wave train span a greater frequency interval.

We take the half-width of the central maximum, or $2\pi/\tau_0$, to indicate in a rough way the range of dominant frequencies required. This criterion at least preserves the important inverse relationship with τ_0. Accordingly, we write, as a measure of the frequency band $\Delta\omega$ centered around ω_0 required to represent the harmonic wave train of frequency ω_0 and lifetime τ_0,

$$\Delta\omega = \frac{2\pi}{\tau_0} \quad \text{or} \quad \Delta\nu = \frac{1}{\tau_0} \tag{9-15}$$

Equation (9-15) shows that if $\tau_0 \rightarrow \infty$, corresponding to a wave train of infinite length, $\Delta\omega \rightarrow 0$, and a single frequency ω_0 or wavelength λ_0 suffices to represent the wave train. In this idealized case we have a perfectly monochromatic beam, as considered previously. On the other hand, as $\tau_0 \rightarrow 0$, approximating a harmonic "spike," $\Delta\omega \rightarrow \infty$. Thus, the sharper or narrower the pulse, the greater is the number of frequencies required to represent it, and so the greater the frequency *bandwidth* of the harmonic wave package.

9-3 TEMPORAL COHERENCE AND LINE WIDTH

Clearly, there are no perfectly monochromatic sources. Sources we call "monochromatic" emit light that can be represented as a sequence of harmonic wave trains of finite length, as suggested in Figure 9-7, each separated from the others by a discontinuous change in phase. These phase changes reflect the erratic process by which excited atoms in a light source undergo transitions between energy levels, producing brief and random radiation wave trains. A given source can be characterized by an average wave train lifetime τ_0, called its *coherence time*. Thus, the physical implications of Eq. (9-15) may be summarized as follows: The frequency width $\Delta\nu$ of a spectral line is inversely proportional to the coherence time of the source. The greater its coherence time, the more monochromatic the source. The *coherence length* l_t of a wave train is the length of its coherent pulse, or

$$l_t = c\tau_0 \tag{9-16}$$

Combining Eqs. (9-15) and (9-16), the coherence length is

$$l_t = \frac{c}{\Delta\nu}$$

The frequency band $\Delta\nu$ can be related to the *line width* $\Delta\lambda$ by taking the differential of the relation $\nu = c/\lambda$. That is, $\Delta\nu = |-(c/\lambda^2)\,\Delta\lambda|$. We note that it

Figure 9-7 Sequence of harmonic wave trains of varying finite lengths or lifetimes τ. The wave train may be characterized by an average lifetime, the coherence time τ_0.

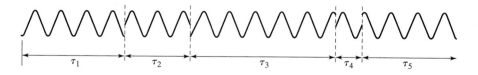

is conventional to take both $\Delta\nu$ and $\Delta\lambda$ to be positive. Then, in terms of the line width $\Delta\lambda$, the coherence length takes the form

$$l_t \cong \frac{\lambda^2}{\Delta\lambda} \qquad (9\text{-}17)$$

Thus, the line width $\Delta\lambda$ is

$$\Delta\lambda \cong \frac{\lambda^2}{l_t} \qquad (9\text{-}18)$$

To digress briefly, it is interesting to note that Eq. (9-18) has a formal similarity to the uncertainty principle of quantum mechanics, where a wave pulse can be used to represent, say, the location of an electron. If the coherence length l_t is interpreted as the interval Δx within which the particle is to be found—that is, its uncertainty in location—and the uncertainty in momentum Δp is expressed by the differential of the deBroglie wavelength in the equation $p = h/\lambda$, the result is $\Delta x\, \Delta p = h$. The inequality associated with the Heisenberg uncertainty relation is consistent with the inequality inherent in Eq. (9-15).

Since the line width of spectral sources can be measured, average coherence times and coherent lengths may be surmised. White light, for example, has a "line width" $\Delta\lambda$ of around 300 nm, extending roughly from 400 to 700 nm. Taking the average wavelength at 550 nm, Eq. (9-17) gives

$$l_t = \frac{(550)^2}{300}\,\text{nm} \cong 1000\,\text{nm} \cong 2\lambda_{\text{av}}$$

a very small coherence length indeed, of around a millionth of a centimeter or two "wavelengths" of white light. Understandably, interference fringes by white light are difficult to obtain since the difference in the path lengths of the interfering beams should not be greater than the coherence length for the light. Sodium or mercury gas-discharge lamp sources are far more monochromatic and coherent. For example, the green line of mercury at 546 nm may have a line width of around 0.025 nm, giving a coherence length of 1.2 cm. One of the most monochromatic gas-discharge sources is a gas of the krypton 86 isotope, whose orange emission line at 606 nm has a line width of only 0.00047 nm. The coherence length of this radiation, by Eq. (9-17), is 78 cm! Laser radiation has far surpassed even the coherence of this gas-discharge source. The short-term stability of commercially available CO_2 lasers, for example, is such that line widths of around 1×10^{-5} nm are attainable at the infrared emission wavelength of 10.6 μm. These numbers give a coherence length of around 11 km! Under carefully controlled conditions, He-Ne lasers can improve this figure by another order of magnitude. Somewhat discouragingly, the common He-Ne laser used in instructional laboratories may not have coherence lengths much greater than its cavity length, due to random temperature fluctuations and mirror vibrations. These spurious effects change the cavity length, lead to multimode oscillations, and adversely affect the coherence length of the laser. Hence the use of these lasers, in holography experiments, for example, still requires some care in equalizing optical-path lengths.

9-4 PARTIAL COHERENCE

As pointed out previously, when the phase difference between two waves is constant, they are mutually coherent waves. In practice, this condition is only approximately met, and we speak of *partial coherence*. The concept is defined

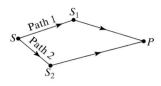

Figure 9-8 Interference at P due to waves from S traveling different paths. The waves are redirected at S_1 and S_2 by various means, including reflection, refraction, and diffraction.

more precisely in what follows. Consider, as in Figure 9-8, a general situation in which interference is produced at P between two beams that originate from a single source S after traveling different paths. A situation like this was discussed in some detail in Chapter 7, but there we considered only the extreme cases of fully coherent and completely incoherent fields. Here we extend that discussion to the case of partially coherent fields. In the present discussion we choose to write the fields $\vec{E}_{1P}$ and $\vec{E}_{2P}$ being superposed at point P in terms of the field $\vec{E}_S$ at the source point S. Further we choose to consider the case in which the fields $\vec{E}_{1P}$ and $\vec{E}_{2P}$ maintain the same polarization as $\vec{E}_S$ so that we can represent the fields by *scalar functions*.

For convenience, we choose to write the source field at point S as

$$E_s(t) = \frac{1}{2}(E(t) + E^*(t)) = \mathrm{Re}(E(t)) \tag{9-19}$$

where,

$$E(t) = E_0 e^{-i\omega t} e^{i\phi(t)} \tag{9-20}$$

Here, $\phi(t)$ models the departure from monochromaticity of the source field. Similarly, for the two fields being superposed at P,

$$E_{1P}(t) = \frac{1}{2}(E_1(t) + E_1^*(t)) = \mathrm{Re}(E_1(t)) \tag{9-21}$$

$$E_{2P}(t) = \frac{1}{2}(E_2(t) + E_2^*(t)) = \mathrm{Re}(E_2(t)) \tag{9-22}$$

Now, we note that the complex superposed fields $E_1(t)$ and $E_2(t)$ are related to the complex source field $E(t)$ via the relations,

$$
\begin{aligned}
E_1(t) &= \beta_1 E(t - T_1) = \beta_1 E_0 e^{-i\omega(t-T_1)} e^{i\phi(t-T_1)} \\
E_2(t) &= \beta_2 E(t - T_2) = \beta_2 E_0 e^{-i\omega(t-T_2)} e^{i\phi(t-T_2)}
\end{aligned}
\tag{9-23}
$$

Here β_1 and β_2 are multiplicative factors resulting from the splitting of the source field and changes in field amplitude due to reflection and transmission in the propagation of the fields from S to P. Further, T_1 is the time of flight for the light field propagating along path 1 and T_2 is the time of flight for the light field propagating along path 2. We choose the forms of the fields shown in the preceding equations in order to lead to a standard expression for the irradiance at point P in terms of the coherence properties of the source field. It is important to note that the fields being superposed are proportional to the source field evaluated at different times. Proceeding, we can form the irradiance at P as,

$$
\begin{aligned}
I_P &= \varepsilon_0 c \langle (E_{1P} + E_{2P})^2 \rangle = \varepsilon_0 c \{ \langle E_{1P}^2 \rangle + \langle E_{2P}^2 \rangle + 2\langle E_{1P} E_{2P} \rangle \} \\
&= I_{1P} + I_{2P} + \frac{\varepsilon_0 c}{2} \langle E_1 E_2 + E_1^* E_2^* + E_1 E_2^* + E_1^* E_2 \rangle
\end{aligned}
\tag{9-24}
$$

where, as before, the brackets denote a time average. In forming the last equality we have made use of the fundamental definition of irradiance to form $I_{1P} = \varepsilon_0 c \langle E_{1P}^2 \rangle$ and $I_{2P} = \varepsilon_0 c \langle E_{2P}^2 \rangle$. In addition we used Eqs. (9-21) and (9-22) to form the last term in Eq. (9-24). This last term is the interference term since its value determines whether the irradiance at P is more than, less than or equal to the sum of the irradiances of the fields being superposed. Note that

$$\langle E_1 E_2 \rangle = 0$$
$$\langle E_1^* E_2^* \rangle = 0$$

since, as one can see from Eq. (9-23) these terms involve the time average of sine and cosine factors that oscillate at 2ω. Thus Eq. (9-24) can be written as,

$$I_P = I_{1P} + I_{2P} + \frac{\varepsilon_0 c}{2}\langle E_1 E_2^* + E_1^* E_2 \rangle = I_{1P} + I_{2P} + \frac{\varepsilon_0 c}{2} 2\,\mathrm{Re}(\langle E_1 E_2^* \rangle)$$

Using the middle members of Eq. (9-23) in the last term in this expression gives,

$$I_P = I_{1P} + I_{2P} + \varepsilon_0 c \beta_1 \beta_2\,\mathrm{Re}\langle E(t - T_1) E^*(t - T_2) \rangle$$

where, for simplicity, we have taken β_1 and β_2 to be real. Shifting the time origin by T_1 and defining the difference in the times of flight for the two paths to be $\tau = T_1 - T_2$, the irradiance at point P can be written as,

$$\begin{aligned} I &= I_{1P} + I_{2P} + \varepsilon_0 c \beta_1 \beta_2\,\mathrm{Re}\langle E(t) E^*(t + (T_1 - T_2)) \rangle \\ &= I_{1P} + I_{2P} + \varepsilon_0 c \beta_1 \beta_2\,\mathrm{Re}\langle E(t) E^*(t + \tau)) \rangle \end{aligned}$$

The remaining time average has the form of a *correlation function*. Accordingly, we define

$$\Gamma(\tau) \equiv \langle E(t) E^*(t + \tau) \rangle \tag{9-25}$$

Note that this correlation function, which determines the size of the interference term, depends on the amount of correlation that exists in the values of the *source field* at two different times. We have achieved one of our objectives: The irradiance at P is dependent on the correlation function $\Gamma(\tau)$ involving the source field. It is convenient to also define the *normalized correlation function*,

$$\gamma(\tau) \equiv \frac{\varepsilon_0 c \beta_1 \beta_2}{2} \frac{\Gamma(\tau)}{\sqrt{I_{1P} I_{2P}}} = \frac{\varepsilon_0 c \beta_1 \beta_2}{2} \frac{\langle E(t) E^*(t + \tau) \rangle}{\sqrt{I_{1P} I_{2P}}} \tag{9-26}$$

so that the irradiance at P may then be expressed as

$$I_P = I_{1P} + I_{2P} + 2\sqrt{I_{1P} I_{2P}}\,\mathrm{Re}[\gamma(\tau)] \tag{9-27}$$

The function $\gamma(\tau)$, now the heart of the interference term, is a function of τ and therefore of the location of point P. We know that the time difference between paths, relative to the average coherence time τ_0 of the source, is crucial to the degree of coherence achieved. We expect that for $\tau > \tau_0$, some coherence between the two beams will be lost. The dependence of $\gamma(\tau)$ on τ_0 is now derived, under the assumption that τ_0 represents a constant coherence time rather than an average. Such a wave train is shown at the top of Figure 9-9a, with regular discontinuities in phase, separated by the time interval τ_0. The normalized correlation function $\gamma(\tau)$, sometimes called the *degree of coherence*, can be simplified by expressing I_{1P} and I_{2P} in terms of the amplitude of the source field. This is most easily accomplished with the help of Eq. (9-23) which indicates that the amplitudes of E_{1P} and E_{2P} are $\beta_1 E_0$ and $\beta_2 E_0$ respectively. Therefore, making use of Eqs. (7-10) and (7-11),

$$I_{1P} = \frac{\varepsilon_0 c}{2}(\beta_1 E_0)^2$$

$$I_{2P} = \frac{\varepsilon_0 c}{2}(\beta_2 E_0)^2$$

Using these relations and Eq. (9-20) in Eq. (9-26) gives,

$$\gamma(\tau) = \frac{\varepsilon_0 c}{2}\beta_1\beta_2 \frac{\langle E_0 e^{-i\omega t} e^{i\phi(t)} E_0 e^{i\omega(t+\tau)} e^{-i\phi(t+\tau)} \rangle}{\sqrt{(\varepsilon_0 c/2)^2 (\beta_1\beta_2 E_0^2)}}$$

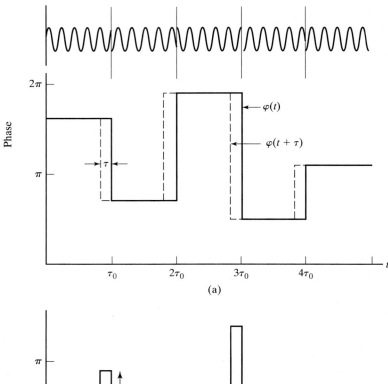

Figure 9-9 (a) Random phase fluctuations $\varphi(t)$ every τ_0 of a wave (solid line) and the same phase fluctuations $\varphi(t + \tau)$ of the wave (dashed line) at a time τ earlier. (b) Difference in the phase between the two waves described in (a).

Simplification gives the important result

$$\gamma(\tau) = e^{i\omega\tau}\langle e^{i[\varphi(t)-\varphi(t+\tau)]}\rangle$$

The time average expressed in this equation may be calculated as

$$\langle e^{i[\varphi(t)-\varphi(t+\tau)]}\rangle = \frac{1}{T}\int_0^T e^{i[\varphi(t)-\varphi(t+\tau)]}\,dt \qquad (9\text{-}28)$$

where T is a sufficiently long time. The function $\varphi(t) - \varphi(t + \tau)$ in the exponent is pictured in Figure 9-9b and is seen to be a series of regularly spaced rectangular pulses with random magnitude falling between -2π and $+2\pi$. Consider the first coherence time interval τ_0, in which the pulse function may be expressed by

$$\varphi(t) - \varphi(t + \tau) = \begin{cases} 0, & 0 < t < (\tau_0 - \tau) \\ H_1, & (\tau_0 - \tau) < t < \tau_0 \end{cases}$$

In successive intervals, the expression is similar, except for the value of H_1. We may then write the normalized coherence function, γ, for a large number, N, of intervals as

$$\gamma = e^{i\omega\tau} \frac{1}{N\tau_0} \left[\underbrace{\int_0^{\tau_0-\tau} e^{i(0)}\, dt + \int_{\tau_0-\tau}^{\tau_0} e^{iH_1}\, dt +}_{\text{interval } N=1} \begin{array}{l} \text{similar terms for } (N-1) \\ \text{successive intervals} \end{array} \right]$$

Integrating over N terms,

$$\gamma = \left(\frac{e^{i\omega\tau}}{N\tau_0} \right) [(\tau_0 - \tau + \tau e^{iH_1}) + (\tau_0 - \tau + \tau e^{iH_2}) + \cdots]$$

Combining the first terms of each interval and summing the rest,

$$\gamma = \left(\frac{e^{i\omega\tau}}{N\tau_0} \right) \left[N(\tau_0 - \tau) + \tau \sum_{j=1}^N e^{iH_j} \right]$$

Because of the random nature of H_j, the terms in the summation average to zero for N sufficiently large. Thus only those times during which the waves coincide—when $\varphi(t) = \varphi(t + \tau)$—contribute to the integral, and we are left with

$$\gamma(\tau) = \left(1 - \frac{\tau}{\tau_0} \right) e^{i\omega\tau} \qquad (9\text{-}29)$$

The real part of γ, required in Eq. (9-27), is given by

$$\mathrm{Re}\,[\gamma(\tau)] = \left(1 - \frac{\tau}{\tau_0} \right) \cos \omega\tau \qquad (9\text{-}30)$$

and so takes on a maximum value of 1 when $\tau = 0$ (equal path lengths), a value of 0 when $\tau = \tau_0$ (path difference equals coherence length), and values between 0 and 1 for τ between τ_0 and 0. The amplitude of the cosine term in Eq. (9-30) is just the magnitude of the degree of coherence γ, that is,

$$|\gamma(\tau)| = 1 - \frac{\tau}{\tau_0} \qquad (9\text{-}31)$$

This quantity sets the limits of the variations in the interference term in Eq. (9-27) and thus controls the contrast or visibility of the fringes as a function of τ. This amplitude, $|\gamma(\tau)|$, is plotted in Figure 9-10. Combining the last three equations,

$$\gamma(\tau) = |\gamma| e^{i\omega\tau} \qquad (9\text{-}32)$$

$$\mathrm{Re}\,\gamma(\tau) = |\gamma| \cos \omega\tau \qquad (9\text{-}33)$$

Recalling the empirical expression for visibility introduced in Chapter 7,

$$V = \frac{I_{\max} - I_{\min}}{I_{\max} + I_{\min}} \qquad (9\text{-}34)$$

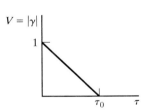

Figure 9-10 Fringe visibility or degree of coherence as a function of the difference in arrival times of two waves with coherence time τ_0.

we may now delineate the following special cases:

1. *Complete incoherence*: $\tau \rightarrow \tau_0$ and $|\gamma| = 0$

$$I_p = I_1 + I_2$$
$$I_p = 2I_0, \quad \text{for equal beams}$$
$$V = \frac{2I_0 - 2I_0}{4I_0} = 0$$

2. *Complete coherence*: $\tau = 0$ and $|\gamma| = 1$

$$I_p = I_1 + I_2 + 2\sqrt{I_1 I_2} \cos \omega\tau$$
$$I_{max} = I_1 + I_2 + 2\sqrt{I_1 I_2} = 4I_0, \quad \text{for equal beams}$$
$$I_{min} = I_1 + I_2 - 2\sqrt{I_1 I_2} = 0, \quad \text{for equal beams}$$
$$V = \frac{4I_0}{4I_0} = 1$$

3. *Partial coherence*: $0 < \tau < \tau_0$ and $1 > |\gamma| > 0$

$$I_p = I_1 + I_2 + 2\sqrt{I_1 I_2} \, \text{Re} \, (\gamma)$$
$$I_p = 2I_0[1 + \text{Re} \, (\gamma)], \quad \text{for equal beams}$$
$$I_{max} = 2I_0(1 + |\gamma|) \quad \text{and} \quad I_{min} = 2I_0(1 - |\gamma|)$$
$$V = \frac{4I_0|\gamma|}{4I_0} = |\gamma|$$

In all cases of equal beams, therefore, the fringe visibility V is equal to the magnitude of the correlation function $|\gamma|$, and either one is a measure of the degree of coherence.

Example 9-1

In an interference experiment, a light beam is split into two equal-amplitude parts. The two parts are superimposed again after traveling along different paths. The light is of wavelength 541 nm with a line width of 1 Å, and the path difference is 1.50 mm. Determine the visibility of the interference fringes. How is the visibility modified if the path difference is doubled?

Solution
The visibility is given by

$$V = 1 - \frac{\tau}{\tau_0} = 1 - \frac{\Delta}{\ell_t}$$

where the ratio of time delay to coherence time is replaced by the corresponding ratio of path difference to coherence length. In this case, $\ell_t = \lambda^2/\Delta\lambda = (5410 \text{ Å})^2/(1 \text{ Å}) = 2.93$ mm. Thus,

$$V = 1 - \frac{1.5}{2.93} = 0.49$$

When the path difference is doubled, $\Delta > \ell_t$ and $\tau > \tau_0$, so that the beams are incoherent and $V = 0$.

9-5 SPATIAL COHERENCE

In speaking of temporal coherence, we have been considering the correlation in phase between temporally distinct points of the radiation field of a source along its line of propagation. For this reason, temporal coherence is also called *longitudinal coherence*. The degree of coherence can be observed by examining the interference fringe contrast in an amplitude-splitting instrument, such as the Michelson interferometer. As we have seen, temporal coherence is a measure of the average length of the constituent harmonic waves, which depends on the radiation properties of the source. In contrast, we now turn our attention to what is referred to as *spatial*, or *lateral, coherence*, the correlation in phase between spatially distinct points of the radiation field. This type of coherence is important when using a wavefront-splitting device, such as the double slit. The quality of the interference pattern in the double-slit experiment depends on the degree of coherence between distinct regions of the wavefield at the two slits.

To sharpen our understanding of the coherence of a wavefield radiating from a source, consider the situation depicted in Figure 9-11. Light from a source S passes through a double slit and is also sampled by a Michelson interferometer located nearby. Spatial coherence between wavefront points A and B at the slits is insured as long as the source S is a true point source. In that case, all rays emanating from S are associated with a single set of spherical waves that have the same phase on any given wavefront. Are clear distinguishable fringes then formed on a screen near point P_1? The answer, of course, depends on whether the light from S, traveling along the two distinct paths SAP_1 and SBP_1, is *temporally* as well as spatially coherent. The matter of temporal coherence requires a comparison between the path difference $\Delta = SAP_1 - SBP_1$ and the coherence length of the radiation. This is equivalent to a comparison of coherence along any radial direction of light propagation from the source at two wavefronts separated by the same path difference. It is this property of temporal coherence that is measured by the Michelson interferometer. If the path difference Δ is much less than the coherence length ($\Delta \ll l_t$), clean interference fringes are formed at P_1; if the path difference is equal to or greater than the coherence length ($\Delta \geq l_t$), interference fringes are poorly defined or absent altogether. In practice, of course, S is always an extended source, so that rays reach A and B from many points of the source. In ordinary (nonlaser) sources, light emitted by different points of a source, well over a wavelength in separation, is not correlated in phase and so lacks coherence. Thus, the spatial coherence of light at

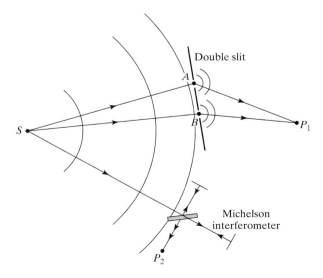

Figure 9-11 Wavefront and amplitude division of radiation from source S, illustrating the practical requirements of spatial and temporal coherence.

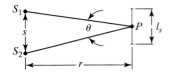

Figure 9-12 Lateral region of coherence l_s, due to two independent point sources.

the slits A and B depends on how closely the source S resembles a point source of light, either in extension or in its actual coherence properties.

We show in the next section that if two source points S_1 and S_2, as in Figure 9-12, are separated by a distance s and if light of wavelength λ from these sources is observed at a distance r away, there will be a region of high spatial coherence of dimension l_s, given by

$$l_s < \frac{\lambda}{\theta} \tag{9-35}$$

where θ is the angle subtended by the point sources at the observation point P. Accepting this result for the moment and combining it with the temporal or longitudinal coherence length l_t, we conclude that there exists at any point in the radiation field of a real light source a region of space in which the light is coherent. This region has lateral dimensions of l_s and longitudinal dimensions of l_t relative to the source and thus occupies a volume of roughly $l_s^2 l_t$ around the point P. It is from this volume that any interferometer must accept radiation if it is to produce observable interference fringes.

9-6 SPATIAL COHERENCE WIDTH

Consider now the spatial coherence at points P_1 and P_2 in the radiation field of a quasi-monochromatic extended source, simply represented by two mutually incoherent emitting points A and B at the edges of the source (Figure 9-13). We may think of P_1 and P_2 as two slits that propagate light to a screen, where interference fringes may be viewed. Each point source, acting alone, then produces a set of double-slit interference fringes on the screen. When both sources act together, however, the fringe systems overlap. If the fringe systems overlap with their maxima and minima falling together, the resulting fringe pattern is highly visible, and the radiation from the two incoherent sources is considered highly coherent! When the fringe systems are relatively displaced, however, so that the maxima of one fall on the minima of the other, the composite pattern is not visible and the radiation is considered incoherent. Suppose that source B is at the position of source A, or that the distance s in Figure 9-13 is zero. The fringe systems at the screen then coincide and correspond to the fringes of a single point source. A maximum in the interference pattern occurs at P if P lies on the perpendicular bisector of the two slits. In this condition,

$$BP_2 - BP_1 = AP_2 - AP_1 = 0$$

If source B is moved below A, the fringe systems separate until, at a certain distance s, where

$$BP_2 - BP_1 = \Delta = \frac{\lambda}{2}$$

Figure 9-13 Light from each of two point sources A and B reach points P_1 and P_2 in the radiation field and are allowed to interfere at the screen. In practice, $s \ll \ell$ and angles θ are approximately equal.

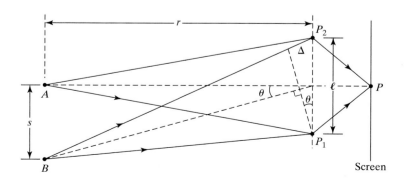

the maximum in the fringe system at P due to source B is replaced by a minimum, and the composite fringe pattern disappears.

If the angle θ represents the angular separation of the sources from the plane of the slits, then from the diagram, $\Delta \cong \ell\theta$, where ℓ is the distance between slits, and $\theta \cong s/r$, where r is the distance to the sources. It follows that

$$\Delta = \frac{\lambda}{2} = \frac{s\ell}{r} \quad \text{or} \quad s = \frac{r\lambda}{2\ell} \tag{9-36}$$

When the distance AB is considered instead to be a continuous array of point sources, the individual fringe systems do not give complete cancellation until the spatial extent AB of the source reaches twice the value of s in Eq. (9-36). If extreme points are separated by an amount $s < r\lambda/\ell$, then fringe definition is assured. Regarding this result as describing instead the maximum slit separation ℓ, given a source dimension s, we have for the *spatial coherence width* ℓ_s,

$$\ell_s < \frac{r\lambda}{s} \cong \frac{\lambda}{\theta} \tag{9-37}$$

As ℓ_s is restricted to smaller fractions of this value, the fringe contrast is correspondingly improved.

According to this argument, moving the source B even farther should bring the fringe system into coincidence again, so that the degree of coherence $|\gamma|$ between P_1 and P_2 is a periodic function. In a more complete mathematical argument, the extended source is represented by a continuous array of elemental emitting areas rather than by two point sources. Results show that outside the coherence width given by Eq. (9-37), the fringe visibility, while oscillatory, is negligible. According to a general theorem, known as the *Van Cittert-Zernike theorem*[1], a plot of the degree of coherence versus spatial separation ℓ of points P_1 and P_2 is the same as a plot of the diffraction pattern due to an aperture of the same size and shape as the extended source. Such patterns for rectangular and circular sources are discussed in Chapter 11.

The significance of Eq. (9-37) is apparent in the case of Young's double-slit experiment, where an extended source is used together with a single slit to render the light striking the double slit reasonably coherent, as in Figure 9-14. We may now use Eq. (9-37) to determine how small the single slit must be to ensure coherence and the production of fringes at the screen. The two slits S_1 and S_2 must fall within the lateral coherence width l_s due to the primary slit of width s.

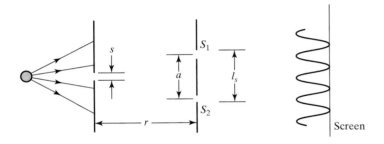

Figure 9-14 Young's double-slit setup. Slits S_1 and S_2 must fall within the lateral coherence width l_s due to the single-slit source.

[1]Born, M. and E. Wolf. *Principals of Optics*, 5th ed., (New York: Pergamon Press, 1975.)

Example 9-2

Let the source-to-slit distance be 20 cm, the slit separation 0.1 mm, and the wavelength 546 nm. Determine the maximum width of the primary or single slit.

Solution

Using Eq. (9-37),

$$s < \frac{r\lambda}{l_s} = \frac{(0.2)(546 \times 10^{-9})}{1 \times 10^{-4}} = 1.1 \text{ mm}$$

Now suppose that the source slit in the example is made exactly 1.1 mm in width and that the separation between slits S_1 and S_2 is adjustable. When the slits are very close together ($a \ll l_s$), they fall within a high coherence region and the fringes in the interference pattern appear sharply defined. As the slits are moved farther apart, the degree of coherence $|\gamma|$ decreases and the fringe contrast begins to degrade. When the slit separation a reaches a value of 0.1 mm, $|\gamma| = 0$ and the fringes disappear. Evidently an experimental determination of this slit separation could be used to deduce the size s of the extended source. This technique was employed by Michelson to measure the angular diameter of stars. Stars are so distant that imaging techniques are unable to resolve their diameters. If a star is regarded as an extended, incoherent source with light emanating from a continuous array of points extending across a diameter s of the star (see Figure 9-15b), then the spatial coherent width l_s in Eq. (9-38) becomes

$$l_s < \frac{1.22\lambda}{\theta} \tag{9-38}$$

Here the factor 1.22 arises from the circular shape of the source, as it does in the Fraunhofer diffraction (Chapter 11) of a circular aperture. Since the angular diameter θ of a star is extremely small, l_s will be correspondingly large. The movable slits were therefore arranged as in Figure 9-15a, using mirrors that direct widely separated portions of the radiation wavefront into a double-slit-telescope instrument. The spacing of the interference fringes depends on the double-slit separation a, whereas their visibility depends on the separation l_s. As l_s is increased, the fringes disappear when equality in Eq. (9-38) is satisfied.

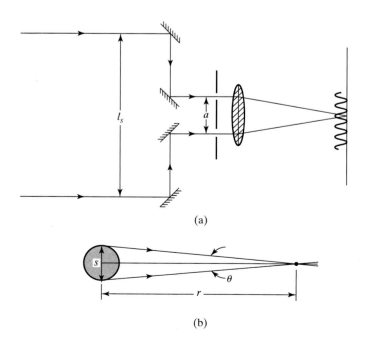

(a)

(b)

Figure 9-15 Michelson stellar interferometer (a) used to determine a stellar diameter (b).

Example 9-3

When Michelson used this technique on the star Betelgeuse in the constellation Orion, he found a first minimum in the fringes at $l_s = 308$ cm. Using an average wavelength of 570 nm, what is the angular diameter of the star?

Solution

Taking Eq. (9-38) as an equality,

$$\theta = \frac{1.22\lambda}{l_s} = \frac{1.22(570 \times 10^{-9})}{3.08} = 2.26 \times 10^{-7} \text{ rad}$$

Since Orion is known to be about 1×10^{15} mi away, the stellar diameter is $s = r\theta = 2.26 \times 10^{8}$ mi, or about 260 solar diameters.

PROBLEMS

9-1 Determine the Fourier series for the function of spatial period L given by

$$f(x) = \begin{cases} -1, & \dfrac{-L}{2} < x < 0 \\ +1, & 0 < x < \dfrac{+L}{2} \end{cases}$$

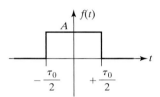

Figure 9-17   Problem 9-4.

9-2 A half-wave rectifier removes the negative half-cycles of a sinusoidal waveform, given by $E = E_0 \cos \omega t$. Find the Fourier series of the resulting wave.

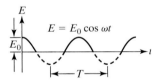

Figure 9-16   Problem 9-2.

9-3 Find the Fourier transform of the Gaussian function given by

$$f(t) = he^{-t^2/2\sigma^2}$$

where h is the height and σ the "width." (*Hint*: Remember how to complete a square? You will also need the definite integral

$$\int_{-\infty}^{+\infty} e^{-x^2} \, dx = \sqrt{\pi}$$

in your calculations.) Does the transform, interpreted as the frequency spectrum, show the proper relationship to the original "pulse" width?

9-4 Using the Fourier transform, determine the power spectrum of a single square pulse of amplitude A and duration τ_0. Sketch the power spectrum, locating its zeros, and show that the frequency bandwidth for the pulse is inversely proportional to its duration.

9-5 Two light filters are used to transmit yellow light centered around a wavelength of 590 nm. One filter has a "broad" transmission width of 100 nm, whereas the other has a "narrow" pass band of 10 nm. Which filter would be better to use for an interference experiment? Compare the coherence lengths of the light from each.

9-6 A continuous He-Ne laser beam (632.8 nm) is "chopped," using a spinning aperture, into 1-μs pulses. Compute the resultant line width $\Delta\lambda$, bandwidth $\Delta\nu$, and coherence length.

9-7 The angular diameter of the sun viewed from the earth is approximately 0.5 degree. Determine the spatial coherence length for "good" coherence, neglecting any variations in brightness across the surface. Let us consider, somewhat arbitrarily, that "good" coherence will exist over an area that is 10% of the maximum area of coherence.

9-8 Michelson found that the cadmium red line (643.8 nm) was one of the most ideal monochromatic sources available, allowing fringes to be discerned up to a path difference of 30 cm in a beam-splitting interference experiment, such as with a Michelson interferometer. Calculate (a) the wavelength spread of the line and (b) the coherence time of the source.

9-9 A narrow band-pass filter transmits wavelengths in the range 5000 ± 0.5 Å. If this filter is placed in front of a source of white light, what is the coherence length of the transmitted light?

9-10 Let a collimated beam of white light fall on one refracting face of a prism and let the light emerging from the second face be focused by a lens onto a screen. Suppose that the linear dispersion at the screen is 20 Å/mm. By introducing a narrow "exit slit" in the screen, one has a type of monochromator that provides a nearly monochromatic beam of light. Sketch the setup. For an exit slit of 0.02 cm, what is the coherence time and coherence length of the light of mean wavelength 5000 Å?

9-11 A pinhole of diameter 0.5 mm is used in front of a sodium lamp (5890 Å) as a source in a Young interference experiment. The distance from pinhole to slits is 1 m. What is the maximum slit space insuring interference fringes that are just visible?

9-12 Determine the linewidth in angstroms and hertz for laser light whose coherence length is 10 km. The mean wavelength is 6328 Å.

9-13 a. A monochromator is used to obtain quasi-monochromatic light from a tungsten lamp. The linear dispersion of the instrument is 20 Å/mm and an exit slit of 200 μm is used. What is the coherence time and length of the light from the monochromator when set to give light of mean wavelength 500 nm?
   b. This light is used to form fringes in an interference experiment in which the light is first amplitude-split into two equal parts and then brought together again. If the optical path difference between the two paths is 0.400 mm, calculate the magnitude of the normalized correlation function and the visibility of the resulting fringes.
   c. If the maximum irradiance produced by the fringes is 100 on an arbitrary scale, what is the difference between maximum irradiance and background irradiance on this scale?

9-14 Determine the length and base area of the cylindrical volume within which light received from the sun is coherent. For this purpose, let us assume "good" spatial coherence occurs within a length that is 25% of the maximum value given by Eq. (9-38). The sun subtends an angle of 0.5° at the earth's surface. The mean value of the visible spectrum may be taken at 550 nm. Express the coherence volume also in terms of number of wavelengths across cylindrical length and diameter.

9-15 a. Show that the fringe visibility may be expressed by

$$V = \frac{2\sqrt{I_1 I_2}\,|\gamma(\tau)|}{(I_1 + I_2)}$$

   b. What irradiance ratio of the interfering beams reduces the fringe visibility by 10% of that for equal-amplitude beams?

9-16 Show that the visibility of double-slit fringes in the mth order is given by

$$V = 1 - \left(m\frac{\Delta\lambda}{\lambda}\right)$$

where λ is the average wavelength of the light and $\Delta\lambda$ is its linewidth.

9-17 A filtered mercury lamp produces green light at 546.1 nm with a linewidth of 0.05 nm. The light illuminates a double slit of spacing 0.1 mm. Determine the visibility of the fringes on a screen 1 m away, in the vicinity of the fringe of order $m = 20$. (See problem 9-16.) If the discharge lamp is replaced with a white light source and a filter of bandwidth 10 nm at 546 nm, how does the visibility change?

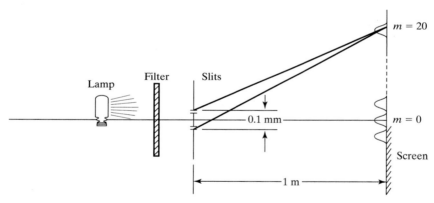

Figure 9-18   Problem 9-17.

9-18 A Michelson interferometer forms fringes with cadmium red light of 643.847 nm and linewidth of 0.0013 nm. What is the visibility of the fringes when one mirror is moved 1 cm from the position of zero path difference between arms? How does this change when the distance moved is 5 cm? At what distance does the visibility go to zero?

9-19 a. Repeat problem 9-18 when the light is the green mercury line of 546.1 nm with a linewidth of 0.025 nm.
   b. How far can the mirror be moved from zero path difference so that fringe visibility is at least 0.85?

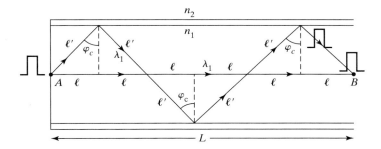

10 *Fiber Optics*

INTRODUCTION

The channeling of light through a transparent conduit has taken on great importance in recent times. This is especially true because of its applications in communications and laser medicine. As long as a transparent solid cylinder, such as a glass fiber, has a refractive index greater than that of its surrounding medium, much of the light launched into one end will emerge from the other end due to a large number of total internal reflections. A comprehensive treatment of fiber optics requires a wave approach in which Maxwell's equations are solved in a dielectric medium, subject to the boundary conditions at the fiber walls. In this chapter, we adopt a simpler and more intuitive approach, describing the propagating wavefronts by their rays, although we appeal in some contexts to wave properties such as phase and interference.

10-1 APPLICATIONS

The simplest use of optical fibers, either singly or in bundles, is as *light pipes*. For example, a flexible bundle of fibers might be used to transport light from inside a vacuum system to the outside, where it can be more easily measured.[1] The bundle might be divided into two or more sections at some point to act as a beam splitter. For such nonimaging applications, the fibers can be randomly distributed within the cable. When imaging is required, however, the fiber ends at the input are coordinated with the fiber ends at the output. To maintain this

---

[1]Interestingly, the rods and cones of the human eye have been shown to function as light pipes, transmitting light along their lengths, as in optical fibers.

coordination, fibers at both ends are bonded together. The fiberscope consists of a bundle of such fibers, end-equipped with objective lens and eyepiece. It is routinely used by physicians to examine regions of the heart, stomach, lungs, and duodenum. Some of the fibers function as light pipes, transporting light from an external source to illuminate inaccessible areas internally. Other fibers return the image.

Fibers can be bound rigidly by fusing their outer coating or *cladding*. In this way, fiber-optic faceplates are made for use as windows in cathode ray tubes. Further, when such fused-fiber bundles are tapered by heating and stretching, images can be magnified or diminished in size, depending on the relative areas of input or output faces. The resolving power of imaging fibers depends on the accuracy of fiber alignment and, as might be expected, on the individual fiber diameter d. A conservative estimate[2] of fiber resolving power, RP, is given by

$$RP \text{ (lines/mm)} = \frac{500}{d(\mu\text{m})} \tag{10-1}$$

Thus, a 5-μm fiber, for example, can produce a high resolution of about 100 lines/mm.

The most far-reaching applications of fiber optics lie in the area of communications, the subject of the following section.

10-2 COMMUNICATIONS SYSTEM OVERVIEW

No application has given greater impetus to the rapid development of fiber optics than has voice or video communication and data transmission. The advantages of fiber-optic conduits, or *waveguides*, over conventional two-wire, coaxial cable or microwave waveguide systems are impressive. The replacement of microwaves and radio waves by light waves is especially attractive, since the information-carrying capacity of the carrier wave increases directly with the width of the frequency band available. Replacement of copper coaxial cable or twisted-pair transmission lines by fiber-optic cable thus offers greater communications capacity with lower loss in a lighter cable that requires less space. Additionally, in contrast with metallic conduction techniques, communication by light offers the possibility of electrical isolation, immunity to electromagnetic interference, and freedom from signal leakage. The latter is especially important where security of information is vital, as in computer networks that handle confidential data.

In Figure 10-1, we give an overview of the essential components and processes involved in a fiber-optic communications system, from message source to message output. At the input end of the fiber-optic cable, the information to be transmitted is converted by some type of transducer from an electrical signal to an optical one. After transmission by the optic fiber, it is reconverted from an optical to an electrical signal. The fiber serves as an *optical waveguide* to propagate the information with as little distortion and loss of power as possible, over a distance that can range from meters to thousands of kilometers.

The message source might be audio, providing an analog electrical signal from a microphone; it might be visual, providing an analog signal from a video camera; or it might be digitally encoded information, like computer data in the form of a train of pulses. Analog and digital formats are convertible into one another, so that the choice of format for transmission through the fiber is always available, regardless of the original nature of the signal.

---

[2]Walter P. Siegmund, "Fiber Optics," in *Handbook of Optics*, edited by Walter G. Driscoll and William Vaughan (New York: McGraw-Hill Book Company, 1978).

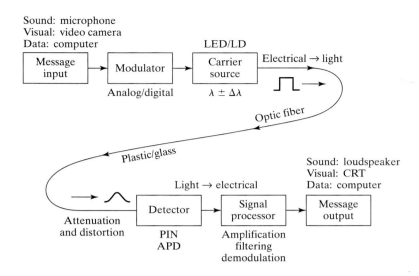

Figure 10-1 Overview of a fiber-optics communication system.

The purpose of the *modulator* is to perform this conversion when desired and to impress this signal onto the carrier wave generated by the *carrier source*. The carrier wave can be modulated to contain the signal information in various ways, usually by amplitude modulation (AM), frequency modulation (FM), or digital modulation.[3] (See Figure 10-2.) In fiber-optics systems, the carrier source is typically either a light-emitting diode (LED) or a laser diode (LD). In Figure 10-1, the carrier source output into the optic fiber is represented by a single, square pulse. As this pulse propagates through the fiber, it suffers both *attenuation* (loss of amplitude) and *distortion* (change in shape) due to several mechanisms to be discussed. The fiber may be, typically, a glass or plastic filament 50 μm in diameter. If the fiber is very long, it may be necessary to amplify the signal at several positions along the fiber. However, while high bit-rate signals carried on copper wire transmission lines may need to be amplified every 300 m, high bit-rate signals carried on optical fibers need amplification only every 100 km or so. At the remote end of the

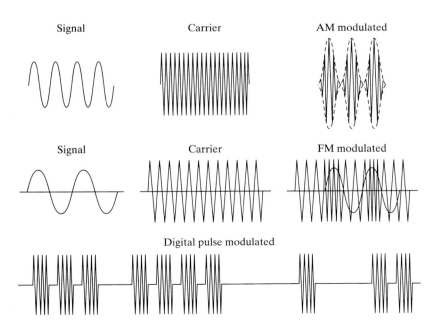

Figure 10-2 Three forms of modulation in which a carrier wave is modified to carry a signal. Top: Amplitude modulation; center: frequency modulation; bottom: digital modulation in which a pulse is either present ("on") or missing ("off").

---

[3]Some of the techniques of light modulation are discussed in Chapter 24.

fiber, the light signal is coupled into a detector that changes the optical signal back into an electrical signal. This service is performed by a semiconductor device, most commonly a *PIN diode*, an *avalanche diode*, or a *photomultiplier* (see Chapter 17). Of course, the response of the detector should be well matched to the optical frequency of the signal received. The detector output is then handled by a signal processor, whose function is to recapture the original electrical signal from the carrier, a process that involves filtering and amplification and, possibly, a digital-to-analog conversion. The message output may then be communicated by loudspeaker (audio), by cathode-ray tube (video), or by computer input (digital).

10-3 BANDWIDTH AND DATA RATE

The more complicated the signal to be communicated, the greater is the range of frequencies required to represent it. The output of a stereo system is more faithful to the original signal than is the output of a telephone receiver because a greater frequency range is devoted to the process of reproduction. The range of frequencies required to modulate a carrier for a single telephone channel is only 4 kHz, whereas the bandwidth of an FM radio broadcasting station is 200 kHz. A commercial TV broadcasting station, which must communicate both sound and video signals, uses a bandwidth of 6 MHz. The great information-carrying potential of a light beam becomes evident when we calculate the ratio of carrier frequency to signal bandwidth, a measure of the number of separate channels that can be impressed on the carrier. For a TV station using a 300-MHz carrier, this ratio is 300 MHz/6 MHz, or 50; for an optical fiber using a carrier of 1-μm wavelength (3×10^8 MHz) to carry the same information, the ratio is $(3 \times 10^8 \text{ MHz})/6 \text{ MHz}$, or 50,000,000! Currently, optical fibers use only a small fraction of the entire bandwidth theoretically available to an optical signal. Still, state-of-the-art fiber-optic systems carry far more information with much lower loss than can copper-wire transmission lines.

More information can be sent by optical fiber when distinct pulses can be transmitted in more rapid succession. This implies higher frequencies or, in the case of digital information, higher *bit rates*. In the latter case, suppose that 8 bits (*on* or *off* pulses) are required to represent the amplitude of an analog signal. According to the *sampling theorem*,[4] an analog signal must be sampled at a rate at least twice as high as its highest-frequency component in order to be faithfully represented. In the case of the TV channel with a bandwidth of 6 MHz, this means that 2×6 MHz or 12×10^6 samples must be taken each second. Since each sample is described using 8 bits, the required data rate is 96 Mbps (megabits per second). As we shall see in discussions to follow, data rates are limited by the modulator capabilities as well as by fiber distortions that prevent distinct identification of neighboring pulses. *Wavelength-division multiplexing* (WDM) is a means of combining 40 or more signals, carried in different wavelength channels, so that they propagate together through the same fiber. The maximum bit rate of transmission through an optical-fiber system using (dense) WDM exceeds 1 terrabit per second (10^{12} bits/s).

10-4 OPTICS OF PROPAGATION

We consider now the manner in which light propagates through an optical fiber. The conditions for successful propagation are developed here mainly

---

[4]The sampling theorem is also discussed in Chapter 21, in connection with Fourier transform spectroscopy.

from the point of view of geometrical optics. In addition, we consider only the meridional rays, which intersect with the fiber's central axis.[5]

Consider a short section of a straight fiber, pictured in Figure 10-3a. The fiber itself has refractive index n_1, the encasing medium (called *cladding*) has index n_2, and the end faces are exposed to a medium of index n_0. Ray A entering the left face of the fiber is refracted there and transmitted to point C on the fiber surface where it is partially refracted out of the fiber and partially reflected internally. The internal ray continues, diminished in amplitude, to D, then to E, and so on. After multiple reflections, the ray will have lost a large part of its energy. Ray A does not meet the conditions for total internal reflection, that is, it strikes the fiber surface at points $C, D, E, \ldots$ such that its angle of incidence φ is less than the critical angle φ_c, or

$$\varphi < \varphi_c = \sin^{-1}(n_2/n_1) \tag{10-2}$$

Ray B, on the other hand, which enters at a smaller angle θ_m with respect to the axis, strikes the fiber surface at F in such a way that it is refracted parallel to the fiber surface. Other rays, as in Figure 10-3b, incident at angles $\theta < \theta_m$, experience total internal reflection at the fiber surface. Such rays are propagated along the fiber by a succession of such reflections, without loss of energy due to refraction out of the cylinder. However, depending upon the degree of transparency of the fiber material to the light, some attenuation occurs by absorption.

Ray B thus represents an extreme ray, defining the slant face of a cone of rays, all of which satisfy the condition for total internal reflection within the fiber. The maximum half-angle θ_m of this cone is evidently related to the critical angle of reflection φ_c. At the input face of the fiber shown in Figure 10-3a,

$$n_0 \sin \theta_m = n_1 \sin \theta'_m \tag{10-3}$$

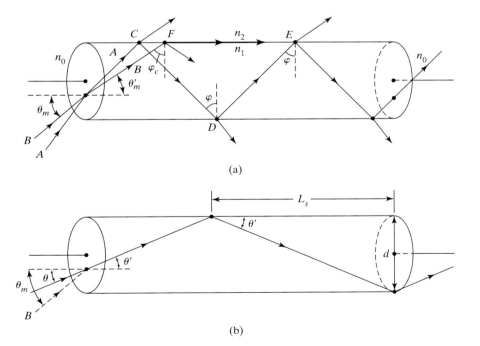

(a)

(b)

Figure 10-3   (a) Propagation of light rays through an optical fiber. Ray B defines the maximum input cone of rays satisfying total internal reflection at the walls of the fiber. (b) Propagation of a typical light ray through an optical fiber.

---

[5]Other rays, the *skew* rays, do not lie in a plane containing the central fiber axis. These rays take a piecewise spiral path through the fiber.

and at a point like F,

$$\sin \varphi_c = \frac{n_2}{n_1}$$

Using the geometrical fact, $\theta'_m = 90° - \varphi_c$, and the trigonometric identity, $\sin^2 \varphi_c + \cos^2 \varphi_c = 1$, these relations combine to give the *numerical aperture*,

$$\text{N. A.} \equiv n_0 \sin \theta_m = n_1 \cos \varphi_c = \sqrt{n_1^2 - n_2^2} \qquad (10\text{-}4)$$

If $n_0 = 1$, the numerical aperture is simply the sine of the half-angle of the largest cone of meridional rays (i.e., rays coplanar with the fiber axis) that are propagated through the fiber by a series of total internal reflections. The numerical aperture clearly cannot be greater than unity, unless $n_0 > 1$. A numerical aperture of 0.6, for example, corresponds to an acceptance cone of 74°. The light-gathering ability of an optical fiber increases with its numerical aperture.

Also from Figure 10-3b, the *skip distance*, L_s, between two successive reflections of a ray of light propagating in the fiber is given by

$$L_s = d \cot \theta' \qquad (10\text{-}5)$$

where d is the fiber diameter. Relating θ' to the entrance angle θ by Snell's law,

$$L_s = d \sqrt{\left(\frac{n_1}{n_0 \sin \theta}\right)^2 - 1} \qquad (10\text{-}6)$$

For example, in the case $n_0 = 1$, $n_1 = 1.60$, $\theta = 30°$, and $d = 50 \ \mu\text{m}$, Eq. (10-6) gives $L_s = 152 \ \mu\text{m}$. Thus, in 1 m of fiber, there are approximately $1/L_s$, or 6580, reflections! Table 10-1 lists various core and cladding possibilities, for which the critical angle, numerical aperture, and skip distances have been calculated. With so many reflections occurring, the condition for total internal reflection must be accurately met over the entire length of the fiber. Surface scratches or irregularities, as well as surface dust, moisture, or grease, become sources of loss that rapidly diminish light energy. If only 0.1% of the light is lost at each reflection, over a length of 1 m, this attenuation would reduce the energy by a factor of about 720. Therefore, to protect the optical quality of the fiber, it is essential that it be coated with a layer of plastic or glass called the *cladding*. Cladding material need not be highly transparent, but must be compatible with the fiber core in terms of expansion coefficients, for example. The index of refraction n_2 of the cladding, where $n_2 < n_1$, influences the critical angle and numerical aperture of the fiber.

The cladding around the fiber cores has another important function, which is to prevent what is called *frustrated total internal reflection* from occurring. When the process of total internal reflection is treated as the interaction of a wave disturbance with the electron oscillators comprising the medium, it becomes apparent that there is some short-range penetration of the wave beyond the boundary. Although the wave amplitude decreases rapidly beyond

TABLE 10-1 CHARACTERIZATION OF SEVERAL OPTICAL FIBERS

Core/cladding	n_0	n_1	n_2	φ_c	θ_{max}	N. A.	$1/L_s$
Glass/air	1	1.50	1.0	41.8°	90.0°	1	8944
Plastic/plastic	1	1.49	1.39	68.9°	32.5°	0.54	3866
Glass/plastic	1	1.46	1.40	73.5°	24.5°	0.41	2962
Glass/glass	1	1.48	1.46	80.6°	14.0°	0.24	1657

Note: The reciprocal of the skip distance ($1/L_s$, or skips per meter) is calculated for a fiber of diameter $100 \ \mu\text{m}$ and at $\theta = \theta_{max}$.

the boundary, a second medium introduced into this region can couple into the wave and provide a means of carrying away energy that otherwise would return into the first medium. Thus, if bare optic fibers are packed closely together in a bundle, there is some leakage between fibers, a phenomenon called *cross talk* in communications applications. The presence of cladding of sufficient thickness prevents leakage, or, to put it more obliquely, negates the frustration of total internal reflection.[6]

The optic-fiber cores are assumed to be homogeneous in composition, characterized by a single index of refraction n_1. Light is propagated through them by multiple total internal reflections. Such fibers are called *step-index fibers* because the refractive index changes discontinuously between core and cladding. They are *multimode* fibers if they permit a discrete number of modes (or ray directions) to propagate. When the fiber is thin enough so that only one mode (a ray in the axial direction) satisfies this condition, the fiber is said to be *single-mode*. Restrictions on possible modes will be described later. Another type of fiber, the *graded-index fiber*, is produced with an index of refraction that decreases continuously from the core axis as a function of radius. All these types are discussed in the sections that follow.

10-5 ALLOWED MODES

Not every ray that enters an optical fiber within its acceptance cone can propagate successfully through the fiber. Only certain ray directions or *modes* are allowed. To see why, we consider the simpler case of a symmetric *planar* or *slab waveguide*, shown in Figure 10-4. The waveguide core of index n_1 has a rectangular (rather than cylindrical) shape and is bounded symmetrically above and below by cladding of index n_2. A sample ray is shown undergoing two total internal reflections from the core-cladding interface at points A and B. Recalling that the ray represents plane waves moving up and down in the waveguide, it is evident that such waves overlap and interfere with one another. Only those waves that satisfy a *resonance condition* are sustained. Notice that points A and C lie on a common wavefront of such waves. If the net phase change that develops between points A and C is some multiple of 2π, then the interfering wavefronts experience constructive interference and corresponding ray directions are allowed. The net phase change is made up of two parts, the optical-path difference Δ and the phase change $2\phi_r$ that occurs due to the two total reflections at points A and B. Thus, the self-sustaining waves must satisfy the condition

$$\frac{\Delta}{\lambda}2\pi + 2\phi_r = 2m\pi$$

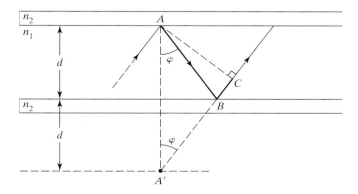

Figure 10-4 Section of a slab waveguide showing a successfully propagating ray or one of the possible modes. The geometry is used to determine the condition for constructive interference.

---

[6]This topic is treated quantitatively in Section 23-5.

where m is an integer. The geometrical extensions denoted by the dashed lines in Figure 10-4, identifying triangle ACA', make it evident that

$$\Delta = AB + BC = A'B + BC = 2n_1 d \cos \varphi \qquad (10\text{-}7)$$

so that the possible modes are given by

$$m = \frac{2n_1 d \cos \varphi}{\lambda} + \frac{\phi_r}{\pi}$$

Now, since $\phi_r \leq \pi$, the second term is 1 at most and is typically negligible compared with the first term. Thus, each successful mode of propagation in the waveguide has an integer mode number m, related to a direction φ_m and given by

$$m \cong \frac{2n_1 d \cos \varphi_m}{\lambda} \qquad (10\text{-}8)$$

For our present purposes, the precise number of allowable modes is not as important as the qualitative dependence of the mode order m on the fiber characteristics. Notice that low-order modes—m small—correspond to $\varphi \cong 90°$, or ray directions that are nearly axial, and high-order modes—m large—correspond to rays that propagate with φ near φ_c, or at steeper ray angles. The total number of propagating modes m_{tot} is the value of m when $\cos \varphi_m$ has its maximum value. This occurs at the critical angle, $\varphi_m = \varphi_c$. Since from Eq. (10-4), $n_1 \cos \varphi_c = \sqrt{n_1^2 - n_2^2} = $ N. A., we can write

$$m_{\text{tot}} \cong \frac{2d}{\lambda} \text{N. A.} + 1 = \frac{2d}{\lambda} \sqrt{n_1^2 - n_2^2} + 1 \qquad (10\text{-}9)$$

We have added 1 to the total number of modes to account for the "straight-through" mode ($m = 0$) at $\varphi = 90°$. Finally, we should point out that, because two independent polarizations are possible for the propagating plane wave, the total number of modes is *twice* that given by Eq. (10-9).

This analysis for the slab waveguide has served to elucidate the physical reasons for mode restriction. The analysis giving the possible modes in a cylindrical fiber is based on the same physical principles but is more complicated and is not developed here. It can be shown[7] that, in this case, the maximum mode number m is the largest integer that is less than the parameter m_{max}, which is given by,

$$m_{\text{max}} = \frac{1}{2} \left(\frac{\pi d}{\lambda} \text{N. A.} \right)^2 \qquad (10\text{-}10)$$

Notice that, as for the slab waveguide, the number of possible modes increases with the ratio d/λ. Thus, larger-diameter fibers are *multimode fibers*. If d/λ is small enough to make $m_{\text{max}} < 2$, the fiber allows only the axial mode to propagate. This is the *monomode* (or *single-mode*) *fiber*. The required diameter for single-mode performance is found by imposing the condition $m_{\text{max}} < 2$ on Eq. (10-10), giving

$$\frac{d}{\lambda} < \frac{2}{\pi (\text{N. A.})}$$

---

[7]Amnon Yariv, *Optical Electronics*, 3d ed. (New York: Holt, Rinehart and Winston, 1985). Peter K. Cheo, *Fiber Optics Devices and Systems* (Englewood Cliffs, N.J.: Prentice-Hall, 1985). Ch. 4.

A more careful analysis indicates that single-mode performance results even when

$$\frac{d}{\lambda} < \frac{2.4}{\pi(\text{N. A.})} \qquad (10\text{-}11)$$

Example 10-1

Suppose an optical fiber (core index of 1.465, cladding index of 1.460) is being used at $\lambda = 1.25\ \mu$m. Determine the diameter for single-mode performance and the number of propagating modes when $d = 50\ \mu$m.

Solution

The N. A. is then $\sqrt{(1.465^2 - 1.46^2)} = 0.121$, and the required diameter for single-mode performance is, using Eq. (10-11),

$$d < \frac{2.4}{\pi(0.121)}(1.25\ \mu\text{m}) \quad \text{or} \quad d < 7.9\ \mu\text{m}$$

On the other hand, if $d = 50\ \mu$m, the fiber is multimode with

$$m_{max} = \frac{1}{2}\left[\pi\frac{50}{1.25}(0.121)\right]^2 = 115$$

giving the number of propagating modes according to Eq. (10.10).

10-6 ATTENUATION

The irradiance of light propagating through a fiber invariably attenuates due to a variety of loss mechanisms that can be classified as *extrinsic* and *intrinsic*. Among the extrinsic losses are inhomogeneities and geometric effects. Inhomogeneities whose dimensions are much greater than the optical wavelength can result, for example, from inadequate mixing of the fiber material before solidification and from an imperfect interface between core and cladding. Extrinsic losses of a geometric nature include sharp bends in the fiber as well as microbends, both of which cause radiation loss because the condition for total internal reflection is no longer satisfied (see Figure 10-5). Other extrinsic losses occur as light is coupled into and out of the fiber. At the fiber input end there are losses due to the restrictions of numerical aperture, as well as losses due to inevitable reflections at the interface, the so-called *Fresnel losses*. The radiation pattern and size of the light source may also be ill-adapted to the fiber end, reducing input efficiency. Of course, such losses also occur at the output end, where the light from the fiber is fed to a detector. Still other losses become important over longer lines wherever connectors, couplers, or splices are necessary. Losses can include mismatch of coupled fiber ends, involving core diameter and lateral and angular alignment. Separation and numerical aperture incompatibility are also possible and can lead to large losses when not properly corrected.

Intrinsic losses are due to absorption, both by the core material and by residual impurities, and by *Rayleigh scattering* from microscopic inhomogeneities, dimensionally smaller than the optical wavelength. The core material—silica, in the case of glass fibers—absorbs in the region of its electronic and molecular transition bands (see Figure 10-6). Strong absorption in the ultraviolet occurs due to electronic and molecular bands. Absorption in the infrared is due to molecular vibrational bands. Both UV and IR absorption decrease as wavelengths approach the visible region. Figure 10-6 shows a minimum of absorption at around 1.3 μm. Residual impurities, such as the transitional metal ions (Fe, Cu, Co, Ni, Mn, Cr, V) and, in particular, the hydroxyl (OH) ion, also contribute to absorption, the last producing significant

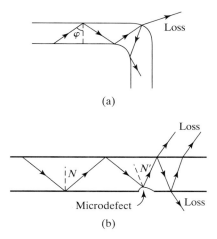

Figure 10-5 Radiation loss from an optical fiber because of (a) a sharp bend and (b) microdefects at the fiber surface. Loss occurs where the condition for total internal reflection fails. Notice that in (b) the defect is also responsible for *mode coupling*, in this case a conversion from a lower to a higher mode.

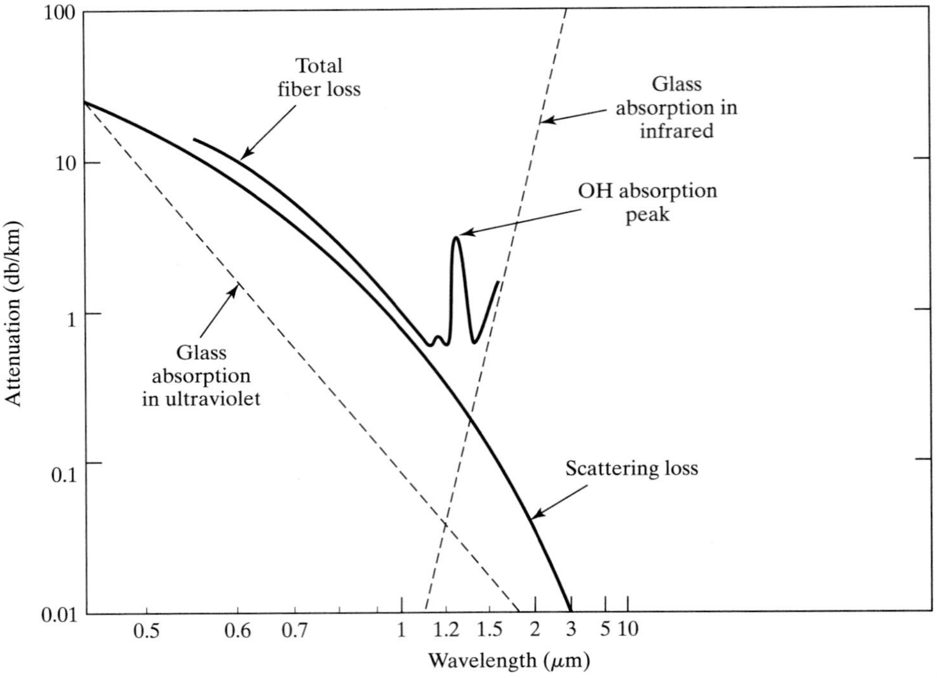

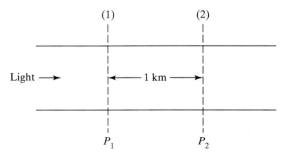

Figure 10-6 Contributions to the net attenuation of a germanium-doped silica glass fiber. (From H. Osanai, T. Shioda, T. Moriyama, S. Araki, M. Horiguchi, T. Izawa, and H. Takara, "Effects of Dopants on Transmission Loss of Low-OH-Content Optical Fibers," *Electronics Letters*, Vol. 12, No. 21 (October 14, 1976): 550. Adapted with permission.)

absorption at 0.95, 1.23, and 1.73 μm. Rayleigh scattering, with its characteristic $1/\lambda^4$ dependence,[8] occurs from localized variations in the density or refractive index of the core material. For example, an optical fiber transmitting at 1.3 μm rather than, say, 800 nm represents a seven-fold reduction in Rayleigh scattering losses.

Absorption losses over a length L of fiber can be described by the usual exponential law for light irradiance I,

$$I = I_0 e^{-\alpha L} \tag{10-12}$$

where α is an *attenuation* or *absorption coefficient* for the fiber, a function of wavelength.[9] For optical fibers, the defining equation for the absorption coefficient in *decibels* per kilometer (db/km) is given by

$$\alpha_{db} \equiv (10 \text{ db/km}) \log_{10}\left(\frac{P_1}{P_2}\right) \tag{10-13}$$

where P_1 and P_2 refer to power levels of the light at two fiber cross sections separated by 1 km, as illustrated in Figure 10-7. For example, if a particular

Figure 10-7 Schematic used to define the absorption coefficient for a glass fiber.

---

[8]Rayleigh scattering is discussed in greater detail in Section 15-3.

[9]Since rays that strike the fiber wall at smaller angles of incidence travel a greater distance through the same axial length L of the absorbing medium, α is also a function of the angle of incidence.

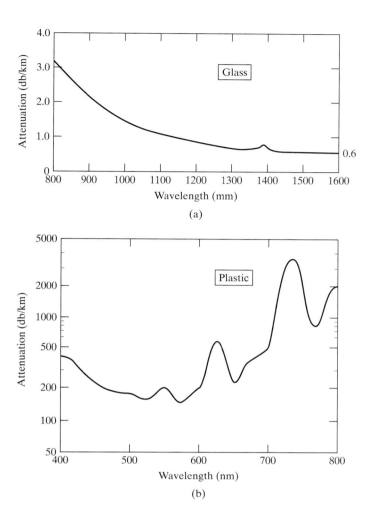

Figure 10-8   (a) Spectral attenuation for all-glass multimode fibers. (Courtesy Corning Glass Works.) (b) Spectral attenuation for all-plastic fiber cable. (Courtesy Mitsubishi Rayon America, Inc.)

fiber experiences a loss given by $\alpha_{db} = 5$ db/km, it means that only 32% of the light energy launched into a 1-km-long fiber arrives at the other end. (Negative values of α_{db} indicate amplification, rather than attenuation!) Dramatic advances have been made in reducing the absorption of fused silica so that today, fibers rated at 0.2 db/km (operating at 1.55 μm) are readily available. Plastic fibers are less expensive but not nearly as transparent. Their overall attenuation is at least an order of magnitude higher than for glass. Glass fibers are therefore preferable in long-distance applications. Figure 10-8 illustrates spectral absorption in silica and plastic fibers.

10-7 DISTORTION

Light transmitted by a fiber may not only lose power by the mechanisms just mentioned; it may also lose information through pulse broadening. When input light is modulated to convey information, the signal waveform becomes distorted due to several mechanisms to be discussed. The major causes of distortion include *modal distortion, material dispersion*, and *waveguide dispersion*, in order of decreasing severity.

Modal Distortion
Figure 10-9 indicates schematically the input of a square wave (a digital signal) into a fiber. The output pulse at the other end suffers, in general, from both attenuation and distortion. Modal distortion occurs because propagating rays

Figure 10-9 Symbolic representation of modal distortion. Portions of a square wave input pulse starting at point A arrive at the fiber end B at different times, depending on the path taken. Shown are the extreme paths: an axial ray and one propagating at the critical angle. The locations of the portions of the pulse taking the extreme paths are shown at the instant the pulse following the axial path reaches point B.

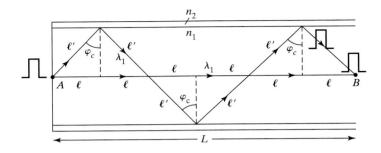

(fiber modes) travel different distances in arriving at the output. Consequently, these rays arrive at different times, broadening the square wave, as shown. The shortest distance L from A to B is taken by the axial ray; the longest distance L' from A to B is taken by the steepest propagating ray that reflects repeatedly at the critical angle φ_c. The distances L and L' are related, as suggested by the geometry in Figure 10-9, by

$$\sin \varphi_c = \frac{n_2}{n_1} = \frac{\ell}{\ell'} = \frac{L}{L'}$$

Thus, the times of flight for the two rays taking the extreme paths between points A and B differ by the time interval $\delta\tau$, given by

$$\delta\tau = \tau_{max} - \tau_{min} = \frac{L'}{v} - \frac{L}{v} = \frac{L}{v}\left(\frac{n_1}{n_2} - 1\right)$$

where v is the speed of light in the fiber core. Since $v = c/n_1$, this result is conveniently expressed as a temporal pulse spread per unit length, in the form

modal distortion (step-index fiber): $\delta\left(\dfrac{\tau}{L}\right) = \dfrac{n_1}{c}\left(\dfrac{n_1 - n_2}{n_2}\right)$ (10-14)

Example 10-2

Suppose the fiber has a core index of 1.46 and a cladding index of 1.45. Determine the modal distortion for this fiber.

Solution

Using Eq. (10-14),

$$\delta\left(\frac{\tau}{L}\right) = \frac{1.46}{3 \times 10^8 \text{ m/s}}\left(\frac{1.46 - 1.45}{1.45}\right) = 3.4 \times 10^{-11} \text{ s/m}$$

$$= 34 \text{ ns/km}$$

The pulse broadens by 34 ns in each km of fiber.[10]

Clearly, this broadening effect limits the possible frequency of distinct pulses. Modal distortion can be lessened by reducing the number of propagating modes. Consequently, the best solution is to use a single-mode fiber,

---

[10]Actual values are somewhat better than predicted by Eq. (10-14) due to *mode coupling* or *mixing* (rays may switch modes in transit due to scattering mechanisms that, on the average, shift power from higher and lower modes to intermediate ones) and due to preferential attenuation (higher modes taking longer paths suffer greater attenuation and so contribute less to overall pulse spreading). For longer distances, this leads to a modified dependence of the form, $\delta\tau \propto \sqrt{L}$.

with only one propagating mode. The next-best solution is to use a *graded index* (GRIN) fiber, which is described next.

The Graded Index (GRIN) Fiber

A GRIN fiber is produced with a refractive index that decreases gradually from the core axis as a function of radius. Figure 10-10 shows the GRIN fiber profile, together with the profile of the ordinary *step-index fiber* for comparison. In the GRIN fiber, a process of continuous refraction bends rays of light, as shown. Notice that at every point of the path, Snell's law is obeyed on a microscopic scale. Ray containment now occurs by a process of continuous refraction, rather than by total reflection. Refraction may not suffice to contain rays making steeper angles with the axis, so GRIN fibers are also characterized by an acceptance cone. When the index profile is suitably adjusted, the rays shown in Figure 10-10c form isochronous loops, an aspect of graded-index fiber that is responsible for reducing modal distortion. Like ordinary fibers, GRIN fibers are also cladded for protection.

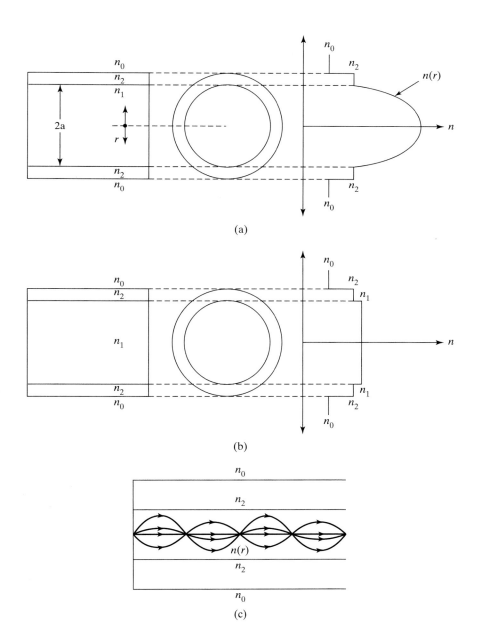

Figure 10-10 (a) Profile of a graded-index (GRIN) fiber, showing a parabolic variation of the refractive index within the core. (b) Profile of a step-index fiber, in which the core index is constant and slightly greater than that of the cladding. (c) Several ray paths within a GRIN fiber, showing their self-confinement due to continuous refraction.

The variation of refractive index with fiber radius is given,[11] in general, by

$$n(r) = n_1 \sqrt{1 - 2\left(\frac{r}{a}\right)^{\alpha_p} \Delta}, \quad 0 \le r \le a \qquad (10\text{-}15)$$

where $\Delta \cong (n_1 - n_2)/n_1$ and $n_1 = [n(r)]_{max}$. The parameter α_p is chosen to minimize modal distortion. For $\alpha_p = 1$, the profile has a triangular shape; for $\alpha_p = 2$, it is parabolic; for higher values of α_p, the profile gradually approaches its limiting case, the step-index profile, as $\alpha_p \to \infty$. Minimizing $\delta\tau$ for all modes requires a value of $\alpha_p = 2$. Thus, the parabolic profile shown in Figure 10-10 is optimum. It can be shown[12] that for this case, pulse broadening is given approximately by

$$\text{modal distortion (GRIN fiber, } \alpha_p = 2): \delta\left(\frac{\tau}{L}\right) = \frac{n_1}{2c}\Delta^2 \qquad (10\text{-}16)$$

Comparing with modal distortion in the step-index fiber, Eq. (10-14), we can write

$$\delta\left(\frac{\tau}{L}\right)_{GRIN} = \frac{\Delta}{2}\left(\frac{n_1}{c}\Delta\right) = \frac{\Delta}{2}\delta\left(\frac{\tau}{L}\right)_{SI}$$

The factor $\Delta/2$ thus represents the improvement offered by a GRIN fiber. For the example used previously, where $n_1 = 1.46$ and $n_2 = 1.45$, we have $\Delta/2 = 1/292$. The GRIN fiber reduces the pulse-broadening effect of modal distortion in this case by a factor of 292.

Material Dispersion

Even if modal distortion is absent, some pulse broadening still occurs because the refractive index is a function of wavelength. Dispersion for a silica fiber is shown in Figure 10-11. Since no light source can be precisely

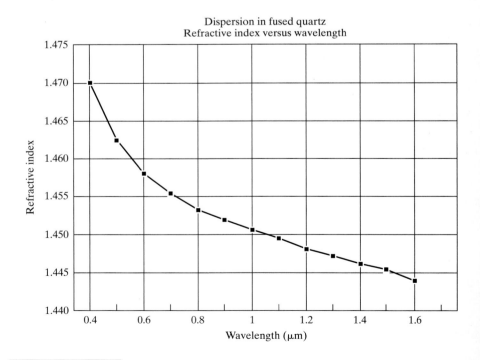

Figure 10-11   Dispersion in fused quartz.

[11]D. Gloge, and E. A. J. Marcatili, "Multimode Theory of Graded-Core Fibers," *Bell Syst. Tech. J.*, Vol. 52 (Nov. 1973), 1563.

[12]Stewart E. Miller, Enrique A. J. Marcatili, and Li Tingye, "Research toward Optical-Fiber Transmission Systems," *Proc. IEEE*, Vol. 61, No. 12 (Dec. 1973): 1703.

monochromatic, the light propagating in the fiber is characterized by a spread of wavelengths determined by the light source. Each wavelength component has a different refractive index and therefore a different speed through the fiber. Pulse broadening occurs, in this case, because each wavelength component arrives at a slightly different time. Light that is more monochromatic suffers less distortion due to material dispersion. To be detected as a single pulse, the output pulse must not spread to the extent of significant overlap with neighboring pulses. Again, this requirement places a limitation on the frequency of input pulses or the rate at which bits of information may be sent.

Figure 10-12 illustrates material dispersion by showing the progress of two square pulses (initially coincident) in a fiber at wavelengths λ_1 and λ_2. If the corresponding refractive indices are n_1 and n_2, the figure implies that $n_1 > n_2$. These wavelengths are but two in a continuum described by the spectral width, $\Delta\lambda$, of the source, usually chosen as the width of the source's spectral output at half-maximum, as shown.

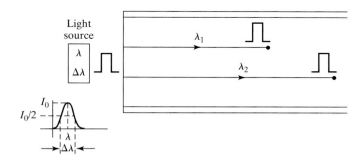

Figure 10-12 Symbolic representation of material dispersion. A square wave input arrives at the fiber end at different times, depending on wavelength. The spectral output of the light source is characterized both by a central wavelength λ and a spectral width $\Delta\lambda$.

Because the optical fiber is dispersive, we describe the speed of propagation of a pulse by its group velocity, v_g (Section 5-6). The time τ required for a signal of angular frequency ω to travel a distance L along the fiber is therefore given by

$$\tau(\omega) = \frac{L}{v_g(\omega)} \quad \text{where} \quad v_g(\omega) = \frac{d\omega}{dk}$$

If the signal bandwidth is $\Delta\omega$, the spread in arrival times per unit distance is expressed by

$$\delta\left(\frac{\tau}{L}\right) = \frac{d}{d\omega}\left(\frac{1}{v_g}\right)\Delta\omega = \frac{d^2k}{d\omega^2}\Delta\omega$$

Now the first derivative $dk/d\omega$ can be calculated from $k = 2\pi/\lambda = n\omega/c$, where n is a function of ω. This gives

$$\frac{dk}{d\omega} = \frac{1}{c}\left(n + \omega\frac{dn}{d\omega}\right) = \frac{1}{c}\left(n - \lambda\frac{dn}{d\lambda}\right) \qquad (10\text{-}17)$$

where we have used the proportion $\omega/d\omega = -\lambda/d\lambda$ in the last step. Progressing to the second derivative, we write

$$\delta\left(\frac{\tau}{L}\right) = \frac{d}{d\omega}\left(\frac{dk}{d\omega}\right)\Delta\omega = \frac{d}{d\lambda}\left(\frac{dk}{d\omega}\right)\Delta\lambda$$

and substitute Eq. (10-17), giving

$$\delta\left(\frac{\tau}{L}\right) = \frac{d}{d\lambda}\left[\frac{1}{c}\left(n - \lambda\frac{dn}{d\lambda}\right)\right]\Delta\lambda = \frac{1}{c}\left(\frac{dn}{d\lambda} - \frac{dn}{d\lambda} - \lambda\frac{d^2n}{d\lambda^2}\right)\Delta\lambda$$

or simply,

$$\text{material dispersion:} \quad \delta\left(\frac{\tau}{L}\right) = -\frac{\lambda}{c}\frac{d^2n}{d\lambda^2}\,\Delta\lambda \equiv -M\Delta\lambda \qquad (10\text{-}18)$$

where M is a property of the core material defined by the prefactor, $(\lambda/c)\,(d^2n/d\lambda^2)$, involving the second derivative of the dispersion. From Eq. (10-18), we see that M has the significance of a temporal pulse spread per unit of spectral width per unit of fiber length. Values of M (in units of ps/nm-km) for pure silica are given in Figure 10-13.

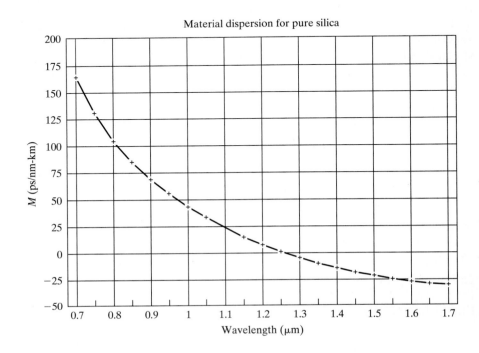

Figure 10-13 Material dispersion in pure silica. The quantity M, representing the pulse broadening (ps) per unit of spectral width (nm) per unit of fiber length (km), is plotted against the wavelength. Pulse broadening becomes zero at 1.27 μm and is negative as wavelength increases further.

Example 10-3

Using Figure 10-13, calculate the pulse spread due to material dispersion in pure silica for both a LED and a LD light source. Consider the source wavelength to be 0.82 μm, with a spectral width of 20 nm for the LED and 1 nm for the more monochromatic LD.

Solution

At 0.82 μm, Figure 10-13 gives a value of near 100 ps/nm-km. Calculation then gives

$$\text{LED:} \quad \delta(\tau/L) = (100\ \text{ps/nm-km})\,(20\ \text{nm}) = 2\ \text{ns/km}$$

$$\text{LD:} \quad \delta(\tau/L) = (100\ \text{ps/nm-km})\,(1\ \text{nm}) = 0.1\ \text{ns/km}$$

At 0.82 μm, the LD is 20 times better than the LED, as a direct result of its superior monochromaticity.

Notice also that pulse broadening due to material dispersion is much smaller than that due to modal distortion. Material dispersion therefore becomes significant only when modal distortion is greatly reduced, in both single-mode and GRIN fibers. Consequently, in the presence of modal distortion, the advantage of superior monochromaticity of a LD over a LED is lost. In applications where fiber lengths are short enough, plastic fibers and LED sources may well represent the best compromise between performance and cost. Finally, notice from Figure 10-13 that M actually passes through zero at around 1.27 μm, so that material dispersion can also be reduced by finding light sources that operate in this spectral region.

We shall extend the previous numerical example to determine the bandwidth limitation due to pulse spreading. Pulse distortion limits transmission frequency and information rate in a way that we can roughly estimate. Let us use as a reasonable criterion for successful discrimination between neighboring pulses that their separation $\delta\tau$ be no less than half their period T:

$$\delta\tau > \frac{T}{2} \quad \text{or} \quad \delta\tau > \frac{1}{2\nu}$$

where ν is the frequency. It follows that the maximum frequency[13]

$$\nu_{max} = \frac{0.5}{\delta\tau} \quad \text{or} \quad \nu_{max}L = \frac{0.5}{\delta(\tau/L)} \qquad (10\text{-}19)$$

For the preceding numerical examples, we calculate an approximate bandwidth, as follows:

$$\text{LED:} \quad \nu_{max}L = \frac{0.5}{2.0 \text{ ns/km}} = 0.25 \text{ GHz-km}$$

$$\text{LD:} \quad \nu_{max}L = \frac{0.5}{0.1 \text{ ns/km}} = 5.0 \text{ GHz-km}$$

Waveguide Dispersion

The last pulse-broadening effect to be discussed is called *waveguide dispersion*, a geometrical effect that depends on waveguide parameters. Compared with modal distortion and material dispersion, waveguide dispersion is a small effect that becomes important only when the other pulse-broadening effects have been essentially eliminated. However, its presence is important in determining the wavelength at which net fiber dispersion is zero, as we shall see.

The variation of the refractive index with wavelength leads to material dispersion, as previously discussed. An *effective refractive index* n_{eff} for the guided wave is defined by $n_{eff} = c/v_g$, where v_g is the group velocity. Waveguide dispersion leads to a variation of n_{eff} with λ for a fixed-diameter fiber, even in the absence of material dispersion. It can be shown[14] that $n_{eff} = n_1 \sin \varphi$. Since φ varies between 90° and φ_c, and $\sin \varphi_c = n_2/n_1$, it follows that n_{eff} varies between n_1 (at $\varphi = 90°$) and n_2 (at $\varphi = \varphi_c$). Thus, n_{eff} for an axial ray depends only on the core index; for a ray at the critical angle, it depends only on the cladding index. The variation of n_{eff} is quite small because $n_1 - n_2$ is, in practice, quite small. Figure 10-14 suggests waveguide dispersion, in the ray representation. For a given mode, the angle between the ray and the fiber axis varies with λ. Thus, the ray paths and times for two different wavelengths also vary with λ, leading to pulse broadening.

The variation of n_{eff} with λ simulates material dispersion and can be handled quantitatively by Eq. (10-18) by simply replacing n by n_{eff}:

$$\text{material dispersion:} \quad \delta\left(\frac{\tau}{L}\right) = -\frac{\lambda}{c}\frac{d^2n}{d\lambda^2}\,\Delta\lambda \equiv -M\,\Delta\lambda$$

$$\text{waveguide dispersion:} \quad \delta\left(\frac{\tau}{L}\right) = -\frac{\lambda}{c}\frac{d^2n_{eff}}{d\lambda^2}\,\Delta\lambda \equiv -M'\Delta\lambda$$

$$(10\text{-}20)$$

---

[13]This value corresponds approximately to the so-called 3-db bandwidth, the modulation frequency at which the signal power is reduced by one-half due to signal distortion.

[14]Joseph C. Palais, *Fiber Optic Communications* (Englewood Cliffs, N.J.: Prentice-Hall, 1988).

Figure 10-14 Symbolic representation of waveguide dispersion. A square wave input arrives at the fiber end at different times, depending on wavelength, even in a dispersionless medium. For any one mode, the angle of propagation is a function of the wavelength.

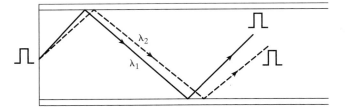

We can appreciate the relative contribution of material and waveguide dispersion by comparing values of M and M'. In Figure 10-13, M ranges from about 165 to -30 ps/nm-km over the spectral range of 0.7 to 1.7 μm. Values of M' for fused quartz over the same range are only about 1 to 4.5 ps/nm-km.[15] For example, the calculation carried out for material dispersion, using a LED source at 0.82 μm, with $M = 100$ ps/nm-km, gave a temporal pulse broadening of 2000 ps/km. For the same wavelength, $M' = 2$ ps/nm-km, giving a pulse broadening of 2/100 times as great, or only 40 ps/km.

Figure 10-13 shows that M for material dispersion becomes zero at around 1.27 μm and then becomes negative for longer wavelengths. Waveguide dispersion, on the other hand, is always positive. Combination of the two thus shifts the wavelength of zero net dispersion toward a longer wavelength of about 1.31 μm in a typical fiber. Sources operating at or near this wavelength are thus ideal in reducing pulse broadening and increasing transmission rates. In discussing attenuation earlier, we pointed out that minimum absorption in silica fibers occurs at around 1.55 μm. The closeness of the wavelengths for minimum absorption and minimum dispersion has motivated attempts to satisfy both conditions by shifting the dispersion curve towards longer wavelengths, so that it passes through zero at 1.55 μm instead of 1.31 μm. Means of modifying the dispersion curve include the use of multiple cladding layers, control of the core/cladding index difference, and variation of the profile parameter α_p in GRIN fibers.

By way of summary, we have discussed three principal ways of reducing pulse broadening in fibers: (1) use a single-mode fiber to eliminate modal distortion, (2) use a light source of small spectral width $\Delta\lambda$ to reduce material dispersion, and (3) use a light source operating in a spectral region where both attenuation and dispersion are as small as possible. Clearly, the required length of fiber and the cost of components play major roles in determining the selection of the best system for a specific application.

10-8 HIGH-BIT-RATE OPTICAL- FIBER COMMUNICATIONS

In previous sections we have discussed some of the limitations to high-bit-rate transmission through optical fibers. We concluded that in order to transfer information at a high rate over long distances, a combination of source wavelength and single-mode fiber that minimizes attenuation and distortion should be used. Most fiber-optic communications systems in use today use semiconductor sources that emit near 1550 nm in order to minimize the attenuation of the optical signal. In this section, we discuss some of the components used in high-bit-rate fiber-optic communication systems.

Wavelength-Division-Multiplexing
The information-carrying capacity of fiber-optic cables has been greatly increased by the combination of *time-division multiplexing* (TDM) and

---

[15]Joseph C. Palais, *Fiber Optic Communications* (Englewood Cliffs, N.J.: Prentice-Hall, 1988).

wavelength-division multiplexing (WDM). Time-division multiplexing is an information encoding scheme that ensures that all time slots in a stream of bits of information are used to full capacity. For example, a normal telephone conversation includes many pauses in which no information is being transmitted. Portions of other conversations can be carried in these pauses in order to increase the carrying capacity of the system. A more dramatic increase in carrying capacity results from wavelength-division multiplexing, in which information is simultaneously carried in different wavelength channels through the same fiber. A schematic of an optical communications system employing WDM is shown in Figure 10-15. As indicated in the figure, carrier signals of different wavelengths originating from different transmitters are combined by a *multiplexer* into a single optical fiber. These signals are then separated back into the different wavelength channels by a *demultiplexer* before reaching separate receivers. Four different wavelength channels are shown in Figure 10-15. Today, fibers carrying 40 different wavelength channels are common, and a combination of TDM and WDM results in optical fibers systems with carrying capacities of more than 10^{12} bits/s.

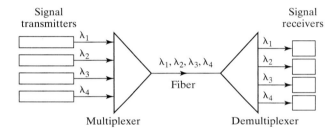

Figure 10-15 Fiber-optic communications system using wavelength-division multiplexing.

   Systems that utilize such a large number of wavelength channels are sometimes said to employ *dense wavelength-division multiplexing* (DWDM). A standard DWDM system might use 40 wavelength channels near 1550 nm. The difference in wavelength between "adjacent" channels is typically about 0.8 nm. Such a system must meet several stringent design considerations. First, sources with a stable wavelength output at each of the wavelengths used in the different channels must be available. As indicated in the summary table of laser systems given at the end of Chapter 6, indium gallium arsenide phosphide (InGaAsP) laser diodes can be engineered to emit near 1550 nm. Tuning these lasers to emit in specific stable wavelengths that correspond to each of the 40 channels is an engineering challenge that has been accomplished by a variety of means.[16] A second key breakthrough enabling the use of DWDM in long-haul fibers was the discovery and development of *erbium-doped fiber amplifiers* (EDFA's). Erbium atoms can be "doped" into sections of silica fiber. When the erbium-doped fiber section is optically pumped, it serves as an amplifer for light near 1550 nm and so can be used to restore attenuated signal strength. The gain bandwidth of an EDFA is about 35 nm, which is enough to amplify some $35/0.8 = 45$ wavelength channels separated by 0.8 nm. Short sections of pumped erbium-doped fiber can be placed at widely spaced (100 km) sections of a long-haul fiber-optic communications system. In Chapter 24 we will discuss briefly another useful fiber-amplifier design based on stimulated Raman scattering. A third DWDM design challenge that has been met is the limiting of dispersion so that separating the wavelength channels by 0.8 nm provides isolation sufficient to prevent cross talk between channels over long hauls. Finally, in order to use DWDM,

---

[16]See, for example, Milorad Cvijetic, *Optical Transmission Systems Engineering* (Norwood, MA: Artech House, Inc., 2004, Ch. 2).

efficient multiplexers and demultiplexers that can discriminate between and combine or separate the different wavelength channels must be used. A variety of multiplexing schemes exist: Here we discuss one based on the Mach-Zehnder interferometer.

Mach-Zehnder Fiber Interferometers

The Mach-Zehnder interferometer was introduced in Section 8-3. We will show here that a Mach-Zehnder fiber interferometer can be used to demultiplex (and multiplex) an optical signal containing a number of different wavelength channels. Before doing so, we will briefly review the operation of a standard (mirror and beam splitter) Mach-Zehnder interferometer emphasizing characteristics of importance to the present discussion. Such a standard interferometer is depicted in Figure 10-16a. The interferometer consists of two 50-50 beam splitters, BS1 and BS2, and two mirrors, M1 and M2. Consider light entering the interferometer through the Input 1 "port" of BS1. This light is split into two beams that travel the different paths, labeled Path 1 and Path 2, respectively, before encountering BS2. To understand the operation of the Mach-Zehnder interferometer, it is helpful to note that beam splitters with real transmission coefficients must have the property that the reflection coefficients from opposite sides of the beam splitter differ by a factor of $-1 = e^{i\pi}$. That is, the phase shifts upon reflection from opposite sides of the beam splitter differ by π. In Figure 10-16a, reflection from the lower surface of BS2 is taken to be $r_2 = 1/\sqrt{2}$, and that from the upper surface of BS2 is taken to be $r_2' = -1/\sqrt{2} = e^{i\pi}/\sqrt{2}$. *The extra π phase shift upon reflection from the upper beam splitter surface ensures that when constructive interference occurs in output port 1, destructive interference will occur in output port 2.* This behavior is required in order that the total energy exiting the interferometer be the same as that entering the interferometer. Indeed, requiring energy conservation in a Mach-Zehnder interferometer is one way to prove that beam splitters with real transmission coefficients have reflection coefficients from opposite surfaces that differ by a factor of -1. (See problem 10-26.)

The Mach-Zehnder fiber interferometer shown in Figure 10-16b operates by the same set of principles as the standard interferometer shown in Figure 10-16a. Light entering Input 1 of the fiber interferometer is split at a four-port fiber coupler, FC1. The separate portions of the beams then travel different paths before being recombined and directed into two different output ports by the second fiber coupler, FC2. The fiber couplers FC1 and FC2 function as beam splitters. If the light from Paths 1 and 2, of Figure 10-16b, happens to constructively interfere upon combination into Output 1, it will destructively interfere upon combination into Output 2. The difference in path lengths can be controlled by a delay line (or a variety of phase shift mechanisms) in one of the "arms" of the fiber interferometer. Figure 10-16b shows a Mach-Zehnder fiber interferometer acting as a demultiplexer. Light composed of two different (free-space) wavelength components λ_1 and λ_2

Figure 10-16 Standard and fiber Mach-Zehnder interferometers. (a) Standard Mach-Zehnder interferometer with two mirrors, M1 and M2, and two 50-50 beam splitters. Reflection coefficients of the beam splitter surfaces are indicated. (b) Mach-Zehnder fiber interferometer used as a wavelength demultiplexer. The four-port fiber couplers FC1 and FC2 act as beam splitters.

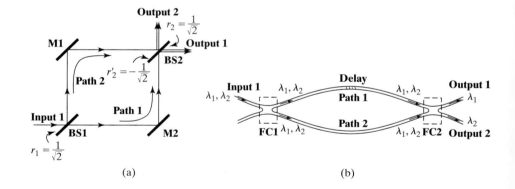

(a)   (b)

enters the interferometer through the port marked Input 1. To operate as a demultiplexer, the difference in path lengths of the arms of the Mach-Zehnder fiber interferometer must be chosen so that light of wavelength λ_1 constructively interferes upon recombination into Output 1 while light of wavelength λ_2 constructively interferes upon recombination into Output 2. Let us assume that light from Path 1 suffers no phase shift upon reflection into Output 1 while light from Path 2 suffers a π phase shift upon reflection into Output 2. Further, let us take the index of refraction of the fiber to be n so that the wavelength of the two components in the fiber are λ_1/n and λ_2/n, respectively. In that case, light of free-space wavelength λ_1 will constructively interfere in the direction of Output 1 if the difference in path lengths of Path 1 and Path 2, ΔL, is given by

$$\Delta L = m\lambda_1/n \quad m = 0, \pm 1, \pm 2 \ldots \tag{10-21}$$

Under this condition, light of wavelength λ_1 will not be present in Output 2. Due to the extra π phase shift upon reflection into Output 2, light of free-space wavelength λ_2 will constructively interfere in the direction of Output 2 if the path length difference ΔL differs by an odd multiple of the half-wavelength of this component in the fiber. That is, for constructive interference of light of free-space wavelength λ_2 in Output 2,

$$\Delta L = (m + 1/2)\lambda_2/n \quad m = 0, \pm 1, \pm 2 \ldots \tag{10-22}$$

Both Eqs. (10-21) and (10-22) must be satisfied for efficient demultiplexing. These demultiplexing relations are explored in Example 10-4 and problem 10-28.

Example 10-4

Consider a Mach-Zehnder interferometer of the type just described and illustrated in Figure 10-16b. Assume that the interferometer is designed to demultiplex two light signals of free-space wavelengths $\lambda_2 = 1550$ nm and $\lambda_1 = 1551$ nm. Assume that when a signal containing these two wavelength components enters the interferometer through Input 1, the λ_2 light component exits the interferometer through Output 2 and the λ_1 light component exits the interferometer through Output 1.

a. Calculate the path length difference ΔL required to perform this task. Assume that the index of refraction of the fiber is $n = 1.500$.

b. Through which output port would light of wavelength $\lambda_3 = 1549$ nm exit?

c. Through which output port would light of wavelength $\lambda_4 = 1548$ nm exit?

Solution

a. Solving Eq. (10-21) for the mode number m gives $m = \dfrac{n\Delta L}{\lambda_1}$. Using this relation in Eq. (10-22) yields

$$\Delta L = \left(\frac{n\Delta L}{\lambda_1} + \frac{1}{2} \right) \frac{\lambda_2}{n}$$

Solving this expression for ΔL gives

$$\Delta L = \frac{1}{2n}\left(\frac{1}{\lambda_2} - \frac{1}{\lambda_1}\right)^{-1} = \frac{1}{2(1.500)}\left(\frac{1}{1550 \text{ nm}} - \frac{1}{1551 \text{ nm}}\right)^{-1}$$

$$= 801350 \text{ nm} = 8.0135 \times 10^{-4} \text{ m}$$

One might be tempted to accept this answer with no further analysis. However, in fact, one must check to see if the mode number m associated with this answer is an integer. For the fortuitous case at hand, this turns out to be the case since Eq. (10-21) gives

$$m = \frac{n\Delta L}{\lambda_1} = \frac{1.500(801350 \text{ nm})}{1551 \text{ nm}} = 775.0$$

In general, the procedure used here to solve for ΔL only ensures that Eqs. (10-21) and (10-22) are both satisfied for some value of m, not necessarily an integer. This complication is explored in problem 10-28.

b. Light of wavelength $\lambda_3 = 1549$ nm satisfies the constructive interference condition (Eq. (10-21)) for Output 1 since

$$m = \frac{\Delta L}{(\lambda_3/n)} = \frac{801350 \text{ nm}}{(1549 \text{ nm})/(1.500)} = 776.0$$

is an integer. Thus, this wavelength, like $\lambda_1 = 1551$ nm, would predominately exit Output 1.

c. Light of wavelength $\lambda_4 = 1548$ nm very nearly satisfies the constructive interference condition (Eq. (10-22)) for Output 2 since

$$m = \frac{\Delta L}{(\lambda_4/n)} = \frac{801350 \text{ nm}}{(1548 \text{ nm})/(1.500)} = 776.502 \approx 776.5$$

is very nearly an integer $+ 1/2$. Thus, this wavelength, like $\lambda_2 = 1552$ nm, would predominately exit Output 2.

The results of Example 10-4 indicate that a large number of equally spaced wavelength channels could be demultiplexed by a system of Mach-Zehnder interferometers. Figure 10-17a shows three Mach-Zehnder fiber interferometers (MZ1, MZ2, and MZ3) arranged to demultiplex a signal containing four different wavelength channels. Adding more Mach-Zehnder interferometers to the chain would allow the system to demultiplex a signal containing a larger number of wavelength channels. The system displayed in Figure 10-17b, in which the directions of all the light fields in Figure 10-17a are simply reversed, shows clearly that an array of Mach-Zehnder interferometers can also serve a multiplexing function.

In this section, we have discussed but a few of the many and varied optical engineering challenges encountered when developing a high-bit-rate fiber-optic communications system. In later chapters, we will have occasion to further discuss optical communications systems and the use of fiber-optic components as optical switches and modulators. Optical fibers play an increasingly important role in a wide array of optical systems.

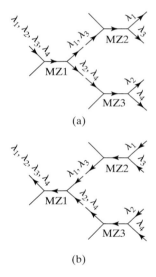

(a)

(b)

Figure 10-17 Array of Mach-Zehnder fiber interferometers used to (a) demultiplex a signal into four wavelength channels and (b) multiplex four wavelength channels. The arrays consist of three Mach-Zehnder interferometers, MZ1, MZ2, and MZ3.

PROBLEMS

10-1 The bandwidth of a single telephone channel is 4 kHz. In a particular telephone system, the transmission rate is 44.7 Mbps. In an actual system, some channels are devoted to housekeeping functions such as synchronization. In this system, 26 channels are so devoted. How many independent telephone channels can the system accommodate?

10-2 Determine the limit to the number of TV station channels that could transmit on a single optical beam of 1.55 μm wavelength if

    a. The entire bandwidth $(\Delta\nu = \nu)$ of the signal was used.

    b. A bandwidth of 4×10^{12} Hz is used. (This corresponds to a typical DWDM system.)

10-3 a. Referring to Figure 10-3, show that, for a guided ray traveling at the steepest angle relative to the fiber axis, the skip distance L_s can be expressed by

$$L_s = \frac{n_2 d}{\sqrt{n_1^2 - n_2^2}}$$

b. How many reflections occur per meter for such a ray in a step-index fiber with $n_1 = 1.460$, $n_2 = 1.457$, and $d = 50\ \mu m$?

10-4 Refractive indices for a step-index fiber are 1.52 for the core and 1.41 for the cladding. Determine (a) the critical angle; (b) the numerical aperture; (c) the maximum incidence angle θ_m for light that is totally internally reflected.

10-5 A step-index fiber 0.0025 in. in diameter has a core of index 1.53 and a cladding of index 1.39. Determine (a) the numerical aperture of the fiber; (b) the *acceptance angle* (or maximum entrance cone angle); (c) the number of reflections in 3 ft of fiber for a ray at the maximum entrance angle, and for one at half this angle.

10-6 a. Show that the actual distance x_s a ray travels during one skip distance is given by

$$x_s = \frac{n_1 d}{\sin\theta}$$

where θ is the entrance angle and the fiber is used in air.

b. Show that the actual total distance x_t a ray with entrance angle θ travels over a total length L of fiber is given by

$$x_t = \frac{n_1 L}{(n_1^2 - \sin^2\theta)^{1/2}}$$

c. Determine x_s, L_s, and x_t for a 10-m-long fiber of diameter 50 μm, core index of 1.50, and a ray entrance angle of $\theta = 10°$.

10-7 How many modes can propagate in a step-index fiber with $n_1 = 1.461$ and $n_2 = 1.456$ at 850 nm? The core radius is 20 μm.

10-8 Determine the maximum core radius of a glass fiber so that it supports only one mode at 1.25 μm wavelength, for which $n_1 = 1.460$ and $n_2 = 1.457$.

10-9 Consider a slab waveguide of AlGaAs for which $n_1 = 3.60$ and $n_2 = 3.55$. How many independent modes can propagate in this waveguide if $d = 5\lambda$ and $d = 50\lambda$? (See Figure 10-4.)

10-10 A signal of power 5 μW exists just inside the entrance of a fiber 100 m long. The power just inside the fiber exit is only 1 μW. What is the absorption coefficient of the fiber in db/km?

10-11 An optic-fiber cable 3 km long is made up of three 1-km lengths, spliced together. Each length has a 5-db loss and each splice contributes a 1-db loss. If the input power is 4 mW, what is the output power?

10-12 The attenuation of a 1-km length of RG-19/U coaxial cable is about 12 db at 50 MHz. Suppose the input power to the cable is 10 mW and the receiver sensitivity is 1 μW. How long can the coaxial cable be under these conditions? If optical fiber is used instead, with a loss rated at 4 db/km, how long can the transmission line be?

10-13 A Ge-doped silica fiber has an attenuation loss of 1.2 db/km due to Rayleigh scattering alone when light of wavelength 0.90 μm is used. Determine the attenuation loss at 1.55 μm.

10-14 a. Show that the attenuation db/km is given by

$$\alpha_{db} = (10\ db/km)\log_{10}(1 - f)$$

where f is the overall fractional power loss from input to output over a 1-km-long fiber.

b. Determine the attenuation in db/km for fibers having an overall fractional power loss of 25%, 75%, 90%, and 99%.

10-15 Determine (a) the length and (b) transit time for the longest and shortest trajectories in a step-index fiber of length 1 km having a core index of 1.46 and a cladding index of 1.45. (See Figure 10-9.)

10-16 Evaluate modal distortion in a fiber by calculating the difference in transit time through a 1-km fiber required by an axial ray and a ray entering at the maximum entrance angle of 35°. Assume a fused silica core index of 1.446. What is the maximum frequency of input pulses that produce nonoverlapping pulses on output due to this case of modal dispersion?

10-17 Calculate the time delay between an axial ray and one that enters a 1-km-long fiber at an angle of 15°. The core index is 1.48.

10-18 Calculate the group delay between the fastest and slowest modes in a 1-km-long step-index fiber with $n_1 = 1.46$ and a *relative index difference* $\Delta = (n_1 - n_2)/n_2 = 0.003$, using a light source at wavelength 0.9 μm.

10-19 Plot the refractive index profile for a GRIN fiber of radius 50 μm and with $n_1 = 1.5$ and $\Delta = 0.01$. Do this for the profile parameter $\alpha_p = 2$ and repeat for $\alpha_p = 10$.

10-20 Calculate the delay due to modal dispersion in a 1-km GRIN fiber with $\alpha_p = 2$. The maximum core index is 1.46 and the cladding index is 1.44. By what factor is this fiber an improvement over a step-index fiber with $n_1 = 1.46$ and $n_2 = 1.44$?

10-21 Equation (10-19) allows calculation of bandwidth for distances less than the *equilibrium length* of fiber (see footnote 10). Assume an equilibrium length of 1 km and determine for this fiber length the 3-db bandwidth of a step-index multimode fiber whose pulse broadening is given by 20 ns/km.

10-22 Determine the material dispersion in a 1-km length of fused silica fiber when the light source is (a) a LED centered at 820 nm with a spectral width of 40 nm and (b) a LD centered at 820 nm with a spectral width of 4 nm.

10-23 The total delay time $\delta\tau$ due to *both* modal distortion and material dispersion is given by

$$(\delta\tau)^2 = (\delta\tau_{mod})^2 + (\delta\tau_{mat})^2$$

Determine the total delay time in a 1-km fiber for which $n_1 = 1.46$, $\Delta = 1\%$, $\lambda = 820$ nm, and $\Delta\lambda = 40$ nm.

10-24 Waveguide dispersion is measured in a silica fiber at various wavelengths using laser diode sources with a spectral width of 2 nm. The results are

$\lambda(\mu m)$	$\delta(\tau/L)(ps/km)$
0.70	1.88
0.90	5.02
1.10	7.08
1.40	8.40
1.70	8.80

  a. Plot the waveguide parameter M' versus λ in the range 0.70 to 1.70 μm.

  b. Determine the waveguide dispersion in ps/km at $\lambda = 1.27$ and 1.55 μm for a source with a spectral width of 1 nm.

10-25 Compare pulse broadening for a silica fiber due to the three principal causes—modal distortion, material dispersion, and waveguide dispersion—in a step-index fiber. The core index is 1.470 and the cladding index is 1.455 at $\lambda = 1$ μm. Assume a LED source with a spectral width of 25 nm. The values of the parameters M and M' are 43 ps/nm-km and 3 ps/nm-km, respectively.

  a. Determine each separately by calculating $\delta\tau$ for a 1-km length of fiber.

  b. Determine an overall broadening $\delta\tau$ for a 1-km length of fiber, using

$$(\delta\tau)^2 = (\delta\tau_{mod})^2 + (\delta\tau_{mat})^2 + (\delta\tau_{wg})^2$$

10-26 Consider the Mach-Zehnder interferometer depicted in Figure 10-16a. Take the transmission coefficients of the beam splitters to be real. Find the irradiance of the light exiting each of the output ports of the interferometer and show that the sum of these output irradiances is equal to the input irradiance.

10-27 a. Find the difference in frequency between two wavelength channels near 1550 nm that differ in wavelength by 0.8 nm.

  b. Find the frequency bandwidth $\Delta\nu$ of a DWDM system utilizing 40 wavelength channels near 1550 nm if the channels are separated by 0.8 nm.

10-28 Consider a Mach-Zehnder fiber demultiplexer depicted in Figure 10-16b. Assume that $n = 1.5000$.

  a. Find a difference in path length between the two arms of the fiber interferometer ΔL that will efficiently demultiplex a signal containing wavelength components $\lambda_1 = 1550.8$ nm and $\lambda_2 = 1550.0$ nm. Assume that the λ_1 component exits through Output 1. (Take care to ensure that your solution nearly satisfies both Eqs. (10-21) and (10-22) with integer m.)

  b. If the path length difference is as found from part (a), through which output port would light of wavelength $\lambda_3 = 1551.6$ nm exit the interferometer?

  c. If the path length difference is as found from part (a), find the ratio of the irradiances exiting through the two output ports if the input signal has a wavelength of 1550.4 nm.

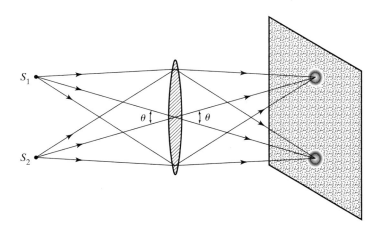

11 *Fraunhofer Diffraction*

INTRODUCTION

The wave character of light has been invoked to explain a number of phenomena, classified as "interference effects" in preceding chapters. In each case, two or more individual coherent beams of light, originating from a single source and separated by amplitude or wavefront division, were brought together again to interfere. Fundamentally, the same effect is involved in the *diffraction* of light. In its simplest description, diffraction is any deviation from geometrical optics that results from the obstruction of a wavefront of light. For example, an opaque screen with a round hole represents such an obstruction. On a viewing screen placed beyond the hole, the circle of light may show complex edge effects. This type of obstruction is typical in many optical instruments that utilize only the portion of a wavefront passing through a round lens. Any obstruction, however, shows detailed structure in its own shadow that is quite unexpected on the basis of geometrical optics.

Diffraction effects are a consequence of the wave character of light. Even if the obstacle is not opaque but causes local variations in the amplitude or phase of the wavefront of the transmitted light, such effects are observed. Tiny bubbles or imperfections in a glass lens, for example, produce undesirable diffraction patterns when transmitting laser light. Because the edges of optical images are blurred by diffraction, the phenomenon leads to a fundamental limitation in instrument resolution. More often, though, the sharpness of optical images is more seriously degraded by optical aberrations due to the imaging components themselves. *Diffraction-limited* optics is good optics indeed.

The double slit studied previously constitutes an obstruction to a wavefront in which light is blocked everywhere except at the two apertures. Recall that the irradiance of the resulting fringe pattern was calculated by treating the two openings as point sources, or long slits whose widths could be treated as points. A more complete analysis of this experiment must take into account the finite size of the slits. When this is done, the problem is treated as a diffraction problem. The results show that the interference pattern determined earlier is modified in a way that accounts for the actual details of the observed fringes.

Adequate agreement with experimental observations is possible through an application of the *Huygens-Fresnel principle*. According to Huygens, every point of a given wavefront of light can be considered a source of secondary spherical wavelets. To this, Fresnel added the assumption that the actual field at any point beyond the wavefront is a superposition of all these wavelets, taking into account both their amplitudes and phases. Thus, in calculating the diffraction pattern of the double slit at some point on a screen, one considers every point of the wavefront emerging from each slit as a source of wavelets whose superposition produces the resultant field. This procedure then takes into account a continuous array of sources across both slits, rather than two isolated point sources, as in the interference calculation. Diffraction is often distinguished from interference on this basis: In diffraction phenomena, the interfering beams originate from a continuous distribution of sources; in interference phenomena, the interfering beams originate from a discrete number of sources. This is not, however, a fundamental *physical* distinction.

A further classification of diffraction effects arises from the mathematical approximations possible when calculating the resultant fields. If both the source of light and observation screen are *effectively* far enough from the diffraction aperture so that wavefronts arriving at the aperture and observation screen may be considered plane, we speak of *Fraunhofer, or far-field, diffraction*, the type treated in this chapter. When this is not the case and the curvature of the wavefront must be taken into account, we speak of *Fresnel, or near-field, diffraction*, the subject of Chapter 13. In the far-field approximation, as the viewing screen is moved relative to the aperture, the *size* of the diffraction pattern scales uniformly, but the *shape* of the diffraction pattern does not change. In the near-field approximation, the situation is more complicated. Both the shape and size of the diffraction pattern depend on the distance between the aperture and the screen. As the screen is moved away from the aperture, the image of the aperture passes through the forms predicted in turn by geometrical optics, near-field diffraction, and far-field diffraction.

It should be stated at the outset that the Huygens-Fresnel principle we shall employ to calculate diffraction patterns is itself an approximation. When no light penetrates an opaque screen, it means that the interaction of the incident radiation with the electronic oscillators, set into motion within the screen, is such as to produce zero net field beyond the screen. This balance is not maintained at the edge of an aperture in the screen, where the distribution of oscillators is interrupted. The Huygens-Fresnel principle does not include the contribution to the diffraction field of the electronic oscillators in the screen material at the edge of the aperture. Such edge effects are important, however, only when the observation point is very near the aperture itself.

11-1 DIFFRACTION FROM A SINGLE SLIT

We first calculate the Fraunhofer diffraction pattern from a *single slit*, a rectangular aperture characterized by a length much larger than its width. For Fraunhofer diffraction, the wavefronts of light reaching the slit must be essentially plane. In practice, this is easily accomplished by placing a source in the focal plane of a positive lens or by simply using a laser beam with a small divergence angle as the source. Similarly, we consider the observation screen to be effectively at infinity by using a lens on the exit side of the slit, as shown

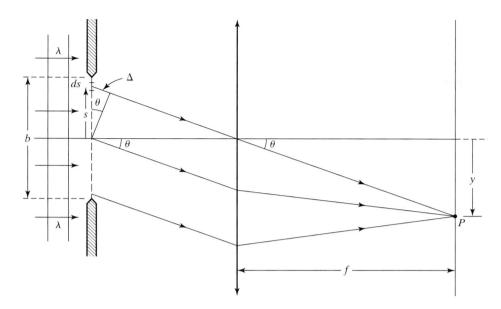

Figure 11-1 Construction for determining irradiance on a screen due to Fraunhofer diffraction by a single slit.

in Figure 11-1. Then the light reaching any point such as P on the screen is due to parallel rays of light from different portions of the wavefront at the slit (dashed line). According to the Huygens-Fresnel principle, we can consider spherical wavelets to be emanating from each point of the wavefront as it reaches the plane of the slit and then calculate the resultant field at P by adding the waves according to the principle of superposition. As shown in Figure 11-1, the waves do not arrive at P in phase. A ray from the center of the slit, for example, has an optical-path length that is an amount Δ shorter than one leaving from a point a vertical distance s above the optical axis.

The plane portion of the wavefront in the slit opening represents a continuous array of Huygens' wavelet sources. We consider each interval of length ds as a source and calculate the result of all such sources by integrating over the entire slit width b. Each interval contributes a spherical wavelet at P whose magnitude is directly proportional to the infinitesimal length ds. Thus,

$$dE_p = \left(\frac{E_L\,ds}{r}\right)e^{i(kr-\omega t)} \tag{11-1}$$

where r is the optical-path length from the interval ds to the point P. The amplitude $(E_L\,ds/r)$ has a $1/r$ dependence because spherical waves decrease in irradiance with distance, in accordance with the inverse square law. That is, for spherical waves the irradiance (which is proportional to the square of the electric field amplitude) is proportional to $1/r^2$ and so the electric field amplitude of a spherical wave is proportional to $1/r$ (see Section 4-6). The proportionality constant E_L, here taken to be constant, determines the strength of the electric field contribution coming from each slit interval ds. Let us set $r = r_0$ for the wave from the center of the slit (at $s = 0$). Then, for any other wave originating at height s, taking the difference in phase into account, the differential field at P is

$$dE_p = \left(\frac{E_L\,ds}{r_0 + \Delta}\right)e^{i[k(r_0+\Delta)-\omega t]} = \left(\frac{E_L\,ds}{r_0 + \Delta}\right)e^{i(kr_0-\omega t)}e^{ik\Delta} \tag{11-2}$$

Note that the quantity $r_0 + \Delta$ appears both in the amplitude factor and in the phase factor. The path difference Δ is much smaller than r_0 and so (to lowest order) can be ignored in the amplitude factor. However, this path difference

Δ cannot be ignored in the phase factor. To understand why this is so, note that $k\Delta = (2\pi/\lambda)\,\Delta$. So as Δ varies by one wavelength, the phase $k\Delta$ varies over an entire cycle of range 2π. Figure 11-1 shows that $\Delta = s\sin\theta$. With these modifications, Eq. 11-2 can be rewritten as

$$dE_P = \left(\frac{E_L ds}{r_0}\right) e^{i(kr_0 - \omega t)} e^{iks\sin\theta} \tag{11-3}$$

The total electric field at the point P is found by integrating over the width of the slit. That is,

$$E_P = \int_{\text{slit}} dE_P = \frac{E_L}{r_0} e^{i(kr_0 - \omega t)} \int_{-b/2}^{b/2} e^{iks\sin\theta}\, ds \tag{11-4}$$

Integration gives

$$E_P = \frac{E_L}{r_0} e^{i(kr_0 - \omega t)} \left(\frac{e^{iks\sin\theta}}{ik\sin\theta}\right)_{-b/2}^{b/2} \tag{11-5}$$

Inserting the limits of integration into Eq. (11-5),

$$E_P = \frac{E_L}{r_0} e^{i(kr_0 - \omega t)} \frac{e^{(ikb\sin\theta)/2} - e^{-(ikb\sin\theta)/2}}{ik\sin\theta} \tag{11-6}$$

The phases of the exponential terms suggest we make a convenient substitution,

$$\beta \equiv \tfrac{1}{2} kb\sin\theta \tag{11-7}$$

Then,

$$E_P = \frac{E_L}{r_0} e^{i(kr_0 - \omega t)} \frac{b(e^{i\beta} - e^{-i\beta})}{2i\beta} = \frac{E_L}{r_0} e^{i(kr_0 - \omega t)} \frac{b(2i\sin\beta)}{2i\beta} \tag{11-8}$$

where we have applied Euler's equation to obtain the last equality. Simplifying, we find

$$E_P = \frac{E_L b}{r_0} \frac{\sin\beta}{\beta} e^{i(kr_0 - \omega t)} \tag{11-9}$$

Thus, the amplitude of the resultant field at P, given by Eq. (11-9), includes the *sinc* function $(\sin\beta)/\beta$, where β varies with θ and thus with the observation point P on the screen. We may give physical significance to β by interpreting it as a *phase difference*. Since a phase difference is given in general by $k\Delta$, Eq. (11-7) indicates a path difference associated with β of $\Delta = (b/2)\sin\theta$, shown in Figure 11-1. Thus $|\beta|$ represents the magnitude of the phase difference, at point P, between waves from the center and either endpoint of the slit, where $|s| = b/2$. In the analysis leading to Eq. (11-9), we assumed that the field strength E_L associated with each slit interval ds was a constant. In Chapter 21, we show that if the field strength is not uniform across the slit, then the Fraunhofer diffraction pattern is the Fourier transform of the function that describes the field strength at various points within the aperture.

The irradiance I at P is proportional to the square of the resultant field amplitude there. The amplitude of the electric field given in Eq. (11-9) is

$$E_0 = \frac{E_L b}{r_0} \frac{\sin\beta}{\beta}$$

Thus, with the help of Eq. (4-42) in Chapter 4 on wave equations, we find the irradiance I to be

$$I = \left(\frac{\varepsilon_0 c}{2}\right)E_0^2 = \frac{\varepsilon_0 c}{2}\left(\frac{E_L b}{r_0}\right)^2 \frac{\sin^2 \beta}{\beta^2}$$

or

$$I = I_0\left(\frac{\sin^2 \beta}{\beta^2}\right) \equiv I_0 \, \text{sinc}^2(\beta) \qquad (11\text{-}10)$$

where I_0 includes all constant factors. Equations (11-9) and (11-10) now permit us to plot the variation of irradiance with vertical displacement y from the symmetry axis at the screen. The sinc function has the property that it approaches 1 as its argument approaches 0:

$$\lim_{\beta \to 0} \text{sinc}(\beta) = \lim_{\beta \to 0}\left(\frac{\sin \beta}{\beta}\right) = 1 \qquad (11\text{-}11)$$

Otherwise, the zeros of $\text{sinc}(\beta)$ occur when $\sin \beta = 0$, that is, when

$$\beta = \tfrac{1}{2}(kb \sin \theta) = m\pi \qquad m = \pm 1, \pm 2, \ldots$$

Equation (11-11) shows that the value $m = 0$ should not be included in this condition. The irradiance is plotted as a function of β in Figure 11-2. Setting $k = 2\pi/\lambda$, the condition for zeros of the sinc function (and so of the irradiance) is

$$m\lambda = b \sin \theta \qquad m = \pm 1, \pm 2, \ldots \qquad (11\text{-}12)$$

Referring to Figure 11-1, note that the distance y from the center of the screen to a point on the screen P located by the angle θ is given approximately by $y \cong f \sin \theta$, where we have made the small angle approximation $\sin \theta \cong \tan \theta$. On the screen, therefore, in accordance with Eqs. (11-11) and (11-12), the irradiance is a maximum at $\theta = 0$ ($y = 0$) and drops to zero at values y_m such that

$$y_m \cong \frac{m\lambda f}{b} \qquad (11\text{-}13)$$

The irradiance pattern is symmetrical about $y = 0$.

The secondary maxima of the single-slit diffraction pattern do not quite fall at the midpoints between zeros, even though this condition is more nearly

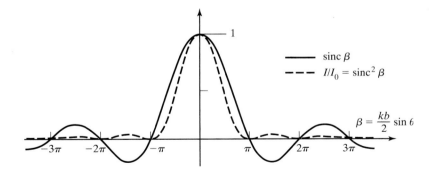

Figure 11-2   Sinc function (solid line) plotted as a function of β. The normalized irradiance function I/I_0 (dashed line) for single-slit Fraunhofer diffraction is the square of $\text{sinc}(\beta)$.

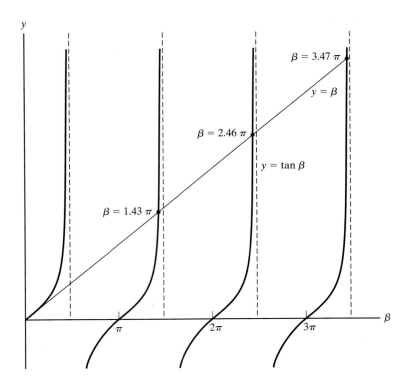

Figure 11-3 Intersections of the curves $y = \beta$ and $y = \tan \beta$ determine the angles β at which the sinc function is a maximum.

approached as β increases. The maxima coincide with maxima of the sinc function, which occur at points satisfying

$$\frac{d}{d\beta}\left(\frac{\sin \beta}{\beta}\right) = \frac{\beta \cos \beta - \sin \beta}{\beta^2} = 0$$

or $\beta = \tan \beta$. An angle equals its tangent at intersections of the curves $y = \beta$ and $y = \tan \beta$, both plotted in Figure 11-3. Intersections, excluding $\beta = 0$, occur at 1.43π (rather than 1.5π), 2.46π (rather than 2.5π), 3.47π (rather than 3.5π), and so on. The plot clearly shows that intersection points approach the vertical lines defining midpoints more closely as β increases. Thus, in the irradiance plot of Figure 11-2, secondary maxima are skewed slightly away from the midpoints toward the central peak. Most of the energy of the diffraction pattern falls under the central maximum, which is much larger than the adjoining maximum on either side.

Example 11-1

What is the ratio of irradiances at the central peak maximum to the first of the secondary maxima?

Solution

The ratio to be calculated is

$$\frac{I_{\beta=0}}{I_{\beta=1.43\pi}} = \frac{(\sin^2 \beta/\beta^2)_{\beta=0}}{(\sin^2 \beta/\beta^2)_{\beta=1.43\pi}} = \frac{1}{(\sin^2 \beta/\beta^2)_{\beta=1.43\pi}}$$

$$= \left(\frac{\beta^2}{\sin^2 \beta}\right)_{1.43\pi} = \frac{20.18}{0.952} = 21.2$$

Thus the maximum irradiance of the nearest secondary peak is only 4.7% that of the central peak.

The central maximum represents essentially the image of the slit on a distant screen. We observe that the edges of the image are not sharp but reveal a series of maxima and minima that tail off into the shadow surrounding the image. These effects are typical of the blurring of images due to diffraction and will be seen again in other cases of diffraction to be considered. The angular width of the central maximum is defined as the angle $\Delta \theta$ between the first minima on either side. Using Eq. (11-12) with $m = \pm 1$ and approximating $\sin \theta$ by θ, we get

$$\Delta \theta = \frac{2\lambda}{b} \tag{11-14}$$

From Eq. (11-14), it follows that the central maximum will spread as the slit width is narrowed. Since the length of the slit is very large compared to its width, the diffraction pattern due to points of the wavefront along the length of the slit has a very small angular width and is not prominent on the screen. Of course, the dimensions of the diffraction pattern also depend on the wavelength, as indicated in Eq. (11-14).

11-2 BEAM SPREADING

According to Eq. (11-14), the angular spread $\Delta \theta$ of the central maximum in the far field is independent of distance between aperture and screen. The linear dimensions of the diffraction pattern thus increase uniformly with distance L, as shown in Figure 11-4, such that the width W of the central maximum is given by

$$W = L\,\Delta\theta = \frac{2L\lambda}{b} \tag{11-15}$$

We may describe the content of Eq. (11-15) as a linear spread of a beam of light, originally constricted to a width b. Indeed, the means by which the beam is originally narrowed is not relevant to the nature of the diffraction pattern that occurs. If one dispenses with the slit in Figure 11-4 and merely assumes an original beam of constant irradiance across a finite width b, all our results follow in the same way. After collimation, a "parallel" beam of light spreads just as if it emerged from a single opening.

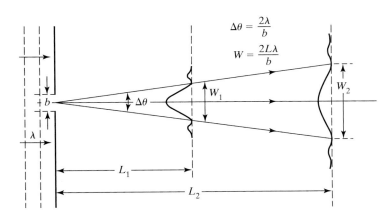

Figure 11-4 Spread of the central maximum in the far-field diffraction pattern of a single slit.

Example 11-2

Imagine a parallel beam of 546-nm light of width $b = 0.5$ mm propagating a distance of 10 m across the laboratory. Estimate the final width W of the beam due to diffraction spreading.

Solution

Using Eq. (11-15),

$$W = \frac{2L\lambda}{b} = \frac{2(10 \text{ m})(546 \times 10^{-9} \text{ m})}{0.5 \times 10^{-3} \text{ m}} = 0.0218 \text{ m} = 21.8 \text{ mm}$$

Even highly collimated laser beams are subject to beam spreading as they propagate, due to diffraction. It is a fundamental consequence of the wave nature of light that beams of finite transverse extent must spread as they propagate.

The beam spreading described by Eq. (11-14) is valid for a rectangular aperture of width much less than its length. As we show in the next section, the spreading due to diffraction from a circular aperture follows a form similar to Eq. (11-14) but with the replacement of the width b of the slit by the diameter D of the circular aperture and with the replacement of the wavelength λ by the factor 1.22λ. Furthermore, one must keep in mind that this treatment assumes a plane wavefront of uniform irradiance.[1] The spreading described by Eq. (11-15) has been deduced on the basis of Fraunhofer, or far-field, diffraction, which means here that L must remain reasonably large. If L is taken small enough, for example, the equation predicts a beam width less than b, contrary to assumption. Evidently L must be larger than some minimum value, $L_{\min}$, which gives a beam width $W = b$, that is,

$$L_{\min} = \frac{b^2}{2\lambda}$$

We may conclude that we are in the *far field* when

$$L \gg \frac{b^2}{\lambda}$$

A more general approach leads to the commonly stated criterion for *far-field diffraction* in the form[2]

$$L \gg \frac{\text{area of aperture}}{\lambda} \tag{11-16}$$

11-3 RECTANGULAR AND CIRCULAR APERTURES

We have been describing diffraction from a slit having a width b much smaller than its length a, as illustrated in Figure 11-5a. When both dimensions of

---

[1]A laser beam usually does not have constant irradiance across its diameter. In its fundamental mode, the transverse profile is a Gaussian function. Still, its spread formula is essentially that of Eq. (11-14) with the beam diameter replacing b and the constant factor of 2 replaced by $4/\pi \cong 1.27$. Laser-beam spreading is mentioned in Chapter 6 and is treated in detail in Chapter 27. In comparing formulas for divergence angles, care must be taken to distinguish between the *full angular spread* illustrated in Figure 11-4 and the *half-angle spread* illustrated, for example, in Figure 6-19.

[2]Many practitioners in the field of high-energy lasers use the far-field criterion, $L > 100$ (area of aperture)$/\lambda$.

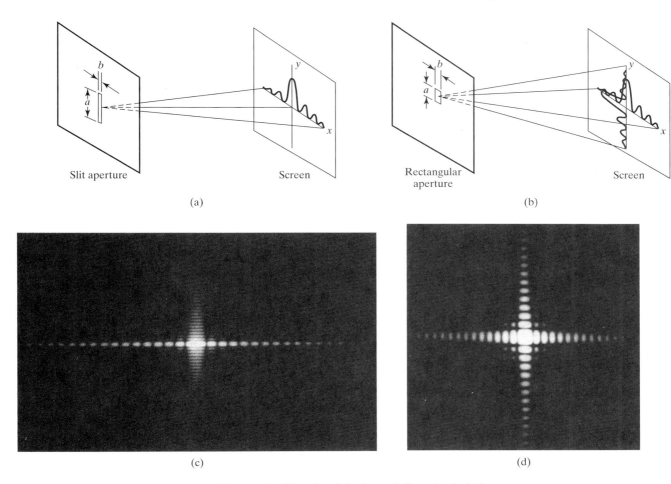

Figure 11-5 (a) Single-slit diffraction. Only the small dimension b of a long, narrow slit causes appreciable spreading of the light along the x-direction on the screen. (b) Rectangular aperture diffraction. Both dimensions of the rectangular aperture are small and a two-dimensional diffraction pattern is discernible on the screen. (c) Photograph of the diffraction image of a rectangular aperture with $b < a$, as in the representation of Figure 11-5a. (d) Photograph of the diffraction image of a rectangular aperture with $b = a$, as in the representation of Figure 11-5b. (Both photos are from M. Cagnet, M. Francon, and J. C. Thrierr, *Atlas of Optical Phenomenon*, Plate 17, Berlin: Springer-Verlag, 1962.)

the slit are comparable and small, each produces appreciable spreading, as illustrated in Figure 11-5b. For the aperture dimension a, we write analogously, for the irradiance, as in Eq. (11-10),

$$I = I_0\left(\frac{\sin \alpha}{\alpha}\right)^2 \quad \text{where } \alpha \equiv \left(\frac{k}{2}\right)a \sin \theta \qquad (11\text{-}17)$$

The two-dimensional pattern now gives zero irradiance for points x, y satisfied by either

$$y_m = \frac{m\lambda f}{b} \quad \text{or} \quad x_n = \frac{n\lambda f}{a}$$

where both m and n represent nonzero integral values. The irradiance over the screen turns out to be just a product of the irradiance functions in each dimension, or

$$I = I_0(\text{sinc}^2 \beta)(\text{sinc}^2 \alpha) \qquad (11\text{-}18)$$

In calculating this result, the single integration over one dimension of the slit is replaced by a double integration over both dimensions of the aperture. Photographs of single-aperture diffraction patterns for rectangular and square apertures are shown in Figure 11-5c and d.

When the aperture is circular, the integration is over the entire area of the aperture since both vertical and horizontal dimensions of the aperture are comparable. Equation (11-4), which describes the total electric field at point P of Figure 11-1 due to single-slit diffraction, can be modified to describe diffraction from a circular aperture. The required modification involves the replacement of the incremental *electric field amplitude* $E_L\,ds/r_0$ by $E_A\,dA/r_0$ and the conversion of the integral over the slit width to an integral over the aperture area. Here, E_A is a constant factor (with "units" of electric field per unit length) that determines the strength of the electric field in the aperture and dA is the elemental area of the aperture. The electric field at P (as in Figure 11-1) due to diffraction through a circular aperture can then be written as

$$E_P = \frac{E_A}{r_0} e^{i(kr_0 - \omega t)} \iint\limits_{\text{Area}} e^{isk\,\sin\theta}\,dA$$

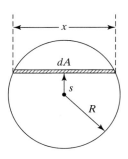

Figure 11-6   Geometry used in the integration over a circular aperture.

We take a rectangular strip of area $dA = x\,ds$ as the elemental area of integration, shown in Figure 11-6. Using the equation of a circle, we calculate the length x at height s to be given by

$$x = 2\sqrt{R^2 - s^2}$$

where R is the aperture radius. The preceding integral can then be rewritten, leading to

$$E_P = \frac{2E_A}{r_0} e^{i(kr_0 - \omega t)} \int_{-R}^{R} e^{isk\,\sin\theta}\sqrt{R^2 - s^2}\,ds$$

The integral takes the form of a standard definite integral upon making the substitutions $v = s/R$ and $\gamma = kR\sin\theta$:

$$E_P = \frac{2E_A R^2}{r_0} e^{i(kr_0 - \omega t)} \int_{-1}^{+1} e^{i\gamma v}\sqrt{1 - v^2}\,dv$$

The integral has the value

$$\int_{-1}^{+1} e^{i\gamma v}\sqrt{1 - v^2}\,dv = \frac{\pi J_1(\gamma)}{\gamma}$$

where $J_1(\gamma)$ is the first-order *Bessel function of the first kind*, expressible by the infinite series

$$J_1(\gamma) = \frac{\gamma}{2} - \frac{(\gamma/2)^3}{1^2 \cdot 2} + \frac{(\gamma/2)^5}{1^2 \cdot 2^2 \cdot 3} - \cdots$$

As can be verified from this series expansion, the ratio $J_1(\gamma)/\gamma$ has the limit $\frac{1}{2}$ as $\gamma \to 0$. Thus, the circular aperture requires, instead of the sine function for the single slit, the Bessel function J_1, which oscillates somewhat like the sine function, as shown in the plot of Figure 11-7. One important difference is that the amplitude of the oscillation of the Bessel function decreases as its argument departs from zero.

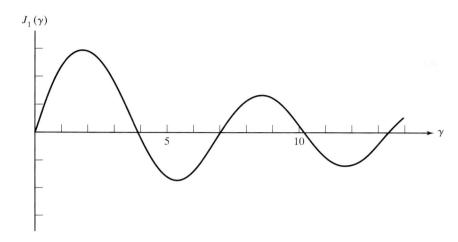

Figure 11-7 A plot of the Bessel function $J_1(\gamma)$ vs. γ. The first few zeroes of the Bessel function occur at $\gamma = 0$, $\gamma = 3.832$, $\gamma = 7.016$, $\gamma = 10.173$, and $\gamma = 13.324$.

The irradiance for a circular aperture of diameter D can now be written as

$$I = I_0\left(\frac{2J_1(\gamma)}{\gamma}\right)^2, \quad \text{where } \gamma \equiv \tfrac{1}{2}kD\sin\theta \qquad (11\text{-}19)$$

where I_0 is the irradiance at $\gamma \to 0$ or at $\theta = 0$. The equations should be compared with those of Eq. (11-17) to appreciate the analogous role played by the Bessel function. Like $(\sin x)/x$, the function $J_1(x)/x$ approaches a maximum as x approaches zero, so that the irradiance is greatest at the center of the pattern $(\theta = 0)$. (In fact, $J_1(x)/x$ tends to $\tfrac{1}{2}$ as x tends to zero, so the irradiance tends to I_0 as γ tends to zero.) The pattern is symmetrical about the optical axis through the center of the circular aperture and has its first zero when $\gamma = 3.832$, as indicated in Figure 11-8a and b. Thus, the irradiance first falls to zero when

$$\gamma = \left(\frac{k}{2}\right)D\sin\theta = 3.832 \quad \text{or} \quad \text{when } D\sin\theta = 1.22\lambda \qquad (11\text{-}20)$$

The irradiance pattern of Eq. (11-19) is plotted in Figure 11-8a. The first few zeroes, and maxima of the normalized irradiance $I/I_0 = (2J_1(\gamma)/\gamma)^2$ are listed in Figure 11-8b. The pattern is similar to that of Figure 11-2 for a slit, except that the pattern for a circular aperture has rotational symmetry about the optical axis. A photograph is shown in Figure 11-8c. The central maximum is a circle of light, the diffracted "image" of the circular aperture, and is called the *Airy disc*. Equation (11-20) should be compared with the analogous equation for the narrow rectangular slit, $m\lambda = b\sin\theta$. We see that $m = 1$ for the first minimum in the slit pattern is replaced by the number 1.22 in the case of the circular aperture. Successive minima are determined in a similar way from other zeros of the Bessel function, as indicated in the table in Figure 11-8b.

Note that the far-field angular *radius* (i.e., the angular half-width) of the Airy disc, according to Eq. (11-20), is very nearly

$$\Delta\theta_{1/2} = \frac{1.22\lambda}{D} \qquad (11\text{-}21)$$

In Example 11-3, the beam spread from a circular aperture is compared with that from a single slit.

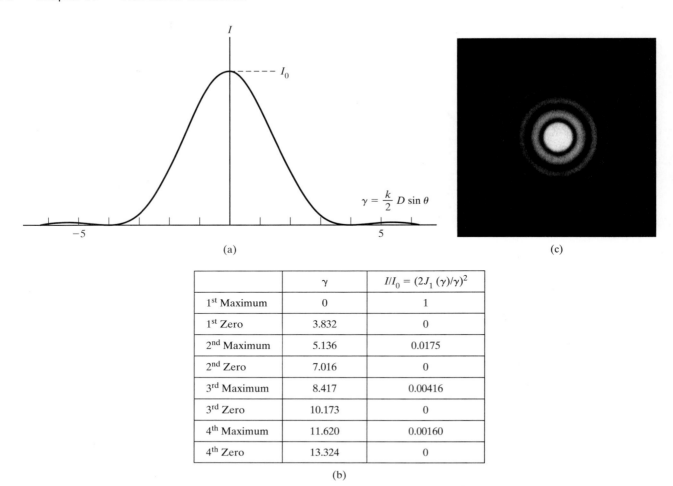

	γ	$I/I_0 = (2J_1(\gamma)/\gamma)^2$
1st Maximum	0	1
1st Zero	3.832	0
2nd Maximum	5.136	0.0175
2nd Zero	7.016	0
3rd Maximum	8.417	0.00416
3rd Zero	10.173	0
4th Maximum	11.620	0.00160
4th Zero	13.324	0

(b)

Figure 11-8 Circular aperture diffraction pattern. (a) Irradiance $I = I_0(2J_1(\gamma)/\gamma)^2$ of the diffraction pattern of a circular aperture. By far the largest amount of light energy is diffracted into the central maximum. (b) The first few zeroes and maxima of the normalized irradiance $I/I_0 = (2J_1(\gamma)/\gamma)^2$. (c) Diffraction image of a circular aperture. The circle of light at the center corresponds to the zeroth order of diffraction and is known as the Airy disc. (From M. Cagnet, M. Francon, and J. C. Thrierr, *Atlas of Optical Phenomenon*, Plate 16, Berlin: Springer-Verlag, 1962.)

Example 11-3

Find the diameter of the Airy disc at the center of the diffraction pattern formed on a wall at a distance $L = 10$ m from a uniformly illuminated circular aperture of diameter $D = 0.5$ mm. Assume that the illuminating light has wavelength of $\lambda = 546$ nm. Compare the beam spread to that from the slit of width $b = 0.5$ mm of Example 11-2.

Solution

The angular radius of the Airy disc is found using Eq. 11-21,

$$\Delta\theta_{1/2} = \frac{1.22\lambda}{D} = \frac{1.22(546 \times 10^{-9}\,\text{m})}{5 \times 10^{-4}\,\text{m}} = 1.33 \times 10^{-3}\,\text{rad}$$

The radius r_d of the Airy disc is then found using an argument similar to that used in Figure 11-4 for single-slit diffraction,

$$r_d = L\,\Delta\theta_{1/2} = (10\,\text{m})(1.33 \times 10^{-3}) = 0.013\,\text{m} = 13\,\text{mm}$$

The diameter D_d of the Airy disc is, then,

$$D_d = 2r_d = 26 \text{ mm}$$

The beam spread is comparable to, but slightly more than, that from the single slit of Example 11-2, where W was found to be near 22 mm.

11-4 RESOLUTION

In forming the Fraunhofer diffraction pattern of a single slit, as in Figure 11-1, we notice that the distance between slit and lens is not crucial to the details of the pattern. The lens merely intercepts a larger solid angle of light when the distance is small. If this distance is allowed to go to zero, aperture and lens coincide, as in the objective of a telescope. Thus, the image formed by a telescope with a round objective is subject to the diffraction effects described by Eq. (11-19) for a circular aperture. The sharpness of the image of a distant point object—a star, for example—is, then, limited by diffraction. The image occupies essentially the region of the Airy disc. An eyepiece viewing the primary image and providing further magnification merely enlarges the details of the diffraction pattern formed by the lens. The limit of resolution is already set in the primary image. The inevitable blur that diffraction produces in the image restricts the resolution of the instrument, that is, its ability to provide distinct images for distinct object points, either physically close together (as in a microscope) or separated by a small angle at the lens (as in a telescope). Figure 11-9a illustrates the diffraction of two point objects S_1 and S_2 formed

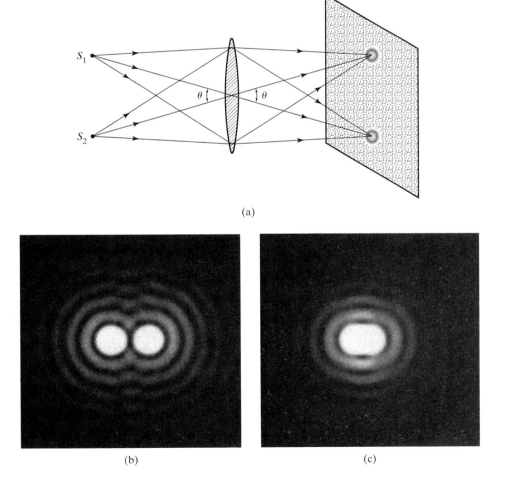

(a)

(b)   (c)

Figure 11-9 (a) Diffraction-limited images of two point objects formed by a lens. As long as the Airy discs are well separated, the images are well resolved. (b) Separated images of two incoherent point sources. In this diffraction pattern, the two images are well resolved. (c) Image of a pair of incoherent point sources at the limit of resolution. (Photos from M. Cagnet, M. Francon, and J. C. Thrierr, *Atlas of Optical Phenomenon*, Plate 16, Berlin: Springer-Verlag, 1962.)

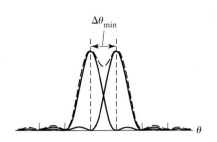

Figure 11-10 Rayleigh's criterion for just-resolvable diffraction patterns. The dashed curve is the observed sum of independent diffraction peaks.

by a single lens. The point objects and the centers of their Airy discs are both separated by the angle θ. If the angle is large enough, two distinct images will be clearly seen, as shown in the photograph of Figure 11-9b. Imagine now that the objects S_1 and S_2 are brought closer together. When their image patterns begin to overlap substantially, it becomes more difficult to discern the patterns as distinct, that is, to resolve them as belonging to distinct object points. A photograph of the two images at the limit of resolution is shown in Figure 11-9c.

Rayleigh's criterion for just-resolvable images—a somewhat arbitrary but useful criterion—requires that the angular separation of the centers of the image patterns be no less than the angular radius of the Airy disc, as in Figure 11-10. In this condition, the maximum of one pattern falls directly over the first minimum of the other. Thus, for the *limit of resolution*, we have, using Eq. (11-21),

$$(\Delta\theta)_{min} = \frac{1.22\lambda}{D} \qquad (11\text{-}22)$$

where D is now the diameter of the lens. In accordance with this result, the minimum resolvable angular separation of two object points may be reduced (the resolution improved) by increasing the lens diameter and decreasing the wavelength.

We consider several applications of Eq. (11-22), beginning with the following example.

Example 11-4

Suppose that each lens on a pair of binoculars has a diameter of 35 mm. How far apart must two stars be before they are theoretically resolvable by either of the lenses in the binoculars?

Solution

According to Eq. (11-22),

$$(\Delta\theta)_{min} = \frac{1.22(550 \times 10^{-9})}{35 \times 10^{-3}} = 1.92 \times 10^{-5} \text{ rad}$$

or about 4″ of arc, using an average wavelength for visible light. If the stars are near the center of our galaxy, a distance, d, of around 30,000 light-years, then their actual separation s is approximately

$$s = d\,\Delta\theta_{min} = (30{,}000)(1.92 \times 10^{-5}) = 0.58 \text{ light-years}$$

To get some appreciation for this distance, consider that the planet Pluto at the edge of our solar system is only about 5.5 light-*hours* distant. If the stars are being detected by their long-wavelength radio waves—the lenses being replaced by dish antennas—the resolution must, by Eq. (11-22), be much less.

If the lens is the objective of a microscope, as indicated in Figure 11-11, the problem of resolving nearby objects is basically the same. Making only rough estimates, we shall ignore the fact that the wavefronts striking the lens from nearby object points A and B are not plane, as required in far-field

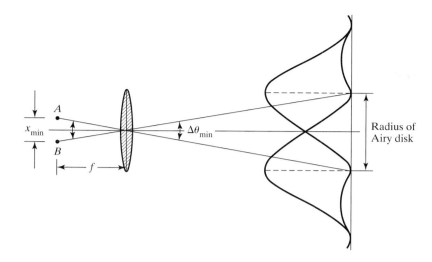

Figure 11-11 Minimum angular resolution of a microscope.

diffraction equations. The minimum separation, x_{min}, of two just-resolved objects near the focal plane of the lens of diameter D is then given by

$$x_{min} = f \, \Delta\theta_{min} = f\left(\frac{1.22\lambda}{D}\right)$$

The ratio D/f is the *numerical aperture*, with a typical value of 1.2 for a good oil-immersion objective. Thus,

$$x_{min} \cong \lambda$$

The resolution of a microscope is roughly equal to the wavelength of light used, a fact that explains the advantage of ultraviolet, X-ray, and electron microscopes in high-resolution applications. In Chapter 28, we will discuss some techniques used in *near-field microscopy* that allow one to surpass the diffraction-limited resolution just discussed.

The limits of resolution due to diffraction also affect the human eye, which may be approximated by a circular aperture (pupil), a lens, and a screen (retina), as in Figure 11-12. Night vision, which takes place with large, adapted pupils of around 8 mm, is capable of higher resolution than daylight vision. Unfortunately, there is not enough light to take advantage of the situation! On a bright day the pupil diameter may be 2 mm. Under these conditions, Eq. (11-22) gives $(\Delta\theta)_{min} = 33.6 \times 10^{-5}$ rad, for an average wavelength of 550 nm. Experimentally, one finds that a separation of 1 mm at a distance of about 2 m is just barely resolvable, giving $(\Delta\theta)_{min} = 50 \times 10^{-5}$ rad, about 1.5 times the theoretical limit. One's own resolution (*visual acuity*) can easily be tested by viewing two lines drawn 1 mm apart at increasing distances until they can no longer be seen as distinct. It is interesting to note that the theoretical resolution just determined for a 2-mm-diameter pupil is consistent with the value of 1′ of arc (29×10^{-5} rad) used by Snellen to characterize normal visual acuity (see Chapter 19).

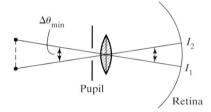

Figure 11-12 Diffraction by the eye with pupil as aperture limits the resolution of objects subtending angle $\Delta\theta_{min}$.

11-5 DOUBLE-SLIT DIFFRACTION

The diffraction pattern of a plane wavefront that is obstructed everywhere except at two narrow slits is calculated in the same manner as for the single slit. The mathematical argument departs from that for the single slit with Eq. (11-4). Here, the limits of integration covering the apertures of the two slits become those indicated in Figure 11-13.

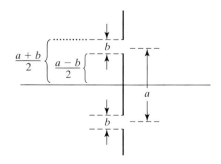

Figure 11-13 Specification of slit width and separation for double-slit diffraction.

We find

$$E_P = \frac{E_L}{r_0} e^{i(kr_0-\omega t)} \int_{-(1/2)(a+b)}^{-(1/2)(a-b)} e^{isk\sin\theta}\, ds \;+\; \frac{E_L}{r_0} e^{i(kr_0-\omega t)} \int_{(1/2)(a-b)}^{(1/2)(a+b)} e^{isk\sin\theta}\, ds$$

$$(11\text{-}23)$$

Integration and substitution of the limits leads to

$$E_P = \frac{E_L}{r_0} e^{i(kr_0-\omega t)} \frac{1}{ik\sin\theta} [e^{(1/2)ik(-a+b)\sin\theta} - e^{(1/2)ik(-a-b)\sin\theta}$$

$$+ e^{(1/2)ik(a+b)\sin\theta} - e^{(1/2)ik(a-b)\sin\theta}]$$

Reintroducing the substitution of Eq. (11-7), involving the slit width b,

$$\beta \equiv \tfrac{1}{2} kb \sin\theta \qquad (11\text{-}24)$$

and a similar one involving the slit separation a,

$$\alpha \equiv \tfrac{1}{2} ka \sin\theta \qquad (11\text{-}25)$$

our equation is written more compactly as

$$E_P = \frac{E_L}{r_0} e^{i(kr_0-\omega t)} \frac{b}{2i\beta} [e^{i\alpha}(e^{i\beta} - e^{-i\beta}) + e^{-i\alpha}(e^{i\beta} - e^{-i\beta})]$$

Employing Euler's equation,

$$E_P = \frac{E_L}{r_0} e^{i(kr_0-\omega t)} \frac{b}{2i\beta} (2i\sin\beta)(2\cos\alpha)$$

Finally,

$$E_P = \frac{E_L}{r_0} e^{i(kr_0-\omega t)} \frac{2b\sin\beta}{\beta}\cos\alpha \qquad (11\text{-}26)$$

The amplitude of this electric field is

$$E_0 = \frac{E_L}{r_0} \frac{2b\sin\beta}{\beta}\cos\alpha$$

so that the irradiance at point P in the double-slit diffraction pattern is

$$I = \left(\frac{\varepsilon_0 c}{2}\right) E_0^2 = \left(\frac{\varepsilon_0 c}{2}\right)\left(\frac{2E_L b}{r_0}\right)^2 \left(\frac{\sin\beta}{\beta}\right)^2 \cos^2\alpha$$

or

$$I = 4I_0 \left(\frac{\sin\beta}{\beta}\right)^2 \cos^2\alpha \qquad (11\text{-}27)$$

where

$$I_0 = \left(\frac{\varepsilon_0 c}{2}\right)\left(\frac{E_L b}{r_0}\right)^2$$

as defined in Eq. (11-10) for the single slit. Since the maximum value of Eq. (11-27) is $4I_0$, we see that the double slit provides four times the maximum irradiance in the pattern center as compared with the single slit. This is exactly what should be expected where the two beams are in phase and amplitudes add.

On closer inspection of Eq. (11-27), we find that the irradiance is just a product of the irradiances found for double-slit interference and single-slit diffraction. The factor $[(\sin\beta)/\beta]^2$ is that of Eq. (11-10) for single-slit diffraction. The $\cos^2\alpha$ factor, when α is written out as in Eq. (11-25), is

$$\cos^2\alpha = \cos^2\left[\frac{ka(\sin\theta)}{2}\right] = \cos^2\left[\frac{\pi a(\sin\theta)}{\lambda}\right]$$

equivalent to the factor in Eq. (7-21) for Young's double slit. The sinc and cosine factors of Eq. (11-27) are plotted in Figure 11-14a for the case $a = 6b$ or $\alpha = 6\beta$. Because $a > b$, the $\cos^2\alpha$ factor varies more rapidly than the $(\sin^2\beta)/\beta^2$ factor. The product of the sine and cosine factors may be considered a modulation of the interference fringe pattern by a single-slit diffraction envelope, as shown in Figure 11-14b. The diffraction envelope has a minimum

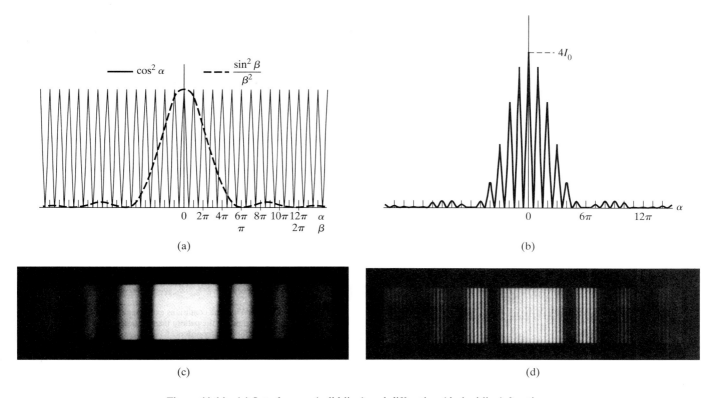

Figure 11-14 (a) Interference (solid line) and diffraction (dashed line) functions plotted for double-slit Fraunhofer diffraction when the slit separation is six times the slit width ($a = 6b$). (b) Irradiance for the double slit of (a). The curve represents the product of the interference and diffraction factors. (c) Diffraction pattern due to a single slit. (d) Diffraction pattern due to a double-slit aperture, with each slit of width b like the one that produced (c), but with a/b unspecified. (Both photos are from M. Cagnet, M. Francon, and J. C. Thrierr, *Atlas of Optical Phenomenon*, Plate 18, Berlin: Springer-Verlag, 1962.)

when $\beta = m\pi$, with $m = \pm1, \pm2, \ldots$, as shown. In terms of the spatial angle θ, this condition is

$$\text{diffraction minima: } m\lambda = b \sin \theta \qquad (11\text{-}28)$$

as in Eq. (11-12). When these minima happen to coincide with interference fringe maxima, the fringe is missing from the pattern. Interference maxima occur for $\alpha = p\pi$, with $p = 0, \pm1, \pm2, \ldots$, or when

$$\text{interference maxima: } p\lambda = a \sin \theta \qquad (11\text{-}29)$$

When the conditions expressed by Eqs. (11-28) and (11-29) are satisfied at the same point in the pattern (same θ), dividing one equation by the other gives the condition for missing orders.

$$\text{condition for missing orders: } a = \left(\frac{p}{m}\right)b \qquad (11\text{-}30)$$

or

$$\alpha = \left(\frac{p}{m}\right)\beta$$

Thus, when the slit separation is an integral multiple of the slit width, the condition for missing order is met exactly. For example, when $a = 2b$, then $p = 2m = \pm2, \pm4, \pm6, \ldots$ gives the missing orders of interference. For the case plotted in Figure 11-14a and b, $a = 6b$, and the missing orders are those for which $p = \pm6, \pm12$, and so on. Figure 11-14c and d contains photographs of a single-slit pattern and a double-slit pattern with the same slit width. (What is the ratio of a/b in this case? Would a ratio of $a/b = 9$ fit the pattern shown?) Evidently, when $a = Nb$ and N is large, the first missing order at $p = \pm N$ is far from the center of the pattern. To produce a simple Young's interference pattern for two slits, one accordingly makes $a \gg b$ so that N is large. A large number of fringes then fall under the central maximum of the diffraction envelope. As a trivial but satisfying case, observe that when $a = b$, Eq. (11-30) requires that all orders (except $p = 0$) are missing. These dimensions cannot be satisfied, however, unless the two slits have merged into one and are unable to produce interference fringes. When $a = b$, the resulting pattern is, of course, that of a single slit.

11-6 DIFFRACTION FROM MANY SLITS

For an aperture of multiple slits (a *grating*), the integrals of Eq. (11-23), together with Figure 11-13, are extended by integrating over N slits. The individual slits are identified by the index j in the following expression for the resultant amplitude:

$$E_P = \frac{E_L}{r_0} e^{i(kr_0 - \omega t)} \sum_{j=1}^{N/2} \left\{ \int_{[-(2j-1)a-b]/2}^{[-(2j-1)a+b]/2} e^{isk \sin \theta}\, ds + \int_{[(2j-1)a-b]/2}^{[(2j-1)a+b]/2} e^{isk \sin \theta}\, ds \right\}$$

$$(11\text{-}31)$$

As j increases, pairs of slits symmetrically placed below (first integral) and above (second integral) the origin are included in the integration. When $j = 1$, for example, Eq. (11-31) reduces to the double-slit case, Eq. (11-23). When $j = 2$, the next two slits are included, whose edges are located at

$\frac{1}{2}(-3a - b)$ and $\frac{1}{2}(-3a + b)$ below the origin and $\frac{1}{2}(3a - b)$ and $\frac{1}{2}(3a + b)$ above the origin.[3] When $j = N/2$, all slits are accounted for.

Let us first concentrate on the integrals contained within the curly brackets, which we shall refer to as K, temporarily. After integration and substitution of limits, we get

$$K = \frac{1}{ik \sin \theta} \{e^{-ik \sin \theta[(2j-1)a-b]/2} - e^{-ik \sin \theta[(2j-1)a+b]/2}\}$$

$$+ \frac{1}{ik \sin \theta} \{e^{ik \sin \theta[(2j-1)a+b]/2} - e^{ik \sin \theta[(2j-1)a-b]/2}\}$$

Using Eqs. (11-24) and (11-25) again for α and β,

$$K = \frac{b}{2i\beta} [e^{-i(2j-1)\alpha}(e^{i\beta} - e^{-i\beta}) + e^{i(2j-1)\alpha}(e^{i\beta} - e^{-i\beta})]$$

With the help of Euler's equation, this can be written as

$$K = \frac{b}{2i\beta} (2i \sin \beta)\{2 \cos[(2j - 1)\alpha]\}$$

or

$$K = 2b \frac{\sin \beta}{\beta} \text{Re} [e^{i(2j-1)\alpha}]$$

where we have expressed the cosine as the real part of the corresponding exponential. Returning to Eq. (11-31), we need next the sum S:

$$S = 2b \frac{\sin \beta}{\beta} \text{Re} \sum_{j=1}^{N/2} e^{i(2j-1)\alpha}$$

Expanding the sum, we find

$$S = 2b \frac{\sin \beta}{\beta} \text{Re} [e^{i\alpha} + e^{i3\alpha} + e^{i5\alpha} + \cdots + e^{i(N-1)\alpha}]$$

The series in brackets is a geometric series whose first term a and ratio r can be used to find its sum, given by

$$a\left(\frac{r^n - 1}{r - 1}\right) = e^{i\alpha} \left[\frac{(e^{2i\alpha})^{N/2} - 1}{e^{2i\alpha} - 1}\right] = \frac{e^{iN\alpha} - 1}{e^{i\alpha} - e^{-i\alpha}}$$

Using Euler's equation, this can be recast into the form

$$\frac{(\cos N\alpha - 1) + i \sin N\alpha}{2i \sin \alpha} = \frac{i(\cos N\alpha - 1) - \sin N\alpha}{-2 \sin \alpha}$$

whose real part is $(\sin N\alpha)/(2 \sin \alpha)$. Then,

$$S = b \frac{\sin \beta}{\beta} \frac{\sin N\alpha}{\sin \alpha}$$

---

[3]This expression is adapted to N even. For N large, one need not be concerned about the parity of N. For N small, however, N odd can be handled by taking the origin at the center of the central slit. This approach is left to the problems.

and

$$E_P = \frac{E_L}{r_0} e^{i(kr_0 - \omega t)} \left\{ \frac{b \sin \beta}{\beta} \frac{\sin N\alpha}{\sin \alpha} \right\}$$

As before, the irradiance is proportional to the square of the field amplitude,

$$I = I_0 \underbrace{\left(\frac{\sin \beta}{\beta} \right)^2}_{\text{diffraction}} \underbrace{\left(\frac{\sin N\alpha}{\sin \alpha} \right)^2}_{\text{interference}} \tag{11-32}$$

where I_0 includes all the constants, the first set of brackets encloses the diffraction factor, and the second set of brackets encloses the interference factor.

Although derived here for an even number N of slits, the result expressed by Eq. (11-32) is valid also for N odd (see problem 11-21). When $N = 1$ and $N = 2$, Eq. (11-32) reduces to the results obtained previously for single- and double-slit diffraction, respectively. By now we are familiar with the factor in β representing the diffraction envelope of the resultant irradiance. Let us examine the factor $(\sin N\alpha/\sin \alpha)^2$, which evidently describes interference between slits. When $\alpha = 0$ or some multiple of π, the expression reduces to an indeterminate form. We can show, in fact, that for such values, the expression is a maximum. Employing L'Hôpital's rule for any $m = 0, \pm 1, \pm 2, \ldots,$

$$\lim_{\alpha \to m\pi} \frac{\sin N\alpha}{\sin \alpha} = \lim_{\alpha \to m\pi} \frac{N \cos N\alpha}{\cos \alpha} = \pm N$$

Thus, the interference factor in Eq. (11-32) describes a series of sharp irradiance peaks (*principal maxima*). The irradiance at a principal maximum is proportional to N^2 and the principal maxima are centered at values for which $\alpha = 0, \pm\pi, \pm 2\pi, \pm 3\pi,$ and so on. For the case $N = 8$, four such peaks, at $\alpha = 0, \pi, 2\pi,$ and 3π are shown in Figure 11-15a. In between successive peaks there are shown $N - 2 = 6$ *secondary* peaks. The diffraction factor in Eq. (11-32) is plotted as the dotted line in Figure 11-15a, and the full irradiance which is proportional to the product of the diffraction and interference factors is plotted in Figure 11-15b. Note that the resulting irradiance in Figure 11-15b reflects the presence of the limiting diffraction envelope.

Let us now develop a more explicit understanding of the formation of the *secondary* peaks. The interference factor $(\sin(N\alpha)/\sin \alpha)^2$ goes to zero when the function in its numerator $(\sin(N\alpha))$ goes to zero but the function in its denominator $(\sin \alpha)$ does not. The numerator is identically zero under the condition $\alpha = p\pi/N$, where p takes on integer values. For the 8-slit case $(N = 8)$ and p from 0 to $N = 8$, the numerator goes to zero for the sequence of values $\alpha = 0, \pi/8, 2\pi/8, 3\pi/8, 4\pi/8, 5\pi/8, 6\pi/8, 7\pi/8,$ and $8\pi/8$. Note that $\alpha = 0$ when $p = 0$ and $\alpha = \pi$ when $p = N = 8$. These values, $\alpha = 0$ and $\alpha = \pi$, correspond to the first two principal maxima in Figure 11-15. For $N = 8$, the function $\sin(N\alpha)$ in the numerator of the interference factor goes to zero for each of the seven intermediate terms in the sequence $(\alpha = \pi/8$ to $\alpha = 7\pi/8)$, but the function in the denominator $\sin \alpha$ does *not* go to zero for the these seven intermediate values. Thus, for the case at hand, there are $N - 1 = 7$ zeroes, and as a consequence $N - 2 = 6$ secondary maxima, between the principal maxima. For the case of arbitrary N, there will be $N - 1$ zeroes and $N - 2$ secondary peaks between principal maxima. We have looked in detail at the behavior as p ranges from 0 to N. This pattern simply repeats for p from N to $2N$ and so on, thereby accounting for all of the principal and secondary peaks. The situation described by Eq. (11-32) and

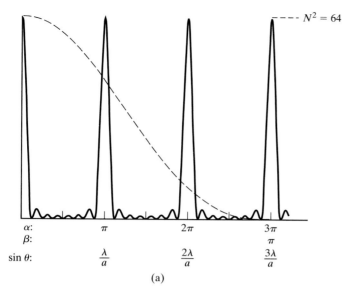

(a)

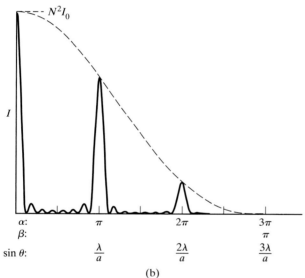

(b)

Figure 11-15 (a) Interference factor $\sin^2(N\alpha)/\sin^2(\alpha)$ (solid line) and diffraction factor $\sin^2\beta/\beta^2$ (dashed line) plotted for multiple-slit Fraunhofer diffraction when $N = 8$ and $a = 3b$. The interference factor peaks at $N^2 = 8^2 = 64$. The diffraction factor has a maximum value of 1 for $\beta = 0$.

(b) Irradiance function $I = I_0 \dfrac{\sin^2\beta}{\beta^2}\dfrac{\sin^2(N\alpha)}{\sin^2\alpha}$ for the multiple slit of (a). The irradiance at the peak of the central principal maximum (at $\alpha = 0$) is $I = N^2 I_0$. Subsequent principal maxima are less bright since they are limited by the diffraction envelope, $\sin^2\beta/\beta^2$ (dashed line).

presented graphically in Figure 11-15 is precisely described by the following set of equations and conditions:

$$\text{for } \alpha = \frac{p\pi}{N}, \qquad p = 0, \pm 1, \pm 2, \ldots \pm N \ldots \pm 2N \ldots$$

principal maxima occur for $p = 0, \pm N, \pm 2N, \ldots$ \hfill (11-33)

secondary minima occur for $p = $ all other integer values

A practical device that makes use of multiple-slit diffraction is the *diffraction grating*. For large N, its principal maxima are bright, distinct, and spatially well separated. According to Eq. (11-33) the principal maxima occur for $p/N = m = 0, \pm 1, \pm 2, \ldots$. Thus the condition for the principal maxima is simply

$$\alpha = m\pi \qquad m = 0, \pm 1, \pm 2 \ldots$$

Recall from Eq. (11-25) that $\alpha = (1/2)ka\sin\theta = \pi a \sin\theta/\lambda$, so that the condition for the existence of a principal maximum can be recast as

$$m\lambda = a\sin\theta \qquad (11\text{-}34)$$

Equation (11-34) is sometimes called the *diffraction grating equation* and m is identified as the *order* of the diffraction.

Now as the number N of slits increases, the brightness of the principal maxima increase as N^2. This increase in irradiance at the peaks of the principal maxima must be accompanied by an overall decrease in irradiance between the peaks of the principal maxima. Thus gratings with more slits direct a greater fraction of the energy emerging from the slits towards the positions of the peaks of the principal maxima than do gratings with fewer slits. Gratings with more slits produce brighter and narrower principal maxima.

Returning to Eq. (11-34), some insight is gained by examining Figure 11-16, which shows representative slits of a grating illuminated by plane wavefronts of monochromatic light. Wavelets emerging from each slit arrive in phase at angular deviation θ from the axis if every path difference like AB ($= a \sin \theta$) equals an integral number m of wavelengths. When $AB = m\lambda$, the grating Eq. (11-34) follows immediately. When all waves arrive in phase, the resulting phasor diagram is formed by adding N phasors all in the same "direction," giving a maximum resultant. At such points, the principal maxima of Figure 11-15 are produced. Secondary maxima result because a uniform phase difference between waves from adjoining slits causes the phase diagram to curl up, with a smaller resultant. At each of the minima, the phasor diagram forms a closed figure, so that cancellation is complete. The phase difference between waves from adjoining slits and in the direction of θ can be found from Figure 11-15a by recalling that the angle α represents half the phase difference between successive slits. Thus, the first principal maximum from the center, at $\alpha = \pi$, occurs when the phase difference between successive waves is precisely 2π.

Photographs of diffraction fringes produced by 2, 3, 4, and 5 slits are shown in Figure 11-17. An examination of the four photographs shows that the principal maxima become narrower and secondary maxima begin to appear as the number of slits increases. For example, notice that the $N - 2 = 3$ secondary maxima appear between the principal maxima for the case $N = 5$. The diffraction grating—for N very large—is discussed further in some detail in the next chapter.

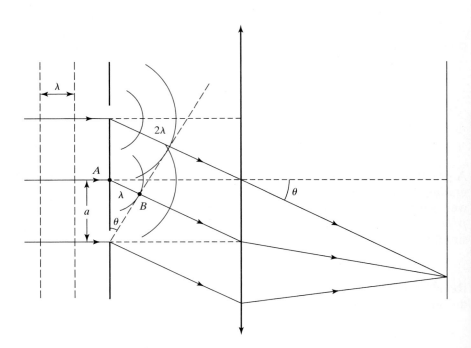

Figure 11-16 Representative grating slits illuminated by collimated monochromatic light. Formation of the first-order diffraction maximum is shown.

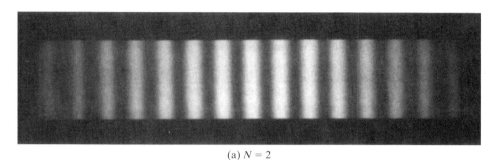

(a) $N = 2$

(b) $N = 3$

(c) $N = 4$

(d) $N = 5$

Figure 11-17   Diffraction fringes produced in turn by two, three, four, and five slits. (From M. Cagnet, M. Francon, and J. C. Thrierr, *Atlas of Optical Phenomenon*, Plate 19, Berlin: Springer-Verlag, 1962.)

PROBLEMS

11-1 A collimated beam of mercury green light at 546.1 nm is normally incident on a slit 0.015 cm wide. A lens of focal length 60 cm is placed behind the slit. A diffraction pattern is formed on a screen placed in the focal plane of the lens. Determine the distance between (a) the central maximum and first minimum and (b) the first and second minima.

Collimated
beam   Slit Lens
f = 60 cm
S c r e e n
60 cm
0.015 cm

Figure 11-18   Problem 11-1.

11-2 Call the irradiance at the center of the central Fraunhofer diffraction maximum of a single slit I_0 and the irradiance at some other point in the pattern I. Obtain the ratio I/I_0 for a point on the screen that is 3/4 of a wavelength farther from one edge of the slit than the other.

11-3 The width of a rectangular slit is measured in the laboratory by means of its diffraction pattern at a distance of 2 m from the slit. When illuminated normally with a parallel beam of laser light (632.8 nm), the distance between the third minima on either side of the principal maximum is measured. An average of several tries gives 5.625 cm.

a. Assuming Fraunhofer diffraction, what is the slit width?
b. Is the assumption of far-field diffraction justified in this case? What is the ratio L/L_{min}?

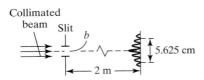

Figure 11-19   Problem 11-3.

11-4 In viewing the far-field diffraction pattern of a single slit illuminated by a discrete-spectrum source with the help of absorption filters, one finds that the fifth minimum of one wavelength component coincides exactly with the fourth minimum of the pattern due to a wavelength of 620 nm. What is the other wavelength?

11-5 Calculate the rectangular slit width that will produce a central maximum in its far-field diffraction pattern having an angular breadth of 30°, 45°, 90°, and 180°. Assume a wavelength of 550 nm.

11-6 Consider the far-field diffraction pattern of a single slit of width 2.125 μm when illuminated normally by a collimated beam of 550-nm light. Determine (a) the angular radius of its central peak and (b) the ratio I/I_0 at points making an angle of $\theta = 5°, 10°, 15°$, and $22.5°$ with the axis.

11-7 a. Find the values of β for which the fourth and fifth secondary maxima of the single-slit diffraction pattern occur. (See the discussion surrounding Figure 11-3.)
  b. Find the ratio of the irradiance of the maxima of part (a) to the irradiance at the central maximum of the single-slit diffraction pattern.

11-8 Compare the relative irradiances of the first two secondary maxima of a circular diffraction pattern to those of a single-slit diffraction pattern.

11-9 The Lick Observatory has one of the largest refracting telescopes, with an aperture diameter of 36 in. and a focal length of 56 ft. Determine the radii of the first and second bright rings surrounding the Airy disc in the diffraction pattern formed by a star on the focal plane of the objective. See Figure 11-8b.

11-10 A telescope objective is 12 cm in diameter and has a focal length of 150 cm. Light of mean wavelength 550 nm from a distant star enters the scope as a nearly collimated beam. Compute the radius of the central disk of light forming the image of the star on the focal plane of the lens.

11-11 Suppose that a CO_2 gas laser emits a diffraction-limited beam at wavelength 10.6 μm, power 2 kW, and diameter 1 mm. Assume that, by multimoding, the laser beam has an essentially uniform irradiance over its cross section. Approximately how large a spot would be produced on the surface of the moon, a distance of 376,000 km away from such a device, neglecting any scattering by the earth's atmosphere? What will be the irradiance at the lunar surface?

11-12 Assume that a 2-mm-diameter laser beam (632.8 nm) is diffraction limited and has a constant irradiance over its cross section. On the basis of spreading due to diffraction alone, how far must it travel to double its diameter?

11-13 Two headlights on an automobile are 45 in. apart. How far away will the lights appear to be if they are just resolvable to a person whose nocturnal pupils are just 5 mm in diameter? Assume an average wavelength of 550 nm.

11-14 Assume that the pupil diameter of a normal eye typically can vary from 2 to 7 mm in response to ambient light variations.

  a. What is the corresponding range of distances over which such an eye can detect the separation of objects 1 mm apart?
  b. Experiment to find the range of distances over which you can detect the separation of lines placed 1 mm apart. Use the results of your experiment to estimate the diameter range of your own pupils.

11-15 A double-slit diffraction pattern is formed using mercury green light at 546.1 nm. Each slit has a width of 0.100 mm. The pattern reveals that the fourth-order interference maxima are missing from the pattern.

  a. What is the slit separation?
  b. What is the irradiance of the first three orders of interference fringes, relative to the zeroth-order maximum?

11-16 a. Show that the number of bright fringes seen under the central diffraction peak in a Fraunhofer double-slit pattern is given by $2(a/b) - 1$, where a/b is the ratio of slit separation to slit width.
  b. If 13 bright fringes are seen in the central diffraction peak when the slit width is 0.30 mm, determine the slit separation.

11-17 a. Show that in a double-slit Fraunhofer diffraction pattern, the ratio of widths of the central diffraction peak to the central interference fringe is $2(a/b)$, where a/b is the ratio of slit separation to slit width. Notice that the result is independent of wavelength.
  b. Determine the peak-to-fringe ratio, in particular when $a = 10b$.

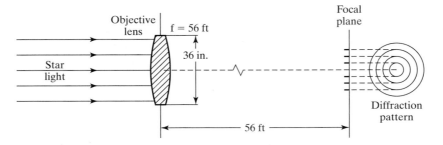

Figure 11-20   Problem 11-9.

11-18 Calculate by integration the irradiance of the diffraction pattern produced by a three-slit aperture, where the slit separation a is three times the slit width b. Make a careful sketch of I versus $\sin\theta$ and describe properties of the pattern. Also show that your results are consistent with the general result for N slits, given by Eq. (11-32).

11-19 Make a rough sketch for the irradiance pattern from seven equally spaced slits having a separation-to-width ratio of 4. Label points on the x-axis with corresponding values of α and β.

11-20 A 10-slit aperture, with slit spacing five times the slit width of 1×10^{-4} cm, is used to produce a Fraunhofer diffraction pattern with light of 435.8 nm. Determine the irradiance of the principal interference maxima of orders 1, 2, 3, 4, and 5 relative to the central fringe of zeroth order.

11-21 Show that one can arrive at Eq. (11-32) by taking the origin of coordinates at the midpoint of the central slit in an array where N is odd.

11-22 A rectangular aperture of dimensions 0.100 mm along the x-axis and 0.200 mm along the y-axis is illuminated by coherent light of wavelength 546 nm. A 1-m focal length lens intercepts the light diffracted by the aperture and projects the diffraction pattern on a screen in its focal plane. See Figure 11-21.

   a. What is the distribution of irradiance on the screen near the pattern center as a function of x and y (in mm) and I_0, the irradiance at the pattern center?
   b. How far from the pattern center are the first minima along the x and y directions?
   c. What fraction of the I_0 irradiance occurs at 1 mm from the pattern center along the x- and y-directions?
   d. What is the irradiance at the point $(x = 2, y = 3)$ mm?

11-23 What is the angular half-width (from central maximum to first minimum) of a diffracted beam for a slit width of (a) λ; (b) 5λ; (c) 10λ?

11-24 A property of the Bessel function $J_1(x)$ is that, for large x, a closed form exists, given by

$$J_1(x) = \frac{\sin x - \cos x}{\sqrt{\pi x}}$$

Find the angular separation of diffraction minima far from the axis of a circular aperture.

11-25 We have shown that the secondary maxima in a single-slit diffraction pattern do not fall exactly halfway between minima, but are quite close. Assuming they are halfway:

   a. Show that the irradiance of the mth secondary peak is given approximately by

   $$I_m \cong I_0 \frac{1}{\left[\left(m + \frac{1}{2}\right)\pi\right]^2}$$

   b. Calculate the percent error involved in this approximation for the first three secondary maxima.

11-26 Three antennas broadcast in phase at a wavelength of 1 km. The antennas are separated by a distance of $\frac{2}{3}$ km and each antenna radiates equally in all horizontal directions. Because of interference, a broadcast "beam" is limited by interference minima. How many well-defined beams are broadcast and what are their angular half-widths?

11-27 A collimated light beam is incident normally on three very narrow, identical slits. At the center of the pattern projected on a screen, the irradiance is $I_{\max}$.

   a. If the irradiance I_P at some point P on the screen is zero, what is the phase difference between light arriving at P from neighboring slits?
   b. If the phase difference between light waves arriving at P from neighboring slits is π, determine the ratio $I_P/I_{\max}$.
   c. What is $I_P/I_{\max}$ at the first principal maximum?
   d. If the average irradiance on the entire screen is I_{av}, what is the ratio I_P/I_{av} at the central maximum?

11-28 Draw phasor diagrams illustrating the principal maxima and zero irradiance points for a four-slit aperture.

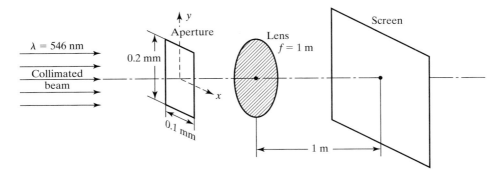

Figure 11-21   Problem 11-22.

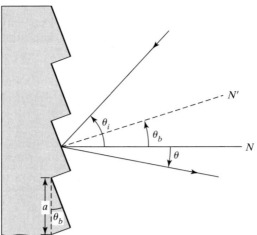

12

The Diffraction Grating

INTRODUCTION

The discussion of the preceding chapter continues here in the formal treatment of diffraction due to a large number of slits or apertures. The diffraction grating equation is first generalized to handle light beams incident on the grating at an arbitrary angle. Performance parameters of practical interest are then developed in discussions of the *spectral range, dispersion, resolution*, and *blaze* of a grating. A brief discussion of interference gratings and several conventional types of grating spectrographs ends the chapter.

12-1 THE GRATING EQUATION

A periodic, multiple-slit device designed to take advantage of the sensitivity of its diffraction pattern to the wavelength of the incident light is called a *diffraction grating*. The grating equation developed in Chapter 11 may be generalized for the case when the incident plane wavefronts of light make an angle θ_i with the plane of the grating, as in Figure 12-1. The net path difference for waves from successive slits is then

$$\Delta = \Delta_1 + \Delta_2 = a \sin \theta_i + a \sin \theta_m \qquad (12\text{-}1)$$

The two sine terms in the path difference may add or subtract, depending on the direction θ_m of the diffracted light. To make Eq. (12-1) correct for all angles of diffraction, we need to adopt a sign convention for the angles. When the incident and diffracted rays are on the same side of the grating normal, as they are in Figure 12-1, θ_m is considered positive. When the diffracted rays are on the side of the grating normal opposite to that of the incident rays, θ_m is

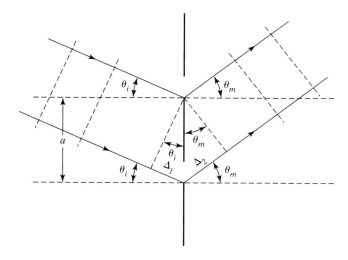

Figure 12-1 Neighboring grating slits illuminated by light incident at angle θ_i with the grating normal. For light diffracted in the direction θ_m, the net path difference from the two slits is $\Delta_1 + \Delta_2$.

considered negative. In the latter case, the net path difference for waves from successive slits is the difference $\Delta_1 - \Delta_2$, as would be evident in a modified sketch of Figure 12-1. In either case, when $\Delta = m\lambda$, all diffracted waves are in phase and the grating equation becomes

$$a(\sin \theta_i + \sin \theta_m) = m\lambda, \qquad m = 0, \pm 1, \pm 2, \ldots \qquad (12\text{-}2)$$

When it is not necessary to distinguish between angles, the subscript on the angle of diffraction, θ_m, is often dropped. For each value of m, monochromatic radiation of wavelength λ is enhanced by the diffractive properties of the grating. By Eq. (12-2), the zeroth order of interference, $m = 0$, occurs at $\theta_m = -\theta_i$, the direction of the incident light, for all λ. Thus, light of all wavelengths appears in the central or zeroth-order peak of the diffraction pattern. Higher orders—both plus and minus—produce spectral *lines* appearing on either side of the zeroth order. For a fixed direction of incidence given by θ_i, the direction θ_m of each principal maximum varies with wavelength. For orders $m \neq 0$, therefore, the grating separates different wavelengths of light present in the incident beam, a feature that accounts for its usefulness in wavelength measurement and spectral analysis. As a dispersing element, the grating is superior to a prism in several ways. Figure 12-2a illustrates the formation of the spectral orders of diffraction for monochromatic light. Figure 12-2b shows the angular spread of the continuous spectrum of visible light for a particular grating. Note that second and third orders in this case partially overlap. Before wavelengths of spectral lines appearing in a region of overlap can be assigned, the actual order of the line must first be ascertained so that the appropriate value of m can be used in Eq. (12-2). Unlike the prism, a grating produces greater deviation from the zeroth-order point for longer wavelengths. Thus, when the spectrum is not a simple one, the overlap ambiguity is often resolved experimentally by using a filter that removes, say, the shorter wavelengths from the incident light. In this way, the spectral range of the incident light is limited by filtering until overlap is removed and each line can be correctly identified. At other times it may be advisable to limit the wavelength range accepted by the grating by first using an instrument of lower dispersion.

12-2 FREE SPECTRAL RANGE OF A GRATING

For diffraction gratings, the nonoverlapping wavelength range in a particular order is called the *free spectral range*, λ_{fsr}. Overlapping occurs because in the grating equation, the product $a \sin \theta$ may be equal to several possible combinations of $m\lambda$ for the light actually incident and processed by the optical system.

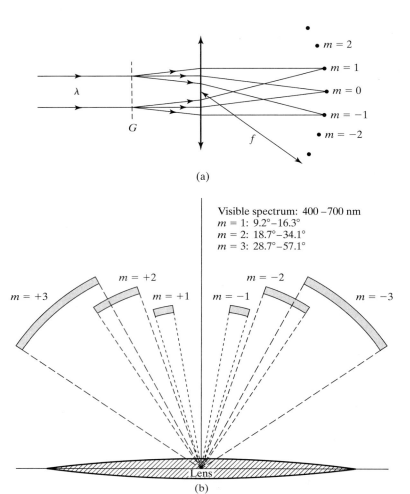

(a)

Visible spectrum: 400–700 nm
$m = 1$: 9.2°–16.3°
$m = 2$: 18.7°–34.1°
$m = 3$: 28.7°–57.1°

(b)

Figure 12-2 (a) Formation of the orders of principal maxima for monochromatic light incident normally on grating G. The grating can replace the prism in a spectroscope. Focused images have the shape of the collimator slit (not shown). (b) Angular spread of the first three orders of the visible spectrum for a diffraction grating with 400 grooves/mm. Orders are shown at different distances from the lens for clarity. In each order, the red end of the spectrum is deviated most. Normal incidence is assumed.

Thus at the position corresponding to λ in the first order, we may also find a spectral line corresponding to $\lambda/2$ in the second order, $\lambda/3$ in the third order, and so on. The free spectral range in order m may be determined by the following argument. If λ_1 is the *shortest detectable wavelength* in the incident light, then the longest nonoverlapping wavelength, λ_2, in order m is coincident with the beginning of the spectrum again in the next higher order $m + 1$, or

$$m\lambda_2 = (m + 1)\lambda_1$$

The free spectral range for order m is then given by

$$\lambda_{fsr} = \lambda_2 - \lambda_1 = \frac{\lambda_1}{m} \tag{12-3}$$

The free spectral range is the maximum wavelength separation, $\Delta\lambda_{max}$, that can be unambiguously resolved in a given order. Notice that this nonoverlapping spectral region is smaller for higher orders.

Example 12-1

The shortest wavelength of light present in a given source is 400 nm. Determine the free spectral range in the first three orders of grating diffraction.

Solution

$$\lambda_{fsr}^m = \frac{\lambda_1}{m}$$

Thus,

$$\lambda^1_{fsr} = \frac{400}{1} = 400 \text{ nm (from 400 to 800 nm in first order)}$$

$$\lambda^2_{fsr} = \frac{400}{2} = 200 \text{ nm (from 400 to 600 nm in second order)}$$

$$\lambda^3_{fsr} = \frac{400}{3} = 133 \text{ nm (from 400 to 533 nm in third order)}$$

12-3 DISPERSION OF A GRATING

As Figure 11-15b in the previous chapter shows, higher diffraction orders grow less intense as they fall more and more under the constraining diffraction envelope. On the other hand, Figure 12-2b shows clearly that wavelengths within an order are better separated as their order increases. This property is precisely described by the *angular dispersion*, $\mathcal{D}$, defined by

$$\mathcal{D} \equiv \frac{d\theta_m}{d\lambda} \tag{12-4}$$

which gives the angular separation per unit range of wavelength. The variation of θ_m with λ is described by the grating Eq. (12-2), from which we may conclude

$$\mathcal{D} = \frac{m}{a \cos \theta_m} \tag{12-5}$$

If a photographic plate is used in the focal plane of the lens to record the spectrum as in Figure 12-2a, it is convenient to describe the spread of wavelengths on the plate in terms of a *linear dispersion*, $dy/d\lambda$, where y is measured along the plate. Since $dy = f \, d\theta$, the linear dispersion is given by

$$\text{linear dispersion} \equiv \frac{dy}{d\lambda} = f \frac{d\theta_m}{d\lambda} = f\mathcal{D} \tag{12-6}$$

The reciprocal of the linear dispersion is known as the *plate factor*.

Example 12-2

Light of wavelength 500 nm is incident normally on a grating with 5000 grooves/cm. Determine its angular and linear dispersion in first order when used with a lens of focal length 0.5 m.

Solution

The *grating constant* or groove separation a is

$$a = \frac{1}{5000 \text{ cm}^{-1}} = 2 \times 10^{-4} \text{ cm}$$

Clearly, for zeroth order, there is no dispersion. For first order, Eq. (12-5) requires knowledge of the the diffraction angle θ_1. This can be obtained from the grating equation (12-2), $a \sin \theta = m\lambda$, so that

$$\sin \theta_1 = \frac{(1)\lambda}{a} = \frac{500 \times 10^{-7}}{2 \times 10^{-4}} = 0.25$$

Thus, $\theta_1 = 14.5°$ and $\cos \theta_1 = 0.968$.

The angular dispersion in the wavelength region around 500 nm can now be calculated:

$$\mathfrak{D} = \frac{m}{a \cos \theta_m} = \frac{1}{(2 \times 10^{-4} \text{ cm})(0.968)} = 5165 \text{ rad/cm}$$

or

$$\mathfrak{D} = 5.165 \times 10^{-4} \frac{\text{rad}}{\text{nm}} \times \frac{180°}{\pi \text{ rad}} = 0.0296°/\text{nm}$$

The linear dispersion is then found from

$$f\mathfrak{D} = (500 \text{ mm})(5.165 \times 10^{-4} \text{ rad/nm}) = 0.258 \text{ mm/nm}$$

and the plate factor is $1/0.258 = 3.88$ nm/mm. One mm of film then spans a range of almost 4 nm, or 40 Å.

At normal incidence, the grating equation can be incorporated with the angular dispersion relation to give

$$\mathfrak{D} = \frac{m}{a \cos \theta_m} = \left(\frac{a \sin \theta_m}{\lambda} \right) \left(\frac{1}{a \cos \theta_m} \right)$$

or

$$\mathfrak{D} = \frac{\tan \theta_m}{\lambda} \tag{12-7}$$

Thus, the dispersion $\mathfrak{D}$ is actually independent of the grating constant a at a given angle of diffraction θ_m and increases rapidly with θ_m. Since $\mathfrak{D}$ is independent of the grating constant, at a given angle of diffraction, the effect of increasing the grating constant is to increase the order m of the diffraction there, as Eq. (12-5) clearly shows.

12-4 RESOLUTION OF A GRATING

Increased dispersion or spread of wavelengths does not by itself make neighboring wavelengths appear more distinctly, unless the peaks are themselves sharp enough. The latter property describes the *resolution* of the recorded spectrum. By the resolution of a grating, we mean its ability to produce distinct peaks for closely spaced wavelengths in a particular order. Recall from the discussion of Fabry-Perot interferometers in Chapter 8 that the *resolving power* $\mathfrak{R}$ is defined in general by

$$\mathfrak{R} \equiv \frac{\lambda}{(\Delta\lambda)_{\min}} \tag{12-8}$$

In the present context, $(\Delta\lambda)_{\min}$ is the minimum wavelength interval of two spectral components that are just resolvable by Rayleigh's criterion (Figure 11-10). For normally incident light of wavelength $\lambda + d\lambda$, and principal maximum of order m, we have by the grating equation (12-2),

$$a \sin \theta_m = m(\lambda + d\lambda) \tag{12-9}$$

To satisfy Rayleigh's criterion, this peak must coincide (same θ) with the first minimum of the neighboring wavelength's peak in the same order, or

$$a \sin \theta_m = \left(m + \frac{1}{N} \right) \lambda \qquad (12\text{-}10)$$

as can be deduced from both Figure 11-15a and Eq. (11-33). Equating the right members of Eqs. (12-9) and (12-10), we obtain $\lambda/d\lambda = mN$. Since $d\lambda$ here is the minimum resolvable wavelength difference, the resolving power of the grating is, by Eq. (12-8),

$$\Re = mN \qquad (12\text{-}11)$$

For a grating of N grooves, the resolving power is simply proportional to the order of the diffraction. In a given order of diffraction, the resolving power increases with the total number of illuminated grooves. It must be remembered that if N is to be increased within a given width W of grating, the grooves must be proportionately closer together. To take advantage of the maximum resolution, the light must cover the entire ruled width of the grating. If the grating in our previous example, with 5000 grooves/cm, has a width of 8 cm, then $N = 40,000$ and the resolving power in the first order is 40,000. This means, by Eq. (12-8), that in the region of $\lambda = 500$ nm, spectral components as close together as 0.0125 nm can be resolved. In the second order, this figure improves to 0.0063 nm, and so on. The best values for the grating resolving power $\Re$ are in the range of 10^5 to 10^6, which is one or two orders of magnitude less than the resolving powers of the Fabry-Perot interferometers discussed in Chapter 8. (When describing theoretical resolution, it must be remembered that the Rayleigh criterion is somewhat arbitrary and that spectral line widths also enter into the actual resolution.) A grating with 10,000 grooves/cm and 20 cm width provides a resolving power of 1 million in fifth order. For normally incident light, however, the grating equation limits the maximum wavelength (at $\theta = 90°$!) under these conditions to 200 nm. As indicated by Eq. (12-2), if the light is not incident along the normal, the maximum diffractable wavelength can be increased; when θ_i nears $90°$, it is twice as much, or 400 nm. Operation in high orders further severely restricts available light because of the diffraction envelope constraint, unless means are taken to *redirect the central diffraction peak* into the desired order. This is achieved through *blazing*, to be discussed presently. Notice that the resolving power, like the dispersion, is independent of groove spacing for a given diffraction angle. If we write $N = W/a$ for a ruled grating width W and incorporate the grating equation for normal incidence, Eq. (12-11) becomes

$$\Re = mN = \left(\frac{a \sin \theta_m}{\lambda} \right) \frac{W}{a}$$

or

$$\Re = \frac{W \sin \theta_m}{\lambda} \qquad (12\text{-}12)$$

According to Eq. (12-12), the resolution of a grating at diffracting angle θ_m depends on the width of the grating rather than on the number of its grooves. For a fixed ratio of $(\sin \theta_m)/\lambda$, however, the grating equation also fixes the ratio m/a. Thus using a grating with fewer grooves and a larger grating constant requires that we work at a higher order m, where there is increased complication due to overlapping orders. Such confusion in high orders is sometimes alleviated by using a second dispersing instrument that spreads the first spectrum again but in a direction orthogonal to the first. One such instrument is described later in this chapter.

TABLE 12-1 FABRY-PEROT INTERFEROMETER AND DIFFRACTION GRATING FIGURES OF MERIT

	Fabry-Perot Interferometer	Diffraction Grating
Resolving power, $\Re$	$m\widetilde{\mathfrak{F}}$	mN
Minimum resolvable wavelength separation, $\Delta\lambda_{min} = \lambda/\Re$	$\dfrac{\lambda}{m\widetilde{\mathfrak{F}}}$	$\dfrac{\lambda}{mN}$
Free spectral range, λ_{fsr}	$\dfrac{\lambda_1}{m}$	$\dfrac{\lambda_1}{m}$
Meaning of parameters	m: Number of half–wavelengths in the Fabry-Perot length. $\widetilde{\mathfrak{F}}$: Cavity finesse	m: Diffraction order N: Number of grooves in grating

In Section 8-6 we discussed the performance characteristics of scanning Fabry-Perot interferometers. Three useful figures of merit that describe these devices as well as diffraction gratings are the resolving power $\Re$, the minimum wavelength separation $\Delta\lambda_{min} = \lambda/\Re$ that can be resolved, and the maximum wavelength separation $\Delta\lambda_{max}$ that can be unambiguously resolved. As discussed in Sections 12-2 and 8-6, the maximum wavelength separation that can be resolved is the free spectral range of the device, so that $\Delta\lambda_{max} = \lambda_{fsr}$. In Table 12-1, these figures of merit are tabulated. Notice that the two types of devices have figures of merit of similar form. The number of grooves N in a diffraction grating plays the role of the finesse $\widetilde{\mathfrak{F}}$ of a Fabry-Perot interferometer. The order number m for a Fabry-Perot interferometer is the number of half-wavelengths that fit into the Fabry-Perot length and so is typically in the range 10^5–10^6. The order number m for a diffraction grating is, of course, much less. Since the free spectral range of both devices is λ_1/m, diffraction gratings typically have a much larger free spectral range than do Fabry-Perot interferometers. (Here, λ_1 is the wavelength of the "short" wavelength end of the free spectral range.) Although the large order number of a typical Fabry-Perot interferometer is a disadvantage in that it leads to a small free spectral range, it is an advantage, as Table 12-1 indicates, in that it allows for higher resolving powers and smaller minimum resolvable wavelength separations. As mentioned, a good Fabry-Perot interferometer may have, overall, a resolving power in the range 10^6–10^7, whereas the resolving power of a good diffraction grating is in the range 10^5–10^6, an order of magnitude smaller.

12-5 TYPES OF GRATINGS

Up to this point we have considered the diffraction grating to be an opaque aperture in which closely spaced slits have been introduced. Fraunhofer's original gratings were, in fact, fine wires wound between closely spaced threads of two parallel screws or parallel lines ruled on smoked glass. Later, Strong used ruled metal coatings on glass blanks. Today the typical grating master is made by diamond-point ruling of grooves into a low-expansion glass base or into a film of aluminum or gold that has been vacuum-evaporated onto the glass base. The base, or *blank*, itself must first be polished to closer than $\lambda/10$ for green light. The development of ruling machines capable of ruling up to 3600 sculptured grooves per millimeter over a width of 10 in. or more, with suitably uniform depth, shape, and spacing, has been an impressive and far-reaching technological achievement. Techniques involving interferometric and electronic servo-control have been used to enhance the precision of the most modern ruling engines. High-quality grating masters ruled over widths as large as 46 cm or more have become feasible.

A grating may be designed to operate either as a *transmission grating* or a *reflection grating*. In a transmission grating, light is periodically transmitted by the clear sections of a glass blank, into which grooves serving as scattering centers have been ruled. Or the light is transmitted by the entire ruled area but periodically retarded in phase due to the varying optical thickness of the grooves. In the first case, the grating is a *transmission amplitude grating*, functioning like the slotted, opaque aperture. In the second case, the grating is called a *transmission phase grating*. In the reflection grating, the groove faces are made highly reflecting, and the periodic reflection of the incident light behaves like the periodic transmission of waves from a transmission grating. Research-quality gratings are usually of the reflection type. A section of a plane reflection grating is shown in Figure 12-3.

The path difference between equivalent reflected rays of light from successive groove reflections is just the difference

$$\Delta = \Delta_1 - \Delta_2 = a \sin \theta_i - a \sin \theta_m$$

where both rays are assumed to have the direction after diffraction specified by the angle θ_m. When $\Delta = m\lambda$, an interference principal maximum results, so that the reflection-grating equation is the same as that for a transmission grating:

$$m\lambda = a \left(\sin \theta_i + \sin \theta_m \right)$$

The same sign convention also applies to the angles θ_m and θ_i: When θ_m is on the opposite side of the grating normal relative to θ_i, as in Figure 12-3, it is considered negative. The zeroth order of interference occurs when $m = 0$ or $\theta_m = -\theta_i$, that is, in the direction of specular reflection from the grating, acting as a mirror for all wavelengths. The metallic coating of the reflection grating should be as highly reflective as possible. In the ultraviolet range of 110 to 160 nm, coatings of magnesium fluoride or lithium fluoride over aluminum are generally used to enhance reflectivity. Below 100 nm, gold and platinum are often used. In the infrared regions, silver and gold coatings are both effective. The light diffracted from a plane grating must be focused by means of a lens or concave mirror. When the absorption of radiation by the focusing elements is severe, as in the vacuum ultraviolet (about 1 to 200 nm), the focusing and diffraction may both be accomplished by using a *concave grating*, that is, a concave mirror that has been ruled to form grooves onto its reflecting surface.

Figure 12-3 Neighboring reflection grating grooves illuminated by light incident at angle θ_i with the grating normal. For light diffracted in the direction θ_m, the net path difference of the two waves is $\Delta_1 - \Delta_2$.

12-6 BLAZED GRATINGS

The *absolute efficiency* of a grating in a given wavelength region and order is the ratio of the diffracted light energy to the incident light energy in the same wavelength region. Increasing the number of rulings on a grating, for example, increases the light energy *throughput*. The zeroth-order diffraction principal maximum, for which there is no dispersion, represents a waste of light energy, reducing grating efficiency. The zeroth order, it will be recalled, contains the most intense interference maximum because it coincides with the maximum of the single-slit diffraction envelope. The technique of shaping individual grooves so that the diffraction envelope maximum shifts into another order is called *blazing* the grating.

To understand the effect of blazing, consider Figure 12-4 for a transmission grating and Figure 12-5 for a reflection grating. For simplicity, light is shown transmitted or reflected from a single groove, even though diffraction involves the cooperative contribution from many grooves. In each figure, (a) illustrates the situation for an unblazed grating and (b) shows the result of shaping the grooves to shift the diffraction envelope maximum ($\beta = 0$) away from the zeroth-order ($m = 0$) interference or principal maximum. Recall from Chapter 11 that the diffraction envelope maximum occurs where $\beta = 0$, that is, where the far-field path difference for light rays

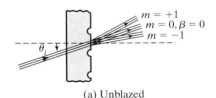

(a) Unblazed

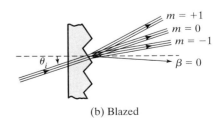

(b) Blazed

Figure 12-4 In an unblazed transmission grating (a), the diffraction envelope maximum at $\beta = 0$ coincides with the zeroth-order interference at $m = 0$. In the blazed grating (b), they are separated.

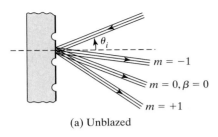

(a) Unblazed

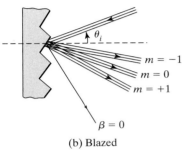

(b) Blazed

Figure 12-5 In an unblazed reflection grating (a), the diffraction envelope maximum at $\beta = 0$ coincides with the zeroth-order interference at $m = 0$. In the blazed grating (b), they are separated.

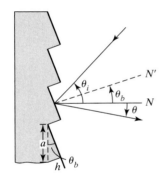

Figure 12-6 Relation of the blaze angle θ_b to the incident and diffracted beams for a reflection grating.

from the center and the edge of any groove is zero. A zero path difference for these rays implies the condition of geometrical optics: For transmitted light, Figure 12-4, the diffraction peak is in the direction of the incident beam; for reflected light, Figure 12-5, it is in the direction of the specularly reflected beam. By introducing prismatic grooves in Figure 12-4 or inclined mirror faces in Figure 12-5, the corresponding zero path difference is shifted into the directions of the refracted beam and the new reflected beam, respectively, which now correspond to the case $\beta = 0$. While the diffraction envelope is thus shifted by the shaping of the individual grooves, the interference maxima remain fixed in position. Their positions are determined by the grating equation, in which angles are measured relative to the plane of the grating. Neither this plane nor the groove separation has been altered in going from (a) to (b) in either Figure 12-4 or 12-5. The result is that the diffraction maximum now favors a principal maximum of a higher order ($|m| > 0$), and the grating redirects the bulk of the light energy where it is most useful.

It remains to determine the proper *blaze angle* of a grating. Consider the reflection grating of Figure 12-6, where a beam is incident on a groove face at angle θ_i and is diffracted at arbitrary angle θ, both measured relative to the grating normal N. The normal N' to the groove face makes an angle θ_b relative to N. This angle is the blaze angle of the grating.

Now let us require that the *diffracted beam* satisfy both the condition of specular reflection from the groove face and the condition for a principal maximum in the mth order, that is, $\theta = \theta_m$. The first condition is satisfied by making the angle of incidence relative to N' equal to the angle of reflection relative to N': $\theta_i - \theta_b = \theta_m + \theta_b$, or

$$\theta_b = \frac{\theta_i - \theta_m}{2} \tag{12-13}$$

The second condition requires that the angle θ_m satisfy the grating equation,

$$m\lambda = a\,(\sin\theta_i + \sin\theta_m) \tag{12-14}$$

Equation (12-13) shows that the blaze angle depends on the angle of incidence, θ_i so that various geometries requiring different blaze angles are possible. In the general case, the equation that must be satisfied by the blaze angle is found by combining Eqs. (12-13) and (12-14). Taking into account the associated sign convention, the grating equation becomes

$$m\lambda = a[\sin\theta_i + \sin(2\theta_b - \theta_i)] \tag{12-15}$$

We consider two special cases of Eq. (12-15). In the *Littrow mount*, incident light is brought in along or close to the groove face normal N', so that $\theta_b = \theta_i$ and $\theta_m = -\theta_i$, as is clear from Figure 12-6 and Eq. (12-13). For this special case, Eq. (12-15) gives

$$\text{Littrow: } m\lambda = 2a\sin\theta_b \quad \text{or} \quad \theta_b = \sin^{-1}\left(\frac{m\lambda}{2a}\right) \tag{12-16}$$

Since the quantity $a\sin\theta_b$ corresponds to the steep-face height h of the groove (Figure 12-6), we see that a grating correctly blazed for wavelength λ and order m in a Littrow mount must have a groove step h of an integral number m of half-wavelengths. Commercial gratings are usually specified by their blaze angles and the corresponding first-order Littrow wavelengths. In another configuration, the light is introduced instead along the normal N to the grating itself. Then $\theta_i = 0$ and, from Eq. (12-13), $\theta_b = -\theta_m/2$. Equation (12-15) now gives

$$\text{normal incidence: } \theta_b = \tfrac{1}{2}\sin^{-1}\left(\frac{m\lambda}{a}\right) \qquad (12\text{-}17)$$

Example 12-3

    a. Consider a 1200-groove/mm grating to be blazed for a wavelength of 600 nm in first order. Determine the proper blaze angle.

    b. *An echelle grating* is a coarsely pitched grating designed to achieve high resolution by operating in high orders. Consider the operation in order $m = 30$ of a commercially available echelle grating with 79 grooves/mm, blazed at an angle of $63°26'$ and ruled over an area of 406×610 mm. Determine its resolution when used in a Littrow mount.

Solution

    a. In a Littrow mount, using Eq. (12-16), the blaze angle must be

$$\theta_b = \sin^{-1}\left[\frac{(1)(600 \times 10^{-6})}{2(1/1200)}\right] = 21.1° = 21°06'$$

On the other hand, if the grating is used in a mount with light incident normal to the plane of the grating, then from Eq. (12-17),

$$\theta_b = \tfrac{1}{2}\sin^{-1}\left[\frac{(1)(600 \times 10^{-6})}{(1/1200)}\right] = 23.03° = 23°02'$$

    b. In a Littrow mount, the grating returns, along the incidence direction, light of wavelength

$$\lambda = \frac{2a\sin\theta_b}{m} = \frac{2(1/79)\sin(63.43)}{30}\text{mm} = 755 \text{ nm}$$

The total number of lines on the grating is $N = (79)(610) = 48,190$ so that the resolving power is $\Re = mN = (30)(48,190) = 1,445,700$ at the blaze wavelength of 755 nm. The minimum resolvable wavelength interval in this region is, then, $\Delta\lambda_{\min} = \lambda/\Re$, or 0.0005 nm. Actual resolutions may be somewhat less than the theoretical value due to grating imperfections. The high resolution is gained at the expense of a contracted spectral range of only $\lambda/m = 755/30 = 25$ nm.

12-7 GRATING REPLICAS

The expense and difficulty of manufacturing gratings prohibit the routine use of grating masters in spectroscopic instruments. Until the technique of making *replicas*—relatively inexpensive copies of the masters—was developed, few research scientists owned a good grating. To make a replica grating, the master is first coated with a layer of nonadherent material, which can be lifted off the master at a later stage. This is followed by a vacuum-evaporated overcoat of aluminum. A layer of resin is then spread over the combination, and a substrate for the future replica is placed on top. After the resin has hardened, the replica grating can be separated from the master. The first good replica grating usually serves as a submaster for the routine production of other replicas. Thin replicas made from a submaster are mounted on a glass or fused silica blank and a highly reflective overcoat of aluminum is added. This is the usual form in which the gratings are made commercially available. Replica gratings can be purchased that are as good as or better than the masters, both in performance and useful life. The efficiency of deep-groove replicas may be better than that of the master because the replication process transfers the smooth parts of the groove faces from bottom to top, improving performance.

12-8 INTERFERENCE GRATINGS

The availability of intense and highly coherent beams of light has made possible the production of gratings apart from the rulings produced by grating engines. As early as 1927, Michelson suggested the possibility of photographing straight interference fringes using an optical system such as that shown in Figure 12-7a. Two coherent, monochromatic beams are made to interfere, producing standing waves in the region between the collimating lens C and a plane mirror M. The resulting straight-line interference maxima are intercepted by a light-sensitive film, inclined at an angle. When the film is developed, straight-line fringes appear.

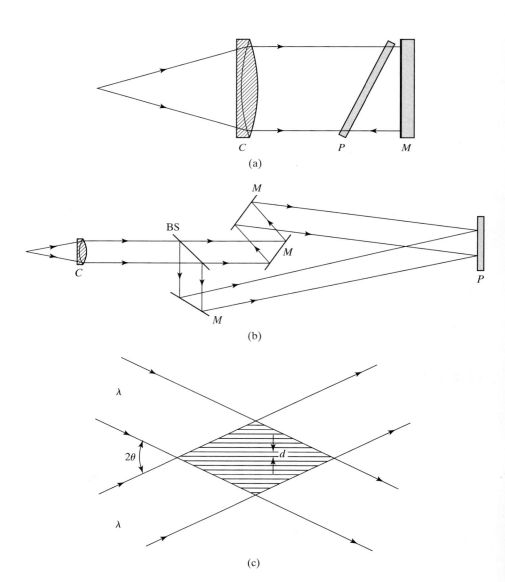

Figure 12-7 (a) Michelson system for producing interference gratings, including collimator C, mirror M, and photographic plate P. (b) Holographic system for producing interference fringes including collimator C, beamsplitter BS, mirrors M, and light-sensitive plate P. (c) Production of interference fringes in the region of superposition of two collimated and coherent beams intersecting at angle 2θ.

Interference gratings produced by such optical techniques are also called *holographic gratings*, since a grating of uniformly spaced, parallel grooves can be considered as a hologram of a point source at infinity. Other interferometric systems, such as that shown in Figure 12-7b, are essentially those used to produce holograms. The interfering wavefronts are photographed on a grainless film of photoresist whose solubility to the etchant is proportional to the irradiance of exposure. The photoresist is spread evenly over the surface of the glass blank to a thickness of 1 μm or less by rapidly spinning the blank. When etched, the interference pattern is preserved in the form of transmission grating grooves whose transmittance varies gradually across the groove in a sine-squared profile. A reflective metallic coating is usually added to the grating by vacuum evaporation. The fringe spacing d, as shown in Figure 12-7c, is determined by the wavelength of the light and by the angle 2θ between the two interfering beams, according to the relation $d = \lambda/(2 \sin \theta)$.

In addition to freedom from the expensive and laborious process of machine ruling, the predominant advantage of the interference grating is the absence of periodic or random errors in groove positions that produce *ghosts* and *grass*, respectively. Thus, interference gratings possess impressive spectral purity and provide a high signal-to-noise advantage. On the other hand, control over groove profile, which affects the blazing and thus the efficiency of the grating, is not easily achieved. The groove profiles of normal interference gratings are sine-squared in form and so symmetrical, rather than sawtooth-shaped, as are the usual blazed gratings. Under normal incidence, a symmetrical groove profile results in an equal distribution of light in the positive and negative orders of diffraction. When used under nonnormal incidence, however, it is possible to disperse light into only one diffracted order (other than the zeroth order), and it has been shown that in this case the distribution of light does not depend to a great extent on groove shape. Efficiencies in this configuration can be comparable to those of blazed gratings. Nevertheless, various efforts are in progress to produce groove shapes more like those of ordinary blazed gratings by exposing the photoresist to two wavelengths of radiation whose Fourier synthesis is more saw-toothed in shape, for example, or by subsequent modification of the symmetrical grooves by argon-ion etching or in a variety of other ways. Interference techniques are not practical in the production of coarse, echellelike gratings.

12-9 GRATING INSTRUMENTS

An instrument that uses a grating as a spectral dispersing element is designed around the type of grating selected for a particular application. An inexpensive transmission grating may be mounted in place of the prism in a *spectroscope*, where the spectrum is viewed with the eye by means of a telescope focused for infinity. The light incident on the grating is rendered parallel by a primary slit and collimating lens. Research-grade instruments, however, make use of reflection gratings. These may be *spectrographs*, which record a portion of the spectrum on a photographic plate, photodiode array, or other image detector, or *spectrometers*, where a narrow portion of the spectrum is allowed to pass through an exit slit onto a photomultiplier or other light detector. In the latter case, the spectrum may be scanned by rotating the grating. There are a number of designs possible; we describe briefly a few of the more common ones.

Figure 12-8 shows the basic *Littrow mount*, where a single focusing element is used both to collimate the light incident on the plane grating and, in the reverse direction, to focus the light onto the photographic plate placed near the slit. Recall that in the Littrow configuration, light is incident along the normal to the groove faces. The Littrow condition is also used in the *echelle spectrograph* (Figure 12-9), which is designed to take advantage of the

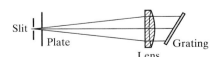

Figure 12-8 Littrow-mounted plane grating. Photographic plate and entrance slit are separated along a direction transverse to the plane of the drawing.

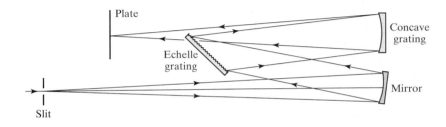

Figure 12-9 Side view of the echelle spectrograph. The echelle is positioned directly over the slit-to-mirror path, but the plate is offset in a horizontal direction.

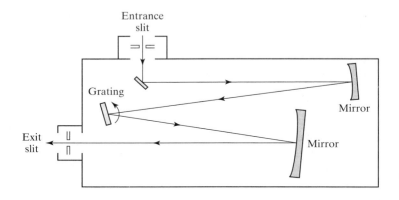

Figure 12-10 Czerny-Turner spectrometer.

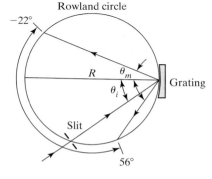

Figure 12-11 Paschen-Runge mounting for a concave grating. Diffracted slit images are formed at the Rowland circle. For a grating of 1200 grooves/mm and $\theta_i = 38°$, the first-order spectrum for wavelengths between 200 and 1200 nm falls between the angles $-22°$ and $56°$, respectively.

high dispersion and resolution attainable with large angles of incidence on a blazed plane grating. As discussed previously, the useful order of diffraction is large and the spectral free range is small, so that a second concave grating is used to disperse the overlapping orders in a direction perpendicular to the dispersion of the echelle grating. In Figure 12-9, a concave mirror collimates the light incident on the echelle, located near the slit and oriented with grooves horizontal. The light diffracted by the echelle is dispersed again by the concave grating, oriented with grooves vertical. The second grating also focuses the two-dimensional spectrum onto the photographic plate. Figure 12-10 shows a *Czerny-Turner* system in a grating spectrometer. Light from an entrance slit is directed by a plane mirror to a first concave mirror, which collimates the light incident on the grating. The diffracted light is incident on a second concave mirror, which then focuses the spectrum at the exit slit. As the grating is rotated, the dispersed spectrum moves across the slit. When the instrument is used specifically to select individual wavelengths from a discrete spectral source or to allow a narrow wavelength range of spectrum through the exit slit, it is called a *monochromator*.

Other instruments dispense with secondary focusing lenses or mirrors and rely on concave gratings both to focus and to disperse the light. The grooves ruled on a concave grating are equally spaced relative to a plane projection of the surface, not relative to the concave surface itself. In this way, spherical aberration and coma are eliminated. Concave-grating instruments are used for wavelengths in the soft X-ray (1 to 25 nm) and ultraviolet regions, extending into the visible. The *Paschen-Runge* design, Figure 12-11, makes use of the *Rowland circle*. This design is used for large concave gratings, whereby the slit, grating, and plate holder all lie on a circle called the Rowland circle that has the following property. If the curved grating surface is tangent at its center to the Rowland circle, which has a diameter equal to the radius of curvature of the concave grating, then a slit source placed anywhere on the circle gives well-focused spectral lines that also fall on the circle. If the light source and slit, grating, and plate holder are placed in a dark room at three stable positions determined by the Rowland circle and the grating equation, the basic requirements of the Paschen-Runge spectrograph are met. Since typical radii of curvature for the grating may be around 6 m, the

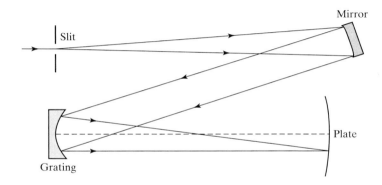

Figure 12-12 Wadsworth mount for a concave grating.

space occupied by this spectrograph can be quite large. The first three orders of diffraction are most commonly used. Typical angles of incidence may vary within the range 30° to 45°, and angles of diffraction may vary between 25° on the opposite side of the grating normal to 85° on the same side of the normal as the slit. Thus, much of the Rowland circle is useful for recording various portions of the spectrum. In Figure 12-11, the first-order spectrum spread (200 to 1200 nm) around the Rowland circle is shown for $\theta_i = 38°$ and a grating of 1200 grooves/mm. Spectral lines formed in this way may suffer rather severely from astigmatism. The *Wadsworth spectrograph* (Figure 12-12) uses a concave mirror, a concave grating, and a plate holder. The plate is mounted normal to the grating. The primary mirror collimates the light incident on the grating. This arrangement eliminates astigmatism and spherical aberration and dispenses with the need for the Rowland circle. Spectra are observed over a range making small angles to the grating normal, perhaps 10° to either side. To record different regions of the spectrum, the grating can be rotated and higher orders can be used. This version of a grating spectrograph allows more compact construction than does the Paschen-Runge design.

The ability of diffraction gratings to direct light of different wavelengths in different directions finds use in several other applications. For example, a Littrow grating can be used as a wavelength-selective mirror to ensure that only one of several laser lines experiences low loss in a laser cavity. Diffraction gratings are also sometimes used in wavelength-division multiplexing and demultiplexing systems in order to combine different-wavelength signals prior to launching them into an optical fiber and then to separate these signals once they have exited the fiber.

PROBLEMS

12-1 What is the angular separation in second order between light of wavelengths 400 nm and 600 nm when diffracted by a grating of 5000 grooves/cm?

12-2 **a.** Describe the dispersion in the red wavelength region around 650 nm (both in °/nm and in nm/mm) for a transmission grating 6 cm wide, containing 3500 grooves/cm, when it is focused in the third-order spectrum on a screen by a lens of focal length 150 cm.

b. Find the resolving power of the grating under these conditions.

12-3 **a.** What is the angular separation between the second-order principal maximum and the neighboring minimum on either side for the Fraunhofer pattern of a 24-groove

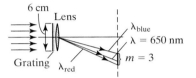

Figure 12-13 Problem 12-2.

grating having a groove separation of 10^{-3} cm and illuminated by light of 600 nm?

b. What slightly longer (or slightly shorter) wavelength would have its second-order maximum on top of the

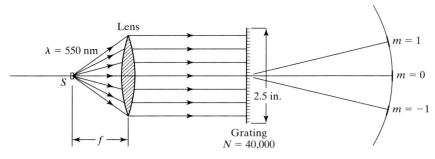

Figure 12-14   Problem 12-6.

minimum adjacent to the second-order maximum of 600-nm light?

c. From your results in parts (a) and (b), calculate the resolving power in second order. Compare this with the resolving power obtained from the theoretical grating resolving power formula, Eq. (12-11).

12-4 How many lines must be ruled on a transmission grating so that it is just capable of resolving the sodium doublet (589.592 nm and 588.995 nm) in the first- and second-order spectra?

12-5 a. A grating spectrograph is to be used in first order. If crown glass optics is used in bringing the light to the entrance slit, what is the first wavelength in the spectrum that may contain second-order lines? If the optics is quartz, how does this change? Assume that the absorption cutoff is 350 nm for crown glass and 180 nm for quartz.

  b. At what angle of diffraction does the beginning of overlap occur in each case for a grating of 1200 grooves/mm?

  c. What is the free spectral range for first and second orders in each case?

12-6 A transmission grating having 16,000 lines/in. is 2.5 in. wide. Operating in the green at about 550 nm, what is the resolving power in the third order? Calculate the minimum resolvable wavelength difference in the second order.

12-7 The two sodium *D* lines at 5893 Å are 6 Å apart. If a grating with only 400 grooves is available, (a) what is the lowest order possible in which the *D* lines are resolved and (b) how wide does the grating have to be?

12-8 A multiple-slit aperture has (1) $N = 2$, (2) $N = 10$, and (3) $N = 15,000$ slits. The aperture is placed directly in front of a lens of focal length 2 m. The distance between slits is 0.005 mm and the slit width is 0.001 mm for each case. The incident plane wavefronts of light are of wavelength 546 nm. Find, *for each case*, (a) the separation on the screen between the zeroth- and first-order maxima; (b) the number of bright fringes (principal maxima) that fall under the central diffraction envelope; (c) the width on the screen of the central interference fringe.

12-9 A reflection grating is required that can resolve wavelengths as close as 0.02 Å in second order for the spectral region around 350 nm. The grating is to be installed in an instrument where light from the entrance slit is incident normally on the grating. If the manufacturer provides rulings over a 10-cm grating width, determine (a) the minimum number of grooves/cm required; (b) the optimum blaze angle for work in this region; (c) the angle of diffraction where irradiance is maximum (show both blaze angle and diffraction angle on a sketch); (d) the dispersion in nanometers per degree.

12-10 A transmission grating is expected to provide an ultimate first-order resolution of at least 1 Å anywhere in the visible spectrum (400 to 700 nm). The ruled width of the grating is to be 2 cm.

  a. Determine the minimum number of grooves required.

  b. If the diffraction pattern is focused by a 50-cm lens, what is the linear separation of a 1-Å interval in the vicinity of 500 nm?

12-11 A concave reflection grating of 2-m radius is ruled with 1000 grooves/mm. Light is incident at an angle of 30° to the central grating normal. Determine, for first-order operation, the (a) angular spread about the grating normal of the visible range of wavelengths (400 to 700 nm); (b) theoretical resolving power if the grating is ruled over a width of 10 cm; (c) plate factor in the vicinity of 550 nm; (d) radius of the Rowland circle in a Paschen-Runge mounting of the grating.

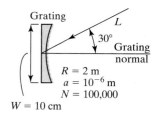

Figure 12-15   Problem 12-11.

12-12 How many grooves per centimeter are required for a 2-m radius, concave grating that is to have a plate factor of around 2 nm/mm in first order?

12-13 A plane reflection grating with 300 grooves/mm is blazed at 10°.

  a. At what wavelength in first order does the grating direct the maximum energy when used with the incident light normal to the groove faces?

  b. What is the plate factor in first order when the grating is used in a Czerny-Turner mounting with mirrors of 3.4-m radius of curvature?

12-14 A reflection grating, ruled over a 15-cm width, is to be blazed for use at 2000 Å in the vacuum ultraviolet. If its theoretical resolving power in first order is to be 300,000, determine the proper blaze angle for use (a) in a Littrow mount and (b) with normal incidence.

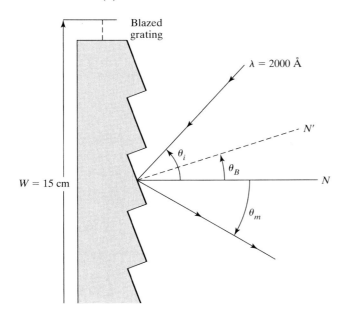

W = 15 cm

Blazed grating

λ = 2000 Å

Figure 12-16   Problem 12-14.

12-15 Show that the spacing d of the fringes in the formation of a holographic grating, as shown in Figure 12-7c, is given by $\lambda/(2\sin\theta)$, where 2θ is the angle between the coherent beams. If the beams are argon-ion laser beams of wavelength 488 nm and the angle between beams is 120°, how many grooves per millimeter are formed in a plane emulsion ($n = 1$) oriented perpendicular to the fringes? What is the effect on the fringe separation d of an emulsion with a high refractive index?

12-16 A grating is needed that is able, working in first order, to resolve the red doublet produced by an electrical discharge in a mixture of hydrogen and deuterium: 1.8 Å at 6563 Å. The grating can be produced with a standard blaze at 6300 Å for use in a Littrow mount. Find (a) the total number of grooves required; (b) the number of grooves per millimeter on the grating with a blaze angle of 22°12′; (c) the minimum width of the grating.

12-17 An echelle grating is ruled over 12 cm of width with 8 grooves/mm and is blazed at 63°. Determine for a Littrow configuration (a) the range of orders in which the visible spectrum (400 to 700 nm) appears; (b) the total number of grooves; (c) the resolving power and minimum resolvable wavelength interval at 550 nm; (d) the dispersion at 550 nm; (e) the free spectral range, assuming the shortest wavelength present is 350 nm.

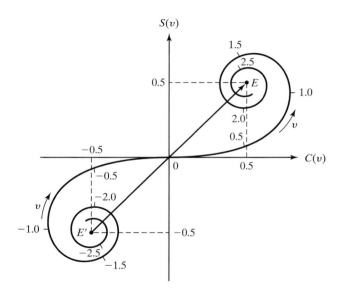

13

Fresnel Diffraction

INTRODUCTION

In the last two chapters, we have dealt with Fraunhofer diffraction, situations in which the wavefront at the diffracting aperture may be considered planar without appreciable error. We turn now to cases where this constitutes an unwarranted approximation, cases in which either or both source and observation screen are close enough to the aperture that wavefront curvature must be taken into account. Collimating lenses are not required, therefore, for the observation of Fresnel, or near-field, diffraction patterns, and in this experimental sense, their study is simpler. The mathematical treatment, however, is more complex and is almost always handled by approximation techniques, as we will see.

Fresnel diffraction patterns form a continuity between the patterns characterizing geometrical optics at one extreme and Fraunhofer diffraction at the other. In geometrical optics, where light waves can be treated as rays propagating along straight lines, we expect to see a sharp image of the aperture. In practice, such images are formed when the observation screen is quite close to the aperture. In cases of Fraunhofer diffraction, where the screen is actually or, through the use of a lens, effectively far from the aperture, the diffraction pattern is a fringed image that bears little resemblance to the aperture. Recall the Fraunhofer double-slit pattern, for example. In the intermediate case of Fresnel diffraction, the diffraction pattern is essentially an image of the aperture, but the edges are fringed.

13-1 FRESNEL-KIRCHHOFF DIFFRACTION INTEGRAL

A typical arrangement is shown in Figure 13-1. Spherical wavefronts emerge from a point source and encounter an aperture. At the aperture, the wavefront

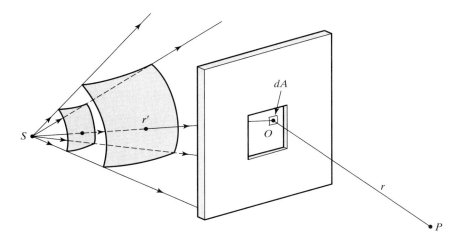

Figure 13-1 Schematic defining the parameters for a typical Fresnel diffraction.

is still substantially spherical, because the aperture is not far from the source. Diffraction effects in the near field on the exit side of the aperture are then of the Fresnel type. The distance from the source S to a representative point O on the wavefront at the aperture is r', and the distance from the point O to a representative point P in the field is r. Compared to Fraunhofer diffraction, this case requires special treatment in several ways. Since the approaching waves are not plane, the distance r' enters into the calculations. Also, the distances r and r' are no longer so much greater than the size of the aperture that Fraunhofer diffraction applies. As a result, the variation of r and r' with different aperture points O and field points P must be taken into account. Finally, because the direction from various aperture points O to a given field point P may no longer be considered approximately constant, the dependence of amplitude on direction of the Huygens wavelets originating at the aperture must be considered. This correction is handled by the *obliquity factor* to be discussed presently.

The electric field at point O in the aperture takes on the usual spherical waveform,

$$E_O = \frac{E_S}{r'} e^{i(kr' - \omega t)} \tag{13-1}$$

Here, $\dfrac{E_S}{r'} e^{ikr'}$ represents the complex amplitude of the electric field at point O.

Employing the *Huygens-Fresnel principle*, as in Fraunhofer diffraction, we seek to find the resultant amplitude of the electric field at P due to a superposition of all the Huygens wavelets from the wavefront at the aperture, each emanating from an infinitesimal region on the wavefront of elemental area dA. The contribution to the resultant field at P due to such an elemental area can be represented by the spherical wave

$$dE_P = \left(\frac{dE_O}{r}\right) e^{i(kr - \omega t)} \tag{13-2}$$

The wave amplitude dE_O/r at the aperture is proportional to the elemental area, so we can write

$$\frac{dE_O}{r} = \frac{E_A}{r} dA \tag{13-3}$$

Here, E_A characterizes the field amplitude per unit area of the Huygens wavelet emanating from the infinitesimal region surrounding point O. As such, E_A should be proportional to the complex amplitude of the electric field originating with the real point source at S. Thus we can write,

$$E_A = \alpha\left(\frac{E_S}{r'}\right)e^{ikr'} \tag{13-4}$$

where α is a proportionality constant with dimensions of inverse length. Combining Eqs. (13-1) through (13-3), we have

$$dE_P = \alpha\left(\frac{E_S}{rr'}\right)e^{ik(r+r')}e^{-i\omega t}\,dA \tag{13-5}$$

The field at P due to the secondary wavelets from the entire aperture is the surface integral of Eq. (13-5),

$$E_P = \alpha E_S e^{-i\omega t}\iint\limits_{\text{Aperture}} \left(\frac{1}{rr'}\right)e^{ik(r+r')}\,dA \tag{13-6}$$

Equation (13-6) is incomplete in two ways. First, it does not take into account the obliquity factor, which attenuates the diffracted waves according to their direction, as described earlier. For the present, we call this factor $F(\theta)$, a function of the angle θ between the directions of the radiation incident and diffracted at the aperture point O. Second, it does not take into account a curious requirement, a $\pi/2$ phase shift of the diffracted waves relative to the primary incident wave. We will return to each of these points in the following discussion. A corrected integral formula was developed by Fresnel and placed on a more rigorous theoretical basis by Kirchhoff. The ad hoc assumptions by Fresnel were shown by Kirchhoff to follow naturally by arguing from Green's integral theorem, whose functions are scalar function solutions to the electromagnetic wave equation.[1] The corrected integral is the *Fresnel-Kirchhoff diffraction formula*, given by

$$E_P = \frac{-ikE_S}{2\pi}e^{-i\omega t}\iint F(\theta)\frac{e^{ik(r+r')}}{rr'}\,dA \tag{13-7}$$

In Eq. (13-7), the factor $-i = e^{-i\pi/2}$ represents the required phase shift, and the obliquity factor

$$F(\theta) = \frac{1+\cos\theta}{2}$$

limits the amplitude, E_S. The result expressed by Eq. (13-7), however, still involves approximations, requiring that the source and screen distances remain large relative to the aperture dimensions and that the aperture dimensions themselves remain large relative to the wavelength of the optical disturbance. The integration specified by Eq. (13-7) is over a closed surface including the aperture but is assumed to make a contribution only over the aperture itself. In arriving at this result, Kirchhoff assumed as boundary conditions that the

---

[1]This derivation requires mathematical ability that is beyond the stated level of this textbook but can be found in many places, for example, Max Born, and Emil Wolf. *Principles of Optics*, 5th ed. (New York: Pergamon Press, 1975, Ch. 8) and Robert Guenther. *Modern Optics* (New York: John Wiley and Sons, 1990, Ch. 9).

wave function and its gradient are zero directly behind the opaque parts of the aperture, and that within the opening itself, they have the same value as they would in the absence of the aperture. These assumptions make the derivation of Eq. (13-7) possible but are not entirely justified. Furthermore, in the theory described here, the $\bar{\mathbf{E}}$-field wave function is a scalar function whose absolute square yields the irradiance. We know that, near the aperture, more rigorous methods must be used that take into account the vector properties of the electromagnetic field, including polarization effects. Nevertheless, the Kirchhoff theory suffices to yield accurate results for most practical diffraction situations.

In the limiting case of Fraunhofer diffraction, Eq. (13-7) is simplified by assuming that (1) the obliquity factor is roughly constant over the aperture due to the small spread in the diffracted light and (2) the variation of distances r and r' remains small relative to that of the exponential function. When all constant (or approximately constant) terms are taken out of the integral and included in an overall constant C_0, Eq. (13-7) is simply

$$E_P = C_0 e^{-i\omega t} \iint e^{ikr} \, dA$$

which is a statement of the Huygens-Fresnel principle and corresponds to the integral used in Chapter 11 to calculate Fraunhofer diffraction patterns there.

For situations in which the assumptions of Fraunhofer diffraction fail, we are left with Eq. (13-7). This integration is, in general, not easy to carry out for a given aperture. Fresnel offered satisfactory methods for simplifying this task, or avoiding it altogether. We apply these methods in the simpler cases to be considered here.

13-2 CRITERION FOR FRESNEL DIFFRACTION

Before dealing with these cases, we wish to establish a practical criterion that determines when we should use Fresnel techniques rather than the simpler Fraunhofer treatment already presented. It will suffice to consider the simple case when both S and P are located on the central axis through the aperture, as in Figure 13-2. Notice that the dimension indicated by Δ is zero when the wavefront is plane. The methods of Fraunhofer diffraction suffice, however, as long as Δ is small, less than the wavelength of the light. From Figure 13-2a we may express this quantity as

$$\Delta = r' - \sqrt{r'^2 - h^2} \tag{13-8}$$

or, equivalently,

$$\Delta = r' - r'\left(1 - \frac{h^2}{r'^2}\right)^{1/2} \cong r' - r'\left(1 - \frac{h^2}{2r'^2}\right)$$

where we have approximated the quantity in parentheses using the first two terms of the binomial expansion, $(1 - x)^{1/2} = 1 - x/2 + \cdots$. Since $p \cong r'$, the condition for significant curvature (near-field case) is

$$\Delta \cong \frac{h^2}{2r'} \cong \frac{h^2}{2p} > \lambda \tag{13-9}$$

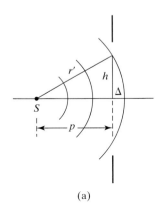

(a)

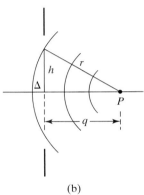

(b)

Figure 13-2 Edge view of Figure 13-1. The curvature of (a) incident and (b) diffracted wavefronts is small when Δ is small.

and similarly, for the diffracted wave curvature in Figure 13-2b,

$$\Delta \cong \frac{h^2}{2q} > \lambda \qquad (13\text{-}10)$$

Combining Eqs. (13-9) and (13-10), the regime of Fresnel, or near-field, diffraction may be expressed by

$$\text{near field: } \frac{1}{2}\left(\frac{1}{p} + \frac{1}{q}\right)h^2 > \lambda \qquad (13\text{-}11)$$

Of course, this condition also applies to the other dimension (transverse to h) of the aperture, not shown in Figure 13-2. When h is taken as the maximum extent of the aperture in either direction or as the radius of a circular aperture, Eq. (13-9) or Eq. (13-10) may also be expressed approximately by the condition

$$\text{near field: } d < \frac{A}{\lambda} \qquad (13\text{-}12)$$

where d represents either p or q and A is the area of the aperture. Note that this condition agrees with Eq. (11-16), which gives the complementary condition under which one can consider the diffraction pattern to be in the *far field*.

13-3 THE OBLIQUITY FACTOR

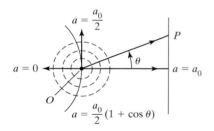

Figure 13-3 Illustration of the obliquity factor.

The effect of the obliquity factor on the secondary wavelets originating at points on the wavefront was introduced by Fresnel. Recall that according to Huygens, a point source of secondary wavelets could radiate with equal effectiveness without regard to direction. This peculiarity would produce new wavefronts in both forward and reverse directions of a propagating wavefront, although the reverse wave does not exist. If point O in Figure 13-3 is the origin of secondary wavelets that arrive at an arbitrary point P in the field, then the correct modification of amplitude a as a function of the angle θ is given by

$$a = \left(\frac{a_0}{2}\right)(1 + \cos\theta) \qquad (13\text{-}13)$$

where a_0 is evidently the amplitude in the forward direction, $\theta = 0$. Notice that $a = 0$ in the reverse direction. The theoretical justification for this relation can also be found in Kirchhoff's derivation.

13-4 FRESNEL DIFFRACTION FROM CIRCULAR APERTURES

Suppose the aperture in Figure 13-1 is circular. Fresnel offered a clever technique for analyzing this special case without having to do the explicit integration of Eq. (13-7). He devised a method for dealing with the contribution from various parts of the wavefront by dividing the aperture into zones with circular symmetry about the axis SOP. The configuration is sketched in Figure 13-4a, which shows an emerging spherical wavefront centered at S. The zones are defined by circles on the wavefront, spaced in such a way that each zone of area S_n is, on the average, $\lambda/2$ farther from the field point P than

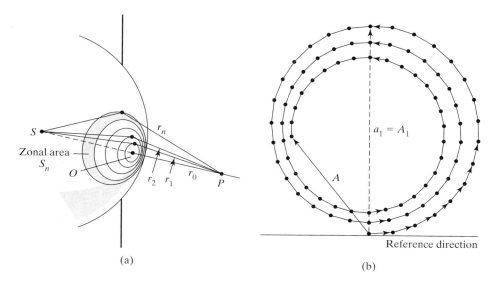

(a)

(b)

Reference direction

Figure 13-4 (a) Fresnel circular half-period zones on a spherical wavefront emerging from an aperture. (b) Phasor diagram for circular Fresnel zones. Each half-period zone is subdivided into 15 subzones. Individual phasors indicate the average phase angle of the subzones and are progressively shorter by 5% to simulate the effect of the obliquity factor. The amplitude a_1 represents contributions of the subzone phazors in the first half-period zone, and the composite amplitude A represents all the zones shown, about $5\frac{1}{2}$ half-period zones.

the preceding zone. In Figure 13-4a, then, $r_1 = r_0 + \lambda/2, r_2 = r_0 + \lambda, \ldots, r_n = r_0 + n\lambda/2, \ldots, r_N = r_0 + N\lambda/2$. This means that each successive zone's contribution is exactly out of phase with that of the preceding one. Of course, each of these *half-period*, or *Fresnel*, zones could be subdivided further into smaller parts–subzones– for which the phase varies from one end of the zone to the other by π. One can show that the resultant contribution from these subzones has an effective phase intermediate between the phases at the zone beginning and end, such that effective phases from successive half-period zones are π, or 180°, apart. This is also clear from Figure 13-4b, a phasor diagram in which each zone is subdivided into 15 subzones. Each of the small phasors represents the contribution from one subzone. The first half-period zone is completed after a number of such phasors culminate in a subzone phasor opposite in direction to the first. The amplitude $a_1 = A_1$ (vertical dashed line) represents the resultant of the subzones in the first half-period zone. Notice that the composite phasor a_1 makes an angle of 90° relative to the reference direction, so that a_1 has a phase of $\pi/2$ relative to the first subzone phasor. For a large number of subzones, the phasor diagram becomes circular and the magnitude of a_1 is the diameter of the circle. The obliquity factor is taken into account in Figure 13-4b by making each succeeding phasor slightly shorter than the preceding one. Thus the circles do not close but spiral inward.

The composite wave amplitudes A_n at P (see Figure 13-4a) from n half-period zones can be expressed as

$$A_n = a_1 + a_2 e^{i\pi} + a_3 e^{i2\pi} + a_4 e^{i3\pi} + \cdots a_n e^{i(n-1)\pi}$$

or

$$A_n = a_1 - a_2 + a_3 - a_4 + \cdots a_n \tag{13-14}$$

The successive zonal amplitudes are affected by three different considerations: (1) a gradual increase with n due to slightly increasing zonal areas, (2) a gradual decrease with n due to the inverse-square law effect as distances from P increase, and (3) a gradual decrease with n due to the obliquity factor. With regard to the first of these, it can be shown that the surface area S_n of the nth Fresnel zone is given by

$$S_n = \frac{\pi r_0' r_0^2}{r_0 + r_0'} \left[\frac{\lambda}{r_0} + (2n - 1)\left(\frac{\lambda}{2r_0}\right)^2 \right] \tag{13-15}$$

The quantity (λ/r_0) is very small in most cases of interest. If the second term in the square brackets is accordingly neglected compared to the first, Eq. (13-15) describes zones with equal areas (independent of n), given by

$$S_n \cong \left[\frac{\pi r_0'}{(r_0 + r_0')}\right] r_0 \lambda \qquad (13\text{-}16)$$

The existence of the second term in Eq. (13-15), however small, indicates increases in zonal areas with n and corresponding small increases in the successive terms of Eq. (13-14). Now one can show that these increases are canceled by the decreases that arise from the second consideration, the effect of the inverse-square law. This leaves only the obliquity factor, which is responsible for systematic decreases in the amplitudes as n increases.

A phasor diagram for the amplitude terms of Eq. (13-14) is shown in Figure 13-5a, as each Fresnel zone contribution a_n is added. The corresponding composite phasors A_n are shown in Figure 13-5b. The individual phasors a_n in Figure 13-5a are separated vertically for clarity. Each phasor is out of phase with its predecessor by 180° and is also shorter, due to the obliquity factor. The composite phasors in Figure 13-5b begin at the start of the phasor a_1 and terminate at the end of the phasor a_n for any number n of contributing Fresnel zones. Notice the large changes in the composite phasor A_n, for small n, as the contribution from each new Fresnel zone is added. For N large, the diagram shows clearly that the resultant amplitude A_R approaches a value of $a_1/2$, or half of that of the first contributing zone. The resultant amplitude A_R is seen to oscillate between magnitudes that are larger and smaller than the limiting value of $a_1/2$, depending on whether it represents an even or an odd number N of contributing zones.

A careful study of Figure 13-5 shows that for N zones, where N is even, the resultant amplitude A_R may be expressed approximately by

$$A_N \cong \frac{a_1}{2} - \frac{a_N}{2}, \ N \text{ even} \qquad (13\text{-}17)$$

and where N is odd by

$$A_N \cong \frac{a_1}{2} + \frac{a_N}{2}, \ N \text{ odd} \qquad (13\text{-}18)$$

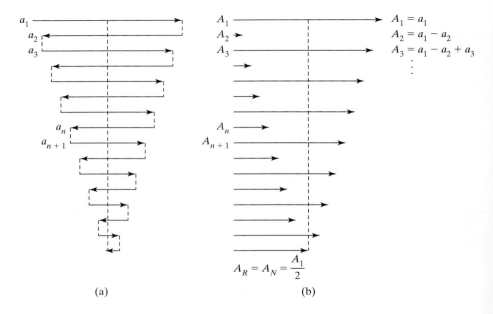

Figure 13-5 Phasor diagram for Fresnel half-period zones. Individual phasors are shown in (a) and the resultant phasors at each step in (b).

(a)   (b)

We may use either Figure 13-5 or Eqs. (13-17) and (13-18) to make the following conclusions:

1. If N is small so that $a_1 \cong a_N$, then for N odd the resultant amplitude is essentially a_1, that of the first zone alone; for N even, the resultant amplitude is near zero.

2. If N is large, as in the case of unlimited aperture, a_N approaches zero, and for either N even or odd, the resultant amplitude is half that of the first contributing zone, or $a_1/2$.

These conclusions produce some curious results, which can be verified experimentally. For example, suppose an amplitude $A_P = a_1$ is measured at P when a circular aperture coincides with the first Fresnel zone. Then by opening the aperture wider to admit the second zone as well, the additional light produces almost zero amplitude at P! Now remove the opaque shield altogether, so that all zones of an unobstructed wavefront contribute. The amplitude at P becomes $a_1/2$, or half that due to the tiny first-zone aperture alone. Since irradiance is proportional to the square of the amplitude, the unobstructed irradiance at P is only $\frac{1}{4}$ that due to the first-zone aperture alone. Such results are surprising because they are not apparent in ordinary experience; yet they necessarily follow once Figure 13-5 is understood.

Another conclusion that is of some historic interest follows from a consideration of the effect at P when a round obstacle or disc just covering the first zone is substituted for the aperture. The light reaching P is now due to all zones *except* the first. The first contributing zone is therefore the second zone, and by the same arguments as those just used, we conclude that light of amplitude $a_2/2$ occurs at P. Thus the irradiance at the center of the shadow of the obstacle should be almost the same as with no disc present! When Fresnel's paper on diffraction was presented to the French Academy, Poisson argued that this prediction was patently absurd and so undermined its theoretical basis. However, Fresnel and Arago showed experimentally that the spot, now known somewhat ironically as *Poisson's spot*, did occur as predicted. The diffraction pattern of an opaque circular disc, including the celebrated Poisson spot, is shown in Figure 13-6. As often happens in such cases, conclusive experimental evidence was already on hand, observed nearly a century before the argument. This sequence of events reminds us of the need to fit experimental results into a successful conceptual framework if they are to make an impact on the scientific world.

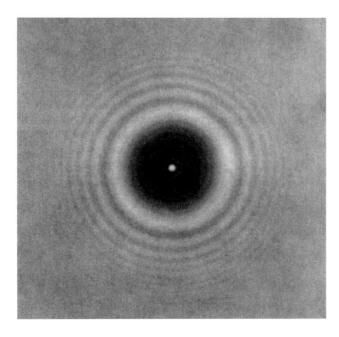

Figure 13-6 Diffraction pattern due to an opaque, circular disc, showing the *Poisson spot* at the center. (From M. Cagnet, M. Francon, and J. C. Thrierr, *Atlas of Optical Phenomenon*, Plate 33, Berlin: Springer-Verlag, 1962.)

13-5 PHASE SHIFT OF THE DIFFRACTED LIGHT

The first phasor a_1 in Figure 13-5 is drawn, rather arbitrarily, in a horizontal direction, and the other phasors are then related to it. As we have seen, however, the phasor a_1, due to the first Fresnel zone, has an effective phase of $\pi/2$ *behind* that of the light reaching P along the axis. The directly propagated light could therefore be represented by a phasor in the vertical direction, making an angle of $\pi/2$ with a_1. The resultant phasor of N zones is also in the direction of a_1. We are forced by these observations to conclude that the phase of the light at P, deduced from the Fresnel zone scheme, is at variance by $\pi/2$ relative to the phase of the light reaching P directly along the axis. To remove this discrepancy and to make the results agree with the phase of the wave without diffraction, Fresnel was forced to assume that the secondary wavelets on diffraction leave with a *gain* in phase of $\pi/2$ relative to the incident wavefront. The factor of i introduced in Eq. (13-7) for this purpose appears naturally in the Kirchhoff derivation of the same equation.

13-6 THE FRESNEL ZONE PLATE

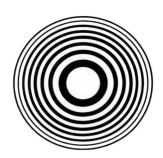

Examination of Eq. (13-14) suggests that if either the negative or the positive terms are eliminated from the sum, the resultant amplitude and irradiance could be quite large. Practically, this means that *every other* Fresnel zone in the wavefront should be blocked. Figure 13-7 shows a drawing of 16 Fresnel zones in which alternate zones are shaded. If such a picture is photographed and a transparency in reduced size is prepared, a *Fresnel zone plate* is produced. Let the light incident on such a zone plate consist of *plane wavefronts*. Then the zone radii required to make the zones half-period zones relative to a fixed field point P can be calculated. From Figure 13-8, the radius R_n of the nth zone must satisfy

$$R_n^2 = \left(r_0 + \frac{n\lambda}{2}\right)^2 - r_0^2 \qquad (13\text{-}19)$$

Figure 13-7 Fresnel zone plate.

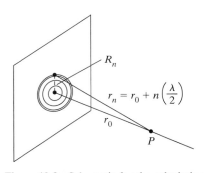

$$R_n$$
$$r_n = r_0 + n\left(\frac{\lambda}{2}\right)$$
$$r_0$$
$$P$$

Figure 13-8 Schematic for the calculation of Fresnel zone plate radii.

which can be written as

$$R_n^2 = r_0^2\left[n\left(\frac{\lambda}{r_0}\right) + \frac{n^2}{4}\left(\frac{\lambda}{r_0}\right)^2\right]$$

We restrict our discussion to cases where $n\lambda/r_0 \ll 1$, so that the second term in square brackets is negligible compared with the first. For example, taking $n = 10{,}000$, $\lambda = 600$ nm, and $r_0 = 30$ cm, one finds $n\lambda/r_0 = 0.02$ and $\frac{n^2}{4}\left(\frac{\lambda}{r_0}\right)^2 = 0.0001$, justifying the neglect of the second term in the square brackets. The zone plate radii are thus given approximately by

$$R_n = \sqrt{nr_0\lambda} \qquad (13\text{-}20)$$

Evidently, the radii of successive zones in Figure 13-7 increase in proportion to $\sqrt{n}$. The radius of the first ($n = 1$) zone determines the magnitude of r_0, or the point P on the axis for which the configuration functions as a zone plate. If the first zone has radius R_1, then successive zones have radii of $1.41R_1$, $1.73R_1$, $2R_1$, and so on.

Example 13-1

If light of wavelength 632.8 nm illuminates a zone plate, what is the first zone radius relative to a point 30 cm from the zone plate on the central axis? How many half-period zones are contained in an aperture with a radius 100 times larger?

Solution

Using Eq. (13-20),

$$R_1 = \sqrt{(1)(30)(632.8 \times 10^{-7})} = 0.0436 \text{ cm}$$

Since $R_n \propto \sqrt{n}$, n increases by a factor of 10^4 when R_n increases by a factor of 10^2. Thus a radius R_n of 4.36 cm encompasses $n = 10{,}000$ Fresnel zones.

If the first, third, fifth, etc., of the 16 zones shown in Figure 13-7 are transmitting, then Eq. (13-14) becomes

$$A_{16} = a_1 + a_3 + a_5 + a_7 + a_9 + a_{11} + a_{13} + a_{15}$$

with 8 zones contributing. When these few zones are reproduced on a smaller scale, the obliquity factor is not very important, and we may approximate $A_{16} = 8a_1$. By comparison, this amplitude at P is 16 times the amplitude $(a_1/2)$ of a wholly unobstructed wavefront. The irradiance at P is, therefore, $(16)^2$, or 256, times as great, even for an aperture encompassing only these 16 zones. If P is 30 cm away, as in the previous example, the radius of this aperture, by Eq. (13-20), is only 1.74 mm for 632.8-nm light. This concentration of light at an axial point shows that the zone plate operates as a lens with P as a focal point. Rearranging Eq. (13-20), we identify the distance r_0 as the first focal length f_1, given by

$$f_1 = \frac{R_1^2}{\lambda}, \qquad n = 1 \tag{13-21}$$

There are other focal points as well. As the field point P approaches the zone plate along the axis, the *same* zonal area of radius R_1 encompasses more half-period zones. In Eq. (13-20), when R_n is fixed, n increases as r_0 decreases. Thus as P is moved toward the plate, $n = 2$ when $r_0 = f_1/2$ for the same zonal radius R_1. At this point, each of the original zones, as in Figure 13-7 for example, now contain two half-period zones. These two half-periods—for each original zone—contribute light at the focal point $r_0 = f_1/2$, out of phase by π with each other. Thus they cancel and so no light is focused by the zone plate at the focal point $r_0 = f_1/2$. Continuing, in this manner, to move the observation point P along the axis toward the Fresnel zone plate, for the *same* zonal radius R_1, we find that $r_0 = f_1/3$ and $n = 3$. Now three half-period zones are contained in each of the original zones in Figure 13-7. Of the three half-period zones, each π out of phase with each other, that contribute light at the focal point $r_0 = f_1/3$, two cancel and only one remains. Suppose then we consider the contribution of all the zones in Figure 13-7, alternately transparent and opaque. The contribution of each original zone, now subdivided into three half-period zones, adds, at the observation point P for $r_0 = f_1/3$, to provide one amplitude A given by

$$A = \underbrace{a_1 - a_2 + a_3}_{a_1} - \underbrace{a_4 + a_5 - a_6}_{\text{removed}} + \underbrace{a_7 - a_8 + a_9}_{a_7} - \cdots \tag{13-22}$$

$$\overset{1^{\text{st}} \text{ zone}}{} \quad \overset{2^{\text{nd}} \text{ zone (opaque)}}{} \quad \overset{3^{\text{rd}} \text{ zone}}{}$$

Comparing the amplitude of Eq. (13-22) at $r_0 = f_1/3$ with the amplitude at $r_0 = f_1$, we see that at $r_0 = f_1$, the entire first zone (made up, effectively, of

a_1, a_2, and a_3) contributes, whereas at $r_0 = f_1/3$, only one of the three does so. Thus the amplitude at $r_0 = f_1/3$, zone by zone, is reduced by a factor of 1/3, so that the irradiance at this point is 1/9 that at $r_0 = f_1$. The argument may, of course, be extended to an observation point at $r_0/5$, when the original zone of radius R_1 includes five half-period zones, and the irradiance is 1/25 that at r_0, and so on. Thus other maximum intensity points along the axis are to be found at

$$f_n = \frac{R_1^2}{n\lambda}, \qquad n \text{ odd} \tag{13-23}$$

Example 13-2

What are the focal lengths for the zone plate described in the preceding example?

Solution

Using Eq. (13-23) together with Eq. (13-20),

$$f_n = \frac{R_1^2}{n\lambda} = \frac{r_0\lambda}{n\lambda} = \frac{r_0}{n}$$

so that $f_1 = 30/1 = 30$ cm, $f_3 = 30/3 = 10$ cm, $f_5 = 30/5 = 6$ cm, and so on.

13-7 FRESNEL DIFFRACTION FROM APERTURES WITH RECTANGULAR SYMMETRY

Diffraction by straight edges, rectangular apertures, and wires are all conveniently handled by another approximation to the Fresnel-Kirchhoff diffraction formula, Eq. (13-7). For this geometry, let the source S in Figures 13-1 and 13-2 represent a slit, so that the wavefronts emerging from S are cylindrical. Recall that cylindrical waves can be expressed mathematically in the same form as spherical waves, except that the amplitude decreases as $1/\sqrt{r}$ so that the irradiance decreases as $1/r$.

Before pursuing Fresnel's quantitative treatment of such cases, consider qualitatively what we might expect by using again the concept of Fresnel half-period zones. This time the zones are rectangular strips along the wavefront, as in Figure 13-9. We wish to show that the sum of all phasors

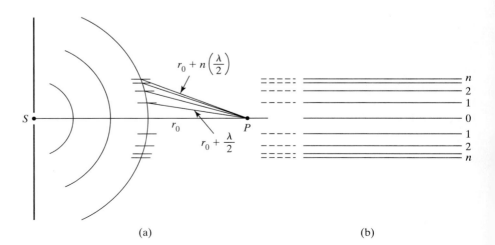

Figure 13-9 Fresnel half-period strip zones on a cylindrical wavefront in an (a) edge view and (b) front view.

(a)   (b)

now gives the endpoints of a curve called the *Cornu spiral.* As before, the average phase at P of the light from each successive zone advances by a half-period, or π. In Figure 13-9b, the rectangular strip zones are shown both above and below the axis SP. Unlike the Fresnel circular zones, the areas of the Fresnel strip zones fall off markedly with n so that successive phasor amplitudes of the zonal contributions are distinctly shorter. A phasor diagram for the complex amplitudes from the Fresnel zones above the axis might look like Figure 13-10. If the first zone is subdivided into smaller segments, which advance by equal phases, the corresponding subzone phasors can be represented by $b_1, b_2, \ldots$, as shown. When the first half-period zone has been included, the last phasor is advanced by π relative to the first and ends at T. The sum of all these contributions is the phasor A_1. In the case of circular zones, Figure 13-4b, the corresponding resultant phasor has a phase angle of $\pi/2$ and the corresponding point, T, would fall on the vertical axis. Because of the rapid decrease in the subzone phasor magnitudes $(b_1, b_2, \ldots)$ the phase angle of A_1 relative to the reference direction is less than $\pi/2$. After advancing through the subzones of the second half-period zone, the phase changes by another π and the last phasor ends at B. The resultant phasor, which includes two full half-period zones, is A_2. By continuing this process, one sees that the phasors approach a smooth curve, which spirals into a limit point E, the eye of the spiral. A phasor A_R from O to E then represents the contributions of half the unimpeded wavefront, the half above the axis SP in Figure 13-9a. A similar argument for the zones below the axis would lead to a twin spiral, represented in the third quadrant and connecting at the origin O. If the coordinates of all points of this Cornu spiral are known, the amplitudes due to contributions from any number of zones can be determined from such a drawing and the relative irradiances compared. The quantitative treatment that allows us to make such calculations follows.

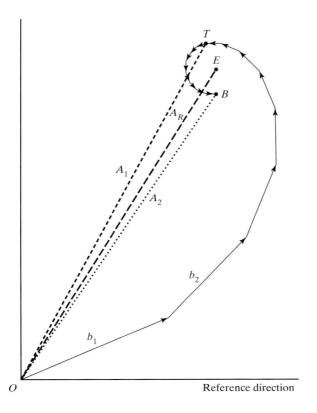

Figure 13-10 Phasor diagram for the first two half-period Fresnel zone strips, each subdivided into smaller zones of equal phase increment.

13-8 THE CORNU SPIRAL

If we neglect the effect of the obliquity factor and the variation of the product rr' in the denominator of Eq. (13-7), the Fresnel-Kirchhoff integral may be approximated by

$$E_P = C_1 e^{-i\omega t} \iint\limits_{A_P} e^{ik(r+r')} \, dA \qquad (13\text{-}24)$$

where all constants are coalesced into C_1. We assume that the surface integral over a closed surface including the aperture is zero everywhere except over the aperture itself, so that we need perform the integration only over the aperture in the yz-plane of Figure 13-11a. A side view, which shows the curvature of the cylindrical wavefront, is drawn in Figure 13-11b. The distance $r + r'$ may be determined approximately from this figure. For $h \ll p$ and $h \ll q$, a binomial expansion approximation gives

$$r' = (p^2 + h^2)^{1/2} = p\left(1 + \frac{h^2}{p^2}\right)^{1/2} \cong p\left(1 + \frac{h^2}{2p^2}\right)$$

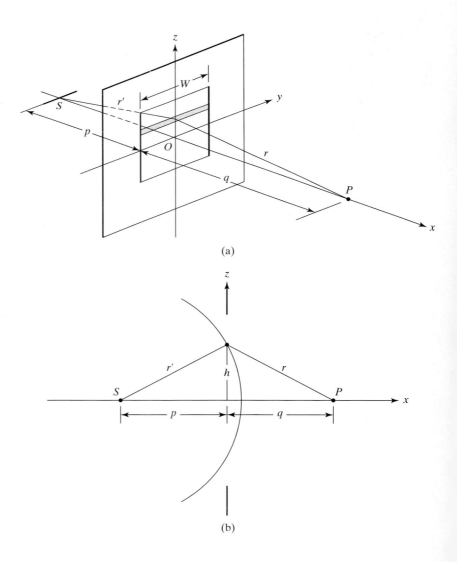

(a)

(b)

Figure 13-11 (a) Cylindrical wavefronts from source slit S are diffracted by a rectangular aperture. (b) Edge view of (a).

Thus,

$$r' \cong p + \frac{1}{2}\left(\frac{h^2}{p}\right)$$

and similarly,

$$r \cong q + \frac{1}{2}\left(\frac{h^2}{q}\right)$$

It follows that

$$r + r' \cong (p + q) + \left(\frac{1}{p} + \frac{1}{q}\right)\frac{h^2}{2}$$

If we abbreviate, using

$$D \equiv p + q \quad \text{and} \quad \frac{1}{L} \equiv \frac{1}{p} + \frac{1}{q} \tag{13-25}$$

we have

$$r + r' \cong D + \frac{h^2}{2L} \tag{13-26}$$

Then Eq. (13-24) becomes

$$E_P = C_1 e^{-i\omega t} \iint e^{ik(D + h^2/2L)} \, dA$$

If the elemental area dA is taken to be the shaded strip in Figure 13-11a, $dA = W\,dz, h = z$, and

$$E_P = C_1 W e^{i(kD - \omega t)} \int_{z_1}^{z_2} e^{ikz^2/2L} \, dz \tag{13-27}$$

The exponent $kz^2/2L = \pi z^2/L\lambda$. Making a change of variable, we let

$$z = v\sqrt{\frac{\lambda L}{2}} \quad \text{or} \quad v = z\sqrt{\frac{2}{L\lambda}} \tag{13-28}$$

whereby E_P is

$$E_P = W\sqrt{\frac{L\lambda}{2}} C_1 e^{i(kD - \omega t)} \int_{v_1}^{v_2} e^{i\pi v^2/2} \, dv \equiv \mathcal{A}_P e^{i(kD - \omega t)} \int_{v_1}^{v_2} e^{i\pi v^2/2} \, dv$$

Here, $\mathcal{A}_P$ is a complex scale factor with dimensions of electric field amplitude. Using Euler's theorem on the integrand, we may write

$$E_P = \mathcal{A}_P e^{i(kD - \omega t)}\left\{ \int_{v_1}^{v_2} \cos\left(\frac{\pi v^2}{2}\right) dv + i \int_{v_1}^{v_2} \sin\left(\frac{\pi v^2}{2}\right) dv \right\} \tag{13-29}$$

The two integrals in this form can be expressed in terms of the *Fresnel integrals*, which we name

$$C(v) \equiv \int_0^v \cos\left(\frac{\pi v^2}{2}\right) dv \tag{13-30}$$

$$S(v) \equiv \int_0^v \sin\left(\frac{\pi v^2}{2}\right) dv \tag{13-31}$$

Using these, Eq. (13-29) may be written as

$$E_P = A_p e^{i(kD - \omega t)}(C(v_2) - C(v_1)) + i(S(v_2) - S(v_1)) \tag{13-32}$$

Now the irradiance at P, since $I_p = \frac{1}{2}\varepsilon_0 c |E_P|^2$, is given by

$$I_P = \frac{1}{2}\varepsilon_0 c |A_p|^2 \{(C(v_2) - C(v_1))^2$$
$$+ (S(v_2) - S(v_1))^2\} \equiv I_0\{(C(v_2) - C(v_1))^2$$
$$+ (S(v_2) - S(v_1))^2\} \tag{13-33}$$

Here we have defined the irradiance scale factor $I_0 = (1/2)\varepsilon_0 c |A_p|^2$. Later we shall show that $2I_0$ is the irradiance at P that results for an *unobstructed wavefront*—that is, for "diffraction" through an aperture of infinite extent. It is useful to note that $C(v)$ and $S(v)$ are both odd functions so that

$$C(-v) = -C(v)$$
$$S(-v) = -S(v)$$

Table 13-1 provides numerical values of these definite integrals for various values of v. As we shall see in several applications, choice of v in the Fresnel integrals is determined by the vertical dimensions of the diffraction aperture.

If the values of the Fresnel integrals are plotted against the variable v, as real and imaginary coordinates on the complex plane, the resulting graph is the *Cornu spiral* (Figure 13-12). According to Eq. (13-33), the square of the length of a straight line drawn between any two points of the spiral must be proportional to the irradiance at point P, since $C(v)$ and $S(v)$ are coordinates in a rectangular coordinate system. For example, consider the spiral points F and G shown in Figure 13-12. The *phasor* $\overrightarrow{FG}$ that connects these points is

$$\overrightarrow{FG} = (C(v_G) - C(v_F)) + i(S(v_G) - S(v_F))$$

The electric field amplitude—see Eq. (13-32)—at point P of Figure 13-11 could then be written as

$$E_p = A_p e^{i(kD - \omega t)}\overrightarrow{FG} \tag{13-34}$$

and the irradiance at point P—see Eq. (13-33)—could be written as

$$I_P = I_0(\overline{FG})^2 \tag{13-35}$$

Here the symbol $\overline{FG}$ is intended to represent the length of the phasor $\overrightarrow{FG}$.

The origin $v = 0$ corresponds to $z = 0$ and therefore to the y-axis through the aperture of Figure 13-11a. The top part of the spiral ($z > 0$ and $v > 0$) represents contributions from strips of the aperture above the y-axis, and the twin spiral below ($z < 0$ and $v < 0$) represents similar contributions from below the y-axis. The limit points or "eyes" of the spiral at E and E'

TABLE 13-1 FRESNEL INTEGRALS

v	$C(v)$	$S(v)$	v	$C(v)$	$S(v)$
0.00	0.0000	0.0000	4.50	0.5261	0.4342
0.10	0.1000	0.0005	4.60	0.5673	0.5162
0.20	0.1999	0.0042	4.70	0.4914	0.5672
0.30	0.2994	0.0141	4.80	0.4338	0.4968
0.40	0.3975	0.0334	4.90	0.5002	0.4350
0.50	0.4923	0.0647	5.00	0.5637	0.4992
0.60	0.5811	0.1105	5.05	0.5450	0.5442
0.70	0.6597	0.1721	5.10	0.4998	0.5624
0.80	0.7230	0.2493	5.15	0.4553	0.5427
0.90	0.7648	0.3398	5.20	0.4389	0.4969
1.00	0.7799	0.4383	5.25	0.4610	0.4536
1.10	0.7638	0.5365	5.30	0.5078	0.4405
1.20	0.7154	0.6234	5.35	0.5490	0.4662
1.30	0.6386	0.6863	5.40	0.5573	0.5140
1.40	0.5431	0.7135	5.45	0.5269	0.5519
1.50	0.4453	0.6975	5.50	0.4784	0.5537
1.60	0.3655	0.6389	5.55	0.4456	0.5181
1.70	0.3238	0.5492	5.60	0.4517	0.4700
1.80	0.3336	0.4508	5.65	0.4926	0.4441
1.90	0.3944	0.3734	5.70	0.5385	0.4595
2.00	0.4882	0.3434	5.75	0.5551	0.5049
2.10	0.5815	0.3743	5.80	0.5298	0.5461
2.20	0.6363	0.4557	5.85	0.4819	0.5513
2.30	0.6266	0.5531	5.90	0.4486	0.5163
2.40	0.5550	0.6197	5.95	0.4566	0.4688
2.50	0.4574	0.6192	6.00	0.4995	0.4470
2.60	0.3890	0.5500	6.05	0.5424	0.4689
2.70	0.3925	0.4529	6.10	0.5495	0.5165
2.80	0.4675	0.3915	6.15	0.5146	0.5496
2.90	0.5624	0.4101	6.20	0.4676	0.5398
3.00	0.6058	0.4963	6.25	0.4493	0.4954
3.10	0.5616	0.5818	6.30	0.4760	0.4555
3.20	0.4664	0.5933	6.35	0.5240	0.4560
3.30	0.4058	0.5192	6.40	0.5496	0.4965
3.40	0.4385	0.4296	6.45	0.5292	0.5398
3.50	0.5326	0.4152	6.50	0.4816	0.5454
3.60	0.5880	0.4923	6.55	0.4520	0.5078
3.70	0.5420	0.5750	6.60	0.4690	0.4631
3.80	0.4481	0.5656	6.65	0.5161	0.4549
3.90	0.4223	0.4752	6.70	0.5467	0.4915
4.00	0.4984	0.4204	6.75	0.5302	0.5362
4.10	0.5738	0.4758	6.80	0.4831	0.5436
4.20	0.5418	0.5633	6.85	0.4539	0.5060
4.30	0.4494	0.5540	6.90	0.4732	0.4624
4.40	0.4383	0.4622	6.95	0.5207	0.4591

represent linear zones at $z = \pm\infty$. Furthermore, the variable v represents the length along the Cornu spiral itself. To see this, recall that the incremental length dl along a curve in the xy-plane is given in general by

$$dl^2 = dx^2 + dy^2$$

In the case at hand, the x- and y-coordinates are the Fresnel integrals $C(v)$ and $S(v)$, respectively. Thus, using Eqs. (13-30) and (13-31), gives

$$dl^2 = (dC(v))^2 + (dS(v))^2 = \left[\cos^2\left(\frac{\pi v^2}{2}\right) + \sin^2\left(\frac{\pi v^2}{2}\right)\right] dv^2$$

or simply,

$$dl = dv \qquad (13\text{-}36)$$

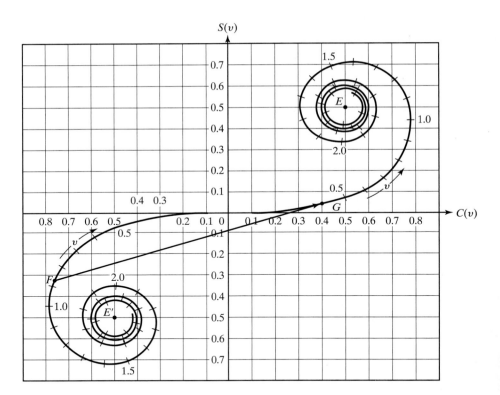

Figure 13-12 The Cornu spiral used to construct the irradiance in a Fresnel diffraction pattern.

13-9 APPLICATIONS OF THE CORNU SPIRAL

Approximate evaluations of the Kirchhoff-Fresnel integral are possible with the help of the Cornu spiral. We examine a few special cases next.

Unobstructed Wavefront

The irradiance in the Fresnel diffraction pattern associated with different apertures are often compared to the irradiance I_u associated with an unobstructed wavefront. An unobstructed wavefront is modeled by passage through an aperture with a vertical dimension 2 that ranges from $-\infty$ to $+\infty$. In this case, the total irradiance I_u at point P is proportional to the square of the length of the phasor drawn from E' to E, as shown in Figure 13-13. The limiting points have the coordinates $(C(v_2), S(v_2)) = (0.5, 0.5)$ and $(C(v_1), S(v_1)) = (-0.5, -0.5)$. These values follow from evaluation of the definite integrals

$$C(\infty) = \int_0^\infty \cos\left(\frac{\pi v^2}{2}\right) dv = 0.5$$

$$S(\infty) = \int_0^\infty \sin\left(\frac{\pi v^2}{2}\right) dv = 0.5$$

and from the previously mentioned fact that $C(v)$ and $S(v)$ are odd functions. Thus, using Eq. (13-33) gives

$$I_u = I_0(\overline{E'E})^2 = I_0\{(C(\infty) - C(-\infty))^2 + (S(\infty) - S(-\infty))^2\}$$
$$= I_0(1^2 + 1^2) = 2I_0 \tag{13-37}$$

Other irradiances may be compared conveniently to this value of $I_u = 2I_0$ for the unobstructed wavefront.

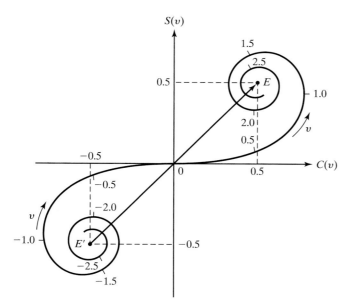

Figure 13-13 The phasor $\overrightarrow{E'E}$ representing the unobstructed wavefront has a length on the Cornu spiral of $\sqrt{2}$.

Straight Edge

Fresnel diffraction by a straight edge is pictured in Figure 13-14a. The Fresnel zones to which we shall refer are long, thin, rectangular regions as pictured earlier in Figure 13-9, rather than annular rings, as pictured in Figure 13-7. Of course, for the straight edge depicted in Figure 13-14a, only those zones above the physical edge contribute light to a given field point. At the field point P on the axis SOP, the edge of the geometric shadow for which $z = v = 0$, the upper half of the zones and Cornu spiral are effective. In this case, the irradiance at point P is proportional to the square of the length of the phasor from O to E. The resulting phasor, shown as OE in Figure 13-14b, has a length of $1/\sqrt{2}$ and, consequently,

$$I_P = I_0(\overline{OE})^2 = \tfrac{1}{2}I_0 = \tfrac{1}{4}I_u \qquad (13\text{-}38)$$

The plot in Figure 13-14c shows the irradiance at point P as well as the irradiance at screen-observation points a vertical displacement y above or below the point P. For a lower point P'' on the screen, we must consider the zones relative to the new axis $SO''P''$, drawn from P'' to the wavefront at the aperture. For P'', the point O'' marks the center of the wavefront, just as the point O marks the center of the wavefront relative to point P. Thus above the axis $SO''P''$, the new "upper half of the wavefront," some of the zones—from O'' to O obstructed by part of the lower half of the edge—do not contribute to the irradiance at point P'' on the screen. Of course, the remaining bottom half of the wavefront is similarly blocked off. Thus contributing zones, relative to the axis $SO''P''$, begin at a finite positive value of z and continue to ∞. These are represented by the amplitude BE on the Cornu spiral. The irradiance at point P'' is thus given by

$$I_{P''} = I_0(\overline{BE})^2 < I_P$$

As indicated in the preceding relation, since $\overline{BE} < \overline{OE}$, the irradiance at P'' is less than that at P. As the observation point P'' moves from P to lower points on the screen, the representative phasor endpoint B slides along the Cornu spiral away from O, with its other end fixed at E. One sees that the amplitude, and so the irradiance, must decrease monotonically, as shown in Figure 13-14c. The edge of the shadow is clearly not sharp. On the other hand, for a point P' above P, we conclude that relative to its axis $SO'P'$, all of the

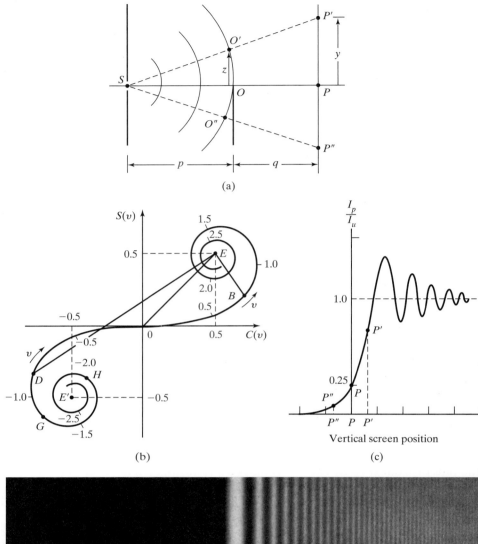

Figure 13-14 (a) Straight-edge diffraction. (b) Use of the Cornu spiral in analyzing straight-edge diffraction. (c) Irradiance pattern due to straight-edge Fresnel diffraction. (d) Diffraction fringes from a straight line. (From M. Cagnet, M. Francon, and J. C. Thrierr, *Atlas of Optical Phenomenon*, Plate 32, Berlin: Springer-Verlag, 1962.)

upper zones (along the wavefront above O') plus some of the first lower zones (between O and O') contribute. Thus, in this case z varies from a certain negative value to ∞. The corresponding field amplitude at P' is then proportional to $\overline{DE} > \overline{OE}$ and so $I_{P'} > I_P$. As P' moves up the screen, D moves down along the spiral shown in Figure 13-14b. In this case, as D winds around the turns of part of the Cornu spiral, the irradiance at P' oscillates with various maxima and minima points, as shown in Figure 13-14c.

Example 13-3

For a straight edge, calculate the irradiance at the first maximum above the edge of the shadow.

Solution

Consider Figure 13-14b. At the first maximum point, the tail of the phasor proportional to the field amplitude is at the extreme point G. Here we read from the curve the value $v \cong -1.2$ at this point. Then from Table 13-1, and the fact that $C(v)$ and $S(v)$ are odd functions we have

$$C(-1.2) = -0.7154 \quad \text{and} \quad S(-1.2) = -0.6234$$

Now the tip of the phasor corresponding to the field maximum is in the positive spiral at point E, which has coordinates $C(\infty) = S(\infty) = 0.5$. The irradiance at such a point on the screen corresponding to this first irradiance maximum is then given by Eq. (13-33) as

$$I_{1st\ max} = I_0\{[0.5 - (-0.7154)]^2 + [0.5 - (-0.6234)]^2\} = 2.74I_0 = 1.37I_u$$

The irradiance at the first maximum is 1.37 times greater than the irradiance I_u for an unobstructed wavefront.

The length $\overline{HE}$, in Figure 13-14b is proportional to the irradiance of the first minimum on the screen. A calculation similar to that carried out in Example 13-3 reveals that in this case $I_{1st\ min} = 0.78I_u$. For observation points P' at very large values of y in Figure 13-14a, the irradiance approaches the value I_u for the unobstructed wavefront. A photograph of the pattern is given in Figure 13-14d. Notice, from Figure 13-14a, that it is possible to relate points at elevation y on the screen to corresponding points at elevation z on the wavefront, such that

$$y = \left(\frac{p + q}{p}\right)z$$

The value of z determines the length v on the Cornu spiral, permitting quantitative calculations of screen irradiance to be made.

Single Slit

Consider a diffracting aperture that is a single slit of width w, as in Figure 13-15a. In this case the contributing zones range from $z = -w/2$ to $z = w/2$ for a field point P along the SOP axis. Thus, for the single slit $\Delta z = w$, and by

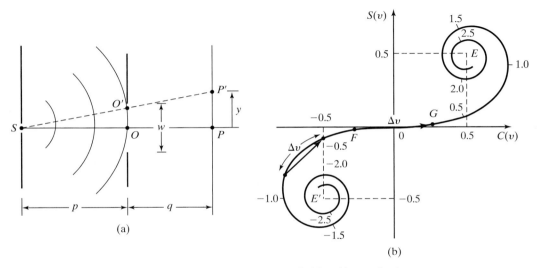

Figure 13-15 Fresnel diffraction from a single slit (a) and its amplitude representation on the Cornu spiral (b).

Eq. (13-28),

$$\Delta v = \Delta z \sqrt{\frac{2}{L\lambda}} = w \sqrt{\frac{2}{L\lambda}} \qquad (13\text{-}39)$$

Once L is calculated from Eq. (13-25), the contributing spiral-length interval Δv on the Cornu spiral can be determined. Note that v plays the role of a universal, dimensionless variable, allowing one Cornu spiral to serve for various combinations of $p, q,$ and λ. For example, if $p = q = 20$ cm, L $= 10$ cm, and $\lambda = 500$ nm, then $\Delta v = 0.632$ for a slit width of 0.01 cm. To calculate the irradiance at the field point P in Figure 13-15a, a length of $\Delta v = 0.632$ symmetrically placed about the origin of the Cornu spiral, as shown in Figure 13-15b, determines the endpoints of the chord FG used to calculate the irradiance at point P. For a field point like P' above P, the contributing zones divide themselves into two parts, with fewer zones above the axis $SO'P'$ (positive z and v) and more zones below $SO'P'$ (negative z and v). Thus the z and v values for the center of the spiral length Δv become increasingly negative. (Note that the center values for point P are $z = 0$ and $v = 0$.) Consequently, as P' moves further above P, Δv slides along the Cornu spiral, toward the lower eye, as shown in Figure 13-15. Although the length Δv along the spiral remains fixed, its placement at different positions along the spiral determines a different chord length and thus a different irradiance. When the observation point is below the axis, Δv is placed along the upper spiral. In this way, the irradiance of the entire pattern can be calculated. From this approach, one can reason that the diffraction pattern of the slit is symmetrical about its center and that the irradiance, while oscillatory, is never zero.

Example 13-4

Let the wavelength of the light be 500 nm and the slit of Figure 13-15a be 1 mm in width. Light emerges from a source slit S, as shown, that is $p = 20$ cm from the diffracting slit. The diffraction pattern is observed $q = 30$ cm from the slit. What is the irradiance at a height of 1 mm above the axis SOP at the screen?

Solution

Using Eq. 13-25 we see that the parameter $L = pq/(p + q) = (20)(30)/(20 + 30) = 12$ cm. We are looking for the irradiance at a point like P' in Figure 13-15a. Contributing zones are like those included in the chord FG of Figure 13-15b, but with the spiral length FG moved somewhat toward the lower end. Fewer zones make a contribution from the upper half of the wavefront relative to P' than they do relative to P. At the screen, point P' corresponds to $y = 1$ mm. The corresponding point z on the wavefront is

$$z = \left(\frac{p}{p+q}\right)y = \frac{20}{50}(1) = 0.4 \text{ mm}$$

The intersection of the dashed line (Figure 13-15a) with the wavefront at the slit thus occurs 0.4 mm above the center of the slit or 0.1 mm below the upper slit edge. The contributing zones as "seen" from P' thus include zones from $z = +0.1$ mm to $z = 0$ in the upper half of its wavefront and from $z = 0$ to $z = -0.9$ mm in the lower half. Contributing zones then span a continuous range from $z_1 = -0.9$ mm to $z_2 = 0.1$ mm relative to the axis $SO'P'$. Corresponding endpoints on the Cornu spiral are v_1 and v_2. For example,

$$v_1 = \sqrt{\frac{2}{L\lambda}} z_1 = \sqrt{\frac{2}{(0.12)(500 \times 10^{-9})}} (-0.9 \times 10^{-3}) = -5.19615$$

Similarly, $v_2 = 0.57735$. The square of the length of chord from v_1 to v_2 on the Cornu spiral is proportional to the irradiance at the point P'. Coordinates of these points are found by interpolation in Table 13-1, giving

For v_1: $C(-5.19615) = -0.44016$ and $S(-5.19615) = -0.50043$

For v_2: $C(0.57735) = 0.56099$ and $S(0.57735)\ 0.10013$

Therefore, the irradiance at point P' is found from Eq. (13-33) to be

$$I_{P'} = I_0\{[0.56099 - (0.44016)]^2 + [0.10013 - (0.50043)]^2\} = 1.36I_0 = 0.68I_u$$

The irradiance at the screen point 1 mm above the axis is 0.68 times the irradiance of an unobstructed wavefront there.

Wire

Suppose now that the narrow slit of Figure 13-15a is replaced by a long, but thin, opaque obstacle such as a wire (Figure 13-16). If the width w of slit and wire are equal, then there is an exact reversal of the transmitting and blocking zones of the wavefront. Now all parts of the Cornu spiral are to be used in calculating the resulting screen irradiance at point P except that portion designated by the spiral-length interval Δv in Figure 13-16b. This situation clearly yields two "vectors," $\overrightarrow{E'F}$ and $\overrightarrow{GE}$, which both contribute to the total field amplitude at point P. In fact, the irradiance at point P is proportional to the square of the length of the *phasor sum* $\overrightarrow{E'F} + \overrightarrow{GE}$. This result can be shown by applying the rules of graphical vector addition in order to subtract the vector associated with the zones blocked by the wire $\overrightarrow{FG}$ from the vector associated with the unobstructed wavefront $\overrightarrow{E'E}$. That is,

$$\overrightarrow{E'F} + \overrightarrow{GE} = \overrightarrow{E'E} - \overrightarrow{FG} \qquad (13\text{-}40)$$

The irradiance at point P is then found as

$$I_P = I_0\big[\overrightarrow{E'F} + \overrightarrow{GE}\big]^2$$

When different field points like P' are considered, the *omitted* spiral interval Δv slides along the spiral as did the *contributing* interval Δv for the case of

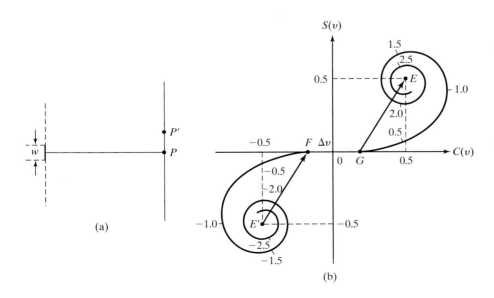

Figure 13-16 Geometry for Fresnel diffraction from a wire of diameter w (a) and its representation on the Cornu spiral (b).

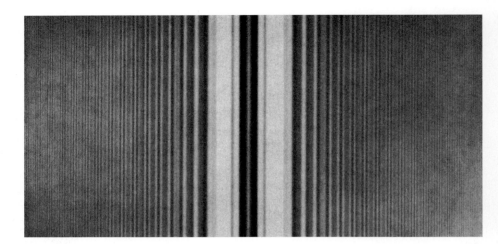

Figure 13-17 Fresnel diffraction pattern of a fine wire. (From M. Cagnet, M. Francon, and J. C. Thrierr, *Atlas of Optical Phenomenon*, Plate 32, Berlin: Springer-Verlag, 1962.)

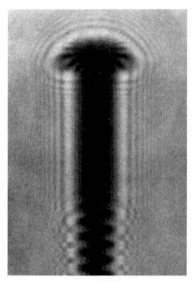

Figure 13-18 Fresnel shadow of a screw. (From M. Cagnet, M. Francon, and J. C. Thrierr, *Atlas of Optical Phenomenon*, Plate 36, Berlin: Springer-Verlag, 1962.)

the rectangular slit considered earlier. The composite diffraction pattern for the wire is shown in the photograph of Figure 13-17. In addition, an example of a more complicated Fresnel pattern than those considered here is given in Figure 13-18.

13-10 BABINET'S PRINCIPLE

Apertures like those of Figures 13-15 and 13-16, in which clear and opaque regions are simply reversed, are called *complementary apertures*. If, in turn, one of the apertures, say A, and then the other, B, are put into place and the amplitude at some point on the screen is determined for each, the sum of these amplitudes must equal the unobstructed amplitude there. This is the content of *Babinet's principle*, which we express as

$$E_A + E_B = E_u \tag{13-41}$$

with A and B representing any two complementary apertures. This result can be demonstrated by analyzing the Cornu spiral results associated with the slit and wire apertures of Figures 13-15 and 13-16. Let the wire be aperture A and the slit be aperture B. By rearranging Eq. (13-40), we may recast Eq. (13-41) as

$$\overrightarrow{E'F} + \overrightarrow{FG} + \overrightarrow{GE} = \overrightarrow{E'E}$$

Now, $\overrightarrow{E'F} + \overrightarrow{GE}$ is proportional to the electric field amplitude E_A at screen point P due to the wire, $\overrightarrow{FG}$ is proportional to the electric amplitude E_B at screen point P due to the slit, and $\overrightarrow{E'E}$ represents the unobstructed amplitude E_u. Thus the wire/aperture complementary pair satisfy Babinet's principle.

It is instructive to apply Babinet's principle at a point where $E_u = 0$. Then, by Eq. (13-41), $E_A = -E_B$ and $I_A = I_B$ at the point. In practice, Fresnel diffraction does not produce amplitudes $E_u = 0$ without an aperture. Fraunhofer diffraction does, however, as in the case of the pattern formed by a point source and a lens. For such a case, in the region outside the small Airy disc, $E_u = 0$, essentially. Complementary apertures introduced into such systems then give, outside the central image, identical diffraction patterns. Thus positive and negative transparencies of the same pattern produce the same diffraction pattern.

PROBLEMS

13-1 A 1-mm-diameter hole is illuminated by plane waves of 546-nm light. According to the usual criterion, which technique (near-field or far-field) may be applied to the diffraction problem when the detector is at 50 cm, 1 m, and 5 m from the aperture?

13-2 A 3-mm-diameter circular hole in an opaque screen is illuminated normally by plane waves of wavelength 550 nm. A small photocell is moved along the central axis, recording the power density of the diffracted beam. Determine the locations of the first three maxima and minima as the photocell approaches the screen.

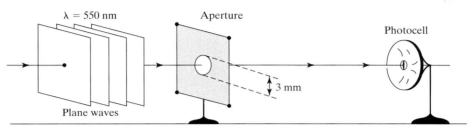

Figure 13-19   Problem 13-2.

13-3 A distant source of sodium light (589.3 nm) illuminates a circular hole. As the hole increases in diameter, the irradiance at an axial point 1.5 m from the hole passes alternately through maxima and minima. What are the diameters of the holes that produce (a) the first two maxima and (b) the first two minima?

13-4 Plane waves of monochromatic (600-nm) light are incident on an aperture. A detector is situated on axis at a distance of 20 cm from the aperture plane.

   a. What is the value of R_1, the radius of the first Fresnel half-period zone, relative to the detector?

   b. If the aperture is a circle of radius 1 cm, centered on axis, how many half-period zones does it contain?

   c. If the aperture is a zone plate with every other zone blocked out and with the radius of the first zone equal to R_1 (found in (a)), determine the first three focal lengths of the zone plate.

13-5 The zone plate radii given by Eq. (13-20) were derived for the case of plane waves incident on the aperture. If instead the incident waves are spherical, from an axial point source at distance p from the aperture, show that the necessary modification yields

$$R_n = \sqrt{nL\lambda}$$

where q is the distance from aperture to the axial point of detection and L is defined by $1/L = 1/p + 1/q$.

13-7 A point source of monochromatic light (500 nm) is 50 cm from an aperture plane. The detection point is located 50 cm on the other side of the aperture plane.

   a. The transmitting portion of the aperture plane is an annular ring of inner radius 0.500 mm and outer radius 0.935 mm. What is the irradiance at the detector relative to the irradiance there for an unobstructed wavefront? The results of problem 13-5 will be helpful.

   b. Answer the same question if the outer radius is 1.00 mm.

   c. How many half-period zones are included in the annular ring in each case?

13-8 By what percentage does the area of the 25th Fresnel half-period zone differ from that of the first, for the case when source and detector are both 50 cm from the aperture and the source supplies light at 500 nm?

13-9 A zone plate is to be produced having a focal length of 2 m for a He-Ne laser of wavelength 632.8 nm. An ink drawing of 20 zones is made with alternate zones shaded in, and a reduced photographic transparency is made of the drawing.

   a. If the radius of the first zone is 11.25 cm in the drawing, what reduction factor is required?

   b. What is the radius of the last zone in the drawing?

13-10 A zone plate has its center half-zone opaque. Find the diameters of the first three clear zones such that the plate focuses parallel light of wavelength 550 nm at 25 cm from the plate.

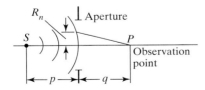

Figure 13-20   Problem 13-5.

13-6 Repeat parts (a) and (b) of problem 13-4 when the source is a point source 10 cm from the aperture. Take into account the results of problem 13-5.

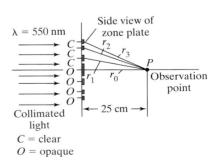

Figure 13-21   Problem 13-10.

13-11 For an incident plane wavefront, show that the areas of the Fresnel half-period zones relative to an observation point at distance x from the wavefront are approximately constant and equal to $\pi\lambda x$. Assume that λ/x is much smaller than 1.

13-12 Light of wavelength 485 nm is incident normally on a screen. How large is a circular opening in an otherwise opaque screen if it transmits four Fresnel zones to a point 2 m away? What, approximately, is the irradiance at the point?

13-13 A single slit of width $\frac{1}{2}$ millimeter is illuminated by a collimated beam of light of wavelength 540 nm. At what observation point on the axis does $\Delta v = 2.5$?

13-14 A source slit at one end of an optical bench is illuminated by monochromatic mercury light of 435.8 nm. The beam diverging from the source slit encounters a second slit 0.5 mm wide at a distance of 30 cm. The diffracted light is observed on a screen at 15 cm farther along the optical bench. Determine the irradiance (in terms of the unobstructed irradiance) at the screen (a) on axis and (b) at one edge of the geometrical shadow of the diffracting slit.

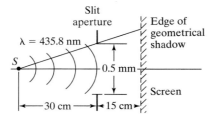

Figure 13-22   Problem 13-14.

13-15 A slit illuminated with sodium light is placed 60 cm from a straight edge and the diffraction pattern is observed using a photoelectric cell, 120 cm beyond the straight edge. Determine the irradiance at (a) 2 mm inside and (b) 1 mm outside the edge of the geometrical shadow.

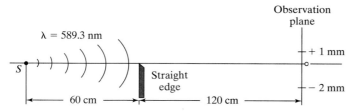

Figure 13-23   Problem 13-15.

13-16 Filtered green mercury light (546.1 nm) emerges from a slit placed 30 cm from a rod 1.5 mm thick. The diffraction pattern formed by the rod is examined in a plane at 60 cm beyond the rod. Calculate the irradiance of the pattern at (a) the center of the geometrical shadow of the rod and (b) the edge of the geometrical shadow.

13-17 For the near-field diffraction pattern of a straight edge, calculate the irradiance of the second maximum and minimum, using the Cornu spiral and the table of Fresnel integral values given.

13-18 Fresnel diffraction is observed behind a wire 0.37 mm thick, which is placed 2 m from the light source and 3 m from the screen. If light of wavelength 630 nm is used, compute, using the Cornu spiral, the irradiance of the diffraction pattern on the axis at the screen. Express the answer as some number times the unobstructed irradiance there.

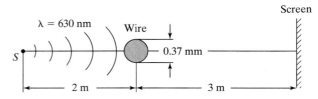

Figure 13-24   Problem 13-18

13-19 Calculate the relative irradiance (compared to the unobstructed irradiance) on the optic axis due to a double-slit aperture that is both 10 cm from a point source of monochromatic light (546 nm) and 10 cm from the observation screen. The slits are 0.04 mm in width and separated (center to center) by 0.25 mm.

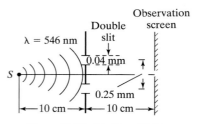

Figure 13-25   Problem 13-19.

13-20 Single-slit diffraction is produced using a monochromatic light source (435.8 nm) at 25 cm from the slit. The slit is 0.75 mm wide. A detector is placed on the axis, 25 cm from the slit.

a. Ensure that far-field diffraction is invalid in this case.

b. Nevertheless, determine the distance above the axis at which single-slit Fraunhofer diffraction predicts the first zero in irradiance.

c. Then calculate the irradiance at the same point, using Fresnel diffraction and the Cornu spiral. Express the result in terms of the unobstructed irradiance.

13-21 A glass plate is sprayed with uniform opaque particles. When a distant point source of light is observed looking through the plate, a diffuse halo is seen whose angular width is about 2°. Estimate the size of the particles. (*Hint*: Use Babinet's principle.)

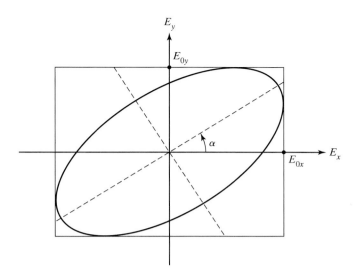

14

Matrix Treatment of Polarization

INTRODUCTION

The polarization of an electromagnetic wave was introduced in Chapter 4. There we noted that the *direction of the electric field vector* $\vec{\mathbf{E}}$ is known as the *polarization* of the electromagnetic wave. In this and the following chapter we extend our discussion of the properties and production of polarized light. As we noted in Chapter 4, the electric field associated with a plane monochromatic electromagnetic wave is perpendicular to the direction of the propagation of the energy carried by the wave. The same can be said of the magnetic field vector, which also maintains an orientation perpendicular to the electric field vector such that the direction of $\vec{\mathbf{E}} \times \vec{\mathbf{B}}$ is everywhere the direction of wave propagation. In general, plane monochromatic waves are *elliptically polarized*, in the sense that, over time, the tip of the electric field vector in a given plane perpendicular to the direction of energy propagation traces out an ellipse. Special cases of electromagnetic waves with elliptical polarization include *linearly polarized* waves in which the electric field vector always oscillates back and forth along a given direction in space and *circularly polarized* waves in which, over time, the tip of the electric field vector traces out a circle. These special cases are shown in Figure 4-12 and are worth reviewing. Monochromatic plane waves are idealized models of the electromagnetic waves produced by, for example, laser sources or a distant single-dipole oscillator. Any electromagnetic wave can be regarded as a superposition of plane electromagnetic waves with various frequencies, amplitudes, phases, and polarizations. "Ordinary" light, such as that produced by a hot filament, is typically produced by a number of independent atomic sources whose radiation is not synchronized. The resultant $\vec{\mathbf{E}}$-field vector consists of many components whose amplitudes, frequencies, polarizations, and phases differ. If the polarizations of the individual fields produced by the independent oscillators are randomly distributed in direction, the field is said to be *randomly polarized* or simply *unpolarized*. If an electromagnetic field consists of the superposition of fields with many different polarizations of which one is (or several are) predominant the field is said to be *partially polarized*.

The possibility of polarizing light is essentially related to its transverse character. If light were a longitudinal wave, the production of polarized light in the ways to be described would simply not be possible. Thus, the polarization of light constitutes experimental proof of its transverse character. In this chapter, we introduce a convenient matrix description of polarization developed by R. Clark Jones.[1] First we develop two-element column matrices or vectors to represent light in various modes of polarization. Then we examine the physical elements that produce polarized light and discover corresponding 2×2 matrices that function as mathematical operators on the Jones vectors. In Chapter 15, we examine in more detail the physical processes that are responsible for producing polarized light.

14-1 MATHEMATICAL REPRESENTATION OF POLARIZED LIGHT: JONES VECTORS

Consider an electromagnetic wave propagating along the z-direction of the coordinate system shown is Figure 14-1. Let the electric field of this wave, at the origin of the axis system, be represented, at a given time, by the vector $\vec{\mathbf{E}}$ shown. Then, in terms of the unit vectors $\hat{\mathbf{x}}$ and $\hat{\mathbf{y}}$,

$$\vec{\mathbf{E}} = E_x\hat{\mathbf{x}} + E_y\hat{\mathbf{y}} \tag{14-1}$$

We write the *complex* field components for waves traveling in the $+z$-direction with amplitudes E_{0x} and E_{0y} and phases φ_x and φ_y as

$$\widetilde{E}_x = E_{0x}e^{i(kz-\omega t+\varphi_x)} \tag{14-2}$$

and

$$\widetilde{E}_y = E_{0y}e^{i(kz-\omega t+\varphi_y)} \tag{14-3}$$

Here, $E_x = \text{Re}\,(\widetilde{E}_x)$ and $E_y = \text{Re}\,(\widetilde{E}_y)$.

Using Eqs. (14-2) and (14-3) in Eq. (14-1) gives, for the complex field $\widetilde{\mathbf{E}}$,

$$\widetilde{\mathbf{E}} = E_{0x}e^{i(kz-\omega t+\varphi_x)}\hat{\mathbf{x}} + E_{0y}e^{i(kz-\omega t+\varphi_y)}\hat{\mathbf{y}}$$

which may also be written

$$\widetilde{\mathbf{E}} = [E_{0x}e^{i\varphi_x}\hat{\mathbf{x}} + E_{0y}e^{i\varphi_y}\hat{\mathbf{y}}]e^{i(kz-\omega t)} = \widetilde{\mathbf{E}}_0 e^{i(kz-\omega t)} \tag{14-4}$$

The bracketed quantity in Eq. (14-4), separated into x- and y-components, is now recognized as the complex amplitude vector $\widetilde{\mathbf{E}}_0$ for the polarized wave. Since the state of polarization of the light is completely determined by the

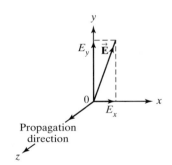

Figure 14-1 Representation of the instantaneous $\vec{\mathbf{E}}$-vector of a light wave traveling in the $+z$-direction.

[1]R. Clark Jones, "A New Calculus for the Treatment of Optical Systems," *Journal of the Optical Society*, Vol. 31 1941; 488.

relative amplitudes and phases of these components, we need concentrate only on the complex amplitude, written as a two-element matrix, or *Jones vector*,

$$\widetilde{\mathbf{E}}_0 = \begin{bmatrix} \widetilde{E}_{0x} \\ \widetilde{E}_{0y} \end{bmatrix} = \begin{bmatrix} E_{0x}e^{i\varphi_x} \\ E_{0y}e^{i\varphi_y} \end{bmatrix} \qquad (14\text{-}5)$$

Let us determine the particular forms for Jones vectors that describe *linear*, *circular*, and *elliptical* polarization. In Figure 14-2a, vertically polarized light travels in the $+z$-direction out of the page with its $\vec{\mathbf{E}}$-oscillations along the y-axis. Since $\vec{\mathbf{E}}$ has a sinusoidally varying magnitude as it progresses, the electric field vector varies between, say, $A\hat{\mathbf{y}}$ and $-A\hat{\mathbf{y}}$. We display this behavior by a double-headed arrow, as shown in Figure 14-2a. As time progresses, the tip of the electric field vector traces out positions along the extent of the double-headed arrow. The field depicted in Figure 14-2a is represented by $E_{0x} = 0$ and $E_{0y} = A$. In the absence of an E_x-component, the phase φ_y may be set equal to zero for convenience. Then, by Eq. (14-5), the corresponding Jones vector is

$$\widetilde{\mathbf{E}}_0 = \begin{bmatrix} E_{0x}e^{i\varphi_x} \\ E_{0y}e^{i\varphi_y} \end{bmatrix} = \begin{bmatrix} 0 \\ A \end{bmatrix} = A\begin{bmatrix} 0 \\ 1 \end{bmatrix} \qquad \text{linear polarization along } y$$

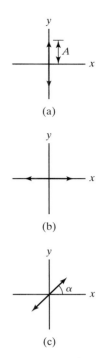

Furthermore, when only the mode of polarization is of interest, the amplitude A may be set equal to 1. The Jones vector for vertically linearly polarized light is then simply $\begin{bmatrix} 0 \\ 1 \end{bmatrix}$. This simplified form is the *normalized* form of the vector. In general, a vector $\begin{bmatrix} a \\ b \end{bmatrix}$ is expressed in normalized form when

$$|a|^2 + |b|^2 = 1$$

Similarly, Figure 14-2b represents horizontally polarized light, for which, letting $E_{0y} = 0$, $\varphi_x = 0$, and $E_{0x} = A$,

$$\widetilde{\mathbf{E}}_0 = \begin{bmatrix} E_{0x}e^{i\varphi_x} \\ E_{0y}e^{i\varphi_y} \end{bmatrix} = \begin{bmatrix} A \\ 0 \end{bmatrix} = A\begin{bmatrix} 1 \\ 0 \end{bmatrix} \qquad \text{linear polarization along } x$$

Figure 14-2 Representation of $\vec{\mathbf{E}}$-vectors of linearly polarized light with various orientations. In each case, the light is propagating in the positive z-direction.

On the other hand, Figure 14-2c represents linearly polarized light whose vibrations occur along a line making an angle α with respect to the x-axis. Both x- and y-components of $\vec{\mathbf{E}}$ are simultaneously present. Evidently this is a general case of linearly polarized light that reduces to the vertically polarized mode when $\alpha = 90°$ and to the horizontally polarized mode when $\alpha = 0°$. Notice that to produce the resultant vibration shown in Figure 14-3a, the two perpendicular vibrations $\widetilde{E}_{0x}$ and $\widetilde{E}_{0y}$ must be in phase. That is, they must pass through the origin together, increase along their respective positive axes together, reach their maximum values together, and then return together to continue the cycle. Figure 14-3a makes this sequence clear. Accordingly, since we require merely a *relative* phase of zero, we set $\varphi_x = \varphi_y = 0$. For a resultant with amplitude A, the perpendicular component amplitudes are $E_{0x} = A\cos\alpha$ and $E_{0y} = A\sin\alpha$. The Jones vector takes the form

$$\widetilde{\mathbf{E}}_0 = \begin{bmatrix} E_{0x}e^{i\varphi_x} \\ E_{0y}e^{i\varphi_y} \end{bmatrix} = \begin{bmatrix} A\cos\alpha \\ A\sin\alpha \end{bmatrix} = A\begin{bmatrix} \cos\alpha \\ \sin\alpha \end{bmatrix} \qquad \text{linear polarization at } \alpha \quad (14\text{-}6)$$

For the normalized form of the vector, we set $A = 1$, since $\cos^2\alpha + \sin^2\alpha = 1$. Notice that this general form does indeed reduce to the Jones vectors found for

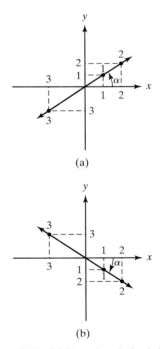

Figure 14-3 (a) Linearly polarized electric field vectors whose x- and y-components are in phase lie in the first and third quadrants. (b) Linearly polarized electric field vectors whose x- and y-components are π out of phase lie in the second and fourth quadrants.

the case $\alpha = 0°$ and $\alpha = 90°$. For other orientations, for example, $\alpha = 60°$,

$$\widetilde{\mathbf{E}}_0 = \begin{bmatrix} \cos(60°) \\ \\ \sin(60°) \end{bmatrix} = \begin{bmatrix} \dfrac{1}{2} \\ \\ \dfrac{\sqrt{3}}{2} \end{bmatrix} = \frac{1}{2}\begin{bmatrix} 1 \\ \\ \sqrt{3} \end{bmatrix}$$

Alternatively, given a vector $\widetilde{\mathbf{E}}_0 = \begin{bmatrix} a \\ b \end{bmatrix}$, where a and b are real numbers, the inclination of the corresponding linearly polarized light is given by

$$\alpha = \tan^{-1}\left(\frac{b}{a}\right) = \tan^{-1}\left(\frac{E_{0y}}{E_{0x}}\right) \tag{14-7}$$

Generalizing a bit, suppose α were a negative angle, as in Figure 14-3b. In this case, E_{0y} is a negative number, since the sine is an odd function, whereas E_{0x} remains positive. The negative sign ensures that the two vibrations are π out of phase, as needed to produce linearly polarized light with $\vec{\mathbf{E}}$-vectors lying in the second and fourth quadrants. Referring to Figure 14-3b again, this means that if the x-vibration is increasing from the origin along its positive direction, the y-vibration must be increasing from the origin along its negative direction. The resultant vibration takes place along a line with negative slope. Summarizing, a Jones vector $\begin{bmatrix} a \\ b \end{bmatrix}$ with both a and b real numbers, not both zero, represents linearly polarized light at inclination angle $\alpha = \tan^{-1}(b/a)$.

By now it may be apparent that in determining the resultant vibration due to two perpendicular components, we are in fact determining the appropriate *Lissajous figure*. If the phase difference between the vibrations is other than 0 or π, the resultant $\vec{\mathbf{E}}$-vector traces out an *ellipse* rather than a straight line. Of course, the straight line can be considered a special case of the ellipse, as can the circle. Figure 14-4 summarizes the sequence of Lissajous figures as a function of relative phase $\Delta\varphi = \varphi_y - \varphi_x$ for the general case $E_{0x} \neq E_{0y}$. Notice the sense of rotation of the tip of the $\vec{\mathbf{E}}$-vector around the ellipses shown in Figure 14-4, which makes the case $\Delta\varphi = \pi/4$, for example, different from the case $\Delta\varphi = 7\pi/4$. When $E_{0x} = E_{0y}$, the ellipses corresponding to $\Delta\varphi = \pi/2$ or $3\pi/2$ reduce to circles.

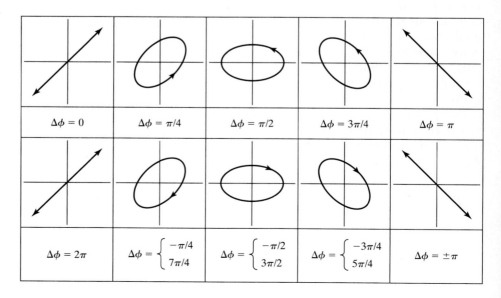

Figure 14-4 Lissajous figures as a function of relative phase for orthogonal vibrations of unequal amplitude. An angle lead greater than π may also be represented as an angle lag of less than π. For all figures we have adopted the phase lag convention $\Delta\varphi = \varphi_y - \varphi_x$.

Now suppose $E_{0x} = E_{0y} = A$ and E_x leads E_y by $\pi/2$. Then at the instant E_x has reached its maximum displacement—$+A$, for example—E_y is zero. A fourth of a period later, E_x is zero and $E_y = +A$, and so on. Figure 14-5 shows a few samples in the process of forming the resultant vibration. For the cases illustrated there, where the x-vibration leads the y-vibration, it is necessary to make $\varphi_y > \varphi_x$. This apparent contradiction results from our choice of phase in the formulation of the $\vec{E}$-field in Eqs. (14-2) and (14-3), where the time-dependent term in the exponent is negative. To show this, let us observe the wave at $z = 0$ and choose $\varphi_x = 0$ and $\varphi_y = \varepsilon$, so that $\varphi_y > \varphi_x$. Equations (14-2) and (14-3) then become

$$\tilde{E}_x = E_{0x}e^{-i\omega t}$$

$$\tilde{E}_y = E_{0y}e^{-i(\omega t - \varepsilon)}$$

The negative sign before ε indicates a lag ε in the y-vibration relative to the x-vibration. To see that these equations represent the sequence in Figure 14-5, we take their real parts and set $E_{0x} = E_{0y} = A$ and $\varepsilon = \pi/2$, giving

$$E_x = A\cos\omega t$$

$$E_y = A\cos\left(\omega t - \frac{\pi}{2}\right) = A\sin\omega t$$

Recalling that $\omega = 2\pi\nu = 2\pi/T$, each of the cases in Figure 14-5 can be easily verified. Also, since

$$E^2 = E_x^2 + E_y^2 = A^2(\cos^2\omega t + \sin^2\omega t) = A^2$$

the tip of the resultant vector traces out a circle of radius A.

We now deduce the Jones vector for this case—where E_x leads E_y—taking $E_{0x} = E_{0y} = A$, $\varphi_x = 0$, and $\varphi_y = \pi/2$. Then,

$$\tilde{\mathbf{E}}_0 = \begin{bmatrix} E_{0x}e^{i\varphi_x} \\ E_{0y}e^{i\varphi_y} \end{bmatrix} = \begin{bmatrix} A \\ Ae^{i\pi/2} \end{bmatrix} = A\begin{bmatrix} 1 \\ i \end{bmatrix} \tag{14-8}$$

To determine the normalized form of the vector, notice that $1^2 + |i|^2 = 1 + 1 = 2$, so that each element must be divided by $\sqrt{2}$ to produce unity. Thus the Jones vector $(1/\sqrt{2})\begin{bmatrix}1\\i\end{bmatrix}$ represents circularly polarized light when $\vec{E}$ rotates *counterclockwise*, viewed head-on. This mode is called *left-circularly polarized* (LCP) light. Thus,

$$\tilde{\mathbf{E}}_0 = \frac{1}{\sqrt{2}}\begin{bmatrix} 1 \\ i \end{bmatrix} \qquad \text{LCP}$$

Similarly, if E_y leads E_x by $\pi/2$, the result will again be circularly polarized light with clockwise rotation leading to *right-circularly polarized* (RCP) light. Replacing $\pi/2$ by $(-\pi/2)$ in Eq. (14-8) gives the normalized Jones vector,

$$\tilde{\mathbf{E}}_0 = \frac{1}{\sqrt{2}}\begin{bmatrix} 1 \\ -i \end{bmatrix} \qquad \text{RCP}$$

Notice that one of the elements in the Jones vector for circularly polarized light is now purely imaginary, and the magnitudes of the elements are the same.

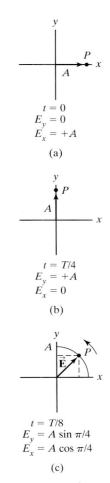

$t = 0$
$E_y = 0$
$E_x = +A$
(a)

$t = T/4$
$E_y = +A$
$E_x = 0$
(b)

$t = T/8$
$E_y = A\sin\pi/4$
$E_x = A\cos\pi/4$
(c)

Figure 14-5 Resultant $\vec{E}$-vibration due to orthogonal component vibrations of equal magnitude and phase difference of $\pi/2$, shown at three different times. The points P represent the position of the resultant. In (c) a sketch of the circular path traced by $\vec{E}$ is also shown. Notice that the $\vec{E}$-vector rotates counterclockwise in this case.

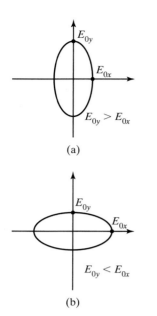

$E_{0y} > E_{0x}$

(a)

$E_{0y} < E_{0x}$

(b)

Figure 14-6 Elliptically polarized light for the case $\Delta\varphi = \pi/2$.

Given a particular mathematical form of the vector, the actual character of the light polarization may not always be immediately apparent. For example, the Jones vector $\begin{bmatrix} 2i \\ 2 \end{bmatrix}$ represents right-circularly polarized light since

$$\begin{bmatrix} 2i \\ 2 \end{bmatrix} = 2 \begin{bmatrix} i \\ 1 \end{bmatrix} = 2i \begin{bmatrix} 1 \\ -i \end{bmatrix}$$

The prefactor of a Jones vector may affect the amplitude and, hence, the irradiance of the light but not the polarization mode. Prefactors such as 2 and $2i$ may therefore be ignored unless information regarding energy is required.

Next suppose that the phase difference between orthogonal vibrations $\widetilde{E}_{0x}$ and $\widetilde{E}_{0y}$ is still $\pi/2$, but $E_{0x} \neq E_{0y}$. In particular, let $E_{0x} = A$ and $E_{0y} = B$, where A and B are positive numbers. In this case, Eq. (14-8) should be modified to give

$$\widetilde{\mathbf{E}}_0 = \begin{bmatrix} A \\ iB \end{bmatrix} \quad \begin{array}{l} \text{counterclockwise} \\ \text{rotation} \end{array} \quad \text{and} \quad \widetilde{\mathbf{E}}_0 = \begin{bmatrix} A \\ -iB \end{bmatrix} \quad \begin{array}{l} \text{clockwise} \\ \text{rotation} \end{array}$$

These instances of elliptical polarization are illustrated in Figure 14-4 for $\Delta\varphi = \pi/2$ and $\Delta\varphi = 3\pi/2$. Notice that a lag of $\pi/2$ is equivalent to a lead of $3\pi/2$. The ellipse is oriented with its major axis along the x- or y-axis, as in Figure 14-6, depending on the relative magnitudes of E_{0x} and E_{0y}. In addition, either case may produce clockwise rotation of $\vec{\mathbf{E}}$ around the ellipse (when E_y leads E_x) or counterclockwise rotation (when E_x leads E_y). Based on these observations, we conclude that a Jones vector with elements of unequal magnitude, one of which is pure imaginary, represents elliptically polarized light oriented along the x,y-axes. The normalized forms of the Jones vectors now must include a prefactor of $1/\sqrt{A^2 + B^2}$.

It is also possible to produce elliptically polarized light with principal axes inclined to the x,y-axes, as evident in Figure 14-4. This situation occurs when the phase difference $\Delta\varphi$ between $\widetilde{E}_{0x}$ and $\widetilde{E}_{0y}$ is some angle other than $\Delta\varphi = 0, \pm\pi, \pm2\pi, \pm m\pi$ (linear polarization) or $\Delta\varphi = \pm\pi/2, \pm3\pi/2, \pm\left(m + \frac{1}{2}\right)\pi$ (circular or elliptical polarization oriented symmetrically about the x,y-axes). Here, $m = 0, \pm1, \pm2, \ldots$. For example, consider the case where E_x leads E_y by some positive angle ε, that is, $\varphi_y - \varphi_x = \varepsilon$. Taking $\varphi_x = 0, \varphi_y = \varepsilon, E_{0x} = A,$ and $E_{0y} = b$ (with A and b positive), the Jones vector is

$$\widetilde{\mathbf{E}}_0 = \begin{bmatrix} E_{0x}e^{i\varphi_x} \\ E_{0y}e^{i\varphi_y} \end{bmatrix} = \begin{bmatrix} A \\ be^{i\varepsilon} \end{bmatrix}$$

Using Euler's theorem, we write

$$be^{i\varepsilon} = b\,(\cos\varepsilon + i\sin\varepsilon) = B + iC$$

The Jones vector for this general case is, then,

$$\widetilde{\mathbf{E}}_0 = \begin{bmatrix} A \\ B + iC \end{bmatrix} \quad \text{counterclockwise rotation, general case} \quad (14\text{-}9)$$

Here the identification of this form with counterclockwise rotation requires that A and C have the same sign. Since multiplying a Jones vector by an overall constant does not change the character of the polarization described by the Jones vector, we shall adopt the convention that A is positive. With that convention a positive imaginary part C of $\widetilde{E}_{0y}$ indicates that the Jones vector represents counterclockwise rotation. Note that one of the elements of the Jones vector in Eq. (14-9) is now a complex number having both real and imaginary parts. The normalized form must be divided by $\sqrt{A^2 + B^2 + C^2}$. The Jones vector of Eq. (14-9) represents an electric field vector whose tip travels in a *counterclockwise* direction as it traces out an ellipse whose symmetry axes are inclined at a general angle relative to the x,y-coordinate system. With the help of analytical geometry, it is possible to show that the ellipse whose Jones vector is given by Eq. (14-9) is inclined at an angle α with respect to the x-axis, as shown in Figure 14-7. The angle of inclination is determined from

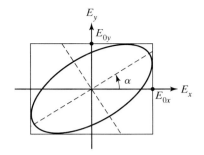

Figure 14-7 Elliptically polarized light oriented at an angle α relative to the x-axis.

$$\tan 2\alpha = \frac{2E_{0x}E_{0y}\cos \varepsilon}{E_{0x}^2 - E_{0y}^2} \qquad (14\text{-}10)$$

The ellipse is situated in a rectangle of sides $2E_{0x}$ and $2E_{0y}$. In terms of the parameters A, B, and C, the derivation of Eq. (14-9) makes clear that

$$E_{0x} = A, \qquad E_{0y} = \sqrt{B^2 + C^2}, \qquad \text{and} \qquad \varepsilon = \tan^{-1}\left(\frac{C}{B}\right) \qquad (14\text{-}11)$$

Example 14-1

Analyze the Jones vector given by

$$\begin{bmatrix} 3 \\ 2 + i \end{bmatrix}$$

to show that it represents elliptically polarized light.

Solution

The light has relative phase between $\widetilde{E}_{0x}$ and $\widetilde{E}_{0y}$ of $\varphi_y - \varphi_x = \varepsilon = \tan^{-1}\left(\frac{1}{2}\right) = 0.148\pi$. Since $E_{0x} = 3$ and $E_{0y} = \sqrt{2^2 + 1^2} = \sqrt{5}$, the inclination angle of the axis is given by

$$\alpha = \frac{1}{2}\tan^{-1}\left(\frac{(2)(3)\left(\sqrt{5}\right)\cos(0.148\pi)}{9 - 5}\right) = 35.8°$$

With this data the ellipse can be sketched as indicated in Figure 14-7. Moreover, from the general equation of an ellipse, we have

$$\left(\frac{E_x}{E_{0x}}\right)^2 + \left(\frac{E_y}{E_{0y}}\right)^2 - 2\left(\frac{E_x}{E_{0x}}\right)\left(\frac{E_y}{E_{0y}}\right)\cos \varepsilon = \sin^2 \varepsilon \qquad (14\text{-}12)$$

For this example, the equation of the ellipse is

$$\frac{E_x^2}{9} + \frac{E_y^2}{5} - 0.267E_xE_y = 0.2$$

When E_x lags E_y, the phase angle ε becomes negative and leads to the Jones vector (with A and C positive numbers) *representing* a clockwise rotation instead:

$$\widetilde{\mathbf{E}}_0 = \begin{bmatrix} A \\ B - iC \end{bmatrix} \qquad \text{clockwise rotation, general case}$$

This form, together with the form representing counterclockwise rotation given in Eq. (14-9), are the most general forms of the Jones vector, including all those discussed previously as special cases.

Table 14-1 provides a convenient summary of the most common Jones vectors in their normalized forms. It should be emphasized that the forms given in Table 14-1 are not unique. First, any Jones vector may be multiplied by a real constant, changing amplitude but not polarization mode. Vectors in Table 14-1 have all been multiplied by prefactors, when necessary, to put them in normalized form. Thus, for example, the vector $\begin{bmatrix} 2 \\ 2 \end{bmatrix} = 2\begin{bmatrix} 1 \\ 1 \end{bmatrix}$ and so represents linearly polarized light making an angle of 45° with the x-axis and with amplitude of $2\sqrt{2}$. Second, each of the vectors in Table 14-1 can be multiplied by a factor of the form $e^{i\varphi}$, which has the effect of promoting the

TABLE 14-1 SUMMARY OF JONES VECTORS $\tilde{\mathbf{E}}_0 = \begin{bmatrix} E_{0x}e^{i\varphi_x} \\ E_{0y}e^{i\varphi_y} \end{bmatrix}$

I. Linear Polarization ($\Delta\varphi = m\pi$)

General:

$$\tilde{\mathbf{E}}_0 = \begin{bmatrix} \cos\alpha \\ \sin\alpha \end{bmatrix}$$

Vertical: $\tilde{\mathbf{E}}_0 = \begin{bmatrix} 0 \\ 1 \end{bmatrix}$

Horizontal: $\tilde{\mathbf{E}}_0 = \begin{bmatrix} 1 \\ 0 \end{bmatrix}$

At $+45°$: $\tilde{\mathbf{E}}_0 = \dfrac{1}{\sqrt{2}}\begin{bmatrix} 1 \\ 1 \end{bmatrix}$

At $-45°$: $\tilde{\mathbf{E}}_0 = \dfrac{1}{\sqrt{2}}\begin{bmatrix} 1 \\ -1 \end{bmatrix}$

II. Circular Polarization $\left(\Delta\phi = \dfrac{\pi}{2}\right)$

Left:

$$\tilde{\mathbf{E}}_0 = \dfrac{1}{\sqrt{2}}\begin{bmatrix} 1 \\ i \end{bmatrix}$$

Right:

$$\tilde{\mathbf{E}}_0 = \dfrac{1}{\sqrt{2}}\begin{bmatrix} 1 \\ -i \end{bmatrix}$$

III. Elliptical Polarization

Left: $A < B$ $A > B$

$$\tilde{\mathbf{E}}_0 = \dfrac{1}{\sqrt{A^2 + B^2}}\begin{bmatrix} A \\ iB \end{bmatrix} \quad A > 0, B > 0$$

$(\Delta\phi = (m + 1/2)\,\pi)$

Right: $A < B$ $A > B$

$$\tilde{\mathbf{E}}_0 = \dfrac{1}{\sqrt{A^2 + B^2}}\begin{bmatrix} A \\ -iB \end{bmatrix} \quad A > 0, B > 0$$

Left:

$$\tilde{\mathbf{E}}_0 = \dfrac{1}{\sqrt{A^2 + B^2 + C^2}}\begin{bmatrix} A \\ B + iC \end{bmatrix} \quad A > 0, C > 0$$

$\left(\Delta\phi \neq \begin{cases} m\pi \\ (m + 1/2)\pi \end{cases}\right)$

Right:

$$\tilde{\mathbf{E}}_0 = \dfrac{1}{\sqrt{A^2 + B^2 + C^2}}\begin{bmatrix} A \\ B - iC \end{bmatrix} \quad A > 0, C > 0$$

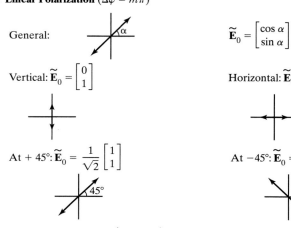
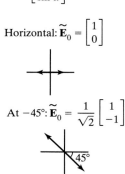
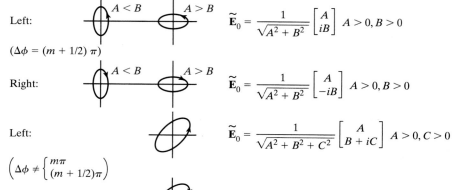

phase of each element by φ, that is, $\varphi_x \rightarrow \varphi_x + \varphi$ and $\varphi_y \rightarrow \varphi_y + \varphi$. Since the phase difference is unchanged in this process, the new vector represents the same polarization mode. Recall that the vectors in Table 14-1 were formulated by choosing, somewhat arbitrarily, $\varphi_x = 0$. Thus, for example, multiplying the vector representing left-circularly polarized light by $e^{i\pi/2} = i$,

$$i\begin{bmatrix} 1 \\ i \end{bmatrix} = \begin{bmatrix} i \\ -1 \end{bmatrix}$$

produces an alternate form of the vector. Clearly, given the second form, one could deduce the standard form in Table 14-1 by extracting the factor i.

The usefulness of these Jones vectors will be demonstrated after Jones matrices representing polarizing elements are also developed. However, at this point it is already possible to calculate the result of the superposition of two or more polarized modes by adding their Jones vectors. The addition of left- and right-circularly polarized light, for example, gives

$$\begin{bmatrix} 1 \\ i \end{bmatrix} + \begin{bmatrix} 1 \\ -i \end{bmatrix} = \begin{bmatrix} 1 + 1 \\ i - i \end{bmatrix} = \begin{bmatrix} 2 \\ 0 \end{bmatrix}$$

or linearly polarized light of twice the amplitude. We conclude that linearly polarized light can be regarded as being made up of left- and right-circularly polarized light in equal proportions. As another example, consider the superposition of vertically and horizontally linearly polarized light in phase:

$$\begin{bmatrix} 0 \\ 1 \end{bmatrix} + \begin{bmatrix} 1 \\ 0 \end{bmatrix} = \begin{bmatrix} 1 \\ 1 \end{bmatrix}$$

The result is linearly polarized light at an inclination of 45°. Notice that the addition of orthogonal components of linearly polarized light is *not* unpolarized light, even though unpolarized light is often symbolized by such components. There is no Jones vector representing unpolarized or partially polarized light.[2]

14-2 MATHEMATICAL REPRESENTATION OF POLARIZERS: JONES MATRICES

Various devices can serve as optical elements that transmit light but modify the state of polarization. The physical mechanisms underlying their operation will be discussed in the next chapter. These polarizers can be generally described by 2×2 Jones matrices,

$$\mathbf{M} = \begin{bmatrix} a & b \\ c & d \end{bmatrix}$$

where the matrix elements a, b, c, and d determine the manner in which the polarizers modify the polarization of the light that they transmit. Here, we will categorize such polarizers in terms of their effects, which are basically three in number.

---

[2]A matrix approach that handles partially polarized light, using 1×4 *Stokes vectors* and 4×4 *Mueller matrices* can be found in M. J. Walker, "Matrix Calculus and the Stokes Parameters of Polarized Radiation," *American Journal of Physics*, Vol. 22, 1954: 170 and W. A. Shurcliff, *Polarized Light: Production and Use* (Cambridge, Mass.: Harvard University Press, 1962).

Linear Polarizer

The linear polarizer selectively removes all or most of the $\vec{\mathbf{E}}$-vibrations in a given direction, while allowing vibrations in the perpendicular direction to be transmitted. In most cases, the selectivity is not 100% efficient, so that the transmitted light is partially polarized. Figure 14-8 illustrates the operation schematically. Unpolarized light traveling in the $+z$-direction passes through a linear polarizer, whose preferential axis of transmission, or transmission axis (TA), is vertical. The unpolarized light is represented by two perpendicular (x and y) vibrations, since any direction of vibration present can be resolved into components along these directions. The light transmitted includes components only along the TA direction and is therefore linearly polarized in the vertical, or y, direction. The horizontal components of the original light have been removed by absorption. In the figure, the process is assumed to be 100% efficient.

Phase Retarder

The phase retarder does not remove either of the orthogonal components of the $\vec{\mathbf{E}}$-vibrations, but rather introduces a phase difference between them. If light corresponding to each orthogonal vibration travels with a different speed through such a *retardation plate*, there will be a cumulative phase difference, $\Delta\varphi$, between the two waves as they emerge.

Symbolically, Figure 14-9 shows the effect of a retardation plate on unpolarized light in a case where the vertical component travels through the plate faster than the horizontal component. This is suggested by the schematic separation of the two components on the optical axis, although of course both

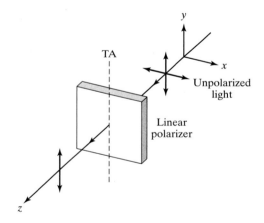

Figure 14-8   Operation of a linear polarizer.

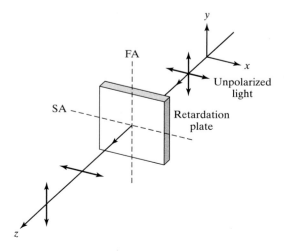

Figure 14-9   Operation of a phase retarder.

waves are simultaneously present at each point along the axis. The fast axis (FA) and slow axis (SA) directions of the plate are also indicated. When the net phase difference $\Delta\varphi = \pi/2$, the retardation plate is called a *quarter-wave plate*; when it is π, it is called a *half-wave plate*.

Rotator

The rotator has the effect of rotating the direction of linearly polarized light incident on it by some particular angle. Vertical linearly polarized light is shown incident on a rotator in Figure 14-10. The effect of the rotator element is to transmit linearly polarized light whose direction of vibration has been, in this case, rotated counterclockwise by an angle θ.

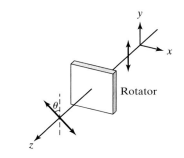

Figure 14-10   Operation of a rotator.

We desire now to create a set of matrices corresponding to these three types of polarizers so that just as the optical element alters the polarization mode of the actual light beam, an element matrix operating on a Jones vector will produce the same result mathematically. We adopt a pragmatic point of view in formulating appropriate matrices. For example, consider a linear polarizer with a transmission axis along the vertical, as in Figure 14-8. Let a 2×2 matrix representing the polarizer operate on vertically polarized light, and let the elements of the matrix to be determined be represented by letters a, b, c, and d. The resultant transmitted or product light in this case must again be vertically linearly polarized light. Symbolically,

$$\begin{bmatrix} a & b \\ c & d \end{bmatrix}\begin{bmatrix} 0 \\ 1 \end{bmatrix} = \begin{bmatrix} 0 \\ 1 \end{bmatrix}$$

This matrix equation—according to the rules of matrix multiplication—is equivalent to the algebraic equations

$$a(0) + b(1) = 0$$

$$c(0) + d(1) = 1$$

from which we conclude $b = 0$ and $d = 1$. To determine elements a and c, let the same polarizer operate on horizontally polarized light. In this case, no light is transmitted, or

$$\begin{bmatrix} a & b \\ c & d \end{bmatrix}\begin{bmatrix} 1 \\ 0 \end{bmatrix} = \begin{bmatrix} 0 \\ 0 \end{bmatrix}$$

The corresponding algebraic equations are now

$$a(1) + b(0) = 0$$

$$c(1) + d(0) = 0$$

from which $a = 0$ and $c = 0$. We conclude here without further proof, then, that the appropriate matrix is

$$\mathbf{M} = \begin{bmatrix} 0 & 0 \\ 0 & 1 \end{bmatrix} \qquad \text{linear polarizer, TA vertical} \qquad (14\text{-}13)$$

The matrix for a linear polarizer, TA horizontal, can be obtained in a similar manner and is included in Table 14-2, near the end of this chapter. Suppose next that the linear polarizer has a TA inclined at $45°$ to the x-axis. To keep matters as simple as possible we consider, in turn, the action of the polarizer on light linearly polarized in the same direction as—and perpendicular to—the TA of the polarizer. Light polarized along the same direction as the TA is represented by the Jones vector $\begin{bmatrix} 1 \\ 1 \end{bmatrix}$, and light with a polarization direction that is perpendicular to the TA is represented by the Jones vector $\begin{bmatrix} 1 \\ -1 \end{bmatrix}$. Then, following the approach used earlier,

$$\begin{bmatrix} a & b \\ c & d \end{bmatrix}\begin{bmatrix} 1 \\ 1 \end{bmatrix} = \begin{bmatrix} 1 \\ 1 \end{bmatrix} \quad \text{and} \quad \begin{bmatrix} a & b \\ c & d \end{bmatrix}\begin{bmatrix} 1 \\ -1 \end{bmatrix} = \begin{bmatrix} 0 \\ 0 \end{bmatrix}$$

Equivalently,

$$a + b = 1$$
$$c + d = 1$$
$$a - b = 0$$
$$c - d = 0$$

or $a = b = c = d = \frac{1}{2}$. Thus, the correct matrix is

$$\mathbf{M} = \frac{1}{2}\begin{bmatrix} 1 & 1 \\ 1 & 1 \end{bmatrix} \qquad \text{linear polarizer, TA at } 45° \qquad (14\text{-}14)$$

In the same way, a general matrix representing a linear polarizer with TA at angle θ can be determined. This is left as an exercise for the student. The result is

$$\mathbf{M} = \begin{bmatrix} \cos^2\theta & \sin\theta\cos\theta \\ \sin\theta\cos\theta & \sin^2\theta \end{bmatrix} \qquad \text{linear polarizer, TA at } \theta \qquad (14\text{-}15)$$

which includes Eqs. (14-13) and (14-14) as special cases, with $\theta = 90°$ and $\theta = 45°$, respectively.

Proceeding to the case of a phase retarder, we desire a matrix that will transform the elements

$$E_{0x}e^{i\varphi_x} \quad \text{into} \quad E_{0x}e^{i(\varphi_x+\varepsilon_x)}$$

and

$$E_{0y}e^{i\varphi_y} \quad \text{into} \quad E_{0y}e^{i(\varphi_y+\varepsilon_y)}$$

where ε_x and ε_y represent the advance in phase of the E_x- and E_y-components of the incident light. Of course, ε_x and ε_y may be negative quantities. Inspection is sufficient to show that this is accomplished by the matrix operation

$$\begin{bmatrix} e^{i\varepsilon_x} & 0 \\ 0 & e^{i\varepsilon_y} \end{bmatrix}\begin{bmatrix} E_{0x}e^{i\varphi_x} \\ E_{0y}e^{i\varphi_y} \end{bmatrix} = \begin{bmatrix} E_{0x}e^{i(\varphi_x+\varepsilon_x)} \\ E_{0y}e^{i(\varphi_y+\varepsilon_y)} \end{bmatrix}$$

Thus, the general form of a matrix representing a phase retarder is

$$\mathbf{M} = \begin{bmatrix} e^{i\varepsilon_x} & 0 \\ 0 & e^{i\varepsilon_y} \end{bmatrix} \qquad \text{phase retarder} \qquad (14\text{-}16)$$

As a special case, consider a quarter-wave plate (QWP) for which $|\varepsilon_x - \varepsilon_y| = \pi/2$. We distinguish the case for which $\varepsilon_y - \varepsilon_x = \pi/2$ (SA vertical) from the case for which $\varepsilon_x - \varepsilon_y = \pi/2$ (SA horizontal). In the former case, then, let $\varepsilon_x = -\pi/4$ and $\varepsilon_y = +\pi/4$. Obviously, other choices—an infinite number of them—are possible, so that Jones matrices, like Jones vectors, are not unique.

This particular choice, however, leads to a common form of the matrix, due to its symmetrical form:

$$\mathbf{M} = \begin{bmatrix} e^{-i\pi/4} & 0 \\ 0 & e^{i\pi/4} \end{bmatrix} = e^{-i\pi/4} \begin{bmatrix} 1 & 0 \\ 0 & i \end{bmatrix} \qquad \text{QWP, SA vertical} \qquad (14\text{-}17)$$

In arriving at the last 2×2 matrix in Eq. (14-17), we used the relationship $e^{i\pi/4} = e^{-i\pi/4} e^{i\pi/2}$ and the indentity $i = e^{i\pi/2}$. Similarly, when $\varepsilon_x > \varepsilon_y$,

$$\mathbf{M} = e^{i\pi/4} \begin{bmatrix} 1 & 0 \\ 0 & -i \end{bmatrix} \qquad \text{QWP, SA horizontal} \qquad (14\text{-}18)$$

Corresponding matrices for half-wave plates (HWP), where $|\varepsilon_x - \varepsilon_y| = \pi$, are given by

$$\mathbf{M} = \begin{bmatrix} e^{-i\pi/2} & 0 \\ 0 & e^{i\pi/2} \end{bmatrix} = e^{-i\pi/2} \begin{bmatrix} 1 & 0 \\ 0 & -1 \end{bmatrix} \qquad \text{HWP, SA vertical} \qquad (14\text{-}19)$$

$$\mathbf{M} = \begin{bmatrix} e^{i\pi/2} & 0 \\ 0 & e^{-i\pi/2} \end{bmatrix} = e^{i\pi/2} \begin{bmatrix} 1 & 0 \\ 0 & -1 \end{bmatrix} \qquad \text{HWP, SA horizontal} \qquad (14\text{-}20)$$

The elements of the matrices are identical in this case, since advancement of phase by π is physically equivalent to retardation by π. The only difference lies in the prefactors that modify the phases of all the elements of the Jones vector in the same way and hence do not affect interpretation of the results.

The requirement for a rotator of angle β is that an $\vec{\mathbf{E}}$-vector, oscillating linearly at angle θ and with normalized components $\cos \theta$ and $\sin \theta$, be converted to one that oscillates linearly at angle $(\theta + \beta)$. That is,

$$\begin{bmatrix} a & b \\ c & d \end{bmatrix} \begin{bmatrix} \cos \theta \\ \sin \theta \end{bmatrix} = \begin{bmatrix} \cos(\theta + \beta) \\ \sin(\theta + \beta) \end{bmatrix}$$

Thus, the matrix elements must satisfy

$$a \cos \theta + b \sin \theta = \cos(\theta + \beta)$$

$$c \cos \theta + d \sin \theta = \sin(\theta + \beta)$$

From the trigonometric identities for the sine and cosine of the sum of two angles,

$$\cos(\theta + \beta) = \cos \theta \cos \beta - \sin \theta \sin \beta$$

$$\sin(\theta + \beta) = \sin \theta \cos \beta + \cos \theta \sin \beta$$

it follows that

$$a = \cos \beta \qquad b = -\sin \beta$$

$$c = \sin \beta \qquad d = \cos \beta$$

so that the desired rotator matrix is

$$\mathbf{M} = \begin{bmatrix} \cos \beta & -\sin \beta \\ \sin \beta & \cos \beta \end{bmatrix} \qquad \text{rotator through angle } +\beta \qquad (14\text{-}21)$$

The Jones matrices derived in this chapter are summarized in Table 14-2.

TABLE 14-2 SUMMARY OF JONES MATRICES

I. Linear polarizers

TA horizontal $\begin{bmatrix} 1 & 0 \\ 0 & 0 \end{bmatrix}$ TA vertical $\begin{bmatrix} 0 & 0 \\ 0 & 1 \end{bmatrix}$ TA at 45° to horizontal $\dfrac{1}{2}\begin{bmatrix} 1 & 1 \\ 1 & 1 \end{bmatrix}$

II. Phase retarders

General $\begin{bmatrix} e^{i\varepsilon_x} & 0 \\ 0 & e^{i\varepsilon_y} \end{bmatrix}$

QWP, SA vertical $e^{-i\pi/4}\begin{bmatrix} 1 & 0 \\ 0 & i \end{bmatrix}$ QWP, SA horizontal $e^{i\pi/4}\begin{bmatrix} 1 & 0 \\ 0 & -i \end{bmatrix}$

HWP, SA vertical $e^{-i\pi/2}\begin{bmatrix} 1 & 0 \\ 0 & -1 \end{bmatrix}$ HWP, SA horizontal $e^{i\pi/2}\begin{bmatrix} 1 & 0 \\ 0 & -1 \end{bmatrix}$

III. Rotator

Rotator $(\theta \rightarrow \theta + \beta)$ $\begin{bmatrix} \cos\beta & -\sin\beta \\ \sin\beta & \cos\beta \end{bmatrix}$

As an important example, consider the production of circularly polarized light by combining a linear polarizer with a QWP. Let the linear polarizer (LP) produce light vibrating at an angle of 45°, as in Figure 14-11, which is then transmitted by a QWP with SA horizontal. In this arrangement, the light incident on the QWP is divided equally between fast and slow axes. On emerging, a phase difference of $\pi/2$ results in circularly polarized light. With the Jones calculus, this process is equivalent to allowing the QWP matrix to operate on the Jones vector for the linearly polarized light,

$$ e^{i\pi/4}\begin{bmatrix} 1 & 0 \\ 0 & -i \end{bmatrix}\frac{1}{\sqrt{2}}\begin{bmatrix} 1 \\ 1 \end{bmatrix} = e^{i\pi/4}\left(\frac{1}{\sqrt{2}}\right)\begin{bmatrix} 1 \\ -i \end{bmatrix} $$

giving right-circularly polarized light (see Table 14-1). If the fast and slow axes of the QWP are interchanged, a similar calculation shows that the result is left-circularly polarized instead.

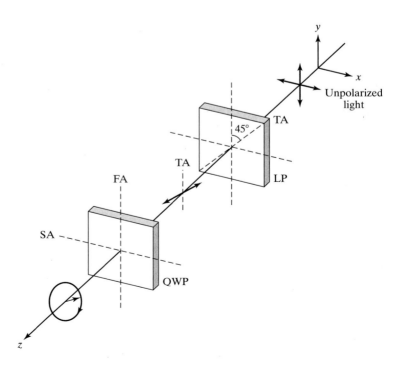

Figure 14-11 Production of right circularly polarized light.

Example 14-2

Consider the result of allowing left-circularly polarized light to pass through an eighth-wave plate.

Solution

We first need a matrix that represents the eighth-wave plate, that is, a phase retarder that introduces a relative phase of $2\pi/8 = \pi/4$. Thus, letting $\varepsilon_x = 0$,

$$\mathbf{M} = \begin{bmatrix} e^{i\varepsilon_x} & 0 \\ 0 & e^{i\varepsilon_y} \end{bmatrix} = \begin{bmatrix} 1 & 0 \\ 0 & e^{i\pi/4} \end{bmatrix}$$

This matrix then operates on the Jones vector representing the left-circularly polarized light:

$$\begin{bmatrix} 1 & 0 \\ 0 & e^{i\pi/4} \end{bmatrix}\begin{bmatrix} 1 \\ i \end{bmatrix} = \begin{bmatrix} 1 \\ ie^{i\pi/4} \end{bmatrix} = \begin{bmatrix} 1 \\ e^{i3\pi/4} \end{bmatrix}$$

The resultant Jones vector indicates that the light is elliptically polarized, and the components are out of phase by $3\pi/4$. Using Euler's equation to expand $e^{i3\pi/4}$, we obtain

$$e^{i3\pi/4} = -\frac{1}{\sqrt{2}} + i\left(\frac{1}{\sqrt{2}}\right)$$

and using our standard notation for this case, we have

$$\begin{bmatrix} \widetilde{E}_{0x} \\ \widetilde{E}_{0y} \end{bmatrix} = \begin{bmatrix} A \\ B + iC \end{bmatrix}, \quad \text{where } A = 1, B = -\frac{1}{\sqrt{2}}, \text{ and } C = \frac{1}{\sqrt{2}}$$

Since A and C have the same sign, the output field vector represents elliptically polarized light with counterclockwise rotation. Comparing this matrix with the general form in Eq. (14-5), we determine that $E_{0x} = A = 1$ and $E_{0y} = \sqrt{B^2 + C^2} = 1$. Making use of Eq. (14-10), we also determine that $\alpha = -45°$.

Of course, the Jones calculus can handle a case where polarized light is transmitted by a series of polarizing elements, since the product of element matrices can represent an overall *system matrix*. If light represented by Jones vector $\mathbf{V}$ passes sequentially through a series of polarizers represented by $\mathbf{M}_1, \mathbf{M}_2, \mathbf{M}_3, \ldots, \mathbf{M}_m$, so that $(\mathbf{M}_m \ldots \mathbf{M}_3\mathbf{M}_2\mathbf{M}_1)\mathbf{V} = \mathbf{M}_s\mathbf{V}$, then the system matrix is given by $\mathbf{M}_s = \mathbf{M}_m \ldots \mathbf{M}_3\mathbf{M}_2\mathbf{M}_1$.

PROBLEMS

14-1 Derive the Jones matrix, Eq. (14-15), representing a linear polarizer whose transmission axis is at an arbitrary angle θ with respect to the horizontal.

14-2 Write the normalized Jones vectors for each of the following waves, and describe completely the state of polarization of each.

a. $\vec{E} = E_0 \cos(kz - \omega t)\hat{x} - E_0 \cos(kz - \omega t)\hat{y}$

b. $\vec{E} = E_0 \sin 2\pi\left(\dfrac{z}{\lambda} - vt\right)\hat{x} + E_0 \sin 2\pi\left(\dfrac{z}{\lambda} - vt\right)\hat{y}$

c. $\vec{E} = E_0 \sin(kz - \omega t)\hat{x} + E_0 \sin\left(kz - \omega t - \dfrac{\pi}{4}\right)\hat{y}$

d. $\vec{E} = E_0 \cos(kz - \omega t)\hat{x} + E_0 \cos\left(kz - \omega t + \dfrac{\pi}{2}\right)\hat{y}$

14-3 Describe as completely as possible amplitude, wave direction, and the state of polarization of each of the following waves.

a. $\vec{E} = 2E_0\hat{x}e^{i(kz-\omega t)}$

b. $\vec{E} = E_0(3\hat{x} + 4\hat{y})e^{i(kz-\omega t)}$

c. $\vec{E} = 5E_0(\hat{x} - i\hat{y})e^{i(kz+\omega t)}$

14-4 Two linearly polarized beams are given by

$$\vec{E}_1 = E_{01}(\hat{x} - \hat{y})\cos(kz - \omega t) \quad \text{and}$$

$$\vec{E}_2 = E_{02}(\sqrt{3}\hat{x} + \hat{y})\cos(kz - \omega t)$$

Determine the angle between their directions of polarization by (a) forming their Jones vectors and finding the vibration direction of each and (b) forming the dot product of their vector amplitudes.

14-5 Find the character of polarized light after passing in turn through (a) a half-wave plate with slow axis at 45°; (b) a linear polarizer with transmission axis at 45°; (c) a quarter-wave plate with slow axis horizontal. Assume the original light to be linearly polarized vertically. Use the matrix approach and analyze the final Jones vector to describe the product light. (*Hint:* First find the effect of the HWP alone on the incident light.)

14-6 Write the equations for the electric fields of the following waves in exponential form:

   a. A linearly polarized wave traveling in the x-direction. The $\vec{E}$-vector makes an angle of 30° relative to the y-axis.

   b. A right-elliptically polarized wave traveling in the y-direction. The major axis of the ellipse is in the z-direction and is twice the minor axis.

   c. A linearly polarized wave traveling in the x,y-plane in a direction making an angle of 45° relative to the x-axis. The direction of polarization is the z-direction.

14-7 Determine the conditions on the elements A, B, and C of the general Jones vector (Eq. 14-9), representing polarized light, that lead to the following special cases: (a) linearly polarized light; (b) elliptically polarized light with major axis aligned along a coordinate axis; (c) circularly polarized light. In each case, from the meanings of A, B, C, deduce the possible values of phase difference between component vibrations.

14-8 Write a computer program that will determine E_y-values of elliptically polarized light from the equation for the ellipse, Eq. (14-12), with input constants A, B, and C and variable input parameter E_x. Plot the ellipse for the example given in the text,

$$\widetilde{\mathbf{E}}_0 = \begin{bmatrix} 3 \\ 2 + i \end{bmatrix}$$

14-9 Specify the polarization mode for each of the following Jones vectors:

   a. $\begin{bmatrix} 3i \\ i \end{bmatrix}$ **b.** $\begin{bmatrix} i \\ 1 \end{bmatrix}$

   c. $\begin{bmatrix} 4i \\ 5 \end{bmatrix}$ **d.** $\begin{bmatrix} 5 \\ 0 \end{bmatrix}$

   e. $\begin{bmatrix} 2 \\ 2i \end{bmatrix}$ **f.** $\begin{bmatrix} 2 \\ 3 \end{bmatrix}$

   g. $\begin{bmatrix} 2 \\ 6 + 8i \end{bmatrix}$

14-10 Linearly polarized light with an electric field $\vec{E}$ is inclined at $+30°$ relative to the x-axis and is transmitted by a QWP with SA horizontal. Describe the polarization mode of the product light.

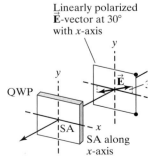

Linearly polarized
$\vec{E}$-vector at 30°
with x-axis

Figure 14-12 Problem 14-10.

14-11 Using the Jones calculus, show that the effect of a HWP on light linearly polarized at inclination angle α is to rotate the polarization through an angle of 2α. The HWP may be used in this way as a "laser-line rotator," allowing the polarization of a laser beam to be rotated without having to rotate the laser.

14-12 An important application of the QWP is its use in an "isolator." For example, to prevent feedback from interferometers into lasers by front-surface, back reflections, the beam is first allowed to pass through a combination of linear polarizer and QWP, with OA of the QWP at 45° to the TA of the polarizer. Consider what happens to such light after reflection from a plane surface and transmission back through this optical device.

14-13 Light linearly polarized with a horizontal transmission axis is sent through another linear polarizer with TA at 45° and then through a QWP with SA horizontal. Use the Jones matrix technique to determine and describe the product light.

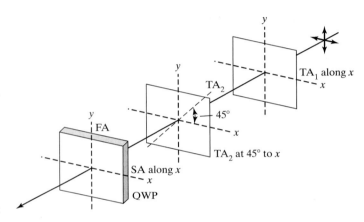

Figure 14-13 Problem 14-13.

14-14 A light beam passes consecutively through (1) a linear polarizer with TA at 45° clockwise from vertical, (2) a QWP with SA vertical, (3) a linear polarizer with TA horizontal, (4) a HWP with FA horizontal, (5) a linear polarizer with TA vertical. What is the nature of the product light?

14-15 Unpolarized light passes through a linear polarizer with TA at 60° from the vertical, then through a QWP with SA horizontal, and finally through another linear polarizer with TA vertical. Determine, using Jones matrices, the character of the light after passing through (a) the QWP and (b) the final linear polarizer.

14-16 Determine the state of polarization of circularly polarized light after it is passed normally through (a) a QWP; (b) an eighth-wave plate. Use the matrix method to support your answer.

14-17 Show that the matrix $\begin{bmatrix} 1 & i \\ -i & 1 \end{bmatrix}$ represents a right-circular polarizer, converting any incident polarized light into right circularly-polarized light. What is the proper matrix to represent a left-circular polarizer?

14-18 Show that elliptical polarization can be regarded as a combination of circular and linear polarizations.

14-19 Derive the equation of the ellipse for polarized light given in Eq. (14-12). (*Hint:* Combine the E_x and E_y equations for the general case of elliptical polarization, eliminating the space and time dependence between them.)

14-20 a. Identify the state of polarization corresponding to the Jones vector

$$\begin{bmatrix} 2 \\ 3e^{i\pi/3} \end{bmatrix}$$

and write it in the standard, normalized form of Table 14-1.

b. Let this light be transmitted through an element that rotates linearly polarized light by $+30°$. Find the new, normalized form and describe the result.

14-21 Determine the nature of the polarization that results from Eq. (14-12) when (a) $\varepsilon = \pi/2$; (b) $E_{0x} = E_{0y} = E_0$; (c) both (a) and (b); (d) $\varepsilon = 0$.

14-22 A quarter-wave plate is placed between crossed polarizers such that the angle between the polarizer TA of the first polarizer and the QWP fast axis is θ. How does the polarization of the emergent light vary as a function of θ?

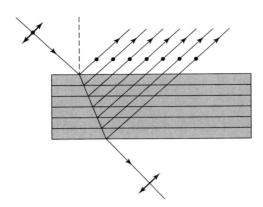

15

Production of Polarized Light

INTRODUCTION

Any interaction of light with matter whose optical properties are asymmetrical along directions transverse to the propagation vector provides a means of polarizing light. Indeed, if light were longitudinal rather than transverse in its nature, transverse material asymmetries along the propagation vector could not alter the sense of the oscillating $\vec{E}$-vector, and the physical mechanisms to be described here would have no polarizing or spatially selective effects on light beams. The experimental observation that light can be polarized is, therefore, clear evidence of its transverse nature. The most important processes that produce polarized light are discussed in this chapter under the following general areas: (1) dichroism, (2) reflection, (3) scattering, and (4) birefringence. Optical activity is described as a mechanism that *modifies* polarized light. Finally, *photoelasticity* is briefly discussed as a useful application.

15-1 DICHROISM: POLARIZATION BY SELECTIVE ABSORPTION

A *dichroic polarizer* selectively absorbs light with $\vec{E}$-vibrations along a unique direction characteristic of the dichroic material. The polarizer easily transmits light with $\vec{E}$-vibrations along a transverse direction orthogonal to the direction of absorption. This preferred direction is called the *transmission axis* (TA) of the polarizer. In the ideal polarizer, the transmitted light is linearly polarized in the same direction as the transmission axis. The state of polarization of the light can most easily be tested by a second dichroic polarizer, which then functions as an *analyzer*, shown in Figure 15-1. When the TA of

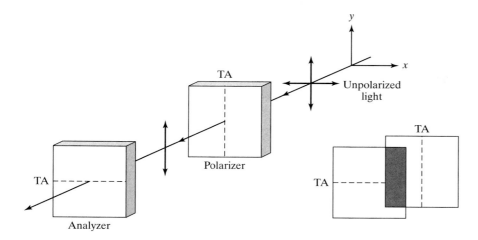

Figure 15-1 Crossed dichroic polarizers functioning as a polarizer-analyzer pair. No light is transmitted through the analyzer.

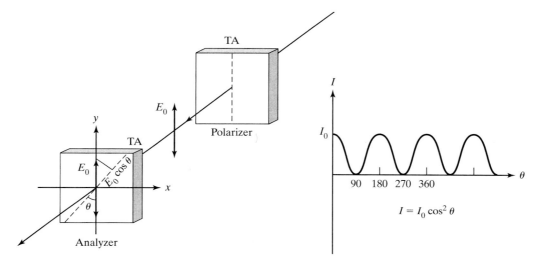

Figure 15-2 Illustration of Malus' law.

the analyzer is oriented at 90° relative to the TA of the polarizer, the light is effectively extinguished. As the analyzer is rotated, the light transmitted by the pair increases, reaching a maximum when their TAs are aligned. If I_0 represents the maximum transmitted irradiance, then *Malus' law* states that the irradiance for any relative angle θ between the TAs is given by

$$I = I_0 \cos^2 \theta \qquad (15\text{-}1)$$

Malus' law is easily understood in conjunction with Figure 15-2. Notice that the amplitude of the light emerging from the analyzer is $E_0 \cos \theta$. The irradiance I (in W/m^2) is then proportional to the square of this result.

The impressive ability of dichroic materials to absorb light strongly with $\vec{\mathbf{E}}$ along one direction and to transmit light easily with $\vec{\mathbf{E}}$ along a perpendicular direction can be understood by reference to a standard experiment with microwaves, illustrated in Figure 15-3. Wavelengths of microwaves range roughly from 1 mm to 1 m. It is found that when a vertical wire grid, whose spacing is much smaller than the wavelength, intercepts microwaves with vertical linear polarization, little or no radiation is transmitted. Conversely, when the grid intercepts waves polarized in a direction perpendicular to the wires, there is efficient transmission of the waves. The explanation of this behavior involves a consideration of the interaction of electromagnetic radiation with

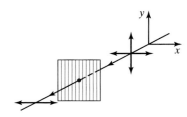

Figure 15-3 Action of a vertical wire grid on microwaves. Effective absorption of the vertical component of the radiation occurs when $\lambda \gg$ grid spacing.

the metal wires that operate as a dichroic polarizer. Within the metal wires, the mobile free electrons are set in oscillatory motion by the oscillations of the electric field of the incident radiation. We know that each electron so oscillating constitutes a dipole source that radiates electromagnetic energy in all directions, except the direction of the electron oscillation itself. Evidently, the superposition of the vertical $\vec{E}$-vibrations of an incident electromagnetic wave with the radiation of these electron oscillators leads to cancellation in the forward direction. It turns out, in fact, that the phase of the electromagnetic wave originating with the oscillating electrons is 180° out of step with that of the incident radiation,[1] so no wave can propagate in the forward direction. In addition, the oscillation of the free electrons is not entirely free. The effective friction due to interaction with lattice imperfections, for example, constitutes some dissipation of energy, which must attenuate the incident wave. The chief reason for the disappearance of the forward wave, however, is destructive interference between the incident and generated waves. Horizontally linearly polarized light incident on the vertical wire grid would suffer the same fate, except that appreciable oscillatory motion of the electrons across the wire is inhibited. As a result, the generated electromagnetic wave is reduced in strength and effective cancellation does not occur. If the grid is rotated by 90°, the vertical $\vec{E}$-vibrations are transmitted and the horizontal $\vec{E}$-vibrations are canceled. The wire grid polarizes microwaves much as a dichroic absorber polarizes optical radiation.

For optical wavelengths, the conduction paths analogous to the grid wires must be much closer together. The most common dichroic absorber for light is Polaroid H-sheet, invented in 1938 by Edwin H. Land. When a sheet of clear, polyvinyl alcohol is heated and stretched along a given direction, its long, hydrocarbon molecules tend to align in the direction of stretching. The stretched material is then impregnated with iodine atoms, which become associated with the linear molecules and provide "conduction" electrons to complete the analogy to the wire grid. Some naturally occurring materials, such as the mineral tourmaline, also possess dichroic properties to some degree. All that is required in principle is that the electrons be much freer to respond to an incident electromagnetic wave in one direction than in an orthogonal direction. In nonmetallic materials, such as Polaroid and tourmaline, the electrons acting as dipole oscillators are not free. Thus, the wave they generate is not out of phase with respect to the incident wave, and complete cancellation of the forward wave does not occur. The energy of the driving wave, however, is gradually dissipated as the wave advances through the absorber, so that the efficiency of the dichroic absorber is a function of the thickness. The absorption follows the usual expression for attenuation,

$$I = I_0 e^{-\alpha x}$$

where I_0 is the incident irradiance and I is the irradiance at depth x of absorber. The constant α is the *absorptivity*, or *absorption coefficient*, characteristic of the material. In a good, practical dichroic absorber, α is relatively independent of wavelength; that is, the material appears transparent and yet behaves as a linear polarizer for all optical wavelengths. This ideal condition is not quite achieved in Polaroid H-sheet, which is less effective at the blue end of the spectrum. Consequently, when a Polaroid H-sheet is crossed with another such sheet acting as an analyzer, the combination contributes a blue tint to the almost canceled transmitted light.

---

[1] The theoretical basis for this statement is presented at the end of Section 25-1, applied to the free electrons of a metal.

15-2 POLARIZATION BY REFLECTION FROM DIELECTRIC SURFACES

Light that is specularly reflected from dielectric surfaces is at least partially polarized. This is most easily confirmed by looking through a piece of polarizing filter while rotating it about the propagation direction of the reflected light. When the preferred $\vec{E}$-direction of the reflected light is perpendicular to the TA of the polarizing filter, regions from which light is specularly reflected into the eye appear reduced in brightness. This is precisely the working principle of Polaroid sunglasses. The $\vec{E}$-vibration in light reflected from a horizontal surface into the eye is preferentially polarized along the horizontal direction. The TA of the Polaroids in a pair of sunglasses is therefore fixed in the vertical direction so as to reduce the partially polarized "glare" from reflection while still blocking only one-half of unpolarized light incident on the sunglasses. To appreciate the physics that underlies this phenomenon, consider Figure 15-4, which shows a narrow beam of light incident at an arbitrary angle on a smooth, flat, dielectric surface. An unpolarized incident beam is conveniently represented by two perpendicular $\vec{E}$-vibrations. One, represented by a dot is perpendicular to the plane of incidence, as in Figure 15-4a. The other, drawn as a double-headed arrow, lies in the plane of incidence, that is, the plane of the page, as in Figure 15-4b. (Recall that the incident ray and the normal to the reflecting surface at the point of incidence define the plane of incidence.) It is common to refer to the two components of the unpolarized incident beam as E_s (perpendicular to the plane of incidence) and E_p (in the plane of incidence). Alternatively, the E_s mode is called the TE (transverse electric) mode, and the E_p mode is called the TM (transverse magnetic) mode, since the $\vec{B}$-component of the wave is transverse to the plane of incidence when the corresponding $\vec{E}$-component is parallel to the plane of incidence.

Consider first the E_s, or TE, component (Figure 15-4a). The action of E_s on the electrons in the surface of the dielectric is to stimulate oscillations along the same direction, that is, perpendicular to the page. The radiation from all these electronic dipole oscillators adds to beams of light in two distinct directions, the direction of the reflected beam and the refracted beam. Each of these beams is made up of light that is linearly polarized perpendicular to the plane of incidence, as indicated by the dots in Figure 15-4a. The reflected and refracted rays are both in a direction corresponding to maximum dipole radiation, perpendicular to the dipole axis.

Consider next the action of the E_p, or TM, component (Figure 15-4b). From the direction of the refracted beam (which may be calculated using Snell's law), we conclude that the $\vec{E}$-field within the isotropic dielectric material, and thus the axis of the dipole oscillations, is oriented perpendicular to the beam direction, as indicated by the double-headed arrows. Notice that the dipole oscillations include a component along the direction of the reflected beam. Recalling that a dipole oscillator radiates only weakly along directions making small angles with the dipole axis ($I \propto \sin^2 \theta$), we conclude that only a fraction of the E_p component of the original light (compared with the E_s component) appears in the reflected beam. Considering both TE and TM modes together, it follows that the reflected light is partially polarized with a predominance of the E_s mode present. Since the energy of the incident beam is equally divided between E_s and E_p components, it also follows that the refracted beam is partially polarized and richer in the E_p component.

This analysis shows that when the dipole axes are in the same direction as the reflected ray, the E_p component is *entirely missing* from the reflected beam. Thus, in this case, the reflected ray is linearly polarized only in the E_s mode. In fact, if the dipoles radiated along the reflected ray, the electromagnetic wave could only be a longitudinal wave! This unique orientation results

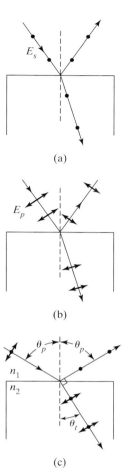

Figure 15-4 Specular reflection of light at a dielectric surface. (a) TE mode. (b) TM mode. (c) Polarization at Brewster's angle.

when the reflected and refracted rays are *perpendicular* to one another (Figure 15-4c). The angle of incidence that produces a linearly polarized beam E_s by reflection is θ_p, the *polarizing angle*, or *Brewster's angle*. Combining Snell's law

$$n_1 \sin \theta_p = n_2 \sin \theta_t$$

with the trigonometric relation $\theta_t = 90 - \theta_p$, we arrive at *Brewster's law*,

$$\theta_p = \tan^{-1}\left(\frac{n_2}{n_1}\right) \tag{15-2}$$

Polarizing angles exist for both external reflection ($n_2 > n_1$) and internal reflection ($n_2 < n_1$) and are clearly not the same. Here, n_1 is the index of refraction of the medium containing the incident beam. For external reflection when light travels from air to glass, with $n_1 = 1$ and $n_2 = 1.5$, for example, $\theta_p = 56.3°$. For internal reflection when light travels in the opposite direction, so that $n_1 = 1.5$ and $n_2 = 1$, $\theta_p = 33.7°$. These angles are seen to be precisely complementary, as required by geometry and the definition of Brewster's angle.

Although reflection at the polarizing angle from a dielectric surface can be used to produce linearly polarized light, the method is relatively inefficient. For reflection from air to glass, as in the example just given, only 15% of the E_s component is found in the reflected beam. (The *Fresnel equations*, which permit calculations of this kind, are treated in Chapter 23.) This deficiency can be remedied to a degree by stepwise intensification of the reflected beam as in a *pile-of-plates* polarizer (Figure 15-5). Repeated reflections by multiple layers of the dielectric at Brewster's angle both increases the irradiance of the E_s component in the integrated, reflected beam and, necessarily, purifies the transmitted beam of this component. If enough plates are assembled, the transmitted beam approaches a linearly polarized condition. Pile-of-plates polarizers are especially helpful in those regions of the infrared and ultraviolet spectrum where dichroic sheet polarizers and calcite prisms are ineffective. Multilayer, thin film coatings that show little absorption in the spectral region of interest behave in a similar manner and can be used as polarization-sensitive reflectors and transmitters. Another interesting application of polarization by reflection is the *Brewster window*. The window (Figure 15-6) operates in the same way as a single plate of the pile-of-plates analyzer. TM linearly polarized light incident at Brewster's angle is fully transmitted at the first surface. The angle of incidence, θ_r, at the second surface also satisfies Brewster's law for internal reflection, so that the light is again fully transmitted. The plate acts as a perfect window for TM polarized light.

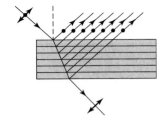

Figure 15-5   Pile-of-plates polarizer.

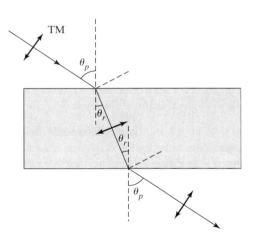

Figure 15-6   Brewster window. Brewster's law is satisfied for the TM mode at both surfaces.

The active medium of a gas laser is often bounded by two Brewster windows, located at the ends of the gas plasma tube. The light in the cavity makes repeated passes through the windows, on its way to and from cavity mirrors positioned beyond alternate ends of the gas tube. Upon each traversal, the TM mode is completely transmitted, whereas the TE mode is partially reflected (rejected). That is, the TE mode will experience more loss per round trip through the cavity than will the TM mode. Typically the extra loss for the TE mode will prevent it from lasing and so the laser output will consist only of the TM mode.

15-3 POLARIZATION BY SCATTERING

Before discussing the polarization of light that occurs in scattering, we make a slight detour to discuss scattering in general, pointing out some familiar consequences of scattering that are in themselves rather interesting. By the *scattering* of light, we mean the removal of energy from an incident wave by a scattering medium and the reemission of some portion of that energy in many directions. We can think of the elemental oscillator or scattering unit as an electronic charge bound to a nucleus (a *dipole oscillator*). The electron is set into forced oscillation by the alternating electric field of incident light and at the same frequency. The response of the electron to this driving force depends on the relationship between the driving frequency ω and the natural or resonant frequency of the oscillator ω_0. In most materials, resonant frequencies lie predominantly in the ultraviolet (due to electronic oscillations) and in the infrared (due to molecular vibrations) rather than in the visible. Because atomic masses are so much larger than the electron mass, amplitudes of induced molecular vibrations are small compared with electronic vibrations and so can be neglected in this discussion. Calculations[2] show that in this case, the induced dipole oscillations have an amplitude that is roughly independent of the frequency ω of the light. The oscillating dipoles, consisting of electrons accelerating in harmonic motion, are tiny radiators—antennas that reradiate or *scatter* energy in all directions except along the dipole axis itself.

Such scattering is most effective when the scattering centers are particles whose dimensions are small compared with the wavelength of the radiation, in which case we speak of *Rayleigh scattering*. The scattering of sunlight from oxygen and nitrogen molecules in the atmosphere, for example, is Rayleigh scattering, whereas the scattering of light from dense scattering centers—like the droplets of water in clouds and fog—is not. In Rayleigh scattering, the well-separated scattering centers act independently (incoherently), so that, according to the conclusions of Section 5-3, their net irradiance is the sum of their individual irradiances. Now, for Rayleigh scattering the radiated power can be shown to be directly proportional to the fourth power of the frequency of the incident radiation. Without deriving this Rayleigh scattering law, we can make the following hand-waving argument: The electric field of a dipole with a charge e accelerating back and forth along a line is proportional to the acceleration. If $d^2\vec{\mathbf{r}}/dt^2 = -\omega^2\vec{\mathbf{r}}$, then the acceleration, $\vec{\mathbf{r}} = \vec{\mathbf{r}_0}\cos(\omega t)$, is proportional to the square of the frequency. Since the power P radiated is in turn proportional to the square of the electric field, it becomes proportional to the fourth power of the frequency. This is the Rayleigh scattering law, which is expressed by[3]

$$P = \frac{e^2\omega^4 r_0^2}{12\pi\varepsilon_0 c^3}$$

---

[2]Such calculations, giving a quantitative treatment to these ideas, are given in Section 25-1.

[3]Richard P. Feynman, Robert B. Leighton, and Matthew Sands, *The Feynman Lectures on Physics*, vol. 1 (Reading, Mass.: Addison-Wesley Publishing Company, 1963), Ch. 32, 33.

Thus the oscillating dipoles radiate more energy in the shorter-wavelength (higher-frequency) region of the visible spectrum than in the longer-wavelength region. The scattered power for violet light of wavelength 400 nm is nearly 10 times as great as for red light of wavelength 700 nm. Rayleigh scattering explains why a clean atmosphere appears blue: Higher-frequency blue light from the sun is scattered by the atmosphere down to the earth more so than is the lower-frequency red light. On the other hand, when we are looking at the sunlight "head-on" at sunrise or sunset, after it has passed through a good deal of atmosphere, we see reddish or yellowish light, that is, white light from which the blues have been preferentially removed by scattering.

Scattering that occurs from larger particles[4] such as those found in clouds, fog, and powdered materials such as sugar appears as white light, in contrast to Rayleigh scattering. Here "larger particles" refers to the size of the scattering particle relative to the wavelength of light. In this case, the scattering centers (particles) are arranged—more or less—in an orderly fashion so that oscillators that are closer together than a wavelength of the incident light become coherent scatterers. The cooperative effect of many oscillators tends to cancel the radiation in all directions but the forward (refraction) direction and the backward (reflection) direction. In other words, the scattering due to these larger particles can be understood in terms of the usual laws of reflection and refraction. However, the usual departures from perfectly ordered atomic arrangement lead to some scattering in other directions as well. The net electric field amplitude of the coherent scattered radiation is now the sum of the individual amplitudes, or the radiated power is proportional to N^2 when there are N coherent oscillators. Although such scattering is much less effective, *per oscillator*, than Rayleigh scattering, the density of oscillators in this case leads to considerable scattering. It can be shown that the number N of such coherent oscillators responsible for the reflected radiation is proportional to λ^2, so that the radiated power is proportional to λ^4, canceling the $1/\lambda^4$ dependence of isolated Rayleigh scatterers. Thus the scattered radiation is essentially wavelength independent, and fog and clouds appear white by scattered light.

Of particular interest in the context of this chapter, however, is the fact that scattered radiation may also be polarized. As an example, consider a vessel of water to which is added one or more drops of milk. The milk molecules quickly diffuse throughout the water and serve as effective scattering centers for a beam of light transmitted across the medium. In pure water the light does not scatter sideways but propagates only in the forward direction. The light scattered in various directions from the milk molecules, when examined with a polarizing filter, is found to be polarized, as shown in Figure 15-7. The unpolarized light incident from the left (along the x-direction) contains $\vec{\mathbf{E}}$-field components oscillating along the y- and z-directions. These perpendicular $\vec{\mathbf{E}}$-field components set the electronic oscillators of a scattering center into forced vibrations along the y- and z-directions. As a result, the electronic oscillators reemit radiation in all directions. This scattered light can include only $\vec{\mathbf{E}}$-field components polarized along the direction of the forced motion executed by the oscillators, that is,

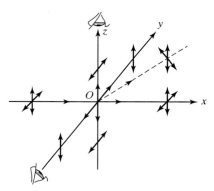

Figure 15-7 Polarization due to scattering. Unpolarized light incident from the left is scattered by a particle at the origin O.

---

[4]The more general theory of scattering, including larger scattering centers, is often called *Mie scattering* after its creator. Mie scattering takes into account the size, shape, refractive index, and absorptivity of the scattering particles and includes Rayleigh scattering as a special case. See Jurgen R. Meyer-Arendt, *Introduction to Classical and Modern Optics*, 3d ed. (Englewood Cliffs, N.J.: Prentice-Hall, 1989), Ch. 4.2.

along the y- and the z-directions. If scattered light is viewed from a point on the y-axis, it will be found to contain $\vec{\mathbf{E}}$-vibrations along the z-direction, but not along the y-direction. Those along the y-direction are absent because they would represent longitudinal $\vec{\mathbf{E}}$-vibrations in an electromagnetic wave. Similarly, viewed from a point on the z-axis, the z-vibrations are missing, and light is linearly polarized along the y-direction. Viewed from off-axis points, the light is partially polarized. The forward beam shows the same polarization as the incident light. In the same way, when the sun is not directly overhead so that its light crosses the atmosphere above us, the light scattered down is found to be partially polarized. The effect is easily seen by viewing the clear sky through a rotating polarizing filter. The polarization is not complete, both because we see light that is multiply scattered into the eye and because not all electronic oscillators in molecules are free to oscillate in exactly the same direction as the incident $\vec{\mathbf{E}}$-vector of the light.

Ordinary polarization by scattering is generally weak and imperfect and so is not used as a practical means of artificially producing polarized light. In the area of nonlinear optics, however, the controlled scattering of light from active media, exemplified by *stimulated Raman, Rayleigh*, and *Brillouin scattering*, provides much vital research in modern optics. In such cases, the scattered light is modified by the resonant frequencies of the medium. An introduction to such nonlinear interactions is given in Chapter 24.

15-4 BIREFRINGENCE: POLARIZATION WITH TWO REFRACTIVE INDICES

Birefringent materials are so named because they are able to cause double refraction, that is, the appearance of two refracted beams due to the existence of two different indices of refraction for a single material. We have already seen that anisotropy in the binding forces affecting the electrons of a material can lead to anisotropy in the amplitudes of their oscillations in response to a stimulating electromagnetic wave and hence to anisotropy of absorption. Such a material displays dichroism. For this to occur, however, the stimulating optical frequencies must fall within the absorption band of the material. Referring to Figure 15-8, we see that the slope of the dispersion curve, $dn/d\omega$, is negative—or "anomalous"—over a certain frequency interval. This interval is an *absorption band* in a given material. Typically, such absorption bands lie in the ultraviolet, above optical frequencies, so that the material is transparent to visible light. In this case, even with anisotropy of electron-binding forces, there is little or no effect on optical absorption, and the material does not appear dichroic. Still, the presence of anisotropic binding forces along the x- and y-directions leads to, for light propagating along the z-direction, different dispersion curves (like that of Figure 15-8) for refractive index n_x corresponding to E_x-vibrations and n_y corresponding to E_y-vibrations. The existence of both an n_x and an n_y for a given optical frequency ω is to be expected, since different binding forces along these directions produce different interactions with the electromagnetic wave and, thus, different velocities of propagation v_x and v_y through the crystal. The result is that such a crystal, although not appreciably dichroic, still manifests the property of birefringence. The critical physical properties here are the refractive index n and the extinction coefficient k (proportional to the absorption coefficient) for a given frequency of light. Each constitutes a part of the *complex refractive index* $\widetilde{n}$, which is given by

$$\widetilde{n} = n + ik$$

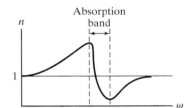

Figure 15-8 Response of refractive index as a function of frequency near an absorption band. The band in which $dn/d\omega < 0$ is said to be a region of anomalous dispersion.

Recapitulating, then, for an ideal dichroic material, $n_x = n_y$ and $k_x \neq k_y$, whereas for an ideal birefringent material, $k_x = k_y$ and $n_x \neq n_y$. Both conditions require anisotropic crystalline structures. The conditions are frequency dependent. Calcite is birefringent in the visible spectrum, for example, and strongly dichroic in certain parts of the infrared spectrum. Other common materials, birefringent in the visible region, are quartz, ice, mica, and even cellophane.

The relationship of crystalline asymmetry with refractive index and the speed of light in the medium may be understood a bit more clearly by considering the case of calcite. The basic molecular unit of calcite is $CaCO_3$, which assumes a tetrahedral or pyramidal structure in the crystal. Figure 15-9a shows one of these molecules, assumed to be surrounded by identical structures that are similarly oriented. The carbon (C) and oxygen (O) atoms form the base of the pyramid, as shown, with carbon lying in the center of the equilateral triangle of oxygen atoms. The calcium (Ca) atom is positioned at some distance above the carbon atom, at the apex of the pyramid. The figure shows unpolarized light propagating through the crystal from two different directions. First consider light entering from below, along the line joining the carbon and calcium atoms. All oscillations of this $\vec{E}$-field are represented by the two transverse vectors, both of which are labeled $E_\perp$ in Figure 15-9a. Since the molecule, and so also the crystal, is symmetric with respect to this direction (from C to Ca), both $\vec{E}$-vibrations interact with the electrons in the *same way* when traveling through the calcite. This direction of symmetry through the crystal is called the *optic axis* (OA) of the crystal. For the light entering from below, then, both $\vec{E}$-components are perpendicular to the OA and "see" no anisotropy. Consider next the light entering the crystal from the left. From this direction the two representative $\vec{E}$-vibrations—labeled $E_\perp$ and $E_\parallel$—have dissimilar effects on the electrons in the oxygen base plane. The component $E_\parallel$, which is parallel to the OA of the crystal, causes electrons in the base plane to oscillate along a direction perpendicular to the plane, whereas its orthogonal counterpart $E_\perp$ causes oscillations within the plane. Oscillations within the plane—where the electrons tend to be confined due to the chemical bonding—take place more

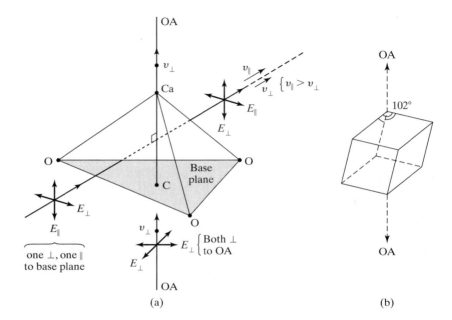

Figure 15-9 (a) Progress of light through a calcite crystal. Three oxygen (O) atoms form the base of a tetrahedron. The optic axis OA is parallel to the line joining the C and Ca atoms. (b) Rhombohedron of calcite, showing the optic axis, which passes symmetrically through a blunt corner where the three face angles equal 102°.

(a)   (b)

easily, that is, with smaller binding forces, than oscillations that are perpendicular to the plane. Since $\vec{E}$-oscillations in the oxygen plane ($\vec{E} \perp OA$) interact more strongly with the electrons, the speed $v_\perp$ of these component waves is reduced most, that is, $v_\perp < v_\parallel$. No interaction at all would make $v = c$. Since $n = c/v$ and $v_\perp < v_\parallel$, we conclude that $n_\perp > n_\parallel$. The measured values for calcite are $n_\perp = 1.658$ and $n_\parallel = 1.486$ for $\lambda = 589.3$ nm. As Table 15-1 indicates, the inequality may be reversed in other materials. In materials that crystallize in the trigonal (like calcite), tetragonal, or hexagonal systems, there is one unique direction through the crystal for which the atoms are arranged symmetrically. For example, the calcite molecule of Figure 15-9a shows a threefold rotational symmetry about the optic axis. Such structures possess a single optic axis and are called *uniaxial birefringent*. Further, when $n_\parallel - n_\perp > 0$, the crystals are said to be *uniaxial positive*, and when this quantity is negative, *uniaxial negative*. Other crystalline systems—the triclinic, monoclinic, and orthorhombic—possess two such directions of symmetry or optic axes and are called *biaxial crystals*.[5] Mica, which crystallizes in monoclinic forms, is a good example. Such materials then possess three distinct indices of refraction. Of course, there are also cubic crystals such as salt (NaCl) or diamond (C) that are optically isotropic and possess one index of refraction. This is the case also for materials that have no large-scale crystalline structure, such as glass or fluids. Naturally occurring calcite crystals are cleavable into rhombohedrons as a result of their crystallization into the trigonal lattice structures. The rhombohedron (Figure 15-9b) has only two corners where all three face angles (each 102°) are obtuse. The OA of calcite is directed through these diagonally opposite corners in such a way that it makes equal angles with the three faces there.

A birefringent crystal can be cut and polished to produce polarizing elements in which the OA may have any desired orientation relative to the incident light. Consider the cases represented in Figure 15-10.

TABLE 15-1   REFRACTIVE INDICES FOR SEVERAL MATERIALS MEASURED AT SODIUM WAVELENGTH OF 589.3 nm

Isotropic (cubic)	Sodium chloride	1.544		
	Diamond	2.417		
	Fluorite	1.392		
Uniaxial (trigonal, tetragonal, hexagonal)	**Positive $n_\parallel > n_\perp$:**	$n_\parallel$	$n_\perp$	
	Ice	1.313	1.309	
	Quartz (SiO_2)	1.5534	1.5443	
	Zircon ($ZrSiO_4$)	1.968	1.923	
	Rutile (TiO_2)	2.903	2.616	
	Negative $n_\parallel < n_\perp$:			
	Calcite ($CaCO_3$)	1.4864	1.6584	
	Tourmaline	1.638	1.669	
	Sodium Nitrate	1.3369	1.5854	
	Beryl ($Be_3Al_2(SiO_3)_6$)	1.590	1.598	
Biaxial (triclinic, monoclinic, orthorhombic)		n_1	n_2	n_3
	Gypsum ($CaSO_4(2\,H_2O)$)	1.520	1.523	1.530
	Feldspar	1.522	1.526	1.530
	Mica	1.552	1.582	1.588
	Topaz	1.619	1.620	1.627

---

[5]For a description of such crystalline systems, see, for example, Charles Kittel, *Introduction to Solid State Physics* (New York: John Wiley & Sons, 1986).

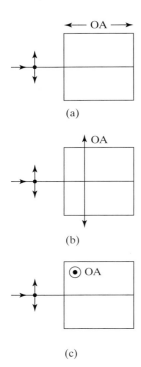

Figure 15-10 Light entering a birefringent plate with its optic axis in various orientations. (a) Light propagation along optic axis. (b) Light propagation perpendicular to optic axis. (c) Light propagation perpendicular to optic axis.

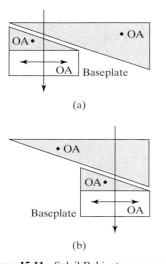

Figure 15-11 Soleil-Babinet compensator. The optic axes are as indicated. The arrow shows the direction of light through the compensator. (a) Zero retardation. (b) Maximum retardation.

In (a), both representative components of the unpolarized light (dot and double-headed arrow) incident from the left are oriented perpendicular to the OA of the crystal. Both propagate at the same speed through the crystal with index of refraction $n_\perp$. In (b) and (c), however, the OA is parallel to one component and perpendicular to the other. In this case, each component propagates through the crystal with a different index of refraction and speed. On emerging, the cumulative relative phase difference can be described in terms of the difference between optical paths for the two components. If the thickness of the crystal is d, the difference in optical paths is

$$\Delta = |n_\perp - n_\parallel| d$$

and the corresponding phase difference is

$$\Delta\varphi = 2\pi\left(\frac{\Delta}{\lambda_0}\right) = \left(\frac{2\pi}{\lambda_0}\right)|n_\perp - n_\parallel| d \qquad (15\text{-}3)$$

where λ_0 is the vacuum wavelength. If the thickness of the plate is such as to make $\Delta\varphi = \pi/2$, it is a *quarter-wave plate* (QWP); if $\Delta\varphi = \pi$, we have a *half-wave plate* (HWP); and so on. These are called *zero-order* (or sometimes *first-order*) plates. Because such plates are extremely thin, it is more practical to make thicker QWPs of higher order m, giving $\Delta\varphi = (2\pi)m + \pi/2$, where $m = 1, 2, 3, \ldots$. A thicker composite of two plates may also be formed, in which one plate compensates for the retardance of all but the desired $\Delta\varphi$ of the other. In this way we can fabricate optical elements that act as *phase retarders*. Mica and quartz are commonly used as retardation plates, usually in the form of thin, flat discs sandwiched between glass layers for added strength. Since the net phase retardation $\Delta\varphi$ is proportional to the thickness d, any device that allows a continuous change in thickness makes possible a continuously adjustable retardation plate.

Such a convenient device is called a *compensator*. Figure 15-11 illustrates the working principle of a *Soleil-Babinet compensator*. Crystalline quartz is used to form a fixed lower baseplate, which is actually a wedge in optical contact with a quartz flat plate. Above is another quartz wedge, with relative motion possible along the inclined face. Notice the direction of the OA in this assembly. In (a) the position of the upper wedge is such that light travels through equal thicknesses of quartz with their optical axes aligned perpendicular to one another. Any retardation due to one thickness is then canceled by the other, yielding zero net retardation. Sliding the upper wedge to the left increases the thickness of the first OA orientation relative to the second, yielding a continuously variable retardation up to a maximum, in position (b), of perhaps two wavelengths, or 4π. Adjustment by a micrometer screw allows small changes in $\Delta\varphi$ to be made.

Birefringence in Optical Fibers

In Chapter 10 some of the advantages of single-mode optical fibers were discussed. A "single-mode" optical fiber actually supports *two* orthogonal linear polarizations. If such a single-mode fiber were perfectly uniform, both polarization modes would travel through the fiber with the same speed. Typical optical fibers have at least a small amount of birefringence due to fiber imperfections and anisotropic stress along the fiber length. As a result, light of orthogonal linear polarizations travels with different speeds through such a fiber. Scattering and other mechanisms lead to a coupling between the orthogonal polarization modes. As a result, linearly polarized light launched into one end of a fiber evolves into light that is a mixture of two orthogonal linear polarizations as the light progresses through the fiber. These components of orthogonal polarization travel with different speeds and so the phase difference between the components changes as the light propagates through the fiber, causing the polarization state of the light field to vary between linear, ellipti-cal, and circular polarizations. In a conventional fiber, the amount of birefrin-gence in the fiber varies randomly along the fiber length and so linearly polarized light entering the fiber quickly attains a state of random polarization that is uncorrelated with the polarization state of the input field. This leads to the phenomenon called *polarization mode dispersion*, which can limit the maximum bit rate of transmission through fibers designed for high-speed communications.

Introducing a high degree of deterministic anisotropy into a fiber re-duces the coupling between orthogonal polarization modes of the fiber. This anisotropy can be introduced by manufacturing fibers with elliptical cores or by applying an anisotropic stress to the fiber. Light that is linearly polarized along one of the symmetry axes of such a fiber can maintain its state of linear polarization over long distances. Such fibers are called *polarization-maintaining fibers*. In addition, through a variety of mechanisms, anisotropic losses can be introduced into the fiber so that one of the orthogonal polarization modes is highly attenuated while the other travels long distances with low loss. When unpolarized light is input into such a fiber, the output light will be linearly polar-ized along the low-loss direction, and the fiber functions as a linear polarizer.

The preceding discussion is but an introduction to the important role that polarization plays in light propagation through optical fibers.[6] Here we simply conclude by noting that birefringent fibers can be used to make quarter- and half-wave "plates" and phase compensators. These devices can be either passive or actively controlled.

15-5 DOUBLE REFRACTION

In the cases depicted in Figure 15-10b and c, the light propagating through the crystal may develop a net phase difference between $\vec{E}$-components perpendicular and parallel to the crystal's OA, but the beam remains a single beam of light. If now the OA is situated so that it makes an arbitrary angle with respect to the beam direction, as in Figure 15-12, the light experiences double refraction[7]; that is, two refracted beams emerge, labeled the *ordinary* and *extraordinary* rays. The extraordinary ray is so named because it does not exhibit ordinary Snell's law behavior on refraction at the crystal surfaces. Thus if a calcite crystal is laid over a black dot on a white piece of paper, or over an il-luminated pinhole, two images are seen while looking into the top surface. If the crystal is rotated about the incident ray direction, the extraordinary image is found to rotate around the ordinary image, which remains fixed in position.

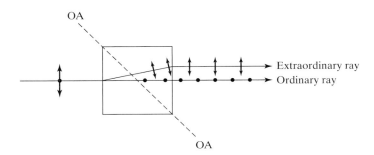

Figure 15-12 Double refraction.

---

[6]For a more complete discussion, see, for example, J-P Goure and I. Verrier, *Optical Fibre De-vices* (Bristol and Philadelphia: Institute of Physics Publishing, 2002).

[7]*Double refraction* is a term used to describe a manifestation of *birefringence* in materials, al-though it has literally the same meaning. Birefringence indicates the possession of two refractive indices, whereas double refraction refers to the splitting of a ray of light into ordinary and extraordinary parts.

Furthermore, the two beams emerge linearly polarized in orthogonal orientations, as shown. Notice that the ordinary ray is polarized perpendicular to the OA and so propagates with a refractive index of $n_o = n_\perp = c/v_\perp$. The extraordinary ray emerges polarized in a direction perpendicular to the polarization of the ordinary ray. Inside the crystal, the extraordinary ray can be described in terms of components polarized in directions both perpendicular and parallel to the optic axis. (This situation is discussed in the following paragraph.) The perpendicular component propagates with speed $v_\perp = c/n_\perp$, as for the ordinary ray. The other component, however, propagates with a refractive index $n_e = n_\parallel = c/v_\parallel$. The net effect of the action of both components is to cause the unusual bending of the extraordinary ray shown in Figure 15-12.

The situation may be clarified somewhat by reference to Figure 15-13a, which shows one Huygens' wavelet created by the extraordinary ray as it contacts the crystal surface at P. The incident $\vec{\mathbf{E}}$-vibration is shown resolved into orthogonal components (*aa*) parallel to the OA and (*bb*) perpendicular to the OA. The parallel component propagates along the direction of $\vec{v}_\parallel$, which must be perpendicular to *aa*, and the perpendicular component propagates along the direction of $\vec{v}_\perp$, which must be perpendicular to *bb*. Since each component travels with a speed determined by the corresponding refractive indices, $n_\parallel$ and $n_\perp$, the speeds are unequal. For calcite, for example, $n_\perp > n_\parallel$, so that $v_\perp < v_\parallel$: The Huygens' wavelet for the extraordinary ray is not spherical as in isotropic media but ellipsoidal as shown, with major axis proportional to $v_\parallel$ and minor axis proportional to $v_\perp$. Figure 15-13b shows several such Huygens' ellipsoidal wavelets and the plane wavefront tangent to the wavelets. This plane wavefront, which constitutes the new surface of constant phase, is perpendicular to the propagation vector $\vec{\mathbf{k}}$ for the wave. The $\vec{\mathbf{E}}$ of the elliptical wavefront is intermediate between $E_\perp$ and $E_\parallel$. Notice that in this case of the extraordinary ray in an anisotropic medium, $\vec{\mathbf{E}}$ is not perpendicular to $\vec{\mathbf{k}}$. Since energy propagates in the direction of the Poynting vector, $\vec{\mathbf{S}} = \varepsilon_0 c^2 \vec{\mathbf{E}} \times \vec{\mathbf{B}}$, and since the ray direction is the same as the direction of energy flow, the extraordinary ray with velocity $\vec{v}$ intermediate between $\vec{v}_\perp$ and $\vec{v}_\parallel$ shows the unusual refraction of Figure 15-12. The extraordinary ray is not perpendicular to the plane wavefront; rather, the ray direction along $\vec{\mathbf{S}}$ is from the wavelet origin O to the point of tangency of the elliptical wavelet with the plane wavefront. For the normal ray, on the other hand, due to the $\vec{\mathbf{E}}$-component perpendicular to the OA, everything is normal; the ray obeys Snell's law, the Huygens' wavelets are spheres, $\vec{\mathbf{k}} \perp \vec{\mathbf{E}}, \vec{\mathbf{k}} \parallel \vec{\mathbf{S}}$, and the ray is perpendicular to its wavefront.

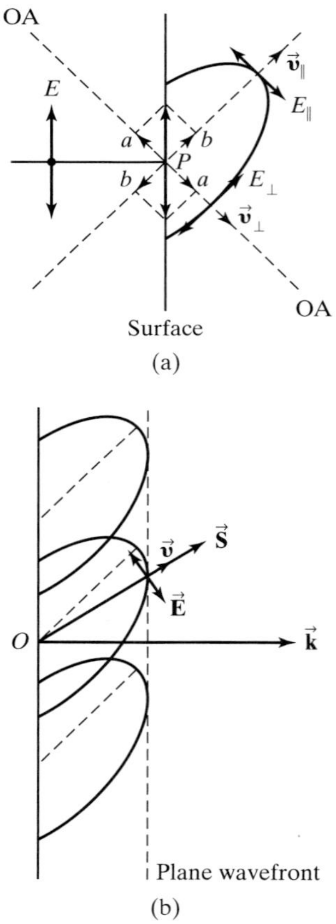

From Figure 15-13a and the preceding discussion, it should be clear that the precise intermediate value of the velocity $\vec{v}$ of the extraordinary ray depends on the relative contributions of $v_\parallel$ and $v_\perp$, that is, on the relative orientations of the incident beam and the OA of the crystal. Thus, both the velocity and index of refraction of the extraordinary ray are continuous functions of direction. On the other hand, the refractive index of the ordinary ray is a constant, independent of direction. Figure 15-14 is a plot of the refractive index versus wavelength for crystalline quartz. At any wavelength, the index for the ordinary ray is a constant, given by the lower curve, whereas the index for the extraordinary ray falls somewhere between the upper and lower curves, depending on the direction of the incident ray relative to the crystal axis. If the two refracted rays, linearly polarized perpendicular to one another, can be physically separated, then double refraction can be used to produce a linearly polarized beam of light. There are various devices that accomplish this. One of the most commonly used is the *Glan-air prism*, shown in Figure 15-15. Two calcite prisms with apex angle θ, as shown, are combined with their long faces opposed and separated by an air space. Their optic axes are parallel,

Figure 15-13 (a) Creation of an elliptical Huygens' wavelet by the extraordinary ray. The material in this case is uniaxial negative, like calcite. (b) Nonalignment of ray direction $\vec{\mathbf{S}}$ and propagation vector $\vec{\mathbf{k}}$ for the extraordinary ray in birefringent material.

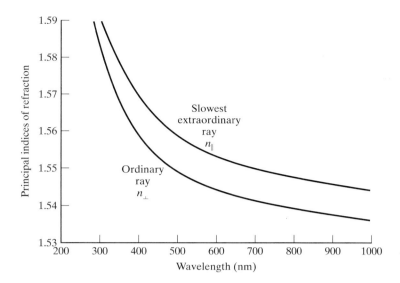

Figure 15-14 Refractive indices of crystalline quartz versus wavelength at 18°C. At a given wavelength, the index for the extraordinary ray may fall anywhere between the two curves, whereas the index for the ordinary ray is fixed. (Adapted from Melles Griot, *Optics Guide 3*, 1985.)

with the orientation perpendicular to the page as shown. At the point of refraction out of the first prism, the angle of incidence is equal to the apex angle θ of the prisms. The critical angle for refraction into air is given as usual by $\sin \theta_c = 1/n$ and so depends on the orientation of the $\vec{E}$-vibration relative to the OA. For $\vec{E} \parallel$ OA, $n_\parallel = 1.4864$ and $\theta_c = 42.3°$, while for $\vec{E} \perp$ OA, $n_\perp = 1.6584$ and $\theta_c = 37.1°$. Thus, by using prisms with apex angles intermediate between these values, the perpendicular component can be totally internally reflected while the parallel component is transmitted. The second prism serves to reorient the transmitted ray along the original beam direction. The entire device constitutes a linear polarizer. When the space between prisms is filled with some other transparent material, such as glycerine, the apex angle must be modified. Several other designs for polarizing prisms constructed from positive uniaxial material (quartz) are illustrated in Figure 15-16. Notice that in these cases, the ordinary and extraordinary rays are separated without the agency of total internal reflection. In each case, the OAs of the two prisms are perpendicular to one another, so that an $E_\perp$-component in the first prism, for instance, may become an $E_\parallel$-component in the second, with a corresponding change in refractive index. Different relative indices for the two components result in different angles of refraction and separation into two polarized beams. We see that birefringent materials are useful in fabricating devices that behave as linear polarizers as well as in producing phase retarders such as QWPs, considered earlier in this chapter.

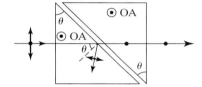

Figure 15-15 Glan-Air prism.

15-6 OPTICAL ACTIVITY

Certain materials possess a property called *optical activity*. When linearly polarized light is incident on an optically active material, it emerges as linearly polarized light but with its direction of vibration rotated from the original. Viewing the beam head-on, some materials produce a clockwise rotation (*dextrorotatory*) of the $\vec{E}$-field, whereas others produce a counterclockwise rotation (*levorotatory*). Optically active materials include both solids (for example, quartz and sugar) and liquids (turpentine and sugar in solution). Some materials, such as crystalline quartz, produce either rotation, traceable to the existence of two forms of the crystalline structure that turn out to be mirror images (*enantiomorphs*) of one another. Optically active materials modify the state of polarization of a beam of polarized light and can be represented mathematically by the Jones rotator matrix given in Table 14-2. Notice that

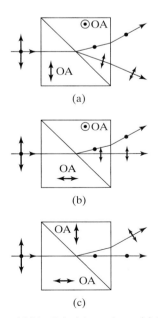

Figure 15-16 Polarizing prisms. (a) Wollaston prism. (b) Rochon prism. (c) Sernamont prism.

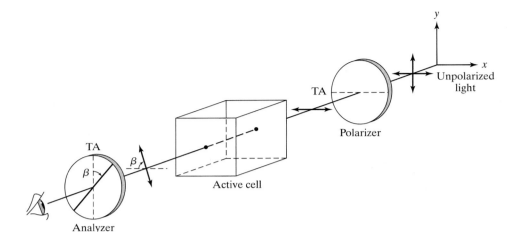

Figure 15-17 Measurement of optical activity. With the active material in place, the optical activity is measured by the angle β required to reestablish extinction.

the rotator mechanism involved in rotating the direction of vibration of linearly polarized light is distinct from the action of phase retarders, such as half-wave plates discussed in Sections 14-2 and 15-4, which may produce the same result.

Optical activity is easily measured using two linear polarizers originally set for extinction, that is, with their TAs crossed in perpendicular orientations (Figure 15-17). When a certain thickness of optically active material is inserted between analyzer and polarizer, the condition of extinction no longer exists because the $\vec{E}$-vector of the light is rotated by the optically active medium. The exact angle of rotation β can be measured by rotating the analyzer until extinction reoccurs, as shown. The rotation so measured depends on both the wavelength of the light and the thickness of the active medium. The rotation (in degrees) produced by a 1-mm plate of optically active solid material is called its *specific rotation*. Table 15-2 gives, in degrees/mm, the specific rotation ρ of quartz for a range of optical wavelengths. The amount of rotation caused by optically active liquids is much less by comparison. In the case of solutions, the specific rotation is defined as the rotation due to a 10-cm thickness and concentration of 1 g of active solute per cubic centimeter of solution. That is, for solutions, ρ has units of degrees/dm · cm^3/g. The net angle of rotation β due to a light path L through a solution of d grams of active solute per cubic centimeter is, then,

$$\beta = \rho L d \qquad (15\text{-}4)$$

where L is in decimeters and d is the concentration in grams per cubic centimeter. For example, 1 dm of turpentine rotates sodium light by $-37°$. The negative sign indicates that turpentine is levorotatory in its optical activity. Measurement of the optical rotation of sugar solutions is often used to determine concentration, via Eq. (15-4).[8] The dependence of specific rotation on wavelength

TABLE 15-2   SPECIFIC ROTATION OF QUARTZ

λ (nm)	ρ (degrees/mm)
226.503	201.9
404.656	48.945
435.834	41.548
546.072	25.535
589.290	21.724
670.786	16.535

[8]Ordinary corn syrup is often used in the optics lab to demonstrate optical activity.

means that if one views white light through an arrangement like that of Figure 15-17, each wavelength is rotated to a slightly different degree. This separation of colors is referred to as *rotatory dispersion*.

Without giving a physical explanation of optical activity, we can, following Fresnel, offer a useful phenomenological description that enables us to relate specific rotation of an active substance to certain physical parameters. This description rests first on the fact, demonstrated in the previous chapter, that linearly polarized light can be assumed to consist of equal amounts of left- and right-circularly polarized light. Second, in using this description one assumes that the left- and right-circularly polarized components move through an optically active material with different velocities, $v_{\mathcal{L}}$ and $v_{\mathcal{R}}$, respectively. Since $v = c/n$, different refractive indices, $n_{\mathcal{L}}$ and $n_{\mathcal{R}}$, may be defined for circularly polarized light.

Consider first the case of an inactive medium for which $v_{\mathcal{L}} = v_{\mathcal{R}}$, or, equivalently, $n_{\mathcal{L}} = n_{\mathcal{R}}$ and $k_{\mathcal{L}} = k_{\mathcal{R}}$. Here $\vec{k}$ is the propagation vector whose magnitude is related to wave speed by $k = \omega/v$. If the incident light is linearly polarized along the x-direction, as in Figure 15-17, it may be resolved into left- and right-circularly polarized light. Figure 15-18 makes this clear by illustrating the vector addition at three different times in an oscillation. The vector sum $\vec{E}$ executes oscillations along the x-axis as the $E_{\mathcal{R}}$- and $E_{\mathcal{L}}$-vectors rotate clockwise and counterclockwise, respectively, at equal rates.

Next, consider the consequences of assuming $n_{\mathcal{L}} \neq n_{\mathcal{R}}$. Now the phases of the $\mathcal{L}$- and $\mathcal{R}$-components, $\vec{E}_{\mathcal{L}}$ and $\vec{E}_{\mathcal{R}}$ respectively, are not equal. In general, their (complex) electric fields may be expressed by

$$\mathbf{E}_{\mathcal{L}} = \widetilde{\mathbf{E}}_{0\mathcal{L}} e^{i(k_{\mathcal{L}}z - \omega t)} \tag{15-5}$$

$$\widetilde{\mathbf{E}}_{\mathcal{R}} = \widetilde{\mathbf{E}}_{0\mathcal{R}} e^{i(k_{\mathcal{R}}z - \omega t)} \tag{15-6}$$

where $k_{\mathcal{L}} = (\omega/c)n$ and $k_{\mathcal{R}} = (\omega/c)n_{\mathcal{R}}$. Of course, the real fields are given by $\vec{E}_{\mathcal{L}} = \text{Re}\,(\widetilde{\mathbf{E}}_{\mathcal{L}})$ and $\vec{E}_{\mathcal{R}} = \text{Re}\,(\widetilde{\mathbf{E}}_{\mathcal{R}})$. The complex vector amplitudes are given by

$$\widetilde{\mathbf{E}}_{0\mathcal{L}} = \left(\frac{E_0}{2}\right)\begin{bmatrix} 1 \\ i \end{bmatrix} \quad \text{and} \quad \widetilde{\mathbf{E}}_{0\mathcal{R}} = \left(\frac{E_0}{2}\right)\begin{bmatrix} 1 \\ -i \end{bmatrix} \tag{15-7}$$

corresponding to the Jones vectors for left- and right-circularly polarized modes, as shown in Table 14-1. The phases of the two components are given by

$$\theta_{\mathcal{L}} = k_{\mathcal{L}}z - \omega t$$

$$\theta_{\mathcal{R}} = k_{\mathcal{R}}z - \omega t \tag{15-8}$$

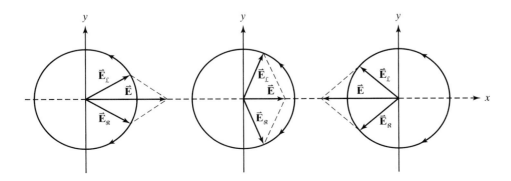

Figure 15-18 Superposition of left- and right-circularly polarized light at different instants. The light is assumed to be emerging from the page.

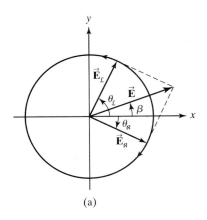

(a)

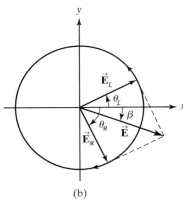

(b)

Figure 15-19 Optical rotation produced by left- and right-circularly polarized light having different speeds through an active medium. (a) Levorotatory: $n_\mathfrak{L} > n_\mathfrak{R}$. (b) Dextrorotatory: $n_\mathfrak{R} > n_\mathfrak{L}$.

Suppose that the active medium is one for which $k_\mathfrak{L} > k_\mathfrak{R}$, which also means that $n_\mathfrak{L} > n_\mathfrak{R}$ and $v_\mathfrak{L} < v_\mathfrak{R}$. Then at some distance z into the medium, $\theta_\mathfrak{L} > \theta_\mathfrak{R}$ for all t. The situation is shown graphically at an arbitrary instant in Figure 15-19a. The vector sum of $\vec{\mathbf{E}}_\mathfrak{L}$ and $\vec{\mathbf{E}}_\mathfrak{R}$ is again linearly polarized light but with an inclination angle $+\beta$ relative to the x-axis. The medium for which $n_\mathfrak{L} > n_\mathfrak{R}$ is therefore levorotatory. In Figure 15-19b, the opposite case is also pictured, for which β is a negative angle and the optical activity is dextrorotatory. The magnitude of β can be determined by noticing that the resultant $\vec{\mathbf{E}}$ that determines the angle β is always the diagonal of an equal-sided parallelogram, so that

$$\theta_\mathfrak{L} - \beta = \theta_\mathfrak{R} + \beta$$

or

$$\beta = \tfrac{1}{2}(\theta_\mathfrak{L} - \theta_\mathfrak{R}) \qquad (15\text{-}9)$$

Using Eq. (15-8) in Eq. (15-9) leads to

$$\beta = \tfrac{1}{2}(k_\mathfrak{L} - k_\mathfrak{R})z$$

Finally, using $k_\mathfrak{L} = k_0 n_\mathfrak{L}$, $k_\mathfrak{R} = k_0 n_\mathfrak{R}$, and $k_0 = 2\pi/\lambda_0$, where λ_0 is the wavelength in vacuum,

$$\beta = \frac{\pi z}{\lambda_0}(n_\mathfrak{L} - n_\mathfrak{R}) \qquad (15\text{-}10)$$

Notice that the linearly polarized light is rotated through an angle that is proportional to the thickness z of the active medium, as verified experimentally. The action of the $\mathfrak{L}$ and $\mathfrak{R}$ modes in producing the resultant light might be visualized in the following way. At incidence, the linearly polarized light is immediately resolved into $\mathfrak{L}$ and $\mathfrak{R}$ circular modes, which, at $z = 0$ and $t = 0$, begin together with $\theta_\mathfrak{L} = \theta_\mathfrak{R} = 0$. If $v_\mathfrak{R} > v_\mathfrak{L}$, the $\mathfrak{R}$ mode reaches some point along its path before the $\mathfrak{L}$ mode. Until the $\mathfrak{L}$ mode arrives, $\vec{\mathbf{E}}$ rotates at this point according to the circular polarization of the $\mathfrak{R}$ mode acting alone. As soon as the $\mathfrak{L}$ mode arrives, however, the two modes superpose to fix the direction of vibration at an angle β in a linear mode. The relative phase between the two modes at this instant determines the angle β, as expressed by Eq. (15-9). Since the frequencies of the two modes are identical, angle β remains constant thereafter.

It should be emphasized that the indices of refraction involved in optical activity characterize *circular birefringence* rather than ordinary birefringence. The indices $n_\mathfrak{R}$ and $n_\mathfrak{L}$ are much closer in value than $n_\perp$ and $n_\parallel$, as can be seen in the case of quartz (Table 15-3).

Example 15-1

Determine the specific rotation produced by a 1-mm-thick quartz plate at a wavelength of 396.8 nm.

TABLE 15-3   REFRACTIVE INDICES FOR QUARTZ

λ (nm)	$n_\parallel$	$n_\perp$	$n_\mathfrak{R}$	$n_\mathfrak{L}$
396.8	1.56771	1.55815	1.55810	1.55821
762.0	1.54811	1.53917	1.53914	1.53920

Solution

From Table 15-3, at $\lambda = 396.8$ nm,

$$n_\mathfrak{L} - n_\mathfrak{R} = 1.55821 - 1.55810 = 0.00011$$

Using Eq. (15-10),

$$\beta = \frac{\pi(10^{-3})}{396.8 \times 10^{-9}}(0.00011) = 0.8709 \text{ rad} = 49.9°$$

in good agreement with Table 15-2 for the neighboring wavelength of 404.6 nm.

The preceding description does not explain why the velocities of the $\mathfrak{L}$ and $\mathfrak{R}$ circularly polarized modes should differ at all. We content ourselves for purposes of this discussion with pointing out that optically active materials possess molecules or crystalline structures that have spiral shapes, with either left-handed or right-handed screw forms. Linearly polarized light transmitted through a collection of such molecules creates forced vibrations of electrons that, in response, move not only along a spiral but necessarily around the spiral. Thus the effect of $\mathfrak{L}$-circularly polarized light on a left-handed spiral would be expected to be different from its effect on a right-handed spiral and should lead to different speeds through the medium. Even if individual spiral-shaped molecules confront the light in random orientations, as in a liquid, there will be a cumulative effect that does not cancel, as long as all or most of the molecules are of the same handedness.

15-7 PHOTOELASTICITY

Consider the following experiment. Two polarizing filters acting as polarizer and analyzer are set up with a white-light source behind the pair. If the TAs of the filters are crossed, no light emerges from the pair. If some birefringent material is inserted between them, light is generally transmitted in beautiful colors. To understand this unusual effect, consider Figure 15-20, where polarizer and analyzer TAs are crossed and at 45° and −45°, respectively, relative to the x-axis. Suppose that the birefringent material introduced in the light beam constitutes a half-wave plate with its fast axis (FA) vertical, as shown. Its action on the incident linearly polarized light is to convert it to linearly polarized light perpendicular to the original direction, or at −45° inclination with the x-axis. This can be understood by resolving the incident light into equal orthogonal components along the FA and SA (slow axis) and with a π phase difference between them. As always, the effect of the HWP on linearly polarized light is to rotate it through 2α, or, in this case, 90°. The same result

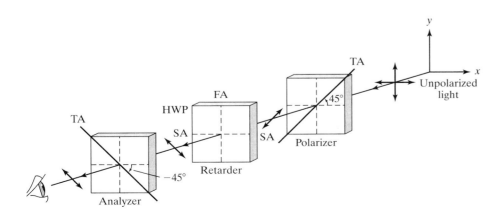

Figure 15-20 Light transmitted by cross polarizers when a birefringent material acting as a half-wave plate is placed between them.

368 Chapter 15 Production of Polarized Light

follows from use of the Jones calculus:

$$\begin{bmatrix} 1 & 0 \\ 0 & -1 \end{bmatrix}\begin{bmatrix} 1 \\ 1 \end{bmatrix} = \begin{bmatrix} 1 \\ -1 \end{bmatrix}$$

HWP   LP   LP
FA vertical at 45°   at −45°

The light emerging from the HWP is now polarized along a direction that is fully transmitted by the analyzer. If the retardation plate introduces phase differences other than π, the light is rendered elliptically polarized, and some portion of the light will still be transmitted by the analyzer. The character of the incident light be will be unmodified by the plate, and so extinguished, if the phase difference introduced by the retarder is 2π or some multiple thereof so that the retardation plate functions as a full-wave plate.

Now recall that the phase difference $\Delta\varphi$ introduced by a retardation plate is wavelength dependent, such that

$$\lambda_0 \, \Delta\varphi = 2\pi d(n_\perp - n_\parallel) \tag{15-11}$$

where d is the thickness of the plate. For a given plate, the right side of Eq. (15-11) is constant throughout the optical region of the spectrum, if the small variation $(n_\perp - n_\parallel)$ is neglected. It follows that the retardation is very nearly inversely proportional to the wavelength. Thus if the retardation plate acts as a HWP for red light, in the arrangement of Figure 15-20, red light will be fully transmitted, whereas shorter visible wavelengths will be only partially transmitted, giving the transmitted light a predominantly reddish hue. If the TA of the analyzer is now rotated by 90°, all components originally blocked are transmitted. Since the sum of the light transmitted under both conditions must be all the incident light, that is, white light, it follows that the colors observed under these two transmission conditions are complementary colors.

Sections of quartz or calcite and thin sheets of mica can be used to demonstrate the production of colors by polarization. Many ordinary materials also show birefringence, either under normal conditions or under stress, as in Figure 15-21. A crumpled piece of cellophane introduced between crossed polarizers shows a striking variety of colors, enhanced by the fact that light must pass through two or more thicknesses at certain points, so that $\Delta\varphi$ varies from point to point due to a

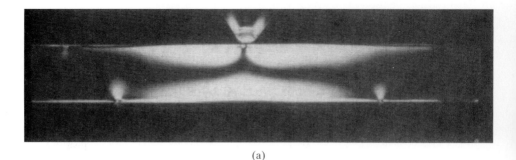

(a)

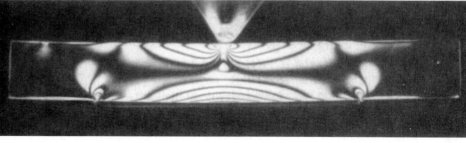

Figure 15-21 Photoelastic stress patterns for a beam resting on two supports and (a) lightly loaded at the center, (b) heavily loaded at the center. (From M. Cagnet, M. Francon, and J. C. Thrierr, *Atlas of Optical Phenomenon*, Plate 40, Berlin: Springer-Verlag, 1962.)

(b)

change in thickness d. A similar effect is produced by wrapping glossy cellophane tape around a microscope slide, allowing for regions of overlap. Finally, $\Delta\varphi$ may also vary from point to point due to local variations in the quantity $n_\perp - n_\parallel$. Formed plastic pieces, such as a drawing triangle or safety glasses, often show such variations due to localized birefringent regions associated with strain. A pair of plastic safety goggles inserted between crossed polarizers shows a higher density of color changes in those regions under greater strain, because the difference in refractive indices changes most rapidly in such regions. The birefringence induced by mechanical stress applied to normally isotropic substances such as plastic or glass is the basis for the method of stress analysis called *photoelasticity*. It is found that in such materials, an optic axis is induced in the direction of the stress, both in tension and in compression. Since the degree of birefringence induced is proportional to the strain, prototypes of mechanical parts may be fabricated from plastic and subjected to stress for analysis. Points of maximum strain are made visible by light transmitted through crossed polarizers when the stressed sample is positioned between the polarizers. Such polarized light patterns for a beam under light and heavy stress is shown in Figure 15-21.

In subsequent chapters we will discuss several applications that rely on the polarization properties of light for their operation. In particular, we will discuss *liquid crystal displays* in Chapter 17 and *electro-optic modulators* and *Faraday rotators* in Chapter 24.

PROBLEMS

15-1 Initially unpolarized light passes in turn through three linear polarizers with transmission axes at $0°, 30°,$ and $60°$, respectively, relative to the horizontal. What is the irradiance of the product light, expressed as a percentage of the unpolarized light irradiance?

15-2 At what angles will light, externally and internally reflected from a diamond-air interface, be completely linearly polarized? For diamond, $n = 2.42$.

15-3 Since a sheet of Polaroid is not an ideal polarizer, not all the energy of the $\bar{E}$-vibrations parallel to the TA are transmitted, nor are all $\bar{E}$-vibrations perpendicular to the TA absorbed. Suppose an energy fraction α is transmitted in the first case and a fraction β is transmitted in the second.

   a. Extend Malus' law by calculating the irradiance transmitted by a pair of such polarizers with angle θ between their TAs. Assume initially unpolarized light of irradiance I_0. Show that Malus' law follows in the ideal case.

   b. Let $\alpha = 0.95$ and $\beta = 0.05$ for a given sheet of Polaroid. Compare the irradiance with that of an ideal polarizer when unpolarized light is passed through two such sheets having a relative angle between TAs of $0°, 30°, 45°,$ and $90°$.

15-4 How thick should a half-wave plate of mica be in an application where laser light of 632.8 nm is being used? Appropriate refractive indices for mica are 1.599 and 1.594.

15-5 Describe what happens to unpolarized light incident on birefringent material when the OA is oriented as shown in each sketch in Figure 15-23. You will want to comment on the following considerations: Single or double refracted rays? Any phase retardation? Any polarization of refracted rays?

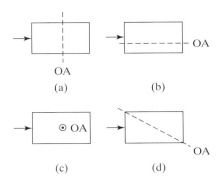

Figure 15-23   Problem 15-5, parts (a)–(d).

   e. Which orientation(s) would you use to make a quarter-wave plate?

15-6 Consider a Soleil-Babinet compensator, as shown in Figure 15-11. Suppose the compensator is constructed of quartz and provides a maximum phase retardation of two full wavelengths of green mercury light (546.1 nm). Refractive indices of quartz at this wavelength are $n_\parallel = 1.555$ and $n_\perp = 1.546$.

   a. How does the total wedge thickness compare with that of the flat plate in the position of maximum retardation?

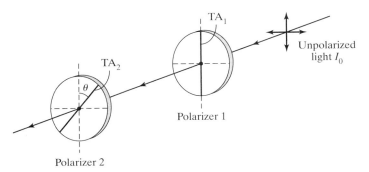

Figure 15-22   Problem 15-3.

b. How do they compare when the emergent light is circularly polarized?

15-7 A number of dichroic polarizers are available, each of which can be assumed perfect, that is, each passes 50% of the incident unpolarized light. Let the irradiance of the incident light on the first polarizer be I_0.

a. Using a sketch, show that if the polarizers have their transmission axes set at angle θ apart, the light transmitted by the pair is given by

$$I = \left(\frac{I_0}{2}\right)\cos^2\theta$$

b. What percentage of the incident light energy is transmitted by the pair when their transmission axes are set at 0° and 90°, respectively?

c. Five additional polarizers of this type are placed between the two just described, with their transmission axes set at 15°, 30°, 45°, 60°, and 75°, in that order, with the 15°-angle polarizer adjacent to the 0° polarizer, and so on. Now what percentage of the incident light energy is transmitted?

15-8 Vertically polarized light of irradiance I_0 is incident on a series of N successive linear polarizers, each with its transmission axis offset from the previous one by a small angle θ. With the help of the Law of Malus, determine the value of N such that the final transmitted irradiance is $I_N = 0.9\,I_0$ when the small angle offsets sum to 90°, that is when the initial vertical polarization is rotated to a horizontal polarization.

15-9 What minimum thickness should a piece of quartz have to act as a quarter-wave plate for a wavelength of 5893 Å in vacuum?

15-10 Determine the angle of deviation between the two emerging beams of a Wollaston prism constructed of calcite and with wedge angle of 45°. Assume sodium light.

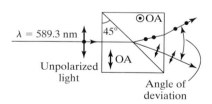

Figure 15-24 Problem 15-10.

15-11 A beam of linearly polarized light is changed into circularly polarized light by passing it through a slice of crystal 0.003 cm thick. Calculate the difference in the refractive indices for the two rays in the crystal, assuming this to be the minimum thickness showing the effect for a wavelength of 600 nm. Sketch the arrangement, showing the OA of the crystal, and explain why this occurs.

15-12 Light is incident on a water surface at such an angle that the reflected light is completely linearly polarized.

a. What is the angle of incidence?

b. The light refracted into the water is intercepted by the top flat surface of a block of glass with index of 1.50. The light reflected from the glass is completely linearly

polarized. What is the angle between the glass and water surfaces?

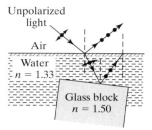

Figure 15-25 Problem 15-12.

15-13 In each of the following cases, deduce the nature of the light that is consistent with the analysis performed. Assume a 100% efficient polarizer.

a. When a polarizer is rotated in the path of the light, there is no intensity variation. With a QWP in front of the rotating polarizer (coming first), one finds a variation in intensity but no angular position of the polarizer that gives zero intensity.

b. When a polarizer is rotated in the path of the light, there is some intensity variation but no position of the polarizer giving zero intensity. The polarizer is set to give maximum intensity. A QWP is allowed to intercept the beam first with its OA parallel to the TA of the polarizer. Rotation of the polarizer now can produce zero intensity.

15-14 Light from a source immersed in oil of refractive index 1.62 is incident on the plane face of a diamond ($n = 2.42$), also immersed in the oil. Determine (a) the angle of incidence at which maximum polarization occurs and (b) the angle of refraction into the diamond.

15-15 The rotation of polarized light in an optically active medium is found to be approximately proportional to the inverse square of the wavelength.

a. The specific rotation of glucose is 20.5°. A glucose solution of unknown concentration is contained in a 12-cm-long tube and is found to rotate linearly polarized light by 1.23°. What is the concentration of the solution?

b. Upon passing through a 1-mm-thick quartz plate, red light is rotated about 15°. What rotation would you expect for violet light?

15-16 a. What thickness of quartz is required to give an optical rotation of 10° for light of 396.8 nm?

b. What is the specific rotation of quartz for this wavelength? The refractive indices for quartz at this wavelength, for left- and right-circularly polarized light, are $n_\mathcal{L} = 1.55821$ and $n_\mathcal{R} = 1.55810$, respectively.

15-17 a. A thin plate of calcite is cut with its OA parallel to the plane of the plate. What minimum thickness is required to produce a quarter-wave path difference for sodium light of 589 nm?

b. What color will be transmitted by a zircon plate, 0.0182 mm thick, when placed in a 45° orientation between crossed polarizers?

15-18 a. Show that polarizing angles for internal and external reflection between the same two media must be complementary.

b. Show that if Brewster's angle is satisfied for a TM light beam entering a parallel plate (a *Brewster window*), it will also be satisfied for the beam as it leaves the plate on the opposite side.

15-19 The indices of refraction for the fast and slow axes of quartz with 546 nm light are 1.5462 and 1.5553, respectively.

a. By what fraction of a wavelength is the e-ray retarded, relative to the o-ray, for every wavelength of travel in the quartz?

b. What is the thickness of a zeroth-order QWP?

c. If a multiple-order quartz plate 0.735 mm thick functions as a QWP, what is its order m?

d. Two quartz plates are optically contacted so that they produce opposing retardations. Sketch the orientation of the OA of the two plates. What should their difference in thickness be such that they function together like a zeroth-order QWP?

15-20 When a plastic triangle is viewed between crossed polarizers and with monochromatic light of 500 nm, a series of alternating transmission and extinction bands is observed. How much does $(n_\perp - n_\parallel)$ vary between transmission bands to satisfy successive conditions for HWP retardation? The plastic triangle is $\frac{1}{16}$ in. thick.

15-21 A plane plate of beryl is cut with the optic axis in the plane of the surfaces. Linearly polarized light is incident on the plate such that the $\bar{E}$-field vibrations are at 45° to the optic axis. Determine the smallest thickness of the plate such that the emergent light is (a) linearly and (b) circularly polarized.

15-22 Find the angle at which a half-wave plate must be set to compensate for the rotation of a 1.15-mm levorotatory quartz plate using 546-nm wavelength light.

15-23 In Chapter 23, the *Fresnel equations* are derived, showing that the fraction r of the incident field that is reflected from a dielectric plane surface for the TE polarization mode is given by Eq. (23-27). Thus, the reflectance $R = r^2$ has the form,

$$R = \left(\frac{\cos\theta - \sqrt{n^2 - \sin^2\theta}}{\cos\theta + \sqrt{n^2 - \sin^2\theta}} \right)^2$$

where θ is the angle of incidence and n is the ratio n_2/n_1.

a. Calculate the reflectance R for the TE mode when the light is incident from air onto glass of $n = 1.50$ at the polarizing angle.

b. The reflectance calculated in part (a) is also valid for an internal reflection as light leaves the glass going into air. This being the case, calculate the net fraction of the TE mode transmitted through a stack of 10 such plates relative to the incident irradiance I_0. Assume that the plates do not absorb light and that there are no multiple reflections within the plates.

c. Calculate the *degree of polarization P* of the transmitted beam, given by

$$P = \frac{I_{TM} - I_{TE}}{I_{TM} + I_{TE}}$$

where I stands for the irradiance of either polarization mode.

15-24 A half-wave plate is placed between crossed polarizer and analyzer such that the angle between the polarizer TA and the FA of the HWP is θ. How does the emergent light vary as a function of θ?

15-25 a. Determine the rotation produced by the optical activity of a 3-mm quartz plate on a linearly polarized beam of light at wavelength 762 nm.

b. What is the rotation due to optical activity by a half-wave plate of quartz using the same light beam?

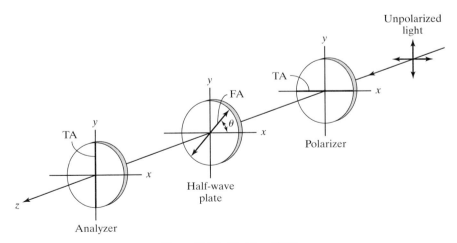

Figure 15-26 Problem 15-24.

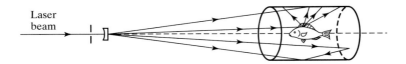

Laser
beam

16 *Holography*

INTRODUCTION

Holography is one of the many flourishing fields that owes its success to the laser. Although the technique was invented in 1948 by the British scientist Dennis Gabor before the advent of coherent laser light, the assurance of success was made possible by the laser. Emmett Leith and Juris Upatnieks at the University of Michigan first applied laser light to holography in 1962 and also introduced an important off-axis technique of illumination that we explain presently.

The spectacular improvement in three-dimensional photography made possible by the hologram has aroused unusual interest in nonscientific circles as well, so that the fast-multiplying applications of holography today also include its use in art and advertising.

16-1 CONVENTIONAL VERSUS HOLOGRAPHIC PHOTOGRAPHY

We are aware that a conventional photograph is a two-dimensional version of a three-dimensional scene, bringing into focus every part of the scene that falls within the depth of field of the lens. As a result, the photograph lacks the perception of depth or the parallax with which we view a real-life scene. In contrast, the hologram provides a record of the scene that preserves these qualities. The hologram succeeds in effectively "freezing" and preserving for later observation the intricate wavefront of light that carries all the visual information of the scene. In viewing a hologram, this wavefront is reconstructed or released, and we view what we would have seen if present at the original scene through the "window" defined by the hologram. The reconstructed wavefront provides depth perception and parallax, allowing us to look around the edge of an object to see what is behind. It may be manipulated by a lens, for example, in the same way as the original wavefront. Thus a "hologram," as its etymology suggests, includes the "whole message."

The real-life qualities of the image provided by a hologram stem from the preservation of information relating to the phase of the wavefront in addition to its amplitude or irradiance. Recording devices like ordinary photographic film and photomultipliers are sensitive only to the radiant energy received. In a developed photograph, for example, the *optical density* of the emulsion at each point is a function of the optical energy received there due to the light-sensitive chemical reaction that reduces silver to its metallic form. When energy alone is recorded, the phase relationships of waves arriving from different directions and distances, and hence the visual lifelikeness of the scene, are lost. To record these phase relationships as well, it is necessary to convert phase information into amplitude information. The *interference of light waves* provides the requisite means. Recall that when waves interfere to produce a large *amplitude*, they must be in *phase*, and when the amplitude is a minimum, the waves are out of phase, so that various contributions effectively cancel one another. If the wavefront of light from a scene is made to interfere with a coherent reference wavefront, then, the resultant interference pattern includes information regarding the phase relationships of each part of the original wavefront with the reference wave and, therefore, with every other part. The situation is sometimes described by referring to the reference wave as a *carrier wave* that is *modulated* by the *signal wave* from the scene. This language allows a fruitful comparison with the techniques of radio wave communication.

In conventional photography, a lens is used to focus the scene onto a film. All the light originating from a single point of the scene and collected by the lens is focused to a single conjugate point in the image. We can say that a one-to-one relationship exists between object and image points. By contrast, a hologram is made, as we shall see, without use of a lens or any other focusing device. The hologram is a complex interference pattern of microscopically spaced fringes, not an image of the scene. Each point of the hologram receives light from every point of the scene or, to put it another way, every object point illuminates the entire hologram. There is no one-to-one correspondence between object points and points in the wavefront before reconstruction occurs. The hologram is a record of the entire signal wave.

16-2 HOLOGRAM OF A POINT SOURCE

To see how the process is realized in practice, both making the hologram and using the hologram to reconstruct the original scene, we begin with a very basic example, the hologram of a point source. In Figure 16-1a, plane wavefronts of coherent, monochromatic radiation (the wavefronts of the reference beam) illuminate a photographic plate. In addition, spherical wavefronts reach the plate after scattering from object point O. The plate, when developed, then shows a series of concentric interference rings about X as a center. Point P falls on such a ring, for example, if the optical path difference $OP - OX$ is an integral number of wavelengths, ensuring that the *reference beam* of plane wavefront light arrives at P in step with the scattered *subject beam* of light. The developed plate is called a *Gabor zone plate*—or zone lens—with circular transmitting zones, whose transmittance is a gradually varying function of radius r. The Gabor zone plate is called a "sinusoidal" circular grating because the optical density, and therefore the transmittance of

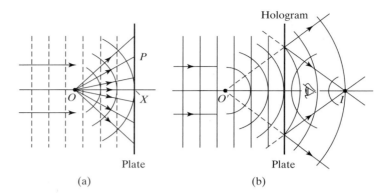

Figure 16-1 Hologram of a point source O is constructed in (a) and used in (b) to reconstruct the wavefront. Two images are formed in reconstruction.

(a)

(b)

the grating, varies as $\cos^2(ar^2)$ along the radius of the zone pattern.[1] Here, a is a constant of dimension m^{-2}. This sinusoidal plate is, in fact, a hologram of the point O. The hologram itself is a series of circular interference fringes that do not resemble the object, but the object may be reconstructed, as in Figure 16-1b, by placing the hologram back into the reference beam without the presence of the object O. Just as light directed from O originally interfered with the reference beam to produce the zone rings, so the same reference beam is now reinforced in diffraction from the rings along directions that diverge from the equivalent point O'. The point O' thus locates a *virtual image* of the original object point O, seen on reconstruction by looking into the hologram. The condition for reinforcement must also be satisfied by a second point on the exit side of the hologram, a point I that is symmetrically placed with O' relative to the plate. Clearly, the set of distances from I to the consecutive zones satisfies the same geometric relationships as the corresponding distances from O' to the zones. Thus the diffracted light also converges to point I, a *real image* of the original object point O that can be focused onto a screen. If, in the making of this hologram, the object point O is moved farther away, the radius of each zone increases. For an off-axis object at infinity, the zones are straight, parallel interference fringes. The hologram is then a *grating hologram*, formed by the intersection of two plane wavefronts of light arriving at the plate along different directions. The grating hologram is discussed briefly in Chapter 12. As explained there, the greater the angle between these wavefronts, the finer the spacing of the interference fringes. The family of circular and straight, parallel fringes we have been discussing can be seen as special cases of two point-source interference, observed in planes perpendicular and parallel, respectively, to the axis joining the points. (See the discussion relating to Figure 7-6.) When object point O is replaced by an extended object or three-dimensional scene, each point of the scene produces its own Gabor zone pattern on the plate. The hologram is now a complex montage of zones in which is coded all the information of the wavefront from the scene. On reconstruction, each set of zones produces its own real and virtual images, and the original scene is reproduced. One usually views the virtual image by looking into the hologram. Figure 16-1b shows that when viewing the virtual image in this way, undesirable light forming the real image is also intercepted. Leith and Upatnieks introduced an off-axis technique, using one or more mirrors to bring in the reference beam from a different angle so that the directions of the reconstructed real and virtual wavefronts are separated.

The two basic types of holograms discussed in the preceding paragraph are the Gabor zone plate and holographic grating, corresponding to point objects at a finite distance and at an infinite distance from the plate, respectively. If the zone plate or grating provides a square wave type of transmittance,

---

[1]More precisely, the transmittance can be expressed as $A + B\cos^2(ar^2)$, where A, B, and a are constants. See problem 16-1.

alternating between minimum and maximum, then multiple diffracted images are possible. The familiar diffraction grating of this type is known to produce orders of diffraction with $m = 0, \pm1, \pm2, \ldots$, limited by the maximum diffraction angle. The zone plate with such transmittance properties is the *Fresnel zone plate*, discussed in Chapter 13, which produces multiple focal points along its axis beyond those discussed here for the Gabor zone plate. It can be shown, however, that when the transmittance profile of the grooves or zones is not sharp but varies continuously, these general remarks concerning orders have to be modified. In particular, when the grating or circular zones are "sinusoidal" in character, that is, their transmittance profiles follow a $\cos^2(bx)$ (grating) or $\cos^2(ar^2)$ (circular zone plate) irradiance, only first-order images appear, in addition to the zeroth order, on reconstruction. Here, b is a constant of dimension m^{-1}. For the circular zones, the two first-order images are the real and virtual images discussed.

In the formation of holograms as shown in Figure 16-1a, the sinusoidal irradiance at the plate can fall to zero at points of destructive interference when the signal and reference beams are equal in amplitude. The emulsion, however, is incapable of responding linearly to all irradiances, varying from zero to maximum, so that the developed plate will show a distorted $\cos^2(ar^2)$ transmittance and higher-order diffractions will not be suppressed. By making the reference beam stronger than the signal beam, the minimum irradiance on the emulsion can be raised to the level of its linear response characteristics. A variation in transmittance of the type

$$T = T_0 + T_m \cos^2(ar^2)$$

is produced, and higher-order images are eliminated. The compromise is that the $\cos^2(ar^2)$ transmittance is now superimposed over a nonzero minimum transmittance T_0, and fringe contrast is somewhat reduced.

As we have just pointed out, the amplitude of the reference beam is made somewhat greater than the average amplitude of the signal or object beam so that the reference wave is modulated by the signal. Even when the signal is zero, the reference beam is of sufficient strength to stimulate the emulsion within its region of linear response to radiant energy. The effect of variations in signal *strength* is then to produce variations in the *contrast* of the interference fringes, whereas variations in *phase* (or *direction*) of the signal waves produce variations in *spacing* of the fringes. Thus it is in the local variations of fringe contrast and spacing across the hologram that the corresponding variations in amplitude and phase of the object waves are encoded. High-resolution plates are used to record this information faithfully.[2]

16-3 HOLOGRAM OF AN EXTENDED OBJECT

One of many holographic techniques for producing an off-axis reference beam in conjunction with the beam of diffusely reflected light from a three-dimensional scene is shown in Figure 16-2a. A combination of pinhole and lens is used to expand the beam from a laser. The expanded beam is then split by a semi-reflecting plate BS to produce two coherent beams. One beam, the *reference beam E_R*, is directed by two plane mirrors $M1$ and $M2$ onto the photographic film, as shown. The other beam, E_S, reflects diffusely from the subject, and

---

[2]Dr. Ramnendra Bahuguna, San Jose State University, has suggested a typical, simple method to process a hologram: 1-minute rinse in developer (Kodak D-19); 1-minute rinse in water; 20-second bleach (4 g potassium dichromate, 4 ml concentrated sulfuric acid, 1 L water); 30-second Kodak foto flo (diluted as on label); 15–20 minute dry vertically on paper towel.

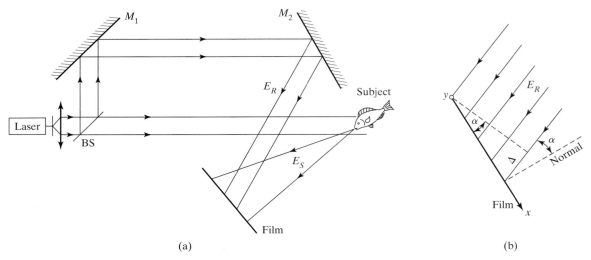

Figure 16-2   (a) Off-axis holographic system. (b) Orientation of film with reference beam in (a).

some of this beam, which we call the *subject beam*, also strikes the film, where it interferes with the reference beam and produces the hologram.

We now make the previous qualitative explanation somewhat quantitative. Let the reference beam be represented by the complex electric field

$$E_R = re^{i(\omega t + \varphi)} \tag{16-1}$$

at the plane of the film. The amplitude $r = r(x, y)$ of the reference beam can be assumed constant over the essentially plane wavefront. The phase angle φ arises from the angle α between the film plane and the plane wavefront of the reference beam, as indicated in Figure 16-2b. If the top edge of the beam strikes the film at $x = 0$, then φ is a linear function of distance x along the film plane, since

$$\varphi = \left(\frac{2\pi}{\lambda}\right)\Delta = \left(\frac{2\pi}{\lambda}\right)x \sin \alpha \tag{16-2}$$

Thus the phase angle φ relates only to the tilt of the film plane relative to the reference beam and appears as an exponential factor in Eq. (16-1):

$$E_R = re^{i\omega t}e^{i\varphi} \tag{16-3}$$

If the reference beam were not present, the film would be illuminated only by the subject beam,

$$E_S = se^{i(\omega t + \theta)} \tag{16-4}$$

where $s(x, y)$ is the amplitude of the reflected light at different points of the film and $\theta = \theta(x, y)$ is a complicated function due to the variations in phase of the light reaching the film from different parts of the subject. If the subject beam alone were present, the film would be darkened in proportion to the irradiance of the subject beam. The irradiance of the subject beam is proportional to the square of the magnitude of the complex field amplitude E_S. So, we define a subject beam scaled irradiance I_S as,

$$I_S = |E_S|^2 = E_S^* E_S = [s(x, y)]^2 \tag{16-5}$$

This scaled irradiance function thus includes no information regarding phase of the subject beam. With the reference beam also present, however, the resultant amplitude E_F at each point of the film—subject to the scalar approximation—is given by

$$E_F = E_R + E_S$$

so that the scaled irradiance on the film is,

$$I_F = |E_F|^2 = (E_R + E_S)(E_R^* + E_S^*)$$

Multiplying the binomials,

$$I_F = E_R E_R^* + E_S E_S^* + E_S E_R^* + E_R E_S^*$$
$$I_F = r^2 + s^2 + E_S E_R^* + E_R E_S^* \tag{16-6}$$

The last two terms now incorporate the important function $\theta(x, y)$. Explicitly,

$$I_F = r^2 + s^2 + rse^{i(\omega t+\theta)} e^{-i(\omega t+\varphi)} + rse^{i(\omega t+\varphi)} e^{-i(\omega t+\theta)}$$
$$I_F = r^2 + s^2 + rse^{i(\theta-\varphi)} + rse^{-i(\theta-\varphi)} \tag{16-7}$$

The scaled irradiance I_F describes the hologram and is a function of x and y and so varies from point to point on the film plane. When the film is developed, its transmittance is determined by I_F.

To reconstruct the image of the scene, the hologram is situated in the reference beam again, as in the formation of the hologram (Figure 16-2b). Of course, the subject is now absent. When illuminated by the reference beam, the hologram, due to its transmittance function, modulates both the amplitude and the phase of the beam. As before,

$$E_R = re^{i(\omega t+\varphi)} \tag{16-8}$$

The resulting emergent beam can then be expressed, except for constants, in terms of the field E_H by

$$E_H \propto I_F E_R = (r^2 + s^2)E_R + r^2 se^{i(\omega t+\theta)} + r^2 e^{i(2\varphi)} se^{i(\omega t-\theta)} \tag{16-9}$$

where we have multiplied together Eqs. (16-7) and (16-8). We now interpret the three terms in Eq. (16-9) as the reconstruction of three distinct beams from the hologram. Each beam is also illustrated in Figure 16-3. The first term,

$$E_{H1} = (r^2 + s^2)E_R = (r^2 + s^2)re^{i(\omega t+\varphi)} \tag{16-10}$$

represents a reference beam modulated in amplitude but not in phase. It therefore appears like the incident beam and passes through the hologram without deviation. In analogy with the holographic grating, it corresponds to zeroth-order diffraction. The second term is

$$E_{H2} = r^2 se^{i(\omega t+\theta)} \tag{16-11}$$

which describes the subject beam, amplitude-modulated by the factor r^2. Thus the beam represents a *reconstructed wavefront* from the subject, making the same angle α relative to the reference beam. Since this beam is essentially the subject beam, it appears to come from the subject. Hence it diverges on emerging from the hologram, as if coming from a *virtual image* behind the hologram. This virtual image is what we customarily view.

The third term is given by

$$E_{H3} = r^2 e^{i(2\varphi)} se^{i(\omega t-\theta)} \tag{16-12}$$

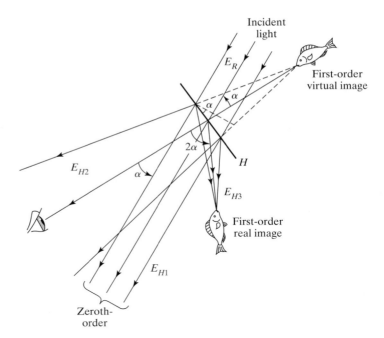

Figure 16-3 Reconstruction of hologram formed in Figure 16-2a.

and represents the subject beam, modulated in both amplitude and phase. This beam reconstructs the subject beam of Eq. (16-4) but with *phase reversal*, that is, with $e^{i\theta}$ replaced by $e^{-i\theta}$. Every delay in phase in E_S now shows up as a phase advance. The image is turned inside out. Because of phase reversal, originally diverging rays—as those in E_{H2}, which form a virtual image—become converging and focus as a *real image* on the viewing side of the hologram. The factor $e^{i(2\varphi)}$, when compared with the phase term in Eq. (16-3), indicates an angular displacement of the image direction by 2α relative to the normal to the film plane. Notice that the off-axis system illustrated in Figure 16-2a produces a hologram in which the two first-order beams are separate in direction from each other and the zeroth-order beam. The virtual image can be observed clearly, without confusion from the other beams.

The hologram made of an extended object shows the same essential features as the hologram of the point object. Photography by holography is a two-step process. Recall that in the making of a hologram, no lens is used, and the presence of the reference beam is essential. The light must have sufficient temporal coherence so that path differences between the two beams do not exceed the coherence length of the light; it must also possess sufficient spatial coherence so that the beam is coherent across that portion of the wavefront needed to encompass the scene. Of course, the holographic system must be vibration-free to within a fraction of the wavelength of the light during the exposure, a condition that is easily satisfied when high-power laser pulses of very short duration are used to freeze undesirable motion. A three-dimensional view of the object from all sides can be produced on a holographic film that is wrapped around the object on a cylindrical form, as shown in Figure 16-4. Light reaches the film both directly and with the help of a mirror at the end of the cylinder (the reference beam) and by light scattered from the object. When viewed under the same conditions, the 360° hologram in Figure 16-4 produces a view of the fish from all sides.

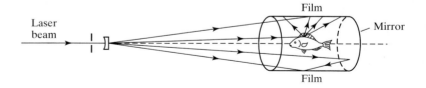

Figure 16-4 Cylindrical film surrounding the subject records a 360° hologram.

16-4 HOLOGRAM PROPERTIES

As stated earlier, the entire hologram receives light from each object point in the scene. As a result, any portion of the hologram contains information of the whole scene. If a hologram is cut up into small squares, each square is a hologram of the whole scene, although the reduction in aperture degrades the resolution of the image. The situation is much the same as when looking through a small, square aperture placed in front of a window. The same scene is viewed, though with slightly varying perspective, as the opening is moved to different parts of the window. Each view is complete, exhibiting both depth and parallax. Another interesting property of a hologram is that a contact print of the hologram, which interchanges the optically dense and transparent regions, has the same properties in use. The "negative" of a hologram alters neither fringe contrast nor spacing and hence does not modify the stored information. Furthermore, the hologram may contain a number of separate exposures, each taken with the film at a different angle relative to the reference beam and with different wavelengths of light. On reconstruction, each scene appears in its own light when viewed along the direction of the original scene, without mutual interference.

16-5 WHITE-LIGHT (RAINBOW) HOLOGRAMS

If the hologram of Figure 16-3 is viewed with a reference beam of color different than that used in its construction, it can be shown that the image of the fish will appear at a different angle. The hologram, like the holographic grating, operates as a dispersing element. If the reference beam is white light, the continuously displaced images due to different spectral regions of the light overlap and produce a colored blur. By producing a hologram that restricts the possible angular views of the subject to one through a horizontal slit, the confusion of images is reduced. In reconstruction, the hologram creates a clearer image in white light. The virtual image now appears colored. The particular color seen depends on the direction along which the hologram is viewed as the head is moved along a vertical line. No parallax is seen and the color of the image sweeps through the colors of the rainbow, from red to blue, thereby giving rise to the name *rainbow hologram*.

Embossed Holograms

Rainbow holograms, sometimes referred to as "embossed holograms," are commonly used on credit cards and in other similar *security* applications. The original hologram is recorded in a photosensitive material called a *photoresist*. When the hologram is developed, it consists of grooves in the surface of the material. A thin layer of nickel is deposited onto the hologram and then is peeled off. This yields a replica of the grooves in a metallic element, which is called a *shim*. The shim is pressed onto a material like mylar by a roller under conditions of high pressure and temperature. This *embosses* the hologram onto the mylar, which is then attached to the credit card. This method allows mass production of many embossed holograms simply and inexpensively.

Embossed holograms are used also in anticounterfeiting applications. The embossing process allows fabrication of a large number of holograms inexpensively. Such holograms are incorporated into the packaging of a product in order to confirm that it is genuine. Anticounterfeiting holograms have been used on many high-value products, including perfumes, automotive parts, and computer software.

Volume Holograms

When the thickness of the film emulsion is large compared with the fringe spacing, the hologram may be considered a three-dimensional, or *volume hologram*. The interference fringes are now interference surfaces within the emulsion that

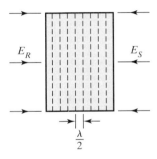

Figure 16-5 Standing wave fringe planes in a volume hologram formed by two plane waves oppositely directed.

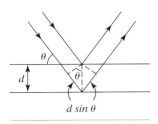

Figure 16-6 Constructive interference of reflected waves from planes of separation d is governed by the Bragg equation, $m\lambda = 2d \sin \theta$.

behave as crystalline planes of atoms in diffracting light, that is, like a three-dimensional grating. Unlike two-dimensional holograms, volume holograms can reproduce images in their original colors when illuminated with white light. To see how this comes about, consider the formation of closely spaced interference surfaces within a thick emulsion by using coherent subject and reference beams with the largest angular separation possible, 180°, as in Figure 16-5. If the two monochromatic beams E_R and E_S have undistorted plane wavefronts, for example, the standing wave pattern produces antinodal planes perpendicular to the beam directions and spaced $\lambda/2$ apart, as shown.

The maximum irradiance in these planes produces, after film development, planes consisting of excess free silver, which function as partially reflecting planes. Of course, the emulsion must itself possess a high-resolution potential to record faithfully such detail. When illuminated from the reference beam direction with white light, for instance, the developed hologram partially reflects light from each silver layer, but only light of the wavelength used in making the hologram is reinforced by such multiple reflections. The physics of the process is, of course, the same as that for X-ray diffraction from crystalline planes, governed by the Bragg equation,

$$m\lambda = 2d \sin \theta$$

and illustrated in Figure 16-6. To apply the equation to the silver layers in an emulsion λ and θ should be taken to be the wavelength and angle in the emulsion. Alternatively, if light of wavelength λ_0 in air is incident on the film at angle θ_0 one can write

$$m\lambda_0 = d \sin \theta_0$$

Since $\lambda = \lambda_0/n$ and $\sin \theta_0 = \sin \theta/n$.

Thus if a volume hologram is illuminated at a given angle θ_0, only the single wavelength that satisfies the Bragg equation locally, where planar spacing is d, is reinforced and appears as a brightly reflected beam. The thicker the emulsion and the greater the number of contributing reflecting planes, the more selective the hologram will be in reinforcing the correct wavelength. If a volume hologram is made by multiple exposures of a scene in each of three primary colors, the reconstruction process with white-light illumination can produce a three-dimensional image in full color.

Example 16-1 and Figure 16-7 illustrate the formation of fringes in a holographic grating by the interference of two argon-ion laser beams.

Example 16-1

Show that the separation d of fringes in the formation of a holographic grating, as in Figure 16-7, is given by $\lambda/(2 \sin \theta)$, where 2θ is the angle between the coherent beams in the film, $\lambda = \lambda_0/n$ is the wavelength of the beams in the film, and λ_0 is the wavelength of the beams in free space. Assume that the beams are incident symmetrically on the film's surface. If the beams are argon-ion laser beams of 488-nm wavelength and the angle between the beams is 120°, how many grooves per centimeter are formed in a plane emulsion oriented perpendicular to the fringes? Let the refractive index n of the emulsion be equal to 1.

Solution

In Figure 16-7, Beam 1 and Beam 2 are shown together with their respective wavefronts or crests $C1$ and $C2$ labeled throughout the emulsion. The angular separation of the two beams is shown as 2θ, and the constructive fringes (vertical dashed lines) formed in the emulsion are labeled as G. Examination

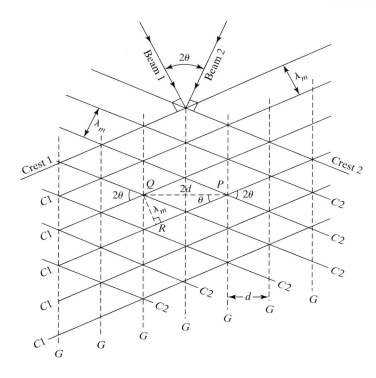

Figure 16-7 Example 16-1. Formation of constructive fringes in a holographic grating.

of the right triangle PQR in the center of the construction leads to the desired relation,

$$\sin \theta = \frac{\lambda}{2d}$$

Thus,

$$d = \frac{\lambda}{2 \sin \theta}$$

Finally, the number of grooves/cm is

$$\frac{1}{d} = \frac{2 \sin \theta}{\lambda} = \frac{2 \sin \theta}{\lambda_0 / n} = \frac{2 \sin(120°/2)}{(488 \times 10^{-7} \text{ cm})/1.6} = 56{,}790 \text{ grooves/cm}$$

16-6 OTHER APPLICATIONS OF HOLOGRAPHY

Holography offers a wide variety of fascinating applications, of which we briefly describe only a few. The hologram, itself a product of the interference of light, has been used as an alternative technique in *interferometry*, the science of using the wavelength of light and interference to measure very small optical-path lengths with precision.

Nondestructive Testing
Suppose that the hologram of the fish in Figure 16-3 and the fish itself are returned exactly to their original positions, and suppose that the same reference beam illuminates the scene. In looking through the hologram, one now sees the virtual image superimposed over the object itself. Both are viewed with the same coherent light. If no change has occurred since recording the hologram, the view appears as if the subject or the hologram *alone* was in place. Suppose, however, that the model of the fish has undergone some small changes in shape, by thermal expansion, for example. Now the direct image of the object and the holographic image are slightly different, and the light forming the two images interferes, producing fringes that measure the extent

of the change at specific locations, as in the case of Newton's rings. The present object can then be compared in *real time* with itself as it existed at an earlier time. This technique—referred to as *nondestructive testing*—is often applied to determine maximum stress points on the subject as pressure is applied, as in the case of an automobile tire, for example. The sensitivity of this technique has been dramatically demonstrated in holographic recordings of convection currents around a hot filament, compressional waves surrounding a speeding bullet, and the wings of a fruit fly in motion.

Time-Average Holographic Interferometry

This type of interferometry is used to study vibrating surfaces, where the object is moving continuously during exposure of the hologram. One makes a single hologram using an exposure time that is long compared to the period of the vibrations being studied. The resulting hologram effectively contains a large number of images mapping the motion of the vibrating surface. The pattern of interference fringes provides information on the relative vibrational amplitude as a function of position on the surface. The vibrational amplitudes of diffusely reflecting surfaces may be measured with high precision, and such measurements can be very useful for determining the modes of vibration of complex structures.

Microscopy

Another useful application of holography is in *microscopy*. When specimens of cells or microscopic particles are viewed conventionally under high magnification, the depth of field is correspondingly small. A photograph that freezes motion of the specimen captures in a focused image a very limited depth of field within the specimen. The disadvantages of this restriction can be overcome if the photograph is a hologram, which in a single snapshot contains potentially all the ordinary photographs that could be made after successive refocusings throughout the depth of the living specimen. A simple interferometric hologram is thus equivalent to many separate observations made with a conventional interferometer. The image provided by the hologram may be viewed by focusing at leisure on any depth of an unchanging field. In making a hologram with a microscope, the specimen is illuminated by laser light, part of which is first split off outside the microscope and routed independently to the photographic plate, where it rejoins the subject beam processed by the microscope optics. Furthermore, it can be shown that if the reconstructing light of wavelength λ_r is longer than the wavelength λ_s used in "holographing" the subject, a magnification given by

$$M = \left(\frac{q}{p}\right)\left(\frac{\lambda_r}{\lambda_s}\right) \tag{16-13}$$

results, where p is the object distance (subject from film) and q is the corresponding image distance (image from hologram). Object and image distances are equal when the reference and reconstructing wavefronts are both plane waves. The content of Eq. (16-13) implies, for example, that if the hologram were made with coherent X-rays and viewed with visible light, magnifications as large as 10^6 could be achieved without deterioration in resolution. This prospect has contributed to interest in developing X-ray lasers. X-ray holograms could provide strikingly detailed three-dimensional images of microscopic objects as small as viruses and DNA molecules. The ability to view a hologram with radiation of a wavelength different than that used in making the hologram offers other interesting possibilities. For example, an ultrasonic wave hologram can be used in place of medical X-rays and a radar hologram can be read with visible wavelengths. In fact, in his original work, Gabor proposed reconstruction of an electron-wave hologram with optical wavelengths in an effort to improve the resolution of electron microscopes.

Ultrasonic Holograms

The mention of *ultrasonic holograms* implies that the waves producing a hologram need not be electromagnetic in nature. Indeed, the principles of holography do not depend on the transverse character of the radiation. Because of the ability of ultrasonic waves to penetrate objects opaque to visible light, holograms formed with such waves can be very useful. Opaque bodies that are promising candidates range from the human body to archeological tombs. Structures and cavities inside can be revealed in three-dimensional images formed by ultrasonic holography. Figure 16-8 illustrates another application of ultrasonic holography that enables one to reveal objects under the surface of the ocean. $G1$ and $G2$ represent two phase-coupled generators radiating coherent ultrasonic waves. The wavefront from $G2$ is deformed by an underwater object and interferes with the undeformed reference beam from $G1$. The deformations of the water surface represent an acoustic hologram. If this region is illuminated with monochromatic light, the light diffracted from the deformations can be photographed and converted into a visual image of the underwater object. The potential offered for submarine detection is an obvious military application.

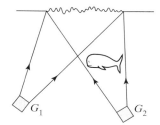

Figure 16-8 Deformations in the surface of the water due to two coherent ultrasonic waves.

Holographic Data Storage

The storage of data in a hologram offers tremendous potential. Because data can be reduced by holographic techniques to dimensions of the order of the wavelength of light, volume holograms can be used to record vast quantities of information. As the hologram is rotated, new exposures can be made. Photosensitive crystals, such as potassium bromide with color centers or lithium niobate can be used in place of thick-layered photoemulsions. Because information can be reduced to such tiny dimensions and the crystal can be repeatedly exposed after small rotations that take the place of turning pages, it is said that all the information in the Library of Congress could theoretically be recorded on a crystal the size of a sugar cube! Information may, of course, be recorded in digital form and thus read by a computer, so holographic storage offers a means of providing computer storage. In conjunction with the optical transport of computer information through optical fibers, information handling, storage, and retrieval can all be done using light. A fascinating aspect of holographic data storage lies in its reliability. Since every data unit is recorded throughout the volume of the hologram, in unique holographic fashion, damage to a portion of the hologram, although affecting the signal-to-noise level of the reconstructed image, does not affect its reliability. Information is not lost, as would be the case in other memory devices, where every bit of information has its own unique storage coordinates.

In a reciprocal sense, computers are used to advantage in the science of holography by making possible the construction of *synthetic holograms* that faithfully represent three-dimensional objects. The object is first defined mathematically by specifying its coordinates and the intensity of all its points. The computer calculates the complex amplitude that is the sum of radiation due to the object and the reference wave and then directs the drawing of the hologram, which can be photographed and reduced to the appropriate fringe spacings required. For example, an ideal aspheric wavefront can be created synthetically to serve as a model against which a mirror may be shaped, using interference between the two surfaces as a guide to making appropriate corrections.

Holocameras

A device called a *holocamera* does not use photographic film, but rather materials like thermoplastics to record a holographic image. The image development is done by *electrical and thermal* means, without the need for wet chemical processing, and can be accomplished in a few seconds without repositioning the recording. Using this process, the hologram can be developed and viewed quickly and so the holocamera can be used for rapid inspection and analysis in an industrial environment. The finished hologram can be erased by heating and the thermoplastic recording material can be reused hundreds of times.

Pattern Recognition

Another area in which holograms may be very useful is in *pattern recognition*. Briefly, the procedure is as follows. A text is scanned, for example, for the presence of a particular letter or word. Light from the text to be searched is passed through a hologram of the letter or word to be identified in an appropriate optical system. The presence of the letter is indicated by the formation of a bright spot in a location that indicates the position of the letter in the text. The hologram acts as a matched filter, recognizing and transmitting only that spatial spectrum similar to the one recorded on it. The technique can be applied to holographic reading of microfilms, for example. Military applications include the use of a memory bank of holograms of particular objects or targets constructed from aerial photographs. Weapons could, by pattern recognition, select proper targets. It has also been suggested that robots could identify and be directed toward appropriate objects in the same way. Pattern recognition is discussed further in Chapter 21.

Holographic Optical Elements

The interference pattern produced by two spherical waves with different radii of curvature can be recorded to produce a holographic device that acts like a lens. Such a recording is a sinusoidal zone plate that—as we have seen in Chapter 13—acts as a positive lens. One can also fabricate holographic optical elements (HOEs) that perform the function of prisms, mirrors, gratings, and so on. Holographic optical elements are generally lighter than the optical components they replace, which proves useful in a number of applications. For example, in one such application, rotating HOEs can be used to scan laser beams on spaceborne platforms in *lidar* (light detection and ranging) systems to monitor atmospheric profiles of wind, aerosols, clouds, temperature, and humidity. Holographic optical elements are also used in optical systems to correct aberrations, in supermarket scanners, and in *heads-up displays* for aircraft pilots. In the latter application, instrument readings are projected so that they seem to be floating in space, allowing the pilot to retain a clear "heads-up" view of the scene in front of the aircraft. High-resolution, holographically recorded gratings have been used—instead of more expensive conventional gratings—in optical spectrometers.

PROBLEMS

16-1 Use Eq. (7-14) for the superposition of two unequal beams to show that the irradiance pattern of a Gabor zone plate (the hologram of a point source) is given approximately by

$$I = A + B \cos^2(ar^2)$$

where $A = I_1 + I_2 - 2\sqrt{I_1 I_2}$, $B = 4\sqrt{I_1 I_2}$, and $a = \pi/(2s\lambda)$. Here, I_1 and I_2 are the irradiances due to the reference and signal beams, respectively, s is the distance of the object point from the film, and λ is the wavelength of the light. For the approximation, assume the path difference between the two beams is much smaller than s, so that we are looking at the inner zones of the hologram.

16-2 **a.** Show that if the local ratio of reference to subject beam irradiances is a factor N at some region of a hologram, then the visibility of the resulting fringes is $2\sqrt{N}/(N + 1)$.

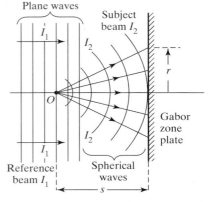

Figure 16-9 Problem 16-1.

b. What is the fringe visibility in a region where the irradiance of the reference beam is three times that of the subject beam?

16-3 A conventional 1-mm-thick compact disc (CD) can store 1 Gb of information in the form of digital data, with all of the data stored in the top 1-μm-thick layer of the CD. How much information could be stored in the 1-mm-thick CD if the data could be stored holographically throughout the entire CD at the same information density?

16-4 The angle between the signal and reference beams during construction of a hologram is 20°. If the light is from a He-Ne laser at 633 nm, what is the fringe spacing? Assume a refractive index of 1 for the emulsion. (See Example 16-1.).

16-5 Suppose a hologram is to be made of a moving object using a 1-ns laser pulse at a wavelength of 633 nm. What is the permissible speed such that the object does not move more than $\lambda/10$ during the exposure?

16-6 During the construction of a hologram, a beam splitter is selected that makes the amplitude of the reference beam eight times that of the signal beam at the emulsion. What is the maximum ratio of beam irradiances there?

16-7 Let us suppose that as a theoretical limit, 1 bit of information can be stored in each λ^3 of hologram volume. At a wavelength of 492 nm and a refractive index of 1.30, determine the storage capacity of 1 mm³ of hologram volume.

16-8 A volume hologram is made using oppositely directed monochromatic beams of coherent, collimated laser light at 500 nm, as in Figure 16-5. Assume a film refractive index of 1.6.

a. Determine the spacing of the developed silver planes within the emulsion.
b. What wavelength is reinforced in reflected light when white light is incident normally on the hologram?
c. Repeat (b) when the angle of incidence from air (relative to the normal) is 30°.

16-9 Two beams of planar wavefront, 633-nm coherent light, whose directions are 120° apart, strike a photographic emulsion of index 1.6.

a. Sketch the arrangement, showing the orientation of the planes of constructive interference within the emulsion.
b. Determine the planar spacing of the developed volume hologram.
c. At what angle of incidence relative to the silver planes is a wavelength of 450 nm reinforced?

16-10 Suppose that the blue component of a white-light hologram is formed as in Figure 16-5, using light of 430-nm wavelength. If emulsion shrinkage is 15% during processing, what wavelength is reinforced by the blue-light fringes on reconstruction? How does this affect the holographic image under white-light viewing?

16-11 A hologram is constructed with ultraviolet laser light of 337 nm and viewed in red laser light at 633 nm.

a. If the original reference beam and the reconstructing beam are both collimated, what is the magnification of the holographic image, compared with the original subject?
b. What magnification would result if coherent X-rays of 1 Å wavelength were available to construct the hologram?

16-12 a. Verify that the reconstructed wavefront from the hologram of a point source produces both the real and virtual images shown in Figure 16-10. First, find the irradiance at the film due to the superposition of a plane and a spherical wave. Then, find the amplitude of the light transmitted by the developed film when irradiated by the reference beam. Interpret the terms as done in the discussion of a hologram of a three-dimensional subject.

b. Show that the phase delay of the diverging subject beam, at a point on the film a distance y from the axis, is given by $\pi y^2/\lambda d$, where d is the distance of the point source from the film. This result follows when $y \ll d$. Show also that reversal of the phase angle produces a converging spherical wavefront associated with the real image on reconstruction.

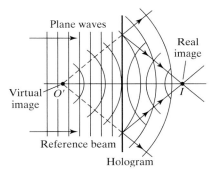

Figure 16-10   Problem 16-12.

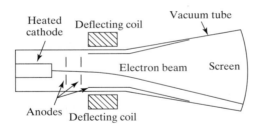

Heated cathode Deflecting coil Vacuum tube Electron beam Screen Anodes Deflecting coil

17 *Optical Detectors and Displays*

INTRODUCTION

In this chapter we give a qualitative description of the more common types of optical detectors and displays. Any device that produces a measurable physical response to incident radiant energy is a detector. The most familiar detector is, of course, the eye. Whereas the eye provides a qualitative and subjective response, the detectors discussed in this chapter provide a quantitative and objective response. In view of the unique role played by the eye in human vision, it is treated separately in Chapter 19. Most detectors may be classified as being either a *thermal* detector or a *quantum* detector. We begin this chapter with a description of thermal and quantum detectors of optical radiation and proceed to an introduction of some of the important figures of merit associated with quantum detectors. We conclude the chapter with an overview of three important types of optical displays—the CRT, the LCD, and the plasma display.

17-1 THERMAL DETECTORS OF RADIATION

When the primary measurable response of a detector to incident radiation is a rise in temperature, the device is a *thermal detector*. The receptor in thermal detectors is typically a blackened surface that efficiently absorbs at all wavelengths. In this section we will describe several basic types of thermal detectors.

Thermocouples and Thermopiles
A thermocouple is a device in which an increase in temperature at a junction of two dissimilar metals or semiconductors generates a voltage (Figure 17-1a). When the effect is enhanced by using an array of such junctions in series, the device is called a *thermopile* (Figure 17-1b).

Bolometers and Thermistors

Thermal detectors also include bulk devices that respond to a rise in temperature with a significant change in resistance. Such an instrument may employ as its sensitive element either a metal (*bolometer*) or, more commonly, a semiconductor (*thermistor*). Typically, two blackened sensitive elements are used in adjacent arms of a bridge circuit, one of which is exposed to the incident radiation. The imbalance in the circuit, due to the change in resistance, is indicated by a change of current in the bridge circuit.

Pyroelectric Detectors

The *pyroelectric* effect can be exploited in order to detect radiation. Certain metals like lithium tantalate or triglycine sulfate (TGS) exhibit the *pyroelectric* effect in that a temperature-dependent charge separation exists between opposite ends of the metal. This charge separation is a result of a permanent macroscopic polarization associated with the metal. The pyroelectric metal behaves like a capacitor whose charge is a function of the temperature. Incident radiation increases the temperature of the pyroelectric material and hence changes the charge on the surface of the metal. This sort of sensor is a common component of motion detectors.

Pneumatic or Golay

A *Golay cell* measures the thermal expansion of a gas induced by radiation incident on the gas enclosure. A schematic of such a cell is shown in Figure 17-2. Radiation is absorbed by a blackened membrane and as a result heat is transmitted to a gas in an airtight chamber. The heat flow into the gas causes an increase in gas pressure, which is typically detected by the deflection of a flexible mirror attached to the cell, as shown in Figure 17-2.

Thermal detectors are generally characterized by a slow response to changes in the incident radiation. If the detector is expected to follow a changing input signal, such as a pulse, thermal detectors are not as desirable as the faster-responding quantum detectors to be discussed next. The speed of response is described by a *time constant*, a measure of the time required to regain equilibrium in output after a change in input. Many of the quantum detectors discussed in the next section are better suited to high-frequency operation.

17-2 QUANTUM DETECTORS OF RADIATION

Quantum detectors respond to the rate of incidence of photons rather than to thermal energy. Photons interact directly with the electrons in the detector material. In this section we describe photoemissive detectors, photoconductive detectors, and photodiodes.

Photoemissive Detectors

When the measurable effect is the release of electrons from an illuminated surface, the device is called a *photoemissive detector*. A photosensitive surface, typically containing alkali metals, absorbs incident photons that transfer enough energy to enable some electrons to overcome the *work function* and escape from the surface. If the photoemitted electrons are simply collected by a positive-biased anode in an evacuated tube, enabling a current to be drawn into an external circuit, the detector is called a *diode phototube*. When the signal is internally amplified by secondary electron emission, the detector is a *photomultiplier*; see Figure 17-3. In this case the primary photoelectrons are accelerated, so that as a result of a sequence of collisions, each multiplying the current by the addition of secondary electrons, an avalanche of electrons becomes available at the output corresponding to each primary photoelectron.

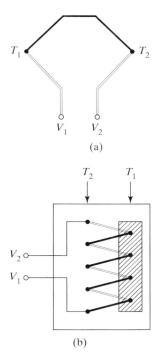

Figure 17-1 (a) Thermocouple made of dissimilar materials (dark and light lines) joined at points T_1 and T_2, where a difference in temperature produces an emf between terminals V_1 and V_2. (b) Thermopile made of couples in series. Radiation is absorbed at the junctions T_1 in thermal contact with a black absorber and thermally insulated from the junctions T_2.

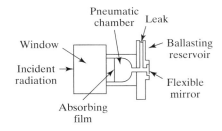

Figure 17-2 Golay pneumatic infrared detector. (Oriel Corp., General Catalogue, Stratford, Conn.)

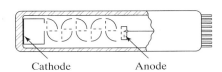

Figure 17-3 One type of photomultiplier tube structure. Electrons photoemitted from the cathode are accelerated along zigzag paths down the tube so as to strike each of the curved dynode surfaces, each time producing additional secondary electrons. The multiplied current is collected at the anode.

Another means of amplification, used in the *gas-filled photocell*, allows the generation of additional electrons by ionization of the residual gas. In the case of energetic photons ($\lambda < 550$ nm), the sensitivity of photoemissive detectors is sufficient to allow the counting of individual photons. Such detectors possess superior sensitivity in the visible and ultraviolet spectral ranges.

Photoconductive Detectors

For wavelengths in the infrared, over 1 μm, photoemitters are not available and *photoconducting detectors* are used. In these detectors, photons absorbed into thin films or bulk material produce additional free charges in the form of electron-hole pairs. Both the negative (electrons) and positive (holes) charges increase the electrical conductivity of the sample. Without illumination, a bias voltage across such a material with high intrinsic resistivity produces a small or "dark" current. The presence of illumination and the extra free-charge carriers so produced effectively lower the resistance of the material, and a larger photocurrent results. Semiconducting compounds cadmium sulfide (CdS) and cadmium selenide (CdSe) are often used in the visible and near-infrared regions; farther out in the near-infrared region, the compounds lead sulfide (PbS) (0.8 to 3 μm) and lead selenide (PbSe) (1 to 5 μm) are popular.

Junction Photodiodes

One of the most common types of photodetectors is the *semiconductor photodiode*. The central component of the simplest type of photodiode is a *p-n* junction, which is a junction between doped *p*-type (rich in positive charge carriers, holes) and doped *n*-type (rich in negative charge carriers, electrons) materials, often silicon. Doping involves adding small amounts of an impurity to the semiconductor to provide either an excess (*n*-type) or deficiency (*p*-type) of conduction electrons. When *p*-type and *n*-type materials are brought into contact, the initially unequal concentration of holes and electrons on opposite sides of the junction induces a diffusion of electrons from the *n*-type material into the *p*-type material and a diffusion of holes from the *p*-type material into the *n*-type material. In this way a voltage difference occurs across the junction that inhibits the diffusion of more carriers. In this situation the *n*-type side of the junction will be at a higher voltage than the *p*-type side of the junction. Thus an electric field directed from the *n*-type to the *p*-type material is created in the junction region. In a photodiode such a *p-n* junction is sandwiched between electrical contacts, as shown in Figure 17-4. An optical window allows electromagnetic radiation to fall on the *p*-type side of the junction.

When photons of energy exceeding the *band-gap energy* of the semiconductor are absorbed in the vicinity of the junction, the created electron-hole pairs are separated by the electric field in the junction region, causing a change in voltage, the *photovoltaic effect*. In this so-called *photovoltaic mode* of operation, depicted in Figure 17-5a, a change in the amount of incoming light leads to a change in the open-circuit voltage across the *p-n* junction. The solar cell and the photographic exposure meter are two well-known applications of the photodiode operating in the photovoltaic mode. More commonly, photodiodes are operated in the *reverse-biased or photoconductive mode*. In this mode, illustrated in Figure 17-5b, the *n*-type side of the junction is connected to the positive terminal of an external voltage source and the *p*-type side of the junction is attached to the negative terminal of the external voltage source. The reverse-biased external voltage increases the electric field in the junction region, which in turn increases the rate at which the electrons and holes generated by absorption of the incident radiation are swept out of the junction region. A strong reverse bias therefore can reduce the response time of the photodiode. In the photoconductive (reverse-biased) mode the current induced in the load resistor has a nearly linear relationship with the incoming irradiance. In the photovoltaic (forward-biased) mode, the open-circuit voltage has a more complicated logarithmic dependence on the incident irradiance.

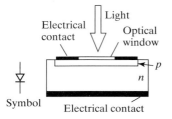

Figure 17-4 Junction photodiode. The diode symbol indicates the orientation of the diode. Note that the arrow in the diode symbol points from the *p*-side to the *n*-side of the diode.

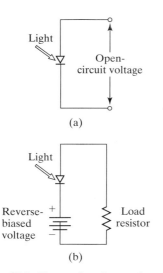

Figure 17-5 Two modes of operation of a junction photodiode. (a) Photovoltaic mode. The open-circuit voltage changes as the optical power incident on the photodiode changes. (b) Photoconductive (reverse-biased) mode. The current through the load resistor changes as the optical power incident on the photodiode changes.

Silicon-based photodiodes are sensitive to optical radiation with wavelengths ranging from 190 to 1100 nm, and Indium-Gallium-Arsenide (InGaAs) *p-n* junction photodiodes can be used to detect radiation with wavelengths ranging from 800 to 1800 nm. A variation of the photovoltaic cell, the *avalanche diode*, provides an internal mechanism of amplification that results in enhanced sensitivity out to around 1.5 μm. In the region of 1 to 8 μm, the semiconductor compounds PbS, PbSe, and PbTe (lead telluride) possess a large photovoltaic effect and greater sensitivity than the thermocouple or the ordinary bolometer. As with other detectors that are designed to operate at longer wavelengths, photodiodes are often cooled to enable operation with greater sensitivity. Many photodiodes used in the visible to near infrared region of the spectrum use *p-i-n* junctions. In this arrangement a region of *intrinsic* (undoped) material is layered between the *p*-type and *n*-type materials. The intrinsic layer is a region of high electric field, and carriers produced by light absorption in this region are quickly swept away to the heavily doped regions. This rapid movement of the carriers produced in the intrinsic region decreases the response time of the detector. Photodiodes using *p-i-n* junctions have response times on the order of 100 ps, allowing for the detection of signals that vary at rates on the order of 10 GHz.

17-3 IMAGE DETECTION

Images can be detected and recorded on a surface sensitive to incoming radiation. Until recently, a *photographic film* or *plate* was the most common image-recording medium. *Charge-coupled devices (CCDs)*, in which the incoming radiation is detected by a two-dimensional array of photodiodes, are becoming the preferred choice for image detection. These two image detection systems are discussed below.[1]

Photographic Film

Until the recent widespread use of digital cameras, photographic film was the most common medium used for image detection and recording. Photographic emulsions are available with spectral sensitivity that extends from the X-ray region into the near infrared at around 1.2 μm. The sensitive material is an emulsion of silver halide crystals or grains. An incident photon imparts energy to the valence electron of a halide ion, which can then combine with the silver ion, producing a neutral silver atom. Even before developing, the emulsion contains a latent image, a distribution of reduced silver atoms determined by the variations in radiant energy received. The latent image is then "amplified," so to speak, by the action of the developer. The resulting chemical action provides further free electrons to continue the reduction process, with the latent image acting as a catalytic agent to further action. The density of the silver atoms, and thus the opacity of the film, is a measure of both the irradiance and the time of exposure, so that photographic film, unlike many other detectors, has the advantage of light-signal integration. Even weak radiation can be detected by the cumulative effect of a long exposure.

CCD Detectors

Two-dimensional arrays or panels of photodiodes can be used to record images. Each photodiode, or *MOS* (metal-oxide-semiconductor) device, responds to the incident radiation to provide one *pixel* (picture element) of output. In a simplistic arrangement, each photodiode could have separate wires for voltage

---

[1] Image detection is a rapidly advancing field. Complementary metal-oxide-semiconductor (CMOS) detectors and charge-injection devices (CID's) are two additional, increasingly important, digital imaging technologies. Readers should consult current opto-electronic trade journals in order to gain an up-to-date understanding of image detection technologies.

Figure 17-6 CCD panel in a wire-bonded package. (Photograph courtesy of NASA.)

supply and signal readout. This sort of arrangement is in fact used for detectors with a relatively small number of pixels. For applications requiring a large number of pixels, the number of wires required by such a design becomes unwieldy. The requirement of a large number of wires is mitigated by the use of a charge-coupled device (CCD). A typical CCD panel in a wire-bonded package is shown in Figure 17-6. When a CCD is exposed to light, each of the discrete devices fabricated on a silicon chip, say, stores photo-induced charge in a potential well created by an applied gate voltage. A CCD panel of area 1 cm^2 might contain 500,000 individual photodiodes. The stored charge contributed by each pixel, a measure of the local irradiance, is electronically scanned to produce an electronic record of the image. Scanning and readout is performed by charge transfer along each pixel row of such a device. The electronic pulses corresponding to the photo-induced charge accumulated near each pixel in a given row arrive sequentially at the end of the row. This pulse sequence provides a record of the local irradiance measured by each pixel in that row. Typically, the record of the irradiance of each pixel is stored digitally. Once each row in a CCD array is scanned in this fashion, a digital record is formed that can be stored in the memory of the device. Digital cameras work in this fashion.

The photodetectors in CCD arrays have a *quantum efficiency* of as much as 80%. That is, as many as 80% of the photons incident on a photodiode in the CCD array are converted to signal electrons. By contrast, the quantum efficiency of photographic film is about 2%. Not surprisingly, the higher quantum efficiency associated with CCD detectors led to their early adoption by astronomers seeking to form and record images of faint astronomical sources. As mentioned above, digital cameras use CCD arrays (or related technologies), instead of photographic film, to record an image. By *repeated* reading of the electronic pulse sequence of the end of each row in the CCD array, the irradiance as a function of position on the array *and of time* can be formed. Thus a CCD array can be (and is) used in digital video cameras, which update the image some 30 times a second. Today digital cameras typically use CCD arrays with resolutions ranging from 1024×768 pixels to 3032×2008 pixels. Even higher resolutions are available for professional use.

A *Bayer mask* can be used in conjunction with a CCD array in order to record color images. A Bayer mask is an array of color filters, with each miniature filter designed to cover one pixel of the CCD array. The color filters in the Bayer mask form a mosaic of 2×2 submasks, each containing four color filters—one red, one blue, and two green. This arrangement is used because the eye is more sensitive to green light than to either red or blue light and doubling the information obtained in the green portion of the spectrum allows for the production of an image that the eye perceives as "true" color. Placing a Bayer mask over a CCD array permits the array to record both irradiance and color information. Since 4 pixels are used to record the color information, the irradiance is recorded with higher resolution than is the color of the image. Better color resolution can be obtained by using three separate CCDs in the camera. A prism or some other device can be used to split an image into red, blue, and green components, which are then imaged on separate CCDs. The digital information from the separate CCDs are then combined to form a color image.

17-4 OPTICAL DETECTORS: NOISE AND SENSITIVITY

In addition to knowledge of the spectral range over which a particular detector is effective, it is important to know the actual sensitivity or, more precisely, the *responsivity, R*, of the detector, defined as the ratio of output to input:

$$R = \frac{\text{output}}{\text{input}} \qquad (17\text{-}1)$$

Input is typically some measure of irradiance and output is almost always a current or voltage. For the responsivity to be a useful specification of a detector, it should be constant over the useful range of the instrument. In other words, the detector, together with its associated amplifier and circuits, should provide a linear response, with output proportional to input. In general, however, responsivity is not independent of wavelength. Curves of responsivity versus wavelength are provided with commercial detectors. When the responsivity is a function of λ, the detector is said to be *selective*. The scaled sensitivity of a CdS photoconducting cell is shown in Figure 17-7. A *nonselective* detector is one that depends only on the radiant flux, not on the wavelength. Thermal detectors using a blackened strip as a receptor may be nonselective; however, entrance windows to such devices may well make them selective.

The *detectivity, D*, of a detector is the reciprocal of the minimum detectable power, called the *noise equivalent power, NEP* of the detector:

$$D = \frac{1}{NEP} \qquad (17\text{-}2)$$

The minimum detectable power is limited by the *noise* inherent in the operation of the detector. The noise is that part of the signal or output not related to the desired input. Many sources of noise in quantum detectors exist,[2] including *shot noise* or *quantum noise*, which arises because of the statistical nature of the conversion of a photon in the incident field to an electron in the detector system. Another type of noise, called *Johnson noise*, due to the thermal agitation of current carriers, is found in all photodetectors. In addition, *generation* and *recombination noise* due to *statistical fluctuations* of current carriers occurs in photoconductors. Mere amplification of a signal is not useful when it does not distinguish between signal and noise and results in the same signal-to-noise ratio, just as the mere magnification of an optical image is not useful since it does not clarify the object details. The fundamental lower limit of the noise inherent in the detection process is set by quantum or shot noise, but in practice the other technical noise sources often determine the noise equivalent power for the detector. (See Problems 17-4 and 17-5.)

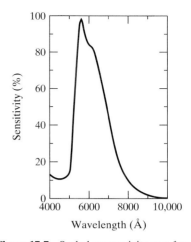

Figure 17-7 Scaled responsivity as a function of wavelength for a CdS photoconducting cell. The peak response at 5500 Å closely matches the response of the human eye.

17-5 OPTICAL DISPLAYS

In this section we describe briefly three different technologies used to display optical images in computer and TV monitors.

Cathode-Ray Tube (CRT) Displays

The *cathode-ray tube* or *CRT* display was until recently used in nearly all televisions, computer displays, and video monitors. A schematic of this device is shown in Figure 17-8. In a CRT display beams of electrons ("cathode rays") emitted from an electron gun into a vacuum tube are steered by electric or magnetic fields and strike a phosphorescent surface that emits light from the point at which the electrons strike the surface.

---

[2]See, for example, Christopher C. Davis, *Lasers and Electro-Optics* (Cambridge, UK: Cambridge University Press, 1996), Ch. 22.

Figure 17-8 CRT display. An electron beam emitted by a heated cathode is steered by electric and magnetic forces to a particular point on a screen covered with a phosphorescent material that emits light when struck by electrons. The beam can be scanned back and forth across the screen, covering the entire screen 30 times each second.

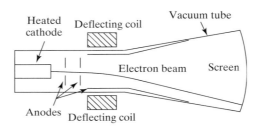

In modern video monitors the entire phosphorescent front of the tube is repetitively scanned in a fixed *raster* pattern. The image is formed by modulating the intensity of the scanning electron beam. Color CRT displays typically use three different phosphors packed closely together in strips or clusters in a repetitive pattern on the display screen. One of these phosphors in a cluster emits blue light, one emits red light, and one emits green light when struck by an electron beam. Three different electron beams are typically used and each beam reaches "dots" of only one of the three types of phosphors on the screen due to a grille or mask that blocks the beams from striking the wrong type of phosphor. In this way, different colors can be displayed by mixing different intensities of the primary red, green, and blue colors. Recently, several different technologies have been developed that allow for displays that are less bulky and/or use less power than CRT displays. Two such technologies are discussed below. Still, many people prefer the quality of the image displayed by the best CRT monitors to those produced by the newer technologies.

Liquid-Crystal Displays (LCDs)

The atoms or molecules in a liquid are arranged in a completely disordered fashion, whereas those in a (solid) crystal are regularly arranged in a periodic fashion. As the name implies, *liquid crystals* are materials with properties intermediate between those of crystals and liquids. A variety of types of liquid crystals exist but here we focus on a description of the *nematic* liquid crystal whose rodlike molecules are randomly positioned but have a common orientation induced by intermolecular interactions. The common orientation of the molecules in a liquid crystal allows the liquid crystal to interact differently with different polarizations of light passing through the crystal. This characteristic is exploited in a *liquid-crystal display (LCD)*.

One means of creating a liquid-crystal display is illustrated in Figure 17-9. A nematic liquid crystal cell is placed between crossed polarizers as shown in Figure 17-9a. The nematic liquid-crystal cell consists of the liquid crystal positioned between two glass plates. As indicated in the figure, the inner surface of each glass plate is prepared with nanometer-wide scratches parallel to the transmission axis of the polarizer adjacent to the glass plate. The molecules of the liquid crystal in contact with the glass sheet tend to align with the scratches in the sheet. The influence of the sheets with perpendicular scratches on either end of the liquid crystal and the long-term order of the liquid crystal leads to the twisted orientation of the molecules shown in Figure 17-9a. A detailed analysis[3] shows that such a *twisted nematic cell* causes the polarization of the light to rotate, as it propagates though the liquid crystal, so as to maintain an electric field polarization directed along the long axis of the liquid-crystal molecules. Thus, vertically polarized light entering the liquid crystal will emerge from the liquid crystal polarized along the horizontal direction and so pass through the second horizontal polarizer on the exit side of the twisted nematic cell. In the configuration of Figure 17-9a, light would be transmitted through the cell. The cell can be made nontransmitting by applying a

---

[3]B. E. A. Saleh, and M. C. Teich, *Fundamentals of Photonics* (New York: John Wiley & Sons, Inc., 1991), Ch. 6.

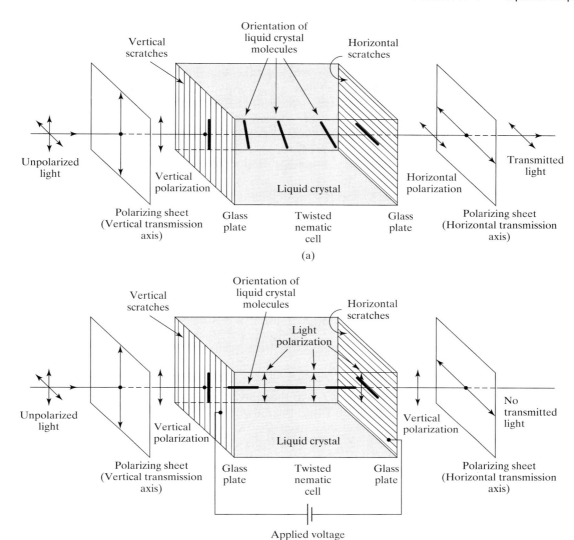

Figure 17-9 Operation of a liquid-crystal display (LCD). (a) With no applied voltage the liquid-crystal molecules change orientation through the twisted nematic cell. The light polarization rotates so as to be aligned along the long axis of the liquid-crystal molecules and so that vertically polarized light entering the twisted nematic cell exits the cell as horizontally polarized light. In this configuration, the LCD transmits light. (b) When sufficient voltage is applied to the cell, the liquid-crystal molecules (except for those adjacent to the scratched glass plates) align with their long axis along the field direction. In this case, the polarization of the light does not change as it progresses across the cell and so vertically polarized light entering the twisted nematic cell is blocked by the second polarizer, which has a horizontal transmission axis.

voltage across the cell. As indicated in Figure 17-9b, such a voltage acts to align the molecules along the direction of the electric field between the glass plates. In this configuration, the liquid crystal does not alter the orientation of the vertical polarization of the light passing through the nematic cell and so the light is blocked by the action of the polarizer with a horizontal transmission axis on the opposite end of the cell. Thus the cell can appear bright or dark depending upon whether or not a voltage is applied across the cell.

In a *passive matrix LCD*, an electronic grid is used to control the voltage across (and so the transmission through) individual pixels in the LCD display. In an *active matrix LCD*, thin-film transistors (TFTs) control the voltage across each pixel in the display. Grayscale images can be formed by either varying the voltage across a pixel, so that the light is neither completely blocked nor transmitted, or by pulsing the voltage at each pixel, so that the length of the transmission pulse leads to differences in perceived brightness of the pixel. Color displays can be made by placing a mask containing an array of red, green, and blue filters arranged in groups of three "subpixels." Controlling the brightness of the light transmitted through the different color subpixels allows for the generation of the complete range of "natural" colors.

Calculator and watch displays often operate with ambient light. In one arrangement for these displays, a mirror is placed after the second polarizing glass sheet in the twisted nematic cell. With no voltage applied across a given pixel, half of the unpolarized ambient light first passes through the vertical polarizer at the input side of the cell, then the polarization of this light is rotated so that it passes the horizontal polarizer at the end of the cell, reflects from the mirror, and retraces its steps so that vertically polarized light emerges from the display panel. When a voltage is applied across a pixel, the liquid crystal does not rotate the polarization of the light and so light is blocked by the action of the crossed polarizers on the ends of the twisted nematic cell. No light is reflected through these pixels and so they appear dark. These dark pixels are used to form the patterns of numbers and letters used in the calculator and watch displays. In addition to those outlined above, many other arrangements involving the liquid crystal and polarizing sheets can be used as elements in displays and as optical modulators and switches.

Flat-panel LCDs are much less bulky and can consume less power than CRT displays with comparably-sized viewing screens. The primary disadvantages of LCDs are that the response time of a nematic LCD is slower than that of a CRT and the angle of view is somewhat limited as the contrast is reduced when the display is not viewed "head on."

Plasma Displays

In order to increase the size of a CRT display, the length of the cathode-ray tube must be correspondingly increased in order for the electron beam to have access to all portions of the CRT screen. In contrast, a *flat-panel plasma display* may have a screen size as large as the largest CRT displays and yet only be a few inches thick. The plasma display essentially consists of hundreds of thousands of tiny neon and xenon gas cells (fluorescent lights) sandwiched between electrodes and glass plates. In a color plasma display, gas cells are grouped into sets of three subpixels. The three subpixels are coated with red, green, or blue phosphor. A voltage placed across an individual subpixel causes the gas in the cell to ionize (that is, causes a *plasma* to be formed) and to subsequently emit ultraviolet light, which in turn strikes the phosphor coating causing the emission of red, green, or blue light. The chief advantages of plasma displays are their thin design, inherent brightness, and wide viewing angle.

PROBLEMS

17-1 Given that the semiconductor germanium has a band-gap energy of 0.67 eV, find the longest wavelength that will be absorbed by a germanium photoconductor.

17-2 A certain photodiode generates one electron for every 10 photons of wavelength 0.9 μm incident on the detector.

    a. What is the quantum efficiency of this detector?

    b. What is the responsivity in A/W (at this wavelength) of this detector?

17-3 The responsivity of an InGaAs *p-i-n* photodiode is 0.8 A/W at a wavelength of 1.5 μm. What photocurrent is

generated by this detector when an electromagnetic field of irradiance 0.1 μW and wavelength 1.5 μm is incident on the detector?

17-4 Due to quantum fluctuations, laser fields have an inherent uncertainty Δn in the number of photons n contained in the field. For an ideal laser field, this inherent uncertainty is given by

$$\Delta n = \sqrt{n}$$

where $\bar{n}$ is the mean number of photons in the field. A signal cannot be extracted from noise unless the mean signal photon number $\bar{n}$ exceeds the uncertainty in photon number $\sqrt{\bar{n}}$. This requirement gives

$$\bar{n} > \sqrt{\bar{n}} \quad \text{or} \quad \bar{n} > 1$$

Thus the minimum optical signal that can be extracted from the inherent quantum noise in an ideal laser field must contain, on average, at least one photon per detector sampling time. Use this relation to estimate the minimum detectable signal power in an ideal laser field of wavelength 1.5 μm using a detector with a detection observation time of 1 μs.

17-5 The noise equivalent power *NEP* (at a wavelength of 1.5 μm) of an InGaAs *p-i-n* photodiode using a detection observation time of 1 μs is about 4×10^{-11} W.

  a. Calculate the number of photons arriving at the detector in one observation time in an optical signal of wavelength 1.5 μm with a power equal to the noise equivalent power of the detector.

  b. Is the noise equivalent power for this detector a result of the intrinsic quantum fluctuations in the incident laser field? (Refer to the discussion in Problem 17-4.)

17-6 Discuss how a twisted nematic liquid crystal between crossed polarizers can be used as a voltage-controlled irradiance modulator.

17-7 Research and describe the manner in which a nematic liquid crystal placed between transparent glass sheets with parallel scratches can be used as a voltage-controlled phase modulator.

17-8 A thin lens is used to image an object 1 m from the lens onto a CCD array 10 cm from the lens. What must the pixel spacing on the CCD array be so that features 1 mm apart on the object can be distinguished in the CCD image? How many pixels would be in such a CCD array of dimensions 2 cm $\times$ 2 cm?

17-9 A photodetector has a saturation photocurrent of 10 μA and a responsivity of 100 mA/W. What is the optical saturation power for this detector?

17-10 Rotate a Polaroid sheet (Polaroid sunglasses will work) in front of a calculator display. Note and carefully explain your observations.

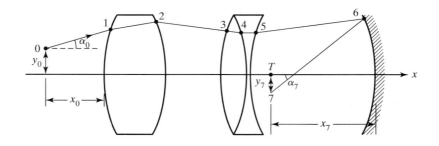

18 *Matrix Methods in Paraxial Optics*

INTRODUCTION

This chapter deals with methods of analyzing optical systems when they become complex, involving a number of refracting and/or reflecting elements in trainlike fashion. Beginning with a description of a single *thick lens* in terms of its *cardinal points*, the discussion proceeds to an analysis of a train of optical elements by means of multiplication of 2×2 matrices representing the elementary refractions or reflections involved in the train. In this way, a *system matrix* for the entire optical system can be found that is related to the same cardinal points characterizing the thick lens. Finally, computer ray-tracing methods for tracing a given ray of light through an optical system are briefly described.

18-1 THE THICK LENS

Consider a spherical *thick lens*, that is, a lens whose thickness along its optical axis cannot be ignored without leading to serious errors in analysis. Just when a lens moves from the category of *thin to thick* clearly depends on the accuracy required. The thick lens can be treated by the methods of Chapter 2. The glass medium is bounded by two spherical refracting surfaces. The image of a given object, formed by refraction at the first surface, becomes the object for refraction at the second surface. The object distance for the second surface takes into account the thickness of the lens. The image formed by the second surface is then the final image due to the action of the composite thick lens.

The thick lens can also be described in a way that allows graphical determination of images corresponding to arbitrary objects, much like the ray rules for a thin lens. This description, in terms of the so-called *cardinal points* of the lens, is useful also because it can be applied to more complex optical

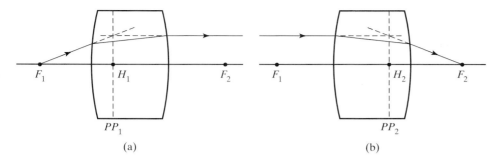

Figure 18-1 Illustration of the (a) first (PP_1) and (b) second (PP_2) principal planes of an optical system. The principal points H_1 and H_2 are also shown.

systems, as will become evident in this chapter. Thus, even though we are at present interested in a single thick lens, the following description is applicable to an arbitrary optical system that we can imagine is contained within the outlines of the thick lens.

There are six cardinal points on the axis of a thick lens, from which its imaging properties can be deduced. Planes[1] normal to the axis at these points are called the *cardinal planes*. The six cardinal points (see Figures 18-1 and 18-2) consist of the first and second *system focal points* (F_1 and F_2), which are already familiar; the first and second *principal points* (H_1 and H_2); and the first and second *nodal points* (N_1 and N_2).

A ray from the first focal point, F_1, is rendered parallel to the axis (Figure 18-1a), and a ray parallel to the axis is refracted by the lens through the second focal point, F_2 (Figure 18-1b). The extensions of the incident and resultant rays in each case intersect, by definition, in the *principal planes*, and these cross the axis at the principal points, H_1 and H_2. If the thick lens were a single thin lens, the two principal planes would coincide at the vertical line that is usually drawn to represent the lens. Principal planes in general do not coincide and may even be located outside the optical system itself. Once the locations of the principal planes are known, accurate ray diagrams can be drawn. The usual rays, determined by the focal points, change direction at their intersections with the principal planes, as in Figure 18-1. The third ray usually drawn for thin-lens diagrams is one through the lens center, undeviated and negligibly displaced. The nodal points of a thick lens, or of any optical system, permit the correction to this ray, as shown in Figure 18-2. Any ray directed toward the first nodal point, N_1, emerges from the optical system parallel to the incident ray, but displaced so that it appears to come from the second nodal point on the axis, N_2.

The positions of all six cardinal points are indicated in Figure 18-3. Distances are *directed*, positive or negative, by a sign convention that makes distances directed to the left negative and distances to the right positive. Notice that for the thick lens, the distances r and s determine the positions of the

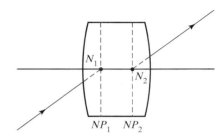

Figure 18-2 Illustration of the nodal points (N_1 and N_2) and nodal planes (NP_1 and NP_2) of an optical system.

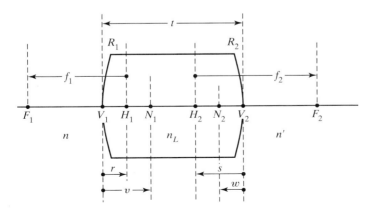

Figure 18-3 Symbols used to signify the cardinal points and locations for a thick lens. Axial points include focal points (F), vertices (V), principal points (H), and nodal points (N). Directed distances separating their corresponding planes are defined in the drawing.

[1]These "planes" are actually slightly curved surfaces that can be considered plane in the paraxial approximation.

principal points relative to the vertices V_1 and V_2, while f_1 and f_2 determine focal point positions relative to the principal points H_1 and H_2, respectively. Note carefully that these focal points are *not* measured from the vertices of the lens.

We summarize the basic equations for the thick lens without proof. Although the derivations involve simple algebra and geometry, they are rather arduous. We shall be content to await the matrix approach later in this chapter as a simpler way to justify these equations, and even then some of the work is relegated to the problems.

Utilizing the symbols defined in Figure 18-3, the focal length f_1 is given by

$$\frac{1}{f_1} = \frac{n_L - n'}{nR_2} - \frac{n_L - n}{nR_1} - \frac{(n_L - n)(n_L - n')}{nn_L} \frac{t}{R_1 R_2} \qquad (18\text{-}1)$$

and the focal length f_2 is conveniently expressed in terms of f_1 by

$$f_2 = -\frac{n'}{n} f_1 \qquad (18\text{-}2)$$

where n, n', and n_L are the refractive indices of the three regions indicated in Figure 18-3.

Notice that the two-focal lengths have the same magnitude if the lens is surrounded by a single refractive medium, so that $n = n'$. The principal planes can be located next using

$$r = \frac{n_L - n'}{n_L R_2} f_1 t \quad \text{and} \quad s = -\frac{n_L - n}{n_L R_1} f_2 t \qquad (18\text{-}3)$$

The positions of the nodal points are given by

$$v = \left(1 - \frac{n'}{n} + \frac{n_L - n'}{n_L R_2} t\right) f_1 \quad \text{and} \quad w = \left(1 - \frac{n}{n'} - \frac{n_L - n}{n_L R_1} t\right) f_2$$

$$(18\text{-}4)$$

Image and object distances and lateral magnification are related by

$$-\frac{f_1}{s_o} + \frac{f_2}{s_i} = 1 \quad \text{and} \quad m = -\frac{ns_i}{n's_o} \qquad (18\text{-}5)$$

as long as the distances s_o and s_i, as well as focal lengths, are measured relative to corresponding principal planes. The signs for s_o and s_i follow the usual sign convention introduced in Chapter 2. In the ordinary case of a lens in air, with $n = n' = 1$, notice that $r = v$ and $s = w$: First and second principal points are superimposed over corresponding nodal points. Also, first and second focal lengths are equal in magnitude, and the usual thin lens equations,

$$\frac{1}{s_o} + \frac{1}{s_i} = \frac{1}{f} \quad \text{and} \quad m = -\frac{s_i}{s_o} \qquad (18\text{-}6)$$

are valid. Here we have noted that $f = f_2 = -f_1$.

Example 18-1

Determine the focal lengths and the principal points for a 4-cm thick, biconvex lens with refractive index of 1.52 and radii of curvature of 25 cm, when the lens caps the end of a long cylinder filled with water ($n = 1.33$).

Solution

Use the equations for the thick lens in the order given:

$$\frac{1}{f_1} = \frac{1.52 - 1.33}{1(-25)} - \frac{1.52 - 1}{1(+25)} - \frac{(1.52 - 1)(1.52 - 1.33)}{1(1.52)} \frac{4}{(+25)(-25)}$$

or $f_1 = -35.74$ cm to the *left* of the first principal plane. Then

$$f_2 = -\left(\frac{1.33}{1}\right)(-35.74) = 47.53 \text{ cm}$$

to the *right* of the second principal plane, and

$$r = \frac{1.52 - 1.33}{(1.52)(-25)}(-35.74)(4) = 0.715 \text{ cm}$$

$$s = -\frac{1.52 - 1}{(1.52)(+25)}(47.53)(4) = -2.60 \text{ cm}$$

Thus the principal point H_1 is situated 0.715 cm to the *right* of the left vertex of the lens, and H_2 is situated 2.60 cm to the *left* of the right vertex V_2.

18-2 THE MATRIX METHOD

When the optical system consists of several elements—for example, the four or five lenses that constitute a photographic lens—we need a systematic approach that facilitates analysis. As long as we restrict our analysis to *paraxial rays*, this systematic approach is well handled by the matrix method. We now present a treatment of image formation that employs matrices to describe changes in the height and angle of a ray as it makes its way by successive reflections and refractions through an optical system. We show that, in the paraxial approximation, changes in height and direction of a ray can be expressed by linear equations that make this matrix approach possible. By combining matrices that represent individual refractions, reflections, and translations, a given optical system may be represented by a single matrix, from which the essential properties of the composite optical system may be deduced. The method lends itself to computer techniques for tracing a ray through an optical system of arbitrary complexity.

Figure 18-4 shows the progress of a single ray through an arbitrary optical system. The ray is described at distance x_0 from the first refracting surface in terms of its height y_0 and slope angle α_0 relative to the optical axis. Changes in angle occur at each *refraction*, such as at points 1 through 5, and at each *reflection*, such as at point 6. The height of the ray changes during *translations* between these points. We look for a procedure that will allow us to calculate the height and slope angle of the ray at any point in the optical system, for example, at point T, a distance x_7 from the mirror. In other words,

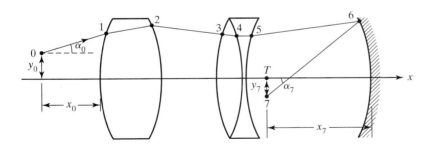

Figure 18-4   Steps in tracing a ray through an optical system. Progress of a ray can be described by changes in its elevation and direction.

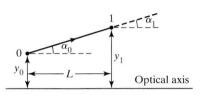

Figure 18-5   Simple translation of a ray.

given the input data (y_0, α_0) at point 0, we wish to predict values of (y_7, α_7) at point 7 as output data.

18-3 THE TRANSLATION MATRIX

Consider a simple translation of the ray in a homogeneous medium, as in Figure 18-5. Let the axial progress of the ray be L, as shown, such that at point 1, the elevation and direction of the ray are given by "coordinates" y_1 and α_1, respectively. Evidently,

$$\alpha_1 = \alpha_0 \quad \text{and} \quad y_1 = y_0 + L \tan \alpha_0$$

These equations may be put into an ordered form,

$$y_1 = (1)y_0 + (L)\alpha_0$$
$$\alpha_1 = (0)y_0 + (1)\alpha_0 \tag{18-7}$$

where the paraxial approximation $\tan \alpha_0 \cong \alpha_0$ has been used. In matrix notation, the two equations are written

$$\begin{bmatrix} y_1 \\ \alpha_1 \end{bmatrix} = \begin{bmatrix} 1 & L \\ 0 & 1 \end{bmatrix} \begin{bmatrix} y_0 \\ \alpha_0 \end{bmatrix} \tag{18-8}$$

The 2×2 *ray-transfer matrix* represents the effect of the translation on a ray. The input data (y_0, α_0) is modified by the ray-transfer matrix to yield the correct output data (y_1, α_1).

18-4 THE REFRACTION MATRIX

Consider next the refraction of a ray at a spherical interface separating media of refractive indices n and n', as shown in Figure 18-6. We need to relate the ray coordinates (y', α') after refraction to those before refraction, (y, α). Since refraction occurs at a point, there is no change in elevation, and $y = y'$.

The angle α', on the other hand, is, by inspection of Figure 18-6 and the use of small angle approximations,

$$\alpha' = \theta' - \phi = \theta' - \frac{y}{R} \quad \text{and} \quad \alpha = \theta - \phi = \theta - \frac{y}{R}$$

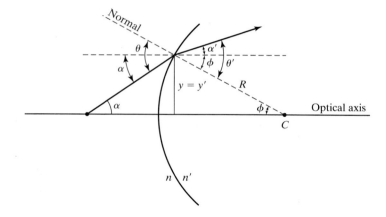

Figure 18-6   Refraction of a ray at a spherical interface.

Incorporating the paraxial form of Snell's law,

$$n\theta = n'\theta'$$

we have

$$\alpha' = \left(\frac{n}{n'}\right)\theta - \frac{y}{R} = \left(\frac{n}{n'}\right)\left(\alpha + \frac{y}{R}\right) - \frac{y}{R}$$

or

$$\alpha' = \left(\frac{1}{R}\right)\left(\frac{n}{n'} - 1\right)y + \left(\frac{n}{n'}\right)\alpha$$

The appropriate linear equations are then

$$y' = (1)y + (0)\alpha$$
$$\alpha' = \left[\left(\frac{1}{R}\right)\left(\frac{n}{n'} - 1\right)\right]y + \left(\frac{n}{n'}\right)\alpha \qquad (18\text{-}9)$$

or, in matrix form,

$$\begin{bmatrix} y' \\ \alpha' \end{bmatrix} = \begin{bmatrix} 1 & 0 \\ \frac{1}{R}\left(\frac{n}{n'} - 1\right) & \frac{n}{n'} \end{bmatrix}\begin{bmatrix} y \\ \alpha \end{bmatrix} \qquad (18\text{-}10)$$

Here, we use the same sign convention for R that was introduced in Chapter 2. If the surface is instead concave, R is negative. Furthermore, allowing $R \rightarrow \infty$ yields the appropriate refraction matrix for a plane interface.

18-5 THE REFLECTION MATRIX

Finally, consider reflection at a spherical surface, illustrated in Figure 18-7. In the case considered, a concave mirror, R, is negative. We need to add a sign convention for the angles that describe the ray directions. Angles are considered positive for all rays pointing upward, either before or after a reflection; angles for rays pointing downward are considered negative. The sign convention is summarized in the inset of Figure 18-7.

From the geometry of Figure 18-7, with both α and α' positive,

$$\alpha = \theta + \phi = \theta + \frac{y}{-R} \quad \text{and} \quad \alpha' = \theta' - \phi = \theta' - \frac{y}{-R}$$

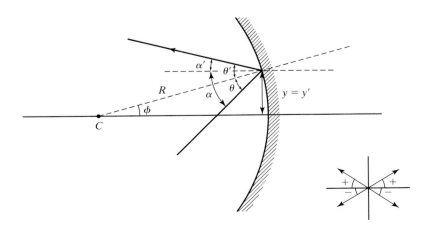

Figure 18-7 Reflection of a ray at a spherical surface. The inset illustrates the sign convention for ray angles.

where we have made the usual small angle approximations. Using these relations together with the law of reflection, $\theta = \theta'$,

$$\alpha' = \theta' + \frac{y}{R} = \theta + \frac{y}{R} = \alpha + \frac{2y}{R}$$

and so the two desired linear equations are

$$y' = (1)y + (0)\alpha$$

$$\alpha' = \left(\frac{2}{R}\right)y + (1)\alpha \tag{18-11}$$

In matrix form,

$$\begin{bmatrix} y' \\ \alpha' \end{bmatrix} = \begin{bmatrix} 1 & 0 \\ \dfrac{2}{R} & 1 \end{bmatrix} \begin{bmatrix} y \\ \alpha \end{bmatrix} \tag{18-12}$$

18-6 THICK-LENS AND THIN-LENS MATRICES

We construct now a matrix that represents the action of a thick lens on a ray of light. For generality, we assume different media on opposite sides of the lens, having refractive indices n and n', as shown in Figure 18-8. In traversing the lens, the ray undergoes two refractions and one translation, steps for which we have already derived matrices. Referring to Figure 18-8, where we have chosen for simplicity a lens with positive radii of curvature, we may write, symbolically,

$$\begin{bmatrix} y_1 \\ \alpha_1 \end{bmatrix} = M_1 \begin{bmatrix} y_0 \\ \alpha_0 \end{bmatrix} \quad \text{for the first refraction}$$

$$\begin{bmatrix} y_2 \\ \alpha_2 \end{bmatrix} = M_2 \begin{bmatrix} y_1 \\ \alpha_1 \end{bmatrix} \quad \text{for the translation}$$

and

$$\begin{bmatrix} y_3 \\ \alpha_3 \end{bmatrix} = M_3 \begin{bmatrix} y_2 \\ \alpha_2 \end{bmatrix} \quad \text{for the second refraction}$$

Telescoping these matrix equations results in

$$\begin{bmatrix} y_3 \\ \alpha_3 \end{bmatrix} = M_3 M_2 M_1 \begin{bmatrix} y_0 \\ \alpha_0 \end{bmatrix}$$

Evidently the entire thick lens can be represented by a matrix $M = M_3 M_2 M_1$. Recalling that the multiplication of matrices is associative but not commutative, the descending order must be maintained. The individual matrices operate on the light ray in the same order in which the corresponding optical actions influence the light ray as it traverses the system. Generalizing, the matrix equation representing any number N of translations, reflections, and refractions is given by

$$\begin{bmatrix} y_f \\ \alpha_f \end{bmatrix} = M_N M_{N-1} \cdots M_2 M_1 \begin{bmatrix} y_0 \\ \alpha_0 \end{bmatrix} \tag{18-13}$$

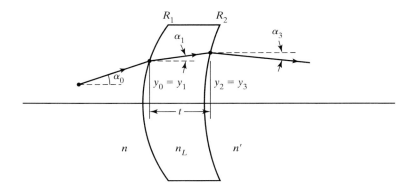

Figure 18-8 Progress of a ray through a thick lens.

and the ray-transfer matrix representing the entire optical system is

$$M = M_N M_{N-1} \cdots M_2 M_1 \qquad (18\text{-}14)$$

We apply this result first to the thick lens of Figure 18-8, whose index is n_L and whose thickness for paraxial rays is t. The correct approximation for a thin lens is then made by allowing $t \to 0$. Letting $\mathfrak{R}$ represent a refraction matrix and $\mathfrak{T}$ represent a translation matrix, the matrix for the thick lens is, by Eq. (18-14), the composite matrix

$$M = \mathfrak{R}_2 \mathfrak{T} \mathfrak{R}_1$$

or

$$M = \begin{bmatrix} 1 & 0 \\ \dfrac{n_L - n'}{n' R_2} & \dfrac{n_L}{n'} \end{bmatrix} \begin{bmatrix} 1 & t \\ 0 & 1 \end{bmatrix} \begin{bmatrix} 1 & 0 \\ \dfrac{n - n_L}{n_L R_1} & \dfrac{n}{n_L} \end{bmatrix} \qquad (18\text{-}15)$$

For the case where t is negligible ($t = 0$) and where the lens is surrounded by the same medium on either side ($n = n'$),

$$M = \begin{bmatrix} 1 & 0 \\ \dfrac{n_L - n}{n R_2} & \dfrac{n_L}{n} \end{bmatrix} \begin{bmatrix} 1 & 0 \\ 0 & 1 \end{bmatrix} \begin{bmatrix} 1 & 0 \\ \dfrac{n - n_L}{n_L R_1} & \dfrac{n}{n_L} \end{bmatrix} \qquad (18\text{-}16)$$

Simplifying Eq. (18-16),

$$M = \begin{bmatrix} 1 & 0 \\ \dfrac{n_L - n}{n}\left(\dfrac{1}{R_2} - \dfrac{1}{R_1} \right) & 1 \end{bmatrix} \qquad (18\text{-}17)$$

The matrix element in the first column, second row, may be expressed in terms of the focal length of the lens, by the lensmaker's formula,

$$\frac{1}{f} = \frac{n_L - n}{n}\left(\frac{1}{R_1} - \frac{1}{R_2} \right)$$

so that the thin-lens ray-transfer matrix is simply

$$M = \begin{bmatrix} 1 & 0 \\ -\dfrac{1}{f} & 1 \end{bmatrix} \qquad (18\text{-}18)$$

As usual, f is taken as positive for a convex lens and negative for a concave lens. This matrix and those previously derived are summarized for quick reference in Table 18-1.

TABLE 18-1 SUMMARY OF SOME SIMPLE RAY-TRANSFER MATRICES

Translation matrix:

$$M = \begin{bmatrix} 1 & L \\ 0 & 1 \end{bmatrix} = \mathfrak{T}$$

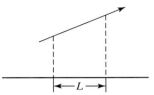

Refraction matrix, spherical interface:

$$M = \begin{bmatrix} 1 & L \\ \dfrac{n-n'}{Rn'} & \dfrac{n}{n'} \end{bmatrix} = \mathfrak{R}$$

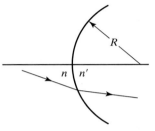

$(+R)$: convex
$(-R)$: concave

Refraction matrix, plane interface:

$$M = \begin{bmatrix} 1 & 0 \\ 0 & \dfrac{n}{n'} \end{bmatrix}$$

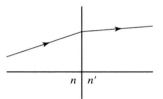

Thin-lens matrix:

$$M = \begin{bmatrix} 1 & 0 \\ -\dfrac{1}{f} & 1 \end{bmatrix}$$

$$\frac{1}{f} = \frac{n'-n}{n}\left(\frac{1}{R_1} - \frac{1}{R_2}\right)$$

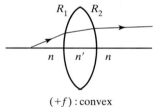

$(+f)$: convex
$(-f)$: concave

Spherical mirror matrix:

$$M = \begin{bmatrix} 1 & 0 \\ \dfrac{2}{R} & 1 \end{bmatrix}$$

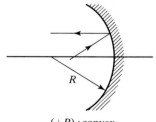

$(+R)$: convex
$(-R)$: concave

18-7 SYSTEM RAY-TRANSFER MATRIX

By combining appropriate individual matrices in the proper order, according to Eq. (18-14), it is possible to express any optical system by a single 2×2 matrix, which we call the *system matrix*.

Example 18-2

Find the system matrix for the thick lens of Figure 18-8, whose matrix before multiplication is expressed by Eq. (18-15), and specify the thick lens exactly by choosing $R_1 = 45$ cm, $R_2 = 30$ cm, $t = 5$ cm, $n_L = 1.60$, and $n = n' = 1$.

Solution

$$M = \begin{bmatrix} 1 & 0 \\ \dfrac{1}{50} & 1.6 \end{bmatrix} \begin{bmatrix} 1 & 5 \\ 0 & 1 \end{bmatrix} \begin{bmatrix} 1 & 0 \\ -\dfrac{1}{120} & \dfrac{1}{1.6} \end{bmatrix} \quad \text{or} \quad M = \begin{bmatrix} \dfrac{23}{24} & \dfrac{25}{8} \\ \dfrac{7}{1200} & \dfrac{17}{16} \end{bmatrix}$$

The elements of this composite ray-transfer matrix, usually referred to in the symbolic form

$$M = \begin{bmatrix} A & B \\ C & D \end{bmatrix}$$

describe the relevant properties of the optical system, as we shall see. Be aware that the particular values of the matrix elements of a system depend on the location of the ray at input and output. In the case of the thick lens just calculated, the *input plane* was chosen at the left surface of the lens, and the *output plane* was chosen at its right surface. If each of these planes is moved some distance from the lens, the system matrix will also include an initial and a final translation matrix incorporating these distances. The matrix elements change and the system matrix now represents this enlarged "system." In any case, the determinant of the system matrix has a very useful property:

$$\text{Det } M = AD - BC = \frac{n_0}{n_f} \tag{18-19}$$

where n_0 and n_f are the refractive indices of the initial and final media of the optical system. The proof of this assertion follows upon noticing first that the determinant of all the individual ray-transfer matrices in Table 18-1 have values of either n/n' or unity and then making use of the theorem[2] that the determinant of a product of matrices is equal to the product of the determinants. Symbolically, if $M = M_1 M_2 M_3 \cdots M_N$, then

$$\text{Det}(M) = (\text{Det } M_1)(\text{Det } M_2)(\text{Det } M_3) \cdots (\text{Det } M_N) \tag{18-20}$$

In forming this product, using determinants of ray-transfer matrices, all intermediate refractive indices cancel, and we are left with the ratio n_0/n_f, as stated in Eq. (18-19). Most often, as in the case of the thick-lens example, n_0 and n_f both refer to air, and Det (M) is unity. The condition expressed by Eq. (18-19) is useful in checking the correctness of the calculations that produce a system matrix.

---

[2]The theorem can easily be verified for the product of two matrices and generalized by induction to the product of any number of matrices. Formal proofs can be found in any standard textbook on matrices and determinants, for example, E. T. Browne, *Introduction to the Theory of Determinants and Matrices* (Chapel Hill: University of North Carolina, 1958).

18-8 SIGNIFICANCE OF SYSTEM MATRIX ELEMENTS

We examine now the implications that follow when each of the matrix elements in turn is zero. In symbolic form, we have, from Eq. (18-13),

$$\begin{bmatrix} y_f \\ \alpha_f \end{bmatrix} = \begin{bmatrix} A & B \\ C & D \end{bmatrix} \begin{bmatrix} y_0 \\ \alpha_0 \end{bmatrix} \tag{18-21}$$

which is equivalent to the algebraic relations

$$y_f = Ay_0 + B\alpha_0$$
$$\alpha_f = Cy_0 + D\alpha_0 \tag{18-22}$$

1. $D = 0$. In this case, $\alpha_f = Cy_0$, independent of α_0. Since y_0 is fixed, this means that all rays leaving a point in the input plane will have the same angle α_f at the output plane, independent of their angles at input. As shown in Figure 18-9a, the input plane thus coincides with the first focal plane of the optical system.
2. $A = 0$. This case is much like the previous one. Here $y_f = B\alpha_0$ implies that y_f is independent of y_0, so that all rays departing the input plane at the same angle, regardless of altitude, arrive at the same altitude y_f at the output plane. As shown in Figure 18-9b, the output plane thus functions as the second focal plane.
3. $B = 0$. Then $y_f = Ay_0$, independent of α_0. Thus, all rays from a point at height y_0 in the input plane arrive at the same point of height y_f in the output plane. The points are then related as object and image points, as shown in Figure 18-9c, and the input and output planes correspond to

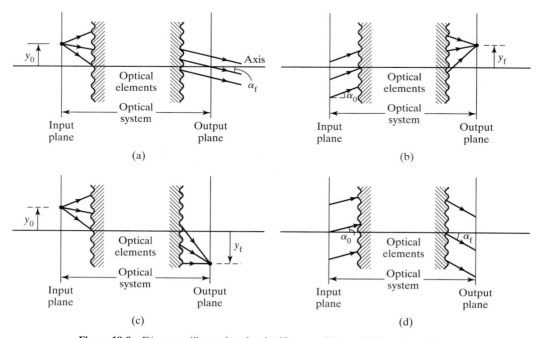

Figure 18-9   Diagrams illustrating the significance of the vanishing of specific system matrix elements. (a) When $D = 0$, the input plane corresponds to the first focal plane of the optical system. (b) When $A = 0$, the output plane corresponds to the second focal plane of the optical system. (c) When $B = 0$, the output plane is the image plane conjugate to the input plane and A is the linear magnification. (d) When $C = 0$, a parallel bundle of rays at the input plane is parallel at the output plane and D is the angular magnification.

conjugate planes for the optical system. Furthermore, since $A = y_f/y_0$, the matrix element A represents the linear magnification.

4. $C = 0$. Now $\alpha_f = D\alpha_0$, independent of y_0. This case is analogous to case 3, with directions replacing ray heights. Input rays, all of one direction, now produce parallel output rays in some other direction. Moreover, $D = \alpha_f/\alpha_0$ is the angular magnification. A system for which $C = 0$ is sometimes called a "telescopic system," because a telescope admits parallel rays into its objective and outputs parallel rays for viewing from its eyepiece.

Example 18-3

We illustrate case 3 in this example. We place a small object on axis at a distance of 16 cm from the left end of a long, plastic rod with a polished spherical end of radius 4 cm, as indicated in Figure 18-10. The refractive index of the plastic is 1.50 and the object is in air. Let the unknown image be formed at the output reference plane, a distance x from the spherical cap. We wish to determine the image distance x and the lateral magnification m. The system matrix connecting the object and image planes consists of the product of three matrices, corresponding to (1) a translation $\mathfrak{T}_1$ in air from object to the rod, (2) a refraction $\mathfrak{R}$ at the spherical surface, and (3) a translation $\mathfrak{T}_2$ in plastic to the image.

Solution

Remembering to take the matrices in "reverse" order and working in cm, we have

$$M = \mathfrak{T}_2 \mathfrak{R} \mathfrak{T}_1 = \begin{bmatrix} 1 & x \\ 0 & 1 \end{bmatrix} \begin{bmatrix} 1 & 0 \\ \dfrac{1-1.50}{4(1.50)} & \dfrac{1}{1.50} \end{bmatrix} \begin{bmatrix} 1 & 16 \\ 0 & 1 \end{bmatrix}$$

or

$$M = \begin{bmatrix} 1 - \dfrac{x}{12} & 16 - \dfrac{2x}{3} \\ -\dfrac{1}{12} & -\dfrac{2}{3} \end{bmatrix}$$

with the unknown quantity x incorporated in the matrix elements. According to this discussion, when $B = 0$, the output plane is the image plane, so that the image distance x is determined by setting

$$16 - \frac{2x}{3} = 0 \quad \text{or} \quad x = 24 \text{ cm}$$

Further, the linear magnification m is then given by the value of element A:

$$m = A = 1 - \frac{x}{12} = -1$$

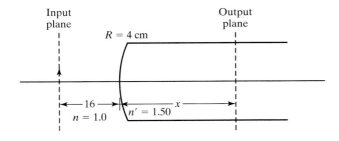

Input plane

Output plane

$R = 4$ cm

16

$n = 1.0$ $n' = 1.50$ x

Figure 18-10 Schematic defining an example for ray-transfer matrix methods.

We conclude that the image occurs 24 cm inside the rod, is inverted, and has the same lateral size as the object. This illustrates how the system matrix can be used to find image locations and sizes, although this may usually be done more quickly by using the Gaussian image formulas derived earlier.

18-9 LOCATION OF CARDINAL POINTS FOR AN OPTICAL SYSTEM

Since the properties of an optical system can be deduced from the elements of the system ray-transfer matrix, it follows that relationships must exist between the matrix elements, A, B, C, and D and the cardinal points of the system. In Figure 18-11, we generalize Figure 18-3 by defining distances locating the six cardinal points relative to the input and output planes that define the limits of an optical system. The focal points F_1 and F_2 are located at distances f_1 and f_2 from the principal points H_1 and H_2 and at distances p and q from the reference input and output planes, respectively. Further, measured from the input and output planes, the distances r and s locate the principal points, and the distances v and w locate the nodal points. Distances measured to the right of their reference planes are considered positive and to the left, negative. The principal points and nodal points often occur outside the optical system, that is, outside the region defined by the input and output planes.

We now derive the relationships between the distances defined in Figure 18-11 and the system matrix elements. Consider Figure 18-12a, which highlights distances p, r, and f_1 as they are determined by the positions of the first focal point and the first principal plane. Input coordinates of the given ray are (y_0, α_0) and output coordinates are $(y_f, 0)$. Thus, the ray equations, Eqs. (18-22), become for this ray

$$y_f = A y_0 + B \alpha_0$$

and

$$0 = C y_0 + D \alpha_0 \quad \text{or} \quad y_0 = -\left(\frac{D}{C}\right)\alpha_0 \tag{18-23}$$

For small angles, Figure 18-12a shows that

$$\alpha_0 = \frac{y_0}{-p}$$

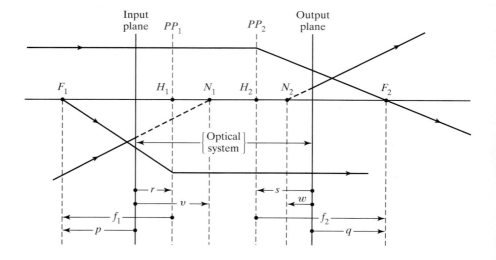

Figure 18-11 Location designations for the six cardinal points of an optical system. Rays associated with the nodal points and principal planes are also shown.

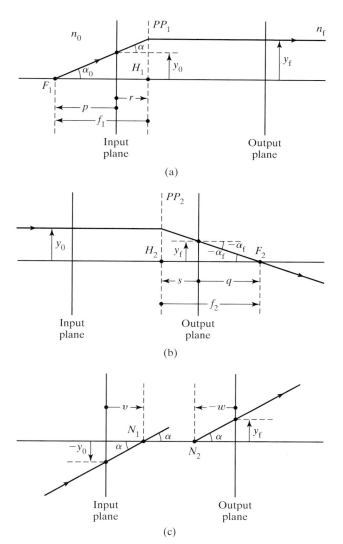

Figure 18-12 (a) Construction used to relate distances p, r, and f_1 to matrix elements. (b) Construction used to relate distances q, s, and f_2 to matrix elements. (c) Construction used to relate distances v and w to matrix elements.

where the negative sign indicates that F_1 is located a distance p to the left of the input plane. Incorporating Eq. (18-23),

$$ p = \frac{-y_0}{\alpha_0} = \frac{D}{C} \tag{18-24} $$

Similarly, $\alpha_0 = y_{\mathrm{f}}/(-f_1)$, and thus

$$ f_1 = \frac{-y_{\mathrm{f}}}{\alpha_0} = \frac{-(Ay_0 + B\alpha_0)}{\alpha_0} = \frac{AD}{C} - B $$

$$ f_1 = \frac{AD - BC}{C} = \frac{\mathrm{Det}(M)}{C} = \left(\frac{n_0}{n_{\mathrm{f}}}\right)\frac{1}{C} \tag{18-25} $$

Finally, using Eqs. (18-24) and (18-25), the positive distance r can be expressed in terms of p and f_1:

$$ r = p - f_1 = \frac{D}{C} - \frac{n_0}{n_{\mathrm{f}}}\frac{1}{C} = \frac{1}{C}\left(D - \frac{n_0}{n_{\mathrm{f}}}\right) \tag{18-26} $$

Using Figure 18-12b, one can similarly discover relations for the output distances q, f_2, and s. The results, together with those just derived for p, f_1,

TABLE 18-2 CARDINAL POINT LOCATIONS IN TERMS OF SYSTEM MATRIX ELEMENTS

$$p = \frac{D}{C} \qquad F_1$$

$$q = -\frac{A}{C} \qquad F_2$$

$$r = \frac{D - n_0/n_f}{C} \qquad H_1$$

$$s = \frac{1 - A}{C} \qquad H_2$$

$$v = \frac{D - 1}{C} \qquad N_1$$

$$w = \frac{n_0/n_f - A}{C} \qquad N_2$$

Located relative to input (1) and output (2) reference planes

$$f_1 = p - r = \frac{n_0/n_f}{C} \qquad F_1$$

$$f_s = q - s = -\frac{1}{C} \qquad F_2$$

Located relative to principal planes

and r, are listed in Table 18-2. With the help of Figure 18-12c, the nodal plane distances v and w may also be determined. For example, for small angle α,

$$\alpha = -\frac{y_0}{v} \tag{18-27}$$

where the negative sign indicates that the ray intersects the input plane below the axis. Input and output rays make the same angle relative to the axis. From Eq. (18-22), with $\alpha_0 = \alpha_f = \alpha$,

$$\alpha = Cy_0 + D\alpha \quad \text{or} \quad \frac{y_0}{\alpha} = \frac{1 - D}{C} \tag{18-28}$$

Combining Eqs. (18-27) and (18-28),

$$v = \frac{D - 1}{C} \tag{18-29}$$

Similarly, one can show that

$$w = \frac{(n_0/n_f) - A}{C} \tag{18-30}$$

again using the fact that Det $(M) = AD - BC = n_0/n_f$. These results are also included in Table 18-2. The relationships listed there can be used to establish the following useful generalizations:

1. Principal points and nodal points coincide, that is, $r = v$ and $s = w$, when the initial and final media have the same refractive indices.
2. First and second focal lengths of an optical system are equal in magnitude when initial and final media have the same refractive indices.
3. The separation of the principal points is the same as the separation of nodal points, that is, $r - s = v - w$.

18-10 EXAMPLES USING THE SYSTEM MATRIX AND CARDINAL POINTS

As an example, consider an optical system that consists of two thin lenses in air, separated by a distance L, as shown in Figure 18-13. The lenses have focal lengths of f_A and f_B, which may be either positive or negative.

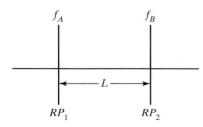

Figure 18-13 Optical system consisting of two thin lenses in air, separated by a distance L.

If input and output reference planes are located at the lenses, the system matrix includes two thin-lens matrices, $\mathcal{L}_A$ and $\mathcal{L}_B$, and a translation matrix $\mathcal{T}$ for the distance L between them. The system matrix is $M = \mathcal{L}_B\mathcal{T}\mathcal{L}_A$, or

$$
M = \begin{bmatrix} 1 & 0 \\ -\dfrac{1}{f_B} & 1 \end{bmatrix}\begin{bmatrix} 1 & L \\ 0 & 1 \end{bmatrix}\begin{bmatrix} 1 & 0 \\ -\dfrac{1}{f_A} & 1 \end{bmatrix}
$$

$$
M = \begin{bmatrix} 1 - \dfrac{L}{f_A} & L \\ \dfrac{1}{f_B}\left(\dfrac{L}{f_A} - 1\right) - \dfrac{1}{f_A} & 1 - \dfrac{L}{f_B} \end{bmatrix} \tag{18-31}
$$

Reference to Table 18-2 shows that the first and second focal lengths of this system are $f_1 = 1/C$ and $f_2 = -1/C$. We shall take the *equivalent focal length* of the two-lens system to be $f_{eq} = f_2 = -1/C$. So,

$$
\frac{1}{f_{eq}} = \frac{1}{f_A} + \frac{1}{f_B} - \frac{L}{f_A f_B} \tag{18-32}
$$

Furthermore, the first principal points and nodal points coincide at a distance given by $r = v = (D - 1)/C$ from the first lens, and the second principal points and nodal points coincide at a distance given by $s = w = (1 - A)/C$ from the second lens. Thus

$$
r = v = \left(\frac{f_{eq}}{f_B}\right)L \quad \text{and} \quad s = w = -\left(\frac{f_{eq}}{f_A}\right)L \tag{18-33}
$$

Example 18-4

Let us apply these results to the case of a Huygens eyepiece, which consists of two positive, thin lenses separated by a distance L equal to the average of their focal lengths. Suppose $f_A = 3.125$ cm and $f_B = 2.083$ cm, giving $L = 2.604$ cm and $f_{eq} = 2.5$ cm, by Eq. (18-32). Incidentally, the magnifying power of this eyepiece, given by $25/f$, is therefore $10\times$. From Eq. (18-33), we conclude that $r = +3.125$ cm and $s = -2.083$ cm. The optical system, together with its cardinal points and sample rays, is shown roughly to scale in Figure 18-14. The converging incident rays 1, 2 and 3 determine an image location between the lenses, which acts as a virtual object VO for the optical system. An enlarged, virtual image (not shown) is formed by the diverging rays leaving the system, as seen by an eye looking into the eyepiece. This eyepiece is also discussed in Chapter 3.

Solution

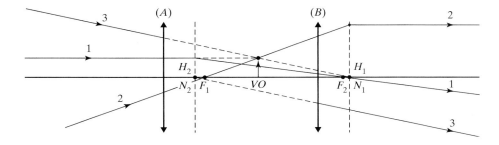

Figure 18-14 Ray construction for a Huygens eyepiece, using cardinal points.

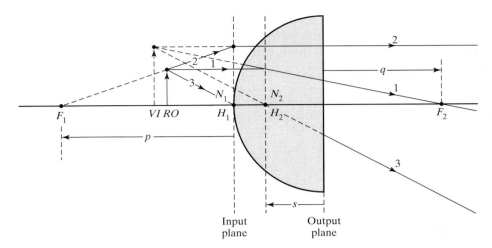

Figure 18-15 Ray construction for a hemi-spherical lens, using cardinal points.

Example 18-5

As a final calculation, let us find the cardinal points and sketch a ray diagram for the hemispherical glass lens shown in Figure 18-15. The radii of curvature are $R_1 = 3$ cm and $R_2 \to \infty$, and the lens in air has a refractive index of 1.50.

Solution

The system matrix, for input and output reference planes at the two surfaces of the lens, is, then,

$$M = \Re_2 \Im \Re_1 = \begin{bmatrix} 1 & 0 \\ 0 & 1.5 \end{bmatrix} \begin{bmatrix} 1 & 3 \\ 0 & 1 \end{bmatrix} \begin{bmatrix} 1 & 0 \\ \dfrac{-0.5}{1.5(3)} & \dfrac{1}{1.5} \end{bmatrix}$$

or

$$M = \begin{bmatrix} \dfrac{2}{3} & 2 \\ -\dfrac{1}{6} & 1 \end{bmatrix}, \quad \text{with } \mathrm{Det}(M) = 1$$

The relations in Table 18-2 then give values of $p = -6$ cm, $q = 4$ cm, $r = 0$, $s = -2$ cm, $f_1 = -6$ cm, and $f_2 = 6$ cm. Principal and nodal points coincide. The cardinal points are located, approximately to scale, in Figure 18-15. Ray diagrams using the principal planes and nodal points are constructed for an arbitrary real object. In this case the emerging rays determine a virtual image *VI* near the object *RO* erect and slightly magnified.

18-11 RAY TRACING

The assumption of paraxial rays greatly simplifies the description of the progress of rays of light through an optical system, because trigonometric terms do not appear in the equations. For many purposes, this treatment is sufficient. In practice, rays of light contributing to an image in an optical system are, in fact, usually rays in the near neighborhood of the optical axis. If the quality of the image is to be improved, however, ways must be found to reduce the ever-present aberrations that arise from the presence of rays deviating, more or less, from this ideal assumption. To determine the actual path of individual rays of light through an optical system, each ray must be *traced*, independent-ly, using only the laws of reflection and refraction together with geometry. This technique is called *ray tracing* because it was formerly done by hand,

graphically, with ruler and compass, in a step-by-step process through an accurate sketch of the optical system. Today, with the help of computers, the
necessary calculations yielding the progressive changes in a ray's altitude and
angle is done more easily and quickly. Graphic techniques are used to actually draw the optical system and to trace the ray's progress through the optical
system on the monitor.[3]

Ray-tracing procedures, such as the one to be described here, are often
limited to *meridional rays*, that is, rays that pass through the optical axis of the
system. Since the law of refraction requires that refracted rays remain in the
plane of incidence, a meridional ray remains within the same meridional
plane throughout its trajectory. Thus the treatment in terms of meridional
rays is a two-dimensional treatment,[4] greatly simplifying the geometrical relationships required. Rays contributing to the image that do not pass through
the optical axis are called *skew rays* and require three-dimensional geometry
in their calculations. The added complexity does not pose a problem for the
computer, once the ray-tracing program is written. Analysis of various aberrations, such as spherical aberration, astigmatism, and coma (described in
Chapter 20), require knowledge of the progress of selected nonparaxial rays
and skew rays. The design of a complex lens system, such as a photographic
lens with four or five elements, is a combination of science and skill. By alternating ray tracing with small changes in the positions, focal lengths, and curvatures of the surfaces involved and in refractive indices of the elements, the
design of the lens system is gradually optimized.

For our present purposes, it will be sufficient to show how the appropriate equations for meridional ray tracing can be developed and how they can
be repeated in stepwise fashion to follow a ray through any number of spherical refracting surfaces that constitute an optical system. The technique is well
adapted to iterative loops handled by computer programs.

Figure 18-16 shows a single, representative step in the ray-tracing analysis. By incorporating a sign convention, the equations developed from this
diagram can be made to apply to any ray and to any spherical refracting surface. The ray selected originates at (or passes through) point A, making an angle
α with the optical axis. The ray passes through the optical axis at O and then

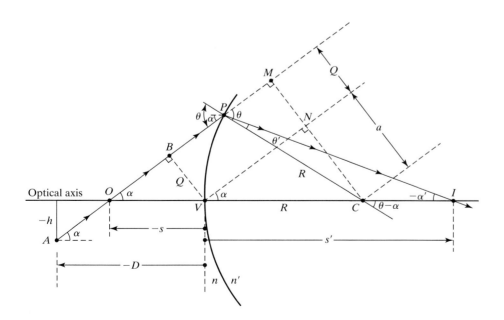

Figure 18-16 Single refraction at a spherical surface. The figure defines the symbols
and shows the geometrical relationships
that lead to ray-tracing equations for a
meridional ray.

---

[3]A search of the Internet will reveal the existence of several free, high-quality, ray-tracing programs.

[4]The two dimensions are those of the page on which we have been drawing our ray diagrams.
Without emphasizing this, we have been using meridional rays in all our diagrams.

intersects the refracting surface at P, where it is refracted into a medium of index n', cutting the axis again at I. The angles of incidence and refraction, θ and θ', are related by Snell's law. Points O and I are conjugate points with distances s and s' from the surface vertex at V. The radius R of the surface is also shown, passing through the center of curvature at C. Other points and lines are added to help in developing the necessary geometrical relationships.

The sign convention is the same as that used previously in this chapter. Distances to the left of the vertex V are negative, and to the right, positive. If we use light rays progressing from left to right, their angles have the same sign as their slopes. Distances measured above the axis are positive and below, negative. An important quantity in the calculations, also subject to this sign convention, is the parameter Q, the perpendicular distance VB from the vertex to the ray, as shown.

The input parameters for the ray are its elevation h, angle α, and distance D. Figure 18-16 shows that the "object distance," s, is related to D by

$$s = D - \frac{h}{\tan \alpha} \tag{18-34}$$

Also, in ΔOBV:

$$\sin \alpha = \frac{Q}{-s} \tag{18-35}$$

In ΔPMC:

$$\sin \theta = \frac{a + Q}{R}$$

In ΔVNC:

$$\sin \alpha = \frac{a}{R}$$

Eliminating the length a from the last two equations, we get

$$\sin \theta = \frac{Q}{R} + \sin \alpha \tag{18-36}$$

Snell's law at P:

$$n \sin \theta = n' \sin \theta' \tag{18-37}$$

In ΔCPI:

$$\theta - \alpha = \theta' - \alpha' \tag{18-38}$$

The Q parameter for the refracted ray is shown in Figure 18-17a as Q'. Analogous to the relations just found, we see that in ΔCMV:

$$\sin(-\alpha') = \frac{a'}{R}$$

in ΔPLC:

$$\sin \theta' = \frac{Q' - a'}{R}$$

As before, when a' is eliminated, there results

$$Q' = R(\sin \theta' - \sin \alpha') \tag{18-39}$$

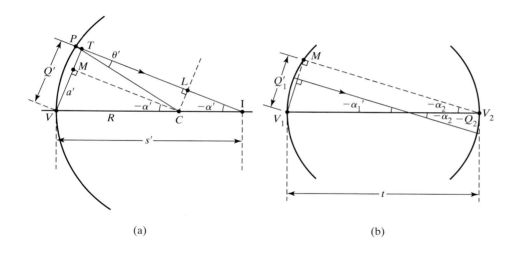

(a)   (b)

Figure 18-17 (a) Geometrical relationship of refracted-ray parameters with the distance Q'. (b) Geometrical relationships illustrating the transfer between Q and α after one refraction and before the next.

In ΔITV:

$$\sin(-\alpha') = \frac{Q'}{s'} \quad \text{or} \quad s' = \frac{-Q'}{\sin \alpha'} \qquad (18\text{-}40)$$

The relevant equations describing the first refraction are included in Table 18-3 under the first column for the general case.

The calculations lead to new values of α, Q, and s (now primed), which prepare for the next refraction in the sequence. The geometrical *transfer* to the next surface, at distance t from the first, is shown in Figure 18-17b, where, in $\Delta V_2 M V_1$,

$$\sin(-\alpha_2) = \frac{V_1 M}{t} = \frac{Q'_1 - Q_2}{t}$$

TABLE 18-3   MERIDIONAL RAY-TRACING EQUATIONS (INPUT: n, n', R, α, h, D)

General case	Ray parallel to axis: $\alpha = 0$	Plane surface: $R \Rightarrow \infty$
$s = D - \dfrac{h}{\tan \alpha}$	—	$s = D - \dfrac{h}{\tan \alpha}$
$Q = -s \sin \alpha$	$Q = h$	$Q = -s \sin \alpha$
$\theta = \sin^{-1}\left(\dfrac{Q}{R} + \sin \alpha\right)$	$\theta = \sin^{-1}\left(\dfrac{Q}{R} + \sin \alpha\right)$	—
$\theta' = \sin^{-1}\left(\dfrac{n \sin \theta}{n'}\right)$	$\theta' = \sin^{-1}\left(\dfrac{n \sin \theta}{n'}\right)$	—
$\alpha' = \theta' - \theta + \alpha$	$\alpha' = \theta' - \theta + \alpha$	$\alpha' = \sin^{-1}\dfrac{n}{n' \sin \alpha}$
$Q' = R(\sin \theta' - \sin \alpha')$	$Q' = R(\sin \theta' - \sin \alpha')$	$Q' = Q\dfrac{\cos \alpha'}{\cos \alpha}$
$s' = \dfrac{-Q'}{\sin \alpha'}$	$s' = \dfrac{-Q'}{\sin \alpha'}$	$s' = \dfrac{-Q'}{\sin \alpha'}$

Transfer: Input: t

$Q = Q' + t \sin \alpha'$

$\alpha = \alpha'$

$n = n'$

Input: new n', R

Return: to calculate θ

or

$$Q_2 = Q_1' + t \sin \alpha_2 \qquad (18\text{-}41)$$

Table 18-3 also shows how the equations must be modified for two special cases: (1) when the incident ray is parallel to the axis and (2) when the surface is plane, with an infinite radius of curvature.

Example 18-6

Do a ray trace for two rays through a *Rapid landscape* photographic lens of three elements. The parallel rays enter the lens from a distant object at altitudes of 1 and 5 mm above the optical axis. The lens specifications (all dimensions in mm) are as follows:

$$R_1 = -120.8$$
$$R_2 = -34.6 \qquad t_1 = 6 \qquad n_1 = 1.521$$
$$R_3 = -96.2 \qquad t_2 = 2 \qquad n_2 = 1.581$$
$$R_4 = -51.2 \qquad t_3 = 3 \qquad n_3 = 1.514$$

Solution

Since the rays are parallel to the axis, the second column of Table 18-3 is used to calculate the progress of the ray. These can be tabulated as follows:

Input	Results: ray at $h = 1$	Results: ray at $h = 5$
First surface:		
$n = 1, n' = 1.521$	$Q = 1$	$Q = 5$
$\alpha = 0$	$\alpha' = 0.1625°$	$\alpha' = 0.8128°$
$h = 1$ or 5	$s' = -352.66$	$s' = -352.53$
$R = -120.8$	$Q' = 1.0000$	$Q' = 5.0010$
Second surface:		
$t = 6$	$Q = 1.0170$	$Q = 5.0861$
$n = 1.581$	$\alpha' = 0.2202°$	$\alpha' = 1.1041°$
$R = -34.6$	$s' = -264.59$	$s' = -264.03$
	$Q' = 1.0170$	$Q' = 5.0876$
Third surface:		
$t = 2$	$Q = 1.0247$	$Q = 5.1261$
$n = 1.514$	$\alpha' = 0.2030°$	$\alpha' = 1.0178°$
$R = -96.2$	$s' = -289.26$	$s' = -288.58$
	$Q' = 1.0247$	$Q' = 5.1260$
Final surface:		
$t = 3$	$Q = 1.0353$	$Q = 5.1793$
$n = 1$	$\alpha' = -0.2883°$	$\alpha' = -1.4520°$
$R = -51.2$	$s' = 205.72$	$s' = 203.91$
	$Q' = 1.0353$	$Q' = 5.1672$

Thus the two rays intersect the optical axis at 205.72 and 203.91 mm beyond the final surface, missing a common focus by 1.8 mm.

PROBLEMS

18-1 A biconvex lens of 5 cm thickness and index 1.60 has surfaces of radius 40 cm. If this lens is used for objects in water, with air on its opposite side, determine its effective focal length and sketch its focal and principal points.

18-2 A double concave lens of glass with $n = 1.53$ has surfaces of 5 D (diopters) and 8 D, respectively. The lens is used in air and has an axial thickness of 3 cm.

a. Determine the position of its focal and principal planes.

b. Also find the position of the image, relative to the lens center, corresponding to an object at 30 cm in front of the first lens vertex.

c. Calculate the paraxial image distance assuming the thin-lens approximation. What is the percent error involved?

18-3 A biconcave lens has radii of curvature of 20 cm and 10 cm. Its refractive index is 1.50 and its central thickness is 5 cm. Describe the image of a 1-in.-tall object, situated 8 cm from the first vertex.

18-4 An equiconvex lens having spherical surfaces of radius 10 cm, a central thickness of 2 cm, and a refractive index of 1.61 is situated between air and water ($n = 1.33$). An object 5 cm high is placed 60 cm in front of the lens surface. Find the cardinal points for the lens and the position and size of the image formed.

18-5 A hollow glass sphere of radius 10 cm is filled with water. Refraction due to the thin glass walls is negligible for paraxial rays.

    a. Determine its cardinal points and make a sketch to scale.

    b. Calculate the position and magnification of a small object 20 cm from the sphere.

    c. Verify your analytical results by drawing appropriate rays on your sketch.

18-6 Light rays enter the plane surface of a glass hemisphere of radius 5 cm and refractive index 1.5.

    a. Using the system matrix representing the hemisphere, determine the exit elevation and angle of a ray that enters parallel to the optical axis and at an elevation of 1 cm.

    b. Enlarge the system to a distance x beyond the hemisphere and find the new system matrix as a function of x.

    c. Using the new system matrix, determine where the ray described above crosses the optical axis.

18-7 Using Figure 18-12b and c, verify the expressions given in Table 18-2 for the distances q, f_2, s, and w.

18-8 A lens has the following specifications:
$R_1 = 1.5$ cm $= R_2$, d(thickness) $= 2.0$ cm,
$n_1 = 1.00, n_2 = 1.60, n_3 = 1.30$.
Find the principal points using the matrix method. Include a sketch, roughly to scale, and do a ray diagram for a finite object of your choice.

18-9 A positive thin lens of focal length 10 cm is separated by 5 cm from a thin negative lens of focal length -10 cm. Find the equivalent focal length of the combination and the position of the foci and principal planes using the matrix approach. Show them in a sketch of the optical system, roughly to scale, and use them to find the image of an arbitrary object placed in front of the system.

18-10 A glass lens 3 cm thick along the axis has one convex face of radius 5 cm and the other, also convex, of radius 2 cm. The former face is on the left in contact with air and the other in contact with a liquid of index 1.4. The refractive index of the glass is 1.50. Find the positions of the foci, principal planes, and focal lengths of the system. Use the matrix approach.

18-11 **a.** Find the matrix for the simple "system" of a thin lens of focal length 10 cm, with input plane at 30 cm in front of the lens and output plane at 15 cm beyond the lens.

    b. Show that the matrix elements predict the locations of the six cardinal points as they would be expected for a thin lens.

    c. Why is $B = 0$ in this case? What is the special meaning of A in this case?

18-12 A gypsy's crystal ball has a refractive index of 1.50 and a diameter of 8 in.

    a. By the matrix approach, determine the location of its principal points.

    b. Where will sunlight be focused by the crystal ball?

18-13 A thick lens presents two concave surfaces, each of radius 5 cm, to incident light. The lens is 1 cm thick and has a refractive index of 1.50. Find (a) the system matrix for the lens when used in air and (b) its cardinal points. Do a ray diagram for some object.

18-14 An achromatic doublet consists of a crown glass positive lens of index 1.52 and of thickness 1 cm, cemented to a flint glass negative lens of index 1.62 and of thickness 0.5 cm. All surfaces have a radius of curvature of magnitude 20 cm. If the doublet is to be used in air, determine (a) the system matrix elements for input and output planes adjacent to the lens surfaces; (b) the cardinal points; (c) the focal length of the combination, using the lensmaker's equation and the equivalent focal length of two lenses in contact. Compare this calculation of f, which assumes thin lenses, with the previous value.

18-15 Enlarge the optical system of Figure 18-15 to include an object space to the left and an image space to the right of the lens. Let the new input plane be located at distance s in object space and the new output plane at distance s' in image space.

    a. Recalculate the system matrix for the enlarged system.

    b. Examine element B to determine the general relationship between object and image distances for the lens. Also determine the general relationship for the lateral magnification.

    c. From the results of (b), calculate the image distance and lateral magnification for an object 20 cm to the left of the lens.

    d. What information can you find for the system by setting matrix elements A and D equal to zero? (See Figure 18-9.)

18-16 Find the system matrix for a Cooke triplet camera lens (see Figure 3-23a). Light entering from the left encounters six spherical surfaces whose radii of curvature are, in turn, r_1 to r_6. The thickness of the three lenses are, in turn, t_1 to t_3, and the refractive indices are n_1 to n_3. The first and second air separations between lens surfaces are d_1 and d_2. Sketch the lens system with its cardinal points. How far behind the last surface must the film plane occur to focus paraxial rays?

Data: $r_1 = 19.4$ mm $t_1 = 4.29$ mm $n_1 = 1.6110$
        $r_2 = -128.3$ mm $t_2 = 0.93$ mm $n_2 = 1.5744$
        $r_3 = -57.8$ mm $t_3 = 3.03$ mm $n_3 = 1.6110$
        $r_4 = 18.9$ mm
        $r_5 = 311.3$ mm $d_1 = 1.63$ mm
        $r_6 = -66.4$ mm $d_2 = 12.90$ mm

18-17 Process the product of matrices for a thick lens, as in Eq. (18-15), without assuming the special conditions, $n = n'$ and $t = 0$. Thus find the general matrix elements $A, B, C,$ and D for a thick lens.

18-18 Using the cardinal point locations (Table 18-2) in terms of the matrix elements for a general thick lens (problem 18-17), verify that f_1 and f_2 are given by Eqs. (18-1) and (18-2).

18-19 Using the cardinal point locations (Table 18-2) in terms of the matrix elements for a general thick lens (problem 18-17), verify that the distances $r, s, v,$ and w are given by Eqs. (18-3) and (18-4).

18-20 Write a computer program that incorporates Eqs. (18-34) to (18-41) for ray tracing through an arbitrary number of refracting, spherical surfaces. The program should allow for the special cases of rays from far-distant objects and for plane surfaces of refraction.

18-21 Trace two rays through the hemispherical lens of Figure 18-15. The rays originate from the same object point, 2 cm above the optical axis and an axial distance of 10 cm from the first surface. One ray is parallel to the axis and the other makes an angle of $-20°$ with the axis.

18-22 Trace a ray originating 7 mm below the optical axis and 100 mm distant from a doublet. The ray makes an angle of $+5°$ relative to the horizontal. The doublet is an equiconvex lens of radius 50 mm, index 1.50, and central thickness 20 mm,

followed by a matched meniscus lens of radii -50 mm and -87 mm, index 1.8, and central thickness 5 mm. Determine the final values of $s, \alpha,$ and Q.

18-23 Trace two rays, both from far-distant objects, through a *Protor photographic lens*, one at altitude of 1 mm and the other at 5 mm. Determine where and at what angle the rays cross the optical axis. The specifications of this four-element lens, including an intermediate air space of 3 mm, are as follows, with distances in mm:

$R_1 = 17.5$	$t_1 = 2.9$	$n_1 = 1.6489$
$R_2 = 5.8$	$t_2 = 1.3$	$n_2 = 1.6031$
$R_3 = 18.6$	$t_3 = 3.0$	$n_3 = 1$
$R_4 = -12.8$	$t_4 = 1.1$	$n_4 = 1.5154$
$R_5 = 18.6$	$t_5 = 1.8$	$n_5 = 1.6112$
$R_6 = -14.3$		

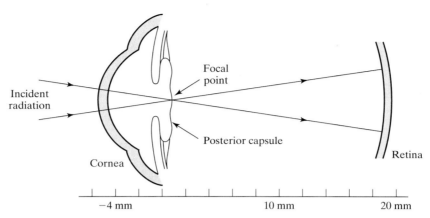

19 *Optics of the Eye*

INTRODUCTION

In this chapter we discuss the optics of the eye. First we examine the structure and functions of the eye. Following this we note the *errors of refraction* in a defective eye and indicate the usual corrective optics. Finally we describe several current surgical procedures that can restore visual acuity in less-than-perfect eyes.

The eyes, in conjunction with the brain, constitute a truly remarkable biooptical system. Consider briefly the distinctive characteristics of this system. It forms images of a continuum of objects, at distances of a foot to infinity. It scans a scene as expansive as the overhead sky or focuses on detail as small as the head of a pin. It adapts itself to an extraordinary range of intensities, from the barely visible flicker of a candle that is miles away on a dark night to sunlight so bright that the optical image on the retina causes serious solar burn. It distinguishes between subtle shades of color, from deep purple to deep red. Most importantly for us, functioning as a unique spatial sense organ, it localizes objects in space, accurately mapping out our three-dimensional world.

19-1 BIOLOGICAL STRUCTURE OF THE EYE

Anatomically, the eyeball is a globe, almost spherically shaped, approximately 22 mm in diameter. It lies buried in fat tissue inside the orbit, or space, in the skull surrounded by bony walls. Optically, the eyeball can be pictured as a positive lens system that refracts incident light onto its rear surface to form a real image, much as does an ordinary camera.

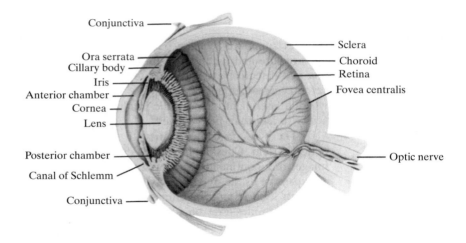

Figure 19-1 Vertical cross section of the eye. (Courtesy of Burroughs Wellcome Co.)

Parts of the Eye

The basic parts of the eye are shown in Figure 19-1. Let us examine the key biological components of the eye along the optical axis, in the same order as they are encountered by light rays in the usual image-forming process. Light first enters the eye through the *cornea*, a transparent tissue devoid of blood vessels but abounding in nerve cells. The cornea is roughly 12 mm in diameter and 0.6 mm thick at its center, thickening somewhat further at its edges, with a refractive index of 1.376. Upon entering the eye at the air-cornea interface, where the refractive index changes abruptly from 1.0 to 1.38, light undergoes a significant degree of bending. The corneal surface provides, in fact, about 73% of the total refractive power of the eye. Immediately behind the cornea is the *anterior chamber*, a small space filled with a watery fluid that provides nutrients for the cornea, the *aqueous humor*. This fluid has a refractive index of 1.336, almost equal to that of water (1.333). Because the refractive indices of the cornea and aqueous humor are nearly alike, little additional bending of rays occurs as light moves from the cornea into the anterior chamber. Situated in the aqueous humor is the *iris*, a diaphragm that gives the eye its characteristic color and controls the amount of light that enters. The amount and location of pigment in the iris determine whether the eye looks blue, green, gray, or brown. The adjustable hole or opening in the iris through which the light passes is called the *pupil*. The iris contains two sets of delicate muscles that change the pupil size in response to light stimuli, adjusting the diameter from a minimum of about 2 mm on a bright day to a maximum of about 8 mm under very dark conditions. While examining the inside of the eye with bright light, doctors often use drugs, such as *atropine*, to maintain the condition of an enlarged or dilated pupil.

Immediately upon passing through the pupil, light falls on the *crystalline lens*, a transparent structure about the size and shape of a small lima bean. The lens provides the fine-tuning in the final light-focusing process, changing its own shape appropriately to transform an external scene into a sharp image on the retina. The shape of the lens is controlled by the *ciliary muscle*, connected by fibers (*zonules*) to the periphery of the lens. When the muscles are relaxed, the lens assumes its flattest shape, providing the least refraction of incident light rays. In this state, the eye forms clear images of distant objects on the retina. When the muscles are tensed, the shape of the lens becomes increasingly curved, providing increased refraction of light. In this "strained" state, the eye is able to form clear images of nearby objects on the retina. The lens is itself a complex, onion-like layered mass of tissue, held intact by an elastic membrane. Due to the rather intricate laminar structure of fibrous tissue, the refractive index of the lens is not homogeneous. Near the center or core of the lens (on axis), the index is about 1.41; near the periphery it falls to about 1.38.

After its final refraction by the crystalline lens, light enters the *posterior chamber* or the *vitreous humor*, a transparent jellylike substance whose refractive index (1.336) is again close to that of water. The vitreous humor, essentially structureless, contains small particles of cellular debris that are referred to as *floaters*. They derive their name from the manner in which they are seen to float in one's field of view, when one looks or squints at a white ceiling, for example.

After traversing the vitreous humor, light rays reach their terminus at the inner rear layer of the eye, the *retina*, literally translated as "net." The retina is that part of the eye, the "screen," that receives light energy and converts it into electrochemical energy. The retina is a complex nerve-related structure that in essence is an outgrowth of the brain. Of the four principal layers that make up the retina—there are in fact ten layers visible with an electron microscope—the second layer below the surface contains the photoreceptors, the light-sensitive *rods* and *cones*. Through successive layers of *neurons* (bipolar cells), the electrical signals induced by the chemical changes in the photoreceptors are led out of the back of the eye along *axons*, which are the fibers of the optic nerve. The two types of photoreceptors, rods and cones, differ in shape, number, location, and function. Both rods and cones are composed of stacks of disks, with the disks of the long rods thinner than those of the shorter cones. The rods are located more densely toward the periphery of the retina. They are exceedingly sensitive to dim light and are unable to distinguish between colors. The cones cluster preferentially near the center of the retina, a 3-mm-diameter region called the *macula*. In sharp contrast to the response of the rods, the cones are *sensitive* to bright light and color but do not function well in dim light. In a human retina there are about 7 million cones and 75 to 150 million rods. These photoreceptors are linked to about 1 million optic nerve fibers, so there is a significant convergence of receptor signals onto the end of the optic nerve. The optic nerve is the main trunk line that carries visual information from the retina to the brain, completing the remarkable process of vision.

In addition to the key optical components encountered by light traveling along the axis of vision, the eye contains other components that should be mentioned. As noted in Figure 19-1, the eye is covered with a tough white coating, the *sclera*, that forms the supporting framework of the eye. Just inside the sclera lies the *choroid*, covering about four-fifths of the eye toward the back and containing most of the blood vessels that nourish the eye. The choroid, in turn, serves as the backing for the retina. At the center of the macula, located somewhat above the optic nerve, is the *fovea centralis*, the region of greatest visual acuity. When it is required that one see sharp and detailed information—while removing a small splinter with a needle, for example— the eyes move continually so that light coming from the area of interest falls precisely on the fovea, a rod-free region about 200 μm in diameter. Quite by contrast, another small region in the retina, located at the point of exit of the optic nerve, is completely insensitive to light. This spot, devoid of any receptors, is appropriately called the *blind spot*.

19-2 PHOTOMETRY

In Chapter 1 we introduced the terms and definitions associated with radiometry, which applies to the measurement of all radiant energy. *Photometry*, on the other hand, applies only to the *visible portion* of the electromagnetic spectrum. Whereas radiometry involves purely physical measurement, photometry takes into account the response of the human eye to radiant energy at various wavelengths and so involves psycho-physical measurements. The distinction rests on the fact that the human eye, as a detector, does not have a "flat" spectral response; that is, it does not respond with equal sensitivity at all wavelengths. If three sources of light of equal radiant power but radiating blue, yellow, and red light, respectively, are observed visually, the yellow source will appear to be far brighter than the others. When we use photometric quantities, then, we are measuring the properties of visible radiation as they appear to the normal eye, rather than as they appear to an "unbiased" detector. Since not all human eyes are identical, a standard response has been determined by the International Commission on Illumination (CIE) and is reproduced in Figure 19-2. The relative response or sensation of brightness for the eye is plotted versus wavelength, showing that peak sensitivity occurs at a "yellow-green" wavelength near 550 nm. Actually, the curve shown is the luminous efficiency of the eye for *photopic vision*, that is, when adapted for day vision. For lower levels of illumination, when adapted for night or *scotopic vision*, the curve shifts toward the green, peaking at 510 nm. It is interesting to note that human color sensation is a function of illumination and is almost totally absent at lower levels of illumination. One way to confirm this is to compare the color of stars, as they appear visually, to their photographic images made on color film using a suitable time exposure. On the other hand, very intense radiation may be visible beyond the limits of the CIE curve. The reflection of an intense laser beam of wavelength 694.3 nm from a ruby laser is easily seen. Even the infrared radiation around 900 nm from a gallium-arsenide semiconductor laser can be seen as a deep red.

A summary of the radiometric quantities and units introduced in Chapter 1 and the corresponding photometric quantities and units are given in Table 19-1. Radiometric quantities are related to photometric quantities through the luminous efficiency curve of Figure 19-2 in the following way: Corresponding to a radiant flux of 1 W at the peak wavelength of 555 nm, where the luminous efficiency is a maximum, the *luminous flux* is defined to be 685 lm. Then, for example, at λ = 610 nm, in the range where the luminous efficiency is 0.5 or

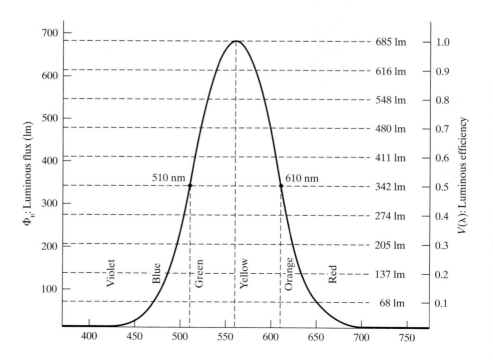

Figure 19-2   CIE luminous efficiency curve. The luminous flux corresponding to 1 W of radiant power at any wavelength is given by the product of 685 lm and the luminous efficiency at the same wavelength: $\Phi_v(\lambda) = 685V(\lambda)$ for each watt of radiant power.

TABLE 19-1   RADIOMETRIC AND PHOTOMETRIC TERMS

Term	Symbol (units)	Defining equation	Term	Symbol (units)	Defining equation
Radiant energy	Q_e (J = W $\cdot$ s)	—	Luminous energy	Q_v (lm $\cdot$ s) (talbot)	—
Radiant energy density	w_e (J/m^3)	$w_e = dQ_e/dV$	Luminous energy density	w_v (lm $\cdot$ s/m^3)	$w_v = dQ_v/dV$
Radiant flux	Φ_e (W)	$\Phi_e = dQ_e/dt$	Luminous flux	Φ_v (lm)	$\Phi_v = dQ_v/dt$
Radiant exitance	M_e (W/m^2)	$M_e = d\Phi_e/dA$	Luminous exitance	M_v (lm/m^2)	$M_v = d\Phi_v/dA$
Irradiance	E_e (W/m^2)	$E_e = d\Phi_e/dA$	Illuminance	E_v (lm/m^2) or (lx)	$E_v = d\Phi_v/dA$
Radiant intensity	I_e (W/sr)	$I_e = d\Phi_e/d\omega$	Luminous intensity (candlepower)	I_v (lm/sr) or (cd)	$I_v = d\Phi_v/d\omega$
Radiance	L_e (W/sr $\cdot$ m^2)	$Le = dI_e/dA \cos\theta$	Luminance	L_v (cd/m^2)	$L_v = dI_v/dA \cos\theta$

Abbreviations: J, joule; W, watt; m, meter; lm, lumen; lx, lux; sr, steradian; cd, candela.

50%, 1 W of radiant flux would produce only 0.5×685 or 342 lm of luminous flux. The curve shows that again at $\lambda = 510$ nm, in the blue-green, the brightness has dropped to 50%.

Photometric units, in terms of their definitions, parallel radiometric units. This is amply demonstrated in the summary and comparison provided in Table 19-1. In general, for monochromatic radiation of wavelength λ, analogous units are related by the following equation:

$$\text{photometric unit} = K(\lambda) \times \text{radiometric unit} \qquad (19\text{-}1)$$

where $K(\lambda)$ is called the *luminous efficacy*. If $V(\lambda)$ is the *luminous efficiency*, as given on the CIE curve, then

$$K(\lambda) = 685V(\lambda) \qquad (19\text{-}2)$$

Photometric terms are preceded by the word *luminous* and the corresponding units are subscripted with the letter v (*visual*); otherwise, the symbols are the same. (Radiometric quantities are sometimes subscripted with the letter e. Also, throughout most of this text we use the symbol I, rather than E_e, to represent irradiance.) Notice that the SI unit of luminous energy is the *talbot*, the unit of luminous incidence is the *lux* (lx), and the unit of luminous intensity is the *candela* (cd). Notice also the distinction between the analogous terms *irradiance* (radiometric) and *illuminance* (photometric).

Example 19-1

A lightbulb emitting 100 W of radiant power is positioned 2 m from a surface. The surface is oriented perpendicular to a line from the bulb to the surface. Calculate the irradiance at the surface. If all 100 W is emitted from a red bulb at $\lambda = 650$ nm, calculate also the illuminance at the surface.

Solution

$$\text{irradiance } E_e = P/A = 100 \text{ W}/4\pi(2 \text{ m})^2 \cong 2 \text{ W/m}^2$$

From the CIE curve, $V(650 \text{ nm}) = 0.1$. Thus,

$$\text{illuminance } E_v = K(\lambda) \times \text{irradiance} = 685V(\lambda) \times E_e$$

$$E_v = 685 \times 0.1 \times 2 = 137 \text{ lm/m}^2 \text{ or lux}$$

Thus, whereas a *radiometer* with aperture at the surface measures 2 W/m^2, a *photometer* in the same position would be calibrated to read 137 lx.

When the radiation consists of a spread of wavelengths, the radiometric and photometric terms may be functions of wavelength. This dependence is noted by preceding the term with the word *spectral* and by using a subscript λ or adding the λ in parentheses. For example, *spectral radiant flux* is denoted by $\Phi_{e\lambda}(\lambda)$ The total radiant flux is then determined by integration over the wavelength region of interest:

$$\Phi_e = \int_{\lambda_1}^{\lambda_2} \Phi_{e\lambda}(\lambda)\, d\lambda$$

19-3 OPTICAL REPRESENTATION OF THE EYE

As we have seen, the normal biological eye is a near spheroid, some 22 mm from cornea to retina. The optical surfaces that provide the bulk of the focusing power are essentially three: the air-cornea interface, the aqueous-lens interface, and the lens-vitreous interface. Overall, the eye can be represented approximately as a thin, positive lens of focal length equal to 17 mm in the relaxed state (distant vision) or 14 mm in the tensed state (near vision). In an attempt to represent the optical powers of the eye more faithfully, *schematic eyes* have been designed. Although still an approximation, a schematic eye presents a fairly valid representation of the true (but complex) biological eye.

A schematic eye (after H. V. Helmholtz and L. Laurance) that represents a living, biological eye with fair accuracy is shown in Figure 19-3. Relative locations of the refracting surfaces are shown, as are the cardinal points (discussed in Chapter 18) of interest for the eye as a whole. The schematic eye shown corresponds to its relaxed state. For the fully tensed eye, the front surface of the lens sharpens its curvature from a radius of $R = 10$ mm to $R = 6$ mm. By way of summary and in conjunction with Figure 19-3, Table 19-2 lists the important optical surfaces, their distances from the corneal vertex on the optical axis, several radii of curvature, indices of refraction, and refracting powers of the optical surfaces related to the cornea and lens. Note carefully that

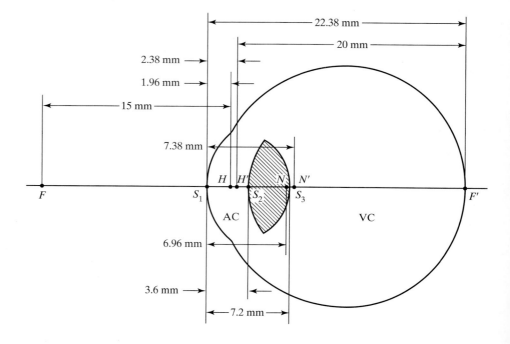

Figure 19-3 Representation of H. V. Helmholtz's schematic eye 1, as modified by L. Laurance. For definition of symbols, refer to Table 19-2. We have included the locations of the cardinal points for the eye as a whole. The meaning of these points is discussed in Chapter 18. (Adapted with permission from Mathew Alpern, "The Eyes and Vision," Section 12 in *Handbook of Optics*, New York: McGraw-Hill, 1978.)

TABLE 19-2   CONSTANTS OF A SCHEMATIC EYE

Optical surface or element	Defining symbol	Distance from corneal vertex (mm)	Radius of curvature of surface (mm)	Refractive index	Refractive power (diopters)
Cornea	S_1	—	$+8^a$	—	+41.6
Lens (unit)	L	—	—	1.45	+30.5
Front surface	S_2	+3.6	$+10^b$	—	+12.3
Back surface	S_3	+7.2	−6	—	+20.5
Eye (unit)	—	—	—	—	+66.6
Front focal plane	F	−13.04	—	—	—
Back focal plane	F'	+22.38	—	—	—
Front principal plane	H	+1.96	—	—	—
Back principal plane	H'	+2.38	—	—	—
Front nodal plane	N	+6.96	—	—	—
Back nodal plane	N'	+7.38	—	—	—
Anterior chamber	AC	—	—	1.333	—
Vitreous chamber	VC	—	—	1.333	—
Entrance pupil	E_nP	+3.04	—	—	—
Exit pupil	E_xP	+3.72	—	—	—

^{a} The cornea is assumed to be infinitely thin.
^{b} Value given is for the relaxed eye. For the tensed or fully accommodated eye, the radius of curvature of the front surface is changed to +6 mm.
Source: Adapted with permission from Mathew Alpern, "The Eyes and Vision," Table 1, Section 12, in *Handbook of Optics* (New York: McGraw-Hill Book Company, 1978).

the values for the refractive indices of various parts of the *schematic* eye, as well as radii of curvature of surfaces, may not agree with values of the *biological* eye itself. When taken as a whole, however, the optical values that describe the schematic eye do faithfully represent the optical performance of a living, biological eye.

19-4 FUNCTIONS OF THE EYE

To operate as an effective optical system, the eye must form a retinal image of an external object or scene, either distant or nearby, in bright as well as dim light. To achieve efficient operation, the eye takes advantage of special functions. To see objects closely and far away, the eye *accommodates*. To process light signals of varying brightness, the eye *adapts*. To sense the spatial orientation of three-dimensional scenes, the eyes make use of *stereoscopic vision*. To form a faithful, detailed image of the external object, the eye relies on its *visual acuity*. In what follows, we discuss each of these visual functions in somewhat more detail.

Accommodation
Depending on the distance of the object or scene from the eye, the lens accommodates—tenses or relaxes—appropriately to "fine-focus" the image on the retina. For a distant object, the ciliary muscle attached to the lens relaxes and the lens assumes a flatter configuration, increasing its radii of curvature and, consequently, its focal length. As the object moves closer to the eye, the ciliary muscle tenses or contracts, squeezing or bulging the lens and resulting in decreased radii of curvature and a shorter focal length. The smaller the radii of curvature and focal length, the higher the refractive or bending power of the lens, precisely the condition needed to bring near objects into sharp focus. In the normal eye—and before the normal aging process robs the lens of its elasticity and ability to reshape itself—accommodation produces faithful retinal images of objects from distant points (infinity) to nearby

points about one foot away. The *near point* (closest point of accommodation) moves further away from the eye with advancing age, starting at a position of 7 to 10 cm from the eye for a teenager, increasing to 20 to 40 cm for a middle-aged adult, and extending to as far as 200 cm in later years. For the average person, *presbyopia* (loss of accommodation) sets in during the early 40s, signaling the need for reading glasses to restore the near point to a comfortable position near 25 cm or so.

Adaptation

The ability of the eye to respond to light signals that range from very dim[1] to very bright, a range of light irradiances that differ by an astonishing factor of about 10^5, is referred to as *adaptation*. The amount of light (flux or photon number) that enters the eye is regulated first of all by the iris, with its adjustable aperture, the pupil. This adjustment of pupil diameter (from 8 mm down to 2 mm) cannot of itself account for the enormous range of intensities processed by the eye. The remarkable adaptivity of the eye can be traced, in fact, to the particular photosensitivity of the rods and cones in the retina of the eye. The key ingredient seems to be a pigment, called *visual pigment*, contained in both the rods and cones. The rods, stimulated by low-level light signals (*scotopic vision*), contain pigment of only one kind, called *visual purple*. The cones, sensitive to light signals of high intensity and variable color composition (*photopic vision*), each contain one of three different kinds of visual pigment. The numerous, thin rods are multiply connected to nerve fibers, making it possible for any one of a hundred rods or so to activate a single nerve fiber. The less numerous, wider cones in the macular region, by contrast, are individually connected to nerve fibers, and thus individually activated.

The activation of nerve fibers—the very heart of the vision process itself—depends on chemical changes that occur in the visual pigment contained in the rods and cones. When light falls on either type of photoreceptor, the visual pigment changes from a dark state to a clearer state, undergoing a sort of bleaching process. The change in state of the visual pigment in the rods or cones is transformed into an electrical output or nerve fiber impulse. These electrical impulses are transmitted to the optic nerve and on to the brain, recording as it were the light intensity of the stimulating signal. When the visual purple is fully bleached out in the rods, the photoreceptor cells become insensitive to further light signals and a regeneration of pigment in the rods must occur before they can respond again. Apparently, the single type of visual pigment in rods is much more sensitive to light than is any of the three pigments in cones. Accordingly, rods bleach out completely at much lower light levels than do cones. A change from low-level light or scotopic vision to high-level light or photopic vision in the process of adaption consists of a rapid bleaching out of rod pigment and a resulting insensitivity of the rod receptors. The bright light is then processed efficiently by the less-sensitive cones. Conversely, adaptation from intensely bright light (handled by the cones) to very dim light involves regeneration of pigment in the rods and a restoration of "night vision." In the full process of adaptation, the scotopic response is active over light levels that range from starlight on a clear, moonless night to lunar light from a quarter-moon. The photopic response (rods completely bleached out and inactive) operates between light levels ranging roughly from twilight to bright sunlight. Between light levels of quarter-moon and twilight, rods and cones both receive light and transmit nerve impulses.

Stereoscopic Vision

The ability to judge depth or position of objects accurately in a three-dimensional field is called *stereoscopic vision*. In humans, the optic nerves

---

[1]Signals containing fewer than 100 photons can trigger a visual response in humans.

from the two eyes come together at the *optic chiasma*, near the brain. From the optic chiasma, nerve fibers originating in the right half of each eye extend to the right half of the brain. Nerve fibers originating in the left half of each eye terminate in the left half of the brain. Thus, even though each half of the brain receives an image from *both* eyes, the brain forms but a single image. The fusion by the brain of two distinct images into a single image is referred to as *binocular vision*. Nevertheless, the slight differences between the two images from the left and right eyes provide the basis for stereoscopic vision in humans. It should be noted that even monocular vision is not without some depth perception. This is due to visual clues like parallax, shadowing, and the particular perspective of familiar objects.

To experience proper binocular vision without *double vision*, the images of an object must fall at corresponding points on each retina. This, of course, is what happens when the eyes move appropriately to focus on an object or scene, causing the image to fall on the fovea centralis of each eye. Most individuals are either right-eyed or left-eyed, indicating a dominance of one eye over the other. To determine which is your dominant eye, try the following simple test. Hold a pencil a foot or so in front of you at eye level. With both eyes open, line the pencil up with the vertical edge of a picture, door, or window across the room. Holding the pencil fixed, close one eye at a time. Whichever eye is open when the pencil remains lined up with the reference object is your dominant eye. The brain records the message seen by the dominant eye, while suppressing the other.

Visual Acuity

The ability to see detail clearly and to perceive real differences in spatial orientation of objects is related to *visual acuity*. This ability depends directly upon the resolving power of the eye or its minimum angle of resolution of two closely spaced objects or points. Technically speaking, visual acuity is defined as the reciprocal of the minimum angle of resolution.

Operationally, assessment of resolving power or visual acuity of the eye is measured in different ways. Two-point discrimination is referred to as *minimum separable* resolution; the smallest resolvable angle subtended by a black bar on a white background is called *minimum visible*, and the smallest angle subtended by block letters that can be read (on an eye chart) is called *minimum legible*. Since most of us, at one time or another, are required to read eye charts in a vision test, we limit our discussion of visual acuity to resolving powers associated with minimum legible resolution.

The nature of the eye chart owes its existence to a Dutch ophthalmologist, Herman Snellen. According to Snellen, the letters on the eye chart are constructed so that the overall block size of a letter, from top to bottom, or side to side, subtends an angle of 5′ of arc at the test distance. The detailed lines within a letter, such as the vertical bar in the letter T or the horizontal bar in the letter H, are all constructed so that the width of each "bar" subtends an angle of 1′ of arc at the test distance. The two choices of angle grew out of the best data available to Snellen on the minimum separable resolution of the eye. For Snellen, the normal eye could just resolve a letter that subtended 5′ of arc at 20 ft, with 1′ of arc contained in the details of the letter. (See Figure 19-4.) In this case the eye is considered "normal," and its visual acuity is referred to as "20/20 vision."

To detect defects in visual acuity, Snellen letters of different sizes are also included on the eye chart. For example, a very large letter may be such that it subtends angles of 5′ and 1′ of arc for a test distance of, say, 300 ft. Other letters are constructed of appropriate size, subtending angles of 5′ and 1′ for other selected distances, such as 200 ft, 100 ft, 80 ft, and so on, down to 15 or even 10 ft. Then, when the letters are read by a test subject at a test distance of 20 ft, visual acuity is measured in terms of the *Snellen fraction*. The numerator of the Snellen fraction expresses the fixed testing distance, and the

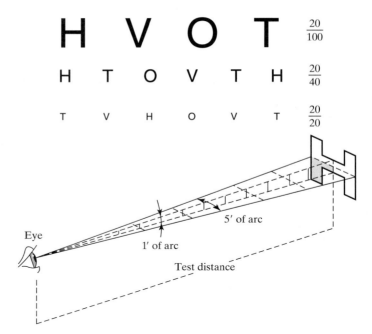

$$\frac{20}{100}$$

$$\frac{20}{40}$$

$$\frac{20}{20}$$

Figure 19-4 Construction of a Snellen eye-chart letter H to measure visual acuity. The top portion of the figure shows a section of an eye chart (reduced) containing the letter H and several other letters.

denominator expresses the distance at which the smallest readable letter subtends 5′ of arc overall. For example, if the large block letter E that subtends an angle of 5′ at 300 ft is just readable by a test subject seated 20 ft from the letter, visual acuity is reported as 20/300. A Snellen fraction of 20/300 means that the test subject sees poorly, reading at a distance of 20 feet what the normal eye reads as well at a distance of 300 ft. While normal vision is 20/20, visual acuity readings as good as 20/15 are not uncommon.

19-5 VISION CORRECTION WITH EXTERNAL LENSES

The errors of refraction of the eye lead to four well-known defects in vision: *myopia* (nearsightedness), *hyperopia* (farsightedness), *presbyopia*, and *astigmatism*. The first two are traceable, for the most part, to an abnormally shaped eyeball, axially too long or too short. Either deviation from normal length impairs the ability of the combined refracting elements, cornea and lens, to form a clear retinal image of objects located at both remote and nearby positions. As mentioned earlier, presbyopia refers to the loss of accommodation that occurs as the lens in an aging eye loses the ability to reshape itself and thus alter its focal length. The last defect, astigmatism, is due to unequal or asymmetric curvatures in the corneal surface, thereby rendering impossible the simultaneous focusing, for example, of all the spokes in a many-spoked wheel. Whether the errors of refraction occur singly or in some combination (as they often do), they are generally correctable with appropriately shaped external optics (eyeglasses or contact lenses).

As a point of reference for judging the departure of defective vision from the norm, refer to the *normal* eye depicted in the left column of Figure 19-5. With accommodation, the normal eye forms a distinct image of objects located anywhere between its far point (F.P.) at infinity and its near point (N.P.), nominally a distance of 25 cm for the young adult. When the normal eye looks at distant objects (that is, objects at "infinity"), parallel light enters the relaxed eye and forms a clear image (Figure 19-5a) on the retina. When the normal eye looks at objects at the near point, diverging light enters the tensed

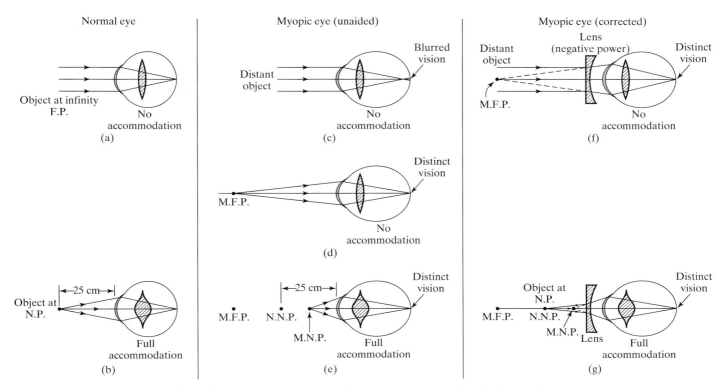

Figure 19-5   A comparison of normal and myopic vision, with optical correction. Note that refraction by the eye lens is not shown. The abbreviations read as follows: M.F.P. = myopic far point; N.N.P. = normal near point; M.N.P. = myopic near point.

eye (fully accommodated) and, again, a clear image is formed on the retina (Figure 19-5b).

If a corrective lens is placed in front of (or on) the eye, the eye no longer views a real object directly. Rather, the eye views a *virtual image* formed by the corrective lens. A good corrective lens forms, therefore, virtual images of real objects located anywhere from about 25 cm from the eye to very far from the eye at positions where the unaided defective eye sees clearly.

Myopia

When compared with the normal eye, a myopic eye or nearsighted eye is commonly found to be longer in axial distance—from cornea to retina—than the usual, accepted span of 22 mm. As a consequence, and as illustrated schematically in Figure 19-5c, the uncorrected myopic eye forms a sharp image of distant objects in *front* of the retina, and, of course, a blurred image results at the retina. Distinct retinal images are not formed with the unaccommodated myopic eye until the object moves inward from infinity and reaches the myopic far point (M.F.P.), the most distant point for clear vision (Figure 19-5d). From the far point inward, with appropriate accommodation, the myopic eye sees quite clearly, even at points *closer* than the normal near point (N.N.P.); see Figure 19-5e. Since angular magnification of detail increases with proximity to the eye, the myopic eye enjoys "superior" vision of objects held close to the eye. (Therefore, it can be an advantage for a watchmaker to be myopic, at least during working hours!) In short, then, the nearsighted person has a contracted, drawn-in field of vision, a less-remote far point and a closer near point than a person with normal vision. While the more proximate near point might serve as an advantage, the less-remote far point is a distinct disadvantage and calls for correction.

Myopic vision is routinely corrected with spectacles (or contact lenses) of *negative* dioptic power (diverging lenses) that effectively move both the myopic far point and near point *outward* to normal positions. Figure 19-5f shows the corrected vision for distant objects. Note that as far as the optics of the eye itself is concerned, light from distant objects appears to originate at its own myopic far point (M.F.P.), ensuring sharp vision. Similarly, Figure 19-5g illustrates the situation for corrected near vision under accommodation. Light from an object at the normal near point (N.N.P.) appears now to originate at the myopic near point (M.N.P.) closer in, thereby again ensuring clear vision for the myopic eye. A close examination of Figures 19-5f and 19-5g shows that the corrective negative lens provides clear vision for objects located anywhere from infinity to the normal near point, N.N.P.

To gain some insight into the degree of negative lens power required to correct myopic vision, consider Example 19-2.

Example 19-2

A myopic person (without astigmatism) has a far point of 100 cm and a near point of 15 cm. (a) What power contact lens should an optometrist prescribe to move the myopic far point out to infinity? (b) With this correction, can the myopic person also read a book held at the normal near point, 25 cm from the eye?

Solution

a. The corrective contact lens of focal length f should form an image (recall the sign convention for image positions) at $s' = -100$ cm of an object at infinity $(s = \infty)$. The image of a faraway object formed by the corrective lens will then be 100 cm in front of the eye so that the relaxed unaided eye can view it. Thus, referring to Figure 19-5f and making use of the thin-lens equation, one has

$$\frac{1}{s} + \frac{1}{s'} = \frac{1}{f} \quad \text{or} \quad \frac{1}{\infty} + \frac{1}{-100} = \frac{1}{f}$$

This gives $f = -100$ cm. Accordingly, the corrective contact lens should have a focal length of -100 cm. The optometrist would prescribe a contact lens with a power of $P = \dfrac{1}{f} = \dfrac{1}{-1.00 \text{ m}} = -1.00$ diopter $(-1.00$ D$)$.

b. Referring to Figure 19-5g and again applying the thin-lens equation with $f = -100$ cm and $s = 25$ cm, the virtual image distance s' is found from

$$\frac{1}{s} + \frac{1}{s'} = \frac{1}{f} \quad \text{or} \quad \frac{1}{25} + \frac{1}{s'} = \frac{1}{-100}$$

Solving gives $s' = -20$ cm. Thus the virtual image of the print held at $s = 25$ cm is formed by the contact lens at a distance of 20 cm in front of the eye. Since this myopic person can see clearly objects brought in as close as 15 cm from the eye, the virtual image of the print formed by the lens at 20 cm is seen without difficulty. In fact, using $(1/s) + (1/s') = 1/f$, with $s' = -15$ cm (myopic near point for person) and $f = -100$ cm, solving for s, one finds that objects can be brought in as close as 17.6 cm from the eye and still be seen clearly.

Hyperopia

The farsighted, or hyperopic, eye is commonly shorter than normal. Whereas the longer-than-normal myopic eye has too much convergence in its "optical system" and requires a diverging lens to correct its over-refraction, the shorter-than-normal hyperopic eye has too little convergence and requires a converging lens to increase refraction. The drawings in Figure 19-6,

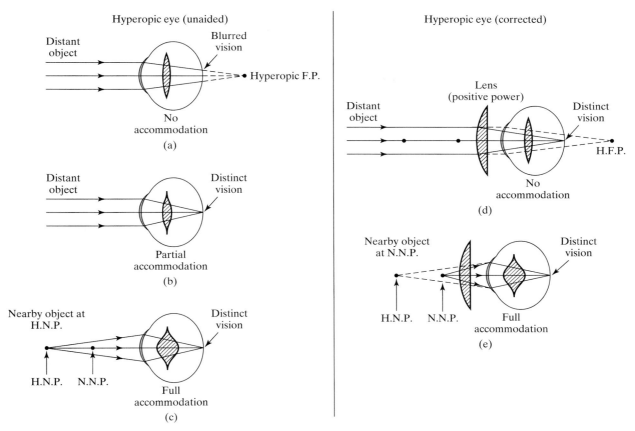

Figure 19-6 Hyperopic vision with correction. The abbreviations read as follows: H.N.P. = hyperopic near point; N.N.P. = normal near point; H.F.P. = hyperopic far point.

in analogy with those in Figure 19-5, illustrate the defects and correction associated with the farsighted eye. In Figure 19-6a, light from a distant object enters the relaxed eye and focuses *behind* the retina, causing blurred vision. The focal point behind the retina is considered as the hyperopic far point. Figure 19-6b shows that the hyperopic eye must (and can) accommodate to see distant objects clearly. In Figure 19-6c, it is clear that for distinct vision, the hyperopic near point is farther away than the normal near point. Consequently, objects located closer than the hyperopic near point would be out of focus, even with full accommodation. The two endpoint corrective measures, with the appropriate positive spectacle lens in place, are indicated in Figure 19-6d and e. The corrected eye now sees distant objects clearly, without accommodation, and at the normal near point it sees objects clearly, with full accommodation.

Let's see how an optometrist might calculate the spectacle power required to correct hyperopic vision. Consider the following example.

Example 19-3

A farsighted person is diagnosed to have a near point at 150 cm. It is found that a corrective lens of power 3.5 diopters placed 1.5 cm from the eye will allow this person to view objects at infinity with a relaxed eye. What is the near point (with the corrective lens) for this eye?

Solution

The near point with the corrective lens will be at the object distance from the lens for which a virtual image is formed at the near point of the unaided eye. Referring to Figure 19-6e and making use of the thin-lens equation, with $f = (1/3.5 \text{ diopters}) = 0.286 \text{ m} = 28.6 \text{ cm}$ and $s' = -148.5 \text{ cm}$, one can solve for the object distance corresponding to the near point of the corrected eye:

$$\frac{1}{s} + \frac{1}{s'} = \frac{1}{f} \quad \text{or} \quad \frac{1}{s} + \frac{1}{-148.5} = \frac{1}{28.6}$$

Calculation gives $s = 24.0 \text{ cm}$. This object distance corresponds to a total distance of 25.5 cm from the eye. So, with the corrective lens, this person can see faraway objects with a relaxed eye and objects as near as 25.5 cm from the eye with a fully accommodated eye.

Presbyopia

As mentioned in the preceding section, human eyes lose ability to accommodate with age. As a result, the near point of an eye tends to move further from the eye as the eye ages. Correction for presbyopia is similar to that for hyperopia in that a converging lens is needed to form clear images of nearby objects on the retina. As we show in Example 19-4, a corrective lens of a single power (as used in *reading glasses*) will not restore clear vision of both near and distant objects for a person who has lost accommodation. Rather, to restore clear vision of objects positioned anywhere from about 25 cm from the eye to very far from the eye, the presbyopic eye requires a lens of multiple powers. *Bifocals*, in which lenses of different powers are situated in the upper and lower half of a spectacle frame, can restore, to the presbyopic eye, clear vision of both near and distant objects. A person with bifocals learns to look through the upper half of the spectacle lenses to view distant objects and uses the lower half of the spectacle lenses to read. *Multifocal* lenses, in which the power gradually increases from the top to the bottom of the lens, are now available. Such a multifocal lens can be used to restore clear vision, to the presbyopic eye, over the full distance range. In Example 19-4 we consider the effect of using reading glasses on the vision of a person with presbyopic eyes.

Example 19-4

A certain presbyopic eye can form clear retinal images of objects that are positioned anywhere from 150 cm from the eye to very far from the eye. (a) What power spectacle lens, placed 2 cm from the eye, will allow this eye to form clear retinal images of the print on a computer screen that is as close as 40 cm to the eye? (b) With the corrective lens of part (a), what will be the far point of this eye?

Solution

a. To move the near point in from 150 cm for the unaided eye to 40 cm for the eye/corrective lens system, the corrective lens must form a virtual image of an object placed 40 cm from the eye at a location of 150 cm from the eye. In that case, the presbyopic, unaided eye can clearly view the virtual image formed by the corrective lens. Using the thin-lens sign conventions and noting that the spectacle lens is placed 2 cm from the eye, this condition implies that the image distance from the corrective lens should be $s' = -148 \text{ cm}$ when the object distance is $s = 38 \text{ cm}$. The thin-lens equation can be used to determine the correct power for the spectacle lens,

$$\frac{1}{s} + \frac{1}{s'} = \frac{1}{f} \quad \text{or} \quad \frac{1}{38} + \frac{1}{-148} = \frac{1}{f}$$

Calculation gives $f = 51$ cm. The power of this lens is $P = \dfrac{1}{f} = \dfrac{1}{0.51 \text{ m}} = 1.96$ dipoters.

b. Assuming that *without* the corrective lens, the *relaxed* eye forms clear retinal images of faraway objects, the far point of the eye/corrective lens system will be at the position of the object whose image formed by the spectacle lens is faraway. The thin lens equation with $s' = -\infty$ gives

$$\frac{1}{s} + \frac{1}{s'} = \frac{1}{f} \quad \text{or} \quad \frac{1}{s} + \frac{1}{-\infty} = \frac{1}{51}$$

Thus, $s = 51$ cm. The far point with the corrective lens is therefore 53 cm from the eye. With the corrective lens, this eye can see clearly print held at any position from 40 to 53 cm from the eye.

Astigmatism

The astigmatic eye suffers from uneven curvature in the surface of the refracting elements, most significantly, the cornea. Generally speaking, the radii of curvature of the corneal surface in two meridional planes (those containing the optical axis) are unequal. Such asymmetry leads to different refractive powers and, consequently, to image formation at different distances from the cornea, resulting, of course, in blurred vision. If the two meridional planes are orthogonal to one another, one horizontal and the other vertical, say, the defect is referred to as *regular astigmatism*, a condition that is correctable with appropriate spectacles. If the two planes are not orthogonal, a rather rare condition called *irregular astigmatism*, the surface anomaly is not so easily corrected. Regular astigmatism can be corrected with a lens that has cylindrical surfaces ground on the back surface of the required spectacle lens. Assume, for example, that the refractive power in the vertical meridian of the cornea is greater by 1 diopter than the power in the horizontal meridian. This situation means that the corneal surface is more sharply curved in the vertical meridian and that vertically oriented details in an object are brought to a focus nearer the cornea than are horizontally oriented details. Consider now a cylindrical surface with a negative power of 1 diopter in the vertical meridian. Since a cylinder has no curvature along its axis of symmetry, the surface has no power in the horizontal meridian. If this surface is included in the spectacle design, it would cancel exactly the distortion introduced by the cornea and equalize powers in both meridians. As a result, the images of vertical and horizontal details in the object scene are formed at the same distance from the cornea, and astigmatic blurring does not occur.

Typically, blurred vision is a result of astigmatism mixed with myopia or hyperopia. If myopic astigmatism is present, for example, vision is faulty on two counts. The myopia itself causes an overall blurring of distant objects; the astigmatism compounds the problem by adding considerably more blurring in one meridian than another. Correcting for both defects is accomplished with sphero-cylindrical lenses, spherical surfaces to correct for myopia and cylindrical surfaces to correct for astigmatism.

When optometrists prescribe corrective eyeglasses for conditions of myopic or hyperopic astigmatism, they generally identify three numbers. For myopic astigmatism, the three numbers, written in prescription format, might be

$$\text{R}_\text{x}: \quad -2.00 \quad -1.00 \times 180$$

For hyperopic astigmatism, the prescription might read

$$\text{R}_\text{x}: \quad +2.00 \quad -1.50 \times 180$$

The first number in the prescription refers to the *sphere power*, the power in diopters of the spherical surfaces on the spectacle lens required to correct for

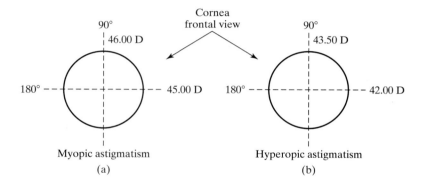

Figure 19-7 Conditions of myopic and hyperopic astigmatism with corrective spectacle prescriptions. (a) Refraction in the 180° meridian yields −2.00 D of myopia. The eyeglass prescription is R$_x$: −2.00 −1.00 × 180. (b) Refraction in the 180° meridian yields +2.00 D of hyperopia. The eyeglass prescription is R$_x$: +2.00 −1.50 × 180.

the overall myopia or hyperopia. The second number refers to the *cylinder power*, the power of the cylindrical surface superimposed on the back surface of the spectacle lens required to correct for astigmatism. The third number refers to the orientation of the *cylinder axis*, specifying whether the axis of the cylinder is to be vertical, horizontal, or somewhere in between. In optometric notation, the horizontal axis is referred to as the 180° axis, or simply "× 180," and the vertical axis as "× 90."

Figure 19-7 indicates the optical conditions associated with the corrective prescriptions just cited for both myopic and hyperopic astigmatism. For the case of myopic astigmatism, Figure 19-7a, the corneal surface is evidently less sharply curved in the horizontal meridian (power = 45.00 D) than in the vertical meridian (power = 46.00 D). The myopic correction, always measured in the meridian of least refractive power, is found in this instance to be −2.00 D, along the horizontal meridian. The astigmatic correction, with cylinder axis horizontal (× 180), is determined to be −1.00 D. With the appropriate cylindrical surface ground on the rear of the spectacle lens, the correction of −1.00 D reduces the power in the vertical meridian from 46.00 D to 45.00 D, thereby equalizing the refracting powers in the two meridians and negating the corneal astigmatism.

Figure 19-7b shows a comparable condition and prescription for hyperopic astigmatism. Note that a sphere power correction of +2.00 D is needed to correct for the hyperopia, and a cylinder power correction of −1.50 D is needed along the vertical meridian (× 180) to equalize the refractive power in the two orthogonal meridians.

19-6 SURGICAL VISION CORRECTION

In the preceding section we discussed the correction of myopia and hyperopia with the help of contact lenses or eye glasses. Alternatively, the refractive power of the eye itself can be altered through surgical procedures that reshape the cornea. In this section we discuss three such procedures; *radial keratotomy, corneal sculpting,* and *conductive keratoplasty.*

Radial Keratotomy
The first eye-shaping procedure used to correct myopia was called radial keratotomy. Radial keratotomy introduces radial cuts in the cornea of the myopic eyeball (see Figure 19-8). After the cuts have healed, the cornea flattens, refractive power decreases, and as a result, normal, or near-normal, vision can be restored. This radical procedure, done mostly with a surgical blade in the hands of a skilled ophthalmologist, had its beginning in the Soviet Union in 1972. As the story is told, a rather myopic Soviet lad, Boris Petrov, was engaged in a schoolyard fight in Moscow. In the usual exchange, a hard punch struck one of the thick lenses he was wearing, shattering it into multiple fragments. Some of the shards of glass, in one of those freak accidents of nature,

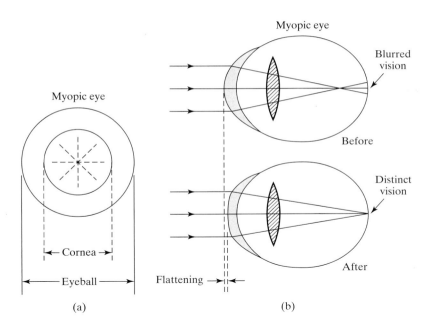

Figure 19-8   Correcting myopia by reshaping the cornea with radial keratotomy. (a) Frontal view showing radial cuts. (b) Side view before and after cuts.

embedded themselves in the lad's cornea in a somewhat regular, radial pattern, deep enough to form cuts but not deep enough to penetrate the cornea. The Soviet ophthalmologist, Svyatoslav N. Fyodorov, who treated the youngster did not hold out much hope for restored vision in the eye, even though the cuts were superficial. Astonishingly, though, as the corneal surface healed with all its scars, the cornea flattened out and most of the myopia disappeared. The lad saw better than ever before. Recognizing the significance of what he had witnessed, Fyodorov replicated, under controlled conditions, what nature and a fist had accomplished so haphazardly. Today, as a remedy for myopia, radial keratotomy has largely been supplanted by the two, more precise, corneal sculpting procedures discussed in the next subsection.

Corneal Sculpting

Laser corneal sculpting is a medical procedure, introduced in the early 1980s, that uses beams of laser energy to reshape the surface of the cornea. Typically, a computer-controlled argon-fluoride excimer laser—of wavelength 193 nm— directs UV radiation onto the cornea in order to remove microscopic amounts of tissue at strategic locations. This careful recontouring of the surface is designed to correct for myopia, hyperopia, astigmatism, and combinations thereof.

Two procedures currently available for corneal sculpting are known as *PRK (photorefractive keratectomy)* and *LASIK (laser in-situ keratomileusis)*. Each procedure modifies the top layer of the 530-μm-thick cornea, while preserving the structural integrity of the layers underneath. An optical topography of the cornea, taken prior to either procedure, identifies the regions of the cornea that need to be reshaped in order to make the needed corrections to the refractive power of the eye. This topography provides the information required for careful control of the positioning, the energy output, the beam diameter, and the pulse length of the computer-driven excimer laser used to reshape the cornea. In PRK, the laser *removes* the corneal epithelium (topmost surface) and then recontours the underlying layers as needed. In LASIK, a microkeratome (sharp blade) is first used to create a corneal flap, 160 to 180 μm thick. The flap, hinged along one edge, opens like the page of a book to reveal the underlying layers. The flap is held open while the underlying layer is reshaped with the excimer laser, and then the flap is replaced when the reshaping is completed. Thus, in contrast to the PRK procedure, in LASIK surgery the corneal epithelium is not removed. In a recent advancement, the

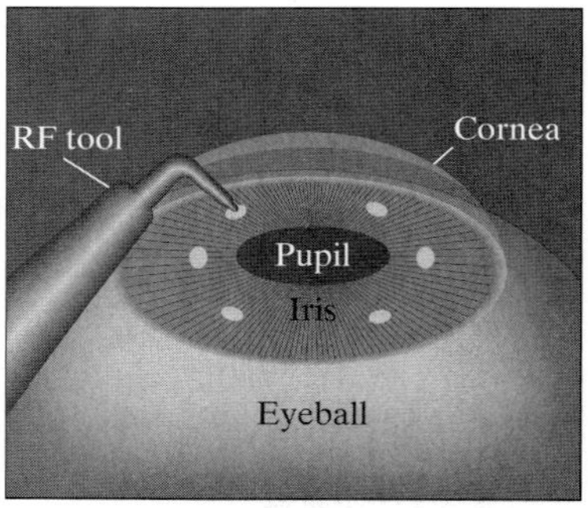

Figure 19-9 Creating corneal lesions with a radio frequency probe in the CK procedure.

flap is created with a precisely controlled laser so that computer-driven lasers perform the entire optical-sculpting procedure.

To correct for myopia, tissue in the center of the cornea is removed in order to effectively flatten the cornea and reduce the refractive power of the eye—allowing the relaxed eye to form images of faraway objects on the retina. For hyperopia, the curvature of the corneal surface is steepened by removing tissue around the periphery of the cornea, thereby increasing the refractive power of the eye. Astigmatism can be corrected by removing tissue in an asymmetric manner, thus producing an eye whose refractive power is the same along any meridian.

Conductive Keratoplasty (CK)

An alternate corrective procedure, known as *conductive keratoplasty* (CK), is currently available to patients over 40 years of age with mild amounts of hyperopia—up to about +2.5 diopters. This *nonlaser* procedure uses the mild heat produced from radio waves to create tiny spots (lesions) along the periphery of the central region of the cornea (see Figure 19-9). These carefully numbered and carefully positioned lesions act like a tightening, contracting belt that causes the cornea to bulge or steepen in the center, increasing the refractive power of the eye. Presbyopia can be remedied in some patients by using CK to correct the refractive power of one of the eyes for near vision while the other eye is left untreated and is used by the patient for middle-to-distant vision.

PROBLEMS

19-1 a. A 50-mW He-Cd laser emits at 441.6 nm. A 4-mW He-Ne laser emits at 632.8 nm. Using Figure 19-2, compare the relative brightness of the two laser beams of equal diameter when projected side by side on a white piece of paper. Assume photopic vision.

    b. What power argon laser emitting at 488 nm is required to match the brightness of a 0.5-mW He-Ne green laser at 543.5 nm under the conditions of (a)?

19-2 A lamp located 3 m directly above a point P on the floor of a room produces at P an illuminance of 100 lm/m^2.

    a. What is the luminous intensity of the lamp?

    b. What is the illuminance produced at another point on the floor, 1 m distant from P?

19-3 A driveway is illuminated at night by identical lamps at the top of two poles 30 ft high and 40 ft apart. Assuming the lamps radiate equally in all directions, compare the illuminance at ground level for points directly under one lamp and midway between them.

19-4 A small source of 100 cd is situated at the focal point of a spherical mirror of 50-cm focal length and 10-cm diameter. What is the average illuminance of the parallel beam reflected from the mirror, assuming an overall reflectance of about 80%?

19-5 a. The sun subtends an angle of 0.5° at the earth's surface, where the illuminance is about 10^5 lx at normal incidence. Determine the luminance of the sun.

    b. Determine the illuminance of a horizontal surface under a hemispherical sky with uniform luminance L.

19-6 A circular disc of radius 20 cm and uniform luminance of 10^5 cd/m^2 illuminates a small plane surface area of 1 cm^2, 1 m distant from the center of the disc. The small surface is oriented such that its normal makes an angle of 45° with the axis joining the centers of the two surfaces. The axis is perpendicular to the circular disc. What is the luminous flux incident on the small surface?

19-7 Reference to Table 19-2 indicates that the corneal radius of curvature for the unaccommodated schematic eye is 8 mm. Treating the cornea as a thin surface (whose own refraction can be neglected), bounded by air on one side and aqueous humor on the other, determine the refractive power (see Section 2-10) of the corneal surface.

19-8 Consider the unaccommodated crystalline lens of the eye as an isolated unit having radii of curvature and effective refractive index as given for the schematic eye in Table 19-2.

    a. Calculate its focal length and refracting power as a thin lens in air.

    b. Calculate its focal length and refracting power in its actual environment, surrounded on both sides with fluid of effective index 1.33. Assume a thin lens.

    c. Calculate its focal length and refracting power again by treating it as a thick lens of thickness 3.6 mm. (The matrix techniques of Chapter 18 may well be applied to this problem).

19-9 Taking values for refractive indices and separation of elements from the schematic of the unaccommodated eye given in Table 19-2 and Figure 19-3, determine the distance behind the cornea where an image is focused for (a) an object at infinity and (b) an object at 25 cm from the eye. Use the Gaussian formula for image formation by a spherical surface in a three-step chain of calculations. In part (b), assume that the fully accommodated eye differs in the following ways: The front surface of the lens is more sharply curved, having a radius of +6 mm, but the back surface remains at −6 mm. As a result, the thickness of the lens along the axis increases to 4.0 mm, and the distance from cornea to the front surface of lens is shortened to 3.2 mm.

19-10 Use the matrix approach of Chapter 18 to find the system matrix for the unaccommodated schematic eye of Table 19-2 and Figure 19-3.

a. Determine the four matrix elements of the system matrix where the system extends from the first refraction at the cornea to the final refraction at the second lens surface.

b. From the matrix elements, determine the first and second focal points and the first and second principal points relative to the corneal surface. Compare with the distances given in Figure 19-3.

19-11 You have been asked to design a Snellen eye chart for a test distance of 5 ft. The chart is to include rows of letters to test for visual acuities of 20/300 (same as 5/75), 20/100, 20/60, 20/20, and 20/15. Determine the size of the block letter and letter detail (in inches) for each row of letters.

19-12 A presbyopic eye has no astigmatism, a near point of 125 cm, and a far point of infinity. Correction, with glasses using a lens placed 1.5 cm from the eye, requires that this person see objects at the normal near point (25 cm) clearly.

a. What is the power of the corrective lens?

b. With the lens of part (a), what is the far point of the corrected eye?

19-13 One eye of a person has a far point of 50 cm and a near point of 15 cm.

a. What power contact lens is needed to correct the far point of this eye?

b. Using the contact lens, what is the new near point of this eye?

c. Repeat parts (a) and (b) if the corrective lens is a spectacle lens placed 2 cm from the eye.

19-14 Consider each of the following spectacle prescriptions and describe the refractive errors that are involved:

a. −1.50, −1.50, axis 180

b. −2.00

c. +2.00

d. +2.00, −1.50, axis 180

19-15 Consider a woman with two myopic eyes. The vision in the woman's left eye is corrected with a contact lens of power −7 diopters, and the vision in this woman's right eye is corrected with a contact lens of power −5 diopters. The *corrected* near point of each eye is 15 cm and the corrected far point of each eye is infinity.

a. Find the near and far points for each unaided eye.

b. Suppose this woman mistakenly puts the right contact lens in her left eye and the left contact lens in her right eye. What are the near and far points of each eye when wearing the wrong corrective lens?

19-16 Consider a myopic, presbyopic eye with a near point of 13 cm and a far point of 15 cm. Design bifocals that will allow this person to see clearly both faraway objects and objects at a comfortable reading distance from the eye.

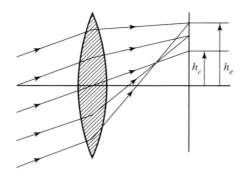

20 Aberration Theory

INTRODUCTION

The paraxial formulas developed earlier for image formation by spherical reflecting and refracting surfaces are, of course, only approximately correct. In deriving those equations, it was necessary to assume paraxial rays, that is, rays both near to the optical axis and making small angles with it. Mathematically, the power expansions for the sine and cosine functions, given by

$$\sin x = x - \frac{x^3}{3!} + \frac{x^5}{5!} - \cdots$$

$$\cos x = 1 - \frac{x^2}{2!} + \frac{x^4}{4!} - \cdots$$

were accordingly approximated by their first terms. To the extent that these first-order approximations are valid, Gaussian optics implies exact imaging. The inclusion of higher-order terms in the derivations, however, predicts increasingly larger departures from "perfect" imaging with increasing angle. These departures are referred to as "aberrations." When the next term involving x^3 is included in the approximation for $\sin x$, a *third-order aberration theory* results. The aberrations have been studied and classified by the German mathematician Ludwig von Seidel and are referred to as third-order or *Seidel aberrations*. For monochromatic light, there are five Seidel aberrations: *spherical aberration, coma, astigmatism, curvature of field*, and *distortion*. An additional aberration, *chromatic aberration*, results from the wavelength dependence of the imaging properties of an optical system. The full details of aberration theory are too formidable to treat in this chapter. We include here a brief, quantitative description of how the various aberrations follow from a third-order treatment and a qualitative description of each aberration, with typical procedures for its elimination.

20-1 RAY AND WAVE ABERRATIONS

The departure from ideal, paraxial imaging may be described quantitatively in several ways. In Figure 20-1 two wavefronts are shown emerging from an optical system. Wavefront W_1 is a spherical wavefront representing the Gaussian, or paraxial, approximation that produces an image at I. Wavefront W_2 is an example of the actual wavefront, an aspherical envelope whose shape represents an exact solution of the optical system. This shape could be deduced by precisely tracing a sufficient number of rays, using the laws of reflection and refraction, through the optical system. Rays from adjacent points A and B, being normal to their respective wavefronts, do not intersect the paraxial image plane at the same point. The "miss" along the optical axis, represented by the distance LI, is called the *longitudinal aberration*, and the miss IS, measured in the image plane, is called the *transverse*, or *lateral, aberration*. These are *ray aberrations*. Alternatively, the aberration may be described in terms of the deviation of the deformed wavefront from the ideal at various distances from the optical axis. At the location of point B, shown in Figure 20-1, the wave aberration is given by the distance AB. Notice that rays from both wavefronts, at their point O of tangency on the optical axis, reach the same image point I. Rays from intermediate points of the actual wavefront between O and B intersect the image screen at other points around I, producing a blurred image, the result of aberration. The maximum ray aberration thus indicates the size of the blurred image. The ultimate goal of optical design is to reduce the ray aberrations until they are comparable to the unavoidable blurring due to diffraction itself.

Lateral ray aberrations corresponding to the wave aberration AB may be calculated once the variation in AB with perpendicular distance from the optical axis is known. Referring to Figure 20-2, the angle α between actual and ideal rays from a point P of the wavefront, at elevation y, is the same as the angle between wavefront tangents at P. The wavefronts, having been shaped by the optical system, exist in image space with refractive index n_2. The offset detail construction of Figure 20-2 then shows that the incremental wave aberration da, expressed as an optical path length in image space, is

$$da = n_2(\alpha \, dy) \tag{20-1}$$

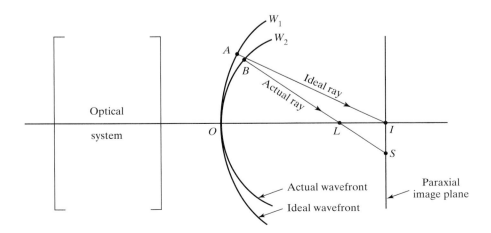

Figure 20-1 Illustration of ray and wave aberrations.

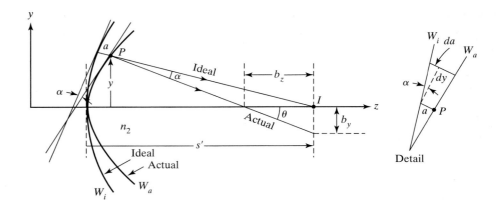

Figure 20-2 Construction used to relate the ray aberrations b_y and b_z to the wave aberration a. The detail shows how to relate a change da in wave aberration to a change dy in the aperture dimension.

The derivative da/dy describes the local curvature of the wavefront at P. The lateral ray aberration b_y due to the rays from the neighborhood of P may then be approximated by

$$b_y = \alpha s' = \frac{s' da}{n_2\, dy} \tag{20-2}$$

where s' is the paraxial image distance from the wavefront and α has been taken from Eq. (20-1). Similarly, along the other transverse direction, perpendicular to the y,z-axes in the plane of the page,

$$b_x = \frac{s' da}{n_2\, dx} \tag{20-3}$$

The longitudinal ray aberration b_z is related to the lateral ray aberration b_y by

$$b_z = \frac{b_y}{\tan \theta} = \frac{s' b_y}{y} \tag{20-4}$$

20-2 THIRD-ORDER TREATMENT OF REFRACTION AT A SPHERICAL INTERFACE

Let us solve now the case of refraction from a single spherical surface, where we improve the approximation to include "third-order" angle effects. In Figure 20-3, an arbitrary ray PQ from an axial object point P is refracted by a

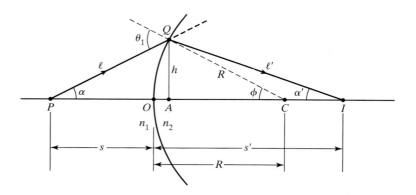

Figure 20-3 Refraction of a ray at a spherical surface.

spherical surface, centered at C, that separates media of refractive indices n_1 and n_2. The refracted ray locates an axial image at I. To a first approximation, the optical path lengths of rays PQI and POI are identical, according to Fermat's principle. Aberration contributes to the formation of the image because, beyond a first approximation, the ray path PQI depends on the position of point Q along the spherical surface. Thus we define the aberration at Q as

$$a(Q) = (PQI - POI)_{opd} \tag{20-5}$$

where opd indicates the optical-path difference. More precisely,

$$a(Q) = (n_1\ell + n_2\ell') - (n_1 s + n_2 s') \tag{20-6}$$

Employing the cosine law in triangles PQC and CQI, the lengths ℓ and ℓ' may be exactly expressed, in terms of the quantities defined in Figure 20-3, by

$$\ell^2 = R^2 + (s + R)^2 - 2R(s + R)\cos\phi \tag{20-7}$$

$$\ell'^2 = R^2 + (s' - R)^2 + 2R(s' - R)\cos\phi \tag{20-8}$$

Now,

$$\cos\phi = (1 - \sin^2\phi)^{1/2} = [1 - (h/R)^2]^{1/2} \tag{20-9}$$

where we have used the fact that $\sin\phi = h/R$. The binomial expansion then permits expansion to the fourth power in h:

$$\cos\phi \cong 1 - \frac{h^2}{2R^2} - \frac{h^4}{8R^4} \tag{20-10}$$

Introducing Eq. (20-10) into Eqs. (20-7) and (20-8) and rearranging terms,

$$\ell = s\left(1 + \left[\frac{h^2(R + s)}{Rs^2} + \frac{h^4(R + s)}{4R^3 s^2}\right]\right)^{1/2} \tag{20-11}$$

$$\ell' = s'\left(1 + \left[\frac{h^2(R - s')}{Rs'^2} + \frac{h^4(R - s')}{4R^3 s'^2}\right]\right)^{1/2} \tag{20-12}$$

Next, representing the quantities enclosed in square brackets by x in Eq. (20-11) and x' in Eq. (20-12), the square roots of the expression in outer parentheses may be approximated, again using the binomial expansion:

$$(1 + x)^{1/2} \cong 1 + \frac{x}{2} - \frac{x^2}{8} \tag{20-13}$$

Thus,

$$\ell \cong s\left(1 + \frac{x}{2} - \frac{x^2}{8}\right) \tag{20-14}$$

$$\ell' \cong s'\left(1 + \frac{x'}{2} - \frac{x'^2}{8}\right) \tag{20-15}$$

When all terms of order higher than h^4 are discarded, there remains

$$\ell = s\left[1 + \frac{h^2(R + s)}{2Rs^2} + \frac{h^4(R + s)}{8R^3s^2} - \frac{h^4(R + s)^2}{8R^2s^4}\right] \tag{20-16}$$

$$\ell' = s'\left[1 + \frac{h^2(R - s')}{2Rs'^2} + \frac{h^4(R - s')}{8R^3s'^2} - \frac{h^4(R - s')^2}{8Rs'^4}\right] \tag{20-17}$$

To find the aberration $a(Q)$ for an axial object point, these expressions for ℓ and ℓ' can be introduced into Eq. (20-6). In the result, there occur terms proportional to both h^2 and h^4. The h^2 term is proportional to the expression

$$\left(\frac{n_1}{s} + \frac{n_2}{s'}\right) - \left(\frac{n_2 - n_1}{R}\right)$$

When this expression is set equal to zero, it reproduces the Gaussian formula for imaging by a spherical surface. Terms proportional to this expression, therefore, vanish, by Fermat's principle. Upon setting this expression equal to zero, there remains in $a(Q)$, then, only the third-order aberration represented by a term in h^4,

$$a(Q) = -\frac{h^4}{8}\left[\frac{n_1}{s}\left(\frac{1}{s} + \frac{1}{R}\right)^2 + \frac{n_2}{s'}\left(\frac{1}{s'} - \frac{1}{R}\right)^2\right] \quad \text{axial object points}$$

$$\tag{20-18}$$

When h is small enough, the rays are essentially paraxial and the aberration represented by this term may be negligible. In any case, since the square brackets include quantities independent of h, we have shown that third-order theory predicts a wave aberration $a(Q)$ that is proportional to the fourth power of the aperture h, measured from the optical axis, or

$$a(Q) = ch^4 \quad \text{axial object points} \tag{20-19}$$

where c represents the constant of proportionality. This is the principal result of our calculation for *axial* object points. We will use this in generalizing the aberration calculation to include *off-axis* imaging. In this way, the other Seidel aberrations will also appear.

The aberration $a(Q)$ we have calculated as a difference in optical-path lengths between ideal and actual rays must correspond to the wave aberration AB of Figure 20-1. The deviation AB of the actual from the ideal spherical wavefront is clearly a function of the distance from the optical axis at which the ray intersects the wavefront and is referred to as *spherical aberration*.

Before examining spherical aberration in more detail, however, we wish to show how the other third-order aberrations arise. To do this, we need to consider the case of an off-axis object point. Shown in Figure 20-4 are two pencils of rays whose limits are determined by an aperture E_nP serving as the entrance pupil. (See Chapter 3.) An axial pencil (shaded) from the on-axis object point O forms an image at and around the paraxial image point I. This image will be affected by spherical aberration, as discussed earlier, to a degree determined by the displacement y of the extreme rays of the pencil. This pencil is symmetrical about the axis OCI, where C is the center of curvature of the refracting surface. Also shown is an oblique pencil of rays originating at the off-axis point O'. This pencil is certainly not symmetrical about the axis OI; in the absence of the limiting aperture E_nP, its axis of symmetry would be

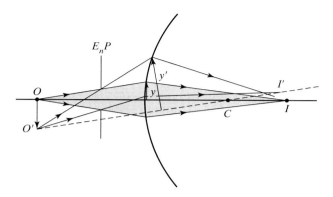

Figure 20-4 Comparison of axial and oblique pencils of rays from an object, defined by passage through entrance aperture E_nP.

the line $O'CI'$. It is from this axis that the displacement y' of the rays of the oblique pencil would have to be measured to determine the degree of aberration described by Eq. (20-19). Notice that such displacement from the axis of symmetry is much greater in the case of the oblique pencil. Thus an oblique pencil of rays due to off-axis object points is far more susceptible to aberration than corresponding axial points. The position of the aperture is critical in determining the magnitude of y' and is least harmful in this respect when placed at the center of curvature, C. (In this regard, one may recall the use of symmetrical lenses or lens combinations, such as the achromatic double meniscus objective, where the aperture is placed midway between them.)

Consider then the off-axis pencil of rays from object point P, as shown in Figure 20-5. The aberration function $a'(Q)$ for the point Q on the wavefront may be expressed as

$$a'(Q) = (PQP' - PBP')_{opd} = c(BQ)^4 = c\rho'^4 \qquad (20\text{-}20)$$

In Eq. (20-20) we relate the elevation of the ray PQP' to the axis PBP' and consider points $B, O,$ and Q to lie in a vertical plane approximating the wavefront at O. It can be shown that this approximation does not affect the results of third-order aberration theory. We have also made use of Eq. (20-19) and identified the distance BQ with a quantity ρ'. A section of the plane that includes the relevant points and defines the distances $\rho', b,$ and r is also shown in Figure 20-5 (detail). In a similar manner, we may write, for the wavefront point O,

$$a'(O) = (POP' - PBP')_{opd} = c(BO)^4 = cb^4 \qquad (20\text{-}21)$$

If the point Q is referred to the optical axis OC, an off-axis aberration function $a(Q)$ may be expressed as the difference between the axial aberrations at Q and O found previously.

$$a(Q) = a'(Q) - a'(O) = c\rho'^4 - cb^4 = c(\rho'^4 - b^4) \qquad (20\text{-}22)$$

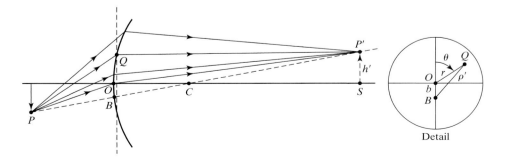

Figure 20-5 Imaging of off-axis point P. Aberration at an arbitrary point Q on the wavefront may be related to the symmetry axis PBP' or the optical axis OCS. The detail shows a frontal view of a portion of a wavefront.

Applying the cosine law to the triangle BOQ in the geometric detail shown in Figure 20-5, we have

$$\rho'^2 = r^2 + b^2 + 2rb \cos \theta$$

and introducing this expression for ρ' into Eq. (20-22) gives

$$a(Q) = c(r^4 + 4r^2b^2 \cos^2 \theta + 2r^2b^2 + 4r^3b \cos \theta + 4rb^3 \cos \theta) \tag{20-23}$$

From similar triangles OBC and SCP' in Figure 20-5, we see that the distance $OB = b$ is proportional to the height h' of the paraxial image P' above the optical axis. This may be expressed by

$$b = kh' \tag{20-24}$$

where k is the appropriate proportionality constant. When b in Eq. (20-23) is replaced by kh', we have, lumping all constants into term-by-term coefficients,

$$a(Q) = {}_0C_{40}r^4 + {}_1C_{31}h'r^3 \cos \theta + {}_2C_{22}h'^2r^2 \cos^2 \theta$$

$$+ {}_2C_{20}h'^2r^2 + {}_3C_{11}h'^3r \cos \theta \tag{20-25}$$

The C coefficients in Eq. (20-25) are subscripted by numbers that specify the powers of the term dependence on h', r, and $\cos \theta$, respectively. For example, the C coefficient ${}_1C_{31}$ accompanies the term $h'r^3 \cos \theta$, where h' is to the first power, r is cubed, and $\cos \theta$ is to the first power. The individual terms describe wavefront aberrations that contribute to the total aberration at the image. These terms comprise the five monochromatic, or Seidel, aberrations, as follows:

r^4	spherical aberration
$h'r^3 \cos \theta$	coma
$h'^2r^2 \cos^2 \theta$	astigmatism
h'^2r^2	curvature of field
$h'^3r \cos \theta$	distortion

Each aberration is characterized by its dependence on h' (departure from axial imaging), r (aperture of refracting surface), and θ (symmetry around the axis). Notice that the first term for spherical aberration agrees with Eq. (20-19), derived for axial imaging, where h represents the aperture.

We now briefly describe each of these aberrations in terms of their visual effects and indicate some means that are employed to reduce them.

20-3 SPHERICAL ABERRATION

The aberration known as *spherical aberration* results from the first term, ${}_0C_{40}r^4$, in Eq. (20-25). It is the only term in the third-order wave aberration $a(Q)$ that does not depend on h'. Thus spherical aberration exists even for axial object and image points, as illustrated for a single lens in Figure 20-6a. The paraxial image point I is distinct from axial image points, such as E, due to rays refracted at lens positions further from the optical axis. The axial miss distance EI, due to rays from the extremities of the lens, provides the usual measure of *longitudinal spherical aberration,* whereas the distance IG in the paraxial image plane measures the corresponding *transverse spherical aberration.* These

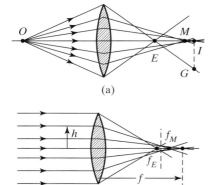

Figure 20-6 Spherical aberration of a lens, producing in (a) different image distances and in (b) different focal lengths, depending on the lens aperture.

quantities also depend on the object distance. When E is to the left of I, as shown for the case of a positive lens, the spherical aberration is positive; for a negative lens, E falls to the right of I, and the spherical aberration is considered negative. At some intermediate point M between E and I, a "best" focus is attained in practice. The broadened image there is called, descriptively, the "circle of least confusion." Using Eqs. (20-2) and (20-4) for lateral aberration b_y and longitudinal aberration b_z, the corresponding spherical aberrations in the yz-plane may be determined with the help of Eq. (20-25) as follows:

$$b_y = \frac{s'}{n_2}\frac{da}{dy} = \frac{s'}{n_2}\frac{da}{dr} = \frac{4_0 C_{40}s'}{n_2}r^3$$

and

$$b_z = \frac{s'b_y}{y} = \frac{s'b_y}{r} = \frac{4_0 C_{40}s'^2}{n_2}r^2$$

Example 20-1

Axially collimated light enters a glass rod through its end, a convex, spherical surface of radius 4 cm. The glass rod has a refractive index of 1.60. Determine the longitudinal and lateral spherical ray aberrations for light entering at an aperture height of $h = 1$ cm.

Solution

According to Eq. (20-18), with object distance s very large, there remains

$$a = -\frac{h^4}{8}\left[\frac{n_2}{s'}\left(\frac{1}{s'} - \frac{1}{R}\right)^2\right]$$

To calculate b_y and then b_z, one needs the derivative da/dh:

$$\frac{da}{dh} = -\frac{h^3}{2}\left[\frac{n_2}{s'}\left(\frac{1}{s'} - \frac{1}{R}\right)^2\right]$$

The image distance s', also the focal length of the surface, is found from the paraxial equation, giving

$$\frac{1}{\infty} + \frac{1.6}{s'} = \frac{0.6}{4} \quad \text{or} \quad s' = 10.667 \text{ cm}$$

Then da/dh and the spherical aberrations are,

$$\frac{da}{dh} = -\frac{1}{2}\left[\frac{1.6}{10.67}\left(\frac{1}{10.67} - \frac{1}{4}\right)^2\right] = -0.001831$$

$$b_y = \frac{s'}{n_2}\frac{da}{dy} = \frac{s'}{n_2}\frac{da}{dh} = \frac{10.667}{1.6}(-0.001831) = -0.0122 \text{ cm}$$

$$b_z = \frac{s'}{r}b_y = \frac{s'}{h}b_y = \frac{10.667}{1}(-0.0122) = -0.130 \text{ cm}$$

Figure 20-6b shows spherical aberration when the object is at infinity. Various circular zones of the lens about the axis produce different focal lengths, so that f is a function of aperture h. The specified focal length of the

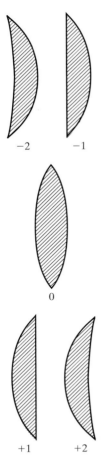

-2   -1

0

+1   +2

Figure 20-7 "Bending" of a single lens into various shapes having the same focal length. The Coddington shape factor below each shape serves to classify them.

lens is due to the intersection of paraxial rays for which $h \to 0$. This focal length is given by the lensmaker's formula,

$$\frac{1}{f} = (n-1)\left(\frac{1}{r_1} - \frac{1}{r_2}\right) \tag{20-26}$$

for a thin lens of refractive index n and radii of curvature r_1 and r_2, when used in air. From Eq. (20-26) it is obvious that a given f may result from different combinations of r_1 and r_2. Various choices of the radii of curvature, while not changing the focal length, may have a large effect on the degree of spherical aberration of the lens. Figure 20-7 illustrates the "bending," or change in shape, of a lens as its radii of curvature vary but its focal length remains fixed. A measure of this bending is the *Coddington shape factor* σ, defined by

$$\sigma = \frac{r_2 + r_1}{r_2 - r_1} \tag{20-27}$$

where the usual sign convention for r_1 and r_2 is assumed. For example, a thin lens of $n = 1.50$ and $f = 10$ cm may result from an equiconvex lens of $\sigma = 0$ ($r_1 = 10, r_2 = -10$ cm); a plano-convex lens of $\sigma = +1$ ($r_1 = 5$ cm); a meniscus lens of $\sigma = +2$ ($r_1 = 3.33, r_2 = 10$ cm). These shapes, as well as their mirror images with negative shape factors, are shown in Figure 20-7.

The spherical aberration of a single, spherical refracting surface is given in Eq. (20-18). A thin lens combines two such surfaces, each of which contributes to the total spherical aberration. The total longitudinal spherical aberration, $s_h' - s_p'$, of a thin lens with focal length f and index n, where s_h' is the image distance for a ray at elevation h, s_p' is the paraxial image distance, and

$$p = \frac{s' - s}{s' + s}$$

is given by[1]

$$\frac{1}{s_h'} - \frac{1}{s_p'} = \frac{h^2}{8f^3}\frac{1}{n(n-1)}\left[\frac{n+2}{n-1}\sigma^2 + 4(n+1)p\sigma\right.$$

$$\left. + (3n+2)(n-1)p^2 + \frac{n^3}{n-1}\right] \tag{20-28}$$

One can show further (problem 20-11), that minimum (but not zero!) spherical aberration results when the bending is such that

$$\sigma = -\frac{2(n^2-1)}{n+2}p \tag{20-29}$$

Notice that for an object at infinity $\sigma \cong 0.7$ for a lens of refractive index $n = 1.50$. This shape factor is close to that of the plano-convex lens with $\sigma = +1$. Accordingly, optical systems often employ plano-convex lenses (with the convex side facing the parallel incident rays) to reduce spherical aberration. In general, a minimum in spherical aberration is associated with the condition of equal refraction by each of the two surfaces, calling to mind the case of minimum deviation in a prism. When lenses are used in combination, the possibility of canceling spherical aberration arises from the fact that

---

[1]See Francis A. Jenkins and Harvey E. White, *Fundamentals of Optics*, 4th ed. (New York: McGraw-Hill Inc. 1976), Ch. 9.

positive and negative lenses produce spherical aberration of opposite sign. A common application of this technique is found in the cemented "doublet" lens.

20-4 COMA

Coma, represented by the term $_1C_{31}h'r^3 \cos \theta$, is an off-axial aberration ($h' \neq 0$) that is not symmetrical about the optical axis ($\cos \theta \neq$ constant) and increases rapidly as the third power of the lens aperture. Because of coma, an off-axis object point images as a blurred shape that resembles a *comet* with a head and a tail—hence the name "coma."

Figure 20-8 illustrates the formation of a comatic image. Figure 20-8a shows four parallel rays—1, 2, 3, and 4—arriving from a distant, below-the-axis object point and being refracted by a convex lens. The rays all pass through a thin annular region of the lens, centered about the optical axis. We will refer to this region as a *zone*. Thus the lens can be considered as a series of concentric

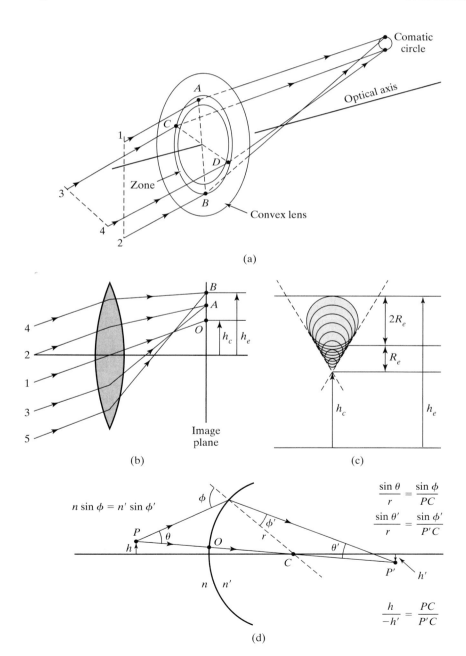

Figure 20-8 (a) and (b) Coma due to fans of parallel rays. The image points due to all of the azimuthal fans that pass through the same annular region lie on a comatic circle. (c) Formation of a comatic image from a series of comatic circles. The shape of the comatic image is such that its maximum extension is three times the radius of the comatic circle formed by rays from the outer zone of the lens. The angle between the dashed lines is 60°. (d) Nonparaxial rays from object point *P* near the axis form an image at *P'*, subject to the Abbe sine condition. The condition follows from Snell's law and the geometric relationships given in the figure.

zones, extending from its center to its outer edge. One such zone ($ABCD$) is shown. All rays from the distant object point that pass through the zone form the *comatic circle* shown. Rays 1 and 2 are in a vertical plane and pass through points A and B of the zone, whereas rays 3 and 4 are in a horizontal plane and pass through points C and D of the zone. The top of the comatic circle is formed by rays 1 and 2; the bottom by rays 3 and 4. Each such zone of the lens produces its own comatic circle, whose diameter increases as the radius of the zone increases.

In Figure 20-8b, a vertical fan of rays (numbered 1, 2, 3, 4, and 5) is shown passing through the center and two outer zones of the lens. The central ray arrives at point O in the image plane. Rays 2 and 3 form the top of the comatic circle at point A for their zone, whereas rays 4 and 5 form the top of the comatic circle at point B for their zone. The height above the optical axis of the bottommost point O is shown as h_c and the distance to the outermost comatic circle at point B is shown as h_e. A sketch of these and several others of the comatic circles is shown in Figure 20-8c. Since the lens zones are continuous, so are their associated comatic circles. Thus, the inner details of these comatic circles are not visible and all that is seen is the cometlike comatic shape. Figure 20-8b shows that each zone produces a different magnification, so that h_c due to the central ray is not equal to h_e due to the extreme rays. Coma, like spherical aberration, may occur as a positive quantity ($h_e > h_c$) or a negative quantity ($h_e < h_c$). Notice, as shown in Figure 20-8c, that the maximum extent of the comatic image ($R_e + 2R_e$) is three times the radius R_e of the largest comatic circle.

Without the usual paraxial approximation—restricting rays to those making small angles with the axis—one can show that for a small object near the axis, any ray from an object point that is refracted at a spherical interface must satisfy the *Abbe sine condition*,

$$nh \sin \theta + n'h' \sin \theta' = 0 \tag{20-30}$$

Here h and h' are object and image size, respectively, and the angles θ and θ' are the slope angles of the rays in optical media n and n', respectively. These quantities are illustrated in Figure 20-8d. When Eq. (20-30) is rearranged to express the lateral magnification, the condition can be written

$$m = \frac{h'}{h} = -\frac{n \sin \theta}{n' \sin \theta'}$$

To prevent coma, the lateral magnification resulting from refraction by all zones of a lens must be the same. Thus coma is absent when, for all values of θ,

$$\frac{\sin \theta}{\sin \theta'} = \text{constant}$$

The bending of a lens, found useful in reducing spherical aberration, is also useful in reducing coma. The Coddington shape factor, Eq. (20-27), which results in minimum spherical aberration, is close to that producing zero coma, so that both aberrations may be significantly reduced in the same lens by proper bending. One can show that coma is absent in a lens when

$$\sigma = \left(\frac{2n^2 - n - 1}{n + 1} \right) \left(\frac{s - s'}{s + s'} \right) \tag{20-31}$$

For the example of the lens considered previously, with $n = 1.50$ and object at infinity, Eq. (20-31) gives a value of $\sigma = 0.8$, quite close to the value of $\sigma = 0.7$, which yielded minimum spherical aberration. A lens or optical system free of both spherical aberration and coma is said to be *aplanatic*.

20-5 ASTIGMATISM AND CURVATURE OF FIELD

Aplanatic optics is still susceptible to two closely related aberrations whose wave aberration terms can be combined to give $h'^2r^2(_2C_{22}\cos^2\theta + {_2C_{20}})$. The first term produces *astigmatism*, and the second, which is symmetrical about the optical axis, is called *curvature of field*. Both aberrations increase similarly with the off-axis distance of the object and with the aperture of the refracting surface.

Figures 20-9a and b illustrate the astigmatic images of an off-axis point P due to a tangential fan of rays through the section tt' and a sagittal fan of

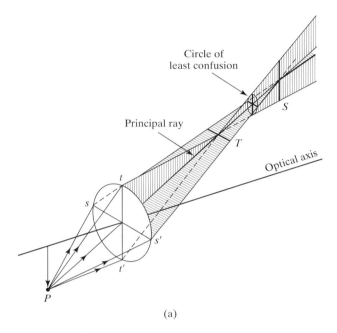

(a)

(b)

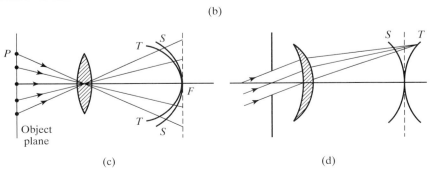

(c)   (d)

Figure 20-9 (a) Astigmatic line images T and S of an off-axial point P due to tangential (tt') and sagittal (ss') fans of light rays through a lens. (b) Photograph of astigmatic images formed by a lens, as illustrated in Figure 20-9a. The separated line images T and S are revealed as sections of the beam by fluorescent screens. (From M. Cagnet, M. Francon, and J. C. Thrierr, *Atlas of Optical Phenomenon*, Plate 4, Berlin: Springer-Verlag, 1962.) (c) Astigmatic surfaces in the field of a lens. (d) Use of a stop to artificially "flatten" the field of a lens. The compromise surface between the S and T surfaces is indicated by the dashed line.

rays through the section ss' of a single lens. Since these perpendicular fans of rays focus at different distances from the lens, the two images are line images, shown as T and S for the tangential and sagittal fans, respectively. The focal line T lies in the sagittal plane, and the focal line S falls in the tangential plane. If a screen held perpendicular to the principal ray is moved from S to T, intermediate images will be elliptical in shape. Approximately midway between S and T, the focus will be circular, the *circle of least confusion*. The locus of the line images S and T for various object points P are paraboloidal surfaces, as illustrated in Figure 20-9c. The deviation between the two surfaces along any principal ray from a given object point measures the magnitude of the astigmatism for this object point, approximately proportional to the square of the distance from the optical axis. When the T surface falls to the left of the S surface, as shown, the astigmatic difference is taken as positive; otherwise it is negative.

If points like P fall along a circle in an object plane perpendicular to the optical axis, the corresponding line images in the T surface merge into a well-focused image circle. In the S surface, however, the image of the circle will not be sharp, having everywhere the width of the S focal line. On the other hand, for object points along radial lines in the object circle, sharp radial images are produced only in the S surface, where the elongated radial images merge to produce well-focused radial lines. Thus if the object plane contains both circular and radial elements, the image distance for a good focus will be different for each type of element, with a compromise image somewhere between.

From the point of view of Figure 20-9c, the elimination of astigmatism requires that the tangential and sagittal surfaces be made to coincide. When the curvatures of these surfaces are changed by altering lens shapes or spacings so that they coincide, the resulting surface is called the *Petzval surface*. In this focal surface, for an aplanatic system, point images are formed. If the surface is curved, then, although astigmatism has been eliminated, the associated aberration called *curvature of field* remains. To record sharp images under these conditions, the film must be shaped to fit the Petzval surface. A Petzval surface can be determined for any optical system, even when the T and S surfaces do not coincide.

Unlike the T and S surfaces, the Petzval surface is unaffected by lens bending or placement and depends only on the refractive indices and focal lengths of the lenses involved. In third-order theory, the Petzval surface is always situated three times farther from the T surface than from the S surface and always lies on the side of the S surface opposite to that of the T surface. For example, two thin lenses (of respective indices of refraction n_1 and n_2 and respective focal lengths f_1 and f_2) will have a flat Petzval surface, eliminating curvature of field, if

$$n_1 f_1 + n_2 f_2 = 0$$

In general, the Petzval surface for a number of thin lenses in air satisfies

$$\sum \frac{1}{n_i f_i} = \frac{1}{R_p} \tag{20-32}$$

where R_p is the radius of curvature of the Petzval surface. Field flattening by this condition cannot be accomplished for a single lens, but artificial field flattening may be accomplished by use of an aperture stop positioned as in Figure 20-9d. In this arrangement, oblique chief rays, now determined by the aperture, do not penetrate the lens center. The S and T astigmatic surfaces then appear oppositely curved, and the surface of least confusion is flat, as shown. This inexpensive method for artificially flattening the field has been used in simple box cameras. In more difficult situations, where the Petzval condition cannot be satisfied without sacrificing other requirements, a low-power lens is sometimes used near the image plane. The lens helps to counteract curvature of field without otherwise seriously compromising image quality. Finally, according to fifth-order aberration theory, the T and S surfaces may actually be made to come together again and intersect at some distance from the optical axis. The result is less average astigmatism over the compromise focal plane. The *anastigmat* camera objective is designed to take advantage of this.

20-6 DISTORTION

The last of the five monochromatic Seidel aberrations, present even if all the others have been eliminated, is *distortion*, represented by the term $_3C_{11}h'^3r\cos\theta$. Even though object points are imaged as points, distortion shows up as a variation in the lateral magnification for object points at different distances from the optical axis. If the magnification increases with distance from the axis, the rectangular grid of Figure 20-10a, serving as object, will have an image as shown in Figure 20-10b. This is descriptively called *pincushion distortion*. On the other hand, if magnification decreases with distance from the axis, the image appears as in Figure 20-10c, with *barrel distortion*. The image in either case is sharp but distorted. Such distortion is often augmented due to the limitation of ray bundles by stops or by elements effectively acting as stops. To see this effect, refer to Figure 20-11a. Shown there is the image of an off-axis point, formed by a single lens. Two pencils of rays are drawn—one shaded, one clear—each limited by an aperture stop located (1) at some distance from the lens and (2) near the lens. As the aperture approaches the lens, it permits a shorter average distance to the lens. Thus, it can be seen in Figure 20-11a that for aperture position (1), the average distance PM for the shaded pencil is greater than the average distance PN for the lower pencil. Similarly, the image distance MP' is less than the distance NP'. Thus, the ratio MP'/PM (magnification for the upper pencil) is less than NP'/PN (magnification for the lower pencil). Therefore, the lateral magnification is less for aperture position (1) than for aperture position (2). This decrease in lateral magnification due to the aperture position is more noticeable as the object point recedes farther from the axis, so that the image suffers from barrel distortion. The effect of placing the aperture stop on the image side of the lens can also be seen from the same figure by reversing all rays and the roles played by object and image. Now the ratio of effective object-to-image distance is smaller, and pincushion distortion appears in the image. When the aperture stop is placed at the position of the lens, such distortion does not occur. Also, a symmetric doublet with a central stop, combining both effects, is free from distortion for unit magnification. Photographs of the effects of stop location on distortion are reproduced in Figure 20-11b, c, and d.

20-7 CHROMATIC ABERRATION

The final aberration to be discussed is not one of the Seidel aberrations, which are all monochromatic aberrations. Neither our first-order (Gaussian or paraxial) approximations nor the third-order theory sketched briefly in the preceding sections took into account an important fact of refraction: the variation of refractive index with wavelength, or the phenomenon of dispersion. Because of dispersion, an additional *chromatic aberration* appears, even for paraxial optics, in which images formed by different colors of light are not coincident. In terms of the monochromatic third-order aberrations of Eq. (20-25), we could introduce chromatic effects by considering the wavelength dependence of each of the coefficients of the terms.

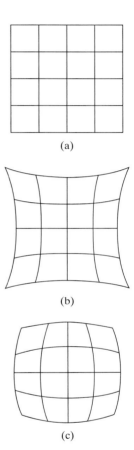

(a)

(b)

(c)

Figure 20-10 Images of a square grid (a) showing pincushion distortion (b) and barrel distortion (c) due to nonuniform magnifications.

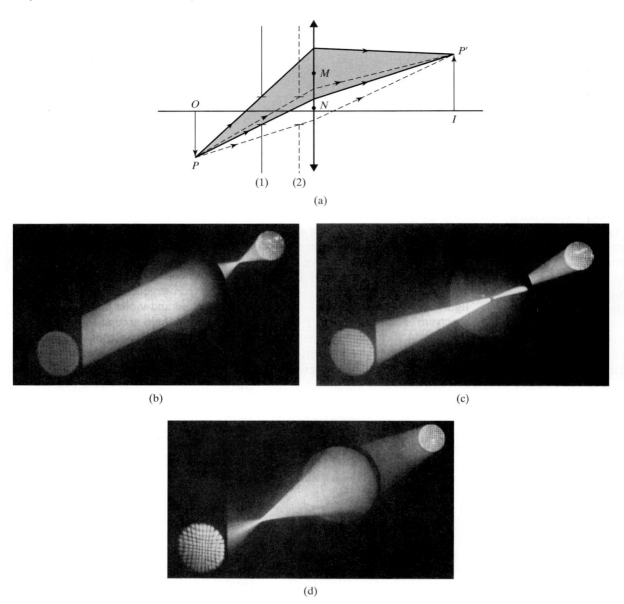

Figure 20-11   (a) Effect of an aperture stop on the distortion of an image by a lens. The aperture in position (1) produces more barrel distortion than it does in position (2). If object and image are interchanged, the same system produces pincushion distortion. (b) Image of a square grid by a positive lens. With the stop located between object (far right) and lens, barrel distortion occurs in the image. (c) Image of a square grid by a positive lens. With the stop located at the lens, the image is free from distortion. (d) Image of a square grid by a positive lens. With the stop located between lens and image, pincushion distortion occurs in the image. (Figures 20-11b, c, and d from M. Cagnet, M. Francon, and J. C. Thrierr, *Atlas of Optical Phenomenon*, Plate 5, Berlin: Springer-Verlag, 1962.)

Chromatic aberration is briefly discussed in Chapter 3. Here we review and extend that discussion. Since the focal length f of a lens depends on the refractive index n of the glass, f is also a function of wavelength. Figure 20-12a shows convergence of parallel incident light rays by a lens to distinct focal points for the red and violet ends of the visible spectrum. Notice that a cone of violet light will form a halo around the red focus at R. If the incident light contains all wavelengths of the visible spectrum, intermediate colors focus between these points on the axis. Just as for a prism, greater refraction of shorter wavelengths brings the violet focus nearer the lens for the positive lens shown.

Figure 20-12b illustrates chromatic aberration for an off-axial object point and displays both *longitudinal chromatic aberration* (LCA) and *lateral* or *transverse chromatic aberration* (TCA). Notice that if longitudinal chromatic aberration were absent, the transverse chromatic aberration could be interpreted as a difference in magnification for different colors. The longitudinal chromatic aberration of a convex lens may easily be comparable to its spherical aberration for rays at widest aperture.

Chromatic aberration is eliminated by making use of multiple refracting elements of opposite power. The most common solution is achieved with the *achromatic doublet* (Figure 20-13), consisting of a convex and concave lens, of different glasses, cemented together. The focal lengths and powers of the lenses differ, through shaping of their surfaces, to produce a net power of the doublet that may be either positive or negative. The dispersing powers of the components are, through appropriate selection of glasses, in inverse proportion to their powers. The result is a compound lens that has a net focal length but reduced dispersion over a significant portion of the visible spectrum.

We consider next the quantitative details of this design. The general shape of an achromatic doublet is shown in Figure 20-13. The powers of the two lenses for the yellow center of the visible spectrum, conveniently represented by the *Fraunhofer wavelength*, $\lambda_D = 587.6$ nm, are

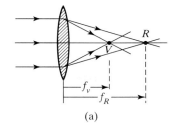

(a)

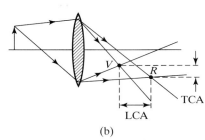

(b)

Figure 20-12   Chromatic aberration (exaggerated) for a thin lens, illustrating the effect on the focal length (a) and the lateral and longitudinal misses (b) for red (R) and violet (V) wavelengths.

$$P_{1D} = \frac{1}{f_{1D}} = (n_{1D} - 1)\left(\frac{1}{r_{11}} - \frac{1}{r_{12}}\right) = (n_{1D} - 1)K_1 \qquad (20\text{-}33)$$

$$P_{2D} = \frac{1}{f_{2D}} = (n_{2D} - 1)\left(\frac{1}{r_{21}} - \frac{1}{r_{22}}\right) = (n_{2D} - 1)K_2 \qquad (20\text{-}34)$$

where the radii of curvature are designated in Figure 20-13. Here, n_D refers to the refractive index of each glass for the D Fraunhofer line, and we have introduced constants K_1 and K_2 as an abbreviation for the curvatures. In Chapter 18 we showed that the focal length f of a thin-lens doublet with lens separation L satisfies the relation

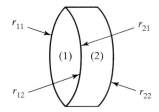

Figure 20-13   Achromatic doublet, consisting of (1) crown glass equiconvex lens cemented to (2) a negative flint glass lens. The four radii of curvature are indicated.

$$\frac{1}{f} = \frac{1}{f_1} + \frac{1}{f_2} - \frac{L}{f_1 f_2} \qquad (20\text{-}35)$$

Here, f_1 and f_2 are the focal lengths of the two lenses in the doublet. Consequently, the power $P = 1/f$ of the doublet is

$$P = P_1 + P_2 - LP_1P_2 \qquad (20\text{-}36)$$

For a cemented doublet of thin lenses, $L = 0$, and, as also discussed in Chapter 2, the powers of the lenses are simply additive:

$$P = P_1 + P_2 \qquad (20\text{-}37)$$

For the case of the cemented doublet, incorporating Eqs. (20-33) and (20-34) into Eq. (20-37) gives

$$P = (n_1 - 1)K_1 + (n_2 - 1)K_2 \qquad (20\text{-}38)$$

Chromatic aberration is absent at the wavelength λ_D if the power is independent of wavelength, or $(\partial P/\partial \lambda)_D = 0$. Applied to Eq. (20-38), this condition is

$$\frac{\partial P}{\partial \lambda} = K_1 \frac{\partial n_1}{\partial \lambda} + K_2 \frac{\partial n_2}{\partial \lambda} = 0 \qquad (20\text{-}39)$$

The variation of n with λ in the neighborhood of λ_D may be approximated using the red and blue Fraunhofer wavelengths, $\lambda_C = 656.3$ nm and $\lambda_F = 486.1$ nm, respectively:

$$\frac{\partial n}{\partial \lambda} \cong \frac{n_F - n_C}{\lambda_F - \lambda_C} \tag{20-40}$$

The dispersion constant for the glasses may be introduced by expressing the terms of Eq. (20-39) as

$$K_1 \frac{\partial n_{1D}}{\partial \lambda} = K_1 \left(\frac{n_{1F} - n_{1C}}{\lambda_F - \lambda_C} \right) \left(\frac{n_{1D} - 1}{n_{1D} - 1} \right) = \frac{P_{1D}}{(\lambda_F - \lambda_C)V_1} \tag{20-41}$$

$$K_2 \frac{\partial n_{2D}}{\partial \lambda} = K_2 \left(\frac{n_{2F} - n_{2C}}{\lambda_F - \lambda_C} \right) \left(\frac{n_{2D} - 1}{n_{2D} - 1} \right) = \frac{P_{2D}}{(\lambda_F - \lambda_C)V_2} \tag{20-42}$$

where we have used Eqs. (20-33) and (20-34) as well as a dispersive constant V, defined as the reciprocal of the *dispersive power* (see Eq. (3-18)) and given by

$$V \equiv \frac{1}{\Delta} = \frac{n_D - 1}{n_F - n_C} \tag{20-43}$$

Substituting Eqs. (20-41) and (20-42) into Eq. (20-39), the condition for the absence of chromatic aberration may be written as

$$V_2 P_{1D} + V_1 P_{2D} = 0 \tag{20-44}$$

Combining Eqs. (20-37) and (20-44), the powers of the individual elements may be expressed in terms of the desired power P_D of the combination:

$$P_{1D} = P_D \frac{-V_1}{V_2 - V_1} \quad \text{and} \quad P_{2D} = P_D \frac{V_2}{V_2 - V_1} \tag{20-45}$$

The K curvature factors expressed in Eqs. (20-33) and (20-34) may then be calculated using

$$K_1 = \frac{P_{1D}}{n_{1D} - 1} \quad \text{and} \quad K_2 = \frac{P_{2D}}{n_{2D} - 1} \tag{20-46}$$

Finally, from the values of K_1 and K_2, the four radii of curvature of the lens faces may be determined. For simplicity of construction, the crown glass lens (1) may be chosen to be equiconvex. In addition, the curvature of the two lenses must match at their interface. The radii of curvature thus satisfy

$$r_{12} = -r_{11}, \quad r_{21} = r_{12}, \quad \text{and} \quad r_{22} = \frac{r_{12}}{1 - K_2 r_{12}} \tag{20-47}$$

In the design of an achromatic doublet, the three indices of refraction for each of the glasses to be used are taken from manufacturer's specifications, like those presented in Table 20-1. One also inputs the desired overall focal length of the achromat. In the series of calculations leading to the four radii of curvature, a calculation that is easily programmed, Eqs. (20-43), (20-45), (20-46), and (20-47) are employed in sequence. For example, if 520/636 crown glass and 617/366 flint glass are used in designing an achromat of focal length 15 cm,

TABLE 20-1   SAMPLE OF OPTICAL GLASSES

Type	Catalog code	V	n_C	n_D	n_F
	$\dfrac{n_D-1}{10V}$	$\dfrac{n_D-1}{n_F-n_C}$	656.3 nm	587.6 nm	486.1 nm
Borosilicate crown	517/645	64.55	1.51461	1.51707	1.52262
Borosilicate crown	520/636	63.59	1.51764	1.52015	1.52582
Light barium crown	573/574	57.43	1.56956	1.57259	1.57953
Dense barium crown	638/555	55.49	1.63461	1.63810	1.64611
Dense flint	617/366	36.60	1.61218	1.61715	1.62904
Flint	620/380	37.97	1.61564	1.62045	1.63198
Dense flint	689/312	31.15	1.68250	1.68893	1.70462
Dense flint	805/255	25.46	1.79608	1.80518	1.82771
Fused silica	458/678	67.83	1.45637	1.45846	1.46313

these equations lead to lenses with radii of curvature given by

$$r_{11} = 6.6218 \text{ cm}$$
$$r_{12} = -6.6218 \text{ cm}$$
$$r_{21} = -6.6218 \text{ cm}$$
$$r_{22} = -223.29 \text{ cm}$$

With these values, Eqs. (20-33) and (20-34) permit the calculation of focal lengths for each of the Fraunhofer wavelengths. In this case, we find

	f_1	f_2	f
λ_D	6.3653 cm	−11.0575 cm	15.0000 cm
λ_C	6.3961 cm	−11.147 cm	15.007 cm
λ_F	6.2966 cm	−10.8485 cm	15.007 cm

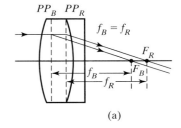

(a)

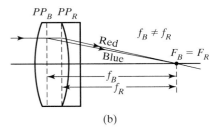

(b)

Figure 20-14   Doublet with second principal planes separated for red and blue light. (a) Equal focal lengths result in residual longitudinal chromatic aberration. (b) Equal foci result in residual lateral chromatic aberration.

For a thin lens, achromatizing renders focal lengths (nearly) equal, eliminating longitudinal and lateral aberration at the same time. In a thick lens or an optical system of lens combinations, the second principal planes for different wavelengths may not coincide as they do in a thin lens. When this is the case, equal focal lengths for two wavelengths, measured as they are from their respective principal planes, do not lead to a single focal point on the axis, and longitudinal chromatic aberration remains (Figure 20-14a). If the focal lengths for red and blue light are made unequal, such that they produce a single focus (Figure 20-14b), the difference in f_B and f_R results in a difference of lateral magnifications, and lateral chromatic aberration remains. Thus the condition for removing lateral chromatic aberration is the coincidence of the principal planes for the two corrected wavelengths.

Another solution for zero longitudinal chromatic aberration results if one uses two separated lenses ($L \neq 0$) of the same glass ($n_1 = n_2 = n$). The condition $\partial P/\partial\lambda = 0$ applied to Eq. (20-36) now gives

$$\frac{\partial P}{\partial\lambda} = \frac{\partial}{\partial\lambda}[(n-1)(K_1 + K_2) - (n-1)^2 K_1 K_2 L] = 0$$

Performing the differentiation and canceling $\partial n/\partial\lambda$, there remains

$$L = \frac{f_1 + f_2}{2} \qquad (20\text{-}48)$$

which is the same result derived for a double-lens eyepiece in Chapter 3. Thus two lenses of the same material, separated by a distance equal to the average of their focal lengths, exhibit zero longitudinal chromatic aberration for the wavelength at which the focal lengths are calculated.

PROBLEMS

20-1 Carry out the "rearranging" called for in arriving at Eq. (20-18).

20-2 If image and object distance for a spherical refracting surface—in addition to satisfying Eq. (2-19)—also satisfy the relation $1/s' = (1/s) + (1/R)$, show that

    a. $s' = -(n_1/n_2)s$, and

    b. $a(Q)$ for spherical aberration in Eq. (20-18) vanishes.

    c. Show that $a(Q)$ also vanishes for $s' = R$ and for rays intersecting with the spherical surface vertex. Such image points are called *aplanatic points*.

    d. Find the aplanatic points for a spherical surface of $+8$ cm separating two media of refractive indices 1.36 and 1.70, respectively.

20-3 A collimated light beam is incident on the plane side of a plano-convex lens of index 1.50, diameter 50 mm, and radius 40 mm. Find the spherical wave aberration and the longitudinal and transverse spherical ray aberrations.

20-4 Show that for a spherical concave mirror, a calculation like that done for a refracting surface gives a third-order aberration of

$$a = \frac{h^4}{4R}\left(\frac{1}{s} - \frac{1}{R}\right)^2$$

where R is the magnitude of the radius of curvature.

20-5 Using the result of problem 20-4, determine the wave aberration, transverse aberration, and longitudinal aberration for a spherical mirror of 2-m focal length and 50-cm diameter, when it forms an image of a distant point object.

20-6 A reflecting telescope uses a spherical mirror with a 3-m focal length and an aperture given by $f/3.75$.

    a. Using the results of problem 20-4, determine the magnitude of the spherical wave aberration for the telescope.

    b. If a Schmidt-type correcting plane of refractive index 1.40 were installed to correct the spherical aberration, what would be the required difference in thickness between the center and edge of the plate?

20-7 In forming an image of an axial point object, a $+4.0$-diopter lens with a diameter of 6.0 cm gives a longitudinal spherical aberration of $+1.0$ cm. If the object is 50 cm from the lens, determine (a) the transverse spherical aberration and (b) the diameter of the blur circle in the paraxial focal plane.

20-8 Determine the longitudinal and lateral spherical ray aberration for a thin lens of $n = 1.50$, $r_1 = +10$ cm, and $r_2 = -10$ cm due to rays parallel to the axis and through a zone of radius $h = 1$ cm.

20-9 Using the equation for spherical aberration of a thin lens, see problem 20-8, find the longitudinal spherical ray aberration of a lens as a function of ray height h. Do this by plotting the longitudinal ray aberration as a function of ray height for $h = 0, 1, 2, 3, 4,$ and 5 cm. The lens has a refractive index of 1.60 and radii $r_1 = 36$ cm and $r_2 = -18$ cm. The incident light rays are parallel to the optical axis.

20-10 An equiconvex thin lens of index 1.50 and radius 15 cm forms an image of an axial object point 25 cm in front of the lens and for rays through a zone of radius $h = 2$ cm. Determine the longitudinal and lateral spherical ray aberration. (See problem 20-8.)

20-11 Show that if $L = (1/s'_h) - (1/s'_p)$, setting $dL/d\sigma = 0$ produces the condition for minimum spherical aberration:

$$\sigma = -\frac{2(n^2 - 1)p}{n + 2}$$

20-12 A positive lens of index 1.50 and focal length 30 cm is "bent" to produce Coddington shape factors of 0.700 and 3.00. Determine the corresponding radii of curvature for the two lenses.

20-13 A positive thin lens of focal length 20 cm is designed to have minimal spherical aberration in its image plane, 30 cm from the lens. If the lens index is 1.60, determine its radii of curvature.

20-14 A thin, plano-convex lens with 1-m focal length and index 1.60 is to be used in an orientation that produces less spherical aberration while focusing a collimated light beam. Prove that the proper orientation is with light incident on the spherical side by comparing the Coddington shape factor for each orientation with the value giving minimum spherical aberration.

20-15 A positive lens is needed to focus a parallel beam of light with minimum spherical aberration. The required focal length is 30 cm. If the glass has a refractive index of 1.50, determine (a) the required Coddington factor and (b) the radii of curvature of the lens. (c) If the lens is to be used instead to produce a collimated beam, how do these answers change?

20-16 Answer problem 20-15 when the lens is designed to reduce coma.

20-17 A 20-cm focal length positive lens is to be used as an inverting lens; that is, it simply inverts an image without altering its size. What radii of curvature lead to minimum spherical aberration in this application? The lens refractive index is 1.50.

20-18 Answer problem 20-17 when the lens is designed to reduce coma.

20-19 It is desired to reduce the curvature of field of a lens of 20-cm focal length made of crown glass ($n = 1.5230$). For this purpose a second lens of flint glass ($n = 1.7200$) is added. What should be its focal length? Refractive indices are given for sodium light of 589.3 nm.

20-20 A doublet telescope objective is made of a cemented positive lens ($n_1 = 1.5736$, $f_1 = 3.543$ cm) and negative lens ($n_2 = 1.6039$, $f_2 = 5.391$ cm).

    a. Determine the radius of their Petzval surface.

    b. What focal length for the negative lens gives a flat Petzval surface?

20-21 Design an achromatic doublet of 517/645 crown and 620/380 flint glasses that has an overall focal length of 20 cm. Assume the crown glass lens to be equiconvex. Determine the radii of curvature of the outer surfaces of the lens, as well as its resultant focal length for the *D, C,* and *F* Fraunhofer lines.

20-22 Design an achromatic doublet of 5-cm focal length using 638/555 crown and 805/255 flint glass. Determine (a) radii of curvature; (b) focal lengths for *D, C,* and *F* Fraunhofer lines; (c) powers and dispersive powers of the individual elements. (d) Is Eq. (20-44) satisfied?

20-23 Design an achromatic doublet of −10-cm focal length, using 573/574 and 689/312 glasses. Assume the crown glass lens to be equiconcave. Determine (a) radii of curvature of the lens surfaces; (b) individual focal lengths for the Fraunhofer *D* line; (c) the overall focal lengths of the lens for the Fraunhofer *D, C,* and *F* lines.

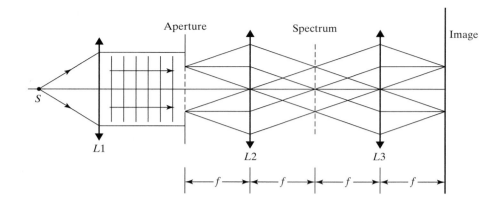

21

Fourier Optics

INTRODUCTION

Two rather extensive areas in which the Fourier transform is central to applications in optics are treated in this chapter, although, necessarily, somewhat cursorily. The first is included under the general heading of *optical data imaging and processing* and the second, *Fourier-transform spectroscopy*. Both are included within a branch of physics referred to generally as *Fourier optics*, in which the *Fourier transform, convolution*, and *correlation* are central concepts of mathematical analysis.

Optical data processing takes advantage of the fact that the simple lens constitutes a Fourier-transform computer, capable of transforming a complex two-dimensional pattern into a two-dimensional transform at very high resolution and at the speed of light. The diffraction pattern of a spatial object formed by the lens is shown to be a two-dimensional Fourier transform, or *spectrum*, of the input. This pattern may be manipulated in turn, using masks or filters to modify the final image produced by a second lens in a process called *spatial filtering*. Since various details of the image can be modified by appropriate filtering, this technique is exploited in such areas as *contrast enhancement* and *image restoration*. If the image is compared directly with a second object, the two may be *optically correlated*. Such correlation is applied, for example, in the problem of *pattern recognition*. By such optical means, two-dimensional pictures or text are processed at once, without the necessity of sequential scanning of the object. Optical data processing represents a fruitful convergence of the fields of optics, information science, and holography. As in many other fields, the availability of the laser as a coherent light source has ensured rapid growth.

Fourier-transform spectroscopy capitalizes on the fact that the spatial or temporal variations of an irradiance pattern due to polychromatic radiation can be Fourier-transformed into a spectral decomposition of the radiation. This technique makes possible another application of interferometry with distinct advantages for spectroscopy. Fourier-transform spectroscopy is the subject of the second part of the present chapter.

21-1 OPTICAL DATA IMAGING AND PROCESSING

Fraunhofer Diffraction and the Fourier Transform

We wish to show that the Fraunhofer diffraction pattern is, within certain approximations, the Fourier transform (Section 9-1) of the $\vec{E}$-field amplitude distribution in the object plane. Recall the one-dimensional Fourier transform pair presented in Chapter 9:

$$f(x) = \frac{1}{2\pi} \int_{-\infty}^{+\infty} g(k)e^{-ikx} \, dk \tag{21-1}$$

$$g(k) = \int_{-\infty}^{+\infty} f(x)e^{ikx} \, dx \tag{21-2}$$

Equation (21-1) states that an arbitrary, nonperiodic function $f(x)$ can be synthesized by summing a continuous distribution of plane waves with amplitude distribution $g(k)$ given by Eq. (21-2). The functions $f(x)$ and $g(k)$ are said to be a Fourier-transform pair. Symbolically,

$$g(k) = \Im\{f(x)\} \tag{21-3}$$

$$f(x) = \Im^{-1}\{g(k)\} \tag{21-4}$$

Here $\Im$ and $\Im^{-1}$ represent, respectively, the Fourier-transform operation and its inverse. The inverse transform of the transform of a function $f(x)$ returns the function $f(x)$. That is,

$$\Im^{-1}\{\Im(f(x))\} = \Im^{-1}\{g(k)\} = f(x) \tag{21-5}$$

in accordance with Eqs. (21-3) and (21-4).

In two dimensions, the transform pair takes the form

$$f(x, y) = \frac{1}{(2\pi)^2} \iint_{-\infty}^{+\infty} g(k_x, k_y)e^{-i(xk_x + yk_y)} \, dk_x \, dk_y \tag{21-6}$$

$$g(k_x, k_y) = \iint_{-\infty}^{+\infty} f(x, y)e^{i(xk_x + yk_y)} \, dx \, dy \tag{21-7}$$

Any nonperiodic function of two variables $f(x, y)$ can thus be synthesized from a distribution of plane waves, each with amplitude $g(k_x, k_y)$ and constant phase, such that

$$xk_x + yk_y = \text{constant} \tag{21-8}$$

The quantities k_x and k_y are the *spatial frequency* components needed in the expansion to represent the desired function $f(x, y)$. The individual plane

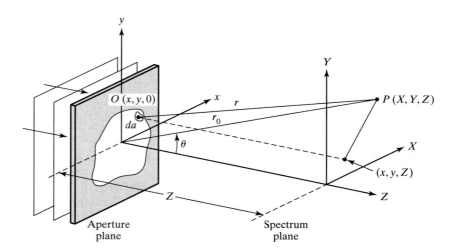

Figure 21-1 Fraunhofer diffraction in the spectrum XY-plane due to an aperture in the xy-plane.

waves in the continuous distribution intersect the xy-plane along the straight lines defined by Eq. (21-8). As k_x and k_y vary, the slopes of these lines vary. Thus the synthesis involves plane waves that vary in direction.

Consider the Fraunhofer diffraction pattern due to an arbitrary aperture situated in an xy-plane, as shown in Figure 21-1. Plane monochromatic waves diffract from the *aperture (xy) plane*. The diffraction pattern is observed in the XY-plane, which we shall call the *spectrum plane*, a distance Z along the axis.

The contribution dE_P at an arbitrary point P due to the light amplitude from an elemental area da surrounding point O in the aperture is given by

$$dE_P = \left(\frac{E_A \, da}{r} \right) e^{i(\omega t - kr)} \qquad (21\text{-}9)$$

where r is the distance from point O to point P. Neglecting the obliquity factor for small angles θ, Eq. (21-9) represents a spherical wave whose amplitude decreases with distance r. The quantity E_A is the *source strength*, or amplitude per unit area of aperture, in the neighborhood of point O. If the aperture is not uniformly illuminated or is not uniformly transparent, then $E_A = E_A(x, y)$ and it is called the *aperture function*. In Eq. (21-9), ω and k refer to the properties of the incident and diffracted radiation. The point P in the spectrum plane is a distance r_0 from the origin of the xy-coordinate system in the aperture plane. The distance r may be referred to the distance r_0 as follows. From the geometry apparent in Figure 21-1,

$$r^2 = (X - x)^2 + (Y - y)^2 + (Z - 0)^2$$

and

$$r_0^2 = X^2 + Y^2 + Z^2$$

so that

$$r^2 = r_0^2 - 2xX - 2yY + (x^2 + y^2) \qquad (21\text{-}10)$$

Although the dimensions X and Y in the spectrum plane may be appreciable, the dimensions x and y are typically negligible in comparison with r_0 for far-field diffraction. Accordingly, the terms x^2 and y^2 are ignored and Eq. (21-10) is rewritten as

$$r = r_0 \left[1 - 2 \frac{(xX + yY)}{r_0^2} \right]^{1/2} \qquad (21\text{-}11)$$

In this form, Eq. (21-11) is immediately adaptable to approximation by the binomial expansion $(1 + u)^{1/2} = 1 + \left(\frac{1}{2}\right)u + \cdots$, so that, retaining only the first two terms. we have,

$$r = r_0\left[1 - \frac{(xX + yY)}{r_0^2}\right] \tag{21-12}$$

In Eq. (21-9), the distance r appears in both the amplitude and the phase. In the amplitude it can be safely approximated by the distance Z between planes, but in the phase we use the approximate expression just derived. Then

$$dE_P = \left(\frac{E_A\,dx\,dy}{Z}\right)e^{i\omega t}e^{-ik[r_0-(xX+yY)/r_0]} \tag{21-13}$$

so that, upon integration over the area of the aperture, we have

$$E_P = \left[\frac{e^{i(\omega t-kr_0)}}{Z}\right]\iint E_A(x,\,y)e^{ik(xX+yY)/r_0}\,dx\,dy \tag{21-14}$$

If we are interested in the relative *amplitude* distribution of the electric field in the spectrum plane, it is convenient to define the relative amplitude function

$$A_P = ZE_Pe^{i(\omega t-kr_0)} = \iint E_A(x,\,y)e^{ik(xX+yY)/r_0}\,dx\,dy \tag{21-15}$$

Next, introducing the *angular spatial frequencies*,

$$k_X \equiv \frac{kX}{r_0} \quad\text{and}\quad k_Y \equiv \frac{kY}{r_0} \tag{21-16}$$

corresponding to each point (X, Y) in the spectrum plane, Eq. (21-15) may be expressed as

$$A_P(k_X, k_Y) = \iint E_A(x,\,y)e^{i(xk_X+yk_Y)}\,dx\,dy \tag{21-17}$$

In this form, Eq. (21-17) may be compared directly with Eq. (21-7), and our goal is established. We see that $A_P(k_X, k_Y)$ and $E_A(x, y)$ are related through a Fourier transformation. The inverse transform, as in Eq. (21-6), is

$$E_A(x,\,y) = \frac{1}{(2\pi)^2}\iint A_P(k_X, k_Y)e^{-i(xk_X+yk_Y)}\,dk_X\,dk_Y \tag{21-18}$$

Within the approximations made, we have shown that the Fraunhofer diffraction pattern described by $A_P(k_X, k_Y)$ is just the two-dimensional Fourier transform of the aperture function described by $E_A(x, y)$. The continuous distribution of constituent multidirectional plane waves is responsible for the redirection of the light into the various regions of the two-dimensional diffraction pattern.

Optical Spectrum Analysis

The Fraunhofer diffraction pattern of a given aperture is most conveniently displayed using a positive lens, as in Figure 21-2. Light from a monochromatic (temporally coherent) point source (spatially coherent) is collimated by lens $L1$ and illuminates, in the *input* or *aperture plane*, a two-dimensional pattern whose transmittance varies across the aperture. Lens $L2$ forms the Fraunhofer pattern in the spectrum plane. We shall neglect lens aberrations and

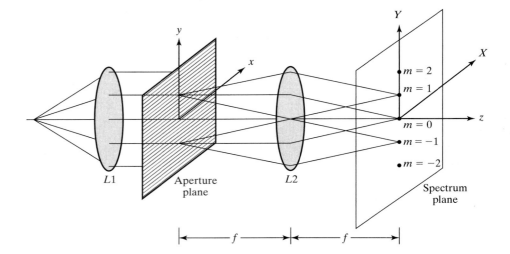

Figure 21-2 Fraunhofer diffraction of a Ronchi ruling.

also assume that the aperture is large enough so that its own boundaries do not appreciably modify the diffraction pattern. The aperture function $E_A(x, y)$ may thus be formed by any photographic negative. For simplicity we shall imagine the aperture function to vary like a square wave, such as would be produced by a *Ronchi ruling*, a grating of parallel straight lines with large grating space, whose opaque and transparent regions are of equal width.

Since the Fourier transform is an amplitude (not an irradiance) transform, we describe the square wave in Figure 21-3 by the amplitude of the transmitted light. We refer to the ratio of transmitted to incident amplitudes E_t/E_0 as the *transmission*, in contrast with the ratio of irradiances I_t/I_0, which we have called the *transmittance*. Transmittance is then just the square of the transmission. The aperture function $E_A(x, y)$, involving amplitudes, may also be called the *transmission function*. Lens $L2$ acts as a *Fourier-transform lens*. With a transmission function $E_A(x, y)$ in its first focal plane, the Fraunhofer diffraction pattern $A_P(k_X, k_Y)$, which is its Fourier transform, is produced in the second focal plane, the *spectrum*, or *output, plane*. The Ronchi ruling acts as a coarse grating, producing a series of bright spots that correspond to the various orders of diffraction. Since the Ronchi rulings are aligned parallel to the x-axis in the aperture plane, the spectrum of bright spots in the output plane occurs along the Y-direction, as shown. Now, according to the grating equation,

$$m\lambda = d \sin \theta = d \frac{Y_m}{f} \qquad (21\text{-}19)$$

where d is the spatial period of the ruling. Spots appear at distances Y_m from the optical axis given by

$$Y_m = m\left(\frac{\lambda f}{d}\right) \qquad (21\text{-}20)$$

Figure 21-3 Transmission function of period d due to a Ronchi ruling, in which opaque and transmitting widths are equal.

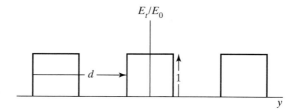

We wish to show now that this series of bright spots is, in fact, the spectrum of frequencies required in a Fourier representation of the aperture or transmission function of Figure 21-3. The angular spatial frequencies required in the Fourier integral were introduced in Eq. (21-16). In the Y-direction these are given by

$$k_Y = \frac{kY}{f} \qquad (21\text{-}21)$$

Since the transmission function for the Ronchi ruling is a periodic square function, it is represented by a discrete set of frequencies in a Fourier series rather than by a continuous distribution of frequencies in a Fourier integral. Let us introduce a wave number or "normalized" form of the spatial frequencies in the Fourier series by

$$\nu_Y \equiv \frac{1}{\lambda_Y} = \frac{k_Y}{2\pi} \qquad (21\text{-}22)$$

Then, substituting for k_Y from Eq. (21-21) and for Y from Eq. (21-20), we have, for the spectrum of spatial frequencies displayed in the diffraction pattern,

$$\nu_Y = \frac{m}{d} \qquad (21\text{-}23)$$

The central spot with $m = 0$ thus corresponds to a normalized spatial frequency $\nu_Y = 0$, the *DC component*, in analogy with electrical frequencies. The first-order ($m = 1$) spots above and below the central spot represent the fundamental frequency $\nu_{Y1} = 1/d$. Higher-order ($m > 1$) spots represent higher harmonics given by $m\nu_{Y1}$. We see that when the frequency of the square wave is larger (more closely spaced rulings with smaller d), the fundamental frequency in the Fourier spectrum is also larger, and the separation $Y_1 = \lambda f/d$ is increased—a fact that should already be familiar from our study of the diffraction grating.

A Fourier analysis of the square function (also calculated in Chapter 9 for a square wave as an even function) gives the Fourier series

$$f(Y) = \frac{1}{2} + \frac{2}{\pi}\left(\cos kY - \frac{1}{3}\cos 3kY + \frac{1}{5}\cos 5kY + \cdots\right) \qquad (21\text{-}24)$$

Here we find a constant ($k = 0$) term of $\frac{1}{2}$ corresponding to the DC component or central spot of the diffraction pattern; a term with fundamental (spatial) frequency, $k_1 = 2\pi/d$; and terms with higher *odd* harmonics, $3k_1, 5k_1, \ldots$. The absence of the even harmonics might at first be puzzling, on the basis of Eq. (21-23), since it would lead us to expect all the higher harmonics in the representation. The even harmonics, however, are just those corresponding to the missing orders in the grating diffraction. These missing orders are expected when the slit separation is twice the width of the slit opening, precisely the case in the Ronchi ruling. The squares of the coefficients in the Fourier series are proportional to the irradiances of the corresponding diffraction spots.

Example 21-1

Consider a Ronchi ruling with slits of width 0.1 mm illuminated by light of wavelength 488 nm. A lens of focal length 40 cm is used in a configuration like that shown in Figure 21-2.

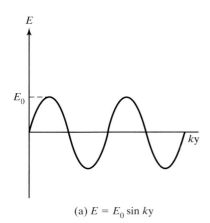

(a) $E = E_0 \sin ky$

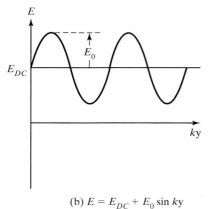

(b) $E = E_{DC} + E_0 \sin ky$

Figure 21-4 Sinusoidal amplitude or transmission functions including negative displacements.

a. Find the distances of the $m = 1$ and $m = 3$ spots from the central DC spot in the diffraction pattern on the screen in the spectrum plane.

b. Find the angular spatial frequencies associated with the $m = 1$ and $m = 3$ spots.

Solution

The ruling period d is twice the slit width, so $d = 0.2$ mm.

a. Using Eq. 21-20, the distances from the central spot are

$$Y_1 = (1)\frac{\lambda f}{d} = (1)\frac{(488 \times 10^{-9})(0.4)}{0.2 \times 10^{-3}}\text{m} = 9.76 \times 10^{-4}\text{ m} = 0.976\text{ mm}$$

$$Y_3 = (3)\frac{\lambda f}{d} = (3)\frac{(488 \times 10^{-9})(0.4)}{0.2 \times 10^{-3}}\text{m} = 2.93 \times 10^{-3}\text{ m} = 2.93\text{ mm}$$

b. From Eq. (21-22) $k_Y = 2\pi\nu_Y$. Using Eq. (21-23) then gives $k_Y = 2\pi$ (m/d).
For $m = 1$ the angular spatial frequency is $k_Y = 2\pi(1/d) = 2\pi/$ $(0.2\text{ mm}) = 31.4/\text{mm}$.
For $m = 3$ the angular spatial frequency is $k_Y = 2\pi(3/d) = 6\pi/$ $(0.2\text{ mm}) = 94.2/\text{mm}$.

Suppose now that the transmission function is not a square wave but a sine wave. If the lines of the Ronchi ruling have gradually changing opacity, such that the amplitude transmitted varies sinusoidally, we have the *sinusoidal grating*. Arguing from the Fourier series required to represent this kind of aperture function, it is clear that orders in the diffraction spectrum higher than $m = 1$ do not appear. Clearly, only one frequency is required to represent a sine wave. Why then does the spectrum also show a central spot, the DC component with $m = 0$? A little thought will make clear that an amplitude aperture function cannot be produced with both positive and negative portions, like the pure sine wave of Figure 21-4a. A photographic negative, at points of ideal opacity, may produce an amplitude $E = 0$ but cannot provide negative values. Thus the sinusoidal grating produces a transmission function like that of Figure 21-4b, in which the sine wave is offset by a DC bias. It is precisely the component E_{DC} in the figure that accounts for the zeroth-order diffraction signal.

Optical Filtering

We have seen that the back focal plane of the transform lens is the spectrum plane in which a Fourier transform of the aperture or transmission function is located. If this spectrum plane now serves in turn as a new aperture function for a second lens $L3$, a focal length away (Figure 21-5), the back focal plane of the second lens $L3$ receives the Fourier transform of the new aperture function. This second Fourier transform is thus the transform of the transform of the original aperture function and so returns the original aperture function; that is, an image of the original aperture is formed there. This conclusion also follows from an application of the laws of geometrical optics, evident from the ray diagram included in Figure 21-5.

Each diffraction spot in the spectrum plane, with coordinates (X, Y), represents spatial frequencies present in the aperture function, as we have pointed out. Each diffraction spot now helps to illuminate the image of the aperture in the image plane. How is this image affected if the light from one or more of these diffraction spots is blocked so that its contribution to the

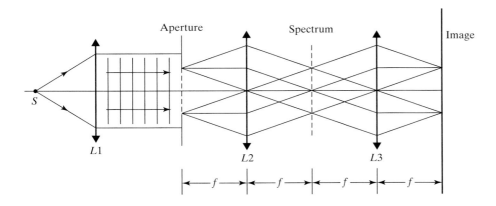

Aperture   Spectrum   Image

S

$L1$ $L2$ $L3$

|←— f —→|←— f —→|←— f —→|←— f —→|

Figure 21-5 Optical filter.

image is subtracted out? From our knowledge of Fourier series, we conclude that the finer features of the image disappear when spots corresponding to the higher spatial frequencies are blocked. If all spots are blocked except the DC component, or undeviated diffraction beam—say by an iris diaphragm centered on the central spot—the image plane is illuminated but no image details appear. As the circular opening of the diaphragm is gradually widened, higher spatial frequencies are admitted and the image gradually sharpens. The physical operation of opening the diaphragm is thus analogous mathematically to the systematic inclusion of higher and higher frequency terms in the Fourier series representing the aperture function.

Optical filtering is the process of intentionally blocking certain portions—that is, certain spatial frequencies—present in the diffraction pattern, to manipulate the image. Suppose, for example, that the aperture function is the superposition of two sine waves that are produced by back-to-back sinusoidal gratings with parallel rulings but different line spacings or spatial frequencies. The diffraction pattern consists, in addition to the direct beam, of two pairs of light spots, each pair due to one of the spatial frequencies present. If one of these pairs is blocked, that frequency is eliminated, or *filtered* from the illumination. The image is a sinusoidal pattern of the other frequency.

This example shows how optical filtering is applied to the extraction of desired periodic signals from background noise or, on the other hand, to the elimination of periodic noise from a desirable signal. As another example, suppose the aperture function is a television picture in which horizontal raster lines are visible. The diffraction pattern due to this function may be quite complicated, but the raster lines, like a Ronchi ruling, produce a series of diffraction spots along the vertical direction in the spectrum plane. If a rectangular-shaped, opaque shield is used to block the contribution of these spots, the raster line frequencies are filtered out and the final image is a reproduction of the TV picture but without the raster lines present. A technique like this was used to remove a sawtooth pattern from the video micrograph of a diatom frustule, as shown in Figure 21-6.

From the point of view of optical filtering, then, it should be clear that a diaphragm, which blocks all but those frequencies near the direct beam, functions as a low-pass optical filter. A diaphragm, which blocks only those frequencies near the direct beam, functions as a high-pass optical filter; and a clear annular ring, which blocks the lowest and the highest frequencies, functions as a band-pass filter. A case in point is the suppression of low spatial frequencies, or high-pass optical filtering, to enhance the contrast in a photograph. (Recall the importance of the high-frequency components in a Fourier series when synthesizing the fine features of a function, like the corners of a square

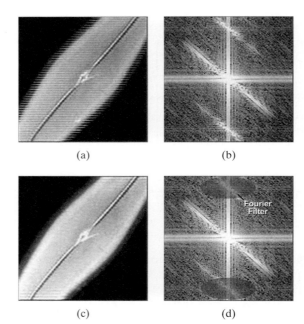

(a)   (b)

(c)   (d)

Fourier
Filter

Figure 21-6 (a) Video image of a diatom frustule imaged in dark field illumination with a superimposed sawtooth pattern. (b) Fourier transform power spectrum for the image. (c) Image of the frustule after applying a spatial filter as indicated in (d).

wave; Section 9-1.) More complex filtering has also been used in image restoration, for example, in the deblurring of lunar photographs.

Optical Correlation

As we have seen, an image of the two-dimensional object situated in the aperture plane is formed in the image plane of the optical filter (Figure 21-5). Suppose now that in the position of the image plane we insert a mask containing a pattern which transmits light at certain positions and blocks light at other positions. The image of the object in the aperture plane is thus superimposed over the pattern on the partially transparent mask. At a given point, the amount of light passed by the mask depends both on the amount of light available in the image and the transparency of the mask at that point. Let the light so transmitted be intercepted by an additional lens $L4$, as shown in Figure 21-7. The transmitted light is then monitored by a light detector placed in the second focal plane of $L4$, labeled Output in Figure 21-7. We have, in effect, added an *optical spectrum analyzer* to the optical filter of Figure 21-5. In the output plane where the detector is placed, we expect to measure the spectrum or Fourier transform of the transmission function represented by the light transmitted through the mask located in the image plane of Figure 21-7. This system provides an experimental means of comparing, or *correlating*, the light pattern in the image of the object in the aperture plane and the pattern contained on the mask. If the two patterns are identical, for example, and so situated that

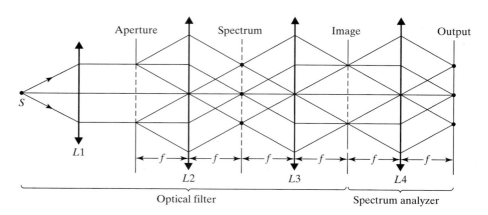

Figure 21-7 Optical correlator formed by the combination of an optical filter and a spectrum analyzer.

the image of the object in the aperture plane coincides with the transmission pattern on the mask, then maximum light throughput occurs, a case of maximum *correlation*. If one pattern is translated relative to the other, however, the bright points of the image no longer all coincide with the transparent regions of the mask, and light throughput and correlation are reduced. If the object in the aperture plane is a photographic image of the block letter *A* and the mask in the image plane contains a pattern of similar shape, a high degree of correlation should be obtained when the objects are properly positioned. On the other hand, if the mask contains a pattern of the letter *B*, the maximum light throughput and correlation should be significantly reduced. This technique of *pattern recognition* is applied, for example, to the recognition and counting of small particles with different shapes, as in the case of blood cells, or to the search for characteristic patterns in aerial photographs, medical X-rays, and fingerprint files.

Let us express the situation more precisely in mathematical terms. Let the object in the aperture plane be illuminated uniformly by light of unit amplitude, and let its transmission function be described by $E_1(x, y)$. The transmitted light, amplitude modulated and imaged at the position of the mask in the image plane of Figure 21-7, is then represented by $E_1(-x, -y)$. The change to negative coordinates is required by the inversion of the real image relative to the object. If the function representing the transmission of the mask alone is $E_2(x, y)$, then the light transmitted through the mask is the product function $E_1(-x, -y)E_2(x, y)$. The Fourier transform or spectrum of this composite transmission function is formed in the output plane, that is, the diffraction pattern there is described by

$$\Im[E_1(-x, -y)E_2(x, y)] = \iint\limits_{-\infty}^{\infty} E_1(-x, -y)E_2(x, y)e^{i(xk_x+yk_y)}\, dx\, dy$$

$$(21\text{-}25)$$

To concentrate only on the direct beam, or DC component, in the pattern, we set the spatial frequencies k_x and k_y equal to zero so that

$$\Im[E_1(-x, -y)E_2(x, y)]_{DC} = \iint\limits_{-\infty}^{\infty} E_1(-x, -y)E_2(x, y)\, dx\, dy$$

$$(21\text{-}26)$$

Both transmission functions $E_1(x, y)$ and $E_2(x, y)$ have been referred to xy-coordinate system origins that differ only by translation along the z-, or optical, axis. If the object in the aperture plane is shifted by an arbitrary translation given by components (q_x, q_y), for instance, its transmission function must reflect a translation of origin within the xy-plane, and Eq. (21-26) is expressed more generally by

$$\Im[E_1(-x, -y)E_2(x, y)]_{DC} = \iint\limits_{-\infty}^{\infty} E_1(q_x - x, q_y - y)E_2(x, y)\, dx\, dy$$

$$(21\text{-}27)$$

The integral in Eq. (21-27) is an example of the two-dimensional *convolution function*,

$$\rho_{12}(q_x, q_y) = \iint\limits_{-\infty}^{\infty} f_1(q_x - x, q_y - y)f_2(x, y)\, dx\, dy \qquad (21\text{-}28)$$

If the transmission function possesses inversion symmetry, that is, if

$$f_1(-x, -y) = f_1(x, y)$$

then the negative signs in the integrand of Eq. (21-28) may be written as positive signs, and the integral is instead the *correlation function*,

$$\Phi_{12}(q_x, q_y) = \Im[f_1(x, y)f_2(x, y)]_{DC}$$

$$\Phi_{12}(q_x, q_y) = \iint\limits_{-\infty}^{\infty} f_1(x + q_x, y + q_y)f_2(x, y) \, dx \, dy$$

(21-29)

Further, when f_1 and f_2 are merely shifted versions of the *same* function, we speak instead of the *autocorrelation function*,

$$\Phi_{11}(q_x, q_y) = \iint\limits_{-\infty}^{\infty} f(x + q_x, y + q_y)f(x, y) \, dx \, dy \qquad (21\text{-}30)$$

Transmission functions with inversion symmetry are imaged in such a way that the actual image inversion due to the lens is not apparent. Let us briefly examine the autocorrelation integral of Eq. (21-30). The integrand is a product of two functions and is nonzero only at those (x, y) points where both functions have nonzero values. With (q_x, q_y) fixed, the integral is the area under a curve representing the product of the two functions. This area, which we call the correlation, clearly depends on the choice of (q_x, q_y). If (q_x, q_y) are large enough so that there is no overlap of the functions, the area and correlation are zero. When q_x and q_y are both zero, the functions coincide, yielding a product curve with the maximum area and correlation. As an example, Figure 21-8, we have chosen as a function the top half of a circle. As one such curve (dashed semicircle) is translated along the x-axis relative to the other, their autocorrelation $\Phi(B)$ varies as a function of the parameter B, the displacement of their y-axes. The example illustrates a one-dimensional correlation.

We see from Eq (21-29) that the correlation is given by the DC component, or zeroth-order spectral point of the Fourier transform, or spectrum. Thus a detector, placed on axis at the output plane in the optical correlation system of Figure 21-7, measures the correlation. More precisely, since it is

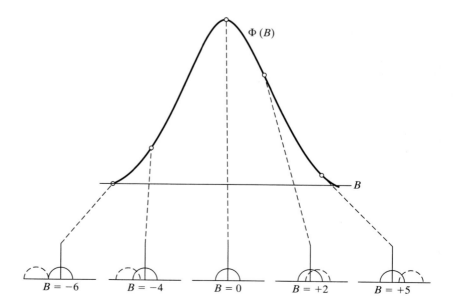

Figure 21-8 One-dimensional autocorrelation $\Phi(B)$ of a semicircle with radius $= 3$ as a function of the displacement parameter B. Several points on the correlation curve are referred to the specific translations that produce them. Note that $\Phi(-6) = 0$.

sensitive only to irradiances, the detector measures a quantity proportional to the square of the correlation. As the object in the aperture plane is translated along its x-axis, the light energy in the direct beam varies, producing the correlation function $\Phi_{12}(q_x)$. The transmission function of a given object in the aperture plane can be simultaneously correlated with many different reference functions by separating the reference functions as horizontal strips, or *channels*, on the mask. The DC spectral components corresponding to each reference channel are kept separated in the output plane by using a cylindrical lens as the final transform lens.

The method of pattern recognition just described is perhaps the simplest to understand, but many other techniques with various advantages have been developed. In this introduction, we briefly describe one other approach that makes use of a hologram as a spatial filter. The technique was introduced by Vander Lugt in 1963. First, a hologram is made of a particular pattern to be "recognized." Let us refer to its amplitude distribution by the function f. The holographic plate is situated in the Fourier-transform plane—the spectrum plane. The resulting hologram is called a *matched filter* or a *Vander Lugt filter*, after its originator. This holographic filter is subsequently used in the Fourier-transform plane, together with various test patterns having amplitude distributions $g_1, g_2, \ldots$, in the object plane. It can be shown[1] that, in general, three angularly distinct beams result so that three distinct spots appear in the image plane. One of these is centered on the optical axis, and the other two, off-axis, represent, respectively, the convolution and the correlation of the f and g pattern functions. When the test pattern g matches the desired pattern f, the correlation image appears with a bright, central spot, and pattern recognition is achieved.

Optical correlation techniques have been developed that allow recognition of a pattern independently of its size or orientation. Matched filtering techniques using incoherent light have also been devised to reduce the signal-to-noise background that is typical in coherent light systems.

Another Model of Imaging: Convolution

In the preceding sections, we have presented imaging as (1) a result of diffraction or Fourier analysis, producing a spectrum of spatial frequencies; and (2) their subsequent recombination, or Fourier synthesis, to form the image. We wish now to introduce some of the mathematical formalism and terminology commonly used to discuss Fourier transformations in another approach to imaging.

Consider a two-dimensional aperture (xy-plane) and its image (XY-plane) formed by some intervening optical system. We assume that the two sets of axes are similarly oriented, as in Figure 21-1. In the case of a perfect optical system, there is established a one-to-one correspondence between conjugate object and image points. For simplicity, we shall assume a lateral magnification of 1. Let the irradiance of such a (hypothetical) perfect image be given by $I_0(X, Y) \equiv I_0(x, y)$. In reality, light from each object point is spread out over its conjugate image point, due to diffraction and aberration. In this model, the resultant image is considered to be the overlapping of such "blurred" image points. In a *linear system*, these elementary irradiance patterns are simply additive. Let the actual irradiance over the image plane be given by $I_i(X, Y)$. The transformation from $I_0(X, Y)$ to $I_i(X, Y)$ clearly characterizes the optical system and is accomplished by a third function, called the *point spread function, G* (x, y, X, Y). For example, in the case of an

---

[1]See, for example, E. G. Steward, *Fourier Optics: An Introduction*, 2d ed. (New York: Halsted Press, 1987). Chaps. 4, 5, and Joseph W. Goodman, *Introduction to Fourier Optics* (New York: McGraw-Hill Book Company, 1968).

aberration-free system, G is just the function describing the Airy pattern (Section 11-3).

Now, if we assume the point spread function to be space-invariant (independent of object point coordinates), it can only depend on the relative displacement of conjugate points:

$$G(x, y, X, Y) = G(X - x, Y - y)$$

Further, if the light from the object plane is incoherent, irradiances add,[2] and we can write for the irradiance at the image point (X, Y) due to all object points (x, y):

$$\underbrace{I_i(X, Y)}_{\substack{\text{image} \\ \text{irradiance}}} = \iint \underbrace{I_0(x, y)}_{\substack{\text{object} \\ \text{irradiance}}} \underbrace{G(X - x, Y - y)}_{\substack{\text{point spread} \\ \text{function}}} dx \, dy \tag{21-31}$$

The integral in Eq. (21-31) is called the *convolution*[3] of the functions I_0 and G, usually abbreviated by

$$I_i = I_0 \otimes G \tag{21-32}$$

Suppose that we calculate the Fourier transform of each of these functions, represented by $\Im(I_0)$, $\Im(I_i)$, and $\Im(G)$. The *convolution theorem* (see problems) states that the Fourier transform of the convolution of two functions is equal to the product of their individual transforms. Symbolically,

$$\Im(I_i) = \Im(I_0 \otimes G) = \Im(I_0) \times \Im(G) \tag{21-33}$$

The content of Eqs. (21-32) and (21-33) can be succinctly summarized by stating that convolution in *real space* corresponds to multiplication in *Fourier space*. Combining this result with our understanding of the equivalence of Fourier transform and spatial frequency spectrum (or Fraunhofer diffraction function) of an aperture function, we can read Eq. (21-33) as follows: The spatial frequency spectrum of image irradiance is equal to the product of the spatial frequency spectrum of object irradiance and the spatial frequency spectrum of the point spread function. The last of these, $\Im(G)$, is called the *optical transfer function* (OTF), because it *transfers* or changes the object spectrum into the image spectrum. Thus the OTF is used to characterize the performance of an optical system.

As an example of the convolution theorem, recall the results given in Eq. (11-32) for the Fraunhofer diffraction of a grating. There we found that the two product functions could be interpreted separately as diffraction from a single slit and interference from multiple (negligible width) slits. Since these functions are the Fourier transforms of their respective aperture functions, we can say either that (1) in Fourier space, the grating diffraction pattern is given by the product of the Fourier transform of the single-aperture function and the transform of the array of line sources defining the grating; or (2) in real space, the grating aperture function is a convolution of the slit-array aperture function with the single-slit aperture function. The second formulation is suggested by Figure 21-9. Practically speaking, if one knows the Fourier transform of simple aperture functions, one can more easily calculate the

---

[2]If the light is coherent, the sum is a vector sum of complex electric field amplitudes.

[3]This integral has other important applications in physics. It requires the multiplication of one function at each point by the whole of another function and then the summation of the results. Hence it is also called a *folding* or *superposition* integral.

Aperture function for
an array of slits

⊛

Aperture function for
a single slit

⇒

Aperture function for
a grating

Figure 21-9 Symbolic representation of the convolution theorem for a grating.

Fraunhofer pattern that results from more complicated aperture functions by using the convolution theorem.

System Evaluation Using the Transfer Function

Characterization of the imaging capacity of an optical system by simply citing its resolving power does not give an adequate assessment of the system's performance. The preferred criterion of performance is the optical transfer function (OTF). To test an optical system properly, objects having both high and low spatial frequencies are required. As usual, low spatial frequencies are sufficient to image the gross details of an object, whereas high spatial frequencies are required to reproduce the finer details.

One technique for testing an optical system is to use a series of test patterns with sinusoidally varying darkness, each at a different spatial frequency K. When illuminated, incoherent imaging of the test pattern takes place. Let us assume a system magnification of 1. The image produced by a linear optical system is also sinusoidal at the same spatial frequency, but with a modification of amplitude and phase, as shown in Figure 21-10a. The OTF encompasses both modifications when it is written in complex form:

$$\text{OTF} = (\text{MTF})e^{i(PTF)} \qquad (21\text{-}34)$$

where MTF is the modulus and PTF is the phase. When either is known as a function of frequency, MTF is the *modulation transfer function* and PTF is the *phase transfer function*. The sinusoids corresponding to object O and image I are described by their *contrast modulation* γ, given by

$$\gamma_0 = \frac{O_{\max} - O_{\min}}{O_{\max} + O_{\min}} \quad \text{and} \quad \gamma_I = \frac{I_{\max} - I_{\min}}{I_{\max} + I_{\min}} \qquad (21\text{-}35)$$

Then the MTF and OTF are given, simply, by

$$\text{MTF} = \gamma_I/\gamma_0 \quad \text{and} \quad \text{PTF} = \Phi \qquad (21\text{-}36)$$

where Φ is illustrated in Figure 21-10a. Determination of MTF at various spatial frequencies, like the curves shown in Figure 21-10b, allows a more complete evaluation of system performance than resolution alone. Three systems are shown characterized. All approach a MTF of 1 as the spatial frequency approaches zero but indicate different resolution limits as MTF becomes zero. System A clearly shows the best performance. System B has a lower frequency limit than system C, but better performance at lower frequencies.

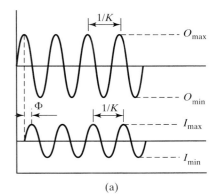

(a)

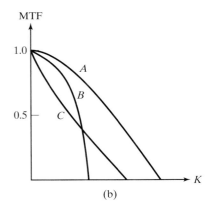

(b)

Figure 21-10 (a) Irradiance sinusoids of object and image, both of spatial frequency K. The optical system has unit magnification. (b) Modulation transfer function (MTF) for three optical systems plotted against spatial frequency.

21-2 FOURIER-TRANSFORM SPECTROSCOPY

Fourier-transform spectroscopy represents an elegant alternative to traditional methods of spectrum analysis. The special advantages of this technique have led to widespread applications in research and industry. Employing as a spectrometer an instrument such as the Michelson interferometer, these advantages derive both from the use of a large aperture at signal input and from the presence of the entire spectrum at signal output. The large energy throughput that results from the use of a large aperture is called the *Jacquinot advantage,* whereas the simultaneous processing of the entire spectral range during a single scan of the instrument is referred to as the *Fellgett,* or *multiplex, advantage.* Thus the Fourier-transform spectrometer is not limited, as are prism and grating spectrometers, by the presence of narrow slits that restrict both the wavelength interval and irradiance available at any one time. In addition, the technique is capable of high resolution, limited in principle only by the sample width of the input data and the wavelength region under analysis.

The large aperture and integrated throughput of the Michelson interferometer make it useful as a Fourier-transform spectrometer. It will be shown in the following treatment that the spectral distribution, or *spectrogram* (irradiance versus wave number), of the light incident on a Michelson interferometer is just the Fourier transform of the irradiance distribution, or *interferogram* (irradiance versus path difference), of its two-beam interference as a function of mirror movement. Figure 21-11 schematically shows the Michelson interferometer, which uses a beam splitter SP to separate equal-amplitude portions of a spectral input beam from source S and reunite them again after reflection from mirrors M_1 and M_2. The interfering beams are collected at detector D. Let the electric fields, E_1 and E_2, of the interfering beams for a particular wave number $k(= 2\pi/\lambda)$ component in the light source, on arrival at the detector, be represented by

$$E_1 = E_0 \cos(kx_1 - \omega t) \tag{21-37}$$

and

$$E_2 = E_0 \cos(kx_2 - \omega t) \tag{21-38}$$

where the two beams have experienced a physical path difference of $x = x_2 - x_1$ between separation and recombination.

The time-averaged irradiance for the k component at the detector is then

$$I_k \propto \langle (E_1 + E_2)^2 \rangle$$

which gives (see Chapter 7),

$$I_k = 2I_0(1 - \cos kx) \tag{21-39}$$

where I_0 represents the time-averaged irradiance of one beam. The minus sign in Eq. (21-39) stems from the fact that the reflection coefficients from opposite sides of the beam splitter differ by a factor of -1. Since there will be a spread of k values in the source, I_k can be interpreted as irradiance $I(k)$ per unit k interval at k, giving an integrated irradiance over all wavelengths of

$$I = \int_0^\infty I(k)\, dk = \int_0^\infty 2I_0(k)\, dk - \int_0^\infty 2I_0(k)\cos(kx)\, dk \tag{21-40}$$

The first term in the result behaves as a bias term, representing the constant integrated irradiance due to all wavelength components in the two noninterfering beams added together. The second term represents interference between the two beams and can be considered as a positive or negative deviation

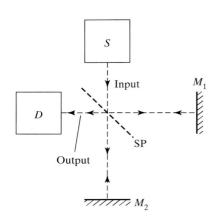

Figure 21-11 Elements of a Michelson interferometer used as a Fourier-transform spectrometer.

from the constant term, dependent upon the path difference x. Irradiance fluctuations about the constant bias comprise the spectral distribution (interferogram) given by

$$I(x) = \int_0^\infty I(k)\cos(kx)\,dk \qquad (21\text{-}41)$$

which is the (cosine) Fourier transform of the spectrogram,

$$I(k) = \left(\frac{2}{\pi}\right)\int_0^\infty I(x)\cos(kx)\,dx \qquad (21\text{-}42)$$

Thus, detection of the interferogram output $I(x)$, as a function of path difference x, at a point on the optical axis of the system enables one to calculate the spectral irradiance distribution $I(k)$ as a function of wavenumber by the Fourier-transform integration indicated in Eq. (21-42). In Figure 21-12, three experimental sample interferograms are shown, produced by a Michelson interferometer using various spectral inputs. Such interferograms are approximated, for the purposes of Fourier-transform calculations, by periodic sampling. When the function $I(x)$ is such a discrete set of sample points, the continuous Fourier transform goes over into sums and is referred to as a *discrete Fourier transform*. The use of finite sampling intervals across a finite total sample width or window leads to limitations both in the resolving power of the instrument and in the minimum wavelength that is unambiguously handled by the transform calculation. It can be shown that the restriction of data to a finite window x_w limits the resolution of the spectral distribution so that the minimum resolvable wavelength interval is given by

$$\Delta\lambda = \frac{\lambda^2}{x_w} \qquad (21\text{-}43)$$

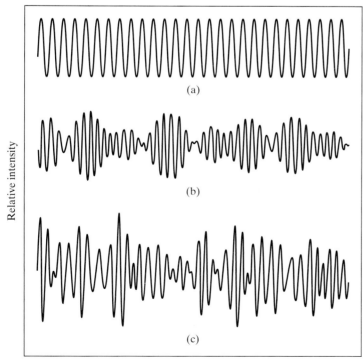

Relative intensity

(a)

(b)

(c)

Path difference

Figure 21-12 Interferograms produced by a Michelson interferometer using different light sources. (a) He-Ne laser. (b) Hg source, violet filter. (c) Hg source, unfiltered.

yielding a resolving power of

$$\mathfrak{R} \equiv \frac{\lambda}{\Delta\lambda} = \frac{x_w}{\lambda} \qquad (21\text{-}44)$$

One sees that the resolution is improved by using large sample widths. For example, a mirror movement of 0.5 cm, producing a total path difference or window of 1 cm, results in a resolving power at 500 nm of 20,000 and a resolution of 0.025 nm. Spectrometers have been built with relative mirror displacements of a meter or more, yielding resolving powers of 10^5 or greater. However, another important limitation must be taken into account. Because the true interferogram is only approximated at a specific sampling interval (nm/reading), a well-known phenomenon in sampling theory called *aliasing* places a limit on the smallest wavelength that can be unambiguously processed by this method. Wavelengths present in the input radiation, which are smaller than a particular λ_{min}, show up as longer wavelengths in the transformed spectrum. Such overlapping of wavelengths can be avoided by observing the *Nyquist criterion* of sampling theory: The signal must be sampled at a rate at least twice as high as its highest-frequency component. It is interesting to note that this criterion is also used in the production of modern digital audio recordings, where an audio signal sampling rate of 50 kHz ensures accurate reproduction of the maximum audio frequency of 20 kHz. Expressed in terms of our experimental parameters, the criterion states equivalently that, to avoid aliasing, the sampling interval must be less than half the smallest wavelength present in the source. Thus the minimum wavelength is given by

$$\lambda_{min} = \frac{2x_w}{N-1} \qquad (21\text{-}45)$$

where N is the total number of samples, giving $N-1$ sampling intervals. One sees now that a large x_w, which is beneficial in producing good resolution, may also be detrimental in limiting the spectral range of the spectrometer, unless N is also suitably large. The maximum number of data points, however, is limited by computer data-storage requirements and by computer time in handling the calculations. The number of operations performed by a computer in calculating the spectral distribution $I(k)$ is roughly equal to N^2. Use of the *Cooley-Tukey algorithm* for carrying out this series of calculations reduces the number of calculations to about $N \log_2 N$ and is known as the *fast Fourier transform*. For example, a transform using 1000 data points would be reduced from 1,000,000 operations to around 10,000, a considerable saving of computer time and expense. In the example just discussed, if the input radiation includes wavelengths in the visible and near ultraviolet, then N could not be less than about 67,000 without jeopardizing the correct analysis of wavelengths as small as 300 nm.

PROBLEMS

21-1 a. Calculate the distances from the axis of the first three bright spots produced by a Ronchi ruling with transmitting slits of width 0.25 mm, as in Figure 21-2. Assume laser irradiation of 632.8 nm and a 50-cm focal length lens.

  b. What is the wavelength corresponding to the fundamental frequency?

  c. Determine the three lowest angular spatial frequencies, apart from the DC component, required in a Fourier representation of the Ronchi aperture function.

  d. What are the ratios of irradiance of the first three spots, relative to the irradiance of the "fundamental"?

21-2 a. When two transmission functions are put together, by physically placing two transparencies back-to-back in the aperture plane, how must the combined transmission function relate to the individual transmission functions?

  b. Consider an aperture function formed by two perpendicularly crossed Ronchi rulings. What would you expect to see in the spectrum plane?

21-3 The optical density of film is defined as the common logarithm of its opacity. The opacity, in turn, is just the reciprocal of the transmittance, T.

    a. Thus, show that optical density is equal to $-\log_{10} T$.

    b. Show that the total optical density of several film layers is just the sum of their individual optical densities.

    c. What is the transmittance of five layers of film, each with an opacity of 1.25? What is the net optical density of the combined layers?

21-4 The sinusoidal transmission of a grating varies as $5\sin(ay)$, in arbitrary units.

    a. To produce faithfully the sinusoidal variation in the transmittance of the grating, what bias is required in the transmission function, assuming 100% maximum transmission?

    b. Sketch the aperture function with and without the bias term.

    c. What is the irradiance function at the detector for unit irradiance incident at the grating?

21-5 Prove the convolution theorem, that is, prove that if

$$h(x) = f(x) \otimes g(x)$$

then

$$\Im[h(x)] = \Im[f(x)]\Im[g(x)]$$

21-6 Plot the convolution in one dimension of two identical square pulses, of unit height and of 6 units length.

21-7 Determine the one-dimensional autocorrelation function $\Phi_{11}(\tau)$ for the sinusoidal function $y = A\sin(\omega t + \alpha)$.

21-8 a. The output of a Michelson spectrometer is fed to a photodetector. The input is mercury green light of 546.1 nm. If one mirror translates at a speed of 5 mm/s, what is the frequency of modulation of the photocurrent?

    b. What is the beat frequency of the photocurrent when the input is the yellow light of sodium, at 5889.95 Å, and 5895.92 Å? [*Hint*: Recall Eq. (8-14).]

21-9 The mirror translation in a Michelson spectrometer is 5 cm. What is the minimum resolvable wavelength at (a) 632.8 nm and at (b) 1 μm?

21-10 Light from a mercury lamp falls on the beam splitter of a student Michelson spectrometer. Wavelengths shorter than 360 nm are filtered from the light. The mirror translation rate is 71.5 nm/s. The rate at which spectrogram data is sampled is 1.28 readings/s. A total of 256 data points is fed to the computer for Fourier-transform analysis. Find the (a) window width x_w; (b) minimum resolvable wavelength interval at 400 nm; (c) minimum wavelength that is not subject to aliasing; (d) minimum sampling rate according to the Nyquist criterion.

21-11 The total path difference executed by a Fourier-transform spectrometer operating in the infrared is 2.78 mm. Its range is from 4400 to 400 cm^{-1}.

    a. What is its resolution in wave number?

    b. How many data points must be taken over the scan to avoid aliasing within this range?

    c. What is the scan rate if one run is completed in 30 s?

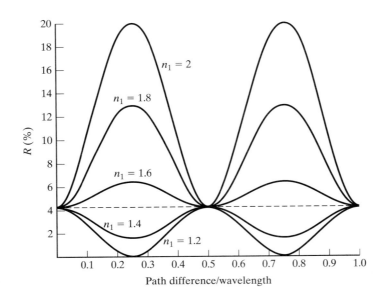

22

Theory of Multilayer Films

INTRODUCTION

The physics of interference in single-layer dielectric films has been treated, in its essentials, in Chapter 7. Many useful and interesting applications of thin films, however, make use of multilayer stacks of films. It is possible to evaporate multiple layers while maintaining control over both refractive index (choice of material) and individual layer thickness. Such techniques provide a great deal of flexibility in designing interference coatings with almost any specified frequency-dependent reflectance or transmittance characteristics. Useful applications of such coatings include antireflecting multilayers for use on the lenses of optical instruments and display windows; multipurpose broad and narrow band-pass filters, available from near ultraviolet to near infrared wavelengths; thermal reflectors and cold mirrors, which reflect and transmit infrared, respectively, and are used in projectors; dichroic mirrors consisting of band-pass filters deposited on the faces of prismatic beam splitters to divide light into red, green, and blue channels in color television cameras; and highly reflecting dielectric mirrors for use in gas lasers and in Fabry-Perot interferometers.

Computer techniques have made routine the rather detailed calculations involved in the analysis of multilayer film performance. The design of a multilayer stack that will meet arbitrary prespecified characteristics, however, remains a formidable task. In this chapter we develop a *transfer matrix* to represent the film and characterize its performance. The approach differs from that used in treating multiple reflections from a thin film in Chapter 7. There we added the amplitudes of all the individual reflected or transmitted beams to find the resultant reflectance or transmittance. It will be more efficient, in the general treatment that follows, to consider all transmitted or reflected beams as already summed in corresponding electric fields that satisfy the general boundary conditions required by Maxwell's equations.

The relationships we require from electromagnetic theory, already presented in Chapter 4, are summarized here. The energy of a plane, electromagnetic wave propagates in the direction of the Poynting vector, given by

$$\vec{S} = \varepsilon_0 c^2 \vec{E} \times \vec{B} \qquad (22\text{-}1)$$

The magnitudes of electric and magnetic fields in the wave are related by

$$E = vB \qquad (22\text{-}2)$$

where the wave speed is related to the the refractive index n by

$$n = \frac{c}{v} \qquad (22\text{-}3)$$

The wave speed in vacuum is a constant, equal to

$$c = \frac{1}{\sqrt{\varepsilon_0 \mu_0}} \qquad (22\text{-}4)$$

where ε_0 and μ_0 are the permittivity and permeability, respectively, of free space. Combining Eqs. (22-2), (22-3), and (22-4), the magnitudes of the magnetic and electric fields can also be related by

$$B = \frac{E}{v} = \left(\frac{n}{c}\right)E = n\sqrt{\varepsilon_0 \mu_0}E \qquad (22\text{-}5)$$

22-1 TRANSFER MATRIX

Our analysis is carried out in terms of the quantities defined in Figure 22-1. An incident beam is shown, with $\vec{E}$ chosen for the moment in a direction perpendicular to the plane of incidence. (Keep in mind, however, that at normal incidence $E_\perp$ and $E_\parallel$ are equivalent since a unique plane of incidence cannot be specified.) The beam undergoes external reflection at the plane interface (a) separating the external medium of index n_0 from the nonmagnetic ($\mu = \mu_0$) film of index n_1. The transmitted portion of the beam undergoes an internal

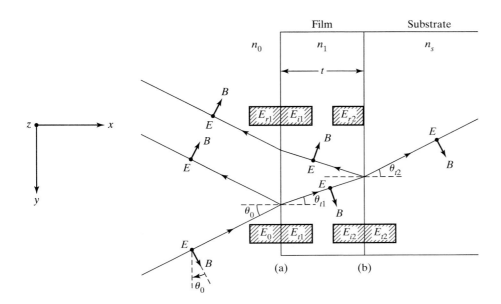

Figure 22-1 Reflection of a beam from a single layer. The diagram defines quantities used in applying boundary conditions to write Eqs. (22-6)–(22-9). Note that a bold dot is used to denote directions perpendicular to the plane of incidence.

reflection and transmission at the plane interface (b) separating the film from the substrate of index n_s. Along each beam the $\vec{E}$-field is shown—by the usual dot notation—to be pointing out of the page ($-z$-direction), and the $\vec{B}$-field is shown in a direction consistent with Eq. (22-1). Notice that the y-component of $\vec{B}$ must reverse on reflection. The insets define a terminology for the magnitudes of the electric fields *at the boundaries* (a) and (b). For example, E_{r1} represents the sum of all the multiply reflected beams at interface (a) in the process of emerging from the film, E_{i2} represents the sum of all the multiple beams incident at interface (b) and directed toward the substrate, and so on. In this way, we account for multiple beams in the interference.

We assume that the film is both homogeneous and isotropic. We assume further that the film thickness is of the order of the wavelength of light, so that the path difference between multiply reflected and transmitted beams remains small compared with the coherence length of the monochromatic light. This ensures that the beams are essentially coherent. The width of the incident beam, finally, is assumed to be large compared with its lateral displacement due to the many reflections that contribute significantly to the resultant reflected and transmitted beams.

Boundary conditions for the electric and magnetic fields of plane waves incident on the interfaces (a) and (b) follow from Maxwell's equations and are simply stated: The tangential components of the resultant $\vec{E}$- and $\vec{B}$-fields are continuous across the interface: that is, their magnitudes on either side are equal. For the case considered in Figure 22-1, $\vec{E}$ is everywhere tangent to the planes at (a) and (b), whereas $\vec{B}$ consists of both a tangential component (y-direction) and a perpendicular component (x-direction). Thus the boundary conditions for the electric field at the two interfaces become

$$E_a = E_0 + E_{r1} = E_{t1} + E_{i1} \tag{22-6}$$

$$E_b = E_{i2} + E_{r2} = E_{t2} \tag{22-7}$$

Corresponding equations for the magnetic field are

$$B_a = B_0 \cos\theta_0 - B_{r1} \cos\theta_0 = B_{t1} \cos\theta_{t1} - B_{i1} \cos\theta_{t1} \tag{22-8}$$

$$B_b = B_{i2} \cos\theta_{t1} - B_{r2} \cos\theta_{t1} = B_{t2} \cos\theta_{t2} \tag{22-9}$$

Rewriting Eqs. (22-8) and (22-9) in terms of electric fields with the help of Eq. (22-5),

$$B_a = \gamma_0(E_0 - E_{r1}) = \gamma_1(E_{t1} - E_{i1}) \tag{22-10}$$

$$B_b = \gamma_1(E_{i2} - E_{r2}) = \gamma_s E_{t2} \tag{22-11}$$

where we have written

$$\gamma_0 \equiv n_0\sqrt{\varepsilon_0\mu_0} \cos\theta_0 \tag{22-12}$$

$$\gamma_1 \equiv n_1\sqrt{\varepsilon_0\mu_0} \cos\theta_{t1} \tag{22-13}$$

$$\gamma_s \equiv n_s\sqrt{\varepsilon_0\mu_0} \cos\theta_{t2} \tag{22-14}$$

Now E_{i2} differs from E_{t1} only because of a phase difference δ that develops due to one traversal of the film. In Chapter 7 [see Eq. (7-35)] we showed that the optical path length Δ associated with *two* traversals of the thin film is $\Delta = 2n_1 t \cos(\theta_{t1})$. Thus the optical path length Δ_1 associated with *one* traversal is $\Delta_1 = \Delta/2 = n_1 t \cos(\theta_{t1})$, and phase difference δ that develops due to one traversal of the film is

$$\delta = k_0 \Delta_1 = \left(\frac{2\pi}{\lambda_0}\right) n_1 t \cos \theta_{t1} \qquad (22\text{-}15)$$

Thus,

$$E_{i2} = E_{t1} e^{-i\delta} \qquad (22\text{-}16)$$

In the same way,

$$E_{i1} = E_{r2} e^{-i\delta} \qquad (22\text{-}17)$$

Using Eqs. (22-16) and (22-17), we may eliminate the fields E_{i2} and E_{r2} in the boundary conditions at (b), expressed by Eqs. (22-7) and (22-11), as follows:

$$E_b = E_{t1} e^{-i\delta} + E_{i1} e^{i\delta} = E_{t2} \qquad (22\text{-}18)$$

$$B_b = \gamma_1(E_{t1} e^{-i\delta} - E_{i1} e^{i\delta}) = \gamma_s E_{t2} \qquad (22\text{-}19)$$

Disregarding for the moment the rightmost members, these equations may be solved simultaneously for E_{t1} and E_{i1} in terms of E_b and B_b, yielding

$$E_{t1} = \left(\frac{\gamma_1 E_b + B_b}{2\gamma_1}\right) e^{i\delta} \qquad (22\text{-}20)$$

$$E_{i1} = \left(\frac{\gamma_1 E_b - B_b}{2\gamma_1}\right) e^{-i\delta} \qquad (22\text{-}21)$$

Finally, substituting the expressions from Eqs. (22-20) and (22-21) into the Eqs. (22-6) and (22-10) for boundary (a), the result is

$$E_a = E_b \cos \delta + B_b\left(\frac{i \sin \delta}{\gamma_1}\right) \qquad (22\text{-}22)$$

$$B_a = E_b(i\gamma_1 \sin \delta) + B_b \cos \delta \qquad (22\text{-}23)$$

where we have used the Euler identities

$$2 \cos \delta \equiv e^{i\delta} + e^{-i\delta} \quad \text{and} \quad 2i \sin \delta \equiv e^{i\delta} - e^{-i\delta}$$

Equations (22-22) and (22-23) relate the net fields at one boundary with those at the other. They may be written in matrix form as

$$\begin{bmatrix} E_a \\ B_a \end{bmatrix} = \begin{bmatrix} \cos \delta & \dfrac{i \sin \delta}{\gamma_1} \\ i\gamma_1 \sin \delta & \cos \delta \end{bmatrix} \begin{bmatrix} E_b \\ B_b \end{bmatrix} \qquad (22\text{-}24)$$

The 2×2 matrix is called the *transfer matrix* of the film, represented in general by

$$M = \begin{bmatrix} m_{11} & m_{12} \\ m_{21} & m_{22} \end{bmatrix} \qquad (22\text{-}25)$$

If boundary (b) is the interface of another film layer, rather than the substrate, Eq. (22-24) is still valid. The fields E_b and B_b are then related to the fields E_c and B_c at the back boundary of the second film layer by a second transfer matrix. Generalizing, then for a multilayer of arbitrary number N of layers,

$$\begin{bmatrix} E_a \\ B_a \end{bmatrix} = M_1 M_2 M_3 \cdots M_N \begin{bmatrix} E_N \\ B_N \end{bmatrix}$$

An overall transfer matrix, M_T, representing the entire multilayer stack is the product of the individual transfer matrices, in the order in which the light encounters them,

$$M_T = M_1 M_2 M_3 \cdots M_N \qquad (22\text{-}26)$$

We return now to Eqs. (22-6), (22-7), (22-10), and (22-11) to make use of those members previously ignored in first finding the transfer matrix. Those remaining equations are

$$E_a = E_0 + E_{r1} \qquad (22\text{-}27)$$

$$E_b = E_{t2} \qquad (22\text{-}28)$$

$$B_a = \gamma_0(E_0 - E_{r1}) \qquad (22\text{-}29)$$

$$B_b = \gamma_s E_{t2} \qquad (22\text{-}30)$$

For the fields represented by Eqs. (22-27) to (22-30), Eq. (22-24) takes the form,

$$\begin{bmatrix} E_0 + E_{r1} \\ \gamma_0(E_0 - E_{r1}) \end{bmatrix} = \begin{bmatrix} \cos\delta & \dfrac{i\sin\delta}{\gamma_1} \\ i\gamma_1\sin\delta & \cos\delta \end{bmatrix} \begin{bmatrix} E_{t2} \\ \gamma_s E_{t2} \end{bmatrix} = \begin{bmatrix} m_{11} & m_{12} \\ m_{21} & m_{22} \end{bmatrix} \begin{bmatrix} E_{t2} \\ \gamma_s E_{t2} \end{bmatrix}$$

$$(22\text{-}31)$$

where in the last equality we have used the generic form of the transfer matrix given in Eq. (22-25). The last equality in Eq. (22-31) defines the transfer matrix elements, m_{11}, m_{12}, m_{21}, and m_{22} for the case at hand.

Equation (22-31) is equivalent to the two equations,

$$E_0 + E_{r1} = m_{11}E_{t2} + m_{12}\gamma_s E_{t2}$$

$$\gamma_0(E_0 - E_{r1}) = m_{21}E_{t2} + m_{22}\gamma_s E_{t2}$$

Dividing through these two equations by E_0 and making use of the reflection coefficient r and transmission coefficient t, defined as

$$r \equiv \frac{E_{r1}}{E_0} \quad \text{and} \quad t \equiv \frac{E_{t2}}{E_0} \qquad (22\text{-}32)$$

we obtain

$$1 + r = m_{11}t + m_{12}\gamma_s t \qquad (22\text{-}33)$$

$$\gamma_0(1 - r) = m_{21}t + m_{22}\gamma_s t \qquad (22\text{-}34)$$

Equations (22-33) and (22-34) can be solved for the transmission and reflection coefficients in terms of the transfer-matrix elements to give

$$t = \frac{2\gamma_0}{\gamma_0 m_{11} + \gamma_0 \gamma_s m_{12} + m_{21} + \gamma_s m_{22}} \qquad (22\text{-}35)$$

$$r = \frac{\gamma_0 m_{11} + \gamma_0 \gamma_s m_{12} - m_{21} - \gamma_s m_{22}}{\gamma_0 m_{11} + \gamma_0 \gamma_s m_{12} + m_{21} + \gamma_s m_{22}} \qquad (22\text{-}36)$$

Equations (22-35) and (22-36), together with the transfer-matrix elements, defined by Eqs. (22-24) and (22-31), now enable one to evaluate the reflective and transmissive properties of a single or multilayer film represented by the transfer matrix.

Before continuing with applications of these equations, we must take into account the necessary modification of the theory that results when the incident electric field of Figure 22-1 has the other polarization, that is, in the plane of incidence. Suppose that $\vec{E}$ is chosen in the original direction of $\vec{B}$ and $\vec{B}$ is rotated accordingly to maintain the same wave direction. If the equations are developed along the same lines, one finds that only a minor alteration of the transfer matrix becomes necessary: In the expression for γ_1, Eq. (22-13), the cosine factor now appears in the denominator rather than in the numerator. Summarizing,

$$\vec{E} \perp \text{plane of incidence:} \quad \gamma_1 = n_1 \sqrt{\varepsilon_0 \mu_0}\, \cos \theta_{t1}$$

$$\vec{E} \parallel \text{plane of incidence:} \quad \gamma_1 = n_1 \frac{\sqrt{\varepsilon_0 \mu_0}}{\cos \theta_{t1}} \qquad (22\text{-}37)$$

Notice that for normal incidence, where $\vec{E}_\perp$ and $\vec{E}_\parallel$ are indistinguishable, we have $\cos \theta_{t1} = 1$, and the expressions are equivalent. For oblique incidence, however, results must be calculated for each polarization. An average can be taken for unpolarized light. For example, the reflectance becomes

$$R = \tfrac{1}{2}(R_\parallel + R_\perp) \qquad (22\text{-}38)$$

22-2 REFLECTANCE AT NORMAL INCIDENCE

We apply the theory now for normally incident light, the case most commonly found in practice. Results apply quite well also to cases of near-normal incidence. The beam remains normal at all interfaces, so that all angles of incidence, reflection, and refraction are zero. In Eqs. (22-12) to (22-14), the cosine factors in the γ-terms are all unity. The matrix elements from Eq. (22-24), appropriately modified to become

$$m_{11} = \cos \delta \qquad m_{12} = \frac{i \sin \delta}{n_1 \sqrt{\varepsilon_0 \mu_0}}$$

$$m_{21} = i n_1 \sqrt{\varepsilon_0 \mu_0}\, \sin \delta \qquad m_{22} = \cos \delta \qquad (22\text{-}39)$$

are substituted into Eq. (22-36). After cancellation of the constant $\sqrt{\varepsilon_0 \mu_0}$ and some simplification, we find

$$r = \frac{n_1(n_0 - n_s)\cos \delta + i(n_0 n_s - n_1^2)\sin \delta}{n_1(n_0 + n_s)\cos \delta + i(n_0 n_s + n_1^2)\sin \delta} \qquad (22\text{-}40)$$

The reflectance R, which determines the reflected irradiance, is defined by

$$R = |r|^2 \qquad (22\text{-}41)$$

To calculate R, first notice that the reflection coefficient r is complex and that it has the general form

$$r = \frac{A + iB}{C + iD}$$

so that

$$|r|^2 = rr^* = \left(\frac{A + iB}{C + iD}\right)\left(\frac{A - iB}{C - iD}\right) = \frac{A^2 + B^2}{C^2 + D^2}$$

By inspection then, we may write

$$\text{normal incidence}$$
$$R = \frac{n_1^2(n_0 - n_s)^2\cos^2 \delta + (n_0 n_s - n_1^2)^2\sin^2 \delta}{n_1^2(n_0 + n_s)^2\cos^2 \delta + (n_0 n_s + n_1^2)^2\sin^2 \delta} \qquad (22\text{-}42)$$

Example 22-1

A 400-Å-thick film of ZrO_2 ($n = 2.10$) is deposited on glass ($n = 1.50$). Determine the normal reflectance for sodium light of wavelength $\lambda_0 = 589.3$ nm.

Solution

The phase difference is given by

$$\delta = \frac{2\pi}{\lambda_0}(n_1 t) = \frac{2\pi}{589.3}(2.1)(40) = 0.8956 \text{ rad}$$

so that $\cos \delta = 0.6250$ and $\sin \delta = 0.7806$. Then, substituting into Eq. (22-42),

$$R = \frac{2.1^2(1 - 1.5)^2(0.6250)^2 + [(1)(1.5) - 2.1^2]^2(0.7806)^2}{2.1^2(1 + 1.5)^2(0.6250)^2 + [(1)(1.5) + 2.1^2]^2(0.7806)^2} = 0.174$$

That is, the irradiance of the reflected beam is 17.4% of the irradiance of the incident beam.

A plot of reflectance versus the optical path difference $\Delta_1 = n_1 t$ associated with one traversal of the film is shown in Figure 22-2, where the abscissa is calibrated in ratios of Δ_1/λ, with $\lambda = \lambda_0/n_1$ being the wavelength in the film. Each curve corresponds to a different film index, but the glass substrate index has been chosen $n_s = 1.52$ in all cases. The magnitude of the film index n_1 evidently determines whether the reflectance is enhanced (for $n_1 > n_s$) or reduced (for $n_1 < n_s$) from that for uncoated glass. The curves show that quarter-wave thicknesses, or odd multiples thereof, lead either to optimum enhancement (high-reflectance coating) or to maximum reduction (antireflection coating). These minima or maxima points in R can be made to occur at various wavelengths by changing Δ_1 through selection of the film thickness. Notice that for $\Delta_1 = \lambda/2$ or any even multiple of a quarter-wavelength, the reflectance is just that from the uncoated glass. An antireflecting single coat, with $n_1 < n_s$, never reflects more than the uncoated glass at any wavelength. The periodic variation in R with Δ_1, which is proportional to the film thickness, provides a practical way of monitoring film thickness in the course of a film deposition.

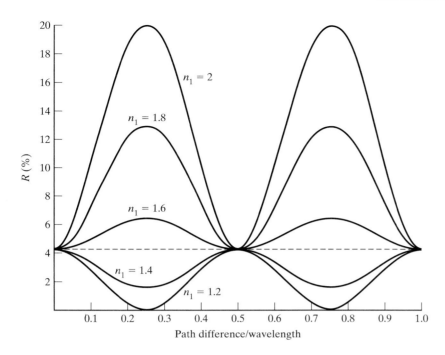

The important case of quarter-wave film thickness,

$$t = \frac{\lambda}{4} = \frac{\lambda_0}{4n_1}$$

makes the phase difference, Eq. (22-15), $\delta = 2\pi n_1 t/\lambda_0 = \pi/2$, so that $\cos \delta = 0$ and $\sin \delta = 1$. In this case, Eq. (22-42) reduces to

normal incidence quarter-wave thickness $\quad R = \left(\dfrac{n_0 n_s - n_1^2}{n_0 n_s + n_1^2}\right)^2$

$$(22\text{-}43)$$

From Eq. (22-43), it follows that a perfectly *antireflecting* film can be fabricated with a coating of $\lambda/4$ thickness and refractive index $n_1 = \sqrt{n_0 n_s}$. If the substrate is glass, with $n_s = 1.52$, the ideal index for a nonreflecting coating is $n_1 = 1.23$, assuming an ambient with $n_0 = 1$. A compromise choice among available coating materials is a film of MgF_2, with $n_1 = 1.38$. For this film, Eq. (22-43) predicts a reflectance of 1.3% in the visible region, where the uncoated glass (set $n_1 = n_0$) would reflect about 4.3%. This difference represents a significant saving of light energy in an optical system where multiple surfaces occur. For example, after only six such interfaces, or three optical components in series, 93% of the incident light survives in the case of MgF_2 coatings, compared with 77% in the case of uncoated glass.

22-3 TWO-LAYER ANTIREFLECTING FILMS

Durable coating materials with arbitrary refractive indices are, of course, not immediately available. Practically speaking then, single films with zero reflectances cannot be fabricated. By using a double layer of quarter-wave-thickness films, however, it is possible to achieve essentially zero reflectance at one wavelength with available coating materials. At normal incidence, the transfer matrix of a single film of quarter-wave thickness is

$$M_1 = \begin{bmatrix} 0 & \dfrac{i}{\gamma_1} \\ i\gamma_1 & 0 \end{bmatrix}$$

The transfer matrix M for two such layers is found, according to Eq. (22-26), by forming the product

$$M = M_1 M_2 = \begin{bmatrix} 0 & \dfrac{i}{\gamma_1} \\ i\gamma_1 & 0 \end{bmatrix}\begin{bmatrix} 0 & \dfrac{i}{\gamma_2} \\ i\gamma_2 & 0 \end{bmatrix} = \begin{bmatrix} -\dfrac{\gamma_2}{\gamma_1} & 0 \\ 0 & -\dfrac{\gamma_1}{\gamma_2} \end{bmatrix}$$

Matrix components are $m_{11} = -\gamma_2/\gamma_1$, $m_{22} = -\gamma_1/\gamma_2$, and $m_{12} = m_{21} = 0$. Using these values in Eq. (22-36), the result is

$$r = \frac{\gamma_2^2 \gamma_0 - \gamma_s \gamma_1^2}{\gamma_2^2 \gamma_0 + \gamma_s \gamma_1^2} \tag{22-44}$$

Incorporating the refractive indices through the use of Eqs. (22-12) to (22-14) and then squaring to get the reflectance,

$$\text{normal incidence quarter-wave thickness} \quad R = \left(\frac{n_0 n_2^2 - n_s n_1^2}{n_0 n_2^2 + n_s n_1^2}\right)^2 \tag{22-45}$$

Zero reflectance is predicted by Eq. (22-45) when $n_0 n_2^2 = n_s n_1^2$, or

$$\frac{n_2}{n_1} = \sqrt{\frac{n_s}{n_0}} \tag{22-46}$$

For a glass substrate ($n_s = 1.52$) and incidence from air ($n_0 = 1$), the ideal ratio for the two films is $n_2/n_1 = 1.23$. The requirement is met quite well using zirconium dioxide ($n_2 = 2.1$) and cerium trifluoride ($n_1 = 1.65$), both good coating materials. The ratio of refractive indices for CeF_3 and ZrO_2 of 1.27 produces a reflectance of only 0.1% according to Eq. (22-45). The arrangement is shown in Figure 22-3 and is plotted as curve (a) in Figure 22-4. Achieving zero reflectance at some wavelength may not satisfy the very common need to reduce reflectance over a broad region of the visible spectrum. Curve (a) is rather steep on both sides of its minimum at 550 nm. Broader regions of low reflectance result for $\lambda/4$–$\lambda/4$ coatings when the substrate index is larger than that of the adjacent film layer, that is, $n_s > n_2$. In such cases, the index is "stepped down" consistently from substrate to ambient. Indices high enough to satisfy this condition are possible in infrared applications where large values of n_s are available, as in the case of germanium with $n_s = 4$. A list of useful refractive indices is given in Table 22-1. Broader regions of low reflectance also become possible in the visible region of the spectrum, once the restriction of using equal $\lambda/4$ coatings is relaxed. For example, curves (b) and (c) of Figure 22-4 show two such solutions to the problem, where the inner coating has a thickness of $\lambda/2$, as illustrated in Figure 22-5. At the wavelength of 550 nm, for which the $\lambda/4$ and $\lambda/2$ thicknesses are determined, the $\lambda/2$ layer

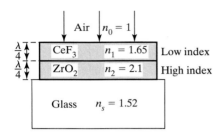

Figure 22-3 Antireflecting double layer, using $\lambda/4$–$\lambda/4$ thickness films.

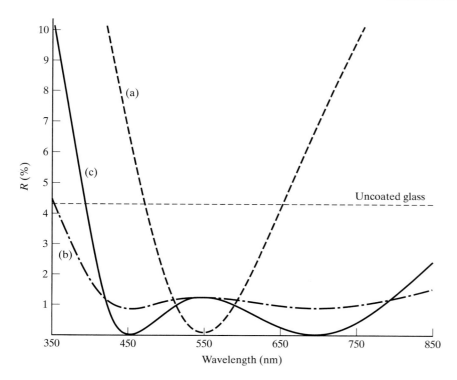

Figure 22-4 Reflectance from a double-layer film versus wavelength. In all cases $n_0 = 1$ and $n_s = 1.52$. Thicknesses are determined at $\lambda = 550$ nm. (a) $\lambda/4 - \lambda/4$; $n_1 = 1.65, n_2 = 2.1$. (b) $\lambda/4 - \lambda/2$: $n_1 = 1.38, n_2 = 1.6$. (c) $\lambda/4 - \lambda/2$: $n_1 = 1.38, n_2 = 1.85$.

TABLE 22-1   REFRACTIVE INDICES FOR SEVERAL COATING MATERIALS

Material	Visible ($\sim$550 nm)	Near infrared ($\sim$2 μm)
Cryolite	1.30–1.33	—
MgF_2	1.38	1.35
SiO_2	1.46	1.44
SiO	1.55–2.0	1.5–1.85
Al_2O_3	1.60	1.55
CeF_3	1.65	1.59
ThO_2	1.8	1.75
Nd_2O_3	2.0	1.95
ZrO_2	2.1	2.0
CeO_2	2.35	2.2
ZnS	2.35	2.2
TiO_2	2.4	—
Si	—	3.3
Ge	—	4.0

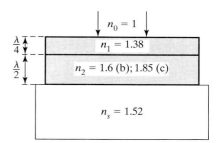

Figure 22-5 Antireflecting double layer using $\lambda/4 - \lambda/2$ thickness films. Reflectance curves are shown in Figure 22-4.

has no effect on the reflectance and the double layer behaves like a single $\lambda/4$ layer with $R = 1.3\%$. At nearby wavelengths, however, the $\lambda/2$ layer helps to keep R below values attained by a single $\lambda/4$ layer alone. For $n = 1.85$ [curve (c)], two minima near $R = 0$ appear. Although reflectance at 550 nm is about 1.3%, greater than for the $\lambda/4 - \lambda/4$ coating of curve (a), it remains at values less than this over the broad range of wavelengths from about 420 to 800 nm. For $n_2 = 1.6$ [curve (b)], the spectral response of the double layer, while more reflective, is flatter over the visible spectrum. Still other practical solutions for double-layer antireflecting films become possible if the thicknesses of the layers are allowed to have values other than multiples of $\lambda/4$.

The curves of Figure 22-4 have been calculated using the theory presented in this chapter. The overall transfer-matrix elements are first determined by forming the product of the transfer matrices of the individual layers. In these elements, the phase difference δ is expressed as a function of λ, and

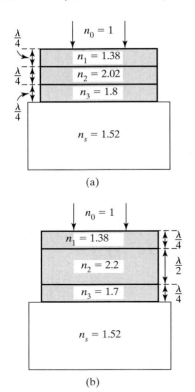

$\frac{\lambda}{4}$

$n_0 = 1$

$n_1 = 1.38$

$\frac{\lambda}{4}$ $n_2 = 2.02$

$n_3 = 1.8$

$\frac{\lambda}{4}$

$n_s = 1.52$

(a)

$n_0 = 1$

$n_1 = 1.38$ $\frac{\lambda}{4}$

$n_2 = 2.2$ $\frac{\lambda}{2}$

$n_3 = 1.7$ $\frac{\lambda}{4}$

$n_s = 1.52$

(b)

Figure 22-6 Antireflecting triple layers. (a) Quarter-quarter-quarter wavelength layers. (b) Quarter-half-quarter wavelength layers. Reflectance curves are shown in Figure 22-7.

the film thickness is determined by the $\lambda/4$ or $\lambda/2$ requirement at a single wavelength. These matrix elements are then used in Eq. (22-36) for the reflection coefficient. When squared, the reflectance as a function of wavelength is determined. Although the calculations can be tedious, they are easily done using a programmable calculator or computer.

22-4 THREE-LAYER ANTIREFLECTING FILMS

The procedure just outlined can be used to calculate the spectral reflectance of three-layer films as well. The use of three or more layer coatings makes possible a broader, low-reflectance region in which the response is flatter. If each of the three layers is of $\lambda/4$ thickness, one can show that a zero reflectance occurs when the refractive indices satisfy

$$\frac{n_1 n_3}{n_2} = \sqrt{n_0 n_s} \qquad (22-47)$$

One such practical solution is shown in Figure 22-6a and plotted as curve (a) in Figure 22-7. Some improvement results when the middle layer is of $\lambda/2$ thickness, as in Figure 22-6b and curve (b) of Figure 22-7.

22-5 HIGH-REFLECTANCE LAYERS

If the order of the layers in a $\lambda/4-\lambda/4$ double-layer film optimized for antireflection is reversed, so that the order is air–high index–low index–substrate, all three reflected beams are in phase on emerging from the structure, and the reflectance is enhanced rather than reduced. A series of such double layers increases the reflectance further, and the structure is called a *high-reflectance stack*, or *dielectric mirror*.

Figure 22-7 Reflectance from triple-layer films versus wavelength. In all cases $n_0 = 1$ and $n_s = 1.52$. Thicknesses are determined at $\lambda = 550$ nm.   (a) $\lambda/4-\lambda/4$: $n_1 = 1.38$, $n_2 = 2.02$, $n_3 = 1.8$.   (b) $\lambda/4-\lambda/2-\lambda/4$: $n_1 = 1.38$, $n_2 = 2.2$, $n_3 = 1.7$.

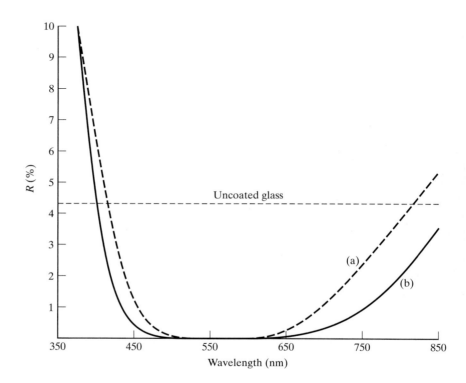

We derive now an expression for the reflectance of this type of structure, shown schematically in Figure 22-8, where High and Low signify high- and low-refractive indices, respectively. The transfer matrix for one double layer of $\lambda/4$-thick coatings at normal incidence is the product of the individual film matrices, just as in the case of the double-layer antireflecting films:

$$M_{HL} = M_H M_L$$

or

$$M_{HL} = \begin{bmatrix} 0 & \dfrac{i}{\gamma_H} \\ i\gamma_H & 0 \end{bmatrix} \begin{bmatrix} 0 & \dfrac{i}{\gamma_L} \\ i\gamma_L & 0 \end{bmatrix} = \begin{bmatrix} \dfrac{-\gamma_L}{\gamma_H} & 0 \\ 0 & \dfrac{-\gamma_H}{\gamma_L} \end{bmatrix} \qquad (22\text{-}48)$$

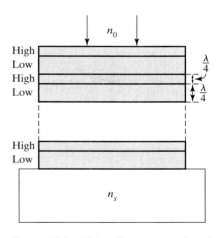

Figure 22-8 High-reflectance stack of double layers with alternating high- and low-refractive indices. Reflectance curves are shown in Figure 22-9.

For N similar double layers in series,

$$M = (M_{H1}M_{L1})(M_{H2}M_{L2})\cdots(M_{HN}M_{LN}) = (M_H M_L)^N = (M_{HL})^N$$

$$(22\text{-}49)$$

Substituting the double-layer matrix, Eq. (22-48),

$$M = \begin{bmatrix} \dfrac{-\gamma_L}{\gamma_H} & 0 \\ 0 & \dfrac{-\gamma_H}{\gamma_L} \end{bmatrix}^N = \begin{bmatrix} \left(\dfrac{-\gamma_L}{\gamma_H}\right)^N & 0 \\ 0 & \left(\dfrac{-\gamma_H}{\gamma_L}\right)^N \end{bmatrix}$$

For normal incidence,

$$\frac{\gamma_L}{\gamma_H} = \frac{n_L}{n_H} \quad \text{and} \quad \frac{\gamma_H}{\gamma_L} = \frac{n_H}{n_L}$$

so that

$$M = \begin{bmatrix} \left(\dfrac{-n_L}{n_H}\right)^N & 0 \\ 0 & \left(\dfrac{-n_H}{n_L}\right)^N \end{bmatrix} \qquad (22\text{-}50)$$

The matrix elements of the transfer matrix representing N high-low double layers of $\lambda/4$-thick coatings in series are thus

$$m_{11} = \left(\frac{-n_L}{n_H}\right)^N, \qquad m_{22} = \left(\frac{-n_H}{n_L}\right)^N, \qquad m_{12} = m_{21} = 0$$

$$(22\text{-}51)$$

Using these matrix elements in the expression for the reflection coefficient, Eq. (22-36), we arrive at

$$r = \frac{n_0(-n_L/n_H)^N - n_s(-n_H/n_L)^N}{n_0(-n_L/n_H)^N + n_s(-n_H/n_L)^N} \qquad (22\text{-}52)$$

When the numerator and denominator of Eq. (22-52) are next multiplied by the factor $(-n_L/n_H)^N/n_s$ and the result is squared to give reflectance, we have

$$R_{max} = \left[\frac{(n_0/n_s)(n_L/n_H)^{2N} - 1}{(n_0/n_s)(n_L/n_H)^{2N} + 1}\right]^2 \qquad (22\text{-}53)$$

Example 22-2

A high-reflectance stack like that of Figure 22-8 incorporates six double layers of SiO_2 ($n = 1.46$) and ZnS ($n = 2.35$) films on a glass ($n = 1.48$) substrate. What is the reflectance for light of 550 nm at normal incidence?

Solution

Substituting directly into Eq. (22-53), we get

$$R = \left[\frac{(1/1.48)(1.46/2.35)^{12} - 1}{(1/1.48)(1.46/2.35)^{12} + 1}\right]^2$$

or $R = 99.1\%$.

Equation (22-53) predicts 100% reflectance when either N approaches infinity or when (n_L/n_H) approaches zero. Some data indicating these tendencies are given in Table 22-2. One sees that the reflectance quickly approaches 100% for several double layers. Since the smallest ratio of n_L/n_H yields best reflectances, high-reflectance stacks may be fabricated from alternating layers of MgF_2 ($n_L = 1.38$) and ZnS ($n_H = 2.35$) or TiO_2 ($n_H = 2.40$).

The reflectance given in Eq. (22-53) represents the maximum reflectance at the wavelength λ_0, for which the layers have *optical* thicknesses of $\lambda_0/4$. For other wavelengths the transfer matrix must be used in its general form, containing the wavelength-dependent phase differences. Spectral reflectance curves for $N = 2$ and $N = 6$ double-layer stacks have been calculated and plotted in Figure 22-9. Curve (c) shows the improvement in the maximum reflectance that results for $N = 2$ stacks when an extra high-index layer is inserted between the substrate and the last low-index layer. The width of the high-reflectance region in these curves is nearly independent of the number of double layers used but increases when the ratio n_L/n_H decreases. This ratio is 0.587 in Figure 22-9, representing alternating MgF_2 and ZnS layers on glass. Outside the central *stopband*—the region of highest reflectance—the

TABLE 22-2   REFLECTANCE OF A HIGH-LOW QUARTER-WAVE STACK

Reflectance for $N = 3$ high-low layers versus n_L/n_H		Reflectance versus N when $n_L/n_H = 0.587$ for alternating double layers of MgF_2 and ZnS	
n_L/n_H	R (%)	N	R (%)
1.0	4.26	1	39.71
0.91	21.01	2	73.08
0.83	40.82	3	89.77
0.77	57.77	4	96.35
0.71	70.44	5	98.72
0.67	79.35	6	99.56
0.625	85.48	7	99.85
0.59	89.67	8	99.95
0.56	92.55		
0.53	94.56		
0.50	95.97		

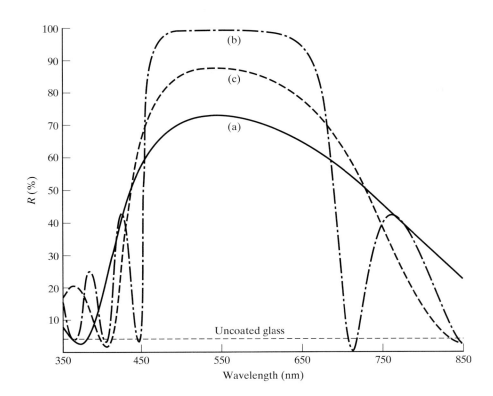

Figure 22-9 Spectral reflectance of a high-low index stack for (a) $N = 2$ and (b) $N = 6$ double layers. Curve (c) represents an $N = 2$ stack with an additional high-index layer adjacent to the substrate. Layers are $\lambda/4$ thick at $\lambda = 550$ nm. In all cases, $n_H = 2.35$, $n_L = 1.38$, $n_s = 1.52$, and $n_0 = 1.00$.

reflectance oscillates between a series of maxima and minima. The center of the stopband can be shifted by depositing layers whose thickness is $\lambda/4$ at another λ. Except for light energy lost by absorption and scattering during passage through the dielectric layers, the percent transmission of the structure is given by $T(\%) = 100 - R(\%)$. Thus such structures can be designed as *band-pass filters* with high spectral transmittance in the wide region of low spectral reflectance. Narrow band-pass filters that behave like Fabry-Perot etalons can be fabricated by separating two dielectric-mirror, multilayer structures with a spacer of, say, MgF_2 film. Narrow wavelength regions that satisfy constructive interference can be produced far enough apart in wavelength so that all but one such region is easily filtered out by a conventional absorption color filter. The result is a filter with a pass-band width of perhaps 15 Å and 40% transmittance.

PROBLEMS

22-1 Show that when the incident $\vec{E}$-field is parallel to the plane of incidence, γ_1 has the form given in Eq. (22-37).

22-2 A transparent film is deposited on glass of refractive index 1.50.

    a. Determine values of film thickness and (hypothetical) refractive index that will produce a nonreflecting film for normally incident light of 500 nm.

    b. What reflectance does the structure have for incident light of 550 nm?

22-3 Show from Eq. (22-42) that the normal reflectance of a single half-wave thick layer deposited on a substrate is the same as the reflectance from the uncoated substrate.

$$R = \frac{(n_0 - n_s)^2}{(n_0 + n_s)^2}$$

22-4 A single layer of SiO_2 ($n = 1.46$) is deposited to a thickness of 137 nm on a glass substrate ($n = 1.52$). Determine the normal reflectance for light of wavelength (a) 800 nm; (b) 600 nm; (c) 400 nm. Verify the reasonableness of your results by comparison with Figure 22-2.

22-5 A 596-Å-thick layer of ZnS ($n = 2.35$) is deposited on glass ($n = 1.52$). Calculate the normal reflectance of 560 nm light.

22-6 Determine the theoretical refractive index and thickness of a single film layer deposited on germanium ($n = 4.0$) such that normal reflectance is zero at a wavelength of 2 μm. What actual material could be used?

22-7 A double layer of quarter-wave layers of Al_2O_3 ($n = 1.60$) and cryolite ($n = 1.30$) are deposited in turn on a glass substrate ($n = 1.52$).

a. Determine the thickness of the layers and the normal reflectance for light of 550 nm.

b. What is the reflectance if the layers are reversed?

22-8 Quarter-wave thin films of ZnS ($n = 2.2$) and MgF$_2$ ($n = 1.35$) are deposited in turn on a substrate of silicon ($n = 3.3$) to produce minimum reflectance at 2 μm.

a. Determine the actual thickness of the layers.

b. By what percentage difference does the ratio of the film indices differ from the ideal?

c. What is the normal reflectance produced?

22-9 By working with the appropriate transfer matrix, show that a quarter-wave/half-wave double layer, as in Figure 22-5, produces the same reflectance as the quarter-wave layer alone.

22-10 Write a computer program that will calculate and/or plot reflectance values for a double layer under normal incidence. Let input parameters include thickness and indices of the layers and the index of the substrate. Check results against Figure 22-4.

22-11 Prove the condition given by Eq. (22-47) for zero reflectance of three-layer, quarter-quarter-quarter-wave films when used with normal incidence. Do this by determining the composite transfer matrix for the three quarter layers and using the matrix elements in the calculation of the reflection coefficient in Eq. (22-36).

22-12 Using the materials given in Table 22-1, design a three-layer multifilm of quarter-wave thicknesses on a substrate of germanium that will give nearly zero reflectance for normal incidence of 2 μm radiation.

22-13 Determine the maximum reflectance in the center of the visible spectrum for a high-reflectance stack of high-low index double layers formed using $n_L = 1.38$ and $n_H = 2.6$ on a substrate of index 1.52. The layers are of equal optical thickness, corresponding to a quarter-wavelength for light of average wavelength 550 nm. The high-index material is encountered first by the incident light, as in Figure 22-8. Assume normal incidence and stacks of (a) 2; (b) 4; (c) 8 double layers.

22-14 A high-reflectance stack of alternating high-low index layers is produced to operate at 2 μm in the near infrared. A stack of four double layers is made of layers of germanium ($n = 4.0$) and MgF$_2$ ($n = 1.35$), each of 0.5-μm optical thickness. Assume a substrate index of 1.50 and normal incidence. What reflectance is produced at 2 μm?

22-15 What theoretical ratio of high-to-low refractive indices is needed to give at least 90% reflectance in a high-reflectance stack of two double layers of quarter-wave layers at normal incidence? Assume a substrate of index 1.52.

22-16 Show that R_{max} in Eq. (22-53) approaches 1 when either N approaches infinity or when the ratio n_L/n_H approaches zero.

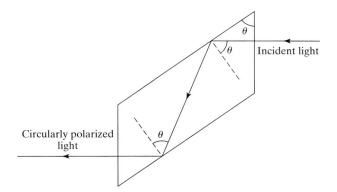

23 *Fresnel Equations*

INTRODUCTION

The basic laws of reflection and refraction in geometrical optics were derived earlier on the basis of either Huygens' or Fermat's principles. In this chapter we regard light as an electromagnetic wave and show that the laws of reflection and refraction can also be deduced from this point of view. More importantly, this approach also leads to the *Fresnel equations*, which describe the fraction of incident energy transmitted or reflected at a plane surface. These quantities will be seen to depend not only on the change in refractive index and the angle of incidence at the surface but also on the polarization of the incident light. Finally, the important differences between internal and external reflection are clarified.

23-1 THE FRESNEL EQUATIONS

Consider Figure 23-1, which shows a ray of light incident at point P on a plane interface—the xy boundary plane—and the resulting reflected and refracted rays. The plane of incidence is the xz-plane. Let us assume the incident light consists of plane harmonic waves, expressed by

$$\vec{E} = \vec{E}_0 e^{i(\vec{k}\cdot\vec{r} - \omega t)} \tag{23-1}$$

where the origin of coordinates is taken to be point O. The wave vector $\vec{E}$ of the incident wave is chosen in the $+y$-direction, so that the wave is linearly

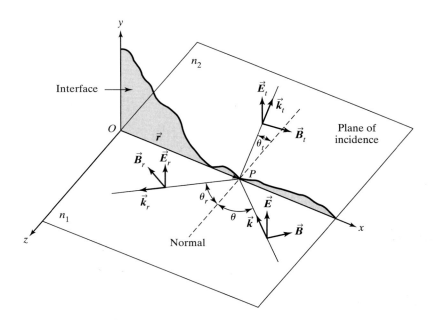

Figure 23-1 Defining diagram for incident, reflected, and transmitted rays at an *xy*-plane interface when the electric field is perpendicular to the plane of incidence, the TE mode.

polarized. The direction of the corresponding magnetic field vector $\vec{\mathbf{B}}$ is then determined to ensure that the direction of $\vec{\mathbf{E}} \times \vec{\mathbf{B}}$ is the direction of wave propagation $\vec{\mathbf{k}}$. This mode of polarization, in which the $\vec{\mathbf{E}}$-field is perpendicular to the plane of incidence and the $\vec{\mathbf{B}}$-field lies in the plane of incidence, is called the *transverse electric* (TE) mode. If instead $\vec{\mathbf{B}}$ is transverse to the plane of incidence, a case to be considered later, the mode is a *transverse magnetic* (TM) mode. An arbitrary polarization direction represents some linear combination of these two special cases. The reflected and transmitted waves in Figure 23-1 can be expressed, respectively, in forms like that of the incident wave of Eq. (23-1):

$$\vec{\mathbf{E}}_r = \vec{\mathbf{E}}_{0r} e^{i(\vec{\mathbf{k}}_r \cdot \vec{\mathbf{r}} - \omega_r t)} \tag{23-2}$$

$$\vec{\mathbf{E}}_t = \vec{\mathbf{E}}_{0t} e^{i(\vec{\mathbf{k}}_t \cdot \vec{\mathbf{r}} - \omega_t t)} \tag{23-3}$$

In the boundary plane *xy*, where all three waves exit simultaneously, there must be a fixed relationship between the three wave amplitudes (and thus their irradiances) that has yet to be determined. Since such a relationship cannot depend on the arbitrary choice of a boundary point $\vec{\mathbf{r}}$ nor a time *t*, it follows that the phases of the three waves, which depend on $\vec{\mathbf{r}}$ and *t*, must themselves be equal:

$$(\vec{\mathbf{k}} \cdot \vec{\mathbf{r}} - \omega t) = (\vec{\mathbf{k}}_r \cdot \vec{\mathbf{r}} - \omega_r t) = (\vec{\mathbf{k}}_t \cdot \vec{\mathbf{r}} - \omega_t t) \tag{23-4}$$

In particular, at the boundary point $\vec{\mathbf{r}} = 0$ of Figure 23-1,

$$-\omega t = -\omega_r t = -\omega_t t$$

or

$$\omega = \omega_r = \omega_t \tag{23-5}$$

so that all frequencies are equal. On the other hand, at *t* = 0 *within the boundary plane*, Eq. (23-4) yields:

$$\vec{\mathbf{k}} \cdot \vec{\mathbf{r}} = \vec{\mathbf{k}}_r \cdot \vec{\mathbf{r}} = \vec{\mathbf{k}}_t \cdot \vec{\mathbf{r}} \qquad (23\text{-}6)$$

Several conclusions can be drawn from the relations of Eq. (23-6). First notice that by subtracting any two members, these relations are equivalent to

$$(\vec{\mathbf{k}} - \vec{\mathbf{k}}_r) \cdot \vec{\mathbf{r}} = (\vec{\mathbf{k}} - \vec{\mathbf{k}}_t) \cdot \vec{\mathbf{r}} = (\vec{\mathbf{k}}_r - \vec{\mathbf{k}}_t) \cdot \vec{\mathbf{r}} = 0 \qquad (23\text{-}7)$$

Equation (23-7) requires that the vectors $\vec{\mathbf{k}}_r$ and $\vec{\mathbf{k}}_t$ lie in the plane determined by the vectors $\vec{\mathbf{k}}$ and $\vec{\mathbf{r}}$. Thus all three propagation vectors are coplanar in the xz-plane, and we conclude that the reflected and refracted waves lie in the plane of incidence. Next, consider the first two members of Eq. (23-6), which govern the relationship between the incident and reflected waves. In terms of the angles designated in Figure 23-1, they are equivalent to

$$kr \sin \theta = k_r r \sin \theta_r$$

Since both waves travel in the same medium, their wavelengths are identical and so $k = k_r$. Therefore, we have the

$$\text{law of reflection:} \quad \theta = \theta_r \qquad (23\text{-}8)$$

Finally, the last two members of Eq. (23-6) are equivalent to

$$k_r r \sin \theta_r = k_t r \sin \theta_t \qquad (23\text{-}9)$$

Writing $k_r = \omega/v_r = n_r \omega/c$ and $k_t = n_t \omega/c$, Eq. (23-9) becomes Snell's

$$\text{law of refraction:} \quad n_r \sin \theta_r = n_t \sin \theta_t \qquad (23\text{-}10)$$

We continue now to specify further the situation at the boundary with the help of boundary conditions arising out of Maxwell's equations and treated in texts on electricity and magnetism. We employ them here without proof. These boundary conditions require that the components of both the electric and magnetic fields parallel to the boundary plane be continuous as the boundary is crossed.

Boundary Conditions for TE Waves

As mentioned earlier, TE waves have electric fields that are perpendicular to the plane of incidence and therefore are parallel to the boundary plane separating the two media. In terms of the choices made for the direction of the electric fields in Figure 23-1, the vector amplitudes of the complex fields of Eqs. (23-1)–(23-3) can be written as

$$\vec{\mathbf{E}}_0 = E\hat{\mathbf{y}} \quad \vec{\mathbf{E}}_{0r} = E_r\hat{\mathbf{y}} \quad \vec{\mathbf{E}}_{0t} = E_t\hat{\mathbf{y}} \qquad (23\text{-}11)$$

Here, E, E_r and E_t are the complex field amplitudes associated, respectively, with the incident, reflected, and transmitted waves. The requirement that the component of the electric field parallel to the boundary plane be continuous at the boundary then gives

$$E + E_r = E_t \qquad (23\text{-}12)$$

The magnetic fields associated with the electric fields of Figure 23-1 have the form,

$$\vec{B} = (B \cos \theta \hat{x} - B \sin \theta \hat{z})e^{i(\vec{k} \cdot \vec{r} - \omega t)}$$

$$\vec{B}_r = (-B_r \cos \theta_r \hat{x} - B_r \sin \theta_r \hat{z})e^{i(\vec{k}_r \cdot \vec{r} - \omega t)} \qquad (23\text{-}13)$$

$$\vec{B}_t = (B_t \cos \theta_t \hat{x} - B_t \sin \theta_t \hat{z})e^{i(\vec{k}_t \cdot \vec{r} - \omega t)}$$

Continuity of the parallel components of the magnetic field requires that the field amplitudes be related by

$$B \cos \theta - B_r \cos \theta = B_t \cos \theta_t \qquad (23\text{-}14)$$

where we have made use of Eq. (23-8). Equations (23-12) and (23-14) are correct for the $\vec{E}$ and $\vec{B}$ vectors as chosen in Figure 23-1. If a different choice is made, for example, by reversing the $\vec{E}$ vector of the incident wave (and also $\vec{B}$ to keep the direction of wave propagation the same), Eqs. (23-12) and (23-14) appear with a change of signs. However, the physical import of these equations is the same when they are interpreted in terms of their original figures.

Boundary Conditions for TM Waves

Before pursuing the significance of Eqs. (23-12) and (23-14) for the TE mode, we parallel their development for the TM mode pictured in Figure 23-2. The electric and magnetic fields in this figure can be written as

$$\vec{E} = (E \cos \theta \hat{x} - E \sin \theta \hat{z})e^{i(\vec{k} \cdot \vec{r} - \omega t)}$$

$$\vec{E}_r = (E_r \cos \theta_r \hat{x} + E_r \sin \theta_r \hat{z})e^{i(\vec{k}_r \cdot \vec{r} - \omega t)} \qquad (23\text{-}15)$$

$$\vec{E}_t = (E_t \cos \theta_t \hat{x} - E_t \sin \theta_t \hat{z})e^{i(\vec{k}_t \cdot \vec{r} - \omega t)}$$

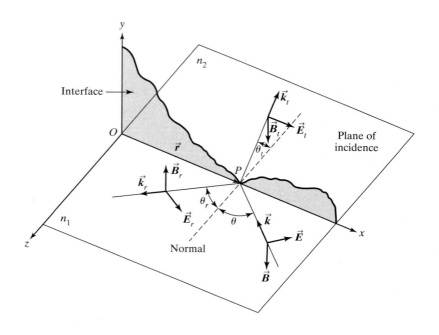

Figure 23-2 Defining diagram for incident, reflected, and transmitted rays at an xy-plane interface when the magnetic field is perpendicular to the plane of incidence, the TM mode.

and

$$\vec{\mathbf{B}} = -B\hat{\mathbf{y}}e^{i(\vec{\mathbf{k}}\cdot\vec{\mathbf{r}}-\omega t)}$$

$$\vec{\mathbf{B}}_r = B_r\hat{\mathbf{y}}e^{i(\vec{\mathbf{k}}\cdot\vec{\mathbf{r}}-\omega t)} \qquad (23\text{-}16)$$

$$\vec{\mathbf{B}}_t = -B_t\hat{\mathbf{y}}e^{i(\vec{\mathbf{k}}_t\cdot\vec{\mathbf{r}}-\omega t)}$$

Requiring continuity of the components of the electric and magnetic fields that are parallel to the boundary gives, in this case,

$$-B + B_r = -B_t \qquad (23\text{-}17)$$

$$E\cos\theta + E_r\cos\theta = E_t\cos\theta_t \qquad (23\text{-}18)$$

Reflection and Transmission Coefficients

The magnetic field amplitudes of Eqs. (23-14) and (23-17) can be expressed in terms of the corresponding electric field amplitudes through the generic relation

$$E = vB = \left(\frac{c}{n}\right)B \qquad (23\text{-}19)$$

Writing the index of refraction for incident and refracting media as n_1 and n_2, respectively, Eqs. (23-12), (23-14), (23-17), and (23-18) can be recast as follows:

$$\text{TE:}\begin{cases} E + E_r = E_t & (23\text{-}20) \\ n_1 E\cos\theta - n_1 E_r\cos\theta = n_2 E_t\cos\theta_t & (23\text{-}21) \end{cases}$$

$$\text{TM:}\begin{cases} -n_1 E + n_1 E_r = -n_2 E_t & (23\text{-}22) \\ E\cos\theta + E_r\cos\theta = E_t\cos\theta_t & (23\text{-}23) \end{cases}$$

Next, eliminating E_t from each pair of equations and solving for the *reflection coefficient* $r = E_r/E$,

$$r_{TE} = \frac{E_r}{E} = \frac{\cos\theta - n\cos\theta_t}{\cos\theta + n\cos\theta_t} \qquad (23\text{-}24)$$

$$r_{TM} = \frac{E_r}{E} = \frac{-n\cos\theta + \cos\theta_t}{n\cos\theta + \cos\theta_t} \qquad (23\text{-}25)$$

where we have introduced a *relative refractive index* $n \equiv n_2/n_1$. Note that we use subscripts to distinguish between the TE and TM cases. Finally, since n and θ_t are related to θ through Snell's law, $\sin\theta = n\sin\theta_t$, θ_t may be eliminated using

$$n\cos\theta_t = n\sqrt{1 - \sin^2\theta_t} = \sqrt{n^2 - \sin^2\theta} \qquad (23\text{-}26)$$

The results are then

$$r_{TE} = \frac{E_r}{E} = \frac{\cos\theta - \sqrt{n^2 - \sin^2\theta}}{\cos\theta + \sqrt{n^2 - \sin^2\theta}} \qquad (23\text{-}27)$$

$$r_{TM} = \frac{E_r}{E} = \frac{-n^2 \cos\theta + \sqrt{n^2 - \sin^2\theta}}{n^2 \cos\theta + \sqrt{n^2 - \sin^2\theta}} \qquad (23\text{-}28)$$

Returning to Eqs. (23-20) through (23-23), if E_r is eliminated instead of E_t, similar steps lead to the following equations describing the *transmission coefficient* $t = E_t/E$:

$$t_{TE} = \frac{E_t}{E} = \frac{2\cos\theta}{\cos\theta + \sqrt{n^2 - \sin^2\theta}} \qquad (23\text{-}29)$$

$$t_{TM} = \frac{E_t}{E} = \frac{2n\cos\theta}{n^2\cos\theta + \sqrt{n^2 - \sin^2\theta}} \qquad (23\text{-}30)$$

Eqs. (23-29) and (23-30) can also be found more quickly by using Eqs. (23-20) and (23-22) written in the form

$$t_{TE} = 1 + r_{TE}$$

$$nt_{TM} = 1 - r_{TM}$$

into which the results expressed by Eqs. (23-27) and (23-28) can be conveniently substituted. Equations (23-27) through (23-30) are the *Fresnel equations*, giving reflection and transmission coefficients, the ratio of both reflected and transmitted $\vec{E}$-field amplitudes to the incident $\vec{E}$-field amplitude. Note that, for normal incidence, the reflection and transmission coefficients for the TE case are identical to those for the TM case. This is sensible since for normal incidence there is no distinction between the two cases.[1] In practice, measured reflection and transmission coefficients also depend on scattering losses from a nonplanar surface.

Example 23-1

Calculate the reflection and transmission coefficients for both TE and TM modes of light incident from air at 30° onto glass of index 1.60.

Solution

Using Eqs. (23-27) and (23-28),

$$r_{TE} = \frac{\cos(30°) - \sqrt{1.6^2 - \sin^2(30°)}}{\cos(30°) + \sqrt{1.6^2 - \sin^2(30°)}} = -0.2740$$

$$r_{TM} = \frac{-1.6^2 \cos(30°) + \sqrt{1.6^2 - \sin^2(30°)}}{1.6^2 \cos(30°) + \sqrt{1.6^2 - \sin^2(30°)}} = -0.1866$$

Using the relations below Eq. (23-30),

$$t_{TE} = 1 + r_{TE} = 1 - 0.2740 = 0.7260$$

$$t_{TM} = \frac{1 - r_{TM}}{n} = \frac{1 + 0.1866}{1.60} = 0.7416$$

---

[1] Some texts use a different convention, in which the positive direction of the reflected electric field for the TM case is opposite to that shown in Figure 23-2, leading to an expression for the reflection coefficient for the TM case that differs from ours by a factor of -1. Of course, both conventions lead to the same physical result since the extra factor of -1 simply reverses the direction of the reflected electric field.

23-2 EXTERNAL AND INTERNAL REFLECTIONS

When interpreting these equations, it is useful to distinguish between two physically different situations:

$$\text{external reflection:} \quad n_1 < n_2 \quad \text{or} \quad n = \frac{n_2}{n_1} > 1$$

$$\text{internal reflection:} \quad n_1 > n_2 \quad \text{or} \quad n = \frac{n_2}{n_1} < 1$$

Figure 23-3 is a plot of Eqs. (23-27) through (23-30) for the case of external reflection with $n = 1.50$. Notice that at both normal and grazing incidence—angles of 0° and 90°, respectively—TE and TM modes have reflection coefficients of the same magnitude and transmission coefficients of the same magnitude. Negative values of r for both the TE and TM modes indicate a phase change of the $\vec{E}$- or $\vec{B}$-field vectors on reflection and will be discussed presently. The fraction of power P in the incident wave that is reflected or transmitted, called the *reflectance* and the *transmittance*, respectively, depends on the ratio of the squares of the amplitudes.

$$\text{reflectance} = R = \frac{P_r}{P_i} = r^2 = \left(\frac{E_r}{E}\right)^2 \tag{23-31}$$

$$\text{transmittance} = T = \frac{P_t}{P_i} = n\left(\frac{\cos\theta_t}{\cos\theta}\right)t^2 \tag{23-32}$$

These expressions are justified later in this chapter.

In Figure 23-4, reflectance is plotted as a function of the angle of incidence θ. The curve for the case of external reflection, TM mode, indicates that no wave energy is reflected when the angle of incidence is near 60°. The angle θ_p at which $R_{TM} = 0$ is known as *Brewster's angle* or the *polarizing* angle and takes the value,

$$\theta_p = \tan^{-1}(n) = \tan^{-1}(n_2/n_1)$$

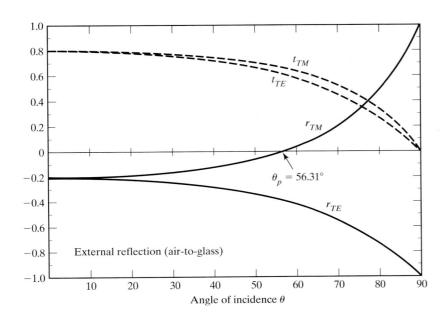

Figure 23-3 Reflection and transmission coefficients for the case of external reflection, with $n = n_2/n_1 = 1.50$.

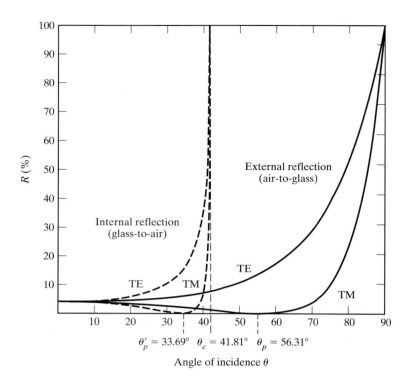

Figure 23-4 Reflectance for both external and internal reflection when $n_1 = 1$ and $n_2 = 1.50$.

This condition is also evident in the vanishing of r_{TM} in Figure 23-3 and the vanishing of the numerator of Eq. (23-28). See problem 23-1. For the case $n = 1.50$ used in Figures 23-3 and 23-4, $\theta_p = 56.31°$. R_{TE} does not go to zero under this condition, so reflected light contains only the TE mode and is linearly polarized, with $R_{TE} = 15\%$. At normal incidence ($\theta = 0°$), for both TE and TM modes, Eqs. (23-24) and (23-25) simplify to give

$$R = r^2 = \left(\frac{1-n}{1+n}\right)^2 \tag{23-33}$$

Equation (23-33) gives a reflectance of 4% from an air/glass interface with $n = 1.5$. Keep in mind, however, that n is a function of wavelength. As the angle of incidence increases to grazing incidence ($\theta = 90°$), both R_{TE} and R_{TM} become unity, although R_{TM} remains quite small until Brewster's angle has been exceeded.

The reflection coefficient for the case of internal reflection is shown in Figure 23-5 with $n = 1/1.50$, as when light encounters a glass/air interface from the glass side. Evidence of phase changes and of a polarizing, or Brewster's, angle may also be seen here. For the case of internal reflection we give Brewster's angle the symbol θ'_p. Examination of Figures 23-4 and 23-5 shows that, for the case of internal reflection, both $R_{TE} = r_{TE}^2$ and $R_{TM} = r_{TM}^2$ reach values of unity before the angle of incidence θ reaches 90°. This is the phenomenon of *total internal reflection*, which occurs at the critical angle $\theta_c = \sin^{-1}(n) = \sin^{-1}(n_2/n_1)$. For the example of glass ($n = 1/1.5$) used in Figure 23-5, $\theta'_p = 33.7°$ and $\theta_c = 41.8°$. When $\sin \theta_c > n$, the radical $\sqrt{n^2 - \sin^2 \theta}$ is negative and both r_{TE} and r_{TM} are complex. Their magnitudes, however, are easily shown to be unity in this range, giving total reflection for $\theta > \theta_c$.

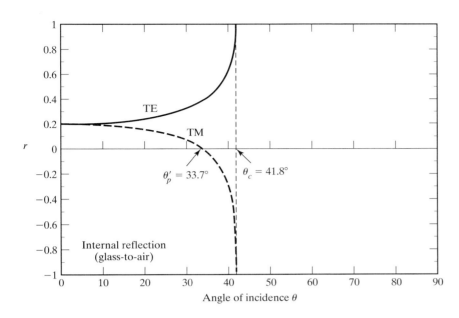

Figure 23-5 Reflection coefficient for the case of internal reflection with $n = n_1/n_2 = 1/1.50$.

23-3 PHASE CHANGES ON REFLECTION

The negative values of the reflection coefficient in Figures 23-3 and 23-5 indicate that $E_r = -|r|E$ in certain situations. Evidently, the electric field vector may reverse direction on reflection. Equivalently, in such cases there is a π-phase shift of E on reflection, as the following mathematical argument demonstrates:

$$E_r = -|r|E = e^{i\pi}|r|E_0 e^{i(\mathbf{k}\cdot\mathbf{r}-\omega t)} = |r|E_0 e^{i(\mathbf{k}\cdot\mathbf{r}-\omega t+\pi)}$$

Thus in the case of external reflection, Figure 23-3, a π-phase shift of E occurs at any angle of incidence for the TE mode and for $\theta < \theta_p$ for the TM mode. When reflection is internal, Figure 23-5, we conclude that a π-phase shift occurs for the TM mode for $\theta'_p < \theta < \theta_c$. However, the situation in the region $\theta > \theta_c$, where r is complex, requires further investigation. When $\theta > \theta_c = \sin^{-1}(n)$, the radical in Eqs. (23-27) and (23-28) becomes imaginary, and the equations may be written in the form

$$r_{TE} = \frac{\cos\theta - i\sqrt{\sin^2\theta - n^2}}{\cos\theta + i\sqrt{\sin^2\theta - n^2}} \tag{23-34}$$

$$r_{TM} = \frac{-n^2\cos\theta + i\sqrt{\sin^2\theta - n^2}}{n^2\cos\theta + i\sqrt{\sin^2\theta - n^2}} \tag{23-35}$$

The reflection coefficients can be written in polar form as $r = |r|e^{i\phi}$ and we shall refer to ϕ as the *phase shift on reflection*. In Eq. (23-34), the reflection coefficient takes the form $r_{TE} = (a - ib)/(a + ib)$. Since the real and imaginary parts of the numerator and denominator are the same, except for a sign, the magnitudes of the numerator and denominator are equal, and r_{TE} has unit amplitude. The phase of r_{TE} may be investigated by expressing Eq. (23-34) in complex polar form, as

$$r_{TE} = \frac{e^{-i\alpha}}{e^{i\alpha}} = e^{-i(2\alpha)}$$

where $\tan\alpha = \sqrt{\sin^2\theta - n^2}/\cos\theta$. So, for the TE case, the phase shift on reflection is $\phi_{TE} = -2\alpha$. A similar analysis (see problem 23-6) can be used to show that r_{TM}, like r_{TE}, has unit magnitude when the angle of incidence exceeds the critical angle and enables one to find the phase shift on total internal reflection ϕ_{TM} for the TM case. The phase shifts on *total internal reflection* for the two cases have the form,

$$\tan\left(\frac{\phi_{TE}}{2}\right) = -\frac{\sqrt{\sin^2\theta - n^2}}{\cos\theta} \tag{23-36}$$

$$\tan\left(\frac{\phi_{TM} - \pi}{2}\right) = -\frac{\sqrt{\sin^2\theta - n^2}}{n^2\cos\theta} \tag{23-37}$$

Clearly, the phase shift on reflection, for total internal reflection, may take on values other than 0 and π, depending on the angle of incidence. The phase shift ϕ, as determined from Eqs. (23-36) and (23-37), is plotted in Figure 23-6.

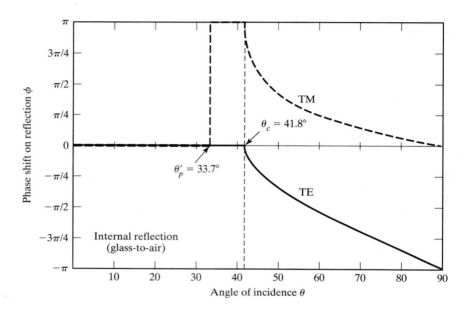

Figure 23-6 Phase shift ϕ on reflection of the electric field for internally reflected rays, with $n = n_1/n_2 = 1/1.5$.

It happens that the relative phase shift $\phi_{TE} - \phi_{TM}$ is about $-3\pi/4$ at an angle of incidence near 53°. Two consecutive internal reflections thus produce a relative phase shift of $2(-3\pi/4) = -3\pi/2$ (equivalently, $+\pi/2$) between the perpendicular components of the $\vec{E}$-field. Recall that circularly polarized light consists of equal amplitude components with phases that differ by $\pm\pi/2$. Thus linearly polarized incident light with equal TM and TE components, after two internal reflections at 53°, will be transformed into circularly polarized light. This technique is utilized in the *Fresnel rhomb* (Figure 23-7). Summarizing these results for the case of internal reflection,

$$\phi_{TM} = \begin{cases} 0, & \theta < \theta'_p \\ \pi, & \theta'_p < \theta < \theta_c \\ -2\arctan\left(\dfrac{\sqrt{\sin^2\theta - n^2}}{n^2\cos\theta}\right) + \pi, & \theta > \theta_c \end{cases} \tag{23-38}$$

$$\phi_{TE} = \begin{cases} 0, & \theta < \theta_c \\ -2\arctan\left(\dfrac{\sqrt{\sin^2\theta - n^2}}{\cos\theta}\right), & \theta > \theta_c \end{cases} \tag{23-39}$$

Phase shifts for both TM and TE modes and for both internal and external reflection are summarized in Figure 23-8.

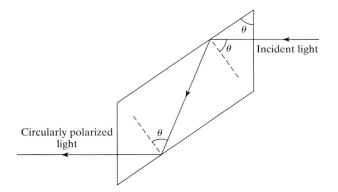

Figure 23-7 The Fresnel rhomb. With the incident light polarized at 45° to the plane of incidence, two internal reflections produce equal-amplitude TE and TM amplitudes with a relative phase of $\pi/2$, or circularly polarized light. For $n = 1.50$, the angle should be $\theta = 53°$. The device is effective over a wide range of wavelengths.

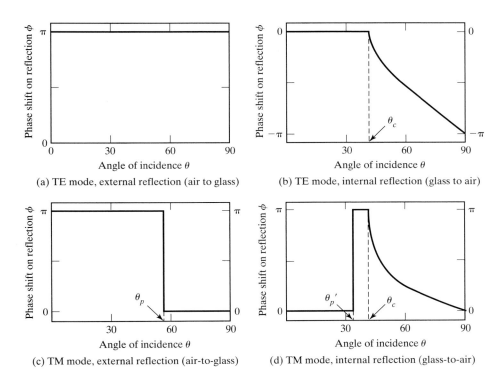

(a) TE mode, external reflection (air to glass)

(b) TE mode, internal reflection (glass to air)

(c) TM mode, external reflection (air-to-glass)

(d) TM mode, internal reflection (glass-to-air)

Figure 23-8 Phase changes on reflection ϕ between incident and reflected rays versus angle of incidence. Discontinuities occur at $\theta_c = 41.8°$, $\theta_p = 56.3°$, and $\theta'_p = 33.7°$ for refractive indices of $n_1 = 1$ and $n_2 = 1.50$.

Example 23-2

What is the phase shift of the TM and TE rays reflected both externally and internally for the situation discussed in Example 23-1?

Solution

For this interface,

$$\theta_c = \sin^{-1}\left(\frac{1}{1.6}\right) = 38.7°$$

$$\theta_p = \tan^{-1}(1.6) = 58.0°$$

$$\theta'_p = \tan^{-1}\left(\frac{1}{1.6}\right) = 32.0°$$

Since the angle of incidence of 30° is less than either θ'_p or θ_c, Eqs. (23-38) and (23-39) or Figure 23-8 require that for internal reflection, $\phi_{TM} = 0$ and $\phi_{TE} = 0$, while Figure 23-8 shows that for external reflection, $\phi_{TM} = \pi$ and $\phi_{TE} = \pi$.

A general conclusion can be drawn from the phase changes for the TE and TM modes under internal and external reflection: Near normal incidence, for both TE and TM modes, the phase shift for an internally reflected beam differs from that of an externally reflected beam by π. For a thin film in air, we are interested in the *relative* phase shift between rays reflected from the first surface (external) and the second surface (internal). Inspection of Figure 23-8 shows that a *relative* phase shift of π occurs in the TE mode for *internal* angles of incidence less than θ_c and in the TM mode for *internal* angles of incidence less than θ'_p. The corresponding ranges of *external* angles of incidence at the first surface are 0° to 90° (TE mode) and 0° to θ_p (TM mode). Thus in the TE mode a relative phase shift of π occurs for *all* external angles of incidence, but in the TM mode this is true only for external angles less than θ_p.

23-4 CONSERVATION OF ENERGY

To conserve energy, at a given boundary, it must be true that the power incident on the boundary be equal to the sum of the power reflected at that boundary and the power transmitted through the boundary. That is,

$$P_i = P_r + P_t \tag{23-40}$$

If we represent the reflectance R as the ratio of reflected to incident power and the transmittance T as the ratio of transmitted to incident power,

$$R = \frac{P_r}{P_i} \quad \text{and} \quad T = \frac{P_t}{P_i} \tag{23-41}$$

then Eq. (23-40) takes the form

$$1 = R + T \tag{23-42}$$

The irradiance I is the power density (W/m^2), so that we may write, in place of Eq. (23-40),

$$I_i A_i = I_r A_r + I_t A_t \tag{23-43}$$

The cross-sectional areas of the three beams (see Figure 23-9) that appear in Eq. (23-43) are all related to the area A intercepted by the beams in the boundary plane through the cosines of the angles of incidence, reflection, and refraction. We may then write

$$I_i(A \cos \theta) = I_r(A \cos \theta_r) + I_t(A \cos \theta_t)$$

Of course, $\theta = \theta_r$ by the law of reflection. Also using the relation between irradiance and electric field amplitude,

$$I = \left(\frac{\varepsilon v}{2}\right)E_0^2$$

and the facts that $v_i = v_r$, and $\varepsilon_i = \varepsilon_r$, since they correspond to the same medium, we arrive at the equation

$$E_0^2 = E_{0r}^2 + \left(\frac{v_t \varepsilon_t}{v_i \varepsilon_i}\right)\left(\frac{\cos \theta_t}{\cos \theta}\right)E_{0t}^2 \tag{23-44}$$

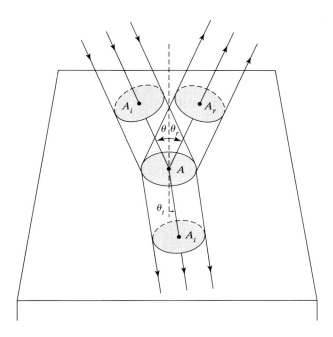

Figure 23-9 Comparison of cross sections of incident, reflected, and transmitted beams.

The quantity $(v_t \varepsilon_t / v_i \varepsilon_i)$ is just a complicated way of expressing the relative refractive index n, which we can show as follows:

$$\frac{v_t \varepsilon_t}{v_i \varepsilon_i} = \frac{v_t}{v_i} \frac{v_i^2 \mu_i}{v_t^2 \mu_t} = \frac{v_i}{v_t} = n \tag{23-45}$$

In arriving at this result we have used

$$\mu_i = \mu_t = \mu_0$$

for nonmagnetic materials and the relation

$$v^2 = \frac{1}{\mu \varepsilon}$$

for the velocity of a plane electromagnetic wave. Incorporating Eq. (23-45) in Eq. (23-44),

$$E_{0i}^2 = E_{0r}^2 + n\left(\frac{\cos \theta_t}{\cos \theta}\right) E_{0t}^2 \tag{23-46}$$

Dividing the equation by the left member, it becomes

$$1 = r^2 + n\left(\frac{\cos \theta_t}{\cos \theta}\right) t^2 \tag{23-47}$$

where the reflection and transmission coefficients r and t have been introduced. Now the quantity r^2 is just the reflectance R:

$$R = \frac{P_r}{P_i} = \frac{I_r}{I_i} = \left(\frac{E_{0r}}{E_{0i}}\right)^2 = r^2$$

Comparing Eq. (23-47) with Eq. (23-42), it follows that the transmittance T is expressed by the relation

$$T = n\left(\frac{\cos \theta_t}{\cos \theta}\right) t^2 \tag{23-48}$$

Notice that T is not simply t^2 since it must take into account a different speed and direction in a new medium. The change in speed modifies the rate of energy propagation and thus the power of the beam; the change in direction modifies the cross section and thus the power density of the beam. However, for normal incidence, Eq. (23-48) reduces to $T = nt^2$ and Eq. (23-47) becomes

$$\text{normal incidence: } 1 = r^2 + nt^2 \qquad (23\text{-}49)$$

Note that throughout this section we have assumed that the reflection and transmission coefficients are real, as they will be for all external reflections and all internal reflections with angles of incidence less than the critical angle.

Example 23-3

Calculate the reflectance R and transmittance T for both TE and TM modes of light incident at $30°$ on glass of index 1.60.

Solution

The reflection and transmission coefficients for this situation are given in the solution to Example 23-1. Using these, the reflectance and transmittance are found to be

$$R_{TE} = r_{TE}^2 = (-0.2740)^2 = 0.075 \quad \text{and} \quad T_{TE} = 1 - R_{TE} = 0.925$$
$$R_{TM} = r_{TM}^2 = (-0.1866)^2 = 0.035 \quad \text{and} \quad T_{TM} = 1 - R_{TM} = 0.965$$

23-5 EVANESCENT WAVES

In discussing the propagation of a light wave by total internal reflection (TIR) through an optical fiber (Chapter 10), we mentioned the phenomenon of *cross talk*, the coupling of wave energy into another medium when it is brought close enough to the reflecting wave. This loss of energy is described as *frustrated total internal reflection*. The theory presented in this chapter allows us to describe this phenomenon quantitatively.

The transmitted wave at a refraction can be represented as

$$E_t = E_{0t} e^{i(\vec{\mathbf{k}}_t \cdot \vec{\mathbf{r}} - \omega t)}$$

where, according to the coordinates chosen in Figure 23-1,

$$\vec{\mathbf{k}}_t \cdot \vec{\mathbf{r}} = k_t \left(-\sin \theta_t \,\hat{\mathbf{x}} - \cos \theta_t \hat{\mathbf{z}} \right) \cdot \left(x\,\hat{\mathbf{x}} + z\,\hat{\mathbf{z}} \right)$$

$$\vec{\mathbf{k}}_t \cdot \vec{\mathbf{r}} = k_t \left(-x \sin \theta_t - z \cos \theta_t \right)$$

We can express $\cos \theta_t$ as

$$\cos \theta_t \equiv \sqrt{1 - \sin^2 \theta_t} = \sqrt{1 - \frac{\sin^2 \theta}{n^2}}$$

where we have used Snell's law, $n \sin \theta_t = \sin \theta$ in writing the last equality. At the critical angle, $\sin \theta = n$ and $\cos \theta_t = \cos(90°) = 0$. For angles such that $\sin \theta > n$, when TIR occurs, $\cos \theta_t$ becomes purely imaginary and we can write

$$\cos \theta_t = i\sqrt{\frac{\sin^2 \theta}{n^2} - 1}$$

Thus the exponential factor

$$\vec{\mathbf{k}}_t \cdot \vec{\mathbf{r}} = -k_t x \frac{\sin \theta}{n} - i k_t z \sqrt{\frac{\sin^2 \theta}{n^2} - 1} = -k_t x \frac{\sin \theta}{n} + i k_t |z| \sqrt{\frac{\sin^2 \theta}{n^2} - 1}$$

In writing the last equality we have noted that, for the situation depicted in Figure 23-1, the transmitted wave exists in the region for which $z < 0$, and so in this region $z = -|z|$. With the definition of the real, positive number,

$$\alpha \equiv k_t \sqrt{\frac{\sin^2 \theta}{n^2} - 1}$$

the transmitted wave may be expressed as

$$E_t = E_{0t} e^{i(\vec{\mathbf{k}}_t \cdot \vec{\mathbf{r}} - \omega t)} = E_{0t} e^{-i\omega t} e^{-ixk_t \sin \theta / n} e^{-\alpha |z|}$$

The last factor on the right-hand side of this relation describes an exponential decrease in the amplitude of the wave as it enters the medium of lesser refractive index along the negative z-direction. When the wave penetrates into the medium of lesser refractive index by an amount

$$|z| = \frac{1}{\alpha} = \frac{\lambda}{2\pi \sqrt{\frac{\sin^2 \theta}{n^2} - 1}} \tag{23-50}$$

the amplitude is decreased by a factor of $1/e$. The energy of this *evanescent wave* returns to its original medium unless a second medium is introduced into its region of penetration. Although detrimental in the case of cross talk in closely bound fibers lacking sufficient thickness of protective cladding, the *frustration* of the total internal reflection is put to good use in devices such as variable output couplers, made of two right-angle prisms whose separation along their diagonal faces can be carefully adjusted to vary the amount of evanescent wave coupled from one prism into the other. Another application involves a prism face brought near to the surface of an optical waveguide so that the evanescent wave emerging from the prism can be coupled into the waveguide at a given angle (mode) of propagation.

Example 23-4

Calculate the penetration depth of an evanescent wave undergoing TIR at a glass- ($n = 1.50$) to-air interface, such that the amplitude is attenuated to $1/e$ of its original value. Assume light of wavelength 500 nm is incident on the interface at an angle of 60°.

Solution

Since $\theta_c = \sin^{-1}(1/1.5) = 41.8°$, TIR occurs at 60°. The penetration depth is given by Eq. (23-50):

$$|z| = \frac{0.500 \ \mu m}{2\pi \sqrt{\frac{\sin^2 60}{(1/1.5)^2} - 1}} = 0.096 \ \mu m$$

23-6 COMPLEX REFRACTIVE INDEX

We wish now to show that when the reflecting surface is metallic, the Fresnel equations we have derived continue to be valid, with one important modification: The index of refraction becomes a complex number, including an imaginary part that is a measure of the absorption of the wave. This subject is treated at greater length in Chapter 25, where we examine the frequency dependence of the real and imaginary parts of the refractive index.

When the reflecting surface is that of a homogeneous dielectric—the case we have been discussing in this chapter—the *conductivity* σ of the material is zero. The conductivity is the proportionality constant in *Ohm's law*,

$$\vec{\mathbf{j}} = \sigma \vec{\mathbf{E}}$$

where $\vec{\mathbf{j}}$ is the *current density* (A/m^2) produced by the field $\vec{\mathbf{E}}$. In such cases, both the $\vec{\mathbf{E}}$- and $\vec{\mathbf{B}}$-fields satisfy a differential wave equation of the form

$$\nabla^2 E = \left(\frac{1}{c^2}\right)\frac{\partial^2 E}{\partial t^2} \tag{23-51}$$

as described in Chapter 4. We have written harmonic waves satisfying Eq. (23-51) in the form

$$E = E_0 e^{i(\vec{\mathbf{k}}\cdot\vec{\mathbf{r}} - \omega t)} \tag{23-52}$$

Now if the material is metallic or has an appreciable conductivity, the fundamental Maxwell equations of electricity and magnetism lead to a modification of Eqs. (23-51) and (23-52). The differential wave equation to be satisfied by the $\vec{\mathbf{E}}$-field is then

$$\nabla^2 E = \left(\frac{1}{c^2}\right)\frac{\partial^2 E}{\partial t^2} + \left(\frac{\sigma}{\varepsilon_0 c^2}\right)\frac{\partial E}{\partial t} \tag{23-53}$$

Note that, compared with Eq. (23-51), the new wave equation, given as Eq. (23-53), includes an additional term involving the conductivity and the first time derivative of E. As a result, when a harmonic wave in the form of Eq. (23-52) is substituted into Eq. (23-53), we find that the propagation vector $\vec{\mathbf{k}}$ must have the complex magnitude

$$\tilde{k} = \frac{\omega}{c}\left[1 + i\left(\frac{\sigma}{\varepsilon_0 \omega}\right)\right]^{1/2} \tag{23-54}$$

Since the refractive index n is related to k by $n = (c/\omega)k$, the refractive index is now the complex number

$$\tilde{n} = \left[1 + i\left(\frac{\sigma}{\varepsilon_0 \omega}\right)\right]^{1/2} \tag{23-55}$$

or we write, in general,

$$\tilde{n} = n_R + in_I \tag{23-56}$$

where $\text{Re}\,(\tilde{n}) = n_R$ and $\text{Im}\,(\tilde{n}) = n_I$. Combing Eqs. (23-55) and (23-56) and equating their real and imaginary parts, the *optical constants* n_R and n_I can be found in terms of the conductivity by the equations

$$n_R^2 - n_I^2 = 1$$

$$2n_R n_I = \frac{\sigma}{\varepsilon_0 \omega}$$

(23-57)

Furthermore, if the complex character of k in the form

$$\widetilde{k} = \left(\frac{\omega}{c}\right)\widetilde{n} = \left(\frac{\omega}{c}\right)[n_R + in_I]$$

(23-58)

is introduced into the harmonic wave, Eq. (23-52), the result is

$$E = E_0 e^{-(\omega n_I s/c)} e^{i\omega(n_R s/c - t)}$$

(23-59)

where s is the directed distance along the propagation direction. We conclude from Eq. (23-59) that the wave propagates in the material at a wave speed c/n_R and is absorbed such that the amplitude decreases at a rate governed by the exponential factor $e^{-(\omega n_I s/c)}$. Thus, Re $(\widetilde{n}) = n_R$ must behave as the ordinary refractive index, and Im $(\widetilde{n}) = n_I$, called the *extinction coefficient*, determines the rate of absorption of the wave in the conductive medium. This absorption, due to the energy contributed to the production of conduction current j in the material, is usually described by the decrease in power density I with distance, given by

$$I = I_0 e^{-\alpha s}$$

(23-60)

By comparison with the power density as determined from Eq. (23-59), where $I \propto |E|^2$,

$$I = I_0 e^{-2\omega n_I s/c}$$

(23-61)

Thus, the *absorption coefficient* α is related to the *extinction coefficient* n_I by

$$\alpha = \frac{2\omega n_I}{c} = \frac{4\pi n_I}{\lambda}$$

(23-62)

23-7 REFLECTION FROM METALS

Replacing n by $\widetilde{n}$ in the Fresnel equations, Eqs. (23-27) and (23-28), we have for metals,

$$\text{TE:} \quad \frac{E_r}{E} = \frac{\cos\theta - \sqrt{\widetilde{n}^2 - \sin^2\theta}}{\cos\theta + \sqrt{\widetilde{n}^2 - \sin^2\theta}}$$

(23-63)

$$\text{TM:} \quad \frac{E_r}{E} = \frac{-\widetilde{n}^2 \cos\theta + \sqrt{\widetilde{n}^2 - \sin^2\theta}}{\widetilde{n}^2 \cos\theta + \sqrt{\widetilde{n}^2 - \sin^2\theta}}$$

(23-64)

Introducing $\widetilde{n}$ as $n_R + in_I$ into Eqs. (23-63) and (23-64) gives

$$\text{TE:} \quad \frac{E_r}{E} = \frac{\cos\theta - \sqrt{(n_R^2 - n_I^2 - \sin^2\theta) + i(2n_R n_I)}}{\cos\theta + \sqrt{(n_R^2 - n_I^2 - \sin^2\theta) + i(2n_R n_I)}}$$

(23-65)

$$\text{TM:}\quad \frac{E_r}{E} = \frac{-[n_R^2 - n_I^2 + i(2n_R n_I)]\cos\theta + \sqrt{(n_R^2 - n_I^2 - \sin^2\theta) + i(2n_R n_I)}}{[n_R^2 - n_I^2 + i(2n_R n_I)]\cos\theta + \sqrt{(n_R^2 - n_I^2 - \sin^2\theta) + i(2n_R n_I)}}$$

$$(23\text{-}66)$$

In calculating the reflectance $R = |E_r/E|^2$, the complex quantity E_r/E can first be reduced to a ratio of complex numbers in the form $(a + ib)/(c + id)$, so that

$$R = (a^2 + b^2)/(c^2 + d^2)$$

In the process, we must take the square root of a complex number, which is done by first putting it into polar form. For example, if $z = A + iB$, then, in polar form,

$$z = (A^2 + B^2)^{1/2} e^{i[\tan^{-1}(B/A)]}$$

and the square root becomes

$$z^{1/2} = (A^2 + B^2)^{1/4} e^{i[(1/2)\tan^{-1}(B/A)]} \qquad (23\text{-}67)$$

The complex expression in Eq. (23-67) can then be returned to the general complex form $C + iD$ using Euler's equation. These mathematical steps are easily performed with a programmable calculator or a computer. In Figure 23-10, the results of such calculations are shown for two metal surfaces, solid sodium and single-crystal gallium. High reflectance in the visible spectrum is characteristic of metallic surfaces, as shown by the curves for solid sodium at a wavelength of 589.3 nm. Strong discrimination between the TE and TM modes in the incident radiation is exhibited by the curves for single-crystal gallium surfaces.

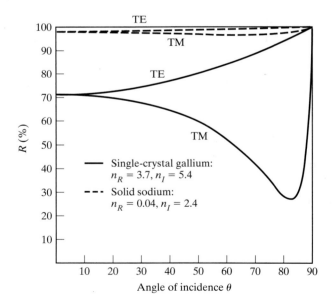

Figure 23-10　Reflectance from metal surfaces by using Fresnel's equations. The values of n_R and n_I are given for sodium light of $\lambda = 589.3$ nm.

PROBLEMS

23-1 Show that the vanishing of the reflection coefficient in the TM mode, Eq. (23-28), occurs at *Brewster's angle*, $\theta_p = \tan^{-1}(n)$.

23-2 The critical angle for a certain oil is found to be 33°33′. What are its Brewster's angles for both external and internal reflections?

23-3 Determine the critical angle and polarizing angles for (a) external and (b) internal reflections from dense flint glass of index $n = 1.84$.

23-4 For what refractive index are the critical angle and (external) Brewster angle equal when the first medium is air?

23-5 Show that the Fresnel equations, Eqs. (23-27) to (23-30), may also be expressed by

$$\text{TE:} \quad r = -\frac{\sin(\theta - \theta_t)}{\sin(\theta + \theta_t)} \qquad t = \frac{2\cos\theta\sin\theta_t}{\sin(\theta + \theta_t)}$$

$$\text{TM:} \quad r = -\frac{\tan(\theta - \theta_t)}{\tan(\theta + \theta_t)} \qquad t = \frac{2\cos\theta\sin\theta_t}{\sin(\theta + \theta_t)\cos(\theta - \theta_t)}$$

23-6 Show that Eq. (23-37) follows from Eq. (23-35).

23-7 Using Eqs. (23-27) through (23-30) and a computer program or a computer algebra system, reproduce Figures 23-3 and 23-5. Also, change the value of n to produce graphs for the case of external and internal reflection from diamond ($n = 2.42$).

23-8 Use a computer to calculate and plot the reflectance curves of Figure 23-4. Also plot the corresponding transmittance.

23-9 Use a computer to calculate and plot the phase shifts on reflection as a function of angle of incidence for $\theta > \theta_c$. Take $n = 1/1.5$ to reproduce Figure 23-6 and then make similar plots for $n = 1/1.3$ and $n = 1/2.42$.

23-10 A film of magnesium fluoride is deposited onto a glass substrate with optical thickness equal to one-fourth the wavelength of the light to be reflected from it. Refractive indices for the film and substrate are 1.38 and 1.52, respectively. Assume that the film is nonabsorbing. For monochromatic light incident normally on the film, determine (a) reflectance from the air–film surface; (b) reflectance from the film–glass surface; (c) reflectance from an air–glass surface without the film; (d) net reflectance from the combination. See Eq. (22-43).

23-11 Calculate the reflectance of water ($n = 1.33$) for both (a) TE and (b) TM polarizations when the angles of incidence are $0°, 10°, 45°$, and $90°$.

23-12 Light is incident upon an air–diamond interface. If the index of diamond is 2.42, calculate the Brewster and critical angles for both (a) external and (b) internal reflections. In each case distinguish between polarization modes.

23-13 Calculate the percent reflectance and transmittance for both (a) TE and (b) TM modes of light incident at 50° on a glass surface of index 1.60.

23-14 Derive Eqs. (23-29) and (23-30) for the transmission coefficients both by (a) eliminating E_r from Eqs. (23-20) to (23-23) and by (b) using the corresponding equations for the reflection coefficients, together with the relationships between reflection and transmission coefficients implied by Eqs. (23-20) and (23-22).

23-15 Unpolarized light is reflected from a plane surface of fused silica glass of index 1.458.
   a. Determine the critical and polarizing angles.
   b. Determine the reflectance and transmittance for the TE mode at normal incidence and at 45°.
   c. Repeat (b) for the TM mode.
   d. Calculate the phase difference between TM and TE modes for internally reflected rays at angles of incidence of $0°, 20°, 40°, 50°, 70°$, and $90°$.

23-16 A Fresnel rhomb is constructed of transparent material of index 1.65.
   a. What should be the apex angle θ, as in Figure 23-7?
   b. What is the phase difference between the TE and TM modes after both reflections, when the angle is 5% below and above the correct value?

23-17 Determine the reflectance for metallic reflection of sodium light (589.3 nm) from steel, for which $n_R = 2.485$ and $n_I = 1.381$. Calculate reflectance for (a) TE and (b) TM modes at angles of incidence of $0°, 30°, 50°, 70°$, and $90°$.

23-18 Determine the reflectance from tin at angles of incidence of $0°, 30°$, and $60°$. Do this for the (a) TE and (b) TM modes of polarization. Real and imaginary parts of the complex refractive index are 1.5 and 5.3, respectively, for light of 589.3 nm.

23-19 a. What is the absorption coefficient for tin, with an imaginary part of the refractive index equal to 5.3 for 589.3-nm light?
   b. At what depth is 99% of normally incident sodium light absorbed in tin?

23-20 a. From the power conservation requirement, as expressed by Eq. (23-47), show that for an external reflection the transmission coefficient t must be less than 1, but for an internal reflection t' may be greater than 1.
   b. Show further, using the Fresnel Eqs. (23-29) and (23-30), that as the angle of incidence approaches the critical angle, t' must approach a value of 2 in the TE mode and $2/n$ in the TM mode.
   c. Plot the transmission coefficient t' for an interface between glass ($n = 1.5$) and air.

23-21 A narrow beam of light ($\lambda = 546$ nm) is rotated through 90° by TIR from the hypotenuse face of a 45°–90°–45° prism made of glass with $n = 1.60$.
   a. What is the penetration depth at which the amplitude of the evanescent wave is reduced to $1/e$ of its value at the surface?
   b. What is the ratio of irradiance of the evanescent wave at 1 μm beyond the surface to that at the surface?

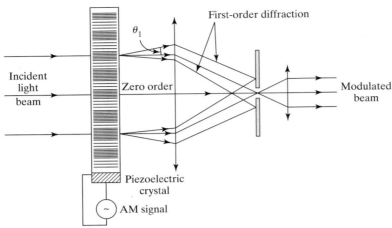

First-order diffraction

Incident
light
beam

θ_1

Zero order

Piezoelectric
crystal

~ AM signal

Modulated
beam

24 *Nonlinear Optics and the Modulation of Light*

INTRODUCTION

The optics treated in most of this text, including the processes of transmission, reflection, refraction, superposition, and birefringence, fall in the category of what is called *linear optics*. When we speak of linear optics, we assume that an optical disturbance propagating through an optical medium can be described by a linear wave equation. As a consequence of this assumption, two harmonic waves in the medium obey the principle of superposition, traveling without distortion due to the medium itself or as a result of the mutual interference of the waves, regardless of the intensity of the light. Only the wavelength and velocity of a light beam in a transparent material are required to describe its behavior.

When the light irradiance becomes great enough, linear optics is not adequate to describe the situation. With the advent of the more intense and coherent light made available by the laser, we find that the optical properties of the medium, such as its refractive index, become a function of the electric field of the light. When two or more light waves interfere within the medium, the principle of superposition no longer holds. The light waves interact with one another and with the medium. These *nonlinear* phenomena require an extension of the linear theory that allows for a nonlinear response of optical materials to the electromagnetic radiation.

In this chapter we define more precisely the area of nonlinear optics, describe and categorize some nonlinear phenomena, and discuss some of their practical applications.

24-1 THE NONLINEAR MEDIUM

Nonlinear phenomena are due ultimately to the inability of the dipoles in the optical medium to respond in a linear fashion to the alternating $\vec{E}$-field associated with a light beam. Atomic nuclei are too massive and inner-core electrons too tightly bound to respond to the alternating $\vec{E}$-field at the frequency of light ($\sim 10^{14}$–10^{15} Hz). Thus the outer electrons of the atoms in a material are primarily responsible for the polarization of the optical medium by the beam's $\vec{E}$-field.[1] When the oscillations of these electrons in response to the field are small, the polarization is proportional to the $\vec{E}$-field, as described later in Section 25-1. However, as the strength of the $\vec{E}$-field increases, strict proportionality begins to fail, just as the harmonic oscillations of a simple spring become increasingly anharmonic as the amplitude of the oscillations increases. Another means of exciting nonlinear behavior without using high beam irradiances is to choose the exciting optical frequency near a resonant frequency of the oscillating dipoles, a technique widely utilized in *nonlinear spectroscopy* and known as *resonance enhancement*.[2]

The polarization of a linear medium by an electric field $\vec{E}$ is usually written in the form

$$\vec{P} = \varepsilon_0 \chi \vec{E} \tag{24-1}$$

where χ is the *susceptibility* and ε_0 is the vacuum permittivity. When departures from linearity are small, it is possible to represent the modification of the susceptibility in a nonlinear medium by a power series in the form

$$\chi = \chi_1 + \chi_2 E + \chi_3 E^2 + \cdots \tag{24-2}$$

When substituted into Eq. (24-1), the polarization strength takes the form

$$P = \varepsilon_0 (\chi_1 E + \chi_2 E^2 + \chi_3 E^3 + \cdots) \tag{24-3}$$

or

$$P = \underbrace{P_1}_{\text{linear}} + \underbrace{(P_2 + P_3 + \cdots)}_{\text{small nonlinear terms}}$$

where the subscripts on χ match the powers of E and reflect the decreasing magnitude of the higher-order terms. The linear and nonlinear susceptibility coefficients characterize the optical properties of the medium, and this relation between P and E completely characterizes the response of the optical medium to the field. We note that in linear media, where only the χ_1 contribution is important, it is common to define the *permittivity* of the medium to be $\varepsilon = \varepsilon_0 \chi_1$ so that $P_1 = \varepsilon E$. The speed of light in a *linear* nonmagnetic material can be written as $v = 1/\sqrt{\varepsilon \mu_0}$ and the index of refraction of such a material is $n = c/v = \sqrt{\varepsilon \mu_0}/\sqrt{\varepsilon_0 \mu_0} = \sqrt{\varepsilon/\varepsilon_0}$.

The first term P_1 in Eq. (24-3) represents linear optics in which the polarization of the medium is simply proportional to the $\vec{E}$-field. Unless the $\vec{E}$-field amplitude is very large, the χ-coefficients of the higher-power E-terms are too small to allow these terms to influence the polarization appreciably. Only with the availability of intense, coherent light have these higher-order terms become important. The high coherence of laser light allows the beam to be focused onto

---

[1]We speak here of the *electric* polarization, rather than the *wave* polarization that was the subject of Chapters 14 and 15.

[2]P. N. Butcher, and D. Cotter, *The Elements of Nonlinear Optics* (New York: Cambridge University Press, 1990), Ch. 6.

small spots with wavelength dimensions, producing electric field strengths exceeding 10^{10} V/m—on the order of the strengths of the fields binding electrons to nuclei in the optical medium. Peter Franken and his associates are credited with the first nonlinear coherent optics experiment,[3] conducted at the University of Michigan in 1961. The team focused the coherent 694.3-nm output from a pulsed ruby laser onto a quartz crystal and detected *second harmonic generation*, the presence in the output of a weak ultraviolet coherent radiation component at 347.15 nm, twice the frequency or half the wavelength of the exciting light. This nonlinear phenomenon is discussed in the next section. Materials used in electro-optic applications typically have second-order nonlinear susceptibility constants χ_2 in the range of 10^{-10} m/V to 10^{-13} m/V, and third-order nonlinear susceptibilities in the range of 10^{-17} m^2/V^2 to 10^{-22} m^2/V^2. In the following example we estimate the field strengths and irradiances required to make the nonlinear contributions to the polarization sizeable compared to the linear contribution.

Example 24-1

a. Estimate the electric field amplitude and irradiance that would cause the second-order nonlinear contribution to the polarization to be 1% of the linear contribution in KDP (KH_2PO_4). KDP has an index of refraction of about 1.5 and a second-order nonlinear susceptibility χ_2 of about 10^{-12} m/V.

b. Estimate the electric field amplitude and irradiance that would cause the third-order nonlinear contribution to the polarization to be 1% of the linear contribution in a certain mix of β-carotene in ethanol. This mix has an index of refraction of about 1.3 and a third-order nonlinear susceptibility χ_3 of about 10^{-20} m^2/V^2.

Solution

a. The linear susceptibility for KDP can be found from the relation given earlier,

$$n = \frac{c}{v} = \sqrt{\frac{\varepsilon}{\varepsilon_0}} = \sqrt{\frac{\varepsilon_0\chi_1}{\varepsilon_0}} = \sqrt{\chi_1}$$

Therefore, $\chi_1 = n^2 = 1.5^2 = 2.25$. Then for the χ_2 term of the polarization to be 1% of the linear term, the electric field amplitude E_0 must satisfy the relation

$$\chi_2 E_0^2 = (0.01)\chi_1 E_0$$

or

$$E_0 = \frac{(0.01)\chi_1}{\chi_2} = \frac{(0.01)(2.25)}{10^{-12} \text{ m/V}} = 2.25 \times 10^{10} \text{ V/m}$$

This electric field strength E_0 corresponds to an irradiance of [see the discussion following Eq. (4-42)]

$$I = \frac{1}{2}\varepsilon_0 c n E_0^2 = \frac{1}{2}(8.85 \times 10^{-12})(3 \times 10^8)(1.5)$$

$$(2.25 \times 10^{10})^2 \text{ W/m}^2 \approx 10^{18} \text{ W/m}^2 = 10^{14} \text{ W/cm}^2$$

---

[3]P. A. Franken, A. E. Hill, C. W. Peters, and G. Weireich. "Generation of Optical Harmonics," *Phys. Rev. Letters*, 7, 1961: 118.

This irradiance would result, for example, from a pulsed laser field of power 100 MW focused to a spot of radius about 5 μm. While achievable, this is a very large irradiance.

b. Following the routine used in part (a), the linear susceptibility of β-carotene in ethanol is $\chi_1 = n^2 = 1.3^2 = 1.69$. Then for the χ_3 term of the polarization to be 1% of the linear term,

$$\chi_3 E_0^3 = (0.01)\chi_1 E_0$$

or,

$$E_0 = \sqrt{(0.01)\frac{\chi_1}{\chi_3}} = \sqrt{(0.01)\frac{1.69}{10^{-20}}} \text{V/m} = 1.3 \times 10^9 \text{ V/m}$$

The corresponding irradiance is about

$$I = \frac{1}{2}\varepsilon_0 cnE_0^2 = \frac{1}{2}(8.85 \times 10^{-12})(3 \times 10^8)(1.3)$$

$$(1.3 \times 10^9)^2 \text{ W/m}^2 \approx 3 \times 10^{15} \text{ W/m}^2 = 3 \times 10^{11} \text{ W/cm}^2$$

24-2 SECOND HARMONIC GENERATION AND FREQUENCY MIXING

In this section we will discuss the manner in which a nonlinear crystal can be used to convert energy in an electromagnetic wave of a given frequency—or from several electromagnetic waves of given frequencies—to energy in an electromagnetic wave at a different frequency. Such processes provide a means of producing intense and coherent electromagnetic radiation at frequencies at which there are no efficient laser transitions.

Second Harmonic Generation

Second harmonic generation results from the contribution of the second-order term in Eq. (24-3):

$$P_2 = \varepsilon_0\chi_2 E^2 \qquad (24\text{-}4)$$

in which the second-order polarization term P_2 of the optical medium is proportional to the square of the electric field. Figure 24-1 shows the polarization as a function of the electric field for the linear case and the deviation from linearity due to this second-order term.

It can be shown that the second-order term makes no contribution to polarization in an isotropic optical material or one having a center of symmetry. A crystal having a center of symmetry is characterized by an inversion center, such that if the radial coordinate r is changed to $-r$, the crystal's atomic arrangement remains unchanged and so the crystal responds in the same way to a physical influence. In such a crystal, reversing the applied field should not—except for a change in sign—change any physical property, such as its polarization. Thus we should have both

$$P_2 = \varepsilon_0\chi_2(+E)^2 \quad \text{and} \quad -P_2 = \varepsilon_0\chi_2(-E)^2$$

Because the E-field is squared, $P_2 = -P_2$, which can only be true if $P_2 = 0$. The quartz crystal used by Franken, and many other crystals as well, do not possess inversion symmetry. They can, therefore, manifest second harmonic generation in addition to other second-order phenomena to be described presently.

The appearance of a *second harmonic* in the polarization is expected from the following mathematical argument. If the applied electric field, or

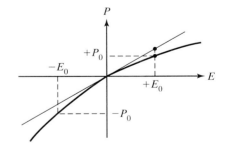

Figure 24-1 Linear and typical nonlinear response of polarization P to an applied electric field E. For equal positive and negative fields, the response of the optical medium is not symmetrical in the case of the nonlinear (curved line) response. In this case, the negative field E_0 produces a greater polarization than a positive field of the same magnitude.

one of its Fourier components, is of the form

$$E = E_0 \cos \omega t$$

substitution into Eq. (24-4) gives

$$P_2 = \varepsilon_0 \chi_2 E_0^2 \cos^2 \omega t = \varepsilon_0 \chi_2 E_0^2 [\tfrac{1}{2}(1+\cos 2\omega t)]$$

where we have substituted the double-angle identity for $\cos^2 \omega t$. Then

$$P_2 = \tfrac{1}{2}\varepsilon_0 \chi_2 E_0^2 + \tfrac{1}{2}\varepsilon_0 \chi_2 E_0^2 \cos 2\omega t \qquad (24\text{-}5)$$

Evidently, the second-order polarization P_2 consists of a term of twice the frequency of the applied optical field as well as a constant or DC component that represents *optical rectification*. Optical rectification results in a time-independent polarization in the medium that manifests itself as a DC voltage, in a direction transverse to the field propagation, across the nonlinear crystal. The component of P_2 oscillating at 2ω corresponds to dipole oscillations in the medium at this same frequency. These dipole oscillations generate electromagnetic radiation of angular frequency 2ω, which is present in the resultant field, together with the stronger first-order field at the fundamental frequency ω. Second harmonic generation is commonly used, for example, to produce light of wavelength 532 nm by passing the 1064-nm light produced by a Nd:YAG laser through a nonlinear crystal.

Phase Matching

The same nonlinear interaction that allows conversion of energy in the fundamental electromagnetic wave of frequency ω into energy in the second harmonic electromagnetic wave of frequency 2ω also allows for energy conversion in the other direction. That is, the nonlinear interaction can convert energy from the second harmonic wave into the fundamental wave. The direction of energy flow depends critically on the phase of the fundamental field, which drives the nonlinear polarization, relative to the phase of the second harmonic field, which is absorbed and emitted by the dipole oscillations comprising the nonlinear polarization. Because of dispersion, the light continuously generated at frequency 2ω travels at a different speed in the optical material than the light at frequency ω. Thus, the two waves are periodically in and out of step as they traverse the crystal. One can show[4] that the irradiance in the second harmonic field is proportional to the irradiance factor

$$\text{sinc}^2\!\left(\frac{\Delta k L}{2}\right)$$

where L is the distance into the crystal and k is the wave propagation constant, equal to $n\omega/c$. Here, $\Delta k = k_{2\omega} - 2k_\omega$. When $\Delta k = 0$ we say the fields are *phase matched* in the crystal and the irradiance factor above is a maximum. Because dispersion is present in materials, Δk is not typically zero and the irradiance factor describes the consequent reduction in the irradiance of the generated second harmonic field due to the phase difference that develops between the two fields as they propagate through the nonlinear crystal. The *coherence length*[5] L_C is defined by the relation $\Delta k L_C = \pi$. Thus,

$$L_C = \pi/\Delta k$$

When $L = L_C$, the sinc-squared intensity factor is reduced to about 0.4 of its maximum value and so represents an estimate of the useful length of a crystal

---

[4]Ammon Yariv, *Optical Electronics*, 3d ed. (New York: Holt, Rinehart and Winston, 1985), Ch. 8.

[5]This *coherence length* is not to be confused with the quantity of the same name presented in Chapter 9.

designed for efficient second harmonic generation using light with a phase mismatch Δk. The coherence length can also be expressed as

$$L_C = \frac{\pi}{\Delta k} = \frac{\pi}{\dfrac{2\omega}{c}\Delta n} = \frac{\pi}{\dfrac{4\pi}{\lambda_0}\Delta n}$$

$$L_C = \frac{\lambda_0}{4\,\Delta n} \qquad\qquad (24\text{-}6)$$

where we have used

$$k = \frac{n\omega}{c} \quad \text{and} \quad \Delta k = \frac{n_{2\omega}(2\omega)}{c} - \frac{2n_\omega(\omega)}{c} = \frac{2\omega}{c}\Delta n$$

In Eq. (24-6), λ_0 is the vacuum wavelength of the fundamental and Δn the amount the index of refraction of the fundamental differs from that of the second harmonic.

Example 24-2

Consider a situation in which an incident fundamental field of wavelength $\lambda_0 = 0.8\ \mu m$ is incident on a KDP crystal with refractive indices 1.4802 for the second harmonic and 1.5019 for the fundamental. What is the maximum crystal thickness useful in generating second harmonic light?

Solution

Substitution into Eq. (24-6) gives

$$L_C = \frac{0.8\ \mu m}{(4)(1.5019 - 1.4802)} = 9.2\ \mu m$$

This calculation shows that the maximum crystal thickness useful in generating second harmonic light is typically quite small, in this case around 10 times the wavelength of the fundamental. Crystals with thicknesses equal to their coherence length are impractically small.

Birefringent Phase Matching

A technique commonly used to circumvent the small coherence length of nonlinear crystals makes use of their birefringence. As described in Section 15-5, the refractive index (and so the velocity) of the extraordinary (E) ray varies with direction through the crystal. If a direction through the crystal is chosen such that $n_{2\omega}$ for the E-ray equals n_ω for the ordinary (O) ray, the fundamental and second harmonic waves remain in step and the crystal can be a centimeter or so thick. This technique is called *index matching* or *birefringent phase matching* and is clarified by Figure 24-2a, which shows how the ellipsoids representing the velocity versus crystal direction for the E- and O-rays intersect along the direction of matching.

Quasi-Phase Matching

An alternative to birefringent phase matching is so-called *quasi-phase matching*, QPM. In this technique a phase mismatch Δk between the fundamental and second harmonic fields can be compensated for by using a nonlinear crystal with a periodic structure such that the sign of χ_2 changes every coherence length L_C of the crystal. Such a periodic structure could be constructed by slicing a crystal into many thin slabs, each having a width equal to L_C and then placing the slabs back together in a manner such that each slab is rotated 180° relative to its neighbors. Since materials with nonzero second-order susceptibilities χ_2 lack inversion symmetry, the resulting crystal will have a second-order susceptibility that changes sign each coherence length of the crystal, as shown in Figure 24-2b.

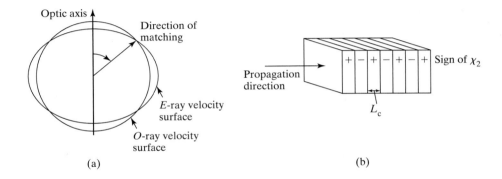

Figure 24-2 Phase matching techniques. (a) Velocity ellipsoids for orthogonally polarized light beams in a birefringent medium. The *O*-ray ellipsoid is spherical and intersects the *E*-ray ellipsoid along a direction (shown relative to the optic axis) for which both rays have the same velocity. (b) QPM crystal.

The coherence length of the crystal is the length over which the direction of energy flow is from the fundamental field to the second harmonic field. Thus, just as the direction of energy flow in the crystal is about to switch so as to cause attenuation of the second harmonic field, the sign of χ_2 changes. This change in sign restores the proper phase relation between the dipoles and the second harmonic field and so ensures continued amplification of the second harmonic field. The small coherence length associated with the phase mismatch in many crystals (see Example 24-2) makes the construction of a QPM crystal by the method just described somewhat impractical. Instead, external fields can be used to perform a *periodic poling* of a nonlinear ferroelectric material or a nonlinear polymer. This permanent periodic poling produces the alternating sign of χ_2 needed for QPM. A common structure of this type is commonly referred to as periodically poled lithium niobate, or PPLN.

Frequency Mixing

When two or more incident beams with different frequencies are allowed to interfere within a nonlinear dielectric material, *frequency mixing* can occur. For example, consider two interfering incident waves of frequencies ω_1 and ω_2 represented in the form

$$E = E_{01} \cos \omega_1 t + E_{02} \cos \omega_2 t$$

or, in the equivalent exponential form,

$$E = \tfrac{1}{2} E_{01}(e^{i\omega_1 t} + e^{-i\omega_1 t}) + \tfrac{1}{2} E_{02}(e^{i\omega_2 t} + e^{-i\omega_2 t})$$

The second-order nonlinear polarization $P_2 = \varepsilon_0 \chi_2 E^2$ is proportional to the square of this incident field and so evidently produces harmonic fields at $2\omega_1, 2\omega_2, \omega_1 - \omega_2$, and $\omega_1 + \omega_2$.

A special case of frequency mixing is the process known as *parametric amplification*. Instead of $2\omega_1 \rightarrow \omega_3$, as in second harmonic generation, it is possible to have sum frequency generation such that $\omega_1 + \omega_2 \rightarrow \omega_3$ or difference frequency generation such that $\omega_1 - \omega_2 \rightarrow \omega_3$. In common parlance, the two wave components input into a nonlinear crystal are called the *pump* and *signal* waves. The generated difference wave is known as the *idler* wave. Suppose, then, that a small signal wave at $\omega_s = \omega_1$ and a powerful *pump* wave at $\omega_p = \omega_2$ interact within a nonlinear medium. A difference or idler frequency $\omega_3 = \omega_i = \omega_p - \omega_s$ can be produced. This idler frequency can in turn beat with the pump frequency to enhance the signal frequency, $\omega_s = \omega_p - \omega_i$. In this process, therefore, both idler and signal waves can be amplified, drawing power from the pump wave. When signal and idler frequencies correspond to resonant frequencies in the nonlinear crystal acting as a tuned Fabry-Perot cavity, the parametric oscillator is a tunable source of coherent radiation. Tuning the cavity is accomplished by varying the refractive index of the cavity through control of temperature or an applied DC field. This important method of producing tunable sources of intense electromagnetic radiation is discussed further in Chapter 28.

TABLE 24-1 LINEAR AND NONLINEAR PROCESSES

Linear first order: $P_1 = \varepsilon_0\chi_1 E$	Nonlinear second order: $P_2 = \varepsilon_0\chi_2 E^2$	Nonlinear third order: $P_3 = \varepsilon_0\chi_3 E^3$
Classical optics:	Materials lacking inversion symmetry:	Materials with inversion symmetry:
Superposition	Second harmonic generation	Third harmonic generation
Reflection	Three-wave mixing	Four-wave mixing
Refraction	Optical rectification	Kerr effect
Birefringence	Parametric amplification	Raman scattering
Absorption	Pockels effect	Brillouin scattering
		Optical phase conjugation

Second harmonic generation is not the only nonlinear phenomenon that results from the quadratic dependence of the polarization on the electric field. Table 24-1 lists others, as well as several that depend on the third-order contribution to the polarization of the medium, $P_3 = \varepsilon_0\chi_3 E^3$. For example, notice that for third-order nonlinear processes, *third harmonic generation* can occur. We describe several of the nonlinear phenomena listed in Table 24-1 in the remaining sections of this chapter.

24-3 ELECTRO-OPTIC EFFECTS

Nonlinear electro-optics effects result from the application of a DC (or low-frequency) electric field to a medium. In this section we discuss two such effects, the *Pockels effect* and the *Kerr effect*, and show that these effects can be used in *light modulators*. By *light modulation*, we mean the modification of the amplitude (AM), frequency (FM), phase, polarization, or direction of a light wave. One purpose of modulation is to render the wave capable of carrying information, as described in the discussion of fiber-optic communication systems appearing in Chapter 10. Of course, light choppers and shutters can accomplish some modulation mechanically. We are interested here in describing the modulation that is accomplished by varying the refractive index of a material through the use of an applied electric field. In subsequent sections of this chapter, we examine so-called magneto-optic and acousto-optic devices.

The basic equation describing nonlinear behavior was given as Eq. (24-3), a relation between the polarization of the medium and the applied electric field. In dealing with crystalline media, which represent most of the useful electro-optic materials, it is customary to express nonlinearity of the refractive index n by an equation[6] analogous to Eq. (24-2) for the susceptibility:

$$\frac{1}{n^2} = \frac{1}{n_0^2} + rE + RE^2 \qquad (24\text{-}7)$$

where r and R are the linear and quadratic *electro-optic coefficients*,[7] respectively, and we assume that there is no other effect present (like crystal strain) that can modify n. The refractive index in the absence of an applied field is n_0. In general, the refractive index depends on the propagation direction and wave polarization relative to the crystal axes. Since $\vec{E}$ is a vector field, the coefficients r and R are tensors that reflect the crystal symmetry. Depending on the degree

---

[6]Ivan P. Kaminov, *An Introduction to Electrooptic Devices* (New York: Academic Press, 1974), Ch. 3.

[7]It is important to distinguish the "order" of an electro-optic process as determined from Eq. (24-3) for the polarization and as determined from Eq. (24-7) for $1/n^2$ or Δn. For example, the Pockels effect is a second-order effect (involving χ_2) by the first criterion and a first-order effect (involving r) by the second. Both criteria are used.

of symmetry, many tensor components may vanish or become equal to others, reducing the total number of independent elements[8] required to represent a particular crystalline material.

The Pockels Effect

The Pockels effect results from the linear term in Eq. (24-7), where E is an applied DC field. This effect can be considered a special case of two-wave mixing, where one of the waves is the incident optical wave and the other a field of zero frequency. The optical electric field can be small, since the DC field is itself large enough to produce nonlinear behavior. In general, the DC field redistributes electrons in such a way that birefringence is induced in an otherwise isotropic material, or new optic axes appear in naturally birefringent crystals. Since the Pockels effect is a second-order effect relative to the polarization [Eq. (24-3)], it is not found in isotropic materials having inversion symmetry. All crystalline materials exhibiting a Pockels effect[9] are also *piezoelectric*; that is, they show induced birefringence due to mechanical strain. Since the effect was discovered in 1893, long before the discovery of the laser, it was well known even before intense optical fields became available.

In one configuration of the Pockels cell, the natural optic axis of the crystal is aligned parallel to the applied field. Fast and slow axes are induced in a plane normal to the applied field, as shown in Figure 24-3. If the Pockels cell crystal is rotated until the FA and SA are at 45° to the x- and y-axes, a vertically polarized light wave $\vec{E}_0$ incident on the crystal along the field direction has equal amplitude components on FA and SA. These components experience different refractive indices and different speeds through the crystal. The

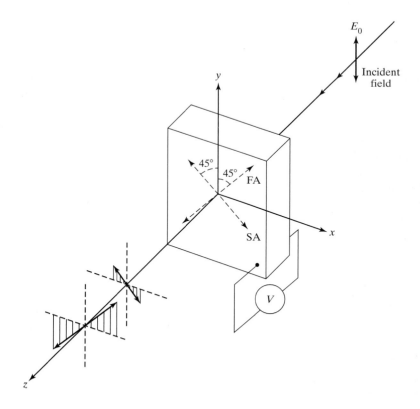

Figure 24-3 Pockels cell schematic. The retardation action due to an applied voltage V is suggested by an on-axis separation of the polarized components transmitted by the fast and slow axes of the crystal.

---

[8]The linear electro-optic tensor p_{ij} is defined by the relation $\Delta(1/n^2)_i = \Sigma_j p_{ij} E_j$, with $i = 1, 2, 3, \ldots, 6$ and $j = x, y, z$. For example, in crystals of triclinic symmetry, all the 18 possible tensor elements are required; in zinc blende (GaAs), only one is required; in centrosymmetric crystals, all elements are zero. See for example, Ammon Yariv, *Optical Electronics*, 3d ed. (New York: Holt, Rinehart and Winston, 1985), Ch. 9.

[9]Single crystals can be divided into 32 symmetry classes; of these, 20 show the Pockels effect.

crystal therefore behaves as a phase retarder (Section 14-2), and the component waves emerge with a phase difference. One component advances in phase by $\Delta\varphi$ and the other lags in phase by $\Delta\varphi$ while traversing the crystal of length L, so that their relative phase on emerging is given by $\Phi = 2\,\Delta\varphi$. Now $\Delta\varphi = (2\pi/\lambda_0)L\Delta n$, where λ_0 is the vacuum wavelength and $L\Delta n$ represents the optical-path difference induced in each component by the applied field. We find Δn from Eq. (24-7), which, for small changes, can be approximated by $d(1/n^2) = rE$. The E^2 term is considered negligible in the Pockels effects. Thus,

$$d\left(\frac{1}{n^2}\right) = -2\frac{dn}{n^3} \cong rE$$

$$|\Delta n| \cong \frac{r}{2}n_0^3 E$$

Substituting into the phase equations, we find

$$\Phi = 2(\Delta\varphi) = 2\frac{2\pi}{\lambda_0}L\Delta n = \frac{2\pi}{\lambda_0}rn_0^3EL = \frac{2\pi}{\lambda_0}rn_0^3V$$

where $V = EL$ is the voltage applied across the length L of the cell. Notice that the phase difference Φ is independent of the crystal length. For example, if the Pockels cell is to behave as a half-wave plate, we need to make $\Phi = \pi$. The half-wave voltage required is then

$$V_{HW} = \frac{\lambda_0}{2rn_0^3} \qquad (24\text{-}8)$$

Example 24-3

Suppose the Pockels cell is made from a KD*P crystal of 1-cm thickness and the optical wave has a wavelength of 633 nm. What half-wave voltage is required?

Solution

From Table 24-2, we find $r = 24.1 \times 10^{-12}$ m/V and a refractive index of 1.51. Then,

$$V_{HW} = \frac{633 \times 10^{-9}}{2(24.1 \times 10^{-12})(1.51)^3} = 3800 \text{ V}$$

Thus an applied voltage of 3.8 kV transforms the crystal into a half-wave plate.

TABLE 24-2 LINEAR ELECTRO-OPTIC COEFFICIENTS FOR REPRESENTATIVE MATERIALS

Material (wavelength if not 633 nm)	Linear electro-optic coefficient[a] r (pm/V)	Refractive index n_0
KH_2PO_4 (KDP)	11	1.51
KD_2PO_4 (KD*P)	24.1	1.51
$(NH_4)H_2PO_4$ (ADP) $\lambda = 0.546$ nm	8.56	1.48
$LiNbO_3$ (lithium niobate)	30.9	2.29
$LiTaO_3$ (lithium tantalate)	30.5	2.18
GaAs (gallium arsenide) $\lambda = 10.6\ \mu$m	1.51	3.3
ZnS (zinc sulfide) $\lambda = 0.6\ \mu$m	2.1	2.36
Quartz	1.4	1.54

[a]Depending on crystalline symmetry, materials have more than one electro-optic coefficient. Only one has been listed here for use in a Pockels cell. These and others may be found in standard references.[10]

---

[10]For example, see Amnon Yariv, *Optical Electronics*, 3d ed. (New York: Holt, Rinehart and Winston, 1985), Ch. 8.

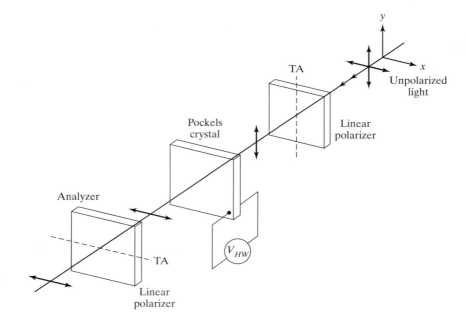

Figure 24-4 The effect of applying a half-wave voltage to a Pockels cell when the system includes a crossed polarizer and analyzer. The vertically polarized beam incident on the Pockels cell is transmitted as a horizontally polarized beam. For other values of applied voltage, the beam incident on the analyzer is elliptically polarized and is only partially transmitted.

Recall (Section 15-7) that for a half-wave plate the linear polarization of the emergent light is rotated by 90° relative to the linear polarization of the incident light. If a linear polarizer with TA along the x-direction intercepts the emergent light, as in Figure 24-4, the transmittance of the system is zero when $V = 0$ and maximum when $V = V_{HW}$. Variations in V therefore modify the polarization state of the emergent light, rendering it elliptical, in general, with an x-component that can be transmitted by the analyzer. In effect, the polarizer-analyzer pair transforms phase modulation into amplitude modulation. Thus we see that the transmittance of the system can be modulated by variations in the applied voltage. Variations of a signal voltage superimposed on V are transformed into variations in light intensity in such a device, known as a *Pockels electro-optic modulator.*

The Pockels cell can be used also with the field oriented orthogonally to the beam direction, an arrangement that simplifies placement of the electrodes. In the geometry we have been describing, the electrodes are usually endrings that allow the light beam to pass through and still provide a reasonably uniform field in the crystal.

The transmittance of the beam can be expressed by the relation[11]

$$I = I_{\max} \sin^2\left(\frac{\pi}{2}\frac{V}{V_{HW}}\right) \qquad (24\text{-}9)$$

and is plotted in Figure 24-5. To take advantage of the more linear region of the transmittance, a quarter-wave plate is often inserted between the initial polarizer and the Pockels crystal. This has the effect of producing 50% transmittance when $V = 0$ so that the operating point is located at P in the figure, rather than at the origin. Variations in modulating voltage, if not too large, then occur with a system response that is linear.

Two other closely related applications of the Pockels cell are illustrated in Figures 24-6 and 24-7. In Figure 24-6, the cell is used as a *Q-switch* that allows

[11]J. Wilson, and J. F. B. Hawkes, *Optoelectronics: An Introduction* (London: Prentice-Hall International, 1983).

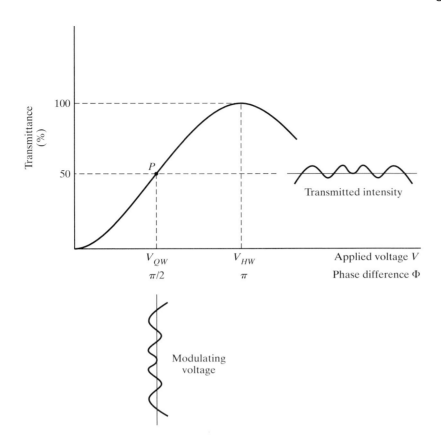

Figure 24-5 Transmittance curve for the Pockels cell modulator. Without a quarter-wave plate, the transmittance is zero when the applied voltage is zero. Using a quarter-wave plate between polarizer and modulator, the transmittance is 50% at operating point P when the applied voltage is zero. Under these conditions, the modulator responds more linearly to an input signal.

sudden dumping of the energy stored in a laser. Without an applied voltage, the Pockels crystal transmits the horizontally polarized beam from a laser cavity without changing its state of polarization. A Glan-laser prism, tuned to pass vertically polarized light, rejects the horizontally polarized beam. With half-wave voltage applied, the horizontally polarized beam from the laser cavity is rotated by 90° and is accordingly passed by the Glan-laser prism, allowing the beam to be backreflected from a high-reflectance mirror. The configuration now permits rapid traversal of the beam back and forth through the laser cavity, and stimulated emission occurs, producing an energetic pulse of laser radiation. In Figure 24-7, the Pockels cell is used to initiate cavity dumping. When half-wave voltage is applied, the polarization state of the laser radiation is rotated by 90°, so that it can be extracted with the help of the polarizing prism, as shown.

The Kerr Effect

When the optical medium is isotropic, as in the case of liquids and glasses, the Pockels effect is absent and the polarization is modified by the third-order [in the expansion of the polarization of the medium, see Eq. (24-3)] electro-optic effect, better known as the *Kerr effect*. This effect, like all third-order effects, occurs whether or not a material possesses inversion symmetry. Actually, the Kerr effect was the first electro-optic effect to be discovered (1875). Kerr cells usually contain nitrobenzene or carbon disulfide in the space between two electrodes across which a voltage is applied, as indicated in Figure 24-8. The applied electric field induces birefringence with an optic axis parallel to the applied field. Light traversing the cell thus encounters two refractive indices, n_e and n_o, for polarizations parallel and perpendicular to the optic axis, and phase retardation results. In this case,

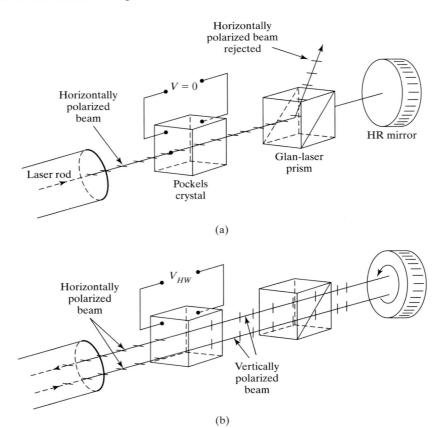

(a)

(b)

Figure 24-6 Light-controlling action of a Pockels cell, used as a Q-switch. The configuration in (a) produces low transmission at zero-cell voltage and in (b) high transmission at half-wave voltage. In (b), the incident and reflected beams are separated for clarity. Repeated reentries of the beam into the laser cavity initiates stimulated emission that produces the laser pulse.

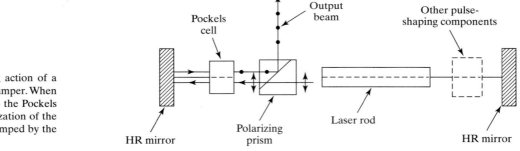

Figure 24-7 Light-controlling action of a Pockels cell, used as a cavity dumper. When half-wave voltage is applied to the Pockels cell, it rotates the linear polarization of the laser beam so that it can be dumped by the polarizing prism.

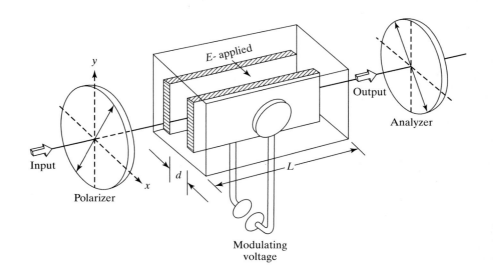

Figure 24-8 Kerr cell. The applied voltage creates a field that is perpendicular to the beam direction. As with the Pockels cell, a modulating voltage produces phase modulation that is converted to amplitude modulation by the polarizer-analyzer pair.

Eq. (24-7) becomes

$$\frac{1}{n^2} = \frac{1}{n_0^2} + RE^2 \quad \text{or} \quad |\Delta n| = \frac{R}{2} n_0^3 E^2 \tag{24-10}$$

Experimentally, the difference between n_e and n_o is found to obey a relation of the form

$$\Delta n = KE^2\lambda \tag{24-11}$$

where K is the *Kerr constant*.

Equating Eqs. (24-10) and (24-11), we find the relationship between K and R to be

$$K = \frac{Rn_0^3}{2\lambda}$$

As explained for the Pockels cell, the relative phase retardation for the ordinary and extraordinary components is

$$\Phi = \frac{2\pi}{\lambda} L \, \Delta n$$

Introducing the Kerr constant through Eq. (24-11),

$$\Phi = \frac{2\pi K V^2 L}{d^2}$$

where we have set $V = Ed$ and d is the interelectrode distance. To function as a half-wave plate, $\Phi = \pi$, and we find that the required voltage V_{HW} is given by

$$V_{HW} = \frac{d}{\sqrt{2KL}} \tag{24-12}$$

Example 24-4

Consider a nitrobenzene Kerr cell for which $K = 2.4 \times 10^{-12}\,\text{m/V}^2$ (see Table 24-3) at room temperature and $\lambda = 589\,\text{nm}$. If $d = 1\,\text{cm}$ and $L = 3\,\text{cm}$, what half-wave voltage is needed?

Solution

Substituting into Eq. (24-12),

$$V_{HW} = \frac{0.01}{\sqrt{(2)(2.4 \times 10^{-12})(0.03)}} = 26.4\,\text{kV}$$

Thus the Kerr cell behaves as a half-wave plate at a voltage of around 26 kV, considerably higher than for a typical Pockels cell.

TABLE 24-3   KERR CONSTANT FOR SELECTED MATERIALS

Material ($\lambda = 589$ nm, room temperature)	K (pm/V^2)
Nitrogen (STP)	4×10^{-6}
Glass (typical)	0.001
Carbon disulfide (CS_2)	0.036
Water (H_2O)	0.052
Nitrotoluene ($C_5H_7NO_2$)	1.4
Nitrobenzene ($C_6H_5NO_2$)	2.4

Kerr cells can be used as modulators in the manner described for Pockels cells. Because of the higher voltages required, and because of the toxic and explosive nature of nitrobenzene, Pockels cells are usually preferred. Nevertheless, Kerr cells find application as high-speed shutters and as a substitute for mechanical light choppers. They are capable of response to frequencies in the range of 10^{10} Hz and they can often be found operating as Q-switches in pulsed lasers.

24-4 THE FARADAY EFFECT

In contrast to the electro-optic effects discussed to this point, the *Faraday effect* is a first-order (i.e., $\Delta n \propto B$) *magneto*-optic interaction.[12] When a transparent material is placed in a magnetic field and linearly polarized light is passed through it along the direction of the magnetic field, the emerging light is found to remain linearly polarized, but with a net rotation β of the plane of polarization that is proportional both to the thickness d of the sample and the strength of the magnetic field B, according to the empirical relation,

$$\beta = VBd \tag{24-13}$$

Here V is the *Verdet constant* for the material, usually expressed in minutes of angle per Gauss-cm (G-cm). The Verdet constant is both temperature and wavelength dependent.

An interesting aspect of the Faraday rotation is that the sense of rotation relative to the magnetic field direction is, for a given material, independent of the propagation direction of the light. Thus, repeated forward and backward traversals of the material by a light beam has a cumulative effect on the angle of rotation β. This behavior contrasts with that exhibited in the closely related phenomenon of *optical activity*, discussed in Section 15-6.

The optical rotation of the polarized light can be understood as *circular birefringence*, the existence of different indices of refraction for left-circularly and right-circularly polarized light components (Section 15-6). Recall that linearly polarized light is equivalent to a combination of right- and left-circularly polarized components. Each component is affected differently by the applied magnetic field and traverses the sample with a different speed, since the refractive index is different for the two components. The end result consists of left- and right-circular components that are out of phase and whose superposition, upon emerging from the Faraday rotator, is linearly polarized light with its plane of polarization rotated relative to its original orientation.

A classical derivation[13] of the angle of rotation β predicts a relation of the form

$$\beta = \left(\frac{e}{2m} \frac{\lambda}{c} \frac{dn}{d\lambda} \right) Bd \tag{24-14}$$

with e and m the electronic charge and mass, c the speed of light, λ the wavelength, and $dn/d\lambda$ the *rotatory dispersion*. By comparison with Eq. (24-13), the theory predicts a dependence of the empirical Verdet constant V given by

$$V = \frac{e}{2mc} \lambda \frac{dn}{d\lambda} \tag{24-15}$$

---

[12]The magnetic field analogs of the Kerr effect (where $\Delta n \propto E^2$) are the *Voigt effect* (in gases) and the *Cotton-Mouton effect* (in liquids), in which a constant magnetic field is applied normal to the light-beam direction. Both are very small effects and will not be discussed further here.

[13]Frank L. Pedrotti, and Peter Bandettini. "Faraday Rotation in the Undergraduate Advanced Laboratory," *Am. J. Phys.*, 58, June 1990: 542.

TABLE 24-4 VERDET CONSTANT FOR SELECTED MATERIALS

Material	V (min/G-cm) $\lambda = 589$ nm
H_2O	0.0131
Crown glass	0.0161
Flint glass	0.0317
CS_2	0.0423
CCl_4	0.0160
NaCl	0.0359
KCl	0.02858
Quartz	0.0166
ZnS	0.225

When the constants are evaluated and V is expressed in the standard units of min/G-cm, Eq. (24-15) becomes

$$V = 1.0083\lambda \frac{dn}{d\lambda}$$

We note that the Verdet constant is proportional to both the wavelength of the light and the induced rotatory dispersion in the medium. Measured values for V at 589 nm are given in Table 24-4.

The Faraday effect can be used for light modulation, although it is difficult, practically speaking, to modulate a magnetic field at very high frequencies. Figure 24-9 shows schematically the Faraday rotator, a crystal or liquid cell whose axis of symmetry is aligned with a magnetic field. The figure shows a field B established by current windings and indicates a rotation of the polarization in the same sense as the current producing the field. This particular geometry defines a *positive* Verdet constant. Figure 24-10 illustrates the principal

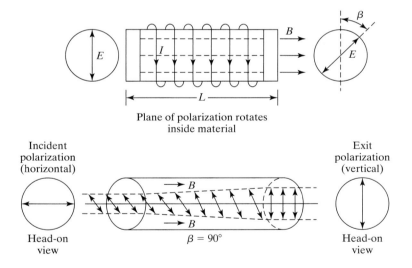

Figure 24-9 Faraday effect producing rotation of the plane of polarization.

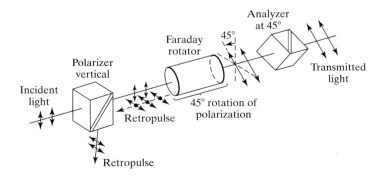

Figure 24-10 Faraday rotator used between a polarizer-analyzer pair to produce optical isolation of the optical system providing the incident light.

application of the Faraday rotator as an *optical isolator*. The isolator consists of a Faraday rotator situated between a polarizer-analyzer pair. As shown, the incident vertically polarized light is rotated 45° counterclockwise by the Faraday rotator and in this orientation is fully transmitted by the analyzer. Optical elements (not shown) farther down the line are responsible for undesirable back reflections (retropulses) of this radiation along the optical axis. In traversing the Faraday rotator a second time, the polarization vector of the reflected light is rotated an additional 45° *in the same rotational sense*, so that it emerges horizontally polarized and encounters the polarizer at an angle of 90° with the polarization direction of the original beam. In this state, it is rejected by the polarizer, preventing it from continuing back into the optical system, where, in high-power laser systems, it can damage optical components. Thus the optical isolator effectively *isolates* the optical system from stray retropulses.

Example 24-5

Let us calculate the required length of SF58 flint glass, having a Verdet constant of 0.112 min/G-cm for 543.5-nm light, if it is to produce the 45° rotation of the polarization vector required in an optical isolator when the magnetic field has a value of 9 kG.

Solution

Using Eq. (24-13), we have

$$d = \frac{\beta}{VB} = \frac{45° \times 60 \text{ min/}°}{(0.112 \text{ min/G-cm}) \times 9000 \text{ G}} = 2.68 \text{ cm}$$

24-5 THE ACOUSTO-OPTIC EFFECT

Photoelasticity (Section 15-7) is the change in refractive index of a crystal due to mechanical stress. This phenomenon makes possible the *AO* or *acousto-optic effect*—the interaction of optical and acoustic waves—in which a longitudinal acoustic wave launched by means of a piezoelectric transducer produces a periodic mechanical stress in the crystal. The acoustic wave (see Figure 24-11) consists of a series of compressions and rarefactions (longitudinal vibrations) in atomic density and so a periodic—although small—variation of the refractive

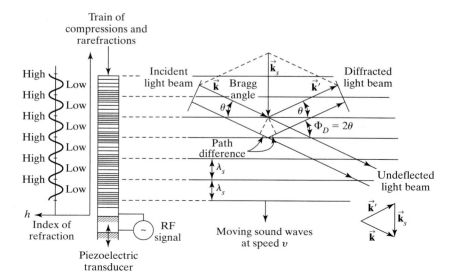

Figure 24-11 Variations in the refractive index of a medium due to the passage of a harmonic acoustic wave (left) and the scattering of an incident optical beam by the induced "planes" (right). The inset shows the relationship of wave vectors required by momentum conservation when the acoustic wave has the propagation direction indicated.

index about its normal value. Light incident on this structure is scattered to a greater extent from regions of higher refractive index. The scattering is called *Brillouin scattering* and is another third-order effect that does not require a medium possessing inversion symmetry.

If the crystal is thin enough, variations in refractive index along its length lead to corresponding variations in the speed of light so that the crystal behaves as a transmission phase grating (Section 12-5). Some of the incident light beam is diffracted into various orders, according to the diffraction grating equation, $m\lambda = d \sin \theta_m$, where the grating constant d in this case should be taken to be the acoustic wavelength λ_S, and θ_m is the angle of diffraction in mth order. This is the so-called *Raman-Nath* regime.[14] If the crystal is thicker, regions of higher refractive index represent planes normal to the direction of the acoustic wave, as suggested in Figure 24-11. In this case, the light wave is diffracted in a way similar to that of X-rays from crystalline planes in Bragg scattering (the *Bragg* regime). The light wave traverses the crystal with a speed that is around five orders of magnitude greater than the speed of the acoustic wave. This means that the induced grating is essentially stationary relative to the light wave.

A diffracted light beam appears in any direction in which portions of the wavefront reflected (1) from different parts of a given plane and (2) from successive planes obey the usual condition for constructive interference; that is, the path difference must be an integral number of wavelengths. The first requirement is satisfied by an angle of scatter that is equal to the angle of incidence. This condition is shown satisfied in Figure 24-11. The second requirement is the *Bragg condition*. The path difference between the incident and diffracted waves is made up of the two segments indicated in the figure. By geometry, each has a magnitude of $\lambda_S \sin \theta$. This leads to the equation

$$m\lambda = 2\lambda_S \sin \theta = 2\lambda_S \sin \frac{\Phi_D}{2} \qquad (24\text{-}16)$$

a relation worked out in this context by Brillouin in 1921 and identical to the Bragg equation for X-ray diffraction.[15] In Eq. (24-16), the angles and optical wavelength are those measured within the medium. (However, see problem 24-15.)

The Bragg condition for diffraction maxima can be found by an alternative argument that makes use of the particle nature of waves (Chapter 1). The incident light beam with wave vector $\vec{k}$ can be considered a flux of photons, each of energy $\hbar\omega$ and momentum $\hbar\vec{k}$, where ω is the angular frequency and $\hbar$ is the Planck constant divided by 2π. In the same way, the acoustic wave with wave vector $\vec{k}_S$ can be considered a flux of quantized particles called *phonons*, each of energy $\hbar\omega_S$ and momentum $\hbar\vec{k}_S$. The acousto-optic interaction then consists of the interaction or collision of these particles in which both energy and momentum are conserved. The two wave vectors $\vec{k}$ and $\vec{k}_S$, as well as the wave vector $\vec{k}'$ of the diffracted light beam, are also shown in Figure 24-11. The acoustic waves are shown propagating downward in the figure. Conservation of momentum in a collision between a photon and a phonon requires that $\vec{k}' = \vec{k} - \vec{k}_S$, as indicated in the inset vector triangle. On the other hand,

---

[14]Robert Guenther, *Modern Optics* (New York: John Wiley and Sons, 1990), Ch. 14.

[15]An important difference between X-ray diffraction and light diffraction in the Bragg regime is that light scattering occurs in a continuous manner from a thick, sinusoidal grating, rather than from discrete planes. As a consequence, only the first-order diffraction, $m = 1$, occurs in the acousto-optic effect. In this connection, see Section 16-2 ("Hologram of a Point Source") and Section 21-1 ("Optical Data Imaging and Processing").

if the acoustic wave is reversed in direction, the corresponding vector triangle is satisfied by $\vec{k}' = \vec{k} + \vec{k}_S$. In general,

$$\vec{k}' = \vec{k} \pm \vec{k}_S \qquad (24\text{-}17)$$

With the plus sign, we interpret this to mean that an incident photon combines with a phonon to produce the diffracted photon; with the negative sign, the incident photon is considered to yield an additional phonon to the acoustic wave, as well as a photon to the diffracted beam. Thus Eq. (24-17) describes both the emission and the absorption of a phonon by the crystal lattice.

We now show that Eq. (24-17) is equivalent to the Bragg diffraction condition. Since light frequencies are of the order of 10^{14} Hz, while acoustic frequencies are generally less than 10^{10} Hz, both ω and ω' are much greater than ω_S, that is,

$$\omega' \cong \omega \quad \text{and accordingly} \quad \vec{k}' \cong \vec{k}$$

The vector triangle (Figure 24-12) for the wave vectors then shows that, $|\vec{k}|_S = 2|\vec{k}| \sin \theta$, or, in terms of wavelength,

$$\lambda = 2\lambda_S \sin \theta \qquad (24\text{-}18)$$

which is Bragg's equation, with $m = 1$.

Conservation of energy of the interacting particles requires that $\hbar\omega' = \hbar\omega \pm \hbar\omega_S$, or

$$\omega' = \omega \pm \omega_S \qquad (24\text{-}19)$$

This result shows that the diffracted photon differs in frequency—however little—from the incident photon by the amount ω_S, greater or less, depending on the direction of the acoustic wave. This turns out to be another way of arriving at the *Doppler effect* for light. When the incident light encounters an approaching wave, the scattered frequency is greater, and when it encounters a receding wave (as in Figure 24-11), the scattered frequency is less.

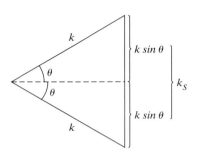

Figure 24-12 Wave vector triangle in the approximation $|\vec{k}'| = |\vec{k}| = k$. The angle θ is the Bragg angle. The geometrical relationship of sides is equivalent to the Bragg diffraction condition.

Example 24-6

To get an idea of the magnitude of the diffraction angle in first order, let us consider a typical case in which the incident light has a wavelength of 550 nm and the acoustic wave has a frequency ν_S of 200 MHz and a speed v_S of 3000 m/s.

Solution

Then,

$$\lambda_S = \frac{v_S}{\nu_S} = \frac{3 \times 10^3}{2 \times 10^8} = 1.5 \times 10^{-5} \text{ m}$$

Using Eq. (24-16),

$$\sin \theta = \frac{\lambda}{2\lambda_S} = \frac{550 \times 10^{-9}}{(2)(1.5 \times 10^{-5})} = 0.0183$$

so that

$$\theta = 1.05°$$

The acousto-optic (AO) effect can be applied to the modulation of a light beam by controlling its amplitude (AM), its frequency (FM), or its direction. Figure 24-13 illustrates one means of achieving AM modulation using

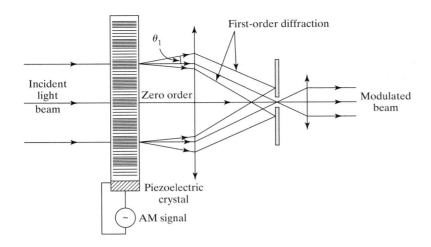

Figure 24-13 Modulation of a light beam by an acousto-optic grating in the Raman-Nath regime. The modulated signal driving the piezoelectric crystal is transferred to the output beam in zero order. Only the zero- and first-order diffracted beams are shown.

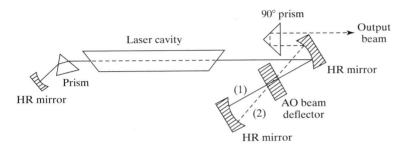

Figure 24-14 Cavity dumping of a laser using an acousto-optic beam deflector. Turning on the acoustic wave deflects the beam outside the laser cavity and initiates cavity dumping.

a thin AO material in the Raman-Nath regime. The fraction of light removed from the zero-order diffracted beam depends on the magnitude of the induced stress and so on the amplitude of the modulating RF signal. The slit allows only the modulated zero-order beam to be transmitted. Other applications make use of the beam deflection capabilities of the AO effect. Referring back to Figure 24-11, it should be evident that both a change in frequency and a change in direction of the acoustic wave cause a change in the direction of the diffracted beam. If the frequency-sensitive position of the output beam is detected by a photodetector array, the AO device can be used as a spectrum analyzer. Again, because the frequency of the diffracted beam is shifted by an amount equal to the acoustic frequency, the beam can be frequency modulated. In this case, the design aims to minimize the angular spread of the diffracted light; when used as a spectrum analyzer, the design aims instead to maximize it.

Another application of the device as a beam deflector to initiate laser-cavity dumping is illustrated in Figure 24-14. When no acoustic wave is applied, the beam (1) bounces back and forth in the laser cavity, building up energy to a maximum value. Turning on the acoustic wave causes a deflection of the beam (2) out of the cavity, thereby dumping the energy stored in the cavity.

24-6 OPTICAL PHASE CONJUGATION

Optical phase conjugation (OPC) represents another third-order, nonlinear phenomenon with some fascinating applications. It was first observed in 1971–1972 by researchers in the Soviet Union. It is so named because the nonlinear interaction produces a light beam that is the *spatial complex conjugate* of one of the waves incident on the nonlinear medium. A process called

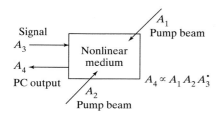

Figure 24-15 Conventional geometry for phase conjugation by four-wave mixing. Pump beams A_1 and A_2 are antiparallel and are much stronger than the signal beam A_3. The phase conjugate output of the signal beam is the beam shown as A_4.

degenerate four-wave mixing, illustrated in Figure 24-15, can be used to produce the phase conjugate beam. In this process, two strong, counterpropagating *pump* beams of complex amplitude A_1 and A_2 interact with a third, weaker *signal* or *probe* beam of complex amplitude A_3 in a nonlinear medium with inversion symmetry. In such a medium, the lowest-order contribution to the nonlinear polarization of Eq. (24-3) will be proportional to the *cube* of the total electric field in the medium. If the two pump beams and the signal beam have the same frequency (that is, are degenerate), a fourth beam of this frequency will be produced. This fourth beam, pictured with complex amplitude A_4 in Figure 24-15, will be the spatial complex conjugate of the signal beam, A_3.

A detailed analysis of this interaction[16] shows that the phase conjugate output beam has an amplitude that is proportional to the product of the amplitudes of the pump beams and the complex conjugate of the amplitude of the signal beam:

$$A_4 \propto A_1 A_2 A_3^* \qquad (24\text{-}20)$$

Now, since the two pump beams are oppositely directed, the complex amplitudes on the right-hand side of Eq. (24-20) can be written as

$$A_1 = |A_1|e^{i\vec{k}_1 \cdot \vec{r}}, \quad A_2 = |A_2|e^{-i\vec{k}_1 \cdot \vec{r}}, \quad A_3 = |A_3|e^{i\vec{k}_3 \cdot \vec{r}}$$

so that

$$A_4 = |A_1||A_2||A_3|e^{-i\vec{k}_3 \cdot \vec{r}}$$

Thus, $\vec{k}_4 = -\vec{k}_3$ and A_4 is proportional to the complex conjugate of A_3. Note that the $\vec{k}$-vectors representing the two pairs of counterpropagating waves comprising the degenerate four-wave mixing process satisfy the phase matching condition,

$$\vec{k}_1 + \vec{k}_2 + \vec{k}_3 + \vec{k}_4 = \vec{k}_1 - \vec{k}_1 + \vec{k}_3 - \vec{k}_3 = 0$$

We have seen that the phase conjugate wave (A_4) exactly reverses the direction and overall phase factor of the signal wave (A_3). Thus the phase conjugate wave precisely retraces the path of the original beam and, at each position, reproduces the exact shape of the original wavefront. Thus optical phase conjugation can be viewed as a unique type of reflection, and we refer to the nonlinear medium that creates the phase conjugate wave as a *phase conjugate mirror* (PCM). To appreciate the uniqueness of the process, consider Figure 24-16, which shows reflections from *ordinary mirrors*. In Figure 24-16a, a plane wave is reflected from an ideal (infinite) plane mirror. The reflected wave is also a plane wave. To express mathematically the reversal in direction of the incident wave, the sign of the kz term is changed from minus to plus. Notice then that, except for the sign of the ωt term, the reflected wave is the complex conjugate of the original wave and has the properties of a phase conjugate wave as described above: It retraces the path of the incident beam and is its phase-reversed replica. Figure 24-16b shows the same process for an incident spherical wave. All that is needed to produce the phase-reversed replica on reflection is a concave spherical mirror whose curvature exactly matches that of the wavefront at incidence. For an arbitrary incident wavefront, as shown in Figure 24-16c, the amplitude $\Psi(r)$ is complex and includes the amplitude and phase factors that describe its deviation from a plane wave. We imagine such a wave as a plane wave that has been shaped by passing through a distorting medium, by diffraction, or by modulation. Its phase-reversed replica is expressed by taking the complex conjugate of both $\Psi(r)$

[16]Amnon Yariv, *Optical Electronics*, 3d ed. (New York: Holt, Rinehart and Winston, 1985), Ch. 16.

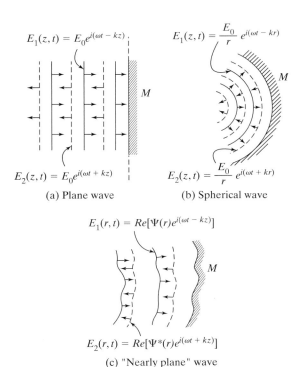

$$E_1(z,t) = E_0 e^{i(\omega t - kz)}$$

$$E_1(z,t) = \frac{E_0}{r} e^{i(\omega t - kr)}$$

M

M

$$E_2(z,t) = E_0 e^{i(\omega t + kz)}$$

$$E_2(z,t) = \frac{E_0}{r} e^{i(\omega t + kr)}$$

(a) Plane wave

(b) Spherical wave

$$E_1(r,t) = Re[\Psi(r)e^{i(\omega t - kz)}]$$

M

$$E_2(r,t) = Re[\Psi^*(r)e^{i(\omega t + kz)}]$$

(c) "Nearly plane" wave

Figure 24-16 Three examples of a phase-reversed replica of an incident wavefront produced by an ordinary mirror. In each case, the rays corresponding to the wavefront are everywhere normal to the mirror surface on reflection. A phase-conjugate mirror handles all cases and also responds to instantaneous changes in the incident wavefront.

and ikz, in other words, the spatial part of the wave equation. To produce this wave with ordinary mirrors, we would need to construct a mirror surface that matched exactly the wavefront of the incident wave at the instant of reflection. The uniqueness of the PCM is that the phase conjugate replica is produced *regardless* of the shape of the incident wavefront, as long as the PCM has an aperture large enough to receive the entire wavefront. Unlike an ordinary mirror, the PCM is able to respond immediately to varying spatial and temporal features of an incident wave such that a phase conjugate wave is continually produced.

Applications of OPC include aberration correction and pointing and tracking. These properties follow from the basic nature of the PC wave, as described above. First, consider a transparent, distorting medium like frosted glass, which is placed in the path of a light wave on its way to a PCM. The wave is modified by the frosted glass in a nonuniform manner, but its phase conjugate—after reflection by the PCM—exactly retraces its path and reverses its modifications, so that the return passage through the distorting medium undoes or "heals" the original distortion. This means that a beam distorted by passing through an optical system with severe aberrations can be recovered by "reflection" from a PCM that sends it back through the same system. It also means that, if the light beam originates from a point source, divergence and diffraction effects are reversed when the beam is returned through the system, so that the PC beam converges to the original point. Furthermore, if the source point moves, the returning beam adjusts so that it continues to point to the source, the pointing and tracking property.

24-7 OPTICAL NONLINEARITIES IN FIBERS

We conclude this chapter with a survey of some aspects of the nonlinear interaction of light waves propagating in optical fibers. We begin by discussing some nonlinear effects that inhibit the successful transmission, at high rates, of information through fibers and then describe a nonlinear fiber amplifier.

Stimulated Raman Scattering

When light of frequency ω is scattered from a molecule, the scattered light consists of a strong component at frequency ω and components of lesser strength at frequencies above or below ω. The scattered light at ω is *elastic or Rayleigh scattering*. An *inelastic* process leading to scattered light at frequencies different from the incident light is known as *spontaneous Raman scattering*. The inelastically scattered light of frequency less than ω is known as the *Stokes* field, and that component with frequency greater than ω is known as the *anti-Stokes* field. In the Raman process the energy lost or gained by the electromagnetic field is accounted for by a change in energy of the molecule. The photons scattered in the spontaneous Raman process can act as seed photons, which in turn *stimulate* additional Raman scattering processes. The predominant frequencies in the Stokes and anti-Stokes fields are characteristic of the molecule and so their detection can be used to determine the composition of, for example, the gases in a particular combustion process. Classically, Raman scattering can be described as arising from the third-order nonlinear polarization of Eq. (24-3). The nonlinearity of silica molecules in optical fibers leads to the generation of, primarily, a Stokes field shifted from the fundamental frequency by an amount in the range 10^{12}–10^{13} Hz. The attenuation, due to stimulated Raman scattering, of the power carried by an optical fiber is typically not significant unless the power carried by the fiber approaches 1 W. Most fiber-optical communication systems use signals of powers significantly less than this and so do not suffer significantly from Raman-induced attenuation. Unfortunately however, the Stokes frequency shift is comparable to the frequency spacing between channels in a wavelength-division-multiplexed (WDM) fiber. The Stokes field can couple the ideally separate channels in the WDM fiber, leading to the degradation of the signal carried by each channel. This effect begins to be important at signal powers of only a few mW.

Stimulated Brillouin Scattering

In Section 24-5 we discussed ordinary Brillouin scattering in which light is scattered by an acoustic wave. *Stimulated Brillouin scattering* is a nonlinear interaction in which light from an incident field is scattered by an acoustic wave, which in turn was generated in the medium by the incident electromagnetic field. The scattered light has a frequency that differs from that of the incident light by an amount equal to the frequency of the acoustic waves in the medium. In optical fibers this frequency shift is typically too small to couple the different channels in a WDM fiber. However, stimulated Brillouin scattering results in a wave that is scattered in the backward direction and so both attenuates and mixes with the forward-traveling signal wave. In contrast to the stimulated Raman process, stimulated Brillouin scattering can lead to significant attenuation and distortion of signal power in a *given* fiber channel when the signal power is just a few mW.

Self-Phase Modulation and Cross-Phase Modulation

In order to carry information in a fiber, the irradiance of the signal wave must vary in time at a given point in the fiber. Due to fiber nonlinearities, this irradiance variation induces a time-varying index of refraction at this given point. This time-varying index of refraction in turn gives rise to a time-dependent phase shift of the signal. This process is known as *self-phase modulation* and leads to what is known as *frequency chirping*. Generally, this phenomenon alters the shape of the information-carrying pulses in a communications system. This effect becomes increasingly important as the pulse width is reduced and the propagation length is increased, and so may be important in long-haul, high-bit-rate systems. Similarly, irradiance variations in one frequency channel of a WDM fiber affect the index of refraction seen by the light signals in the other frequency channels, leading to *cross-phase modulation*.

Raman Amplification in Fibers

The stimulated Raman effect can be used to amplify the signal in an optical fiber, thus counteracting attenuation and increasing the length of fiber that can carry a usable signal. This can be accomplished by propagating a pump laser field through the fiber. The pump field will induce a Stokes field at a frequency less than that of the pump field. In fact, the Stokes field is generated in a range of frequencies known as the *Raman bandwidth* of the fiber. If the signal field input into the fiber has a frequency that is near the peak of the Raman bandwidth, it will strongly stimulate the Raman process and transfer energy from the pump field into the signal field. Raman amplification in fibers is attractive in part because it utilizes the entire unaltered fiber in the amplification process. Competing amplification technologies require either that the signal exit the fiber and be amplified before being inserted into a second fiber or that portions of the fiber be doped with an amplifying medium like erbium. The gain in the Raman process is sufficient to produce a Raman laser in which the Raman gain compensates for the attenuation in a fiber loop, and a self-sustained oscillation at the overlap between a fiber ring mode frequency and the peak of the Raman gain bandwidth occurs.

Nonlinear effects in optical fibers can lead also to *optical pulse compression* and to the propagation of pulses of unchanging shape, *solitons*, through an optical fiber. Nonlinear optics is a burgeoning field driven by the availability of high-power/short-pulse electromagnetic fields. In this chapter we have discussed but a few of the many important effects and applications associated with the nonlinear interaction of light and matter.

PROBLEMS

24-1 Write out the third-order terms of the polarization for a single beam described by a plane wave with amplitude E_0 and frequency ω. What frequencies appear in the polarization wave?

24-2 Write out the third-order terms of the polarization for two-beam interaction, where the beams are plane waves having amplitudes E_{01} and E_{02} and frequencies ω_1 and ω_2, respectively. What frequencies are radiated by the polarization wave?

24-3 Write out the second-order terms of the polarization for three-beam interaction, where the beams are plane waves having amplitudes E_{01}, E_{02}, and E_{03} and frequencies ω_1, ω_2, and ω_3, respectively. What frequencies are radiated by the polarization wave?

24-4 Arguing from Eq. (24-7), show that the linear electro-optic effect is found only in crystals lacking inversion symmetry.

24-5 a. Determine the coherence length for second harmonic generation in KDP when subjected to pulsed ruby laser light at $\lambda_0 = 694$ nm. Appropriate refractive indices are $n(694 \text{ nm}) = 1.505$ and $n(347 \text{ nm}) = 1.534$.

b. The measured coherence length of barium titanate at $\lambda_0 = 1.06$ μm is 5.8 μm. Calculate the expected change in refractive index at $\lambda = 0.53$ μm.

24-6 Determine the half-wave voltage for a longitudinal Pockels cell made of ADP (ammonium dihydrogen phosphate) at $\lambda = 546$ nm. What is its length?

24-7 A longitudinal Pockels cell is made from lithium niobate. Determine the change in refractive index and the phase difference produced by an applied voltage of 426 V when the light beam is from a He-Ne laser at 632.8 nm. The length of the crystal is 1 cm.

24-8 Using Eq. (24-9), show that the transmittance of a Pockels cell can also be written as $I = I_{\max} \sin^2(\Phi/2)$.

a. At what values of V and Φ (greater than zero) is the transmittance zero?

b. If the Pockels cell is preceded by an ordinary half-wave plate, what is the irradiance when $V = 0$ and when $V = V_{HW}$?

24-9 In what kinds of media are both longitudinal Pockels and Kerr effects present? To get some idea of their relative strengths, compare them by calculating the ratio of retardations produced by an appropriately applied 10 kV. Derive an expression for this ratio. Then do a numerical calculation by assuming a hypothetical medium with "typical" values of $r = 10$ pm/V, $K = 1$ pm/V^2, $L = 2$ cm, $d = 1$ cm, and $n_0 = 2$. Take $\lambda = 550$ nm.

24-10 Calculate the length of a Kerr cell using carbon disulfide required to produce half-wave retardation for an applied voltage of 30 kV. The electrodes of the cell have a separation of 1.5 cm. Is this cell practical?

24-11 Show that Eq. (24-19) is equivalent to the Doppler effect for light. Use the fact that the Doppler frequency shift $\Delta \nu$ for light reflected from a moving object is twice that of light emanating from a moving object, or $\Delta \nu = 2 \nu u_p / v$, where ν is the light frequency, v its velocity in the medium, and u_p is the component of the object velocity parallel to the light wave's propagation direction. Use the geometry of Figure 24-11 and the Bragg condition.

24-12 The speed of sound in glass is 3 km/s. For a sound wave having a width of 1 cm, calculate the advance of the sound wave while it is traversed by a light wave. Take $n = 1.50$ for the glass. What is the significance of this result?

24-13   **a.** Show that a small change in angle $\Delta\theta$ around the direction of the diffracted beam in Figure 24-11 can be expressed approximately by $\Delta\theta = \Delta k_S / k$.

   b. Show that this result can be expressed as

$$\Delta\theta = (\lambda/v_S)\, \Delta\nu_S$$

   where λ is the wavelength in the medium.

   c. The factor by which $\Delta\theta$ exceeds the beam divergence is a practically useful number N called "number of resolvable spots." This serves as a figure of merit, giving the number of resolvable positions that can be addressed by the beam deflector. If the beam divergence is expressed by the diffraction angle $\theta_D = \lambda/D$, with D the beam diameter, show that

$$N = \frac{\Delta\theta}{\theta_D} = \tau\Delta\nu_S$$

   where τ is the time for the sound to cross the optical beam diameter.

   d. As a numerical example, consider modulation of the sound frequency in the range 80–120 MHz in fused quartz, where $v_S = 5.95 \times 10^5$ cm/s. If the beam diameter is 1 cm, determine the number of resolvable spots.

24-14 What acoustic frequency is required of a plane acoustic wave, launched in an acousto-optic crystal, so that a He-Ne laser beam is deflected by 1°? The speed of sound in the crystal is 2500 m/s and its refractive index at 632.8 nm is 1.6.

24-15 In Bragg's equation (24-18), the wavelength of the light and the angle are those measured within the medium. Show that, if the medium is isotropic and its sides are parallel to the direction of a plane acoustic wave, the equation also holds for the wavelength and angle of diffraction measured *outside* the medium.

24-16 Determine the difference in deflection angle for a He-Ne laser beam that is Bragg-scattered by an acoustic plane wave when the frequencies are 50 MHz and 80 MHz. The acoustic crystal is sapphire, with $n = 1.76$ and a sound speed of 11 km/s.

24-17 Design an optical isolator, as in Figure 24-10, that uses ZnS as the active medium. Let the magnetic field be produced by winding a solenoid directly onto the ZnS crystal at a turn density of 60 turns/cm. Assume $\lambda = 589$ nm.

24-18 A sample of SF57 glass with polished, parallel sides and 2.73 cm in length is placed between the tapered poles of an electromagnet. A small, central hole is drilled through the pole pieces to allow passage of a linearly polarized He-Ne laser beam through the sample and parallel to the magnetic field direction. The magnetic field is set at 5.098 kG.

   a. When red He-Ne laser light (632.8 nm) is used, the measured rotation is 900 min. Determine the Verdet constant for the glass.

   b. When green He-Ne laser light (543.5 nm) is used, the measured rotation is 1330 min. Determine the Verdet constant for the glass.

24-19 A 5-cm-long liquid cell is situated in a magnetic field of 4 kG. The cell is filled with carbon disulfide and linearly polarized sodium light is transmitted through the cell, along the B-field direction. Determine both the net rotation of the light and the circular dispersion of CS_2 at this wavelength.

24-20 Sketch the shape of a nonsymmetrical pulse before and after reflection from an ordinary mirror and before and after reflection from a PCM. In the latter case, assume that the PCM is "turned on" by initiating the pump beams at the instant the entire pulse has moved inside the PC medium. Show how this effect might be used to correct dispersion broadening in an optical fiber. (If necessary, consult Vladimir V. Shkunov, and Boris Ya. Zel'dovich, "Optical Phase Conjugation," *Scientific American*, Dec. 1985: 54.)

24-21 Sketch an arrangement using a PCM to project a sharp, high-intensity image of a mask onto the photo-resist layer on a semiconducting chip without using lenses. This provides a means of doing photolithography without placing a mask in direct contact with the chip. (If necessary, consult Vladimir V. Shkunov, and Boris Ya. Zel'dovich, "Optical Phase Conjugation," *Scientific American*, Dec. 1985: 54.)

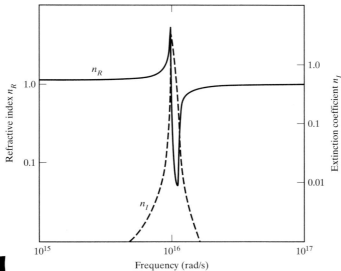

25 Optical Properties of Materials

INTRODUCTION

Electromagnetic waves that encounter materials create a complex of interactions with the charged particles of the medium. Forces are exerted on the charges by the electric field of the waves and, because of the motions of the charges, also by the magnetic field of the waves. In responding to these oscillating fields, the charges themselves oscillate and act as radiators of secondary electromagnetic waves. Thus, in determining the net field at some point, the fields of both the source waves and the waves emitted by the charged oscillators must be taken into account. In the case of ordinary fields, smaller than those now attainable with high-energy lasers, the net fields are assumed to be a linear superposition of the constituent fields. The complicated effects of all the microscopic contributions to the resultant field by the charges in the material can, for certain purposes, be simply described by macroscopic material parameters, the *optical constants* of the material. In this chapter, we show in particular how the *refractive index* and the *absorption coefficient* for isotropic conducting (metals) and nonconducting (insulators or dielectrics) materials can be understood. In order to do this we use Maxwell's equations and the mathematical techniques of vector calculus.

25-1 POLARIZATION OF A DIELECTRIC MEDIUM

We take as our model a *simple dielectric*, that is, a nonconducting material whose properties are isotropic. By *nonconducting* we mean that the medium, unlike a metal, contains no free charges. Positive charges are associated with the constituent nuclei and negative charges with the electrons bound to such nuclei. By *isotropic* we mean that the relevant physical properties we consider are independent of direction in the medium, so that we may treat the physical constants as scalar quantities. Application of an electric field to such a medium causes charge displacement, in which the negative charge distribution bound to the nuclei shifts in a direction opposite to the electric field. The shift may occur in a *polar molecule*, like H_2O, because the molecule has a *permanent electric dipole*, that is, the effective centers of its positive and negative charge distributions do not coincide. In this case, application of the field produces some reorientation of the molecules so that, on the average, the positive end of the dipole is in the direction of the field. The tendency toward alignment is counteracted by the thermal motions of the molecules. The shift in charge distribution may also occur in *nonpolar molecules*, such as O_2, in which positive and negative charge distributions normally have the same effective center. Application of the field results in a slight shift of the electron cloud relative to its nucleus, producing an *induced dipole*. In either case, the *dipole moment* $\vec{p}$ due to each atom or molecule is given by the product of the magnitude of the displaced charge q and the vector $\vec{r}$ that locates the effective negative charge center relative to the effective positive charge center in the dipole, or

$$\vec{p} = -q\vec{r} \tag{25-1}$$

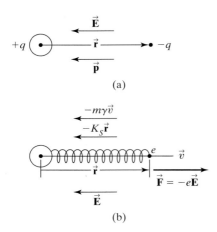

Figure 25-1 The elementary electric dipole. (a) Alignment with the field. (b) Forces acting on a dipole when the electric field has the direction indicated.

as indicated in Figure 25-1a. The direction of the dipole moment is from the negative toward the positive charge. The magnitude of the dipole moment for a given material depends on how easily charge is displaced under the influence of a given electric field. The *polarization* $\vec{P}$ of the medium is then said to be the collective dipole moment per unit volume, the sum of dipole moments given by

$$\vec{P} = -Ne\vec{r} \tag{25-2}$$

where N is the number of elementary dipoles per unit volume and e is the magnitude of the electronic charge.

Electrons behave as though the forces binding them to the nuclei are elastic forces given by Hooke's law, where the restoring force is proportional to the displacement and oppositely directed. The more massive nuclei can be considered stationary since they are unable to respond to the rapid changes in the field representing an electromagnetic wave in the optical region of the spectrum. A simple model in which electrons are held by springlike forces to a fixed nucleus is therefore applicable. In an alternating electric field, however, forced oscillations of electrons remove a certain amount of energy from the incident radiation, the energy that the electrons radiate in turn and the energy of interaction with neighboring atoms that shows up as thermal energy. The model of the oscillating electron is therefore that of a damped, harmonic oscillator, with a frictional force proportional to the velocity. Newton's second law, applied to the electron in the model of Figure 25-1b, then leads to the equation of motion,

$$-K_S\vec{r} - m\gamma\frac{d\vec{r}}{dt} - e\vec{E} = m\frac{d^2\vec{r}}{dt^2} \tag{25-3}$$

In Eq. (25-3), K_S is the force constant of the effective spring, m is the electronic mass, and γ is a frictional constant with dimensions of reciprocal time. Notice that the force $(e\vec{v} \times \vec{B})$ on the electron due to the magnetic field of the radiation is omitted; it is, in fact, negligible compared with the force $(e\vec{E})$ due to the electric field.

When the applied $\vec{\mathbf{E}}$-field is static, there is no oscillation of the dipoles, so that both velocity and acceleration of the electron vanishes. In this special case, Eq. (25-3) reduces to

$$-K_S\vec{\mathbf{r}} = e\vec{\mathbf{E}}$$

or, eliminating $\vec{\mathbf{r}}$ with the help of Eq. (25-2), the *static polarization* is given by

$$\vec{\mathbf{P}} = \frac{Ne^2\vec{\mathbf{E}}}{K_S} \qquad (25\text{-}4)$$

Suppose now that $\vec{\mathbf{E}}$ is a harmonic field with a time dependence given by $\vec{\mathbf{E}} = \vec{\mathbf{E}}_0e^{-i\omega t}$ and that the oscillations respond with a similar dependence, $\vec{\mathbf{r}} = \vec{\mathbf{r}}_0e^{-i\omega t}$. Note that we are representing the electric field $\vec{\mathbf{E}}$ and the dipole displacement vector $\vec{\mathbf{r}}$ by complex vectors. The real field and displacement vectors are formed by taking the real part of these complex vectors. All of the equations used in this chapter are linear in these quantities and so hold for both the real fields and displacements and their complex representations. Inserting the derivatives of the complex displacement vector $d\vec{\mathbf{r}}/dt = -i\omega\vec{\mathbf{r}}$ and $d^2\vec{\mathbf{r}}/dt^2 = -\omega^2\vec{\mathbf{r}}$ into Eq. (25-3) leads to

$$\vec{\mathbf{r}} = \frac{-e\vec{\mathbf{E}}}{-m\omega^2 - im\omega\gamma + K_S} \qquad (25\text{-}5)$$

which, when substituted into Eq. (25-2), yields a time-dependent complex polarization $\vec{\mathbf{P}}$ given by

$$\vec{\mathbf{P}} = \left(\frac{Ne^2}{-m\omega^2 - im\omega\gamma + K_S}\right)\vec{\mathbf{E}} \qquad (25\text{-}6)$$

Note that Eq. (25-6) agrees with Eq. (25-4) in the case of a static ($\omega = 0$) $\vec{\mathbf{E}}$-field. In general, the polarization is a function of the radiation frequency ω and, because the coefficient of $\vec{\mathbf{E}}$ in Eq. (25-6) is complex, the polarization may possess a frequency-dependent phase relative to $\vec{\mathbf{E}}$. If we were to associate the symbol $\vec{\mathbf{E}}$ with the externally applied field, our analysis, thus far, would apply only to the case of a single dipole oscillator. The electric field $\vec{\mathbf{E}}$ in Eq. (25-6) should represent the actual field at the dipole, in the interior of the medium. This local field $\vec{\mathbf{E}}_{\text{loc}}$ is a superposition of the applied field $\vec{\mathbf{E}}_{\text{app}}$ and the field that results from all the other dipoles aligned in a polarized medium. Therefore, in order to treat a macroscopic assembly of dipoles we must make use of a result given in standard texts on electricity and magnetism. That is, the contribution to the total electric field at the position of a given dipole, due to all the other dipoles in the medium, is given by $\vec{\mathbf{P}}/3\varepsilon_0$, where ε_0 is the permittivity of free space. Thus,

$$\vec{\mathbf{E}}_{\text{loc}} = \frac{\vec{\mathbf{P}}}{3\varepsilon_0} + \vec{\mathbf{E}}_{\text{app}} \qquad (25\text{-}7)$$

Choosing to retain the symbol $\vec{\mathbf{E}}$ for the applied field leads to

$$\vec{\mathbf{E}}_{\text{loc}} = \frac{\vec{\mathbf{P}}}{3\varepsilon_0} + \vec{\mathbf{E}}$$

Substituting this expression for the local electric field into the right-hand side of Eq. (25-6) gives

$$\vec{\mathbf{P}} = \left(\frac{Ne^2}{-m\omega^2 - im\omega\gamma + K_S}\right)\left(\vec{\mathbf{E}} + \frac{\vec{\mathbf{P}}}{3\varepsilon_0}\right) \tag{25-8}$$

We can solve for $\vec{\mathbf{P}}$ explicitly by, for the moment, setting the prefactor in Eq. (25-8) equal to F. We conclude

$$\vec{\mathbf{P}} = \left(\frac{F}{1 - F/3\varepsilon_0}\right)\vec{\mathbf{E}} \tag{25-9}$$

The multiplier of $\vec{\mathbf{E}}$ is

$$\frac{F}{1 - F/3\varepsilon_0} = \frac{Ne^2/m}{(K_S/m - Ne^2/3m\varepsilon_0) - \omega^2 - i\omega\gamma}$$

Let us define ω_0^2 as the quantity in parentheses, that is,

$$\omega_0^2 \equiv \frac{K_S}{m} - \frac{Ne^2}{3m\varepsilon_0} \tag{25-10}$$

Then, Eq. (25-9) becomes

$$\vec{\mathbf{P}} = \frac{Ne^2/m}{\omega_0^2 - \omega^2 - i\omega\gamma}\vec{\mathbf{E}} \tag{25-11}$$

Forming the magnitudes of both sides of this relation shows that the magnitude of the polarization is related to the magnitude of the applied field by the relation

$$|\vec{\mathbf{P}}| = \frac{Ne^2/m}{\sqrt{(\omega_0^2 - \omega^2)^2 + \omega^2\gamma^2}}|\vec{\mathbf{E}}|$$

Clearly, $|\vec{\mathbf{P}}|$ can increase dramatically as $\omega \to \omega_0$, so that ω_0 represents a *resonance frequency* for the dipoles of the medium. Equation (25-11) has the same form as the equation of motion of a driven harmonic oscillator with damping. As the driving frequency approaches the resonance frequency ω_0 of the oscillator, the amplitude of the vibrations becomes very large and subsides again as the frequency increases beyond ω_0. For a dielectric medium, the increase of dipole moments at resonance results in a large maximum polarization. Equation (25-11) also indicates that there is a frequency-dependent phase shift between the applied field and the polarization. Far from resonance, $\omega\gamma \ll \omega_0^2 - \omega^2$, and so the damping term in the denominator of the expression on the right-hand side of Eq. (25-11) can be ignored. Then for $\omega \ll \omega_0$, $\vec{\mathbf{P}}$ and $\vec{\mathbf{E}}$ have the same sign and the dipoles are oscillating in phase with the field. Beyond resonance, however, when $\omega \gg \omega_0$, $\vec{\mathbf{P}}$ and $\vec{\mathbf{E}}$ have opposite signs, indicating a phase difference of π. Free electrons respond in this manner. When $\omega \cong \omega_0$, near resonance, the vibrations are large. The damping term in the denominator in this case is not negligible, and the division by $-i$, equivalent to multiplication by i, indicates a $\pi/2$ phase shift between $\vec{\mathbf{E}}$ and $\vec{\mathbf{P}}$.

The dependence of $\vec{\mathbf{P}}$ on $\vec{\mathbf{E}}$, as given in Eq. (25-11), can now be used to discover the conditions under which plane waves are able to propagate in a dielectric. The fundamental wave equation for electromagnetic waves in the dielectric is a consequence of the Maxwell equations.

25-2 PROPAGATION OF LIGHT WAVES IN A DIELECTRIC

The four Maxwell equations may be written in the general form

$$\nabla \cdot \vec{\mathbf{E}} = \frac{\rho}{\varepsilon_0} \tag{25-12}$$

$$\nabla \times \vec{\mathbf{E}} = -\frac{\partial \vec{\mathbf{B}}}{\partial t} \tag{25-13}$$

$$\nabla \cdot \vec{\mathbf{B}} = 0 \tag{25-14}$$

$$c^2 \nabla \times \vec{\mathbf{B}} = \frac{\partial \vec{\mathbf{E}}}{\partial t} + \frac{\vec{\mathbf{J}}}{\varepsilon_0} \tag{25-15}$$

In these equations ρ is the charge density, which in general includes both the free charge density ρ_f and the bound charge density ρ_b, so that $\rho = \rho_b + \rho_f$. In a dielectric, however, $\rho_f = 0$. It is standard practice in a course in electricity and magnetism to show that the bound-charge density is related to the polarization by

$$\rho_b = -\nabla \cdot \vec{\mathbf{P}} \tag{25-16}$$

The quantity $\vec{\mathbf{J}}$ similarly represents the current density and can arise from both free and bound charge, as indicated by $\vec{\mathbf{J}} = \vec{\mathbf{J}}_b + \vec{\mathbf{J}}_f$. In a dielectric where $\rho_f = 0$, $\vec{\mathbf{J}}_f = 0$ also. Furthermore, it can be shown that

$$\vec{\mathbf{J}}_b = \frac{\partial \vec{\mathbf{P}}}{\partial t} \tag{25-17}$$

With these constraints, the four Maxwell equations for a dielectric can be written

$$\nabla \cdot \vec{\mathbf{E}} = \frac{-\nabla \cdot \vec{\mathbf{P}}}{\varepsilon_0} \tag{25-18}$$

$$\nabla \times \vec{\mathbf{E}} = -\frac{\partial \vec{\mathbf{B}}}{\partial t} \tag{25-19}$$

$$\nabla \cdot \vec{\mathbf{B}} = 0 \tag{25-20}$$

$$c^2 \nabla \times \vec{\mathbf{B}} = \frac{\partial \vec{\mathbf{E}}}{\partial t} + \frac{1}{\varepsilon_0} \frac{\partial \vec{\mathbf{P}}}{\partial t} \tag{25-21}$$

Now we take the curl of both sides of Eq. (25-19), giving

$$\nabla \times (\nabla \times \vec{E}) = \nabla \times \left(-\frac{\partial \vec{B}}{\partial t}\right) = -\frac{\partial}{\partial t}(\nabla \times \vec{B}) \tag{25-22}$$

where we have interchanged the order of differentiation with respect to space and time in the last step. The left member of Eq. (25-22) can be reexpressed by the identity

$$\nabla \times (\nabla \times \vec{E}) \equiv \nabla(\nabla \cdot \vec{E}) - \nabla^2 \vec{E} \tag{25-23}$$

In a *homogeneous* dielectric, the effect of polarization is to produce a net surface charge density, while leaving the internal charge density $\rho_b = 0$ unchanged. The internal charge density is zero because, in any internal closed surface, every bit of charge that moves into the enclosed volume in response to a polarizing field is balanced by an equal bit of charge that moves out. The surface charge density appears because such balancing is not possible there. Thus by Eqs. (25-16) and (25-18) we conclude that $\nabla \cdot \vec{E} = 0$ and substitute the remainder of Eq. (25-23) into Eq. (25-22), giving

$$\nabla^2 \vec{E} = \frac{\partial}{\partial t}(\nabla \times \vec{B}) \tag{25-24}$$

For the right member we may make use of Maxwell's equation (25-21) and write

$$c^2 \nabla^2 \vec{E} = \frac{\partial^2 \vec{E}}{\partial t^2} + \frac{1}{\varepsilon_0}\frac{\partial^2 \vec{P}}{\partial t^2} \tag{25-25}$$

The last term is expressible in terms of $\vec{E}$ using Eq. (25-11), so we have

$$c^2 \nabla^2 \vec{E} = \left[1 + \frac{Ne^2}{m\varepsilon_0(\omega_0^2 - \omega^2 - i\omega\gamma)}\right]\frac{\partial^2 \vec{E}}{\partial t^2} \tag{25-26}$$

For a harmonic wave expressed as $\vec{E} = \vec{E}_0 e^{i(kz-\omega t)}$, in which case $\nabla^2 \vec{E} = -k^2\vec{E}$ and $\partial^2 \vec{E}/\partial t^2 = -\omega^2\vec{E}$, Eq. (25-26) solved for k^2 becomes

$$k^2 = \frac{\omega^2}{c^2}\left[1 + \frac{Ne^2}{m\varepsilon_0}\frac{1}{(\omega_0^2 - \omega^2 - i\omega\gamma)}\right] \tag{25-27}$$

We conclude that the analysis of plane waves propagating in a homogeneous dielectric requires in general that the propagation constant k be a complex number. Consequently, we write

$$k = k_R + ik_I \tag{25-28}$$

Inserting this form into the expression for a harmonic wave, we have

$$\vec{E} = \vec{E}_0 e^{i(k_R z + ik_I z - \omega t)} = \vec{E}_0 e^{-k_I z} e^{i(k_R z - \omega t)} \tag{25-29}$$

The exponential factor in k_I represents a depth-dependent absorption of an otherwise harmonic wave, and k_I measures the *amplitude attenuation* of the wave. By taking the square of the magnitude of both sides of Eq. (25-29), the result describes instead the energy flux density, giving

$$I = I_0 e^{-\alpha z}$$

where $\alpha = 2k_I$ is the *absorption coefficient* of the medium. If the propagation constant is complex, so must be the refractive index, since

$$k = \frac{2\pi}{\lambda} = \frac{2\pi\nu}{\upsilon} = \left(\frac{\omega}{c}\right)n \qquad (25\text{-}30)$$

If we identify the real and imaginary parts of the complex refractive index by

$$n = n_R + in_I \qquad (25\text{-}31)$$

where n_R is the usual refractive index and n_I is called the *extinction coefficient*, it follows from Eqs. (25-28) and (25-30) that

$$k_R + ik_I = \left(\frac{\omega}{c}\right)(n_R + in_I)$$

yielding the relations

$$k_R = \left(\frac{\omega}{c}\right)n_R \qquad (25\text{-}32)$$

and

$$k_I = \left(\frac{\omega}{c}\right)n_I \qquad (25\text{-}33)$$

Writing n^2 as

$$n^2 = (n_R + in_I)^2 = \left(\frac{ck}{\omega}\right)^2$$

and relating this equation to Eq. (25-27) gives

$$n^2 = (n_R + in_I)^2 = 1 + \left(\frac{Ne^2}{m\varepsilon_0}\right)\frac{1}{\omega_0^2 - \omega^2 - i\omega\gamma} \qquad (25\text{-}34)$$

Expressions for the real and imaginary parts of the refractive index can be found by equating real and imaginary parts in Eq. (25-34). The left member is

$$(n_R + in_I)^2 = (n_R^2 - n_I^2) + i(2n_R n_I) \qquad (25\text{-}35)$$

The right member can also be written as the sum of a real and imaginary part. The complex term is first rewritten by multiplying numerator and denominator by the complex conjugate of the denominator. The result, after simplification, is

$$(n_R + in_I)^2 = 1 + \frac{Ne^2}{m\varepsilon_0}\left(\frac{\omega_0^2 - \omega^2}{(\omega_0^2 - \omega^2)^2 + \omega^2\gamma^2} + i\frac{\omega\gamma}{(\omega_0^2 - \omega^2)^2 + \omega^2\gamma^2}\right)$$

$$(25\text{-}36)$$

Now by comparing the right members of Eqs. (25-35) and (25-36),

$$n_R^2 - n_I^2 = 1 + \frac{Ne^2}{m\varepsilon_0}\left[\frac{\omega_0^2 - \omega^2}{(\omega_0^2 - \omega^2)^2 + \gamma^2\omega^2}\right] \qquad (25\text{-}37)$$

and

$$2n_I n_R = \frac{Ne^2}{m\varepsilon_0}\left[\frac{\gamma\omega}{(\omega_0^2 - \omega^2)^2 + \gamma^2\omega^2}\right] \qquad (25\text{-}38)$$

The equations can be solved simultaneously for n_I and n_R. The appearance of the mass m in the denominator of these equations shows that electronic oscillations are more important than ionic oscillations in determining the index of refraction. Ionic polarization may be significant in the region of resonance, however, where the large-bracketed terms in Eqs. (25-37) and (25-38) balance the small prefactors containing the mass.

Figure 25-2 shows both n_R and n_I calculated from Eqs. (25-37) and (25-38) as a function of driving frequency ω. The absorption described by the extinction coefficient n_I is seen to peak at the resonant frequency ω_0. The real refractive index experiences a sharp rise and then a fall as ω increases toward and then passes through resonance, after which it increases again, approaching the value $n_R = 1$ at high frequencies. The narrow region where n_R decreases with frequency is contrary to the usual dispersion of transparent media and is called the region of *anomalous dispersion*.

A resonance frequency such as ω_0 for the dielectric means that, for incident photons of frequency ω_0, there is a high probability of absorption. Absorption of such a photon corresponds to a transition between states differing in energy by $\hbar\omega_0 = h\nu_0$ in the energy-band structure of the material. As ω is varied, there will be a series of resonance frequencies characteristic of the material. If such a resonance occurs in the visible range of frequencies, for example, the material absorbs a portion of the spectrum and appears colored, while transmitting the remainder. Transparent materials like glass have resonance frequencies in the infrared and ultraviolet regions but not in the visible. In terms of our simplified model of a dielectric, we interpret the existence

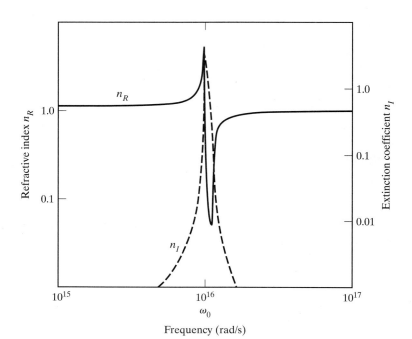

Figure 25-2 Angular frequency dependence of the refractive index n_R and the extinction coefficient n_I for a dielectric. Assumed values are $\omega_0 = 1 \times 10^{16}\,\text{s}^{-1}$, $\gamma = 10^{14}\,\text{s}^{-1}$, and $N = 1 \times 10^{28}\,\text{m}^{-3}$.

of a number of resonance frequencies to mean that electrons experience different degrees of freedom in response to the applied field. To take this into account formally, Eq. (25-34) is usually generalized to include a number of terms summed over the resonant frequencies ω_j, given by

$$n^2 = 1 + \frac{Ne^2}{m\varepsilon_0} \sum_j \frac{f_j}{\omega_j^2 - \omega^2 - i\gamma_j\omega} \qquad (25\text{-}39)$$

where f_j, called the *oscillator strength* for the resonance ω_j, represents the fraction of dipoles having this resonant frequency. The determination of the oscillator strength for a given resonant transition requires the application of quantum theory.

The Dispersion Equation

The variation in refractive index with frequency, described by Eq. (25-39), is what we mean by *dispersion*. We wish to show that the Cauchy dispersion equation, introduced in Eq. (3-17) as an empirical formula, can be deduced from Eq. (25-39) under certain simplifying assumptions. We shall assume a single resonant frequency ω_0 in the ultraviolet, such that frequencies in the visible obey the inequality, $\omega \ll \omega_0$ We shall also assume that, in this regime $\gamma\omega \ll (\omega_0^2 - \omega^2)$. In this case, the index of refraction is real and Eq. (25-34) takes the form

$$n^2 = 1 + \frac{Ne^2}{m\varepsilon_0}\left(\frac{1}{\omega_0^2 - \omega^2}\right)$$

Notice that, for $\omega \ll \omega_0$, as for a gas, the refractive index is nearly constant. As ω increases toward ω_0, the refractive index increases slightly, as shown in Figure 25-2. The slowly increasing index with frequency (decreasing with wavelength) is characteristic of normal dispersion.

To derive the Cauchy dispersion equation, let us first expand the frequency factor in a binomial series:

$$\frac{1}{\omega_0^2 - \omega^2} = \frac{1}{\omega_0^2}\left(1 - \frac{\omega^2}{\omega_0^2}\right)^{-1} = \frac{1}{\omega_0^2}\left(1 + \frac{\omega^2}{\omega_0^2} + \frac{\omega^4}{\omega_0^4} + \cdots\right)$$

so that

$$n^2 = 1 + \frac{Ne^2}{m\varepsilon_0\omega_0^2}\left(1 + \frac{\omega^2}{\omega_0^2} + \frac{\omega^4}{\omega_0^4} + \cdots\right)$$

Writing $\omega = 2\pi c/\lambda$ and gathering constants A', B', and C' appropriately, we can express

$$n^2 = A' + \frac{B'}{\lambda^2} + \frac{C'}{\lambda^4} + \cdots$$

We may take the square root of each side and, since each higher-order term in the expression is less than A', use the binomial expansion again on the right member. After re-collecting constants, we get

$$n = A + \frac{B}{\lambda^2} + \frac{C}{\lambda^4} + \cdots$$

This is the Cauchy relation introduced earlier to describe normal dispersion.

25-3 CONDUCTION CURRENT IN A METAL

In metals, the existence of "free" electrons, not bound to particular nuclei, modifies the treatment outlined above for dielectrics. Although there are also bound electrons, the response of the free electrons dominates the electrical and optical properties of the medium. So, in Eq. (25-3), we set $K_S = 0$, and the equation of motion becomes

$$m\frac{d\vec{\mathbf{v}}}{dt} + m\gamma\vec{\mathbf{v}} = -e\vec{\mathbf{E}} \tag{25-40}$$

The equation may be conveniently expressed in terms of the *conduction current density* $\vec{\mathbf{J}}$, defined by

$$\vec{\mathbf{J}} = -Ne\vec{\mathbf{v}} \tag{25-41}$$

where $\vec{\mathbf{J}}$ has (SI) units of amperes per square meter. Writing Eq. (25-40) in terms of $\vec{\mathbf{J}}$ rather than $\vec{\mathbf{v}}$,

$$\frac{d\vec{\mathbf{J}}}{dt} + \gamma\vec{\mathbf{J}} = \left(\frac{Ne^2}{m}\right)\vec{\mathbf{E}} \tag{25-42}$$

In the case where the applied field is the harmonic wave $\vec{\mathbf{E}} = \vec{\mathbf{E}}_0 e^{-i\omega t}$, we expect the current density to vary at the same rate and write $\vec{\mathbf{J}} = \vec{\mathbf{J}}_0 e^{-i\omega t}$. Equation (25-42) then takes the form

$$(-i\omega + \gamma)\vec{\mathbf{J}} = \left(\frac{Ne^2}{m}\right)\vec{\mathbf{E}} \tag{25-43}$$

In the static, or DC, case specified by $\omega = 0$,

$$\vec{\mathbf{J}} = \left(\frac{Ne^2}{m\gamma}\right)\vec{\mathbf{E}} \tag{25-44}$$

The static *conductivity* σ, defined by Ohm's law,

$$\vec{\mathbf{J}} = \sigma\vec{\mathbf{E}} \tag{25-45}$$

then takes the theoretical form

$$\sigma = \frac{Ne^2}{m\gamma} \tag{25-46}$$

Since conductivities are usually measured, we rewrite Eq. (25-43) in terms of σ, giving

$$\vec{\mathbf{J}} = \left(\frac{\sigma}{1 - i\omega/\gamma}\right)\vec{\mathbf{E}} \tag{25-47}$$

25-4 PROPAGATION OF LIGHT WAVES IN A METAL

An electromagnetic wave propagating in the conducting medium satisfies Maxwell's equations (25-12) through (25-15). Although free charge exists in the metal, the internal free-charge volume density ρ_f is zero. The free charge is so mobile that it quickly redistributes in response to an applied field, preventing the buildup of local charge densities. The appropriate Maxwell equations are then

$$\nabla \cdot \vec{\mathbf{E}} = 0 \tag{25-48}$$

$$\nabla \times \vec{\mathbf{E}} = -\frac{\partial \vec{\mathbf{B}}}{\partial t} \tag{25-49}$$

$$\nabla \cdot \vec{\mathbf{B}} = 0 \tag{25-50}$$

$$c^2 \nabla \times \vec{\mathbf{B}} = \frac{\partial \vec{\mathbf{E}}}{\partial t} + \frac{\vec{\mathbf{J}}}{\varepsilon_0} \tag{25-51}$$

As before, $\nabla \times (\nabla \times \vec{\mathbf{E}}) = -\nabla^2 \vec{\mathbf{E}}$ because $\nabla \cdot \vec{\mathbf{E}} = 0$ in the identity of Eq. (25-23). So, taking the curl of Eq. (25-49), we have

$$-\nabla^2 \vec{\mathbf{E}} = \nabla \times \left(-\frac{\partial \vec{\mathbf{B}}}{\partial t}\right) = -\frac{\partial}{\partial t}(\nabla \times \vec{\mathbf{B}}) = -\frac{1}{c^2}\frac{\partial^2 \vec{\mathbf{E}}}{\partial t^2} - \frac{1}{\varepsilon_0 c^2}\left(\frac{\partial \vec{\mathbf{J}}}{\partial t}\right)$$

where we have used Eq. (25-51) in the last step. Representing $\vec{\mathbf{J}}$ with the help of Eq. (25-47), we conclude

$$\nabla^2 \vec{\mathbf{E}} = \frac{1}{c^2}\left(\frac{\partial^2 \vec{\mathbf{E}}}{\partial t^2}\right) + \frac{1}{\varepsilon_0 c^2}\left(\frac{\sigma}{1 - i\omega/\gamma}\right)\frac{\partial \vec{\mathbf{E}}}{\partial t} \tag{25-52}$$

For plane, harmonic waves given by $\vec{\mathbf{E}} = \vec{\mathbf{E}}_0 e^{i(kz-\omega t)}$, the appropriate space and time derivatives required by Eq. (25-52) can be calculated to give

$$k^2 = \frac{\omega^2}{c^2} + i\left(\frac{\sigma \omega \mu_0}{1 - i\omega/\gamma}\right) \tag{25-53}$$

where we have also made use of the fact that $\varepsilon_0 \mu_0 = 1/c^2$ with μ_0 the permeability of vacuum. Again, we find that the propagation constant must be a complex number to properly describe the propagation of the wave in a metal.

25-5 SKIN DEPTH

Before proceeding with the general case described by Eq. (25-53), we pause to consider the special case in which the frequency ω of the incident radiation is small enough to allow as a good approximation to Eq. (25-53)

$$k^2 = i\omega\sigma\mu_0$$

Expressing i as $e^{i\pi/2}$ and taking the square root of each side,

$$k = (1 + i)\left(\frac{\sigma\mu_0\omega}{2}\right)^{1/2} \tag{25-54}$$

Writing k in the complex form $k = k_R + ik_I$, as before, we can identify the real and imaginary coefficients by

$$k_R = k_I = \left(\frac{\sigma\mu_0\omega}{2}\right)^{1/2} \tag{25-55}$$

and the real and imaginary refractive indices by

$$n_R = \frac{c}{\omega}k_R = \left(\frac{c^2\sigma\mu_0}{2\omega}\right)^{1/2} = \left(\frac{\sigma}{2\omega\varepsilon_0}\right)^{1/2} \tag{25-56}$$

and

$$n_I = \frac{c}{\omega}k_I = \left(\frac{\sigma}{2\omega e_0}\right)^{1/2} \tag{25-57}$$

The complex character of k, when introduced into the plane, harmonic wave equation, leads as in Eq. (25-29) to

$$\vec{E} = \vec{E}_0 e^{-k_I z} e^{i(k_R z - \omega t)}$$

The real exponential factor $e^{-k_I z}$ describes absorption. When the radiation has penetrated a depth of $z = 1/k_I$, therefore, the amplitude has decreased to $1/e$ of its surface value. This particular distance is called the *skin depth*, δ, where

$$\delta \equiv \frac{1}{k_I} = \sqrt{\frac{2}{\sigma\mu_0\omega}} \tag{25-58}$$

and is evidently smaller for better conductors with larger σ. For 3-cm microwaves, for example, the skin depth in copper, with conductivity of $5.8 \times 10^7/\Omega\text{-m}$, is only about 6.6×10^{-5} cm.

25-6 PLASMA FREQUENCY

Returning to the general case of Eq. (25-53) and introducing there the complex refractive index, we write

$$n^2 = \left(\frac{c}{\omega}k\right)^2 = 1 + \frac{i\sigma c^2\mu_0}{\omega(1 - i\omega/\gamma)}$$

After multiplying the second term on the right-hand side of this equation by $i\gamma/i\gamma$, this becomes

$$n^2 = 1 - \frac{\mu_0\sigma c^2\gamma}{\omega^2 + i\omega\gamma} \tag{25-59}$$

The numerator in the second term must have the same dimensions as ω^2 and is identified as the square of a *plasma frequency* given by

$$\omega_p^2 = \mu_0 c^2 \gamma \sigma = \mu_0 c^2 \gamma \left(\frac{Ne^2}{m\gamma} \right) = \frac{Ne^2}{m\varepsilon_0} \qquad (25\text{-}60)$$

where we have made use of both Eq. (25-46) and the relation $\varepsilon_0 \mu_0 = 1/c^2$. The plasma frequency is a resonant frequency for the free oscillations of the electrons about their equilibrium positions. Inserting it into Eq. (25-59),

$$n^2 = 1 - \frac{\omega_p^2}{\omega^2 + i\omega\gamma} \qquad (25\text{-}61)$$

The plasma frequency turns out to be a critical frequency whose value determines whether the refractive index is real or imaginary. This can be seen by neglecting the γ-term, valid for high-enough frequency ($\omega \gg \gamma$), in which case Eq. (25-61) is simply

$$n^2 = 1 - \frac{\omega_p^2}{\omega^2} \qquad (25\text{-}62)$$

Equation (25-62) now shows that for $\omega < \omega_p$, the refractive index of the metal is complex and radiation is attenuated, whereas for $\omega > \omega_p$, the index is real and the metal is transparent to the radiation.

Returning to Eq. (25-61), we find, as before, two equations from which the real and imaginary parts of the refractive index can be calculated. Equating real and imaginary parts in

$$n^2 = (n_R + in_I)^2 = (n_R^2 - n_I^2) + i(2n_R n_I) = 1 - \left(\frac{\omega_p^2}{\omega^2 + i\omega\gamma} \right) \qquad (25\text{-}63)$$

we find

$$n_R^2 - n_I^2 = 1 - \left(\frac{\omega_p^2}{\omega^2 + \gamma^2} \right) \qquad (25\text{-}64)$$

$$2n_R n_I = \frac{\gamma}{\omega} \left(\frac{\omega_p^2}{\omega^2 + \gamma^2} \right) \qquad (25\text{-}65)$$

These equations, solved simultaneously, permit calculation of curves such as those in Figure 25-3. The curves cross at $\omega = (\omega_p^2 - \gamma^2)^{1/2}$, as is evident from Eq. (25-64). Since typically $\omega_p \gg \gamma$, the crossover occurs at $\omega \cong \omega_p$, dividing the transparent and the opaque (and highly reflecting) regions. The plasma frequency for metals falls in the visible to near-ultraviolet regions, so that they are opaque to visible and transparent to ultraviolet radiation at sufficiently high frequency.

Intermediate to the good insulator and good conductor we have treated separately are materials, like semiconductors, for which neither of these extreme cases suffices to explain the properties. Such materials manifest appreciable contributions to their optical properties from both free and bound charges and accordingly must be treated by allowing for both types of behavior.

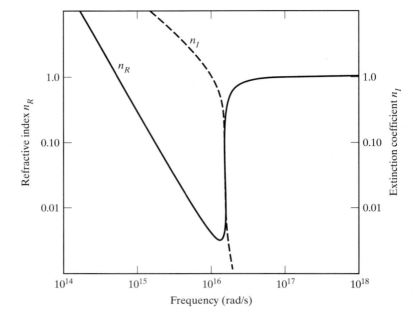

Figure 25-3 Angular frequency dependence of the refractive index n_R and the extinction coefficient n_I for copper. Values assumed are $\omega_p = 1.63 \times 10^{16}\,\text{s}^{-1}$ and $\gamma = 4.1 \times 10^{13}\,\text{s}^{-1}$. The crossover point of the curves coincides with the plasma frequency.

PROBLEMS

25-1 In general, the "electrical constant" K, the *dielectric constant*, is related to the refractive index by

$$K = n^2$$

a. Show that if K_R and K_I are the real and imaginary parts of the dielectric constant, then

$$n_R = \left[\frac{K_R + (K_R^2 + K_I^2)^{1/2}}{2}\right]^{1/2}$$

and

$$n_I = \left[\frac{-K_R + (K_R^2 + K_I^2)^{1/2}}{2}\right]^{1/2}$$

b. Calculate n_R and n_I for a dielectric, in terms of K_I, at frequencies high enough such that $K_I = K_R$.

25-2 Show that in a nearly transparent medium, the absorption coefficient is related to the conductivity and refractive index by

$$\alpha = \frac{(377\ \Omega)\sigma}{n_R}$$

25-3 Calculate and/or plot real and imaginary parts of the refractive index for a dielectric given the frictional parameter γ, the resonant frequency ω_0, and the dipole density N. Check your calculations against Figure 25-2.

25-4 Assume that aluminum has one free electron per atom and a static conductivity given by $3.54 \times 10^7/\Omega$-m. Determine (a) the frictional constant γ, (b) the plasma frequency ω_p, (c) the real and imaginary parts of the refractive index at 550 nm.

25-5 Show that Eq. (25-58) for the skin depth at low frequency is an adequate approximation when $\omega \ll \gamma$ and $\omega \ll \sigma/\varepsilon_0$.

25-6 Calculate the skin depth in copper for radiation of (a) 60 Hz and (b) 3 m. First ensure that the approximations of problem 25-5 are satisfied. (Handbook data for copper: $\sigma = 5.76 \times 10^7/\Omega$-m.)

25-7 Compare the skin depth of (a) aluminum, with conductivity of $3.54 \times 10^7/\Omega$-m and (b) seawater, with conductivity of $4.3/\Omega$-m, for radio waves of 60 kHz.

25-8 Calculate the skin depth of a solid silver waveguide component for 10-cm microwaves. Silver has a conductivity of $3 \times 10^7/\Omega$-m. Explain why a more economical silver-plated brass component will work as well.

25-9 The energy density of red light of wavelength 660 nm is reduced to one-quarter of its original value by passage through 342 cm of seawater.

a. What is the absorption coefficient of seawater for red light of this wavelength?

b. At what depth is red light reduced to 1% of its original energy density?

25-10 Calculate and/or plot the real and imaginary parts of the refractive index for a metal, given the frictional parameter γ and the plasma frequency. Check your results against Figure 25-3.

25-11 Determine the theoretical content of the constants A, B, and C used to express the Cauchy dispersion equation.

25-12 In writing Eq. (25-3) we neglected to include a contribution due to the magnetic force on the electron. Under what condition is the magnitude of the magnetic force exerted by a harmonic electromagnetic wave on an electron much less than the magnitude of the electric force exerted by the same harmonic electromagnetic wave acting on the electron? (*Hint*: see Eq. (4-30)).

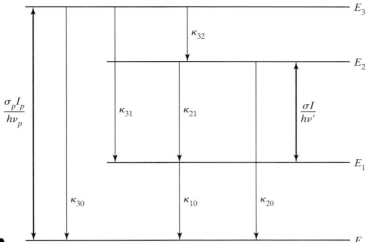

26 *Laser Operation*

INTRODUCTION

In this chapter we extend the largely qualitative discussion of laser systems given in Chapter 6 with a more quantitative treatment of laser operation.[1] We begin by developing the *rate equations* governing the population densities in a medium interacting with an electromagnetic field. These rate equations are developed following an approach taken by Albert Einstein in 1916. The population-density rate equation, together with equations representing the effect of a cavity on an electromagnetic field, are used to develop a relation that predicts the output irradiance from a laser given the characteristics of the pump, gain medium, and cavity that comprise the laser system. We then discuss the gain bandwidth of laser gain media describing both homogeneous and inhomogeneous broadening. The use of Q-switching and mode-locking to produce pulsed output fields is then considered. Finally, we give a qualitative description of the operation and characteristics of diode lasers.

26-1 RATE EQUATIONS

Rate equations relate the population densities (number of atoms or molecules per unit volume in a given energy state) to the properties of an electromagnetic field incident on the atoms or molecules. Recall from the discussion in Chapter 6 that the different energy states of interest in an atom correspond to different configurations of the charge cloud associated with one

---

[1]It may be helpful to review Sections 6-4 through 6-8 before working through the current chapter.

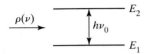

Figure 26-1 Electromagnetic field spectral energy density $\rho(\nu)$ incident on a two-level atom. Level 1 is the ground state and the nominal energy difference between the levels is $E_2 - E_1 = h\nu_0$.

of the outermost electrons in the atom. In molecules, the energy states are also distinguished by the vibrational and rotational state of the molecule. For the moment we will concentrate on but two of the many energy states in an atomic system. We denote the energies of these levels E_1 and E_2 with the nominal difference in energy $E_2 - E_1 = h\nu_0$ and take, again for the moment, the lower energy state to be the *ground state* of the system, shown in Figure 26-1.

We denote the population densities of these states N_1 and N_2. In Chapter 6 we noted that the interaction of an electromagnetic field with a pair of atomic or molecular energy states can be described by three processes: *stimulated emission, stimulated absorption,* and *spontaneous emission.* The rates associated with these processes, given in Eqs. (6-9) through (6-11), are valid for the case of interaction with a nearly monochromatic laser field of frequency ν'. In the more general case of the interaction with an electromagnetic field with spectral energy density $\rho(\nu)$, these relations must be generalized as

$$R_{\text{St.Em.}} = B_{21}N_2 \int_0^{\infty} g(\nu)\rho(\nu)\, d\nu \tag{26-1}$$

$$R_{\text{St.Abs.}} = B_{12}N_1 \int_0^{\infty} g(\nu)\rho(\nu)\, d\nu \tag{26-2}$$

and

$$R_{\text{Sp.Em.}} = A_{21}N_2 \int_0^{\infty} g(\nu)\, d\nu = A_{21}N_2 \tag{26-3}$$

In the last relation we have used the fact that the lineshape function $g(\nu)$ is normalized so that its integral over all frequencies is unity. Note that the rate of spontaneous emission is independent of the spectral energy density of the electromagnetic field. The spectral energy density $\rho(\nu)$ has a dimension of energy per volume per frequency interval. The integration of the spectral energy density over all frequencies gives the time-averaged energy density $\langle u \rangle$ of the electromagnetic field,

$$\int_0^{\infty} \rho(\nu)\, d\nu = \langle u \rangle$$

Evaluation of the integrals in Eqs. (26-1) and (26-2) requires knowledge of the form of the spectral energy density $\rho(\nu)$ and lineshape function $g(\nu)$. These integrals simplify in two general cases that are of particular interest. These cases are described in the following subsections.

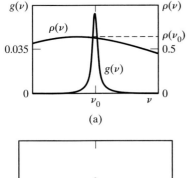

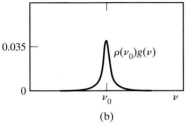

Figure 26-2 Broadband spectral energy density. (a) A lineshape function $g(\nu)$ (left axis) and a broadband spectral energy density $\rho(\nu)$ (right axis) are shown. (b) The product of the two functions plotted in (a) is $\rho(\nu)g(\nu) \approx \rho(\nu_0)g(\nu)$.

Broadband Electromagnetic Energy Density

Consider the case in which the spectral energy density function $\rho(\nu)$ is much broader than the lineshape function $g(\nu)$. The behavior of these functions is illustrated in Figure 26-2. Note that, for such a situation, the product function $\rho(\nu)g(\nu)$ is significant only over the range of frequencies near ν_0 for which $g(\nu)$ is significant. Over this range, the energy density function is nearly constant and can be taken to be $\rho(\nu_0)$. As a consequence the energy density function can be "pulled out" of the integral so that, to a good approximation,

$$\int_0^{\infty} g(\nu)\rho(\nu)\, d\nu = \rho(\nu_0)\int_0^{\infty} g(\nu)\, d\nu = \rho(\nu_0)$$

Thus, for a broadband electromagnetic field, the stimulated rates are

$$R_{\text{St.Em.}} = B_{21}\rho(\nu_0)N_2$$

$$R_{\text{St.Abs.}} = B_{12}\rho(\nu_0)N_1$$

Here, ν_0 is the center frequency of the $2 \to 1$ atomic transition. The rate of spontaneous emission is, of course, still given by Eq. (26-3).

The rate equations describing the interaction of the two-state atomic system with a broadband electromagnetic field then have the form

$$\frac{dN_2}{dt} = -A_{21}N_2 - B_{21}\rho(\nu_0)N_2 + B_{12}\rho(\nu_0)N_1 \qquad (26\text{-}4)$$

and

$$\frac{dN_1}{dt} = +A_{21}N_2 + B_{21}\rho(\nu_0)N_2 - B_{12}\rho(\nu_0)N_1 \qquad (26\text{-}5)$$

Note that

$$\frac{dN_1}{dt} = -\frac{dN_2}{dt}$$

since we are accounting only for processes that *couple* the energy levels 2 and 1. Thus a reduction in the population density of state 2 must be accompanied by an increase in the population density of state 1.

The various parameters in these equations are defined in Section 6-4. The coefficients A_{21}, B_{21}, and B_{12} are characteristic of the two energy states. Their form can be determined using a fully quantum-mechanical treatment[2] of the interaction of the atom with an electromagnetic field, but such a treatment is beyond the scope of this text. However, in 1916 Einstein was able to develop *relationships* between these so-called A and B coefficients, without relying on a fully quantum treatment, by considering the situation in which the atoms in the system come to thermal equilibrium with an electromagnetic field. We summarize his approach in what follows.

Relationships Between *A* and *B* Coefficients

The spectral energy density in an electromagnetic field in thermal equilibrium with its surroundings at temperature T takes the form

$$\rho(\nu) = \frac{8\pi h \nu^3}{c^3} \frac{1}{e^{h\nu/k_B T} - 1} \qquad (26\text{-}6)$$

[See Eq. (6-5) and problem 26-1.] In thermal equilibrium at temperature T the population densities N_1 and N_2 of the atoms satisfy a Boltzmann relation[3] [see Eq. (6-4)],

$$\frac{N_2}{N_1} = e^{-(E_2 - E_1)/k_B T} = e^{-h\nu_0/k_B T} \qquad (26\text{-}7)$$

These relations can be compared to the requirements imposed by the rate equations, Eqs. (26-4) and (26-5). In thermal equilibrium the rates of change

---

[2]See, for example, M. O. Scully and M. S. Zubairy, *Quantum Optics* (Cambridge, UK: Cambridge University Press, 1997).

[3]See footnote 3 of Chapter 6.

of the population densities should be zero. Solving either Eq. (26-4) or (26-5) in steady state (that is, setting the left-hand side of either of these equations to 0) results in the relation

$$\rho(\nu_0) = \frac{A_{21}}{B_{12}(N_1/N_2) - B_{21}}$$

In thermal equilibrium, Eqs. (26-6) and (26-7) can be used in this relation to give

$$\frac{8\pi h\nu_0^3}{c^3} \frac{1}{e^{h\nu_0/k_B T} - 1} = \frac{A_{21}}{B_{12}e^{h\nu_0/k_B T} - B_{21}}$$

Rearranging to isolate multipliers of the term $e^{h\nu_0/k_B T}$,

$$\left(\frac{A_{21}}{B_{21}} - \frac{8\pi h\nu_0^3}{c^3}\frac{B_{12}}{B_{21}}\right)e^{h\nu_0/k_B T} - \left(\frac{A_{21}}{B_{21}} - \frac{8\pi h\nu_0^3}{c^3}\right) = 0$$

In order for this relation to be true at all temperatures T, the term that multiplies $e^{h\nu_0/k_B T}$ and the remaining term in parentheses must each be identically zero. Then, it follows that

$$\frac{A_{21}}{B_{21}} = \frac{8\pi h\nu_0^3}{c^3} \tag{26-8}$$

and

$$B_{12} = B_{21} \tag{26-9}$$

Thus we have derived Eqs. (6-12) and (6-13) by considering a thermal equilibrium situation. However, the A and B coefficients are properties of the atom alone[4] and so Eqs. (26-8) and (26-9) are true whether or not the atom is in thermal equilibrium.

Rate Equations for Monochromatic Light

When the frequency width of the spectral energy density $\rho(\nu)$ is much narrower than the frequency width of the lineshape function $g(\nu)$, as is typically the case for laser light, the stimulated emission and absorption rates appearing in Eqs. (26-1) and (26-2) can be approximated as,

$$R_{\text{St.Em.}} = B_{21}N_2g(\nu')\int_0^\infty \rho(\nu)\,d\nu = B_{21}N_2g(\nu')\langle u\rangle = B_{21}N_2g(\nu')(I/c)$$

$$\tag{26-10}$$

and

$$R_{\text{St.Abs.}} = B_{12}N_1g(\nu')\int_0^\infty \rho(\nu)\,d\nu = B_{12}N_1g(\nu')\langle u\rangle = B_{12}N_1g(\nu')(I/c)$$

$$\tag{26-11}$$

---

[4]Actually, these coefficients depend also on the structure of the enclosure surrounding the atom, but this dependence typically becomes important only for enclosures of dimension not too much larger than the wavelength of light.

Here ν' is the center frequency of the spectral energy density of the electromagnetic field. For nearly monochromatic fields we typically say simply that ν' is the frequency of the field. Finally we have noted [see Eq. (4-37) and problem 26-2] that for nearly monochromatic fields, $\langle u \rangle = I/c$, where I is the irradiance of the electromagnetic field. Using Eqs. (26-10) and (26-11) allows the rate equations governing the population densities in the states 2 and 1 to be written as

$$\frac{dN_2}{dt} = -A_{21}N_2 - B_{21}g(\nu')(I/c)N_2 + B_{12}g(\nu')(I/c)N_1 \quad (26\text{-}12)$$

and

$$\frac{dN_1}{dt} = +A_{21}N_2 + B_{21}g(\nu')(I/c)N_2 - B_{12}g(\nu')(I/c)N_1 \quad (26\text{-}13)$$

Following a particular convention, we choose to define the *stimulated emission cross section* σ as

$$\sigma = B_{21}g(\nu')h\nu'/c$$

For an atomic system in which the degeneracy of the upper and lower states each is 1, σ is also the *stimulated absorption cross section*[5] since, in that case, $B_{12} = B_{21}$. With this definition, the rate equations can be written as

$$\frac{dN_2}{dt} = -A_{21}N_2 - \frac{\sigma I}{h\nu'}(N_2 - N_1) \quad (26\text{-}14)$$

and

$$\frac{dN_1}{dt} = +A_{21}N_2 + \frac{\sigma I}{h\nu'}(N_2 - N_1) \quad (26\text{-}15)$$

Now σ has a dimension of area which accounts for its designation as a cross section. Further, the factor $\sigma I/h\nu'$ has a dimension of inverse time and so is a rate. In fact, it is the stimulated emission rate and can be compared to the spontaneous emission rate A_{21}. As we see in following sections, the ratio of the stimulated emission rate and the spontaneous emission rate governs the behavior of the interaction of light with an atomic system.

26-2 ABSORPTION

Light may either be attenuated or amplified as it propagates through a medium. The population densities of the different energy states in the medium determine the amount of attenuation or amplification that a given electromagnetic field will undergo. In this section we derive steady-state population densities for a two-level atomic system, in which the lower energy level is the ground state of the system. We let laser light of irradiance I and frequency ν' be incident on the atomic medium. This system is described by Eqs. (26-14) and (26-15). As shown in Example 6-3, at room temperature nearly all the atoms in a medium will be in the atomic ground state. When laser light is incident on such a medium a significant population density can accumulate in an excited state only if the

---

[5]More generally, $\sigma_{abs} = (g_2/g_1)\sigma$, where g_2 and g_1 are the degeneracies of the upper and lower levels, respectively.

transition to the excited state is *resonant* with the electromagnetic field so that $E_2 - E_1 \approx h\nu'$. Let us assume that only such an excited state 2 accumulates a significant population density as a result of the interaction with the electromagnetic field. In this case we can set

$$N_1 + N_2 = N_T \tag{26-16}$$

where N_T is the total population density of the atoms in the medium. Equation (26-14), considered in steady state ($dN_2/dt = 0$), and Eq. (26-16) can be solved jointly to give the steady-state values of the population densities N_1 and N_2. The result (see problem 26-3) is

$$N_2 = N_T\left(\frac{\sigma I/(h\nu' A_{21})}{1 + 2\sigma I/(h\nu' A_{21})}\right) \tag{26-17}$$

$$N_1 = N_T\left(\frac{1 + \sigma I/(h\nu' A_{21})}{1 + 2\sigma I/(h\nu' A_{21})}\right) \tag{26-18}$$

We shall see that the *population inversion* $N_{\mathrm{inv}} \equiv N_2 - N_1$ is of physical interest. For the case at hand, the population inversion is always negative since

$$N_{\mathrm{inv}} = N_2 - N_1 = -N_T\left(\frac{1}{1 + 2\sigma I/(h\nu' A_{21})}\right) \tag{26-19}$$

Note that the key parameter governing the distribution of population between the two levels is the ratio of the stimulated emission rate $\sigma I/h\nu'$ and the spontaneous emission rate A_{21}. Plots of the population densities of the two levels and the population inversion as functions of light irradiance are given in Figure 26-3. Note that the inversion is always negative and decreases in magnitude as the irradiance of the light incident on the medium increases. In the limit of very large irradiances, the inversion tends to zero. This behavior can be understood by noting that two processes, stimulated emission and spontaneous emission, drive an atom from level 2 to level 1. Only one process, stimulated absorption, drives the atom from level 1 to level 2. The rates of stimulated emission and absorption are equal, and so the "downward" (2 to 1) rate always exceeds the "upward" (1 to 2) rate; thus a steady-state population inversion cannot occur in a two-level atomic system. In the limit of very large irradiances, the spontaneous emission process becomes negligible, leading to a near equalization of population densities.

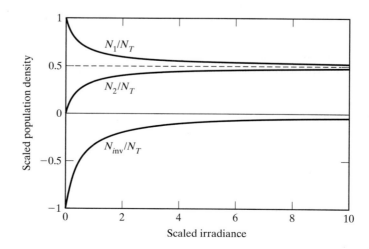

Figure 26-3 Scaled population densities of the excited (2) and ground (1) states of a two-level atomic system as a function of the scaled irradiance $I/(A_{21}h\nu'/\sigma)$. The scaled population inversion density $N_{\mathrm{inv}}/N_T = (N_2 - N_1)/N_T$ is also shown.

Absorption Coefficient and Beer's Law

Thus far we have concentrated on the change in population density induced by an incident electromagnetic field. In doing so we have implicitly assumed that the irradiance of the field is constant temporally and spatially. Consider the situation depicted in Figure 26-4. Light of irradiance I_0 is incident on an atomic medium from the left as shown. As the light propagates in the z-direction across the medium, there can be an exchange of energy between the light and the atoms in the medium. The *absorption or loss coefficient* α characterizes the spatial rate of change of the irradiance, that is,

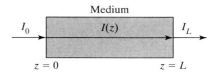

Figure 26-4 Light propagating through a medium of length L. The irradiance I changes due to interaction with the atoms in the medium.

$$\frac{dI}{dz} = -\alpha I \qquad (26\text{-}20)$$

Let us now form the left-hand side of Eq. (26-20). Let $\Delta n / \Delta t$ be the incremental rate of change of the number n of photons (traveling in the $+z$-direction) through a small volume ΔV of cross-sectional area ΔA and length Δz due to interaction with a gain medium. The incremental change in irradiance ΔI across this small volume then can be written as

$$\Delta I = \frac{h\nu'}{\Delta A} \frac{\Delta n}{\Delta t}$$

Dividing each side of this relation by Δz gives

$$\frac{\Delta I}{\Delta z} = \frac{h\nu'}{\Delta V} \frac{\Delta n}{\Delta t} \qquad (26\text{-}21)$$

Now, the net rate of photon production or loss in the small volume ΔV is the result of spontaneous emission, stimulated emission, and stimulated absorption. For sizable incident irradiances, the rate of spontaneous emission into the direction of the beam is much less than the stimulated emission and stimulated absorption rates, and so spontaneous emission makes a negligible contribution to the net rate of photon production. Then, since each stimulated emission event creates one photon and each stimulated absorption event removes one photon from the field, the rate of change of the number of photons in the volume ΔV (see the last term in Eq. 26-14) is

$$\frac{\Delta n}{\Delta t} = \frac{\sigma I}{h\nu'}(N_2 - N_1)\,\Delta V$$

Using this in Eq. (26-21) leads to

$$\frac{\Delta I}{\Delta z} = \sigma I(N_2 - N_1)$$

Letting the increments pass into differentials gives the seminal relation

$$\frac{dI}{dz} = \sigma(N_2 - N_1)I \qquad (26\text{-}22)$$

Thus we have discovered the form of the loss coefficient α appearing in Eq. (26-20):

$$\alpha = -\sigma(N_2 - N_1) \qquad (26\text{-}23)$$

Note that the loss coefficient has a dimension of inverse length.

As is evident from Eq. (26-18) and Figure 26-3, for sufficiently small input irradiance (or sufficiently small cross section σ) the population density remains concentrated in the ground state, $N_1 \approx N_T$. In this case,

$$\alpha \approx \sigma N_T \equiv \alpha_0 \qquad (26\text{-}24)$$

is a constant, independent of the irradiance of the input field. Here we have introduced α_0 as the *small-signal* ($I \rightarrow 0$) loss coefficient. For this case, Eq. (26-22) takes the form

$$\frac{dI}{dz} = -\alpha_0 I$$

and can be directly integrated as

$$\frac{dI}{I} = -\alpha_0\, dz$$

$$\int_{I_0}^{I_L} \frac{dI}{I} = -\alpha_0 \int_0^L dz$$

$$\ln(I_L/I_0) = -\alpha_0 L$$

$$I_L = I_0 e^{-\alpha_0 L} \qquad (26\text{-}25)$$

This result is known as Beer's law. The simple exponential-decay nature of this expression follows only for the case of a weakly absorbing system for which nearly all of the population remains in the ground state. The treatment of a system in which a significant population is transferred to an excited state is mathematically more complex. In such a situation, which is the subject of problem 26-4, the loss coefficient decreases (*saturates*) with increased incident irradiance.

Example 26-1

The cross section σ, for a transition from the ground state to an excited state that is resonant with an electromagnetic field of wavelength 808 nm, for a neodymium (Nd) atom doped into a YAG (yttrium aluminum garnet) crystal[6] is about 3×10^{-20} cm^2. Assume that the dopant density (number of atoms per cm^3) of Nd in the YAG crystal is 10^{20} atoms/cm^3 and that the YAG crystal itself is transparent to 808-nm light. Assume that a diode laser with an emission at a wavelength of 808 nm is to be used to pump an Nd:YAG laser rod. (Pump energy absorbed by the crystal, as we shall see in the next section, can be converted to Nd:YAG laser output.)

   a. Estimate the small-signal absorption coefficient for the Nd:YAG crystal.
   b. Estimate the depth to which the diode laser beam would penetrate significantly into the Nd:YAG crystal.

Solution

   a. The small-signal absorption coefficient is

$$\alpha_0 = \sigma N_T = (3 \times 10^{-20})(10^{20})\ \text{cm}^{-1} = 3\ \text{cm}^{-1}$$

---

[6]Modeling absorption in the the Nd:YAG crystal by a two-level system with no degeneracies is a rather drastic oversimplification of the real situation. Example 26-1 should be regarded as an "order-of-magnitude, back of the envelope" estimation.

b. Let us take the depth of "significant penetration" to be the depth L at which the irradiance of the diode laser has decreased to $1/e \approx 0.368$ of its initial value. Then assuming that it is appropriate to use the small-signal result given in Eq. (26-25),

$$I_L = I_0 e^{-1} = I_0 e^{-\alpha_0 L}$$

$$L = 1/\alpha_0 = 0.33 \text{ cm}$$

Thus Nd atoms deeper than about 0.33 cm would not absorb much of the pump energy.

26-3 GAIN MEDIA

We have seen that a simple two-level system with the lower energy state being the ground state always acts as an absorber. Laser systems require a *gain medium* that provides energy to a field that propagates through it. As discussed in Chapter 6 and alluded to in Example 26-1, gain media must be pumped. That is, the pump energy stored in the gain medium is converted to irradiance by interaction of the electromagnetic field with the gain medium. A common atomic-level structure that serves as a gain medium is illustrated in Figure 26-5. In this figure *four* important levels are shown,[7] with the level of energy E_0 being the ground state. The light field to be amplified is taken to have irradiance I and frequency ν' and should be nearly resonant with the 2 to 1 transition. Pump energy is provided to the system via the 0 to 3 transition. For concreteness, let us assume that the pump is a laser of frequency $\nu_p = (E_3 - E_0)/h$ and irradiance I_p. The rate equations governing population densities of the four levels shown in Figure 26-5, due to the indicated processes, take the form

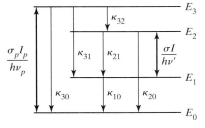

Figure 26-5 Level structure of a four-level gain medium. The thicker, double-headed arrows indicate stimulated processes, and the lighter, single-headed arrows indicate decay processes. Gain is provided to a field of irradiance I and frequency $\nu' \approx (E_2 - E_1)/h$ as a result of an optical pump of irradiance I_p and frequency $\nu_p \approx (E_3 - E_0)/h$.

$$\frac{dN_3}{dt} = -\kappa_3 N_3 - \frac{\sigma_p I_p}{h\nu_p}(N_3 - N_0) \tag{26-26}$$

$$\frac{dN_2}{dt} = \kappa_{32} N_3 - \kappa_2 N_2 - \frac{\sigma I}{h\nu'}(N_2 - N_1) \tag{26-27}$$

$$\frac{dN_1}{dt} = \kappa_{31} N_3 + \kappa_{21} N_2 - \kappa_{10} N_1 + \frac{\sigma I}{h\nu'}(N_2 - N_1) \tag{26-28}$$

$$\frac{dN_0}{dt} = \kappa_{30} N_3 + \kappa_{20} N_2 + \kappa_{10} N_0 + \frac{\sigma_p I_p}{h\nu_p}(N_3 - N_0) \tag{26-29}$$

Here we have replaced the spontaneous emission A coefficients with decay rates κ in order to account for any *decay* process that causes a downward transition in the atom, including spontaneous emission and, for example, inelastic collisions with other atoms, molecules, or particles. These latter processes are examples of what is called *nonradiative decay*. Also, σ_p is the stimulated emission cross section for the 3-to-0 transition and we have again ignored, for simplicity, level degeneracies. Finally, we will assume that the four-level system is *closed*, in the sense that all atoms in the sample exist in one or the other of these four levels so that

$$N_T = N_0 + N_1 + N_3 + N_4 \tag{26-30}$$

---

[7]See problem 26-7 for another common arrangement involving three levels.

If this is to be true, the right-hand sides of the Eqs. (26-26) though (26-29) must sum to zero. This is so provided that $\kappa_3 = \kappa_{32} + \kappa_{31} + \kappa_{30}$ and $\kappa_2 = \kappa_{21} + \kappa_{20}$. The *lifetime* τ of an energy level is defined to be the inverse of the total decay rate from the level so that, in the present case,

$$\tau_3 = \frac{1}{\kappa_3}, \quad \tau_2 = \frac{1}{\kappa_2}, \quad \text{and} \quad \tau_1 = \frac{1}{\kappa_{10}}$$

The lifetime of a level (see problem 6-7) is the time for the population density of a given level to decay to $1/e$ of its initial value, when the decay process is the only process that occurs.

The analysis leading to Eq. (26-22) holds as well for the four-level gain medium being presently considered. Thus gain will occur provided that a *population inversion* $(N_2 > N_1)$ exists in steady state. Under this condition, stimulated emission will exceed stimulated absorption and a net production of photons will occur. Any three of the four-level rate equations, together with Eq. (26-30), can be solved simultaneously to give the steady-state population densities of the four levels and so to find the steady-state population inversion $N_{\text{inv}} = N_2 - N_1$. This full analysis is left as an exercise (problem 26-5). It is instructive to make a series of simplifying assumptions that sometimes apply in real atomic gain media and which dramatically reduce the complexity of the system of equations. We do so in the following subsections.

Undepleted Pump Approximation   It is often true that the laser pump does not significantly empty the ground state of the four-level system so that $N_0 \gg N_3$ and $N_0 \approx N_T$. In this case, Eq. (26-26) can be solved in steady state to give

$$N_3 \approx \frac{1}{\kappa_3} \frac{\sigma_p I_p}{h\nu_p} N_T \tag{26-31}$$

It is important to note that, in the undepleted pump approximation, N_3 is approximately constant and although $N_3 \ll N_0$, the rate $\kappa_3 N_3$ is not negligible. Using Eq. (26-31) in Eqs. (26-27) and (26-28) leads to, in steady state,

$$\frac{dN_2}{dt} = 0 = R_{p2} - \kappa_2 N_2 - \frac{\sigma I}{h\nu'}(N_2 - N_1) \tag{26-32}$$

$$\frac{dN_1}{dt} = 0 = R_{p1} + \kappa_{21} N_2 - \kappa_{10} N_1 + \frac{\sigma I}{h\nu'}(N_2 - N_1) \tag{26-33}$$

where we have defined the effective pump rate densities

$$R_{p2} = \frac{\kappa_{32}}{\kappa_3}\left(\frac{\sigma_p I_p}{h\nu_p} N_T\right)$$

$$R_{p1} = \frac{\kappa_{31}}{\kappa_3}\left(\frac{\sigma_p I_p}{h\nu_p} N_T\right)$$

Equations (26-32) and (26-33) can be solved for the population inversion, giving

$$N_{\text{inv}} = N_2 - N_1 = \frac{(1 - \kappa_{21}/\kappa_{10})R_{p2} - (\kappa_2/\kappa_{10})R_{p1}}{\kappa_2 + (1 + \kappa_{20}/\kappa_{10})(\sigma I/h\nu')} \tag{26-34}$$

When describing gain media, Eq. (26-22) is often recast by defining a *gain coefficient* γ as

$$\gamma = \sigma(N_2 - N_1) = \sigma N_{\text{inv}} \qquad (26\text{-}35)$$

so that

$$\frac{dI}{dz} = \gamma I \qquad (26\text{-}36)$$

It is important to note that the gain coefficient depends on the irradiance I, since the population inversion is, in general, dependent on the irradiance.

Evidently, maximizing the population inversion maximizes the gain coefficient. Examination of Eqs. (26-34) and (26-35) indicates that population inversion and so the gain coefficient are made larger under the following conditions:

1. The gain coefficient becomes larger as κ_{10} is increased relative to the decay rates associated with level 2. This is sensible since stimulated absorption attenuates the irradiance, and a large decay rate from level 1 indicates that the atoms do no not dwell a long time in level 1 and so the chances of stimulated absorption events occurring are reduced.
2. The gain coefficient becomes larger by reducing R_{p1}. This is again sensible since the effective pump rate R_{p1} feeds the population of level 1 and so increases the likelihood of stimulated absorption. The effective pump rate density R_{p1} is less in an atom for which κ_{31} is small.
3. The gain coefficient is increased by reducing the decay rate from level 2. By doing so the dwell time of the atoms in the upper lasing level 2 increases and so the likelihood of the occurrence of stimulated emission, which leads to gain, is increased. Energy levels with such long lifetimes are often referred to as *metastable* states.
4. The gain coefficient is increased by increasing the effective pump rate density R_{p2} that feeds the upper lasing level and so increases the likelihood of occurrence of stimulated emission events. This rate can be increased by increasing the irradiance I_p of the pump laser.

Ideal Four-Level Gain Medium
In the ideal case, $R_{p1} = 0$ and $\kappa_{10} \to \infty$. Then, the gain coefficient takes the simple form

$$\gamma = \frac{\sigma R_{p2}/\kappa_2}{1 + (\sigma I/h\nu')/\kappa_2} \equiv \frac{\gamma_0}{1 + I/I_S} \qquad (26\text{-}37)$$

Here we have introduced two important parameters that describe gain media, the *small-signal gain coefficient* γ_0 and the *saturation irradiance* I_S. For the ideal four-level gain medium, these parameters take the form

$$\gamma_0 = \sigma R_{p2}/\kappa_2 = \sigma R_{p2}\tau_2 \qquad (26\text{-}38)$$

and

$$I_S = h\nu'\kappa_2/\sigma = h\nu'/\sigma\tau_2 \qquad (26\text{-}39)$$

Note that the small-signal gain coefficient is the approximate gain coefficient when the irradiance I in the medium is much less than the saturation irradiance I_S. As illustrated in Figure 26-6, when $I = I_S$ the gain coefficient is reduced by a factor of 2 from its small-signal value. Further, when $I \gg I_S$ the gain coefficient becomes inversely proportional to the irradiance, and the spatial rate of

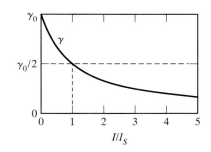

Figure 26-6 Gain coefficient γ as a function of scaled irradiance I/I_S. Note that $\gamma = \gamma_0/2$ when $I = I_S$.

change of the irradiance γI given in Eq. (26-36) becomes constant. Although the middle member of Eq. (26-37) is valid only for an ideal four-level gain medium, the general form of the gain coefficient given as the rightmost member of that equation holds for a wide range of *homogeneously* broadened gain media (see Section 26-5), but with definitions of γ_0 and I_S different from those given in Eqs. (26-38) and (26-39).

That the gain coefficient is less for fields of larger irradiance is said to be due to *gain saturation*. Gain saturation occurs because large irradiances cause more stimulated emission, which depletes the steady-state population density of the upper lasing level, thus reducing the likelihood of a stimulated emission event.

Integrated Gain

Using Eq. (26-37) in Eq. (26-36) results in a differential equation that can be integrated to provide a relation between the irradiance I_0 input into a gain medium and the irradiance I_L output from that gain medium. Proceeding,

$$\frac{dI}{dz} = \gamma I = \frac{\gamma_0}{1 + I/I_S} I \tag{26-40}$$

Separating variables and integrating both sides of the resulting relation gives

$$\int_{I_0}^{I_L} \left(\frac{1}{I} + \frac{1}{I_S} \right) dI = \gamma_0 \int_0^L dz$$

Integration gives

$$\ln\left(\frac{I_L}{I_0}\right) + \frac{1}{I_S}(I_L - I_0) = \gamma_0 L \tag{26-41}$$

The important result given in Eq. (26-41) is a transcendental relation that can be numerically solved for the output irradiance I_L given the input irradiance and the characteristics of the gain medium, γ_0, L and I_S. Important and interesting features of this relationship are illustrated in a series of problems at the end of this chapter.

Figure 26-7 indicates the effects that varying the small-signal gain coefficient γ_0 and the saturation irradiance I_S have on the output irradiance I_L from a gain cell. Note that for all three curves, the irradiance input to the cell is $I_0 = 1 \ \text{W/cm}^2$. The lower curves have the same small-signal gain coefficient

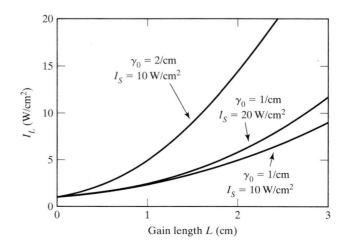

Figure 26-7 Irradiance output I_L from a gain cell as a function of the length L of the gain cell. The plots correspond to the different small-signal gain coefficients γ_0 and saturation irradiances I_S indicated. In each case, the irradiance input into the gain cell is 1 W/cm².

but different saturation irradiances. These two lower curves show that, for gain cell lengths L short enough that the irradiance in the cell remains well below the saturation value, the output irradiance from the cell is nearly independent of the saturation irradiance. However, for gain cells that are long enough to allow the irradiance to grow to an appreciable fraction of the saturation irradiance, the output irradiance is less for cells with smaller saturation irradiances. The gain cell represented by the upper curve has twice the small-signal gain coefficient of the cells represented by the lower two curves, and as a consequence the output irradiance shown in the upper curve is significantly larger, even for short gain lengths, than those shown in the lower curves.

In closing this section, we point out that both the saturation irradiance I_S and the small-signal gain coefficient γ_0 are functions of the electromagnetic field frequency ν', since they each depend on the cross section σ, which in turn is proportional to the lineshape function $g(\nu')$. In fact,

$$\gamma_0 \propto g(\nu')$$

and

$$I_S \propto 1/g(\nu')$$

The import of these relations for the operation of a gain cell is explored in problem 26-13.

26-4 STEADY-STATE LASER OUTPUT

Equation (26-41) allows for the prediction of the steady-state or *continuous wave* (CW) output irradiance of a laser given the natures of the pump, gain medium, and cavity that constitute the laser. We will develop first an expression for the CW output irradiance for a *ring laser* and then write down the corresponding result for the more common case of a laser system that uses a two-mirror *linear cavity*.

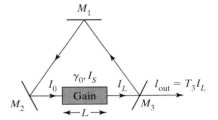

Figure 26-8 Ring laser. The field is constrained to oscillate in the counterclockwise direction by means not shown. The gain cell has length L and is characterized by small signal gain coefficient γ_0 and saturation irradiance I_S. The output mirror has transmittance T_3.

Ring Laser

We choose to illustrate the development of an expression for the steady-state irradiance output from a laser cavity by considering a ring laser consisting of three mirrors as shown in Figure 26-8. Take the reflectances of the three mirrors M_1, M_2, and M_3 to be R_1, R_2, and R_3, respectively. Further, let mirror M_3 be the output mirror with a transmittance T_3. The gain medium is characterized by its saturation irradiance I_S, length L, and small-signal gain coefficient γ_0. We assume that lasing action occurs only in the counterclockwise direction around the ring.

In steady state, the losses due to imperfect reflection and transmission at the mirrors must be, in each round-trip, offset by the increase in irradiance that results from interaction with the gain medium. As indicated in Figure 26-8, we will take the steady-state irradiance at the input end of the gain cell to be I_0 and that at the output end to be I_L. Tracking the irradiance from the output end around the ring to the input end leads to the relation

$$I_0 = SI_L \qquad (26\text{-}42)$$

Here, S is the fraction of the irradiance that *survives* the trip around the cavity from the output to the input of the gain cell. For the simple cavity considered here, $S = R_1R_2R_3$. More generally, the *survival factor* S should include factors describing each process that reduces the irradiance as the beam traverses the ring. Equation (26-42) can be used in Eq. (26-41) to give the desired expression for the steady-state irradiance at the output end of the gain cell. After a bit of rearrangement, we find

$$I_L = I_S\left(\frac{\gamma_0 L - \ln(1/S)}{1 - S}\right)$$

The irradiance of the laser beam exiting the ring cavity is, then,

$$I_{\text{out}} = T_3 I_L = T_3 I_S\left(\frac{\gamma_0 L - \ln(1/S)}{1 - S}\right) \qquad (26\text{-}43)$$

Let us examine the features of this expression. Larger irradiances occur for gain media with larger saturation irradiances. However, the saturation irradiance is a property of the gain media and cannot be manipulated. With a given gain medium one can increase the output irradiance by increasing the pump density R_{p2}, since this increases the small signal gain coefficient γ_0. A bit of analysis (see problem 26-17) indicates that decreasing the cavity losses not associated with the laser output also increases the laser output irradiance.

Note that Eq. (26-43) is only meaningful when the small-signal gain coefficient exceeds a certain *threshold* value,

$$\gamma_{\text{th}} = \frac{1}{L}\ln(1/S) \qquad \text{(Ring cavity)} \qquad (26\text{-}44)$$

since it predicts a negative irradiance when $\gamma_0 < \gamma_{\text{th}}$. Of course, there are no negative irradiances. Rather when this condition holds, the cavity loss per round-trip exceeds the gain per round-trip and no irradiance builds up in the cavity. Note that γ_{th} characterizes the cavity *losses*. Using Eq. (26-44) in Eq. (26-43) yields the indicative relation

$$I_{\text{out}} = T_3 I_S\left(\frac{(\gamma_0 - \gamma_{\text{th}})L}{1 - S}\right) \qquad (26\text{-}45)$$

Let us obtain the expression for the threshold small-signal gain coefficient γ_{th} in another fashion. As discussed in some detail in Chapter 6, the field in a laser cavity builds from spontaneous emission events occurring into a laser cavity mode. These spontaneous emission events are then amplified by the action of stimulated emission within the gain medium. However, the gain provided must exceed the losses encountered per round-trip due to imperfect mirrors, transmission from the output mirror, and other cavity loss mechanisms. The irradiance due to the spontaneous emission is typically much less than the saturation irradiance of the gain medium. Thus, during the initiation of the laser field the gain coefficient γ has its small-signal value γ_0. In this case, the irradiance I_0 input into the gain cell and the irradiance I_L output from the gain cell are simply related (see problem 26-8) by

$$I_L/I_0 = e^{\gamma_0 L}$$

In order for the irradiance to grow during each round-trip, this single-pass small-signal gain factor $e^{\gamma_0 L}$ must more than offset the fractional loss in a round-trip. That is,

$$Se^{\gamma_0 L} > 1 \qquad (26\text{-}46)$$

This condition implies that for the laser field to grow,

$$\gamma_0 > \frac{\ln(1/S)}{L} = \gamma_{\text{th}}$$

If $\gamma_0 > \gamma_{th}$, the field grows with each round-trip through the cavity, but as it does so the gain coefficient γ decreases due to saturation. When the gain coefficient γ is reduced so that the gain per pass through the gain cell just offsets the cavity loss per round-trip, steady state is reached. In summary, the small-signal gain coefficient γ_0 must exceed the threshold gain coefficient γ_{th} for the laser field to grow in the cavity. This field will then continue to grow until the gain coefficient γ saturates to the threshold value γ_{th}.

Example 26-2

A ring laser cavity system like the one shown in Figure 26-8 uses mirrors with reflectances $R_1 = R_2 = 0.99$ and $R_3 = 0.95$. The transmittance of the output mirror is $T_3 = 0.04$. The gain medium is a 10-cm-long Nd:YAG crystal with a saturation irradiance of 2300 W/cm^2. The crystal is optically pumped at a rate that leads to a small-signal gain coefficient of $\gamma_0 = 0.05$/cm.

   a. Find the threshold gain coefficient for this cavity.
   b. Find the irradiance of the laser field that exits this cavity.
   c. Assuming that the laser beam has cross sectional area $A = 0.1$ cm^2, find the output power of the laser.
   d. Assuming that the overall efficiency of this laser system is 3%, find the pump power required to operate this laser system.

Solution

   a. Using Eq. (26-44),

$$\gamma_{th} = \frac{1}{L}\ln(1/S) = \frac{1}{L}\ln\left(\frac{1}{R_1 R_2 R_3}\right) = \frac{1}{10\ \text{cm}}\ln\left(\frac{1}{0.99^2 \cdot 0.95}\right) = 0.0071/\text{cm}$$

   b. According to Eq. (26-45),

$$I_{out} = T_3 I_S\left(\frac{(\gamma_0 - \gamma_{th})L}{1 - S}\right) = (0.04)(2300\ \text{W/cm}^2)\frac{(0.05 - 0.0071)10}{1 - 0.99^2(0.95)}$$

$$I_{out} = 570\ \text{W/cm}^2$$

   c. The output power is, then, approximately

$$P_{out} \approx I_{out}A = (570\ \text{W/cm}^2)(0.1\ \text{cm}^2) = 57\ \text{W}$$

   d. The required pump power would be

$$P_{pump} = \frac{P_{out}}{\text{efficiency}} = \frac{57\ \text{W}}{0.03} = 1900\ \text{W}$$

Two-Mirror Linear Cavity

The analysis leading to the output irradiance for the more common two-mirror linear laser is somewhat more complicated than that given for the ring laser, but nevertheless can be carried out in a similar fashion.[8] The result for a two-mirror cavity like the one shown in Figure 26-9 (see problem 26-19) is

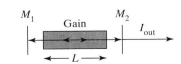

Figure 26-9 Linear laser cavity.

$$I_{out} = \frac{T_2 I_S}{2}\frac{\gamma_0(2L) - \ln\left(\frac{1}{R_1 R_2}\right)}{\left(1 - \sqrt{R_1 R_2}\right)\left(1 + \sqrt{R_2/R_1}\right)} \quad (26\text{-}47)$$

---

[8]See, for example, Joseph T. Verdeyen, *Laser Electronics*, 3d ed. (Prentice-Hall: Englewood Cliffs, New Jersey, 1995), Ch. 9.

Here, T_2 is the transmittance of the laser cavity output mirror, R_1 and R_2 are the reflectances of the cavity mirrors, and the gain cell is characterized by length L, small-signal gain coefficient γ_0, and saturation irradiance I_S. Equation (26-47) implies that for a two-mirror linear cavity, the threshold gain coefficient is

$$\gamma_{\text{th}} = \frac{1}{2L} \ln\left(\frac{1}{R_1 R_2}\right)$$

More generally, an analysis like that leading to Eq. (26-46) leads to

$$\gamma_{\text{th}} = \frac{1}{2L} \ln\left(\frac{1}{S}\right) \qquad \text{(Linear cavity)} \qquad (26\text{-}48)$$

Example 26-3

Estimate the minimum length of a linear cavity with mirror reflectances $R_1 = 0.99$ and $R_2 = 0.98$ that can be used in a He-Ne laser system given a small-signal gain coefficient of 0.001/cm.

Solution

The small-signal gain coefficient must exceed the threshold gain coefficient for lasing action to occur. Thus, using Eq. (26-48),

$$\gamma_{\text{th}} = \frac{1}{2L} \ln(1/S) < \gamma_0$$

so

$$L > \frac{1}{2\gamma_0} \ln(1/S) = \frac{1}{2(0.001/\text{cm})} \ln\left(\frac{1}{(0.99)(0.98)}\right) = 15 \text{ cm}$$

Since the gain cell must fit into the laser cavity, the laser cavity must be longer than 15 cm. One rarely finds a He-Ne laser shorter than 15 cm.

26-5 HOMOGENEOUS BROADENING

In Chapter 6 we introduced the notion of the lineshape factor $g(\nu')$ that governs the strength of the interaction of an atomic system with an incident field of frequency ν'. In this section we discuss the underlying broadening mechanisms that determine the width $\Delta\nu$ of the lineshape function and distinguish between *homogeneous* and *inhomogeneous* broadening mechanisms. Briefly, homogeneous broadening mechanisms are those physical influences that broaden the linewidth of the frequency response of each atom in the medium in the same manner, whereas inhomogeneous broadening mechanisms affect different groups of atoms in different ways—typically making the central frequency of the lineshape function different for different atoms.

Two important homogeneous broadening mechanisms are called *lifetime broadening* and *pressure broadening*. These effects can be shown to lead to a *Lorentzian* lineshape function,

$$g(\nu) = \frac{\Delta\nu_H}{2\pi[(\nu - \nu_0)^2 + (\Delta\nu_H/2)^2]} \qquad (26\text{-}49)$$

This function peaks at $\nu = \nu_0$ and has a full-width at half-maximum of $\Delta\nu_H$. The full-width at half-maximum, sometimes called the *linewidth* of the transition or the *gain bandwidth*, can be shown to be well approximated by

$$\Delta\nu_H = \frac{1}{2\pi}\left(\frac{1}{\tau_2} + \frac{1}{\tau_1} + 2r_{col}\right) \qquad (26\text{-}50)$$

Here τ_2 and τ_1 are the lifetime of the upper and lower levels, respectively, participating in the transition. The terms involving these lifetimes arise from the process called *lifetime broadening*. The remaining term on the right-hand side of Eq. (26-50) is due to *pressure broadening*, which contributes a term proportional to the rate of collisions, r_{col}, of the gain atoms with each other and with other species in the gain medium. This term is important primarily in gas media. We now describe briefly each of these homogeneous-broadening mechanisms.

Lifetime Broadening
Classically, an atom driven by a nearly resonant electromagnetic field oscillates at the frequency of the driving field. This oscillation appears in a quantum-mechanical treatment of the atom as an alternating transition between the upper and lower laser states. Spontaneous decay from either of these levels removes that atom from the coherent interaction with the field. The net effect, in an assembly of oscillating atoms, is to introduce an additional time dependence related to the random deletion of the contribution of one of the oscillators from the net charge oscillation in the medium. As a result, the charge oscillation in the gain medium cannot be represented by a function oscillating with a single frequency but rather has a broadened Fourier spectrum. The lifetime broadening contribution to the lineshape function $g(\nu)$ is related to this broadened spectrum. The finite lifetime of the atomic levels is due both to the fundamental process of spontaneous emission and to transitions induced by *inelastic* collisions with other atoms. In these inelastic interactions, the collisions induce a change in internal state of one or both of the atoms participating in the collisions.

Pressure Broadening
Pressure broadening is a result of *elastic collisions* between atoms in the gain medium. Although these elastic collisions induce no change in the internal state of the colliding atoms, they act to interrupt the regular oscillation of the charge in the atom. These interruptions introduce additional Fourier components into the spectrum of the oscillations of the gain medium and hence contribute to the broadening of the atomic response. The pressure-broadening contribution to the gain bandwidth is, of course, pressure and temperature dependent. For many gas gain media, pressure broadening makes the dominant contribution to the homogeneous bandwidth.

Example 26-4

Consider a homogeneously broadened transition in a carbon dioxide (CO_2) laser.

a. Find the lifetime-broadening contribution to the homogeneous linewidth for this transition if the lifetime of the upper lasing level is 10 μs and that of the lower lasing level is 0.1 μs.

b. The linewidth of this transition is measured to be 1 GHz. Estimate the pressure-broadening contribution to the linewidth of this transition.

c. Gain bandwidths are often expressed as wavelength spreads $\Delta\lambda$ rather than frequency spreads $\Delta\nu$. Express the linewidth of this transition as a wavelength spread given that the light resonant with this transition has a wavelength of 10.6 μm.

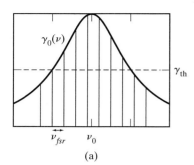

(a)

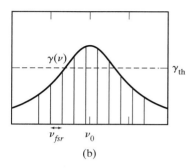

(b)

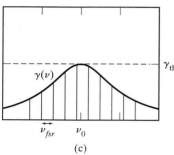

(c)

Figure 26-10   Gain saturation in a homogeneously broadened laser. Cavity mode frequencies are separated by $\nu_{fsr} = 0.15$ GHz. For all three plots the gain bandwidth is 1 GHz. The small-signal gain coefficient at linecenter $\gamma_0(\nu_0)$ is twice the threshold gain coefficient γ_{th}. (a) At laser turn on the gain coefficient has its small-signal value $\gamma_0(\nu)$ and six cavity modes are above threshold. (b) As the irradiance in the cavity modes above threshold grows, the gain coefficient is reduced due to gain saturation. At the instant shown, four cavity modes are above threshold. (c) Gain saturation has reduced the gain coefficient so that only the mode nearest linecenter survives.

Solution

a. According to Eq. (26-50), the lifetime broadening contribution $\Delta\nu_{H,\tau}$ to the linewidth is

$$\Delta\nu_{H,\tau} = \frac{1}{2\pi}\left(\frac{1}{\tau_1} + \frac{1}{\tau_2}\right) = \frac{1}{2\pi}\left(\frac{1}{10\times10^{-6}} + \frac{1}{0.1\times10^{-6}}\right)\text{Hz}$$

$$= 1.61 \times 10^6 \text{ Hz}$$

b. The contribution due to pressure broadening $\Delta\nu_{H,p}$ is evidently

$$\Delta\nu_{H,p} = \Delta\nu_H - \Delta\nu_{H,\tau} = 1\text{ GHz} - 0.00161\text{ GHz} \approx 1\text{ GHz}$$

Under the operating conditions leading to a linewidth of 1 GHz, this transition is predominantly pressure broadened.

c. The wavelength spread associated with the linewidth is found by noting that

$$d\nu = d(c/\lambda) = -(c/\lambda^2)\,d\lambda$$

For small spreads, then, the magnitude of the wavelength spread is

$$\Delta\lambda = \lambda^2\,\Delta\nu/c = (10.6\times10^{-6}\text{ m})^2(10^9\text{ Hz})/(3\times10^8\text{ m/s})$$

$$= 3.75\times10^{-10}\text{ m} = 0.375\text{ nm}$$

Gain Saturation in Homogeneously Broadened Media

It is important to note that the expression for the gain coefficient given in Eq. (26-37) is valid only for homogeneously broadened media since the derivation of this relation implicitly relies on the assumption that all atoms participating in the interaction with the light are of the same type. Further, recall from Chapter 6 [see Eq. (6-15), for example] that the low-loss "cavity modes" of a linear laser cavity are separated by $c/2d$, where d is the cavity length. (For a ring cavity, the mode separation is c/P, where P is the cavity perimeter.) In Chapter 8 we noted that this mode separation is sometimes called the free spectral range, ν_{fsr}, of the cavity. In general, several different cavity modes might have frequencies within the bandwidth of the gain medium and so see significant gain. Let us be a bit more specific. Consider a 1-meter-long linear cavity with a mode spacing $\nu_{fsr} = 1.5 \times 10^8$ Hz. Let this cavity contain a gain medium of bandwidth 1 GHz pumped so that at the frequency of the transition line center ν_0 the small-signal gain coefficient $\gamma_0(\nu_0)$ is twice the threshold gain coefficient γ_{th}. Recall that the small-signal gain coefficient is proportional to the lineshape function so that the width of the small-signal gain coefficient is $\Delta\nu_H = 1$ GHz. The gain coefficient and cavity modes for this system are shown in Figure 26-10.

When the laser is turned on, the irradiance is very small and so the gain coefficient $\gamma(\nu)$ takes on its small-signal value $\gamma_0(\nu)$. This situation is illustrated in Figure 26-10a. Notice that in this case six cavity modes are at frequencies that have gain larger than the threshold value. The irradiance in each of these six modes, therefore, begins to grow. As the irradiance grows, the gain saturates and the gain coefficient is reduced. As the gain coefficient is reduced, the number of cavity modes above threshold decreases. Figure 26-10b shows the gain coefficient at a time in the buildup to steady state when only four cavity modes remain above threshold. These four modes continue to grow while all others die out. As the irradiance in these four modes grows, gain saturation further reduces the gain coefficient, causing still more modes to drop below threshold. Steady state is reached when only the cavity mode with the frequency closest to the line center frequency ν_0 experiences a gain coefficient equal to the threshold value. This situation is shown in Figure 26-10c.

Note that this scenario indicates that only a *single cavity mode* can be present in the CW output of a laser containing a *homogeneously broadened* gain medium. In fact, it is possible that more cavity modes could be present in the output, but only if these modes utilize different spatial portions of the gain medium. In contrast, as we shall see in the next section, many different cavity modes can lase in a system that uses an inhomogeneously broadened gain medium.

26-6 INHOMOGENEOUS BROADENING

The homogeneous gain bandwidth results from the broadening of the frequency response of *each* atom in the medium. In addition, there can be environmental factors that cause the *center* frequencies of each of the atoms in an assembly of atoms to differ from one another. Then, the frequency response of the gain medium reflects both the homogeneous bandwidth of each atom and the distribution of the center frequencies of the atoms. In gas lasers the most important inhomogeneous broadening mechanism is *Doppler broadening*, due to the Doppler effect discussed in Chapter 4. Recall that the frequency measured by a detector depends on the velocity of the source of the wave relative to the velocity of the detector. Since the atoms in a gas gain medium have a distribution of velocities (the Maxwell Boltzmann distribution), the radiation from different atoms, each of which emits light at frequency, say, ν_0 in its own rest frame, will be perceived in the laboratory as radiation with a spread of frequencies related to the spread of velocities of the atoms in the gain medium. Without proof we state that the lineshape function, resulting from this process, is well described by a *Gaussian* function,

$$g(\nu) = \left(\frac{4\ln(2)}{\pi\Delta\nu_D^2}\right)^{1/2} e^{-4\ln(2)[(\nu-\nu_0)/\Delta\nu_D]^2} \qquad (26\text{-}51)$$

Here, ν_0 is the center frequency of the emission from the atoms in their rest frame and $\Delta\nu_D$ is the full width at half-maximum of the Doppler-broadened lineshape function. For an assembly of gain atoms each of mass M at temperature T,

$$\Delta\nu_D = \left(\frac{8k_BT}{Mc^2}\ln(2)\right)^{1/2}\nu_0 \qquad (26\text{-}52)$$

Figure 26-11 shows the lineshape function $g(\nu)$ for a Doppler-broadened gain medium consisting of atoms each of which has a homogeneous linewidth $\Delta\nu_H$. In general, both homogeneous and inhomogeneous broadening occur

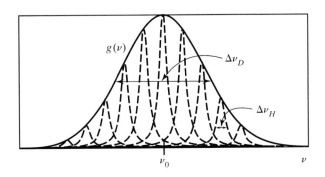

Figure 26-11 A Doppler-broadened lineshape function $g(\nu)$ (solid curve) is shown as a function of ν. The lineshape functions of ten of the continuum of "groups" of atoms of homogeneous linewidth $\Delta\nu_H$ and differing center frequencies that contribute to the overall lineshape function are shown by the dotted curves.

within the same gain system. The forms of the lineshape functions given in the preceding paragraphs are appropriate for those relatively common cases in which either homogeneous broadening or Doppler broadening is dominant. In the intermediate case, the lineshape function does not have a simple closed form. The bandwidth of many gas gain media, like He-Ne and Ar$^+$, for example, are primarily due to Doppler broadening.

Example 26-5

Estimate the linewidth of an Ar$^+$ gain medium. Consider a transition wavelength of $\lambda_0 = 488$ nm and take the temperature of the gas under the operating conditions to be 3000 K.

Solution

The atomic mass of an argon atom is $M = 6.64 \times 10^{-26}$ kg. Then, from Eq. (26-52),

$$\Delta \nu_D = \left(\frac{8 \times 1.38 \times 10^{-23} \times 3000}{6.64 \times 10^{-26}(3 \times 10^8)^2} \ln(2) \right)^{1/2} \frac{3 \times 10^8}{488 \times 10^{-9}} \text{ Hz}$$

$$= 3.8 \times 10^9 \text{ Hz} = 3.8 \text{ GHz}$$

There are many other important mechanisms leading to inhomogeneous broadening. For example, in solid-state lasers in which the active atom is *doped* into a transparent host, inhomogeneities in the host lead to broadening. In many real gain media the gain broadening is due to a complicated mix of many physical processes and is difficult to model accurately.

Gain Saturation in Inhomogeneously Broadened Media

Let us now turn to a description of gain saturation in an inhomogeneously broadened medium. Consider Figure 26-12, which shows the behavior of the gain coefficient during the buildup to steady state in such a medium. As in the corresponding figure for a medium that is homogeneously broadened (Figure 26-10), let us consider a case in which the small-signal gain coefficient at line center $\gamma_0(\nu_0)$ is twice the threshold gain coefficient γ_{th}. Once again we consider a case in which several (for the case shown, three) cavity modes have frequencies at which the small-signal gain coefficient exceeds the threshold gain coefficient, as shown in Figure 26-12a. Irradiance grows at each of these frequencies. The gain coefficient in an inhomogeneously broadened medium has contributions from groups of atoms with different center frequencies and relatively narrow homogeneous bandwidths. Consequently, the growing irradiance only reduces the population inversion, and so the gain coefficient, in atoms within, roughly, one homogeneous gain bandwidth of the corresponding cavity-mode frequency. Thus, as shown in Figure 26-12b, the

Figure 26-12 Gain saturation in an inhomogeneously broadened laser. (a) At laser turn on, the gain coefficient has its small-signal value and three cavity modes are above threshold, and so the irradiance grows at these frequencies. (b) In steady state, the irradiance in each of the three cavity modes has grown large enough to reduce the gain coefficient $\gamma(\nu)$ at these frequencies to the threshold value γ_{th}.

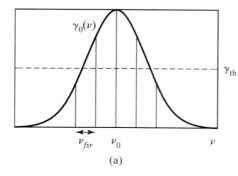

(a)

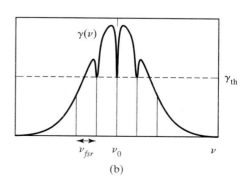

(b)

laser output can consist of frequencies corresponding to *several different cavity modes* in a laser using an *inhomogeneously broadened* gain medium. The dips in the gain curve have a frequency width roughly equal to the *homogeneous gain bandwidth* $\Delta\nu_H$ of the atoms whose center frequency corresponds to the cavity-mode frequency. If the cavity-mode separation is less than the homogeneous gain bandwidth of the atoms, the cavity modes will "compete" for the same group of atoms and not all cavity modes initially above threshold will lase.

The gain coefficient, near a cavity mode above threshold, in an inhomogeneously broadened gain medium saturates according to a law that is different from that given in Eq. (26-37), which is appropriate for homogeneously broadened media. For an inhomogeneous medium with a Gaussian lineshape function, as is appropriate for Doppler-broadened media, the gain saturation takes the form

$$\gamma = \frac{\gamma_0}{\sqrt{1 + I/I_S}} \qquad (26\text{-}53)$$

In Eq. (26-53), the saturation irradiance I_S is that for the group of atoms with a center frequency at the cavity-mode frequency. Note that the gain coefficient for an inhomogeneously broadened gain medium saturates less, for a given increase in irradiance, than does the corresponding gain coefficient given in Eq. (26-37) for a homogeneously broadened medium. For example, according to Eq. (26-53), when $I = I_S$, $\gamma = \gamma_0/\sqrt{2} \approx 0.71\gamma_0$ rather than $0.5\gamma_0$ as in a homogeneously broadened medium. This slower saturation results because, although irradiance at a frequency equal to that of a given cavity mode interacts primarily with the group of atoms whose center frequency is at the cavity-mode frequency, the irradiance also can use other groups of atoms with "nearby" center frequencies. As the irradiance at a given cavity-mode frequency grows, it "reaches" groups of atoms with center frequencies further and further from the cavity-mode frequency. This ability to extract energy from new groups of atoms as the irradiance grows reduces, somewhat, the rate of saturation of the gain coefficient. Equation (26-53) can be used to relate the output irradiance to the input irradiance of an inhomogeneously broadened gain cell using the same procedure that led to Eq. (26-41). This result can then be used together with the cavity survival factor to develop an expression for the output irradiance from a laser system using an inhomogeneously broadened medium. This procedure is cumbersome and the resulting expression inelegant, and so we leave this as an unlisted problem for the ambitious student.

26-7 TIME-DEPENDENT PHENOMENA

In the previous sections of this chapter, we have developed a means for predicting the steady-state output irradiance from a laser. In this section, we write down the equations that govern the time-dependent exchange of energy between the energy stored in a gain medium and that stored in the intracavity field. For simplicity we will treat the ideal, homogeneously broadened, four-level system discussed earlier. Recall that in this case the population of the lower lasing level N_1 is negligible since we take $\tau_1 \approx 0$. Equation (26-32), describing the atomic population density N_2, then becomes

$$\frac{dN_2}{dt} = R_{p2} - \kappa_2 N_2 - \frac{\sigma I}{h\nu'} N_2$$

We wish to develop a rate equation governing the photon number density N_p in the cavity. Let us adopt again the ring cavity model used earlier. In that case the photon number density in the cavity can be related to the field irradiance I in the cavity by the relation

$$I = h\nu'cN_p$$

Using this in the rate equation for N_2 gives,

$$\frac{dN_2}{dt} = R_{p2} - \kappa_2 N_2 - \sigma c N_p N_2 \qquad (26\text{-}54)$$

The rate equation for the photon number density can be formed by adding the contribution originating from cavity losses detailed in Chapter 8 to the contribution originating from interaction with the gain medium. For the case at hand, the gain contribution is found by noting that each stimulated emission event decreases the population of level 2 by 1 and increases the number of photons in the cavity by 1. Recalling that N_2 and N_p are densities and that the atomic population is spread through the volume V_g of the gain medium, whereas the photons are distributed throughout the field in the cavity occupying an effective volume V_c, we write

$$\frac{dN_p}{dt} = -\Gamma N_p + (V_g/V_c)\sigma c N_p N_2 \qquad (26\text{-}55)$$

Here, Γ is the cavity loss rate and the V_g/V_c is approximately the ratio of the length of the gain cell to the perimeter P of the ring cavity. Using the method leading to Eq. (8-45) that is valid for a linear cavity, one can show (see problem 26-23) that the cavity loss rate Γ for the ring cavity is

$$\Gamma = \frac{c}{P}(1 - S)$$

Note that Eq. (26-55) predicts a zero-growth rate if there are no photons in the cavity. This results because we have ignored spontaneous emission in the derivation of this rate equation. Whereas, as noted earlier, spontaneous emission is typically unimportant in steady-state laser operation, it is essential in initiating the growth of the laser field. To treat the initial growth of the laser field, Eq. (26-55) can be rectified by replacing the factor N_p in the last term on the right with the factor $(N_p V_c + 1)/V_c$. This follows from a fully quantum-mechanical treatment of the interaction between the atomic system and the electromagnetic field and indicates that the ratio of stimulated emission to spontaneous emission *into a given cavity mode* is the same as the ratio of the number of photons in the cavity to 1. In practice, one often follows the evolution of a cavity photon number density by using Eqs. (26-54) and (26-55), starting from a nonzero "seed" photon number density. The former approach was used in producing the curves, in Figure 26-13, showing the population inversion $N_{\text{inv}} = N_2$ and photon number density N_p as functions of time after laser turn on. Note that, for the parameters listed in the figure caption, the inversion population density grows to a value which is roughly a factor of 2 larger than its steady-state, threshold value. As a result the initial gain exceeds the cavity losses and the photon number density begins to grow rapidly. When the photon number density grows to near its steady-state value, however, the rate of stimulated emission, which reduces the population inversion, exceeds the pump rate that feeds the inversion, and so the population inversion decreases rapidly. This *decrease* in population density, of course, is accompanied by a rapid increase in photon number density. However, this large photon number density initiates sufficient stimulated emission events

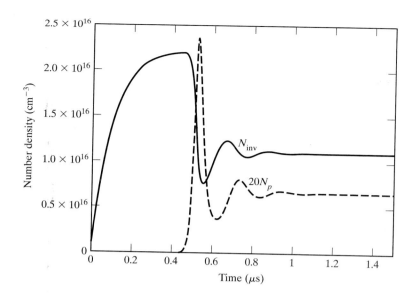

Figure 26-13 Laser turn-on and approach to steady state. The solid curve represents the population inversion N_{inv} as a function of time. The dotted curve shows the photon number density N_p in the cavity (multiplied by 20 so that both curves have similar vertical scales) as a function of time. The parameters used to produce this curve are $\sigma = 10^{-18}$ cm^2, $\Gamma = 10^8$ s^{-1}, $\kappa_2 = 10^7$ s^{-1}, $\gamma_0/\gamma_{th} = 2$, and $V_g/V_c = 0.3$.

to drive the population inversion below its threshold value, at which time cavity loss exceeds gain and the photon number density begins to decrease. The decrease in photon number density allows the population inversion to once again grow and there follows a back-and-forth trading of energy between the atomic population and the electromagnetic field as the system settles into steady state. The oscillations in population inversion in the approach to steady state are known as *relaxation oscillations*. The strength and duration of the relaxation oscillations are sensitive functions of the parameters governing the laser system. (See problems 26-24 and 26-25.)

Note that, as shown in Figure 26-13, in the approach to steady state a laser system can produce pulses of intracavity photon number densities N_p that far exceed the steady-state photon number density. Consequently, in the approach to steady state, the laser system produces pulses that have peak irradiances far larger than the steady-state irradiance. We describe other means of producing pulsed laser output in the next section.

26-8 PULSED OPERATION

In this section we turn to a discussion of the means of producing laser output pulses. The ability to control the temporal delivery of laser energy is important in a wide variety of applications, including materials processing, characterization of fast processes, and laser fusion technology. Perhaps the simplest means of producing pulsed energy output from a laser would be to pulse the pump—that is, turn the source of laser energy on and off. Such a pulsed-pump or *gain-switched* system can produce usable pulses as suggested by Figure 26-13. In general, however, there is a complex exchange of energy between the field and atomic population in such a system that makes the pulse characteristics difficult to control. Two important methods used to control the characteristics of laser light pulses are *Q-switching* and *mode-locking*. These are discussed in the following subsections.

Q-switching

In Chapter 8 we introduced the quality factor Q as a measure of the loss rate of the cavity. There we noted that cavities with higher loss rates have lower quality factors. *Q-switching* refers to the periodic change of the loss rate of a cavity in order to pulse the output of a laser. Q-switching proceeds in the following time sequence:

1. A laser gain medium is pumped while the cavity has high loss (low Q). The low Q of the cavity prevents growth of the intracavity irradiance and so spontaneous emission and incoherent decay processes are the only drains on the population of the upper lasing level. As a result, a large population inversion grows in the gain medium. But since the cavity loss is high, the small-signal gain coefficient is less than the threshold value needed for lasing. That is, $\gamma_0 = \sigma(N_2 - N_1) < \gamma_{th}^{low\,Q}$.

2. The Q of the cavity is rapidly switched to a high value so that the loss rate of the cavity is reduced to a low value. The small-signal gain coefficient, built while the cavity had a high loss rate, now exceeds the high-Q threshold value, $\gamma_0 > \gamma_{th}^{high\,Q}$. As a result, a large irradiance builds rapidly within the cavity, leading to output pulses of high peak irradiance.

3. The large irradiance in the cavity induces a high stimulated-emission rate, which depletes the population inversion, driving the laser system below threshold.

4. Before the population inversion can grow again to sustain a small-signal gain coefficient larger than the threshold value, the cavity Q is switched back to the low value. This prevents the reinitiation of lasing action and allows the population inversion to grow, once again to a large value, storing a large amount of energy in the gain medium.

5. Steps 2–4 are repeated.

Note that this description of the production of a Q-switched pulse has much in common with the description of the first pulse shown in Figure 26-13. There are, however, several important differences between the underlying systems and their behaviors. In a Q-switched system, the population inversion grows to a value larger than that in the equivalent gain-switched system because the low-Q, high-loss cavity prevents the growth of the cavity field until the population inversion grows to the maximum value allowed by the pump-rate/atomic-decay-rate dynamics. Further, when the cavity in a Q-switched system is made to have low loss (high Q), the system is far above the lasing threshold and so the irradiance grows more rapidly and to a larger value than in the gain-switched system. Finally, in a Q-switched system, the cavity is switched to the high-loss state after the pulse has driven the population inversion below threshold and before the inversion can begin to grow again. That is, switching to the low-Q state prevents additional relaxation oscillations characteristic of the approach to steady state. A Q-switched laser system, then, produces a series of irradiance pulses similar in character to the first pulse in a gain-switched system. The Q-switched pulses are, in general, of higher peak irradiance and narrower pulse width than the initial pulse in a similar gain-switched system. A Q-switched laser system is an optical analogue to a capacitor. The high loss rate in the low-Q cavity allows for large energy storage in the gain medium, which can then be released rapidly as light energy when the cavity is switched to the low-loss, high-Q state.

Q-switching can be accomplished by mechanical means, such as rotating a cavity mirror into and out of alignment or passing a miniature fan blade through the cavity. More commonly, Q-switching is accomplished through electro-optic or acousto-optic means, as discussed in Chapter 24 and illustrated in Figures (24-6) and (24-14). A particularly elegant means of Q-switching involves the insertion of a *saturable absorber* into the cavity. A saturable absorber is a system that is absorbing for low light irradiances but transmitting for high light irradiances. In the high-loss state, the population inversion grows, which in turn leads to an increase in spontaneous emission. If the spontaneous emission grows to a value large enough to saturate the absorber, the cavity will suddenly become low loss. Thus, the saturable-absorber Q-switch requires no external intervention and the cavity is naturally Q-switched when the irradiance due to spontaneous emission reaches a value large enough to saturate the absorber.

To a first approximation, Q-switching a laser system does not change the overall efficiency of the system. As a result, the *average power* output P_{av} from a Q-switched system is roughly the same as the CW power of the same system. Q-switching simply redistributes the energy so that it comes out in large bursts separated by periods of nearly zero energy output. Q-switched systems typically have pulse widths on the order of a cavity lifetime $\tau_p = 1/\Gamma$ since the pulse must leak out of the cavity. This leads to pulse widths on the order of 1 μs or less. The *pulse repetition time* (time between pulses) can be no less than the time needed for an inversion to build, which is governed by the lifetime τ_2 of the upper lasing level. The pulse repetition time in a Q-switched system is typically on the order of 1 ms. As a result, the *peak power* P_{peak} in a Q-switched pulse can be a factor of 1000 or more larger than the CW output power from the same laser system.

Example 26-6

A certain Nd:YAG laser is reported to have a CW power output of $P_{CW} = 10$ W. If the system is to be Q-switched with a *pulse repetition rate* of 1 kHz and a pulse width of $\Delta t_p = 0.25$ μs, (a) estimate the peak power in a Q-switched pulse and (b) estimate the energy in each pulse.

Solution

To estimate the peak power, we shall model the pulses as rectangles of width equal to the pulse width and height equal to the peak power P_p, as shown in Figure 26-14. Since the pulse repetition rate is 1 kHz, the time between pulses is $T = 1$ ms. The average power in the Q-switched system is the power averaged over the time T. The energy in each cycle is the energy contained in a single pulse:

$$E_p = P_p \Delta t_p$$

a. The average power is thus $P_{av} = E_p/T$, so that

$$P_{av} = \frac{P_p \Delta t_p}{T}$$

Finally, setting $P_{av} = P_{CW}$ and solving for the peak power gives

$$P_p = \frac{T}{\Delta t_p} P_{CW} = \frac{0.001}{0.25 \times 10^{-6}} (10\text{ W}) = 4 \times 10^4 \text{ W}$$

b. Using the expression given earlier,

$$E_p = P_p \Delta t_p = (4 \times 10^4 \text{ W})(0.25 \times 10^{-6} \text{ s}) = 0.01 \text{ J} = 10 \text{ mJ}$$

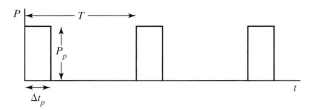

Figure 26-14 Construction showing a rectangular pulse approximation to Q-switched pulses used in Example 26-6. Typically, the pulse width Δt_p is a much smaller fraction of the pulse repetition time T than is shown in this diagram.

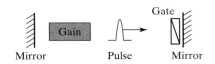

Figure 26-15 A mode-locked laser system. The gate opens for a brief time during each round-trip to let the pulse pass.

Mode-Locking

The physical arrangement of a mode-locked system, shown in Figure 26-15, is similar to that for a Q-switched system. Both systems require the insertion into the laser cavity of an element that performs loss modulation. However, as noted, in a Q-switched system the cavity is left in the high-loss state for a time long enough to allow the population inversion to build to a large value. This process requires many cavity round-trips. In contrast, in a mode-locked system the loss modulation occurs *once each round-trip* in the following fashion. In a linear cavity, the loss modulation "gate" is placed near one of the mirrors and is switched to the low-loss state for a brief period once each round-trip. Here, "a brief period" means a fraction of a round-trip time. As a result, only pulses timed to pass through the gate when it is open can survive in the cavity. All other field shapes see a high loss and so do not lase. In a mode-locked system, the loss modulation occurs too rapidly for the population inversion in the gain medium to respond and so the population inversion, more or less, retains the same value that it would have in a CW system. The mode-locking "gate" can be any of the devices, described earlier, that could be used as Q-switches, provided that the gate can be opened and closed over a time period shorter than a cavity round-trip time.

Inhomogeneously broadened laser systems in which many different cavity modes lase are commonly mode-locked. The total field in the cavity is a superposition of these fields with frequencies corresponding to the cavity-mode frequencies that are separated by the cavity free spectral range $c/2d$, where d is the cavity length. In order that these modes add to a pulse, they must be *locked* to a common phase at a particular place that moves back and forth through the cavity at the speed of light. The opening and closing of the gate, just described, provides such a phase-locking mechanism. The initiation of such a pulse can be viewed as the fortuitous spontaneous emission into the different cavity modes in such a way that the fields in these modes happen to constructively interfere at the gate when the gate is open. Once formed, this pulse passes through the gate, bounces off the mirror, and passes through the gate again before the gate closes. The pulse is then amplified by the gain medium and returns as a larger pulse just as the gate opens again. In this way the pulse builds from phase-matched spontaneous emission events. Field components not phase-locked to form constructive interference, at the position of the gate when the gate is opened, are blocked by the closed gate. *The mode-locked pulse is the low-loss mode of a cavity that is loss modulated once each round-trip.*

Phase-locking the fields in the cavity allows for the formation of a narrow pulse centered on the positions of common phase. Between positions of common phase, since the fields have different frequencies, the fields tend to destructively interfere. At a given place in the cavity, the condition for a common phase for all of the fields recurs at times equal to the inverse of the constant frequency separation of the modes. (See problem 26-26.) As noted earlier, the frequency difference between modes is $\nu_{fsr} = c/2d$ so that the recurrence time for the pulses is the round-trip cavity time $2d/c$.

Let us compare the peak power in a mode-locked, multimode laser to the average power in the same multimode laser system with CW output. Let the number of nearly equal-amplitude modes each of power P_0 in the system be N. In CW operation the phases of these modes are independent from each other and wander randomly since the fields are not perfectly coherent. Thus the total average power, for the CW laser, is simply the sum of the powers in the individual fields. That is,

$$P_{CW} = NP_0 \qquad (26\text{-}56)$$

In the mode-locked pulse, at the positions of equal phase, the fields *constructively* interfere so that the power P_p in the total field at these positions is given by

$$P_p = N^2 P_0 \qquad (26\text{-}57)$$

Now the *average* power in the mode-locked laser is roughly the same as the average power in the same CW system. Thus the relation given in part (a) of Example 26-6, derived there for a Q-switched system, also applies for the mode-locked case. That is, with $P_{av} = P_{CW}$,

$$P_{CW} = \frac{P_p \, \Delta t_p}{T}$$

Using Eqs. (26-56) and (26-57), together with the relation for the pulse repetition time $T = 2d/c$, leads to an approximate expression for the temporal width Δt_p of a mode-locked pulse,

$$\Delta t_p = \frac{P_{CW} T}{P_p} = \frac{N P_0 (2d/c)}{N^2 P_0} = \frac{1}{N(c/2d)} = \frac{1}{N \nu_{fsr}}$$

The number of modes that are above threshold depends on the bandwidth $\Delta \nu$ of the gain medium and on the ratio of the small-signal gain coefficient to the threshold gain coefficient. In an inhomogeneously broadened system pumped so that $\gamma_0 = 2\gamma_{th}$, all modes within $\Delta \nu/2$ of linecenter will be above threshold (see, for example, Figure 26-12a). In that case, $N \nu_{fsr} = \Delta \nu$ and the expression for the pulse width becomes

$$\Delta t_p \approx \frac{1}{\Delta \nu} \qquad (26\text{-}58)$$

This important relation is an example of the more general *bandwidth theorem*, which states that the width of a pulse is inversely proportional to the range of frequencies in the fields that constitute the pulse. So gain media with larger bandwidths can be used to form narrower mode-locked pulses. The relationships given in Eqs. (26-57) and (26-58) are explored in problems 26-27 through 26-31.

26-9 SOME IMPORTANT LASER SYSTEMS

The features of many important laser systems are given in Table 6-1. Detailed descriptions of the characteristics of different laser systems are beyond the scope of this text. In this section, we describe briefly the properties of the different types of atomic and molecular gain media and list a few examples of each. In Section 26-10 we discuss, in a bit more detail, the operation of diode lasers.

Gas Atomic Lasers

By an atomic laser we mean one in which the upper and lower lasing levels are different electronic states of one of the atoms in the gain medium. Helium-neon (He-Ne) and argon-ion (Ar$^+$) lasers are two important gas atomic lasers. In the He-Ne laser (described in some detail in Chapter 6) neon atoms are the lasing species and the helium atoms aid in the pumping process. The lasing levels in the Ar$^+$ laser are two different electronic energy states of a singly ionized argon atom. Gas atomic lasers are typically Doppler broadened and have low efficiencies (less than 1%). The two types mentioned here are known for their coherence. Gas atomic lasers typically emit laser light with frequencies in the visible to near-infrared range. These lasers are usually pumped by an electric discharge.

Gas Molecular Lasers

As mentioned in Chapter 6, molecules have energy levels corresponding to different rotational and vibrational states of the molecule, in addition to the states associated with different electronic configurations. The upper and lower laser levels of gas-molecular gain media are typically different vibrational-rotational states associated with the ground electronic state. These vibrational-rotational states typically differ in energies such that the emitted light is in the mid-infrared range. The most important gas-molecular laser is the carbon dioxide (CO_2) laser that emits radiation of wavelength 10.6 μm. The CO_2 laser is one of the most versatile and efficient laser systems, capable of CW power outputs ranging from mW to MW with efficiencies ranging from 10% to 40%. CO_2 lasers are typically pumped by an electric discharge. High-power CO_2 lasers are sometimes pumped by electron beams or gas dynamic processes.

Excimer Lasers

The gain medium in an *excimer* laser is a rare-gas halogen mixture. Rare-gas atoms have filled outer shells and so tend not to bond to other atoms. However, an excited rare-gas atom (like Kr*) has a single electron in its outer shell and so can form an ionic bond with a halogen atom (like F), which is one electron short of a filled outer shell. A typical reaction might be

$$Kr^* + F_2 \rightarrow (Kr^+F^-)^* + F$$

The production of the excited rare-gas species is accomplished by electric-discharge or e-beam pumping. The molecule $(Kr^+F^-)^*$ is the upper state of the lasing transition. If the Kr is not excited, it repels the F atom and so the lower lasing level is unstable with an extremely short lifetime ($\cong 10^{-13}$ s). Thus this system approximates an ideal four-level laser system. Excimer lasers are typically pulsed systems distinguished by their relatively high average power (~50 W) and, importantly, lasing wavelengths in the UV.

Liquid-Dye Lasers

The gain media in liquid-dye lasers are solutions of certain long-chain organic molecules in alcohol or other solvents. Dye lasers have very large small-signal gain coefficients (4/cm or more) and very large gain bandwidths (50–100 nm). Dye lasers emit radiation in the visible region of the spectrum, and a few different dyes are sufficient to provide coherent radiation tunable across the entire visible spectrum. The broad gain bandwidth of a given dye allows for the production of extremely short (ps-width) mode-locked pulses.

Solid-State Lasers

Solid-state lasers are a class of lasers in which the lasing atomic species is doped into a transparent host material. The most common solid-state laser is the Nd:YAG laser in which neodymium (Nd) atoms replace about 1% of the yttrium (Y) atoms in the host crystal, yttrium aluminum garnet (YAG). Other important solid-state lasers are listed in Table 6-1. These lasers typically have outputs with wavelengths in the near infrared. The Nd:YAG system provides laser output at 1.064 μm. An Nd:YAG system can provide a high power, good beam quality CW output, and can be either mode-locked or Q-switched. The output of an Nd:YAG laser is commonly frequency doubled to produce coherent radiation at 532 nm. Solid-state lasers are typically flash-lamp pulsed or optically pumped by another laser. Efficient, portable Nd:YAG lasers, pumped by an array of diode lasers, are available.

26-10 DIODE LASERS

Semiconductor or diode lasers have gain media that differ significantly from the atomic and molecular gain media discussed in the preceding sections of this chapter. A diode laser is essentially a *p-n* junction whose cleaved edges act as reflecting surfaces that supply the cavity feedback. From a packaging perspective, the main distinguishing feature of the diode laser is its size. The photograph in Figure 26-16 allows one to appreciate the difference in scale between diode lasers and conventional table-top laser systems. In this section we provide a qualitative description of the operating principles of diode lasers and begin with a very brief review of semiconductors and *p-n* junctions.

Each of the electrons in a solid *insulator* is tightly bound to a given atom in the lattice that forms the solid. These tightly bound electrons are said to occupy states in the *valence band* of electronic energy states of the material. In a conductor, on the other hand, one or more electrons per atom in the solid are not tightly bound to a given atom but rather are relatively free to roam about the solid as a whole. These electrons are said to be in the *conduction band* of electron energy states in the solid. In an *intrinsic* (pure) semiconductor, each electron in the solid is bound to a given atom (that is, it is in the valence band). However, the outermost electrons in a semiconductor are not as tightly bound as those in an insulator. The strength of the binding of the outermost valence electrons is described by the *band-gap energy*, which is the energy required to free an electron from its host nucleus. That is, the band-gap energy is the energy required to promote an electron from the valence band to the conduction band. In an insulator, the band-gap energy is typically about 4 eV. In a semiconductor, the band-gap energy is between 1 and 2 eV. (In a conductor, the conduction and valence bands overlap.)

In an *intrinsic* semiconductor, the atoms bond so as to "fill" the outer shells of the atoms. For example, in gallium arsenide (GaAs), gallium has three outer-shell electrons and arsenic has five outer-shell electrons. These atoms bond in the solid lattice in such a way that each atom sees a full outer shell (eight electrons) with some of these electrons shared with neighboring atoms. It is this sharing of the electrons that forms the crystal bond. In a *p*-type semiconductor, the material is *doped* with a small amount of an impurity that has one less electron in its outer shell than does the atom that it replaces in the lattice. For example, zinc has two outer-shell electrons. In GaAs doped with zinc, the zinc replaces gallium (three outer-shell electrons) in the lattice. As a result, for every zinc atom added, there are empty spaces in the lattice reserved for an electron. These empty spaces are called *holes* and act as positive charge carriers. In an *n*-type semiconductor, the material is doped with an impurity whose outer shell has one electron more than the outer shell of the atom that it replaces in the lattice. For example, if selenium (six outer-shell electrons) is doped into GaAs, it replaces As (five outer-shell electrons) in the lattice. This results in one electron per selenium atom that does not have a slot in the charge clouds surrounding the nuclei of the atoms in the lattice. These "extra" electrons act as negative charge carriers. Of course, both *n*-type and *p*-type materials have zero net charge since the dopant atoms are neutral. Laser diodes use forward-biased *p-n* junctions, in which the voltage source drives the holes in the *p*-type material and the conduction electrons in the *n*-type material toward the junction (Figure 26-17).

Figure 26-16 Two views of a packaged laser diode. The packaged diode is about 6 mm in diameter. The laser diode chip itself is located in the small rectangular depression visible in the mount shown in the top image. Courtesy of John Sohl. (Photo by Sheri Trbovich.)

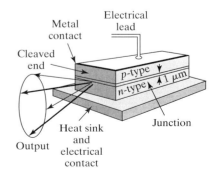

Figure 26-17 Simple *p-n* junction, laser diode pumped by an injection current.

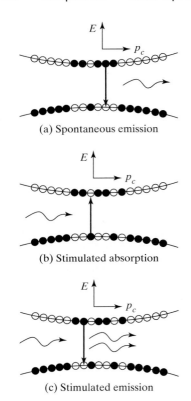

(a) Spontaneous emission

(b) Stimulated absorption

(c) Stimulated emission

Figure 26-18 The interaction of an electromagnetic field with a semiconductor gain medium. The upper energy band is the conduction band and the lower energy band is the valence band. Filled circles indicate electrons and empty circles indicate holes. The thick, straight arrows indicate transitions of an electron from one band to another, and the lighter, curved arrows represent photons. The energy E of the charge carriers varies across the band since the carriers in a given band vary in momentum, p_c.

Figure 26-18 shows the band structure in the junction region. Note that the bands are curved and contain states of differing energy E since the carriers in the bands can have different momenta p_c. A population inversion is shown near the center of the bands where there is an excess of electrons in the conduction band. In the junction region the electrons can "fall" into the available holes, releasing energy in the form of a photon as they do so. This *spontaneous emission* process is responsible for the output of light-emitting diodes and is illustrated in Figure 26-18a. The cleaved edges of these *p-n* junctions, of submillimeter dimension, act as mirrors, providing feedback for the laser system. The spontaneously emitted photons that are reflected back into the thin junction region can cause stimulated absorption (the creation of a conduction electron/valence hole pair and the loss of a photon, shown in Figure 26-18b) or stimulated emission (the creation of a twin photon and the demotion of an electron from the conduction band into a hole in the valence band, shown in Figure 26-18c).

Diode Laser Operating Characteristics

Important favorable characteristics of diode lasers are that they are relatively inexpensive, are small and efficient, and can be engineered to have a variety of wavelengths of interest in many applications. In addition, the output irradiance from laser diodes can be easily modulated by varying the injection current that pumps the diode. Thus, it is relatively easy to encode information into laser-diode light fields. Efficient diodes have a more complicated layered structure than does the simple *p-n* junction device illustrated in Figure 26-17. A more typical stripe-heterojunction device is shown in Figure 26-19. *Arrays* of diode lasers allow for relatively high average power devices. Diode lasers do have several unfavorable characteristics. The small asymmetric output aperture leads to highly divergent nonsymmetrical output beams, as indicated in Figure 26-17. As a result, the output of a diode laser is typically coupled directly into a fiber or collimated by a short focal length lens. Further, it is difficult to limit a diode laser to stable single-mode output, and so diode devices generally have shorter coherent lengths than, for example, do Nd:YAG, He-Ne, or Ar$^+$ laser systems. Still, the advantages of the diode laser design make them the preferred choice in an ever-increasing array of devices, including laser pointers, optical pumps, and optical data storage and read-out.

Semiconductor materials can be engineered to have a variety of band-gap energies resulting in a variety of operating wavelengths. Some of these are listed in Table 26-1 below.

Figure 26-19 Stripe-heterojunction semiconductor laser. Note the dimensions of the cavity region.

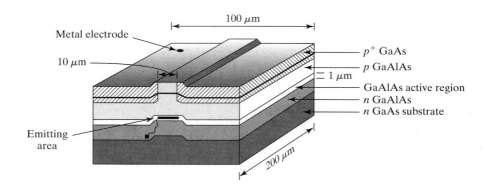

TABLE 26-1 LASER DIODE WAVELENGTHS

Material	GaN	AlGaInP	GaAlAs	InGaAsP	Sb mixtures
Wavelength (nm)	400–480	630–690	750–900	1200–2000	2000–4000

PROBLEMS

26-1 The spectral energy density $\rho(\nu)$ in an electromagnetic field in thermal equilibrium at temperature T is given by Eq. (26-6). Recall that $\rho(\nu)$ is the energy per unit volume per unit frequency interval in the field.

  a. Show that $\rho(\lambda)$, which we define to be the energy per volume per unit *wavelength* interval in the field, is

$$\rho(\lambda) = \frac{8\pi h}{\lambda^5}\left(\frac{1}{e^{hc/\lambda k_B T}-1}\right)$$

  b. Confirm that the relation found in (a) is in agreement with the spectral exitance associated with a blackbody given in Eq. (6-5) as

$$M_\lambda = \frac{2\pi h c^2}{\lambda^5}\left(\frac{1}{e^{hc/\lambda k_B T}-1}\right)$$

Note that M_λ is the the exitance per wavelength interval emitted by a blackbody source. (*Hint*: The spectral energy density in (a) includes the energy in field components moving in all directions while the spectral exitance accounts for the power moving normally away from the blackbody surface.)

26-2 Consider a monochromatic electromagnetic field traveling with speed c in a given direction. Use a conservation of energy argument to show that the time-averaged energy density $\langle u \rangle$ associated with this field is related to the irradiance I of the field by $\langle u \rangle = I/c$.

26-3 Show that Eqs. (26-17), (26-18), and (26-19) follow, in steady state, from Eq. (26-16) and Eq. (26-14).

26-4 One can define a saturation irradiance $I_{S,abs}$ for an absorptive medium as the irradiance for which the loss coefficient α is reduced by a factor of 2 from its small-signal value.

  a. Show that for the two-level absorptive medium considered in Section 26-2,

$$I_{S,abs} = \frac{h\nu'}{2\sigma\tau_2}$$

where $\tau_2 = 1/A_{21}$.

  b. Compare this relation to the saturation irradiance for the ideal four-level gain medium given in Eq. (26-39) and account, with a conceptual argument, for the factor of two difference between the two saturation irradiances.

26-5 Consider an amplifying medium composed of homogeneously broadened four-level atoms like the one depicted in Figure 26-5. Amplification is to occur on the 2-to-1 transition. The medium is pumped by a laser of intensity I_p, which is resonant with the 3-to-0 transition. The spontaneous decay processes are as indicated on the diagram. The total number density of gain atoms is $N_T = N_0 + N_1 + N_2 + N_3$. The various parameters are

$$\kappa_{32} = 10^8/s, \kappa_{21} = 1000/s, \kappa_{10} = 10^8/s, \kappa_{30} = \kappa_{31} = \kappa_{20} \approx 0$$
$$\sigma_p = 3\cdot10^{-19}\text{ cm}^2, \sigma = 10^{-18}\text{cm}^2$$
$$\lambda_{30} = 400\text{ nm}, \lambda_{21} = 600\text{ nm}\quad\text{(in free space)}$$
$$N_T = 1.5\cdot10^{26}/\text{m}^3$$

  a. Write down the rate equations for the population densities of the levels.

  b. Find and plot the steady-state small-signal population inversion $N_2 - N_1$ as a function of the pump irradiance. (Recall that "small signal" is code for "set $I=0$.")

  c. Find the pump irradiance I_p required to sustain a steady-state population inversion.

  d. Find the pump irradiance I_p required to sustain a small-signal gain coefficient of 0.01/cm.

  e. Find the pump irradiance I_p required to sustain a small-signal gain coefficient of 1/cm.

  f. Compare N_0 to N_1, N_2, and N_3 for the pump irradiances of parts (d) and (e). Is it reasonable to set $N_0 \approx N_T$ for either of these irradiances?

  g. Use the ideal four-level gain medium relation given as Eq. (26-38) together with the definition of the effective pump rate density given following Eq. (26-33) to estimate the pump irradiance required to sustain a small-signal gain coefficient of 0.01/cm and 1/cm. Compare these results to those obtained in parts (d) and (e).

26-6 Show that Eq. (26-34) follows from Eqs. (26-32) and (26-33).

26-7 Consider an amplifying medium composed of homogeneously broadened *three*-level atoms. Amplification is to occur on the 2-to-1 transition. The medium is pumped by a laser, of irradiance I_p, that is resonant with the 3-to-1 transition. Level 3 decays spontaneously only to level 2 and level 2 decays spontaneously to level 1, which is the ground state of the system. The total number density of gain atoms is $N_T = N_1 + N_2 + N_3$. The various parameters are

$$\kappa_{32} = 10^8/s, \kappa_{21} = 10^3/s$$
$$\sigma_p = \sigma_{31} = 3\cdot10^{-19}\text{ cm}^2,\quad \sigma_{21} = 10^{-18}\text{ cm}^2$$
$$\lambda_{31} = 400\text{ nm}, \lambda_{21} = 600\text{ nm}\quad\text{(in free space)}$$
$$N_T = 1.5\cdot10^{26}/\text{m}^3$$

  a. Sketch a level diagram like Figure 26-5 appropriate for this case and indicate the various stimulated and decay processes with arrows on the level diagram.

  b. Write down the rate equations for the population densities of the levels. Include the presence of a field of irradiance I resonant with the 2-to-1 transition, the pump interaction, and the decay processes.

  c. Find and plot the steady-state small-signal population inversion $N_2 - N_1$ as a function of pump irradiance. (Recall that "small signal" is code for "set $I=0$.")

  d. Find the pump irradiance I_p required to sustain a steady-state population inversion. Compare this result to the answer obtained for part (c) of problem 26-5.

  e. Find the pump irradiance I_p required to sustain a small-signal gain coefficient of 0.01/cm. Compare this result to the answer obtained for part (d) of problem 26-5.

  f. Find the pump irradiance I_p required to sustain a small-signal gain coefficient of 1/cm. Compare this result to the answer obtained for part (e) of problem 26-5.

  g. Summarize the important differences in the behavior of the three-level gain medium considered in this problem and the four-level gain medium considered in problem 26-5.

26-8 Show that if the irradiance throughout a gain cell described by Eq. (26-41) is much less than the saturation irradiance I_S, the output irradiance I_L is related to the input irradiance I_0 by the simple relation

$$I_L = I_0 e^{\gamma_0 L}$$

That is, show that, in the small-signal regime, the irradiance exhibits *exponential* growth.

26-9 Show that if the irradiance throughout a gain cell described by Eq. (26-41) is much greater than the saturation irradiance I_S, the output irradiance I_L is related to the input irradiance I_0 by the simple relation

$$I_L = I_0 + I_S \gamma_0 L$$

That is, show that for a very large input irradiance, the irradiance exhibits *linear* growth. [It may be somewhat simpler to implement the relation $I \gg I_S$ in Eq. (26-40) and then integrate, than to use Eq. (26-41) directly.]

26-10 Consider the limit described in problem 26-9.

a. Show that in this limit and for an ideal four-level gain medium,

$$I_L - I_0 = h\nu' R_{2p} L$$

b. Argue that the relation in part (a) implies that for the *large* input-irradiance case of problem 26-9 every pump event leads to one photon added to the electromagnetic field being amplified.

c. For the *small* input-irradiance of problem 26-8, even for an ideal four-level gain medium, it is not true that every pump event leads to one photon added to the electromagnetic field being amplified. Conceptually, account for the missing pump events.

26-11 A homogeneously broadened gain medium has a length of $L = 2$ cm, a small-signal gain coefficient at the transition linecenter of $\gamma_0(\nu_0) = 1/$cm, and a saturation irradiance at the transition linecenter of $I_S(\nu_0) = 100$ W/cm^2. Assume that light of frequency $\nu' = \nu_0$ is input into the cell. Find the irradiance I_L exiting the gain cell when the irradiance I_0 input to the cell is (a) 1 W/cm^2, (b) 10 W/cm^2, (c) 100 W/cm^2, (d) 1000 W/cm^2, and (e) 10,000 W/cm^2.

26-12 For each case of problem 26-11, find the irradiance added by passage through the gain cell $I_L - I_0$ and describe how this added irradiance changes with increasing input irradiance.

26-13 Repeat problems 26-11 and 26-12 for the case in which the field input into the cell has a frequency $\nu' = \nu_0 + \Delta\nu/2$, where $\Delta\nu$ is the homogeneous linewidth of the gain medium.

26-14 Reproduce the curves shown in Figure 26-7 but extend the maximum length of the gain cell on the plot to 10 cm.

26-15 Consider an ideal four-level gain medium in a ring cavity like the one of Figure 26-8 but with $R_1 = R_2 = 1$ and $R_3 = 1 - T_3$.

a. Show that, for this case,

$$I_{out} = I_{sat}(\gamma_0 - \gamma_{th})L$$

b. Show that, for this case and for $\gamma_0 \gg \gamma_{th}$, essentially every pump event leads to an output photon.

c. Explain why, even when every pump event leads to an output photon, the efficiency of the laser system is less than 100%.

26-16 In this problem and the following two problems, consider a ring cavity like the one depicted in Figure 26-8. Let the cavity mirrors M_1 and M_2 have reflectances $R_1 = R_2$ and let mirror M_3 have reflectance $R_3 = 1 - T_3 - A_3$, where A_3 characterizes the output mirror absorption. Let the gain medium be homogeneously broadened and have length $L = 10$ cm and a saturation irradiance (at the lasing frequency) of $I_S = 2000$ W/cm^2.

a. Find the threshold gain coefficient if $R_1 = R_2 = 1$, $R_3 = 0.95$, and $T_3 = 0.05$.

b. If the small-signal gain coefficient is twice the threshold value, find the irradiance of the laser output field.

26-17 Consider again the ring laser described in problem 26-16 but now take the small-signal gain coefficient to be 0.01/cm, $R_3 = 0.95$, and $T_3 = 0.05$. Plot the laser output irradiance as a function of the variable reflectance $R = R_1 = R_2$ of the other two cavity mirrors.

26-18 Consider again the ring laser described in problem 26-16. Let $R_1 = R_2 = 0.99$ and $A_3 = 0.01$. Let the small-signal gain coefficient be 0.01/cm.

a. Plot the laser output irradiance as a function of the variable transmittance T_3 of the output mirror.

b. Using the plot produced in part (a), determine the value of T_3 that maximizes the laser output irradiance. Explain why, for this system in which there are unavoidable losses, the output irradiance is reduced from its maximum value if T_3 is either too large or too small.

26-19 Derive Eq. (26-47) by a procedure similar to that leading to Eq. (26-43). The linear cavity case is complicated by the fact that the field encounters the gain medium twice in each round-trip with the losses encountered at the mirrors interspersed between passes through the gain medium. It may be useful to research and then summarize the solution to this problem.

26-20 Show that Eq. (26-47) for a linear cavity reduces to

$$I_{out} = \frac{T_2 I_S}{2}\left(\frac{\gamma_0(2L) - \ln(1/S)}{1 - S}\right)$$

for a cavity with $R_1 = 1$. Here, S is the survival fraction in the linear cavity without gain ($S = R_2$). Compare this result with the similar result given in Eq. (26-43) for the ring cavity and account for the differences between the two results.

26-21 Consider the CO_2 transition described in Example 26-4. In addition to the information given in the example note that the spontaneous emission rate for the transition is $A_{21} = 0.34/$s.

a. What is the stimulated emission cross section σ for this transition?

b. What must be the population inversion in the gain medium to produce a small-signal gain coefficient (at linecenter, $\nu' = \nu_0$) of 0.03/cm?

c. Treating this system as an ideal four-level system, estimate the saturation irradiance for this transition.

26-22 Find the Doppler-broadened gain bandwidth of the 633-nm He-Ne transition. Assume that the operating temperature is 400 K and recall that neon is the lasing species.

26-23 Use the method leading to Eq. (8-45) to show that the loss rate Γ for a ring cavity with round-trip survival factor S and perimeter P is

$$\Gamma = \frac{c}{P}(1 - S)$$

26-24 Reproduce the curves shown in Figure 26-13 using the parameters given in the figure caption. Note that the effective pump rate can be found from the listed condition $\gamma_0 = 2\gamma_{th}$.

26-25 Produce curves like those shown in Figure 26-13 for the parameters given in the figure caption except let (a) $\kappa = 10^{-8} \, s^{-1}$, (b) $\kappa = 10^{-6} \, s^{-1}$, (c) $\gamma_0/\gamma_{th} = 1.1$, and (d) $\gamma_0/\gamma_{th} = 4$. In each case describe how changing the indicated parameter changes the curves.

26-26 Consider the electromagnetic fields given by

$$E_1 = E_0 \cos(2\pi\nu_1 t - k_1 z)$$

$$E_2 = E_0 \cos[2\pi(\nu_1 + \delta\nu)t - k_2 z]$$

    a. Show that at $z = 0$, $E_1 + E_2 = 2E_0$ at times given by $t = n/\delta\nu$, where n is an integer.

    b. Discuss the relevance of the result shown in (a) for a mode-locked laser.

26-27 The gain bandwidth (in nm) and the transition wavelength for three different laser systems are given below. Estimate the pulse width attainable with these laser systems if they are mode-locked.

Ar^+	$\lambda = 488$ nm	$\Delta\lambda = 0.004$ nm
He-Ne	$\lambda = 633$ nm	$\Delta\lambda = 0.002$ nm
Rhodamine 6G dye	$\lambda = 590$ nm	$\Delta\lambda = 80$ nm

26-28 In order to investigate the *bandwidth theorem*, plot the given function F as a function of time t for (a) $N = 5$, (b) $N = 10$, and (c) $N = 50$. In each case estimate the pulse width from the plot and compare the pulse width to the range of frequencies in the superposition.

$$F = \sum_{j=0}^{N} \cos[2\pi((10 + 0.2j)t/s]$$

26-29 Show that the sum E of the electric fields associated with N mode-locked cavity modes of equal amplitude and with frequencies $\nu_j = \nu_0 + j\nu_{fsr}$ can be written as

$$E = \sum_{j=-(N-1)/2}^{(N-1)/2} E_0 \cos(2\pi\nu_j t + \varphi_0)$$

$$= E_0 \cos(2\pi\nu_0 t + \varphi_0) \frac{\sin(N\pi\nu_{fsr}t)}{\sin(\pi\nu_{fsr}t)}$$

You may wish to review the mathematical procedure used to describe multislit diffraction in Section 11-6 as a guide for carrying out the indicated summation.

26-30 Use the relation in problem 26-29 to verify Eqs. (26-57) and (26-58).

26-31 Estimate the peak power and pulse repetition rate in a mode-locked Nd:YAG laser pulse of pulse width 70 ps if the Nd:YAG laser cavity is 1.5 m long, and the CW output power of the Nd:YAG laser system is 10 W.

26-32 Estimate the diffraction-limited far-field divergence angles of a beam output from the heterojunction laser diode illustrated in Figure 26-19.

26-33 What is the band-gap energy of an AlGaAs semiconductor used in a laser diode device that emits light of wavelength 800 nm?

26-34 What must the reflectance of the cleaved ends of the laser diode illustrated in Figure 26-19 be if the small-signal gain coefficient of the medium is 40/cm?

26-35 **a.** Show that solving Eqs. (26-54) and (26-55) for the *steady-state* photon number density N_p and population inversion N_2 gives,

$$N_p = \frac{\sigma c(V_g/V_c)R_{p2} - \kappa_2\Gamma}{\sigma c\Gamma}$$

$$N_2 = \frac{\Gamma}{\sigma c} \frac{V_c}{V_g}$$

    b. Use the result for N_p in (a) to form the following expression for the steady-state output irradiance from a ring laser like the one discussed in connection with Figure 26-8:

$$I_{out} = T_3 \frac{\kappa_2 h\nu'}{\sigma} \frac{(\sigma R_{p2}L/\kappa_2) - (1 - S)}{1 - S}$$

    c. Show that the relation from part (b) agrees with Eq. (26-43) only if the survival fraction is close to 1. (*Hint:* $\ln(1 - x) \approx x$, for small x.)

    d. Which relation, the one from part (b) or the one given in Eq. (26-43), is correct when S is not close to 1? Explain.

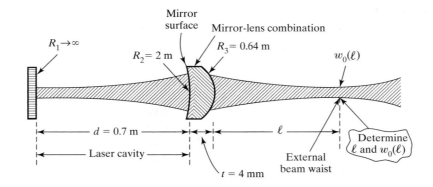

27

Characteristics of Laser Beams

INTRODUCTION

We now turn our attention to the nature of the optical beam generated by a laser. We shall see that a laser beam generated in a spherical mirror cavity has characteristics of both plane and spherical waves. In its simplest mode—the *fundamental TEM$_{00}$ Gaussian mode*—the laser beam takes the form of nearly spherical wavefronts, with the electric field exhibiting a transverse *Gaussian* irradiance distribution localized near the propagation axis. In other, more complex forms, referred to as *higher-order modes* or *Hermite-Gaussians*, the electric field takes on transverse irradiance distributions that depart from the simple Gaussian variation and exhibit an ordered pattern of "hot spots." In many cases, the output laser beam consists of a mixture of modes: the fundamental and several higher-order modes.

In this chapter we describe the general characteristics of Hermite-Gaussian laser beams and treat the propagation of these beams through general optical systems. After a rather thorough study of the fundamental mode of the laser beam, we shall examine the higher-order transverse modes and their transverse irradiance distributions.

27-1 THREE-DIMENSIONAL WAVE EQUATION AND ELECTROMAGNETIC WAVES

Consider a general electromagnetic field of the form

$$\vec{E} = E_x(x, y, z, t)\hat{x} + E_y(x, y, z, t)\hat{y} + E_z(x, y, z, t)\hat{z} \qquad (27\text{-}1)$$

In homogeneous media devoid of free charges or currents, each component of this electromagnetic field satisfies a wave equation like that given as Eq. (4-25). Since that wave equation is linear, one can solve it for each field component separately and then use these components to form the full field of Eq. (27-1). Consequently, without loss of generality, we shall consider an electromagnetic field of the form

$$\vec{E} = E(x, y, z, t)\hat{x} \tag{27-2}$$

We shall find it convenient to represent this field by the complex field $\widetilde{E}$ such that

$$E = \mathrm{Re}(\widetilde{E})$$

This field component $\widetilde{E}$ satisfies the wave equation

$$\nabla^2 \widetilde{E} - \frac{n^2}{c^2} \frac{\partial^2 \widetilde{E}}{\partial t^2} = 0 \tag{27-3}$$

Here,

$$\nabla^2 = \frac{\partial^2}{\partial x^2} + \frac{\partial^2}{\partial y^2} + \frac{\partial^2}{\partial z^2}$$

is the *Laplacian* operator, n is the refractive index of the medium through which the field is propagating, and c is the wave speed in vacuum. In Chapter 4 we described two important harmonic (single-frequency) solutions to Eq. (27-3): plane waves and spherical waves. These waves can be represented by the complex fields

$$\widetilde{E} = E_0 e^{i(kz - \omega t + \phi)} \qquad \text{Plane wave propagating in } +z\text{-direction} \tag{27-4}$$

$$\widetilde{E} = \frac{A}{r} e^{i(kr - \omega t + \phi)} \qquad \text{Spherical wave} \tag{27-5}$$

Here, $r = \sqrt{x^2 + y^2 + z^2}$ is the distance from the source of the spherical waves, and in each waveform $k = n\omega/c$. In Section 4-7 and in Figure 4-9 we also introduced, qualitatively, Gaussian beams. In the remainder of this chapter we will develop and characterize a mathematical description of the Gaussian beam solution and compare this useful representation of a laser field to plane and spherical waves.

27-2 GAUSSIAN BEAMS

If we were to examine the electromagnetic character of a typical laser beam, we would find that its wavefronts are essentially spherical surfaces with long radii of curvature that increase as the beam advances along the propagation axis. The combined wavefront and irradiance variation of such a typical laser beam passing through a converging lens might appear as shown in Figure 27-1. The solid guidelines, above and below the z-axis, represent the locus of points for which the beam's electric field irradiance in a transverse direction is equal to $1/e^2$ of its value on-axis. Thus these lines are used to define a continuously changing *beam width*. The dashed arcs transverse to the z-axis indicate the wavefronts of the beam. We desire solutions to the wave equation, Eq. (27-3),

Figure 27-1 An external laser beam, confined essentially to regions within the $1/e^2$-irradiance guidelines, is focused by a converging lens. The incoming beam is highly collimated so that its wavefronts are very nearly planar. The converging lens reshapes the wavefronts and focuses the beam—forming a beam waist—to the right of the lens. The beam diverges strongly as it propagates on past the beam waist.

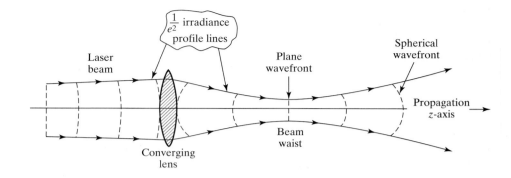

that resemble the field distribution shown in Figure 27-1. We choose, therefore, as a trial solution for the laser beam's electric field $\widetilde{E}(x, y, z, t)$ the form

$$\widetilde{E}(x, y, z, t) = U(x, y, z)e^{i(kz - \omega t + \phi)} \tag{27-6}$$

The term $U(x, y, z)$ describes a departure from a "pure" plane wave and, when determined, provides the details that accurately specify the irradiance and phase variations of the wave. The exponential term in Eq. (27-6) merely reflects the "more-or-less" plane wave nature of the solution.

Since our trial solution in Eq. (27-6) must satisfy the wave equation, we substitute it into Eq. (27-3) and obtain a defining equation for the yet unspecified function $U(x, y, z)$:

$$e^{i(kz - \omega t + \phi)}\left[\frac{\partial^2 U}{\partial x^2} + \frac{\partial^2 U}{\partial y^2} + \frac{\partial^2 U}{\partial z^2} + 2ik\frac{\partial U}{\partial z} - \left(k^2 - n^2\frac{\omega^2}{c^2}\right)U\right] = 0$$

The last term in the brackets on the left side of this equation vanishes outright, since $\omega \equiv kc/n$. That is,

$$\frac{\partial^2 U}{\partial x^2} + \frac{\partial^2 U}{\partial y^2} + \frac{\partial^2 U}{\partial z^2} + 2ik\frac{\partial U}{\partial z} = 0 \tag{27-7}$$

Equation (27-7) is exact, in the sense that it is just a representation of the full wave equation for fields taking the general form given in Eq. (27-6). Unfortunately, this equation is, in general, difficult to solve. However, we intend to describe fields $\widetilde{E}$ whose primary dependence on z is described by the rapidly varying factor e^{ikz}. Thus, it is appropriate to seek approximate solutions, to Eq. (27-7), that vary slowly with z so that the following condition holds:

$$\left|\frac{\partial^2 U}{\partial z^2}\right| \ll 2k\left|i\frac{\partial U}{\partial z}\right| \tag{27-8}$$

That is, we are motivated to drop the term $\partial^2 U/\partial z^2$ from the left side of Eq. (27-7). However, we wish to describe beams of finite transverse extent, so we cannot similarly neglect the transverse variation in the beam described by the second spatial derivatives of U with respect to x and y. Thus we seek solutions to the more tractable equation

$$\frac{\partial^2 U}{\partial x^2} + \frac{\partial^2 U}{\partial y^2} + 2ik\frac{\partial U}{\partial z} = 0 \tag{27-9}$$

Equation (27-9) is a nontrivial partial differential equation with complex terms. To solve it, we make an "educated" guess[1] at a solution, motivated in part by the cylindrical symmetry (about the propagation direction) that we expect in the electric field and in part by the complex nature that the solution $U(x, y, z)$ must exhibit. Thus we "guess" the form

$$U(x, y, z) = E_0 e^{i\{p(z)+[k(x^2+y^2)/2q(z)]\}} = E_0 e^{i\{p(z)+[k\rho^2/2q(z)]\}}$$
(27-10)

In the last equality we have introduced the cylindrical coordinate $\rho = \sqrt{x^2 + y^2}$. Since U depends only on ρ and z, it is cylindrically symmetric about the z-axis. Now, $p(z)$ and $q(z)$ are general functions, as yet undetermined, that are subject to constraints imposed by Eq. (27-9). After substituting Eq. (27-10) into Eq. (27-9), we obtain

$$\frac{2ikU}{q} - \frac{k^2}{q^2}\rho^2 U - 2kU\frac{\partial p}{\partial z} + \frac{k^2}{q^2}\rho^2 U\frac{\partial q}{\partial z} = 0$$

which, when rearranged in terms of powers of ρ^2, becomes

$$\left[\left(\frac{2ik}{q} - 2k\frac{\partial p}{\partial z}\right)(\rho^2)^0 + \left(\frac{k^2}{q^2}\frac{\partial q}{\partial z} - \frac{k^2}{q^2}\right)(\rho^2)^1\right]U = 0 \qquad (27\text{-}11)$$

If the function given in Eq. (27-10) is to be a solution for all ρ, then each factor multiplying a power of ρ^2 must equal zero separately. The coefficients of $(\rho^2)^0$ and $(\rho^2)^1$, when set equal to zero, yield

$$\frac{\partial p}{\partial z} = \frac{i}{q} \qquad (27\text{-}12)$$

and

$$\frac{\partial q}{\partial z} = 1 \qquad (27\text{-}13)$$

Equation (27-13) can be integrated easily to give

$$q(z) = q(0) + z$$

Now, $q(z)$ is, in general, a complex function. The coordinate z is, of course, real so $q(0)$ must be complex. The real and imaginary parts of $q(0)$ are two of the parameters that distinguish Gaussian beams one from another. Noting that q has the dimension of a length, let us take the real and imaginary parts of $q(0)$ to be, respectively, z_{0R} and z_{0I} so that $q(0) = z_{0R} + iz_{0I}$. Then $q(z) = z + z_{0R} + iz_{0I}$. It is convenient to choose the $z = 0$ plane to be the plane in which $q(z)$ is purely imaginary. This choice sets $z_{0R} = 0$ and so $q(0) = iz_{0I}$. An examination of Eq. (27-10), for $z = 0$, indicates that the imaginary part of $q(0)$ must be negative so that the field amplitude does not grow without bound as x and y tend to infinity. To emphasize this requirement, we write $z_0 = |z_{0I}|$ and

$$q(z) = z - iz_0 \qquad (27\text{-}14)$$

---

[1]A more detailed derivation is given in Anthony E. Siegman, *Lasers* (Mill Valley, Calif.: University Science Books, 1986), Ch. 16.

We have seen that the real part of $q(0)$ is related to the position of the plane $z = 0$. In the next section we examine the import of Eq. (27-14) and interpret the meaning of the imaginary part of $q(0)$, z_0, called the *confocal parameter* or *Rayleigh range*.

Before turning to that discussion we complete our general solution by using Eq. (27-14) in Eq. (27-12). This gives

$$\frac{\partial p}{\partial z} = \frac{i}{z - iz_0} \tag{27-15}$$

This relation can be integrated (see problem 27-3) and the solution manipulated to give

$$e^{ip(z)} = \sqrt{\frac{z_0^2}{z_0^2 + z^2}}\, e^{-i(\tan^{-1}z/z_0)} \tag{27-16}$$

Equations (27-14) and (27-16) can be substituted into Eq. (27-10) to form our completed Gaussian beam solution. Before analyzing the resulting expression, we will recast the solution into a more easily interpreted form.

27-3 SPOT SIZE AND RADIUS OF CURVATURE OF A GAUSSIAN BEAM

Motivated more by hindsight than by foresight, we write $q(z)$ in the form[2]

$$\frac{1}{q(z)} = \frac{1}{R(z)} + i\frac{\lambda}{\pi w(z)^2} \tag{27-17}$$

Here, $R(z)$ and $w(z)$ are *real* functions. We shall see that R is the *radius of curvature* of the wavefront and w is related to the transverse dimension of the beam. Following convention, we call $w(z)$ the *spot size* of the beam. The parameter $q(z)$ is often called the *complex radius of curvature* of the beam.

Using Equation (27-17) in the trial solution given in Eq. (27-10) leads to

$$U(\rho, z) = E_0 e^{ip(z)} e^{ik\rho^2/2R(z)} e^{-\rho^2/w^2(z)} \tag{27-18}$$

We see that the spot size $w(z)$ parameterizes the exponential drop-off of the electric field strength in the transverse (ρ) direction. The functions $R(z)$ and $w(z)$ can be determined by using Eq. (27-14) in Eq. (27-17) and performing some complex algebra. Proceeding,

$$\frac{1}{z - iz_0} = \frac{1}{R(z)} + i\frac{\lambda}{\pi w^2(z)}$$

Manipulating the left side of this relation leads to

$$\frac{1}{z - iz_0}\left(\frac{z + iz_0}{z + iz_0}\right) = \frac{z}{z^2 + z_0^2} + i\frac{z_0}{z^2 + z_0^2} = \frac{1}{R(z)} + i\frac{\lambda}{\pi w^2(z)}$$

---

[2]A word of caution is in order here. If we had chosen to write $\widetilde{E} = Ue^{-i(kz - wt + \phi)}$ in Eq. (27-6), then we would have written $q(0) = z + iz_0$. As a result, a *negative sign* between the two terms on the right of Eq. (27-17) would have been an appropriate choice.

Equating real and imaginary parts gives

$$R(z) = z\left(1 + \frac{z_0^2}{z^2}\right) \tag{27-19}$$

and

$$w^2(z) = \frac{\lambda z_0}{\pi}\left(1 + \frac{z^2}{z_0^2}\right) \tag{27-20}$$

Note that the spot size at $z = 0$, which we denote w_0, has the value $w_0 = w(0) = \sqrt{\lambda z_0/\pi}$, so the Rayleigh range z_0 can be written as

$$z_0 = \frac{\pi w_0^2}{\lambda} \tag{27-21}$$

and Eq. (27-20) can be rewritten as

$$w^2(z) = w_0^2\left(1 + \frac{z^2}{z_0^2}\right) \tag{27-22}$$

Equation (27-22) can be used to recast Eq. (27-16) in the somewhat simpler form

$$e^{ip(z)} = \frac{w_0}{w(z)}e^{-i\,\tan^{-1}(z/z_0)} \tag{27-23}$$

This relation and Eq. (27-20) can be used to specify completely the function $U(x, y, z)$ in Eq. (27-10). The final form of $U(x, y, z)$ can then be used in Eq. (27-6) to give a manageable form for the Gaussian beam solution to the wave equation,

$$\widetilde{E} = E_0\left(\frac{w_0}{w(z)}\right)e^{-\rho^2/w^2(z)}e^{ik\rho^2/2R(z)}e^{-i\,\tan^{-1}(z/z_0)}e^{i(kz-\omega t+\phi)} \tag{27-24}$$

where $R(z)$, z_0, and $w(z)$ are given by Eqs. (27-19), (27-21), and (27-22), respectively. Note that a beam traveling in the z-direction described by Eq. (27-24) is uniquely specified if the amplitude E_0, phase constant ϕ, wavelength λ, spot size at the beam waist w_0, and location of the beam waist ($z = 0$ plane) are known. In contrast, a plane wave traveling in the z-direction is parameterized by just E_0, ϕ, and λ. Of course, a complete specification of the full electric field vectors of plane waves and Gaussian beams also requires knowledge of the polarization of the fields.

27-4 CHARACTERISTICS
OF GAUSSIAN BEAMS

Let us now try to make sense of the Gaussian beam form given in Eq. (27-24)—along the way justifying the names "radius of curvature" for $R(z)$ and "spot size" for $w(z)$. We begin with an analysis of the irradiance profile of the Gaussian beam and then turn to an examination of the shape of the phase fronts associated with the beam.

Irradiance Profile

The irradiance I carried by a harmonic electromagnetic wave is proportional to the square of the magnitude of the complex electric field strength $\widetilde{E}$. Thus, using Eq. (27-24) we can write

$$I(\rho, z) = I_0\left(\frac{w_0}{w(z)}\right)^2 e^{-2\rho^2/w^2(z)} \tag{27-25}$$

Here, I_0 is the maximum irradiance, which occurs at the center of the *beam waist*. That is, $I_0 = I(\rho = 0, z = 0)$ For reasons to be elucidated shortly and illustrated in Figure 27-2, we have referred to the $z = 0$ plane as the plane containing the beam waist. Important features of Eq. (27-25) to be discussed in the following paragraphs and the next subsection are shown in Figure 27-2. According to Eq. (27-25), the irradiance along the z-axis (center of the beam) has the form

$$I(\rho = 0, z) = I_0\left(\frac{z_0^2}{z^2 + z_0^2}\right) \qquad \text{Axial irradiance} \tag{27-26}$$

indicating that the axial irradiance is less at axial positions further from the beam waist, a behavior that is consistent with a beam that is wider at points further from the beam waist. This widening of the beam follows from a further examination of the irradiance profile given in Eq. (27-25), which indicates that, at a given z, the irradiance $I(\rho, z)$ a transverse distance ρ from the center of the beam is related to the axial irradiance $I(\rho = 0, z)$ by

$$\frac{I(\rho, z)}{I(\rho = 0, z)} = e^{-2\rho^2/w^2(z)} \tag{27-27}$$

Evidently, the irradiance is reduced from its axial value by a factor of $1/e^2 \simeq 0.135$, where the transverse distance ρ from the axis is equal to the spot size, that is, where $\rho = w(z)$. Now according to Eq. (27-22), the spot size $w(z)$ increases as z^2 increases, accounting for the spreading of the beam depicted in Figure 27-2. The transverse dimension of a Gaussian beam, described by the spot-size parameter $w(z)$ thus changes at it propagates. Note that $w(z)$ is a measure of the radius (not diameter) of the beam.

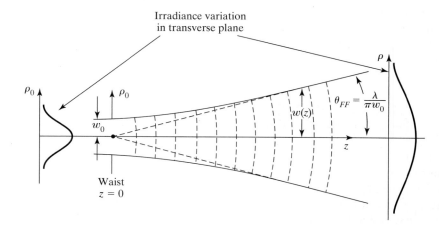

Figure 27-2 Gaussian spherical beam propagating in the z-direction. The spot size $w(z)$ at the beam waist ($z = 0$) is defined as w_0. The half-angle beam divergence $\theta_{FF} = \lambda/(\pi w_0)$ is valid only in the far field. Note the change in transverse irradiance as the beam propagates to the right.

Beam Spreading

Let us further investigate the spreading of a Gaussian beam as it propagates from the beam waist. Equation (27-22) can be used to obtain spot sizes at various transverse planes:

$$w(z = 0) = w_0 \qquad \text{Spot size at the beam waist}$$

$$w(z = z_0) = \sqrt{2}w_0 \qquad \text{Spot size at } z = z_0$$

$$w(z \gg z_0) = \frac{w_0 z}{z_0} = \frac{\lambda}{\pi w_0} z \qquad \text{Spot size in the far field } (z \gg z_0)$$

Here we have used Eq. (27-21) to obtain the final expression for the *far field* (that is, for $z \gg z_0$) spot size. Far-field expressions are strictly valid only in the limit that z/z_0 tends to infinity. In practice, far-field expressions can often be used safely when the distance from the beam waist, z, exceeds the Rayleigh range z_0 by a factor of 20 to 50. Note that the Rayleigh range z_0 is also a convenient measure of the distance over which a beam spreads appreciably in the *near field*. That is, z_0 is related to the *depth of focus* of the beam.

Now, since in the far field the spot size grows linearly with z, the far-field divergence angle[3] θ_{FF}, shown in Figure 27-2, satisfies the relation

$$\theta_{FF} \approx \tan\theta_{FF} = \frac{w(z \gg z_0)}{z} = \frac{\lambda}{\pi w_0} \qquad (27\text{-}28)$$

From Eq. (27-28) we draw the important conclusion that a beam with a smaller beam-waist spreads more rapidly than a beam with a larger beam-waist. This behavior is illustrated in Figure 27-3. Equation (27-28) also indicates that shorter-wavelength Gaussian beams tend to spread less than longer-wavelength beams.

Gaussian Beam Phase Fronts

Having discussed the changing irradiance of a Gaussian beam as it propagates, let us now turn to an investigation of the nature of the wavefronts (that is, the surfaces of constant phase) associated with a Gaussian beam. The nature of these wavefronts that we are about to investigate is illustrated in Figures 27-2 and 27-3. Before turning our full attention to a Gaussian beam, it is useful, for comparison purposes, to consider the nature of the surfaces of constant phase for a spherical wave, such as the one described by Eq. (27-5) and shown in Figure 27-4. At a given instant, the surfaces of constant phase for a spherical wave are given by the simple relation

$$kr = \text{constant} \qquad \text{Spherical wave} \qquad (27\text{-}29)$$

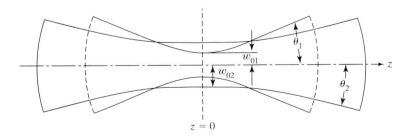

Figure 27-3 Two laser beams with a beam waist at $z = 0$. The beam with the smaller spot size at the beam waist spreads more rapidly than the beam with the larger spot size at the beam waist.

---

[3]Note that the far-field divergence angle, as we define it, is the half-angle spread of the beam.

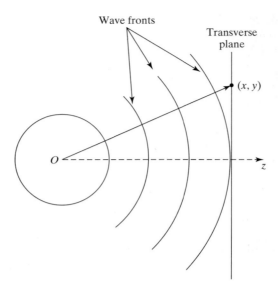

Figure 27-4 The surfaces of constant phase for spherical waves are concentric spheres radiating outwardly from a point source O.

This relation, for different values of the constant, describes concentric spherical surfaces with different radii. To prepare for a comparison with Gaussian beams that are confined near the z-axis, we will develop an approximation to Eq. (27-29) that is valid near the z-axis and away from the beam waist, that is, where $\rho \ll |z|$. Proceeding from Eq. (27-29),

$$kr = k(x^2 + y^2 + z^2)^{1/2} = k(\rho^2 + z^2)^{1/2}$$
$$= kz(1 + \rho^2/z^2)^{1/2} \approx kz(1 + \rho^2/2z^2)$$

Here, in order to form the approximate equality, we used a Taylor series expansion in the small quantity ρ^2/z^2. Without affecting the order of validity of this approximation, we may replace, in the last expression, the term $\rho^2/2z^2$ by $\rho^2/2r^2$, leading to the following expression describing the phase fronts of a spherical wave:

$$kr \approx kz + k\rho^2/2r \approx \text{constant} \qquad \text{Spherical wave near } z\text{-axis}$$

$$(27\text{-}30)$$

Now a Gaussian beam described by Eq. (27-24) has phase fronts that satisfy the condition

$$kz + k\rho^2/2R(z) - \tan^{-1}(z/z_0) = \text{constant} \qquad (27\text{-}31)$$

After neglecting the smaller third term on the left side of this relation, comparison with Eq. (27-30) leads to the conclusion that $R(z)$ does indeed play the role of the radius of curvature of the phase fronts of a Gaussian beam. The justification of this identification makes clear that it is only reasonable near the beam axis. However, since most of the power carried by the beam is concentrated near the axis, referring to the function $R(z)$ as the radius of curvature of the beam is appropriate. However, as suggested by Figures 27-1, 27-2, and 27-3 and in contrast to a spherical wave, the center of curvature of a Gaussian-beam wavefront changes as the beam propagates along the z-axis. In Figure 27-5 the expression given in Eq. (27-19) for the radius of curvature of the Gaussian-beam wavefronts is plotted. It is interesting to note that, at the beam waist, the wavefront has an infinite radius of curvature. That is, the wavefront is *planar* at the beam waist. Wavefronts propagating away from (or toward) the beam waist have reduced radii of curvature with the minimum

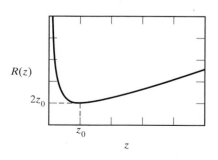

Figure 27-5 Wavefront radius of curvature $R(z)$ for a Gaussian beam. $R(z) = \infty$ at $z = 0$ and $z = \infty$. The minimum wavefront radius of curvature occurs at $z = z_0$ and has the value $R(z_0) = 2z_0$.

radius of curvature occurring at $z = \pm z_0$. In the far field ($z \gg z_0$), the phase fronts again become planar.

27-5 MODES OF SPHERICAL MIRROR CAVITIES

A mode of an optical cavity is a self-replicating field distribution. That is, for an electromagnetic field waveform to be a mode of an optical cavity, it must have the same spatial form after one round-trip that it had at the start of the round-trip. If this is the case, as the field is generated in the cavity, it continually adds in step to the field circulating through the cavity. Since Gaussian beams are confined in the transverse direction and have nearly spherical phase fronts, they "fit" into spherical mirror cavities, as indicated in Figure 27-6. Thus Gaussian beams are modes of spherical mirror cavities. As suggested by Figure 27-6, the mirror radii of curvature R_{M1} and R_{M2} and cavity length d constrain the nature of the Gaussian-beam mode of a spherical mirror cavity. Consider a wavefront moving towards Mirror 2 in Figure 27-6. If, when it reaches Mirror 2, its wavefront radius of curvature $R(z_2)$ matches the radius of curvature R_{M2} of Mirror 2, the wave will reflect back on itself, precisely retracing its path and shape through the beam waist and on to Mirror 1. Similarly, if the wavefront radius of curvature $R(z_1)$ at Mirror 1 is equal to R_{M1}, the left-going beam will reflect from Mirror 1 without distortion and retrace its path to Mirror 2. Thus, such a beam is a mode of the cavity. The constraints placed upon the Gaussian beam by the cavity are, therefore,

$$R(z_1) = z_1(1 + z_0^2/z_1^2) = R_{M1} \tag{27-32}$$

$$R(z_2) = z_2(1 + z_0^2/z_2^2) = R_{M2} \tag{27-33}$$

and

$$z_2 - z_1 = d \tag{27-34}$$

Here, z_1 and z_2 are the coordinates of the two cavity mirrors. (Recall that the beam waist is at $z = 0$.) Because of the back-and-forth nature of a beam propagating in a cavity, some care must be taken when using Eqs. (27-32) through (27-34). Equation (27-24) describes a beam propagating in the $+z$-direction. Let us consider the direction to the right across the page to be the positive z-direction. Then positions to the left of the beam waist (like Mirror 1 in Figure 27-6) have negative z-coordinates, and positions to the right of the beam waist (like Mirror 2 in Figure 27-6) have positive z-coordinates. Further, for Eqs. (27-32) and (27-33) to be sensible, a mirror to the *left* of the

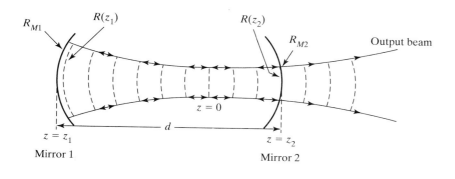

Figure 27-6   The Gaussian-beam modes of a spherical mirror optical cavity have wavefront radii of curvature that match the radii of curvature of the cavity mirrors.

beam waist should be taken to have a *negative* radius of curvature if its reflecting surface faces the beam waist, and a mirror to the *right* of the beam waist should be taken to have a *positive* radius of curvature if its reflecting surface faces the beam waist. Example 27-1 illustrates the manner in which the cavity structure determines the parameters and behavior of a Gaussian beam generated in the cavity.

Example 27-1

Consider the 4-mW, TEM_{00} helium-neon (He-Ne) laser ($\lambda = 632.8$ nm) with cavity length $d = 1$ m shown in Figure 27-7. The left mirror ($|R_{M1}| = 2$ m) is 100% reflecting. The right mirror ($R_{M2} \to \infty$) is a partially reflecting, plane, output mirror. The dashed profiles represent the wavefronts in the cavity.

    a. Determine the location of the beam waist.
    b. Determine the Rayleigh range z_0 for the Gaussian beam generated by this cavity.
    c. Determine the spot size w_0 at the beam waist.
    d. Determine the laser-beam spot size w on Mirror 1.
    e. What is the far-field beam-divergence angle θ_{FF} for this laser?

Solution

    a. The radii of curvature of the beam wavefronts must match the mirror radii of curvature. Thus, $R(z_2) = \infty$. For this to be true, as indicated by Eq. (27-33), $z_2 = 0$. Thus the beam waist is at Mirror 2.
    b. According to Eq. (27-34), $z_1 = z_2 - d = -d$. Using this in Eq. (27-32) gives $R_{M1} = -d(1 + z_0^2/d^2)$. Solving this expression for z_0 gives $z_0 = d\sqrt{-1 - R_{M1}/d} = (1\text{ m})\sqrt{-1 - (-2/1)} = 1$ m.
    c. Using Eq. (27-21), $w_0 = \sqrt{\lambda z_0/\pi} = \sqrt{(632.8 \times 10^{-9})(1)/\pi}$ m $= 4.49 \times 10^{-4}$ m $= 0.449$ mm.
    d. Using Eq. (27-22) with $w_0 = 4.49 \times 10^{-4}$ m, $z = -1$ m, and $z_0 = 1$,

$$w^2(z) = (4.49 \times 10^{-4})^2[1 + (-1/1)^2]\text{ m}$$

giving $w(z) = 6.35 \times 10^{-4}$ m $\cong 0.64$ mm. Thus the spot size increases from a radius of $w = 0.45$ mm at the waist to $w = 0.64$ mm at Mirror 1.
    e. Using Eq. (27-28),

$$\theta_{FF} = \frac{\lambda}{\pi w_0} = \frac{632.8 \times 10^{-9}}{(3.14)(4.49 \times 10^{-4})} = 4.49 \times 10^{-4}\text{ rad} \cong 0.45\text{ mrad}$$

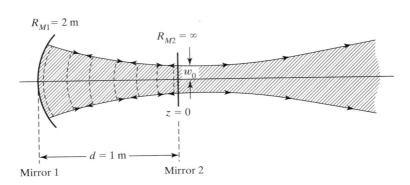

Figure 27-7   Sketch for Example 27-1.

27-6 LASER PROPAGATION THROUGH ARBITRARY OPTICAL SYSTEMS

With the basic law of propagation for the complex radius of curvature $q(z)$ in Eq. (27-14) and the defining equations for the real radius of curvature $R(z)$ and beam width $w(z)$ in Eqs. (27-19) and (27-22), we are able to characterize the beam parameters for laser propagation in any homogeneous medium of refractive index n. We now wish to address the question of how the beam changes when it is modified by an arbitrary optical system, one that contains lenses, mirrors, prisms, and so forth.

It is helpful to note the similarity between the behavior of ordinary spherical waves encountered in geometrical optics and Gaussian spherical waves encountered here. In Figure 27-8, the basic law of propagation for each type of wave and the effect of a lens on reshaping the propagating wavefront is illustrated. One is struck with the correspondence between $R(z)$ for ordinary spherical waves and $q(z)$ for Gaussian spherical waves in the defining equations. For example, the law of propagation of ordinary spherical waves, along the z-axis, is given by

$$R_2 = R_1 + (z_2 - z_1) \qquad (27\text{-}35)$$

Similarly, the basic law of propagation for the laser beam follows from Eq. (27-14) and is given by

$$q_2 = q_1 + (z_2 - z_1) \qquad (27\text{-}36)$$

an equation identical to Eq. (27-35) except that R has been replaced by q.

Seizing the initiative, we claim that this correspondence[4] can be extended to other basic laws, for example, the simple lens law. For ordinary spherical waves, a common form of the law is

$$\frac{1}{R_2} = \frac{1}{R_1} - \frac{1}{f} \qquad (27\text{-}37)$$

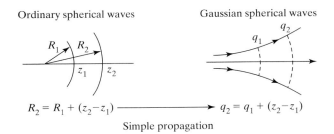

Ordinary spherical waves   Gaussian spherical waves

$$R_2 = R_1 + (z_2-z_1) \longrightarrow q_2 = q_1 + (z_2-z_1)$$

Simple propagation

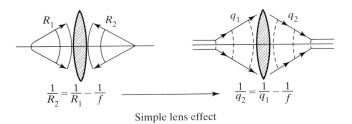

$$\frac{1}{R_2} = \frac{1}{R_1} - \frac{1}{f} \longrightarrow \frac{1}{q_2} = \frac{1}{q_1} - \frac{1}{f}$$

Simple lens effect

Figure 27-8 Correspondence between ordinary spherical waves and Gaussian spherical waves. If one knows the law for ordinary spherical waves, one can infer a similar law for Gaussian spherical waves by simply replacing $R(z)$ with $q(z)$.

---

[4] The correspondence we refer to here can be shown to be rigorous, not just analogous.

where R_1 and R_2 are the radii of curvature at incidence and refraction, respectively, and f is the focal length (see Figure 27-8). Replacing R by q for the laser beam, we obtain at once

$$\frac{1}{q_2} = \frac{1}{q_1} - \frac{1}{f} \tag{27-38}$$

This relationship predicts accurately the reshaping of the incident laser beam after refraction by a simple thin lens.

ABCD Law

With the correspondence between $R(z)$ and $q(z)$ established, we can, with the help of the matrix methods outlined in Chapter 18, develop a simple, yet powerful, recipe for laser-beam propagation through an arbitrary optical system. Consider Figure 27-9. Recall from Chapter 18 that, at a given plane, a ray can be described by its height y and slope angle α relative to the optical axis.

Consider a ray with parameters (y_1, α_1) incident on the entrance plane of an arbitrary optical system described by the overall system matrix

$$\begin{bmatrix} A & B \\ C & D \end{bmatrix}$$

Upon emerging from the system, the same ray has parameters (y_2, α_2). The radius of curvature R of an ordinary spherical wave can be related to its appropriate paraxial ray parameters y and α by

$$R = \frac{y}{\alpha} \tag{27-39}$$

We know from matrix optics that when ray 1 is changed into ray 2 by an optical system, the change can be represented by the $ABCD$ system matrix as follows:

$$\begin{bmatrix} y_2 \\ \alpha_2 \end{bmatrix} = \begin{bmatrix} A & B \\ C & D \end{bmatrix} \begin{bmatrix} y_1 \\ \alpha_1 \end{bmatrix} \tag{27-40}$$

Then it must be true that

$$y_2 = Ay_1 + B\alpha_1 \quad \text{and} \quad \alpha_2 = Cy_1 + D\alpha_1 \tag{27-41}$$

Dividing the first equation by the second, and using Eq. (27-39), we obtain

$$R_2 = \frac{AR_1 + B}{CR_1 + D} \tag{27-42}$$

Generalizing the basic result in Eq. (27-42) to a Gaussian spherical wave, by replacing $R(z)$ with $q(z)$, we obtain at once

$$q_2 = \frac{Aq_1 + B}{Cq_1 + D} \qquad ABCD \text{ propagation law} \tag{27-43}$$

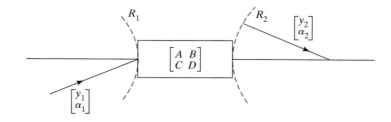

Figure 27-9 Propagation of ordinary spherical waves through an arbitrary optical system via matrix formulation.

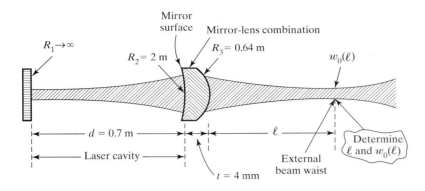

Figure 27-10 Geometry for a He-Ne laser system. With the parameters for the laser system specified, one can use the $ABCD$ propagation law to determine the location ℓ and size of the beam waist $w_0(\ell)$ outside of the laser.

The heuristic approach used here to develop this relation can be shown to result from a general solution to the wave equation for propagation through media that can be represented by an $ABCD$ matrix.[5] The relationship between q_2 and q_1 in Eq. (27-43) enables one to describe the new shape of the laser beam after it passes through an arbitrary optical system. One need only know the incident beam parameter (q_1) and the overall 2×2 matrix that characterizes the optical system. Equation (27-43) is a powerful result. It is often referred to as the $ABCD$ *propagation law* for Gaussian laser beams. For example, Figure 27-10 illustrates a typical laser system to which the $ABCD$ propagation law can be applied.

The problem depicted in Figure 27-10 is to determine the location ℓ of the external beam waist and its spot size w_0. Clearly, Eq. (27-43) can be used if one knows a value for q_1 somewhere in the cavity—say, at the left mirror—and the $ABCD$ matrix for the optical system that must, consequently, extend from the left mirror though the right mirror-lens combination and on to the transverse plane containing the external beam waist. The value of q_2 derived with the aid of Eq. (27-43) then predicts both ℓ and w_0 for the externally focused beam waist. The outline of the solution is given in the example that follows.

Example 27-2

For the He-Ne laser geometry given in Figure 27-10, use the $ABCD$ propagation law to determine the spot size w_0 of the external beam waist and its distance from the outer surface R_3 of the mirror-lens combination.

    a. Determine the complex radius of curvature q_1 at the plane mirror in the cavity.

    b. Then develop an $ABCD$ matrix for propagation to the external beam waist.

    c. Determine q_2, the complex radius of curvature at the external beam waist by using the $ABCD$ propagation law.

    d. From q_2, determine the spot size w_0 and location of the beam waist.

Solution

We sketch the solution in outline form only. Details are to be worked out in a series of problems at the end of this chapter.

    a. Since $R_1 \rightarrow \infty$ at the plane mirror,

$$\frac{1}{q_1} = \frac{1}{R_1} + \frac{i\lambda}{\pi w_1^2} = \frac{i\lambda}{\pi w_1^2} \quad \text{or} \quad q_1 = -\frac{i\pi w_1^2}{\lambda}$$

---

[5]Amnon Yariv, *Quantum Electronics* 3d ed. (New York: John Wiley & Sons, 1989), Ch. 6.

One can evaluate q_1 as follows. Relate the wavefront curvature at R_2—where the wavefront curvature matches the mirror curvature—to the spot size w_1 at the plane mirror and the distance $z_2 = d$ from plane to curved mirror R_2. Use Eqs. (27-19) and (27-21) to write

$$R_2 = z_2\left[1 + \left(\frac{\pi w_1^2}{\lambda z_2}\right)^2\right]$$

where $\lambda = 0.633 \times 10^{-6}$ m, $z_2 = 0.7$ m, and $R_2 = 2.0$ m. Solve for w_1 to get $w_1 = 4.38 \times 10^{-4}$ m. Then $q_1 = -i\pi w_1^2/\lambda = -i\,0.952$ m.

b. From the plane mirror to the external beam waist, form the system matrix

$$\begin{bmatrix} A & B \\ C & D \end{bmatrix} = \begin{bmatrix} 1 & \ell \\ 0 & 1 \end{bmatrix}\begin{bmatrix} 1 & 0 \\ \dfrac{-0.5}{0.64} & 1.5 \end{bmatrix}\begin{bmatrix} 1 & 0 \\ \dfrac{0.5}{3} & \dfrac{1}{1.5} \end{bmatrix}\begin{bmatrix} 1 & 0.7 \\ 0 & 1 \end{bmatrix}$$

$$\underbrace{}_{\text{transfer}} \quad \underbrace{}_{\text{refraction}} \quad \underbrace{}_{\text{refraction}} \quad \underbrace{}_{\text{transfer}}$$

Carry out the indicated matrix multiplication to obtain

$$\begin{bmatrix} A & B \\ C & D \end{bmatrix} = \begin{bmatrix} 1 - 0.53\ell & 0.7 + 0.63\ell \\ -0.53 & 0.63 \end{bmatrix}$$

Note that we have ignored the transfer matrix for propagation through the mirror-lens combination. This neglect is justified in problem 27-7.

c. Use Eq. (27-43) with results from parts (a) and (b) to obtain an expression for $q_2(\ell)$.

d. Obtain a second expression for $q_2(\ell)$ by using Eq. (27-17),

$$\frac{1}{q_2(\ell)} = \frac{1}{R_2(\ell)} + \frac{i\lambda}{\pi w_2(\ell)^2}, \text{ with } R_2(\ell) \longrightarrow \infty.$$

e. Equate the real and imaginary parts of the two expressions for $q_2(\ell)$ to find $\ell = 0.06$ m and $w_2(\ell) = 5.4 \times 10^{-4}$ m. Thus, the external beam waist has a spot size $w_2(\ell)$ of 0.54 mm located 6 cm from the R_3 surface of the mirror-lens combination.

Collimation of a Gaussian Beam

A collimated laser beam is a Gaussian beam with a long waist, as shown in Figure 27-11. The *collimated beam length* is arbitrarily defined as the distance between two symmetrical, transverse planes on either side of the beam waist, the two planes being those in which the spot size $w(z)$ has increased by a factor of $\sqrt{2}$ over the spot size w_0 at the waist. We have shown earlier that this

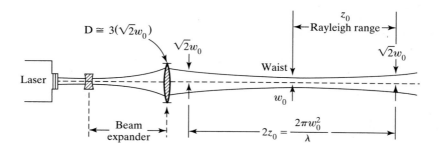

Figure 27-11 A collimated laser beam and the corresponding Rayleigh range z_0.

distance is the Rayleigh range z_0. The collimated region extends over two Rayleigh ranges, one on either side of the beam waist. As illustrated earlier in Figure 27-3, decreasing the spot size w_0 of the beam at the waist causes the beam to diverge more rapidly as it leaves the waist, leading to a smaller Rayleigh range. As demonstrated in Example 27-2, beam-shaping optics can control the size of the beam waist and thus control the length of the collimated region.

Beam-Shaping Optics for Optimum Beam Propagation

Gaussian beams are often passed through apertures such as mirrors, lenses, beam expanders, and telescopes. To find the fraction of incident power that passes through a circular aperture (lens, diaphragm, etc.) of radius a, we revisit the expression for the Gaussian-beam irradiance I given in Eq. (27-25). The total power Φ_{tot} in the beam is then obtained by evaluating the integral

$$\Phi_{\text{tot}} = \iint_A I \, dA = \frac{w_0^2}{w^2(z)} \iint_A I_0 e^{-2\rho^2/w^2(z)} \, dA$$

Here the integration is over the entire—infinite in extent—transverse cross-sectional area of the beam. When this is done, (see problem 27-13) one obtains for the total beam power

$$\Phi_{\text{tot}} = I_0 \left(\frac{\pi w_0^2}{2} \right)$$

Thus the total power carried by the beam is, as it must be, independent of z and has the form of the irradiance I_0 at the center of the beam ($\rho = 0, z = 0$) multiplied by an *effective* area of the beam, $\pi w_0^2/2$. The fractional power $\Phi_{\text{frac}} = \Phi(\rho = a)/\Phi_{\text{tot}}$ passing through a circular aperture of radius a, has the form

$$\Phi_{\text{frac}} = \frac{\Phi(\rho = a)}{\Phi_{\text{tot}}} = \frac{1}{\Phi_{\text{tot}}} \iint_{\text{Aperture}} I \, dA$$

$$\Phi_{\text{frac}} = \frac{(w_0^2/w^2(z))I_0}{\Phi_{\text{tot}}} \int_0^a e^{-2\rho^2/w^2} (2\pi\rho \, d\rho)$$

Carrying out the integration and using the result just found for Φ_{tot}, we find

$$\Phi_{\text{frac}} = 1 - e^{-2a^2/w^2}$$

Figure 27-12 shows the fractional power $\Phi(\rho = a)/\Phi_{\text{tot}}$ that passes through a circular aperture of radius a, versus the ratio a/w, where w is the

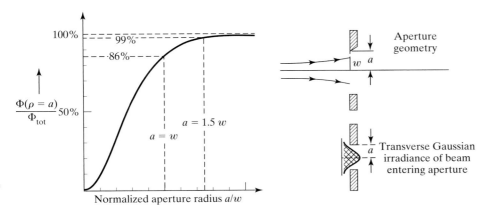

Figure 27-12 Fractional power transmission of a Gaussian beam of spot size w through a circular aperture of radius a.

spot size of the beam at the aperture location. Notice that 86% of the beam gets through (14% blocked) when the aperture radius equals the spot size, while just under 99% gets through when the aperture radius is increased to 1.5 times the spot size. Thus, if each beam-shaping element that passes the laser beam in a given optical system has a *diameter* D equal to 3 times the beam spot size $(D = 3w)$, *nearly* 99% of the beam gets through. Even for this case, we should note[6] that diffraction effects caused by sharp-edged circular apertures produce ripple effects on irradiance patterns in the near field and a reduction of on-axis irradiance in the far field of about 17%. To negate the diffraction effects, one can enlarge the aperture so that $D \cong 4.5w$. Beam-shaping optics with $D \geq 4.5w$ transmit essentially 100% of the beam power without superimposing *additional* diffraction effects on the beam.

With Eq. (27-21) for the Rayleigh range and the aperture diameter criterion $D = 4.5w$, one can calculate the Rayleigh range z_0 for typical lasers as a function of aperture diameter. Results are shown in Figure 27-13 for a He-Ne, HF (hydrogen fluoride), and CO_2 laser on a log-log plot. For example, if the aperture diameter $4.5w = 1$ cm, the collimated beam length $2z_0$ is equal to 24.5 m for He-Ne light at $\lambda = 632.8$ nm and 1.46 m for CO_2 light at 10.6 μm. For an aperture diameter $4.5w = 2$ cm, the collimated beam length becomes 98 m for the He-Ne laser and 5.8 m for the CO_2 laser, increasing as the square of the aperture diameter.

Focusing a Gaussian Beam

Let us now address the problem of focusing a Gaussian laser beam. We consider a rather general case, that of a beam with waist w_{01} located a distance Z_1 to the left of the lens, incident upon a positive thin lens of focal length f and

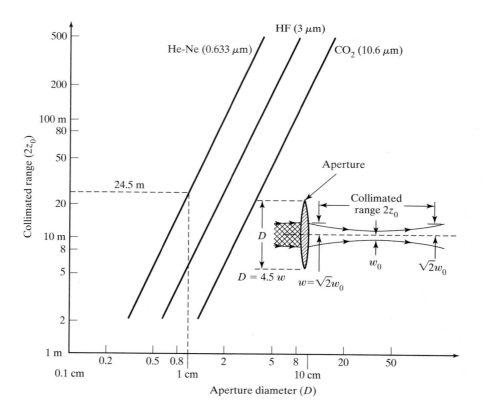

Figure 27-13 Collimated range $2z_0$ versus aperture diameter D, for a laser beam focused to a waist w_0, assuming that the aperture diameter $D = 4.5w$, where $w = \sqrt{2}w_0$.

[6]Anthony E. Siegman, *Lasers* (Mill Valley, Calif.: University Science Books, 1986), Ch. 17.

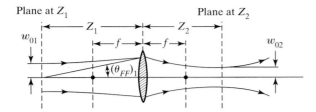

brought to a focus with beam waist w_{02} located a distance Z_2 to the right of the lens. The pertinent geometry is shown in Figure 27-14.

The problem before us is as follows: Given the beam waist w_{01} and distance Z_1 on the incident side of the lens, determine the beam waist w_{02} and the distance Z_2 on the output side of the lens. We shall use the $ABCD$ propagation law, Eq. (27-43), which involves the $ABCD$ matrix from transverse plane Z_1 to transverse plane Z_2. This matrix is

$$
\begin{bmatrix} A & B \\ C & D \end{bmatrix} = \begin{bmatrix} 1 & Z_2 \\ 0 & 1 \end{bmatrix} \begin{bmatrix} 1 & 0 \\ -\dfrac{1}{f} & 1 \end{bmatrix} \begin{bmatrix} 1 & Z_1 \\ 0 & 1 \end{bmatrix}
$$

$$
= \begin{bmatrix} 1 - \dfrac{Z_2}{f} & Z_1 + Z_2 - \dfrac{Z_1 Z_2}{f} \\ -\dfrac{1}{f} & 1 - \dfrac{Z_1}{f} \end{bmatrix}
\tag{27-44}
$$

Then the $ABCD$ propagation law gives

$$
q_2 = \frac{\left(1 - \dfrac{Z_2}{f}\right) q_1 + \left(Z_1 + Z_2 - \dfrac{Z_1 Z_2}{f}\right)}{-\dfrac{1}{f} q_1 + \left(1 - \dfrac{Z_1}{f}\right)}
\tag{27-45}
$$

Making use of Eq. (27-17) and noting that the radius of curvature of the beam at the beam waist is infinite, we find

$$
q_1 = -i \frac{\pi w_{01}^2}{\lambda} \text{ at } Z_1 \quad \text{and} \quad q_2 = -i \frac{\pi w_{02}^2}{\lambda} \text{ at } Z_2
$$

Using these relations in Eq. (27-45), we obtain a rather complicated equation that contains the desired unknowns, Z_2 and w_{02}. After equating real parts and imaginary parts, we sort out the unknowns w_{02} and Z_2 in the following form:

$$
\frac{1}{w_{02}^2} = \frac{1}{w_{01}^2} \left(1 - \frac{Z_1}{f}\right)^2 + \frac{1}{f^2} \left(\frac{\pi w_{01}}{\lambda}\right)^2
\tag{27-46}
$$

$$
Z_2 = f + \frac{f^2 (Z_1 - f)}{(Z_1 - f)^2 + (\pi w_{01}^2/\lambda)^2}
\tag{27-47}
$$

We note from Eq. (27-47) that for the general case, $Z_2 \neq f$. Apparently then, the Gaussian beam is not brought necessarily to a focus in the focal plane of the lens, that is, at a distance f to the right of the lens.

We can make some practical assumptions that will simplify Eqs. (27-46) and (27-47) considerably. First, consider Eq. (27-46). If $w_{01} \gg w_{02}$—that is, a strong positive lens is used in the focusing process—then the first term on the right side of Eq. (27-46) can be neglected, and the focused beam waist w_{02} is given approximately by

$$w_{02} \cong \frac{f\lambda}{\pi w_{01}} = f(\theta_{FF})_1 \tag{27-48}$$

Next, in Eq. (27-47), if we have a physical system for which $\pi w_{01}^2/\lambda \gg (Z_1 - f)^2$ —not an uncommon situation[7]—then it follows that Eq. (27-47) reduces to

$$Z_2 \cong f \tag{27-49}$$

The important special case in which the lens is placed at the beam waist of the input beam ($Z_1 = 0$) is explored in problem 27-12.

For a beam such that w_{01} is approximately equal to the radius of the lens, multiplying the numerator and denominator of Eq. (27-48) by 2 and using the definition $f\# \equiv f/D$ for the f-number of the lens, where D is the lens diameter, leads to

$$w_{02} = \frac{2f\lambda}{\pi(2w_{01})} = \frac{2\lambda f\#}{\pi} \tag{27-50}$$

In general, then, a smaller $f\#$ of the focusing lens produces a smaller beam waist at the focused spot.

27-7 HIGHER-ORDER GAUSSIAN BEAMS

The solution for the cylindrically symmetric Gaussian beam, derived earlier and displayed in Eq. (27-24), represents the lowest order—that is, fundamental—transverse electromagnetic mode that exists in the open-side-wall laser cavity. Other modes—higher-order modes—exist that do not have a pure Gaussian profile for their irradiance variation in the transverse plane. Let us return to our earlier development in this chapter and generalize Eq. (27-10), which we first guessed to have the form

$$U(x, y, z) = E_0 e^{i[p(z)+k(x^2+y^2)/2q(z)]}$$

as

$$U(x, y, z) = E_0 g\left(\frac{x}{w}\right) h\left(\frac{y}{w}\right) e^{i[p(z)+k(x^2+y^2)/2q(z)]} \tag{27-51}$$

The presence of the functions $g(x/w(z))$ and $h(y/w(z))$ admit waveforms that do not have cylindrical symmetry. Substitution of Eq. (27-51) into Eq. (27-9),

---

[7]In many cases, laser beams are first expanded in cross section before being focused to a small spot, as in the case of laser welding. In these instances, the beam waist w_{01} may be of the order of a centimeter or more, thereby ensuring that the inequality $\pi w_{01}^2/\lambda \gg (z_1 - f)^2$ is satisfied.

the defining equation for U, leads to

$$
\underbrace{\frac{1}{w^2}\frac{g''}{g} + \frac{2ikx}{w}\left(\frac{1}{q} - \frac{1}{w}\frac{\partial w}{\partial z}\right)\frac{g'}{g}}_{1} + \underbrace{\frac{1}{w^2}\frac{h''}{h} + \frac{2iky}{w}\left(\frac{1}{q} - \frac{1}{w}\frac{\partial w}{\partial z}\right)\frac{h'}{h}}_{2}
$$

$$
+ \underbrace{\frac{2ik}{q} - 2k\frac{\partial p}{\partial z}}_{3} - \underbrace{\frac{k^2}{q^2}(x^2 + y^2)\left(1 - \frac{\partial q}{\partial z}\right)}_{4} = 0 \tag{27-52}
$$

where the primes denote differentiation with respect to the arguments of g and h.

Inspection of Eq. (27-52) indicates that for a given z, that is, for a given transverse plane, the first bracketed series of terms, the g-expression, is a function of x alone; the second, the h-expression, is a function of y alone; the third is independent of both x and y; and the fourth is identically zero, since $q = q_0 + z$ is the propagation law that we continue to take to be valid. Thus, Eq. (27-52) can be satisfied for all x, y at arbitrary z only if (1) the g-expression equals a constant, say $-C_1$, (2) the h-expression equals a constant, say $-C_2$, thereby (3) leaving the third term equal to $C_1 + C_2$. For the g-expression, we write

$$
\frac{1}{w^2}\frac{g''}{g} + \frac{2ikx}{w}\left(\frac{1}{q} - \frac{1}{w}\frac{\partial w}{\partial z}\right)\frac{g'}{g} = -C_1 \tag{27-53}
$$

We can show, with the help of Eqs. (27-17), (27-21), and (27-22), that the term in parentheses is given by

$$
\frac{1}{q} - \frac{1}{w}\frac{\partial w}{\partial z} = \frac{i\lambda}{\pi w^2} \tag{27-54}
$$

If we use Eq. (27-54) and a change in variable, $\xi = \sqrt{2}x/w$, in Eq. (27-53), we obtain immediately the *Hermite* differential equation,

$$
\frac{\partial^2 q}{\partial \xi^2} - 2\xi\frac{\partial q}{\partial \xi} + \frac{C_1 w^2}{2}q = 0 \tag{27-55}
$$

This equation is known to have a solution only if

$$
\frac{C_1 w^2}{2} = 2m \qquad \text{where } m = 0, 1, 2, \ldots \tag{27-56}
$$

The solutions to Eq. (27-55) are the well-known *Hermite polynomials*,

$$
g(\xi) = H_m(\xi) = H_m\left(\frac{\sqrt{2}x}{w}\right) \tag{27-57}
$$

where the $H_m(\xi)$ can be obtained for each appropriate m from the generating function

$$
H_m(\xi) = (-1)^m e^{\xi^2}\frac{d^m}{d\xi^m}(e^{-\xi^2}) \tag{27-58}
$$

Solution of Eq. (27-58) for $m = 0, 1, 2, \ldots$ gives

$$m = 0: \quad H_0(\xi) = 1$$

$$m = 1: \quad H_1(\xi) = 2\xi = \frac{2\sqrt{2}x}{w}$$

$$m = 2: \quad H_2(\xi) = 4\xi^2 - 2 = \frac{8x^2}{w^2} - 2$$

(27-59)

Applying the second condition and solving for the h-expression in Eq. (27-52), in a manner identical to that for the g-expression, we obtain, using the change of variable $\eta = \sqrt{2}y/w$,

$$\frac{\partial^2 h}{\partial \eta^2} - 2\eta \frac{\partial h}{\partial \eta} + \frac{\lambda_2 w^2}{2} h = 0 \qquad (27\text{-}60)$$

for which the solutions are again the Hermite polynomials:

$$h(\eta) = H_n(\eta) = H_n\!\left(\frac{\sqrt{2}y}{w}\right), \quad \text{where } n = 0, 1, 2, \ldots \qquad (27\text{-}61)$$

obtained for all n from the generating expression, Eq. (27-58).

Before investigating the effects that the Hermite functions $g(\xi)$ and $h(\eta)$ have on the transverse nature of the beam irradiance, let us examine the consequences of the third condition imposed on Eq. (27-52),

$$\frac{2ik}{q} - 2k\frac{\partial p}{\partial z} = C_1 + C_2 \qquad (27\text{-}62)$$

where $C_1 = 4m/w^2$ and $C_2 = 4n/w^2$. Note that if $C_1 = C_2 = 0$, which is necessarily true when $m = n = 0$ (see Eq. 27-56), then $g(\xi) = h(\eta) = 1$ and we recover the cylindrically symmetric Gaussian-beam solution. In this case, Eq. (27-62) reduces to $\partial p/\partial z = i/q$, the defining equation for p obtained earlier. However, in the general case, for which C_1 and C_2 are *not* equal to zero, Eq. (27-62) remains the defining equation for the function $p(z)$. So, we must solve

$$\frac{\partial p}{\partial z} = \frac{i}{q(z)} - \frac{\lambda(m + n)}{\pi w^2(z)} \qquad (27\text{-}63)$$

to obtain an expression for $p(z)$. Substitution of Eq. (27-17) into Eq. (27-63) for $1/q(z)$ and then Eqs. (27-19) and (27-22) for $R(z)$ and $w(z)$, we can integrate the resulting equation to obtain

$$p(z) = \frac{i}{2} \ln\!\left[\frac{\lambda^2 z^2 + (\pi w_0^2)^2}{(\pi w_0^2)^2}\right] - (m + n + 1)\arctan\!\left[\frac{\lambda z}{\pi w_0^2}\right]$$

(27-64)

Here we have also used Eq. (27-21) to eliminate z_0 in favor of w_0 and λ. With the help of Eq. (27-64), we can form the factor $e^{ip(z)}$ as

$$e^{ip(z)} = \frac{w_0}{w(z)} e^{-i(m+n+1)\arctan(\lambda z/\pi w_0^2)} = \frac{w_0}{w(z)} e^{-i(m+n+1)\arctan(z/z_0)}$$

(27-65)

where we have used Eqs. (27-21) for z_0 and (27-22) for $w(z)$ to simplify the form of the expression.

The Hermite-Gaussian Beam Solutions

Now, collecting our results for p, q, $g(\xi)$, and $h(\eta)$, and putting them all together in the general expression for the electric field $\widetilde{E}(x, y, z, t)$ as initially introduced in Eq. (27-6), we obtain finally

$$\widetilde{E}_{mn}(x, y, z, t) = \underbrace{\left[E_0 \frac{w_0}{w(z)} H_m\left(\frac{\sqrt{2}x}{w(z)}\right) H_n\left(\frac{\sqrt{2}y}{w(z)}\right) e^{(-x^2+y^2)/w^2(z)} \right]}_{\text{amplitude}} \times$$

(27-66)

$$\underbrace{\left[e^{ik(x^2+y^2)/2R(z)} e^{i[kz - (m+n+1)\arctan(z/z_0)]} e^{-i\omega t} e^{i\phi} \right]}_{\text{phase}}$$

It is standard to refer to the various waveforms distinguished by the different values of m and n as TEM$_{mn}$ modes. Here TEM stands for **t**ransverse **e**lectric and **m**agnetic. Since these modes have different dependences on the transverse coordinates x and y, they are referred to as different transverse modes. The expression in Eq. (27-66) includes the higher-order transverse modes as well as the TEM$_{00}$ wave derived earlier. The first set of terms describes the *amplitude* variation of the electric field at any transverse plane z. The variations in transverse amplitude deviate more and more from a pure Gaussian form with higher integers m and n.

The second set of terms, a complicated *phase* factor, describes the nature of the wavefront as a function of m and n. Again we see that for $m = n = 0$, the phase term reduces to the form characteristic of the pure Gaussian spherical wave derived earlier. It can be shown that the frequency of a TEM$_{mn}$ beam that is a resonant mode of a spherical mirror cavity of length d must satisfy the relationship[8]

$$\nu_{mnq} = \left[q + \frac{m + n + 1}{\pi} \left(\arctan \frac{z_2}{z_0} - \arctan \frac{z_1}{z_0} \right) \right] \frac{c}{2d} \qquad (27\text{-}67)$$

where q is the *axial mode number* (not the complex radius of curvature); m and n are integers associated with the Hermite polynomials H_m and H_n; z_1 and z_2 are the coordinates ($z = 0$ is at the beam waist) of the cavity mirrors; and z_0 is the Rayleigh range.

The expression for $\widetilde{E}_{mn}(x, y, z, t)$ given in Eq. (27-66) with $\omega = 2\pi\nu_{mnq}$ describes the *Hermite-Gaussian* modes of a laser cavity with two spherical mirrors. Any arbitrary beam can be expanded as a linear combination of Hermite-Gaussian beams, each of which has the same propagation law, $q = q_0 + z$.

Field and Irradiance Patterns for Hermite-Gaussian Beams

With the help of Eq. (27-66) we can sketch the transverse electric field and irradiance variation for several of the lower-order modes—small m and n—and thus predict the nature of the "burn" pattern.

Let us use the scaled coordinates $x_s = \sqrt{2}x/w(z)$ and $y_s = \sqrt{2}y/w(z)$. With these relations and since the beam irradiance is proportional to the square of the magnitude of $\widetilde{E}_{mn}$,

$$I_{mn} = I_0 \left(\frac{w_0^2}{w^2(z)} \right) H_m^2(x_s) H_n^2(y_s) e^{-(x_s^2 + y_s^2)} \qquad (27\text{-}68)$$

---

[8]Amnon Yariv, *Quantum Electronics* 3d ed. (New York: John Wiley & Sons, 1989), Ch. 7.

Figure 27-15 shows the development of the irradiance patterns I_{mn} for the x_s-variation in the transverse plane at $z = z_0$. From symmetry, it is clear that identical results would be obtained for the y_s-variation. Thus, for two cases, $m = 0$ and $m = 1$, Figure 27-15 shows separate sketches for $H_m(x_s)$, for $H_m(x_s)\exp(-x_s^2/2)$, and, finally, for $[H_m(x_s)\exp(-x_s^2/2)]^2$, the sketch that gives direct evidence of the burn pattern to be expected.[9]

After studying Figure 27-15, we see that the irradiance variation has m zeros along the x-axis, discounting the zeros at the distant tails of the Gaussian envelope. Thus, for $m = 1$, there is a single zero at $x_s = 0$ and peaks on either side. The same type of variation is to be expected for the irradiance variation in the y-direction. Some of the irradiance patterns for low values of the Hermite integers m and n—predictable from Eq. (27-68) and Figure 27-15—are reproduced in Figure 27-16.

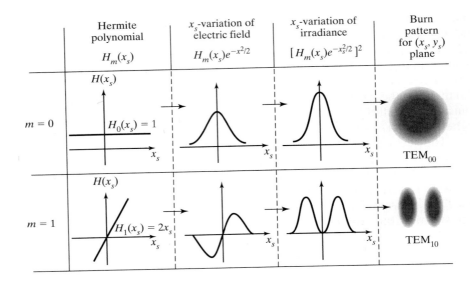

Figure 27-15 Laser-beam electric field and irradiance variations in the x_s-direction for two values of the Hermite integer m. Corresponding burn patterns for $m = 0, n = 0$ and $m = 1, n = 0$ are shown.

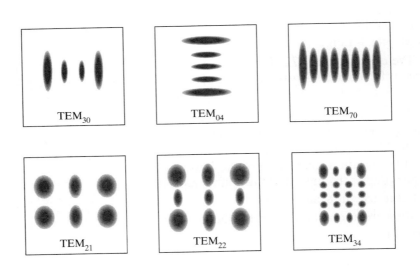

Figure 27-16 Representative sketches of irradiance or burn pattern for several different orders of Hermite-Gaussian optical resonator modes as they might be photographed in the output beam of a laser oscillator.

[9]To ensure understanding of the sketch patterns shown in Figure 27-15, the reader is invited, in problem 27-19, to compute the next row, for $m = 2$.

PROBLEMS

27-1 Describe the ways that the TEM_{00} Gaussian beam is similar to and different from (a) plane waves and (b) spherical waves.

27-2 Show that substitution of $\widetilde{E}(x, y, z, t) = U(x, y, z)e^{i(kz - \omega t + \phi)}$ in the wave equation (27-3) leads to Eq. (27-7).

27-3 Show that Eq. (27-16) follows from Eq. (27-15).

27-4 A TEM_{00} He-Ne laser ($\lambda = 632.8$ nm) has a beam waist w_0 (at $z = 0$) of 0.5 mm and a beam divergence of $\theta_{FF} = 0.4$ mrad.

    a. Determine the value of the complex radius of curvature q at the beam waist.

    b. Determine a numerical expression for the complex radius of curvature q, at a distance of 50 m from the beam waist, using each of the following expressions:

$$\frac{1}{q} = \frac{1}{R} + i\frac{\lambda}{\pi w^2(z)} \quad \text{and} \quad q = z - iz_0$$

    (*Hint*: Is the transverse plane, 50 m from the beam waist, far enough away to be in the far field? If so, what does this say about R and z?)

27-5 a. In problem 27-4, use Eqs. (27-19), (27-21), and (27-22) to determine $R(z)$ and $w(z)$ at $z = 50$ m.

    b. Is it true that $R(z) \cong z$ in the far field? Is it true that $\tan \theta_{FF} \cong w(z)/z$, where θ_{FF} is the beam divergence angle, can be used as a good approximation to determine $w(z)$ at $z = 50$ m?

27-6 A TEM_{00} He-Ne laser ($\lambda = 0.6328 \ \mu m$) has a cavity that is 0.34 m long, a fully reflecting mirror of radius $R = 10$ m (concave inward), and an output mirror of radius $R = 10$ m (also concave inward).

    a. Determine the location of the beam waist in the cavity.

    b. Determine the spot size at the beam waist, w_0.

    c. Determine the beam spot size $w(z)$ at the left and right cavity mirrors.

    d. Determine the beam divergence angle θ_{FF} for this laser.

    e. Where is the far field for this laser if one uses the criterion $z_{FF} \geq 50z_0$?

    f. If the laser emits a constant beam of power 5 mW, what is the on-axis irradiance at the position where $z_{FF} = 50z_0$?

27-7 Refer to Figure 27-10, where the *output element* of the laser is a mirror-lens combination with thickness 0.004 m, mirror surface curvature of $|R_2| = 2$ m, lens surface curvature of $|R_3| = 0.64$ m, and lens refractive index of 1.50.

    a. Using the definitions given for the refraction and translation matrices in Chapter 18, set up the $ABCD$ matrix for this element as follows:

$$\begin{bmatrix} A & B \\ C & D \end{bmatrix} = \begin{bmatrix} 1 & 0 \\ \dfrac{n - n'}{R_3 n'} & \dfrac{n}{n'} \end{bmatrix} \begin{bmatrix} 1 & t \\ 0 & 1 \end{bmatrix} \begin{bmatrix} 1 & 0 \\ \dfrac{n - n'}{R_2 n'} & \dfrac{n}{n'} \end{bmatrix}$$

Pay particular attention to the changing meaning of n and n' for the two refractions and to the sign conventions for R_2 and R_3 in the matrix formulations. Within

rounding approximations, you should find

$$\begin{bmatrix} A & B \\ C & D \end{bmatrix} = \begin{bmatrix} 1.0007 & 0.0027 \\ -0.5318 & 0.9979 \end{bmatrix}$$

    b. Since $L = 0.004$ m is a very small dimension compared with $|R_2| = 2$ m or $|R_3| = 0.64$ m, repeat the $ABCD$ calculation, replacing the translation matrix with the unit matrix

$$\begin{bmatrix} 1 & t \\ 0 & 1 \end{bmatrix} \longrightarrow \begin{bmatrix} 1 & 0 \\ 0 & 1 \end{bmatrix}$$

What then is the result of the $ABCD$ matrix for this "thin lens"?

27-8 a. Since the output element described in problem 27-7 is essentially a thin lens, compare the $ABCD$ matrix obtained in problem 27-7(b) with the thin-lens matrix, namely,

$$\begin{bmatrix} 1 & 0 \\ -\dfrac{1}{f} & 1 \end{bmatrix}$$

and deduce the focal length of the output element.

    b. Use the expression for the focal length of a thin lens, $\dfrac{1}{f} = \dfrac{n_2 - n_1}{n_1}\left(\dfrac{1}{R_1} - \dfrac{1}{R_2}\right)$, with careful attention to thin-lens sign conventions to obtain the focal length of the thin-lens output element. How do the results for parts (a) and (b) compare?

27-9 Referring to Example 27-2 and Figure 27-10, (a) determine an expression for q_1 at the plane mirror; (b) solve Eq. 27-22 for the spot-size value w_1; (c) obtain a numerical value for q_1; (d) multiply q_1 by the $ABCD$ matrix to obtain q_2; (e) use Eq. (27-17) and q_2 from part (d) to obtain ℓ and $w(\ell)$.

27-10 a. Referring to Example 27-2 in which the external beam waist is focused at $\ell \cong 0.06$ m with a waist size $w(\ell) = 0.54$ mm, use Eqs. (27-46) and (27-47) together with Figure 27-14 to obtain values for w_{02} and z_2. How do these results compare with those for $w_0(\ell)$ and ℓ obtained in the example?

    b. Explain why one *cannot* use the approximations

$$w_{02} \cong \frac{f\lambda}{\pi w_{01}} = f\theta \quad \text{and} \quad z_2 \cong f$$

in this instance.

27-11 Refer to the externally focused TEM_{00} laser beam shown in Figure 27-10, with beam waist $w_0(\ell) = 0.54$ mm, located at 0.06 m from the output element.

    a. Calculate the far-field distance $z_{FF} = 50z_0$ for the externally focused beam waist.

    b. Calculate the far-field beam divergence angle for the laser beam that emerges from the focused beam waist.

    c. Insert a 10 $\times$ beam expander in the beam at a distance $z = 30$ m past the focused beam waist. Calculate the beam spot size w at the entrance and exit faces of the beam expander.

d. Now place a thin lens of focal length 10 cm and appropriate diameter at a distance of 20 cm from the output face of the beam expander. With reference to Figure 27-14 and Eqs. (27-46) and (27-47), calculate z_2 and w_{02} for the newly focused beam. Could you have used the approximate formulas $w_{02} \cong f\lambda/\pi w_{01}$ and $z_2 = f$ in this instance? Why? How do the calculations for the exact formulas and approximate formulas compare?

27-12 a. Specialize Eqs. (27-46) and (27-47) for the case in which the lens is placed at the waist of the incident beam.

b. For the case described in (a), show that the location of the beam waist can be written as

$$Z_2 = \frac{f}{1 + f^2/z_{01}^2}$$

where $z_{01} = \pi w_{01}^2/\lambda$.

c. Investigate, for the case described in (a) and (b), whether $Z_2 \approx f$ for a variety of reasonable choices for lens focal lengths and beam parameters.

27-13 Carry out the integration necessary to verify the claim that the total power carried in a TEM$_{00}$ beam is

$$\Phi_{\text{tot}} = I_0 \frac{\pi w_0^2}{2}$$

27-14 Carry out the integrations necessary to show that the fraction of the power in a TEM$_{00}$ beam that is transmitted through a circular aperture of radius a is $1 - e^{-2a^2/w^2}$.

27-15 Explain how you can use an adjustable circular aperture (iris) and a power meter to determine the spot size w of a TEM$_{00}$ laser beam at any position along the beam.

27-16 Determine collimated beam lengths $2z_0$ for a TEM$_{00}$ Nd:YAG laser beam ($\lambda = 1.064\ \mu$m) focused by lenses of aperture diameters $D = 1$ cm, 2 cm, 3 cm, and 5 cm, respectively. Assume that the lens diameter D is related to the focused beam waist w_0 by the equation $D = 4.5\left(\sqrt{2}w_0\right)$. Refer to Figure 27-13 for geometry and similar calculations made for He-Ne, HF, and CO$_2$ TEM$_{00}$ lasers.

27-17 Given the generating function, Eq. (27-58), for Hermite polynomials $H_m(\xi)$, where $\xi = \sqrt{2}x/w$, verify the particular cases for $m = 0, 1, 2, \ldots$ given in Eq. (27-59).

27-18 Fill in the steps to show how Eq. (27-65) follows from Eq. (27-64).

27-19 Refer to Figure 27-15. Extend the "table" to include the case $m = 2, n = 0$. Thus, in a third row, sketch in curves for column 1, $H_m(x_s)$, column 2 for the x-variation curves of the electric field, column 3 for the x_s-variation curves of the irradiance, and column 4 for the expected burn pattern.

27-20 Find expressions for the fraction of the total power in a beam of spot size $w(z)$ that is transmitted through a circular aperture, centered on the beam, of radius a for a (a) TEM$_{00}$, (b) TEM$_{01}$, (c) TEM$_{11}$, and (d) TEM$_{02}$ beam.

27-21 Plot each of the transmittance functions found in problem 27-20 as a function of $a/w(z)$. Plot the four curves on the same set of axes.

27-22 Based on the plots obtained in problem 27-21, describe how an adjustable aperture can be used in a laser cavity to ensure that only the TEM$_{00}$ cavity mode would be present in the laser output.

27-23 The output from a single-mode TEM$_{00}$ Ar$^+$ laser ($\lambda = 488$ nm) has a far-field divergence angle of 0.001 rad and an output power of 5 W.

a. What is the spot size at the beam waist for this laser field?

b. What is the irradiance at the center of the beam waist ($\rho = 0, z = 0$) for this field?

c. What is the irradiance at the center of the beam 10 m from the beam waist?

27-24 Consider a laser cavity consisting of two spherical concave mirrors that are facing each other. Let the mirrors be separated by 20 cm and let each mirror have a radius of curvature of 100 cm. Find the mode-frequency separations: (a) $\nu_{0,0,q+1} - \nu_{0,0,q}$, (b) $\nu_{m,n,q+1} - \nu_{m,n,q}$, (c) $\nu_{0,1,q} - \nu_{0,0,q}$, (d) $\nu_{1,0,q} - \nu_{0,0,q}$, and (e) $\nu_{1,1,q} - \nu_{0,0,q}$.

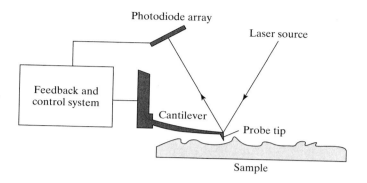

28

Selected Modern Applications

INTRODUCTION

When one endeavors to chronicle current applications in a rapidly evolving field such as optical technology, one accepts the inevitable fact that such information becomes outdated almost as rapidly as it is written. So, rather than attempt to provide an exhaustive summary of the current state of applications dependent upon optical technology, we limit ourselves to an overview of areas in which lasers and optics play an important role and then discuss a few selected applications of lasers and other optical systems. To gain a full understanding of some aspects of these selected applications requires an integration of topics discussed in some of the previous 27 chapters of this text. In the present chapter our intent, in part, is to provide stepping-off points for in-depth student research projects. We begin with a brief description of the breadth of laser applications and proceed to summarize in some detail four particular application areas; medicine, remote sensing, ultrashort pulses, and laser cooling and trapping. We then turn to a discussion of an increasingly important, alternative, source of coherent radiation, optical parametric amplifiers and oscillators. We conclude with an overview of near-field microscopy.

28-1 OVERVIEW OF LASER APPLICATIONS

Most laser applications can be organized into two general classes: (1) lasers and interactions and (2) lasers and information, outlined in Table 28-1. In the former, lasers interact with matter and cause desirable changes, either permanent or temporary. In the latter, lasers are used to send, detect, store, and process information.

TABLE 28-1 A CLASSIFICATION OF LASER APPLICATIONS

Lasers and interactions	Lasers and information
Materials processing	Communication
Cutting and drilling	Information processing
Welding	Optical sensing
Heat treating	Optical ranging
Medical	Optical storage
Marking and scribing	Printers and copiers
Microfabrication	Metrology
Laser spectroscopy	Holography
Laser-driven energy	Alignment
Weaponry	Point-of-sale scanners
Cooling and trapping	Entertainment and display

The broad class of applications that involve the interaction of lasers with matter includes the industrial machining and processing of materials, therapeutic and surgical uses in medicine, laser-driven energy sources, scribing, and microfabrication in semiconductor and computer technologies. The equally broad class that involves lasers and information includes information processing, optical sensing and ranging, optical communication, entertainment, printers and copiers, metrology, and alignment facilitators in construction and agriculture. In the next three sections of this chapter we examine but three of the many application of lasers detailed in Table 28-1.

28-2 LASERS IN MEDICINE

The laser beam—a scalpel of light—can cut, vaporize, rupture, weld, seal, cauterize, coagulate, open, and heal human tissue. In Chapter 19, for example, we discussed in some detail the use of lasers to reshape the cornea of the eye in order to correct defects in vision. An increasingly important addition to hospital operating rooms and outpatient treatment centers, the laser beam has emerged as an effective, noninvasive surgical tool. The suitability of a given laser for a particular medical application depends on several factors. For example, whether or not the radiation emitted by the laser is absorbed by water and/or blood determines the suitability of a given laser for use as a scalpel or for photocoagulation. With these requirements in mind, the absorption coefficients for water and hemoglobin (a primary component of blood) as a function of wavelength are shown in Figure 28-1. In addition, the ability to deliver laser energy in a pulsed form of a particular duration is important in many medical applications—for example, in the treatment of age spots and dental drilling. Finally the output radiation from the laser must propagate with relatively low loss through an optical fiber in order to noninvasively deliver laser energy to specific places in the interior of the body. We now provide brief descriptions of some laser systems commonly used in medical procedures.

Carbon Dioxide Laser

The CO_2 laser, one of the first to be used extensively by medical practitioners, emits mid-infrared (IR) energy at 10.6 μm that, as indicated in Figure 28-1, is strongly absorbed by water. Since body cells and tissues are 70% to 90% water, a CO_2 laser can make a clean incision or, if focused more sharply, can vaporize malignant tissue. Because a laser cauterizes and seals capillaries as it cuts, surgery is essentially bloodless. The laser energy can also seal nerve endings, significantly reducing postoperative pain. A drawback of the CO_2 laser remains its inability to be used readily with fiber optics; transmission of the

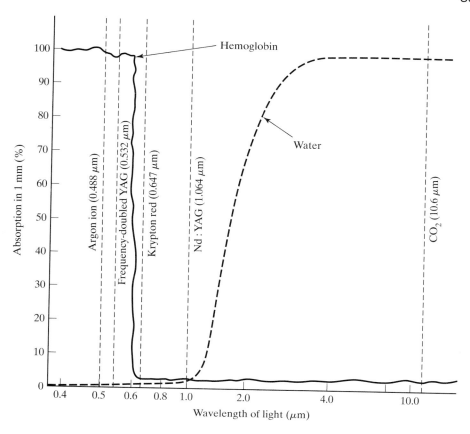

Figure 28-1 The percentage of light irradiance absorbed, as a function of wavelength, due to passage through 1 mm of water (dashed curve) and 1 mm of hemoglobin (solid curve).

beam from the laser to the target area is often accomplished with cumbersome, articulated optical arms. When defocused, the CO_2 laser can be used to ablate, shave, or vaporize soft tissue. Applications in this mode of operation include removal of tumors, treatment for gingivitis, removal of skin lesions, and a variety of cosmetic treatments of the skin. The CO_2 laser is the most common laser scalpel in use today.

Nd:YAG Laser

The Nd:YAG laser, with an operating wavelength of 1064 nm, mates well with an optical fiber. In addition, as is evident from Figure 28-1, Nd:YAG laser radiation is not strongly absorbed by either water or hemoglobin and so can penetrate through the skin and blood in an unfocused state, cutting or coagulating tissue only in the focal region of the beam. (See, for example, the description of its use in posterior capsulotomy in a subsequent subsection.) Passing the output from a Nd:YAG laser through a frequency-doubling crystal—typically, in medical applications composed of potassium-titanyl-phosphate (KTP)—produces a beam of wavelength 532 nm. This wavelength is strongly absorbed by hemoglobin but not by water and so can be used for photocoagulation. "KTP-lasers" find use in the repair of vascular lesions and in the vaporization of prostatic obstructions. The latter procedure can be performed, with a fiber inserted through the urethra, in about 30 minutes as an outpatient procedure.

Argon-Ion Laser

The blue-green argon-ion laser beam (488 nm and 514 nm) is selectively absorbed by red and brown substances, such as hemoglobin and melanin pigment spots. It can therefore stop retinal bleeding (photocoagulation) and weld detached retinas. It can fade port-wine birthmarks and tattoos, penetrating the nonpigmented upper layer of skin without harm.

Er:YAG and Ho:YAG

Erbium doped into the YAG crystal (Er:YAG) emits radiation at a wavelength of 2940 nm and so is strongly absorbed by water. It can be used to ablate soft tissue in cosmetic surgery and to ablate hard tissue, for example, in use as a dental drill. Holmium doped into a YAG crystal (Ho:YAG) emits radiation of wavelength 2070 nm and is used as an alternative to the frequency-doubled Nd:YAG laser to perform photoselective vaporization of prostatic obstructions, to precisely shave bone and cartilage, and to eliminate kidney stones.

Diode Lasers

Semiconductor diode lasers are attractive to medical practitioners because of their simplicity, relatively low cost, compact size, ability to propagate through fibers, low power consumption, ease of intensity modulation, and low cooling requirements—that is, for the same reasons that diode lasers are popular in a wealth of nonmedical applications. Currently diode lasers are used for hair removal and cosmetic surgery. It is likely that diode lasers will find increasing application in the medical arena.

Excimer Lasers

Excimer lasers emit ultraviolet radiation. The argon-fluoride (ArF) laser, which emits radiation of wavelength 193 nm, and the krypton-fluoride (KrF) laser, which emits radiation of wavelength 248 nm, are commonly used in medical procedures. As we discussed in detail in Chapter 19, excimer lasers are used to perform LASIK surgery for vision correction. In addition, these lasers are used in laser angioplasty (the removal of plaque and fatty deposits blocking arterial blood flow) and laser-assisted balloon angioplasty. Excimer lasers are ideal for microsurgical procedures that require precise positioning of the laser beam.

Other lasers used in medical procedures include the alexandrite laser (tattoo removal), tunable dye lasers (vascular lesions, tumor treatments), and copper vapor lasers (vascular lesions). Additional medical procedures, not mentioned in the preceding subsections, performed with the aid of laser light include gall bladder removal, neurosurgery, treatment of nose, ear, and throat disorders, a variety of orthopedic procedures, treatment of gastrointestinal disorders, many cosmetic procedures, and veterinary procedures of all sorts.

Posterior Capsulotomy: A Case Study

We conclude this section with the examination of a particular medical procedure—posterior capsulotomy—that highlights the unique capabilities that the laser brings to medical procedures. Posterior capsulotomy is used to treat the most common complication resulting from cataract surgery, wherein a clouded lens is replaced by a clear plastic lens implant. In a significant number of cases, the back lens covering that remains after the surgery is opacified (clouded). In posterior capsulotomy, a Nd:YAG laser is used to rupture these opacified membranes along the optical axis of the eye, creating an unobstructed path for light to reach the retina. The geometry of the eye and the laser beam used in this procedure are illustrated in Figure 28-2. The pulsed Nd:YAG laser beam can pass through the cornea, aqueous humor, and lens with essentially zero absorption, (see Figure 28-1). The high-intensity beam penetrates the front part of the eye and comes to a sharp focus just beyond the implanted lens, some 7 mm from the front corneal surface. Located there is the opacified membrane, the posterior segment of a weblike pillow (capsule) that previously contained the cataractous lens and now encases the implant.

Prior to the use of lasers in this procedure, the opacified membrane was removed invariably with invasive intravention by surgical instruments. In addition to the trauma involved with the operation, invasive surgery of the

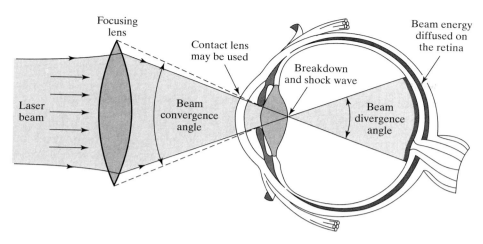

Figure 28-2   Focusing laser light to rupture membranes in posterior capsulotomies. Larger beam convergence angles (less than about ~17°)—limited by the pupillary aperture—lead to larger beam divergence angles and therefore, less irradiance at the retina.

eyeball is always attended by a possible introduction of foreign bodies and an increased risk of infection. By contrast, the Nd:YAG laser surgery, performed on an outpatient basis in a matter of minutes, is neither traumatic nor infectious.

In posterior capsulotomy, a short pulse of 1.06-μm laser radiation is focused at the target just beyond the implanted lens. The typical pulse delivered by a Q-switched Nd:YAG laser has an energy of 1 to 4 mJ and a pulse length of several nanoseconds. With such short times, the laser power, even for pulse energies as small as millijoules, reaches the megawatt range and higher.

Example 28-1

Determine the irradiance (power per unit area) delivered by a Nd:YAG laser pulse of energy 4 mJ and pulse length 1 ns when the pulse is focused on target to a tiny spot of diameter 30 μm.

Solution

The irradiance I is calculated directly as follows:

$$I = \frac{P}{A} \quad \text{where } P = \frac{4 \times 10^{-3}\,\text{J}}{1 \times 10^{-9}\,\text{s}} = 4 \times 10^6\,\text{W}$$

and

$$A = \frac{\pi D^2}{4} = \frac{(3.14)(30 \times 10^{-6})^2}{4} = 7.065 \times 10^{-10}\,\text{m}^2$$

Thus,

$$I = \frac{4 \times 10^6\,\text{W}}{7.065 \times 10^{-10}\,\text{m}^2} = 5.7 \times 10^{15}\,\text{W/m}^2$$

or

$$I = 5.7 \times 10^{11}\,\text{W/cm}^2$$

As the example shows, when such high powers are focused sharply at the target in tiny spots ranging from 5 to 50 μm in diameter, irradiances, or power densities, of 10^{12} W/cm^2 are developed. Power densities of this magnitude are accompanied by very high electric fields that cause, first, a dielectric breakdown of optical tissue and, second, the formation of a plasma. The explosive growth of the plasma gives rise to a strong shock wave that travels radially outward, mechanically rupturing the nearby taut, opacified membrane. Neither the laser radiation, which moves on toward the retina, or the expanding shock wave causes damage elsewhere in the eye.

28-3 REMOTE SENSING

The availability of satellites orbiting the Earth, combined with advances in laser technology, set the stage for laser remote sensing. Simply stated, remote sensing involves the detection of laser-scattered light from objects that are located remotely relative to active laser interrogating systems. The laser systems—sources and detectors—are frequently located on space platforms, whereas the remote objects, far below, may be the atmosphere, terrestrial vegetation, or the Earth's crust. Ground-based systems, also prominent but not necessarily remote, are used to detect pollution and to monitor global winds.

The technique used, often referred to as *laser radar*, or LIDAR (**li**ght **d**etection **a**nd **r**anging), depends on the presence of a laser source, a suitable target for interrogation, and sensitive detectors. For any particular remote-sensing application, an ideal laser must have an appropriate wavelength, tunability, output power, pulse repetition rate, and amplitude and frequency stability. It should, in addition, have high operating efficiency, be mechanically rigid, and possess a long operating lifetime. Current laser sources include carbon dioxide lasers, pulsed Nd:YAG and Nd-glass lasers, alexandrite lasers, Nd:YAG-pumped dye lasers, excimer lasers, AlGaAs-pumped Nd:YAG lasers, and a class of continuously tunable, solid-state lasers, such as the Ti:sapphire laser. Taken together, the family of special lasers provides wavelength coverage from 0.66 μm to 10.6 μm, sufficient for most remote-sensing applications.

The targets, whatever their nature, must interact with the interrogating laser light and produce a measurable change that can be detected. Such changes, returned to the detection system by the backscattered laser radiation, include fluorescence, Raman-shifted wavelengths, and Doppler effects, each of which provides important information on species identity, concentration, or movement. Highly specialized LIDAR systems, such as differential absorption LIDAR, Raman LIDAR, and Doppler LIDAR, are designed to detect the specific information content locked in the backscattered radiation.

The state of the art in remote sensing is immediately evident from a review of two applications involving the measurement of wind speed and shifts in the Earth's crust. The detailed movement and speed of tropospheric winds is of considerable interest in the study of global weather patterns. To ferret out such information, a space-based CO_2 laser system can operate from an orbiting spacecraft at distances from 250 km to 800 km above the Earth. From such altitudes, high initial laser power is needed in order to detect the weak backscattered radiation. The incident laser radiation is backscattered by dust particles carried along by the wind. Calculations of wind speed and direction are based on the Doppler frequency shifts in the reflected radiation. A CO_2 laser, designed to measure wind speeds with an accuracy of 1 m/s, requires 10-J pulses every second or so, with a frequency stability of several hundred kHz or better. Large power requirements of several kilowatts are needed to operate such laser systems. The availability of high power, stored and accessible during the space mission, poses a challenge.

The use of a space-borne LIDAR system to detect movement of the Earth's crust is of value in determining strains near earthquake zones, ground swell prior to volcanic activity, and large-area land settling. Based on time-of-flight measurements of short (nanosecond) laser pulses from space to ground retroflectors and back, and making use of simple methods of triangulation, distances to specific parts of the Earth's crust are determined to within 1 or 2 cm! By repeating the measurements at appropriate time intervals, significant earth movement is detected and monitored. Systems under consideration for studies of the Earth's crust include a Nd:YAG laser with pulse widths less than 0.5 ns and pulse repetition rates of the order of 10 per second. Backscattered signal detection on board the spacecraft can be accomplished with existing 6-in-diameter collecting telescopes and photomultiplier electronics.

LIDAR techniques, coupled with remote-sensing laser systems, are useful in the detection of pollution, identification of toxic-particle concentrations, or global measurements of specific atmospheric gases and aerosols. Optical-sensing technology has been used successfully to monitor smokestack emissions from industrial plants and to detect gas leaks in pipelines. On a broader scale, atmospheric concentrations of ozone, water vapor, and sulfuric acid can be measured with high precision. In particular, laser absorption spectrometers, relying on the principles of differential absorption and resonance fluorescence, have already demonstrated their usefulness. Global measurements that determine atmospheric aerosol concentrations are of much interest and are within the capability of remote-sensing systems. Such determinations clarify the role that aerosols play in the warming or cooling of our planet as well as the formation and evolution of haze layers in the troposphere. LIDAR measurements from a worldwide network of Earth-based observatories use time-of-flight measurements for beams reflected from mirrors placed on the moon to measure the Earth-to-moon distance with millimeter precision. Such measurements can be used to test predictions of general relativity.

28-4 ULTRASHORT PULSE PRODUCTION AND APPLICATIONS

In Chapter 26 we noted that mode-locked pulses could have a temporal pulse width Δt_p as short as the inverse of the gain bandwidth $\Delta \nu$ of the laser medium in the mode-locked system. In Example 28-2 we estimate the *bandwidth-limited* pulse width of the Ti:sapphire laser.

Example 28-2

Estimate the bandwidth-limited pulse width of a Ti:sapphire laser that has a bandwidth extending from 650 to 1100 nm.

Solution

The frequency bandwidth of the Ti:sapphire laser is bounded by the frequencies

$$\nu_{high} = c/\lambda_{low} = (3 \times 10^8 \text{ m/s})/(650 \times 10^{-9} \text{ m}) = 4.62 \times 10^{14} \text{ Hz}$$

$$\nu_{low} = c/\lambda_{high} = (3 \times 10^8 \text{ m/s})/(1100 \times 10^{-9} \text{ m}) = 2.73 \times 10^{14} \text{ Hz}$$

The gain bandwidth expressed as a frequency difference is, therefore, $\Delta \nu = \nu_{high} - \nu_{low} = 1.89 \times 10^{14}$ Hz. The bandwidth-limited pulse width is, then,

$$\Delta t_p = 1/\Delta \nu = 5.3 \times 10^{-15} \text{ s} = 5.3 \text{ fs}$$

Indeed, mode-locked laser pulses from a Ti:sapphire laser with a duration of 5 fs (and less) have been produced. In fact, to date the Ti:sapphire laser produces the shortest pulse of any laser system. (Pulse-shaping postprocessing and harmonic generation can produce narrower pulses of width less than 1 fs.)

This extremely short pulse duration, on the order of the period of an optical wave, can be used to follow fast processes in time-resolved detail. For example, mode-locked pulses of femtosecond duration can be timed to interrogate a chemical reaction at times separated by just a few femtoseconds. Chemical reactions proceed on a time scale of tens of femtoseconds, so with ultrashort pulse technology chemical bonds can be watched as they form. This has led to a new field called *femtochemistry*. Short (ps) and ultrashort (fs) pulses are used to obtain time-resolved information about biological processes as well as to track electron transport phenomena in semiconductors. In addition to interrogating very fast processes, ultrashort pulses have other unique uses. Thermal processes occur on time scales longer than the duration of an ultrashort pulse and so these pulses can cut and treat materials (organic and inorganic) with less thermal shock damage to the surrounding material. This allows for smaller and cleaner cuts in micromachining applications, for example. Finally, ultrashort pulses can have extraordinarily high peak powers, 50 GW or more. These enormous peak powers can initiate intense nonlinear reactions leading to novel states of matter heretofore unseen in laboratories.

Kerr-Lens Mode-Locking—A Case Study

Mode-locking in a femtosecond Ti:sapphire laser is accomplished through a process called Kerr-lens mode-locking, based, as one might guess, on the optical Kerr effect discussed in Chapter 24. When radiation of high irradiance with a TEM_{00} Gaussian spatial profile passes through a Ti:sapphire crystal, it induces a transverse index of refraction profile in the crystal as a result of the nonlinear Kerr effect. This index of refraction profile serves to focus the beam. The self-lensing effect can be exploited for use as a passive, very fast mode-locker by simply inserting an aperture close to the Ti:sapphire crystal. If the aperture diameter is large enough to allow a high-irradiance beam focused in the crystal by the Kerr self-lensing effect to pass through with low loss, but small enough to block a significant fraction of a lower irradiance beam that does not induce significant self-lensing, the system will self mode-lock. That is, high-irradiance mode-locked pulses will be the low-loss field mode in the cavity and so will form naturally from spontaneous emission. The formation of the Kerr-effect-induced index of refraction profile in the crystal occurs fast enough that the full bandwidth of the Ti:sapphire gain medium can be utilized in the pulse formation.

28-5 LASER COOLING AND TRAPPING

Using laser radiation to reduce the speed of atoms in a gas sample is known as *laser cooling*. An *optical trap* is a device that uses laser light to hold atoms in a confined region. Recall that the temperature of a sample of gas is related to the average kinetic energy of the atoms or molecules in the sample. Atoms and molecules moving around at room temperature have speeds on the order of several hundred meters per second. If one wishes to manipulate individual atoms, one first must slow them down. Laser fields can be used to slow atoms to speeds of a few meters per second or less and then to hold these atoms in an optical trap, so that they can be manipulated.

Laser Cooling

Photons carry momentum (see Chapter 1) and energy, so it is not surprising that light can be used to slow an atom through a collision process. An application of the law of conservation of momentum indicates that a "head-on" collision between a photon and an atom will slow the atom. In fact, in an inelastic collision, in which the photon is absorbed by the atom, the momentum of the atom is reduced by precisely the photon momentum, h/λ. Subsequently, of course, the atom recoils as a result of spontaneous emission. However, the spontaneous photon is emitted in a random direction and so the recoil of the atom is also randomly directed. The net effect, then, in collisions between *counterpropagating* photons and atoms, if the photons are resonant with an electronic transition in the atoms, is to slow the atoms. However, if the photons and an assembly of resonant atoms are moving in the same direction, the net effect is to increase the speed of the atoms in the direction of the photon motion. If the photons in a light field are more likely to be absorbed by atoms moving in a direction opposite to the direction of photon propagation than they are to be absorbed by atoms moving in the direction of photon propagation, then the light field can preferentially *slow* an assembly of randomly moving atoms. Just such a situation will occur if the light field is tuned to a ("redshifted") frequency slightly less than the resonant frequency of the atomic transition. In this case, due to the Doppler shift (see Chapter 4), atoms will absorb more photons if they move toward the laser source. (This situation is shown in Figure 28-3.) That is, the Doppler shift will bring the atoms into resonance with the field if the atoms are moving toward the light source. In contrast, atoms moving away from the light source will experience an electromagnetic field that is Doppler-shifted out of resonance with the atomic transition and so will not interact strongly with that light field. Laser fields, incident from all directions on an assembly of atoms, therefore, will slow the atoms if the laser field has a frequency slightly less than the resonant frequency of the atoms. The resulting slowed assembly of atoms is referred to as an *optical molasses*. The minimum temperature (related to the minimum average speed) obtainable by this method is limited by the atomic recoil kinetic energy resulting from the random spontaneous emission events that follow absorption of the photons. Extensions of this *Doppler-cooling* technique have led to the cooling of an assembly of atoms to a small fraction of a millikelvin above absolute zero. The slow-moving atoms can then be manipulated in a variety of interesting applications.

Laser Trapping

Even slow-moving atoms will tend to wander away from each other. In order to hold them in place they must be trapped. Several different trapping schemes exist. In one kind of trap, illustrated in Figure 28-4, magnetic coils surround the slow-moving cooled atoms at the center of the three pairs of counterpropagating laser beams. The transition frequency of an atom is affected by magnetic fields. Thus it is possible to create a spatial magnetic field gradient in the trap region such that atoms wandering out of the trap region have shifted resonant frequencies that cause them to interact more strongly with the incident laser fields and so be preferentially returned to the trap region. Atoms that are cooled (slowed) and trapped exhibit interesting and novel behaviors. Recently, for example, Bose-Einstein condensates have been formed in atom traps. This unique state of matter, predicted to exist by Einstein and Bose in 1925, had not been formed in the laboratory until it was localized in optical traps. In a Bose-Einstein condensate, all of the atoms "condense" into the ground state of the trap system. In addition to allowing for the study of this exotic state of matter, cooled atoms can be used to make better atomic clocks and can be used in "atom-interferometers." (See problem 28-18.) Relatively simple optical traps can be constructed using diode lasers to trap rubidium atoms, which have a suitable resonant transition near 800 nm.

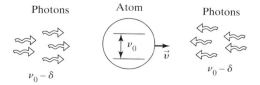

Photons   Atom   Photons

$\nu_0 - \delta$

Figure 28-3 Doppler slowing of an atom. An atom with a transition frequency ν_0 will absorb more photons with a frequency $\nu_0 - \delta$ (slightly less than ν_0) from a counterpropagating light field than it will from a copropagating light field. In this case, the atom will be slowed.

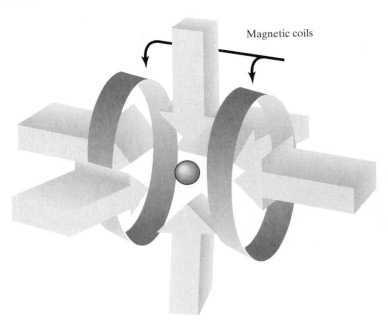

Magnetic coils

Figure 28-4 Schematic of a magneto-optical atom trap. The small sphere at the center of the converging laser beams, represented by the thick arrows, is the trapping region.

Optical Tweezers

A second technique—first performed with a single laser beam in 1986[1]—that can be used to trap tiny bits of matter ranging from atomic size to 1 μm involves so-called *optical tweezers*. This technique differs from the Doppler-cooling technique described in the previous section in that it uses a single focused laser beam as the trap. Optical tweezer devices are often modified microscopes; in a common arrangement, the laser beam is passed through the microscope objective and focused in the "specimen plane" of the microscope. The standard TEM_{00} Gaussian beam intensity profile can lead to a gradient force that tends to trap dielectric particles in the focal region. Other beam profiles also have this property. The full details of this process are to be researched in problem 28-19, but here we note that the trapping force, directed toward the center of the beam focus, arises as a natural consequence of conservation of momentum as light is reflected and refracted by objects in the beam path. Optical tweezers have been used to trap and manipulate bacteria, viruses, dielectric and metal spheres, single strands of DNA, and an assortment of other microscopic objects. The study of molecular motors,—the family of biological molecules responsible for movement in the cellular world—is of keen interest. Molecular motors can be attached by biochemical means to a micron-sized dielectric sphere (glass, for example) and then trapped, pushed around, stretched, and tracked using optical tweezers. In addition to gaining a better understanding of cell motion and dynamics, scientists hope to mimic molecular motors to construct *nanomachines* designed to carry out specific tasks related to atomic and cellular movement. The range of application of optical tweezers is broad—ranging from cell sorting to in vitro fertilization.

28-6 OPTICAL PARAMETRIC OSCILLATORS

The ability to produce coherent radiation at any specified wavelength has long been a goal of laser scientists and engineers. There are several existing

---

[1]A. Ashkin, J. M. Dziedzic, J. E. Bjorkholm and S. Chu. 1986. "Observation of a Single-Beam Gradient Force Optical Trap for Dielectric Particles," *Opt. Lett.*, 11 (5), 288–290.

laser gain media with wide gain bandwidths that serve as sources of tunable electromagnetic radiation. For example, the liquid dye Rhodamine 6G has a gain bandwidth of 50 nm or more centered around 600 nm, and a selection of dyes can provide coherent radiation over the entire visible range of wavelengths. As noted previously, the Ti-sapphire laser has a bandwidth extending from 650 nm to 1100 nm. Another approach to producing coherent radiation at a particular wavelength involves band-gap engineering of semiconductor gain media. That is, constructing layered *p-n* junctions made from particular combinations and concentrations of elements can lead to emission at specific wavelengths. A third approach involves the use of an optical parametric oscillator (OPO) and is the subject of this section. Of course, even if a given method can produce coherent radiation at a particular wavelength, the radiation produced may have undesirable properties (power output in the wrong range, cannot produce pulses with the needed properties, coherence length too short, system too expensive or too difficult to maintain or operate, and so on) for a given application.

An OPO uses a nonlinear medium placed inside an optical cavity. Figure 28-5 shows one possible arrangement that uses an external *pump* beam and a periodically-poled lithium-niobate (PPLN) crystal. A three-wave mixing process, described in Chapter 24, results in the parametric conversion of some pump power at frequency ω_p into a *signal* of frequency ω_s and an *idler* of frequency ω_i, one or both of which might be resonant with the optical cavity. If the rate of production of radiation resonant with the optical cavity exceeds the cavity loss rate, the produced radiation grows, saturates, and reaches steady state much like the radiation produced by a laser gain medium. In the OPO case, however, the nonlinear parametric interaction between the radiation and the gain medium allows for the tuning of the radiation generated inside the optical cavity. Many different configurations are possible. In a system utilizing a lithium-niobate nonlinear crystal, the pump beam might, for example, be the output from either a Nd:YAG laser or a frequency-doubled Nd:YAG laser. Recall from Chapter 24 that in order for an efficient transfer of energy to take place from the pump beam to the signal and idler beams, two conditions must be met. The first of these is a direct consequence of conservation of energy and takes the form

$$\hbar\omega_p = \hbar\omega_s + \hbar\omega_i \tag{28-1}$$

The second condition is the phase-matching condition. If the three interacting beams are colinear, the phase-matching condition can be expressed as

$$k_p = k_i + k_s \quad \text{or} \quad n_p\omega_p/c = n_i\omega_i/c + n_s\omega_s/c \tag{28-2}$$

Due to dispersion, of course, the three indices of refraction are not, in general, equal, and so Eqs. (28-1) and (28-2) can be solved for the frequencies of the signal and idler beams that will be generated by a pump beam of a given frequency.[2] The signal beam or the idler beam (or both) should have a frequency

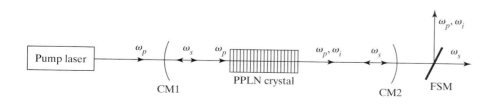

Figure 28-5 Optical parametric oscillator. The pump beam generates the signal and idler beams in the nonlinear PPLN crystal. In the configuration shown, only the signal beam is resonant with the cavity formed by the two mirrors CM1 and CM2, which are nonreflecting for the pump and idler beams. An additional frequency selective mirror (FSM) can be used to separate the signal, idler, and pump beams.

---

[2]Be aware, however, that the indices of refraction are themselves functions of frequency.

that matches a resonant frequency of the cavity. For a sufficiently strong pump beam, the rate of production of the signal or idler beam can be high enough to offset the cavity loss introduced by transmission and absorption at the cavity mirrors (and other loss mechanisms).

OPO Tuning

As noted, Eqs. (28-1) and (28-2) constrain the signal and idler frequencies for a given pump frequency. Some tuning to produce different signal and idler frequencies can be accomplished by changing the temperature of the crystal, which changes the indices of refraction n_p, n_s, and n_i and leads to different signal and idler frequencies. More dramatic tuning capability is provided by the use of a quasi-phase-matched nonlinear crystal. Quasi-phase matching (QPM), as described in Chapter 24, is accomplished through the use of a crystal with an alternating positive and negative nonlinear contribution to the index of refraction of the material. In Chapter 24 we noted that the alternating nonlinear index of refraction can be used to ensure that, in the nonlinear parametric interaction, energy flows from the pump beam into the signal and idler beams, rather than in the other direction. This alternating index of refraction can be accomplished by a *periodic poling* of a nonlinear medium such as lithium-niobate or KTP. Such a poled crystal was discussed in Chapter 24 and displayed in Figure 24-2b. In the present context it is useful to note that QPM leads to an efficient transfer of energy from the pump to the signal and idler when the spatial frequency $K_C = 2\pi/\lambda_C$ associated with the alternating contribution to the nonlinear index of refraction satisfies the condition

$$ k_p = k_i + k_s + K_C \quad \text{or} \quad \frac{n_p\omega_p}{c} = \frac{n_i\omega_i}{c} + \frac{n_s\omega_s}{c} + \frac{2\pi}{\lambda_C} \qquad (28\text{-}3) $$

Figure 24-2b indicates that $\lambda_C = 2L_C$ is the spatial period associated with the periodic poling of the nonlinear crystal. The poling-period λ_C thus becomes a free parameter in Eq. (28-3), which together with Eq. (28-1) then determines the frequency of the signal and idler beams. Different choices of the poling period lead to different signal and idler frequencies. (See problem 28-9.) Periodically poled crystals can be designed to produce specific signal and idler frequencies. In fact, a single crystal can be poled such that the poling period differs gradually across the transverse extent of the crystal. Translating such a crystal so that the optical axis of the cavity intercepts different portions of the crystal leads to the production of different signal and idler frequencies. OPOs using periodically-poled lithium-niobate can produce coherent radiation of wavelengths in the range from 0.7 μm to 4.5 μm and with CW output powers (or pulsed average powers) of several watts. The maximum power output of the OPO is limited by inefficiencies induced by the nonlinear interaction when the pump power is too high. Conversion efficiencies of pump power into the signal and idler of better than 50% are common.

Singly Resonant OPO

A general OPO arrangement is illustrated in Figure 28-5. In a singly resonant OPO the cavity is tuned (by changing the cavity length, for example) to be resonant with either the signal or idler beam, but not both. Noting that which is called the signal beam and which the idler beam is an arbitrary choice, let us say that the signal beam is resonant with the cavity. In this arrangement, then, only the signal beam will experience more "gain" (transfer of energy from the pump beam) than loss per round-trip and so will grow to a large value. As with a laser gain medium, in an OPO the initial gain must exceed a certain gain threshold set by the cavity losses. The rapidly growing intracavity field saturates the nonlinear conversion process. The gain saturates until the round-trip gain is equal to the round-trip loss and a steady-state oscillation is established. The off-resonance idler beam grows, essentially, only due to single-pass amplification through the nonlinear crystal. This commonly used arrangement has the advantage of simplicity but has a relatively high pump threshold. (The pump threshold is the pump power required for the signal-beam gain to exceed the cavity losses.) For a system using a PPLN crystal, the pump threshold is typically a few watts.

Doubly Resonant OPO

It is also possible to arrange (see problem 28-10) for both the signal and idler beam to satisfy a cavity resonance condition so that both signal and idler beams can grow to a large value and oscillate in steady state. The pump threshold in the doubly resonant arrangement is less, perhaps by a factor of 100, than in an otherwise similar singly resonant arrangement. Since the signal and idler waves have different frequencies, they resonate with different cavity modes. It is difficult to ensure that both frequencies are initially matched and remain matched to cavity mode frequencies. As Eq. (28-1) indicates, the signal and idler frequencies are anticorrelated, so that as the signal frequency drifts to higher frequencies, the idler frequency must decrease. Consequently, it is difficult (but not impossible) to develop a feedback mechanism that ensures that both signal and idler remain locked to a cavity mode frequency.

Pump-Resonant OPO

If the pump is resonant with a cavity mode frequency, the pump power experienced by the crystal can be many times the pump power outside the cavity. (See problem 28-11.) In this arrangement, one can lock the pump frequency to a cavity mode frequency. The amplified signal beam will naturally resonate at a low-loss cavity mode frequency, with the nonresonant idler beam choosing a frequency that satisfies both the energy-conservation and phase-matching conditions. This arrangement is thus simpler to maintain than the doubly resonant cavity. The resonant-cavity enhancement of the pump power reduces the extracavity pump threshold.

Optical parametric oscillators are becoming increasingly common and sophisticated in design. Currently they are a source of moderate-power, tunable, coherent radiation that supplements the range of wavelengths, powers, and pulse forms available with laser sources alone.

28-7 NEAR-FIELD MICROSCOPY

We conclude this chapter with a discussion of another optical probe into the nano world—near-field microscopy (NFM). There are myriad NFM techniques; we discuss two: *atomic force microscopy* (AFM) and *near-field scanning optical microscopy*, NSOM. Both of these *scanning probe* microscopes map out and measure properties of a surface by physically moving a probe across the surface to be "imaged." Unlike traditional optical microscopes that cannot resolve details with dimensions less than a wavelength of light, the resolution of a scanning probe microscope is limited by the dimension of the probe-sample interaction volume. Scanning probe microscopes can resolve surface details of linear dimension 10 pm or less and can be used to image the surfaces of nearly any kind of solid object. Typically, scanning probe microscopes are used to image samples with horizontal surface areas less than $100~\mu$m $\times~100~\mu$m and with height variations up to $1~\mu$m.

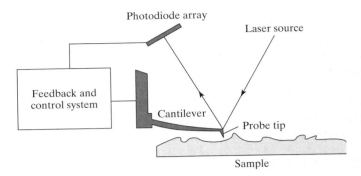

Figure 28-6 A typical atomic force microscope.

Atomic Force Microscope

The first atomic force microscope was developed in 1986. A schematic of a typical AFM is shown in Figure 28-6. In this arrangement a probe tip is attached to a cantilever, which can be scanned back and forth across the surface of the sample to be imaged. To measure the vertical motion of the probe tip, a laser beam is reflected from the top of the cantilever beam supporting the probe tip. The reflected beam is detected at a photodiode array. Motion of the beam across the array indicates vertical motion of the tip. A variety of alternative methods for measuring the vertical motion of the probe tip also exist. (See problem 28-21.)

There are three basic modes of operation of the AFM: *contact mode*, *noncontact mode*, and *tapping* or *dynamic-contact mode*. In the contact mode, which was developed first, the probe tip is dragged across the sample surface and rises and falls as it is repelled by surface contact forces. The tip deflection thus provides a topographical map as it is scanned across the horizontal surface of the sample. Although simple to employ, correlating the deflection of the probe tip with the true topography of the sample is complicated by secondary interactions induced by contact of the probe with the sample. Drag forces reduce the accuracy to which the deflection can be measured, and other secondary probe-sample interactions alter the true topography of the sample. The effects of these secondary probe-sample interactions can be reduced by immersing the sample and the probe tip in a fluid. However, immersion in a fluid can damage or change the properties of some samples. The probe-sample secondary interactions are mitigated when the noncontact mode is used. In this arrangement, the tip is held a distance of 5 to 15 nm from the surface. At this range Van der Waals forces *attract* the tip to the surface and the deflection of the tip toward the surface can be correlated to the sample-surface topography. However, the Van der Waals forces are generally weak and the deflection signal generated in this mode can be quite small. In the third common AFM mode of operation, the tapping mode, the probe tip is caused to oscillate so that it makes intermittent contact with the surface. The oscillation frequency of the tip in the tapping mode is typically in the range from 50 to 500 kHz. This high-frequency tapping reduces the probe-sample secondary interactions, which, in turn, reduce the efficacy of the contact mode of operation. Changes in the amplitude of the tapping-mode oscillations can be correlated to the topography of the sample. An image of the surface of a glass slide, taken with an atomic force microscope is shown in Figure 28-7.

Near-Field Scanning Optical Microscopy

In near-field scanning optical microscopy the probe tip is the tapered end of an optical waveguide (typically, an optical fiber) through which laser light can be guided to the sample surface. The probe tip is of subwavelength dimension and is placed close to the surface to be imaged, so that the surface is in the near field of the light source. For an opaque surface, the signal to be correlated to the surface topography can be the amount of light reflected through the

Figure 28-7 Atomic force microscope topographical scan of a glass surface. The micro- and nano-scale features of the glass can be observed, portraying the roughness of the material. Courtesy of Chytra Pawashe.

fiber tip back through the fiber. For a transparent surface, the signal can be the amount of light transmitted through the surface. Other modes of operation also exist. NSOM devices can resolve surface features of dimension 50 nm or less. In addition to providing a topographical image of the surface of the sample, a near-field scanning optical microscope can be used to probe the optical properties of the surface of opaque samples and the surface and bulk layers close to the surface of transparent samples.

Electron microscopes can also produce high-resolution images of surfaces. Whereas electron microscopes form only two-dimensional images and typically require that the sample be placed in near vacuum in order to be imaged, scanning probe microscopes give three-dimensional images of surfaces, and the samples can be conveniently immersed in gases or liquid ambient mediums. However, *scanning electron microscopes* can produce three-dimensional images of larger surface area (mm by mm) and greater maximum height (mm) than can scanning probe microscopes.

PROBLEMS

28-1 a. Why are excimer lasers particularly useful in microsurgical applications?

   b. The absorption curves shown in Figure 28-1 begin at a wavelength of 0.4 μm, which is longer than the wavelengths of the excimer lasers. Do some research to find the absorption characteristics of water and hemoglobin at the wavelengths of the excimer lasers.

   c. Considering your findings in part (b), discuss the utility of the excimer laser as a scalpel and for photocoagulation.

28-2 After doing some research, summarize the current state of the art with regard to the development of fibers through which excimer lasers can propagate with low loss.

28-3 Consider the medical procedure called posterior capsulotomy, illustrated in Figure 28-2 and discussed in the text surrounding that figure. Using the analytical tools developed in Chapter 27, choose a lens focal length, a beam-waist spot size at the lens, and a lens-to-cornea distance for use in the procedure. Be sure to respect the 17° convergence angle requirement indicated in the caption to Figure 28-2. For the system you design, calculate the beam-waist spot size at the focused spot, and the Rayleigh range parameter (z_0) for the focused beam. Comment on the importance of the Rayleigh range in this design problem. It may be useful to refer to the schematic eye shown in Figure 19-3.

28-4 For the system designed in problem 28-3, find the irradiance at the center of the focused beam waist for an Nd:YAG laser of pulse energy 4 mJ and pulse width 1 ns. Compare the answer obtained with the estimate given in Example 28-1.

28-5 Estimate the electric field strength at the center of the focused spot of Example 28-1.

28-6 A laser system, fixed on a geosynchronous satellite, uses a Nd:YAG laser that emits 1.06 μm radiation in a highly collimated beam of 5-μrad beam divergence.

   a. If the laser is 36,205 km above Earth, what is the minimum diameter of the laser-beam "footprint" on the surface of Earth?

   b. If the laser emits pulses of 200-MW power, what is the average electric field per pulse in the laser beam at Earth's surface?

28-7 Personnel who work with high-power lasers or with unshielded laser beams must take measures to protect themselves from laser damage to the skin and eyes. Toward that end, nominal hazard zones (NHZ) for given lasers can be defined. A NHZ outlines the conical space within which the level of direct, reflected, or scattered radiation—during normal operation—exceeds the assigned levels of maximum permissible exposure (MPE). Exposure levels outside the conical region are below the designated MPE level. The conical region, with tip emanating from the laser, extends to a range R_{NHZ}, with sides spreading at the full-angle beam divergence ϕ. The range for a nonfocused laser beam is given by

$$R_{NHZ} = \frac{1}{\phi}\left[\left(\frac{4P}{\pi \text{ MPE}}\right)^{1/2} - d\right]$$

where P is the laser power, ϕ is the beam divergence, MPE is the maximum permissible exposure in power/area, and d is the aperture diameter of the laser exit port.

   a. Draw a sketch of the NHZ (conical shape), labeling the extent of the cone, R_{NHZ}, and the full-angle beam divergence ϕ.

   b. Determine R_{NHZ} for a continuous wave Nd:YAG laser ($\lambda = 1.06 \mu$m) of 50-W power, 3.0-mrad beam divergence, MPE of 5.1×10^{-3} W/cm², a 10-s exposure, and an exit aperture diameter of 3.0 mm.

28-8 The range R_{NHZ} (refer to problem 28-7) for a laser that directs its exiting laser beam immediately onto a lens is given by

$$R_{NHZ} = \left(\frac{f}{b}\right)\left(\frac{4P}{\pi \text{ MPE}}\right)^{1/2}$$

where R_{NHZ} defines the extent of the cone emanating from the point of beam focus, f is the focal length of the lens, b is the diameter of the beam exiting the laser, P is the power of the laser, and MPE is the maximum permissible exposure.

a. Draw a sketch of the NHZ (conical space) for this focused laser beam.

b. Determine R_{NHZ} for a continuous wave Nd:YAG laser of 50-W power, exit beam diameter of 5.0 mm, and capped by a lens of focal length 7.5 cm, if the MPE of this laser for an 8-hr exposure is 1.6×10^{-3} W/cm^2.

28-9 Investigate the tuning of the signal and idler frequencies of an OPO using a PPLN crystal. In particular, find a PPLN spatial frequency K_C that will produce a signal of wavelength 2400 nm when the pump is (a) an Nd:YAG laser and (b) a frequency-doubled Nd:YAG laser. Assume that the operating temperature is room temperature. (To solve this problem you will have to do some research. The free software program called SNLO, available from the Sandia National Laboratory web site, would prove very useful.)

28-10 Consider an OPO system pumped with Nd:YAG laser radiation. For a certain configuration, the signal beam is found to have a wavelength of 1500 nm.

a. Find the wavelength of the idler beam.

b. Find a cavity length d, near 1 m, that will allow both signal and idler to resonate by modeling the cavity as having two flat mirrors and ignoring the influence of the PPLN crystal on the cavity mode frequencies. That is, take the cavity mode frequencies to be given by $\nu_q = qc/(2d)$.

c. Refine the calculation of part (b) by finding a cavity length near 1 m that will allow both the signal and idler to resonate, in a TEM$_{00}$ mode, in a cavity with one flat mirror and one mirror of radius of curvature 2 m. In this case, the cavity mode frequencies are given by Eq. (27-67).

28-11 Consider a two-mirror cavity and take the reflectance of each mirror to be $R = 0.97$. Let resonant light of power 100 mW be incident into the cavity. Find the (one-way) power in the intracavity beam.

28-12 Who won the Nobel prizes for physics in 1997 and 2001. Summarize the work that led to the awards.

28-13 Who won the Nobel Prize for chemistry in 1999. Summarize the work that led to the award.

28-14 Summarize the contents of a journal article describing a femtosecond dye laser experimental setup.

28-15 Find the aperture size and position used in a real Kerr-lens, mode-locked, Ti:sapphire laser system.

28-16 a. Calculate the root-mean-square room-temperature speed of an oxygen molecule.

b. Find the root-mean-square speed of an atom in a gas sample of rubidium at a temperature of 2 mK.

28-17 Estimate the amount of Doppler detuning δ (see Figure 28-3) that should be used to cool a gas sample of rubidium atoms at a temperature of 3 mK. Take the transition wavelength to be 800 nm. (Equation (4-44) might prove useful.)

28-18 After doing research, answer the following questions.

a. What is an atom-interferometer?

b. Why is there an interest in atom-interferometers?

28-19 After doing research, describe the nature of the trapping force that exists in the focal region of an optical tweezers arrangement.

28-20 After doing research, describe some experiments designed to study DNA with optical tweezers.

28-21 After doing research, describe how the tip deflection in an atomic force microscope can be measured by piezoresistive means.

28-22 Summarize the contents of a journal article detailing the use of a scanning probe microscope to image a biological sample.

28-23 Summarize the contents of a journal article detailing the use of a NSOM to measure the optical properties of a surface.

References

GENERAL REFERENCES

Born, Max, and Emil Wolf. *Principles of Optics,* 5th ed. New York: Pergamon Press, 1975.

Dichtburn, R. W. *Light,* Vol. 2. New York: Academic Press, 1976.

Fincham, W. H. A., and M. H. Freeman. *Optics,* 9th ed. Boston: Butterworth Publishers, 1980.

Fowles, Grant R. *Introduction to Modern Optics.* New York: Holt, Rinehart and Winston, 1968.

Ghatak, Ajoy K. *An Introduction to Modern Optics.* New York: McGraw-Hill Book Company, 1972.

Goodman, Joseph W. *Introduction to Fourier Optics,* 3d ed. Greenwood Village, CO: Roberts & Company Publishers, 2004.

Guenther, Robert D. *Modern Optics.* New York: John Wiley and Sons, 1990.

Hecht, Eugene, *Optics,* 4th ed. Reading, MA: Addison-Wesley, 2002.

Klein, Miles V. *Optics.* New York: John Wiley and Sons, 1970.

Longhurst, R. S. *Geometrical and Physical Optics,* 2d ed. New York: John Wiley and Sons, 1967.

Meyer-Arendt, Jurgen R. *Introduction to Classical and Modern Optics,* 3d ed. Englewood Cliffs, NJ: Prentice-Hall, 1989.

Möller, K. D. *Optics.* Mill Valley, CA: University Science Books, 1988.

Nussbaum, Allen. *Geometric Optics: An Introduction.* Reading, MA: Addison-Wesley Publishing Company, 1968.

Nussbaum, Allen, and Richard A. Phillips. *Contemporary Optics for Scientists and Engineers.* Englewood Cliffs, NJ: Prentice-Hall, 1976.

Rossi, Bruno. *Optics.* Reading, MA: Addison-Wesley Publishing Company, 1957.

Saleh, B. E. A., and M. C. Teich. *Fundamentals of Photonics.* New York: John Wiley and Sons, 1991.

Siegman, Anthony E. *Lasers.* Mill Valley, CA: University Science Books, 1986.

Steward, E. G. *Fourier Optics: An Introduction,* 2d ed. New York: Halsted Press, 1987.

Strong, John. *Concepts of Classical Optics.* San Francisco: W. H. Freeman and Company Publishers, 1958.

Welford, W. T. *Optics,* 3d ed. New York: Oxford University Press, 1988.

Young, M. *Optics and Lasers: Including Fibers and Optical Waveguides,* 5th ed. New York: Springer-Verlag, 2000.

CHAPTER 1 NATURE OF LIGHT

Feinberg, Gerald. "Light." In *Lasers and Light,* pp. 4–13. San Francisco: W. H. Freeman and Company Publishers, 1969.

OPN Trends, "The Nature of Light: What is a Photon?" Vol. 3., No. 1, October 2003.

Ronchi, Vasco. *The Nature of Light.* Cambridge: Harvard University Press, 1970.

CHAPTER 2 GEOMETRICAL OPTICS

Feynman, Richard P., Robert B. Leighton, and Matthew Sands. *The Feynman Lectures on Physics,* Vol. 1. Reading, MA: Addison-Wesley Publishing Company, 1975. Chs. 26, 27.

Longhurst, R. S. *Geometrical and Physical Optics,* 2d ed. New York: John Wiley and Sons, 1967. Chs. 1, 2.

Rossi, Bruno. *Optics.* Reading, MA: Addison-Wesley Publishing Company, 1957. Chs. 1, 2.

CHAPTER 3 OPTICAL INSTRUMENTATION

Cox, A. *Photographic Optics,* 15th ed. New York: Focal Press, 1974.

Goodman, Douglas S. "Basic Optical Instruments." In *Geometrical and Instrumental Optics,* edited by Daniel Malacara. Boston: Academic Press, 1988.

Horne, D. F. *Optical Instruments and Their Applications.* Bristol, England: Adam Hilger Ltd., 1980.

Klein, Miles V. *Optics.* New York: John Wiley and Sons, 1970. Ch. 4.1: "Radiometry and Photometry."

McLaughlin, R. B. *Special Methods in Light Microscopy.* London: Microscope Publications Ltd., 1977.

Smith, W. J. *Modern Optical Engineering.* New York: McGraw-Hill Book Company, 1966.

CHAPTER 4 WAVE EQUATIONS

Ghatak, Ajoy K. *An Introduction to Modern Optics.* New York: McGraw-Hill Book Company, 1972. Ch. 1.

Hecht, Eugene, and Alfred Zajac. *Optics*. Reading, MA: Addison-Wesley Publishing Company, 1974. Ch. 2.

Resnick, Robert. *Basic Concepts in Relativity and Early Quantum Mechanics*. New York: John Wiley and Sons, 1972. Ch. 2.

CHAPTER 5 SUPERPOSITION OF WAVES

Ghatak, Ajoy K. *An Introduction to Modern Optics*. New York: McGraw-Hill Book Company, 1972. Ch. 1.

Hecht, Eugene, and Alfred Zajac. *Optics*. Reading, MA: Addison-Wesley Publishing Company, 1974. Ch. 7.

CHAPTER 6 PROPERTIES OF LASERS

O'Shea, D. C., W. R. Callen, and W. T. Rhodes. *Introduction to Lasers and Their Applications*. Reading MA: Addison-Wesley Publishing Company, 1978.

Saleh, B. E. A., and M. C. Teich. *Fundamentals of Photonics*. New York: John Wiley and Sons, 1991.

Scully, M. O., and M. S. Zubairy. *Quantum Optics*. Cambridge: Cambridge University Press, 1997.

Siegman, A. E. *Lasers*. Mill Valley, CA: University Science Books, 1986.

Verdeyen, J. T. *Laser Electronics*. Englewood Cliffs, NJ: Prentice-Hall, 1995.

CHAPTER 7 INTERFERENCE OF LIGHT

Feynman, Richard P., Robert B. Leighton, and Matthew Sands. *The Feynman Lectures on Physics,* Vol. 1. Reading, MA: Addison-Wesley Publishing Company, 1975. Chs. 28, 29.

Fincham, W. H. A., and M. H. Freeman. *Optics,* 9th ed. London: Butterworths, 1980. Ch. 14.

Ghatak, Ajoy K. *An Introduction to Modern Optics*. New York: McGraw-Hill Book Company, 1972. Ch. 4.

Hecht, Eugene, and Alfred Zajac. *Optics*. Reading, MA: Addison-Wesley Publishing Company, 1974. Ch. 9.

Longhurst, R. S. *Geometrical and Physical Optics,* 2d ed. New York: John Wiley and Sons, 1967. Chs. 7, 8.

CHAPTER 8 OPTICAL INTERFEROMETRY

Francon, Maurice. *Optical Interferometry*. New York: Academic Press. 1966.

Hariharan, P. *Optical Interferometry*. Boston: Academic Press, 2003.

Hernandez, G. *Fabry-Perot Interferometers*. New York: Cambridge University Press, 1986.

James, J. F., and R. S. Sternberg. *The Design of Optical Spectrometers*. London: Chapman and Hall Ltd., 1969. Ch. 7.

Reynolds, George O., John B. DeVelis, George B. Parrent, Jr., and Brian J. Thompson. *Physical Optics Notebook: Tutorials in Fourier Optics*. Bellingham, WA: SPIE Optical Engineering Press, 1989. Chs. 22–24.

Robinson, Glen M., David M. Perry, and Richard W. Peterson. "Optical Interferometry of Surfaces." *Scientific American,* July 1991: 66.

Tolansky, Samuel. *An Introduction to Interferometry*. New York: John Wiley and Sons, 1973.

CHAPTER 9 COHERENCE

Feynman, Richard P., Robert B. Leighton, and Matthew Sands. *The Feynman Lectures on Physics,* Vol. 1. Reading, MA: Addison-Wesley Publishing Company, 1975. Ch. 50.

Parrent, Mark J., and George B. Parrent, Jr. *Theory of Partial Coherence*. Englewood Cliffs, NJ: Prentice-Hall, 1964.

Peřina, Jan. *Coherence of Light*. New York: Van Nostrand Reinhold Co., 1971.

Reynolds, George O., John B. DeVelis, George B. Parrent, Jr., and Brian J. Thompson. *Physical Optics Notebook: Tutorials in Fourier Optics*. Bellingham, WA: SPIE Optical Engineering Press, 1989. Chs. 11, 17, 18.

CHAPTER 10 FIBER OPTICS

Boyd, Waldo T. *Fiber Optics Communications, Experiments and Projects*. Indianapolis: Howard W. Sams and Co., 1982.

Cheo, Peter K. *Fiber Optics Devices and Systems*. Englewood Cliffs, NJ: Prentice-Hall, 1985. Ch. 4.

Cvijetic, Milorad. *Optical Transmission Systems Engineering,* Boston: Artech House, Inc., 2004. Ch. 2.

Karim, Mohammad A. *Electro-Optical Devices and Systems*. Boston: PWS-Kent Publishing Company, 1990. Ch. 9.

Nerou, Jean Pierre. *Introduction to Fiber Optics*. Sainte-Foy, Quebec: Les Editions Le Griffon D'Argile, 1988.

Palais, Joseph C. *Fiber Optic Communications*. Englewood Cliffs, NJ: Prentice-Hall, 1988.

Yariv, Amnon. *Optical Electronics,* 3d ed. New York: Holt, Rinehart and Winston, 1985. Ch. 3.

CHAPTER 11 FRAUNHOFER DIFFRACTION

Ball, C. J. *An Introduction to the Theory of Diffraction*. New York: Pergamon Press, 1971.

Born, Max, and Emil Wolf. *Principles of Optics,* 5th ed. New York: Pergamon Press, 1975. Ch. 8.

Goodman, Joseph W. *Introduction to Fourier Optics*. New York: McGraw-Hill Book Company, 1968. Ch. 3.

Longhurst, R. S. *Geometrical and Physical Optics,* 2d ed. New York: John Wiley and Sons, 1967. Chs. 10, 11.

CHAPTER 12 THE DIFFRACTION GRATING

Davis, Sumner P. *Diffraction Grating Spectrographs*. New York: Holt, Rinehart and Winston, 1970.

Feynman, Richard P., Robert B. Leighton, and Matthew Sands. *The Feynman Lectures on Physics,* Vol. 1. Reading, MA: Addison-Wesley Publishing Company, 1975. Ch. 30.

Hutley, M. C. *Diffraction Gratings*. New York: Academic Press, 1982.

Ingalls, Albert G. "Ruling Engines." *Scientific American,* June 1952: 45.

James, J. F., and R. S. Sternberg. *The Design of Optical Spectrometers*. London: Chapman and Hall Ltd., 1969. Chs. 5, 6.

CHAPTER 13 FRESNEL DIFFRACTION

Born, Max, and Emil Wolf. *Principles of Optics,* 5th ed. New York: Pergamon Press, 1975. Ch. 8.

Guenther, Robert. *Modern Optics*. New York: John Wiley and Sons, 1990. Ch. 9.

Longhurst, R. S. *Geometrical and Physical Optics,* 2d ed. New York: John Wiley and Sons, 1967. Ch. 13.

Reynolds, George O., John B. DeVelis, George B. Parrent, Jr., and Brian J. Thompson. *Physical Optics Notebook: Tutorials in Fourier Optics*. Bellingham, WA: SPIE Optical Engineering Press, 1989. Ch. 9.

CHAPTER 14 MATRIX TREATMENT OF POLARIZATION

Gerrard, A., and J. M. Burch. *Introduction to Matrix Methods in Optics.* New York: John Wiley and Sons, 1975.

Jones, R. Clark. "A New Calculus for the Treatment of Optical Systems." *Journal of the Optical Society,* Vol. 31, 1941: 488.

Shurcliff, W. A. *Polarized Light: Production and Use.* Cambridge: Harvard University Press, 1962.

CHAPTER 15 PRODUCTION OF POLARIZED LIGHT

Feynman, Richard P., Robert B. Leighton, and Matthew Sands. *The Feynman Lectures on Physics,* Vol. 1. Reading, MA: Addison-Wesley Publishing Company, 1963. Chs. 32, 33.

Goure, J-P, and I. Verrier. *Optical Fibre Devices,* Bristol and Philadelphia: Institute of Physics Publishing, 2002.

Kittel, Charles, *Introduction to Solid State Physics.* New York: John Wiley & Sons, 1986.

Kliger, David S. *Polarized Light in Optics and Spectroscopy.* Boston: Academic Press, 1990.

Meyer-Arendt, Jurgen R. *Introduction to Classical and Modern Optics,* 3d ed. Englewood Cliffs, NJ: Prentice-Hall, 1989. Ch. 4.2.

Shurcliff, W. A., and S. S. Ballard. *Polarized Light.* Princeton, NJ: D. Van Nostrand Company, 1964.

Weisskopf, Richard F. "How Light Interacts with Matter." In *Lasers and Light,* pp. 14–16 San Francisco: W. H. Freeman and Company Publishers, 1969.

CHAPTER 16 HOLOGRAPHY

Francon, Maurice. *Holography.* New York: Academic Press, 1974.

Reynolds, George O., John B. DeVelis, George B. Parrent, Jr., and Brian J. Thompson. *Physical Optics Notebook: Tutorials in Fourier Optics.* Bellingham, WA.: SPIE Optical Engineering Press, 1989. Chs. 25–27.

Smith, Howard Michael. *Principles of Holography.* New York: John Wiley and Sons, 1975.

Stroke, George W. *An Introduction to Coherent Optics and Holography,* 2d ed. New York: Academic Press, 1969.

Vest, C. M. *Holographic Interferometry.* New York: John Wiley and Sons, 1979.

CHAPTER 17 OPTICAL DETECTORS AND DISPLAYS

Budde, W. *Physical Detectors of Optical Radiation.* Optical Radiation Measurements Series, Vol. 4. New York: Academic Press, 1983.

Davis, Christopher C. *Lasers and Electro-Optics,* Cambridge: Cambridge University Press, 1996. Ch. 22.

Grum, Franc, and Richard J. Becherer. *Radiometry.* Optical Radiation Measurements Series, Vol. 1. New York: Academic Press, 1979.

Kingston, R. H. *Detection of Optical and Infrared Radiation.* New York: Springer-Verlag, 1978.

Malacara, Zacarias H., and Morales R. Arquimedes. "Light Sources." In *Geometrical and Instrumental Optics,* edited by Daniel Malacara. Boston: Academic Press, 1988.

Nussbaum, Allen. *Geometric Optics: An Introduction.* Reading, MA: Addison-Wesley Publishing Company, 1968. Ch. 6.

Saleh, B. E. A., and M. C. Teich, *Fundamentals of Photonics.* New York: John Wiley & Sons, Inc., 1991. Ch. 6.

Stimson, A. *Photometry and Radiometry for Engineers.* New York: Wiley-Interscience, 1974.

CHAPTER 18 MATRIX METHODS IN PARAXIAL OPTICS

Brouwer, William. *Matrix Methods in Optical Instrument Design.* New York: W. A. Benjamin, 1964.

Fincham, W. H. A., and M. H. Freeman. *Optics,* 9th ed. Boston: Butterworth Publishers, 1980. Chs. 8, 9, 19.

Gerard, A., and J. M. Burch. *Introduction to Matrix Methods in Optics.* New York: John Wiley and Sons, 1975.

Kingslake, Rudolf. *Lens Design Fundamentals.* New York: Academic Press, 1978. Chs. 2, 3, 7.

Nussbaum, Allen, and Richard A. Phillips. *Contemporary Optics for Scientists and Engineers.* Englewood Cliffs, NJ: Prentice-Hall, 1976. Ch. 1.

CHAPTER 19 OPTICS OF THE EYE

Feynman, Richard P., Robert B. Leighton, and Matthew Sands. *The Feynman Lectures on Physics,* Vol. 1. Reading, MA: Addison-Wesley Publishing Company, 1975. Chs. 35, 36.

Fincham, W. H. A., and M. H. Freeman. *Optics,* 9th ed. Boston: Butterworth Publishers, 1980. Ch. 20.

Michaels, D. D. *Visual Optics and Refraction,* 2d ed. St. Louis: C. V. Mosby Company, 1980.

Rubin, M. L. *Optics for Clinicians,* 2d ed. Gainesville, FL: Triad Scientific Publishers, 1974.

Sliney, D. H., and M. L. Wolbarsht. *Safety with Lasers and Other Optical Sources: A Comprehensive Handbook.* New York: Plenum Press, 1980.

CHAPTER 20 ABERRATION THEORY

Fincham, W. H. A., and M. H. Freeman. *Optics,* 9th ed. Boston: Butterworth Publishers, 1980. Ch. 18.

Guenther, Robert D. *Modern Optics.* New York: John Wiley and Sons, 1990. Appendix 5-B.

Kingslake, Rudolf. *Lens Design Fundamentals.* New York: Academic Press, 1978.

Nussbaum, Allen. *Geometric Optics: An Introduction.* Reading, MA: Addison-Wesley Publishing Company, 1968. Chs. 7, 8.

Smith, F. Dow. "How Images Are Formed." *Scientific American,* Sept. 1968: 59–70.

Welford, W. T. *Aberrations of Optical Systems.* Boston: Adam Hilger Ltd., 1986.

CHAPTER 21 FOURIER OPTICS

Bell, Robert John. *Introductory Fourier Transform Spectroscopy.* New York: Academic Press, 1972.

Duffieux, P. M. *The Fourier Transform and Its Applications to Optics,* 2d ed. New York: John Wiley and Sons, 1983.

Françon, M. *Optical Image Formation and Processing.* New York: Academic Press, 1979.

Goodman, Joseph W. *Introduction to Fourier Optics.* New York: McGraw-Hill Book Company, 1968.

Lee, S. H., ed. *Optical Information Processing, Fundamentals.* New York: Springer-Verlag, 1981.

Matthys, D. R., and F. L. Pedrotti. "Fourier Transforms and the Use of a Microcomputer in the Advanced Undergraduate Laboratory." *American Journal of Physics,* Vol. 50, 1982: 990.

Steward, E. G. *Fourier Optics: An Introduction,* 2d ed. New York: Halsted Press, 1987. Chs. 4, 5.

Strong, John. *Concepts of Classical Optics.* San Francisco: W. H. Freeman and Company Publishers, 1958. Appendix F.

CHAPTER 22 THEORY OF MULTILAYER FILMS

Chopra, Kasturi L. *Thin Film Phenomena.* New York: McGraw-Hill Book Company, 1969.

Heavens, O. S. *Thin Film Physics.* New York: Barnes and Noble, 1970.

Knittl, Z. *Optics of Thin Films, an Optical Multilayer Theory.* New York: John Wiley and Sons, 1976.

Macleod, H. A. *Thin Film Optical Filters.* New York: American Elsevier Publishing Company, 1969.

CHAPTER 23 FRESNEL EQUATIONS

Dichtburn, R. W. *Light,* Vol. 2. New York: Academic Press, 1976. Ch. 14.

Longhurst, R. S. *Geometrical and Physical Optics,* 2d ed. New York: John Wiley and Sons, 1967. Ch. 21.

Rossi, Bruno. *Optics.* Reading, MA: Addison-Wesley Publishing Company, 1957. Ch. 8.

CHAPTER 24 NONLINEAR OPTICS AND THE MODULATION OF LIGHT

Butcher, P. N., and D. Cotter. *The Elements of Nonlinear Optics.* New York: Cambridge University Press, 1990. Ch. 6.

Kaminov, Ivan P. *An Introduction to Electrooptic Devices.* New York: Academic Press, 1974. Ch. 3.

Pedrotti, Frank L., and Peter Bandettini. "Faraday Rotation in the Undergraduate Advanced Laboratory." *Am. J. Phys.,* Vol. 58, June 1990: 542.

Yariv, Amnon. *Optical Electronics,* 3d ed. New York: Holt, Rinehart and Winston, 1985. Ch. 8.

CHAPTER 25 OPTICAL PROPERTIES OF MATERIALS

Feynman, Richard P., Robert B. Leighton, and Matthew Sands. *The Feynman Lectures on Physics,* Vol. 1. Reading, MA: Addison-Wesley Publishing Company, 1975. Ch. 31.

Javan, Ali. "The Optical Properties of Materials." *Scientific American,* Sept. 1967: 238.

Sommerfeld, Arnold. *Optics.* New York: Academic Press, 1964. Ch. 3.

CHAPTER 26 LASER OPERATION

Davis, Christopher C. *Lasers and Electro-Optics.* Cambridge: Cambridge University Press, 1996. Ch. 22.

Scully, M. O., and M. S. Zubairy. *Quantum Optics.* Cambridge: Cambridge University Press, 1997.

Siegman, A. E. *Lasers.* Mill Valley, CA: University Science Books, 1986.

Verdeyen, J. T. *Laser Electronics.* Englewood Cliffs, NJ: Prentice-Hall, 1995.

CHAPTER 27 CHARACTERISTICS OF LASER BEAMS

Gerrard, A., and J. M. Burch. *Introduction to Matrix Methods in Optics.* New York: John Wiley and Sons, 1975.

Saleh, B. E. A., and M. C. Teich. *Fundamentals of Photonics.* New York: John Wiley and Sons, 1991. Chs. 3, 9.

Siegman, Anthony E. *Lasers.* Mill Valley, CA: University Science Books, 1986. Chs. 16, 17.

Verdeyen, J. T. *Laser Electronics.* Englewood Cliffs, NJ: Prentice-Hall, 1995.

CHAPTER 28 SELECTED MODERN APPLICATIONS

Berns, Michael H., and Donald E. Rounds. "Cell Surgery by Laser." *Scientific American,* Feb. 1970: 98.

Faller, James E., and E. Joseph Wampler. "The Lunar Laser Reflector." *Scientific American,* Mar. 1970: 38.

Lasers and Optronics. Morris Plains, NJ: Elsevier Communications. Trade journal, published monthly.

Optics and Photonics. Washington, DC: Optical Society of America. Published monthly.

Photonics Spectra. Pittsfield, MA: Laurin Publishing Co. Trade journal, published monthly.

Saleh, B. E. A., and M. C. Teich. *Fundamentals of Photonics.* New York: John Wiley and Sons, 1991. Chs. 3, 9.

Vali, Victor. "Measuring Earth Strains by Laser." *Scientific American,* Dec. 1969: 88.

Yariv, Amnon. *Optical Electronics,* 3d ed. New York: Holt, Rinehart and Winston, 1985.

Answers to Selected Problems

Chapter 1   **1-1** (a) 6.6×10^{-34} m   (b) 3.9 Å

1-2 3.6×10^{-17} W

1-3 3.27 and 1.61 eV

1-4 0.024 Å; 2.7×10^{-22} kg-m/s

1-5 0.511 MeV

1-6 1.422 MeV/c

1-9 (a) 1.49×10^{-18} kg-m/s   (b) 4.45×10^{-16} m   (c) 4.22×10^{-16} m

1-10 2.77×10^{17}

1-12 3.9×10^{14} Hz to 7.9×10^{14} Hz

1-13 1.5 m

1-14 0.83 m

1-15 (a) 0.3 ms   (b) 0.1 m

1-16 (a) 39.8 W/sr   (b) 10^6 W/m^2 (c) 9.95 W/m^2 (d) 0.0195 W

1-17 (a) 0.0096°   (b) 8.73×10^{-8} sr   (c) 76.4 W/m^2 (d) 8.75×10^{10} W/(m$^2 \cdot$ sr)

Chapter 2   **2-1** $t = (\Sigma_i n_i x_i)/c$

2-2 $1.25(x^2 + y^2) + 70(x^2 + y^2)^{1/2} - 135x + 800 = 0$

2-3 4.00 mm

2-4 3 ft, with top edge of mirror at a height halfway between the person's eye level and the top of the person's head

2-5 The ray emerges from the bottom at 45°.

2-6 Reflection from the bottom surface; 1.60

2-7 1.55

2-8 1.153 cm

2-9 8 cm

2-10 Light from the bubble is refracted through the plane surface, both directly and after reflection from the spherical mirror; 3.33 cm and 10 cm.

2-11 12.5 cm; 75 cm

2-12 10 cm behind the near surface; $3 \times$

2-13 (a) $f = n_1 R/(n_2 - n_1)$ (b) $R > 0$ (convex) and $R < 0$ (concave), respectively

2-14 (a) center, $\frac{4}{3}$ actual size   (b) 6.4 cm behind the glass, $\frac{8}{7}$ actual size

2-15 Virtual, inverted, 15 cm from the window, twice the object size

2-16 13.0 cm

2-17 +20 cm or −20 cm

2-18 22.5 cm behind the lens; 1.50 times the actual size

2-19 (a) −6.7 cm   (b) −10 cm or −60 cm

2-20 −50 cm

2-21 (a) 3.33 mm in front of the objective   (b) erect and magnified

2-22 Final image between lens and mirror at 21/34 f from lens, virtual, inverted, and $\frac{1}{17}$ the original size

2-23 (a) 33.3 cm, $2 \times$ (b) 86.67 cm, $2 \times$ (c) 7.37 cm, $-0.316 \times$

2-24 1.63

2-25 150 cm and 600 cm; inverted

2-26 (a) 10, 5, −2.5 diopters; 12.5 diopters   (b) 8.33 m^{-1}, 4.17 m^{-1};   24 cm

2-30 $f_1/f_2 = n/(n-1)$

2-31 At $s = s' = 2f$

2-32 (b) $\Sigma_i \, (t_i \tan \theta_i)$

2-33 Incident on plane side: 8 cm beyond lens; on curved side: 5.33 cm beyond lens

2-34 $+40$ cm, $+30$ cm; -30 cm, -40 cm

2-35 25,000 ft

2-36 The line image is real, 18.75 cm past the lens and 15.75 cm long.

2-37 The line image is virtual at a distance of 65.2 cm on the object side of the lens and 5.65 cm long.

2-38 The line image is virtual at a distance of 11.11 cm on the object side of the lens and 2.80 cm long.

2-39 The line image is virtual at a distance of 13.33 cm on the object side of the lens and 0.67 cm long.

2-40 The line image is virtual at a distance of 21.4 cm on the object side of the lens and 17.97 cm long.

Chapter 3

3-1 Entrance pupil is the stop; exit pupil is 3.33 cm in front of the lens, with an aperture of 3.33 cm; image is 10 cm behind the lens, inverted and 2 cm long.

3-2 Exit pupil is the stop; entrance pupil is 4.29 cm behind the lens, with an aperture of 3.43 cm; image is 10.5 cm behind the lens, inverted, and 3 cm long.

3-3 Entrance pupil is the stop; exit pupil is 12 cm in front of the lens, with an aperture of 6 cm; image is 10.5 cm behind the lens, inverted, and 1.5 cm long.

3-4 (b) 20 cm right of L_2 (c) both at L_1 (d) 8.57 cm beyond L_2 and $\frac{4}{7}$ cm in diameter   (e) field stop at A, entrance window in object plane with 1 cm diameter, exit window in image plane with 1 cm diameter   (f) 2.86°

3-5 (a) Lens L_2 of diameter 6 cm   (b) The entrance window is 12 cm to the right of L_1 and has radius 9 cm. The exit window is lens L_2 of diameter 6 cm.

3-7 53′

3-8 (a) crown: $A = 1.511$, $B = 4240$ nm^2, $n_D = 1.523$; flint: $A = 1.677$, $B = 13{,}190$ nm^2, $n_D = 1.715$
(b) crown: -4.146×10^{-5} nm^{-1}; flint: -1.290×10^{-4} nm^{-1} (c) crown: 3110, 1.9 Å; flint: 9675, 0.61 Å

3-9 (a) 50.0°   (b) 1/55.5   (c) $A = 1.6205$, $B = 6073.7$ nm^2; 4.297×10^{-5} nm^{-1} (d) 1.12 m

3-10 0.01909

3-11 5.99°, 2.16′

3-12 4.82°, 4.37°

3-13 (a) $M_e = 10^4$ W/m^2, $I_e(\theta = 0) = 7.96$ W/sr, $L_e = 3180$ W/m^2-sr   (b) 1.56×10^{-4} W   (c) 35.9 W/m^2

3-15 5.7 cm

3-16 $f = 53.3$ cm, 13.33 cm, 1.86 cm

3-17 5.3 to 7.0 ft

3-18 1.3×10^5 W/cm^2

3-19 (a) 0.90 cm   (b) 5.45 cm, 3 ×

3-20 (a) 27.8 mm   (b) $f/3.1$, $f/5.4$, $f/9.4$ (c) 16.0, 9.26, 5.35 mm   (d) 0.03, 0.09, 0.27 s

3-22 (a) 2.8 cm   (b) 10 ×

3-23 (a) 320 ×   (b) 0.516 cm

3-24 (a) 46.7 ×   (b) 8.68 cm

3-25 5 cm

3-26 14.9 cm

3-27 (a) 7 ×   (b) 2 cm   (c) 5 mm   (d) 2.3 cm   (e) 337 ft

3-28 (b) 7.50 ×; 8.70 ×

3-29 1.05 cm

3-30 (a) 8 cm, 3 ×   (b) 7.38 cm, 2.6 ×

3-31 1.25 cm farther from the objective

3-32 (a) 12.5 ×   (b) 15 ×   (c) 0.13 cm, 3 mm   (d) 3.8°

3-34 -2.5 ft; -180 ×

Chapter 4

4-1 $y = ae^{-b(x+10t)^2}$

4-2 (a) $y = \dfrac{4\text{m}^3}{(x + (2.5 \text{ m/s})t)^2 + 2 \text{ m}^2}$

4-3 (a) (1) and (2) qualify because they satisfy the wave equation; more simply, if $w = z + vt$, they are functions of w: $y = A\sin^2(4\pi w)$ and $y = Aw^2$.   (b) (i) $v = 1$ m/s in $-z$-direction; (ii) $v = 1$ m/s in $+x$-direction.

4-4 10 m/s in $+x$-direction

4-5 (a) $\psi = 2\sin[2\pi(z/5\text{ m} + t/3\text{ s})]$ (b) $\psi = 2\sin(2\pi/5)\left(z/\text{m} + \frac{5}{3}t/\text{s}\right)$ (c) $\psi = 2\exp[(2\pi i(z/5\text{ m} + t/3\text{ s})]$

4-6 (a) $y = (5\text{ m})\sin(\pi x/25\text{ m})$ (b) $y = (5\text{ m})\sin[(\pi/25)(x/\text{m} + 8)]$

4-7 (a) 0.01 cm   (b) 1000 Hz   (c) 628.3 cm^{-1} (d) 6283 s^{-1} (e) 1 ms   (f) 10 cm/s   (g) 10 cm

4-8 (a) $+1$ in y-direction   (b) $-C/B$ in x-direction   (c) C in z-direction

4-9 $y = 15\sin(kx + \pi/3)$

4-10 (b) $\pi/2$, $\pi/3$, 0, $-\pi/2$, 0.6π (c) Subtract $\pi/2$ from each.

4-11 (a) $A\sin(2\pi/\lambda)(z - vt)$ (b) $A\sin(2\pi/\lambda)\left(\sqrt{2}x \pm vt\right)$ (c) $A\sin(2\pi/\lambda)\left[\left(\sqrt{3}/3\right)(x + y + z) \pm vt\right]$

4-15 $E = 870$ V/m, $B = 2.90 \times 10^{-6}$ T

4-16 (a) 5×10^{-7} T   (b) 19.88 W/m^2

4-17 (a) 1.01×10^3 V/m, 3.37×10^{-6} T   (b) 4.76×10^{21}/m^2-s   (c) $E = 1010\sin 2\pi(1.43 \times 10^6 r + 4.28 \times 10^{14}t)$, r in m, t in s

4-18 (a) 8.75×10^{-3} W/m^2, 2.57 V/m   (b) 2×10^{13} W/m^2, 1.23×10^8 V/m, 0.410 T

4-21 $v = 0.168c$

4-22 $v = -0.917c$

4-23 $2\,\Delta\lambda = 0.12$ Å

Chapter 5

5-1 (a) The waves move in opposite directions along the x-axis, E_1 to the right, E_2 to the left, with equal speeds of $\frac{4}{3}$ m/s.   (b) $t = \frac{3}{4}$ s   (c) $x = 1$ m

5-2 (b) $E_R = 8.53\cos(0.20\pi - \omega t)$

5-3 $E_R = 6.08\cos(0.36\pi - 2\pi t/\text{s})$

5-4 $y = 11.6\sin(\omega t + 0.402\pi)$

5-5 $E = 0.695\cos(0.349 - \pi t/\text{s})$

5-6 (a) 2 V/m   (b) 0.2 V/m

5-7 $\psi(t) = (2.48\text{ cm})\cos(2.51 - (20/\text{s})t)$

5-8 (a) $v_g = v_p[1 - (\omega/n)(dn/d\omega)]$ (b) $v_g < v_p$

5-9 $c/1.56$

5-10 $v_p = c/1.5; v_g = c/1.73$

5-12 $v_g = A = $ constant

5-14 $2(v/c)v_0$

5-15 14 cm; 1.57 cm; 0.785 cm; 0 cm/s; T seconds

5-16 (a) 1.5 cm; 25 Hz; 20 cm; 5 m/s; opposite directions   (b) 10 cm   (c) -3 cm; 0 cm/s; 7.40×10^4 cm/s^2

5-18 40

Chapter 6

6-1 (a) 122 nm, 103 nm, 97.3 nm; ultraviolet   (b) 656 nm 486 nm, 434 nm; visible.

6-2 (a) no   (b) less than 91.2 nm   (b) less than 365 nm

6-3 (a) 2.4×10^{-21} J $= 0.015$ eV   (b) 0.55

6-4 (a) 8.6×10^{-20} J $= 0.54$ eV   (c) 5×10^{-10}

6-6 e^{-76}

6-10 (a) $0.4830 \,\mu$m   (b) 0.0756 W

6-11 6266 K; 462.5 nm

6-12 6105 K

6-15 0.45 nm

6-16 10^{-5} s; 3000 m   (b) 5×10^{-10} s; 15 cm

6-17 6

6-18 (a) half angle spread: 0.4 mrad   (b) 80 cm.

6-19 3.6 mm

6-20 (a) $0.81 \,\mu$m; $0.75 \,\mu$m; $0.585 \,\mu$m; $0.525 \,\mu$m   (b) 76%; 70%; 55%; 49%

6-21 (a) 31.5 W   (b) 1.26%

Chapter 7

7-1 (a) 11,945 and 21,235 W/m^2 (b) 18,723 W/m^2 (c) 51,903 W/m^2 (d) 0.96

7-2 0.86, 0

7-3 0.8; 3.73/1

7-4 (b) 1.78, 2.55, 4.00, 13.9

7-5 Lloyd's mirror interference fringes are produced, aligned parallel to the slit, and separated by 0.273 mm. The irradiance of the pattern is given by $I = 4I_0 \sin^2(115y)$, with y measured in cm from the mirror surface.

7-6 509 nm

7-7 514.5 nm

7-8 To acquire coherent beams; 560 nm

7-9 (a) 83.3 cm   (b) 83.3 fringes   (c) 150 nm

7-10 556 nm, 455 nm

7-11 20.3′

7-12 6′5″

7-13 35′40″

7-14 9.09×10^{-5} cm; orders 4 and 3, respectively

7-15 498 nm

7-16 1.33; 103 nm

7-17 (a) 2.78%   (b) 89.3 nm   (c) 1%

7-18 Soap film becomes wedge-shaped under gravity; the angle of the wedge is 1′14″.

7-19 15

7-20 1.16×10^{-3} cm

7-21 1.09 mm; 184

7-23 3 m

7-25 603.5 nm; 2.39 mm; 2.87×10^{-4} cm

7-26 928 nm

7-27 (a) 980 V/m   (b) 30°   (c) $r' = 0.28; tt' = 0.9216$ (d) 274, 253, 19.8 V/m; 7.8%, 6.7%, 0.041%   (e) 903, 70.8 V/m; 85%, 0.52%   (f) 258 nm

Chapter 8

8-1 436 nm

8-2 One mirror makes a wedge angle of 0.0172° with the image of the other, reflected through the beam splitter. Fizeau fringes result.

8-3 $23.75 \,\mu$m

8-4 (a) 80,000;   (b) 79,994

8-5 (a) $n = 1 + N\lambda/2L$ (b) 153

8-6 (a) 11.2°   (b) 45.9°

8-7 79.1 nm or $\lambda/8$

8-8 (a) 48,260   (b) 0.01013 cm

8-9 (a) 3.996×10^6 (b) 3.16×10^6 (c) 0.318 mm   (d) 6.29 Å   (e) 0.002 Å

8-10 (a) 329,670   (b) 361   (c) 9.8×10^6

8-11 2.18 cm

8-12 0.161 mm

8-13 (a) 360°   (b) 180°   (c) 2

8-14 1; 0.47

8-15 (a) $R = \dfrac{4r^2 \sin^2 \delta/2}{(1 - r^2)^2 + 4r^2 \sin^2 \delta/2}$

8-16 16

8-17 For lossless mirrors with $R_{1,2} \equiv |r_{1,2}|^2$, $T = \dfrac{(1 - R_1)(1 - R_2)}{\left(1 - \sqrt{R_1 R_2}\right)^2 + 4\sqrt{R_1 R_2} \sin^2 \delta/2}$; $R = 1 - T$

8-18 (a) 70   (b) 1.5 GHz   (c) 21 MHz   (d) 2.2×10^7 (e) 8 ns

8-20 9.99×10^5; 1570; 3.14×10^8; 8.4×10^{-8}; 3.14×10^8; 1.6×10^{-6} nm; 250 nm; 0.16 nm; 3 GHz; 1.91 MHz

8-21 (a) $5.9 \,\mu$m   (b) $1.2 \times 10^{-6} \,\mu$m   (c) $3.5 \times 10^{-4} \,\mu$m

8-22 (a) 10.6 GHz   (b) 0.83 GHz; 0.62 GHz

Chapter 9   **9-1** $f(x) = (4/\pi)\left(\sin kx + \frac{1}{3}\sin 3kx + \frac{1}{5}\sin 5kx + \cdots\right)$

9-2 $f(t) = \dfrac{E_0}{\pi} + \dfrac{E_0}{2}\cos \omega t + \dfrac{2E_0}{3\pi}\cos 2\omega t - \dfrac{2E_0}{15\pi}\cos 4\pi t + \cdots$

9-3 $g(\omega) = \dfrac{\sigma h}{\sqrt{2\pi}}e^{(-\sigma^2\omega^2)/2}$

If the width of the first is σ, the width of the second is $1/\sigma$. Thus the spectrum broadens as the original Gaussian narrows, and vice versa.

9-4 $|g(\omega)|^2 = (A^2\tau_0^2/4\pi^2)(\sin u/u)^2$, where $u = \omega\tau_0/2$

9-5 The narrow-band filter has a coherence length better by one order of magnitude: 3.48×10^{-5} m

9-6 1.3 fm; 10^6 Hz; 300 m

9-7 0.0243 mm

9-8 (a) 0.00138 nm   (b) 1 ns

9-9 2.5 mm

9-10 0.0625 cm; 2.08×10^{-12} s

9-11 0.144 cm

9-12 4×10^{-7} Å; 3×10^4 Hz

9-13 (a) 2.08×10^{-12} s, 0.0625 cm   (b) 0.36, 0.36   (c) 53

9-14 1.01×10^{-4} cm, 2.90×10^{-6} cm^2; 1.8, 35

9-15 (b) 2.55

9-17 0.998, 0.63

9-18 0.937, 0.686, 15.95 cm

9-19 (a) 0, 0, 0.596 cm   (b) 0.895 mm

Chapter 10   **10-1** 672

10-2 (a) 32 million   (b) 0.67 million

10-3 (b) 1284

10-4 (a) 68.1°   (b) 0.567   (c) 34.5°

10-5 (a) 0.64   (b) 79.5°   (c) 6624, 3281

10-6 (c) 432 μm; 429 μm; 10.07 m

10-7 159

10-8 10.2 μm

10-9 12 and 120, counting both polarizations

10-10 -70 db/km

10-11 0.080 mW

10-12 3.33 km; 10 km

10-13 0.136 db/km

10-14 (b) -1.25 db, -6.02 db, -10 db, -20 db

10-15 (a) 1.0069 km; 1 km   (b) 4.900 μs; 4.867 μs

10-16 431 ns; 2.32 MHz

10-17 77.2 ns

10-18 14.6 ns/km

10-20 457 ps; 1/146

10-21 25 MHz

10-22 (a) 4 ns   (b) 0.4 ns

10-23 48.9 ns

10-24 (b) 3.9 ps/km; 4.3 ps/km

10-25 (a) 50.5 ns; 1.075 ns; 0.075 ns   (b) 50.5 ns

10-27 (a) 100 GHz   (b) 4 THz

10-28 (a) No ΔL satisfies Eqs. (10-21) and (10-22) with *exactly* integer m, but many approximate solutions exist. One such solution is $\Delta L = 950.01\lambda_1 = 950.500\lambda_2$.   (b) Output 2.   (c) 0.94

Chapter 11   **11-1** (a) 0.218 cm   (b) 0.218 cm

11-2 0.090

11-3 (a) 0.135 mm   (b) 139

11-4 496 nm

11-5 2.125, 1.44, 0.778, and 0.55 μm

11-6 (a) 15°   (b) 0.678, 0.166, 0, 0.0461

11-7 (a) 4.477π; 5.482π (b) 1/199; 1/298

11-8 Single slit: 0.047, 0.017; Circular Aperture: 0.018, 0.0042

11-9 1.68×10^{-3} cm; 2.75×10^{-3} cm

11-10 8.4×10^{-4} cm

11-11 9725 km in diameter; 2.69×10^{-11} W/m^2

11-12 5.2 m

11-13 5.3 miles

11-14 3 to 10.4 m

11-15 (a) 0.400 mm   (b) 0.8106, 0.4053, 0.09006

11-16 (b) 2.10 mm

11-17 (b) 20

11-20 0.875, 0.573, 0.255, 0.0547, 0

11-22 (a) $I = I_0\dfrac{\sin^2(1.151y)\,\sin^2(0.575x)}{0.438x^2y^2}$ (b) 5.46 mm along x; 2.73 mm along y (c) 0.895 along x; 0.629 along y (d) 0.005

11-23 (a) 90°   (b) 11.5°   (c) 5.7°

11-24 $\Delta\theta = (\lambda/D)(1/\cos\theta)$

11-25 (b) 4.7%, 1.8%, 0.84% for $m = 1, 2, 3$, respectively

11-26 $m = 0; \theta_{1/2} = 30°$

11-27 (a) 120°   (b) $I_p = \left(\frac{1}{9}\right)I_{max}$ (c) $I_p = I_{max}$ (d) $I_p = 3I_{av}$

Chapter 12 **12-1** 13°18′

12-2 (a) 0.0823°/nm; 0.464 nm/mm   (b) 63,000

12-3 (a) 8.66′   (b) 612.5 nm (or 587.5 nm)   (c) 48; 48

12-4 987; 494

12-5 (a) 700 nm, 360 nm   (b) 57.1°, 25.6°   (c) 350 nm and 175 nm for crown glass; 180 nm and 90 nm for quartz

12-6 120,000; 0.069 Å

12-7 (a) third order   (b) any width smaller than light beam

12-8 (a) 21.8 cm, in each case   (b) 9, in each case   (c) 21.8 cm, 4.37 cm, and 0.0029 cm, respectively

12-9 (a) 8750 grooves/cm   (b) 18.89°   (c) 37.77°   (d) 7.88 nm/deg

12-10 (a) 7000   (b) 0.018 mm

12-11 (a) $-5.7°$ to $+11.5°$ (b) 100,000   (c) 10 Å/mm   (d) 1 m

12-12 about 5000 grooves/cm

12-13 (a) 1.16 μm   (b) 18.4 Å/mm

12-14 (a) 11.5°   (b) 11.8°

12-15 3550 grooves/mm; reduces it

12-16 (a) 3647   (b) 1200 grooves/mm   (c) 3.04 mm

12-17 (a) 557 to 318   (b) 960   (c) 388,800; 0.014 Å   (d) 0.41°/nm   (e) 5.5 Å

Chapter 13 **13-1** near, near, far

13-2 maxima: 409, 136, 81.8 cm; minima: 204.5, 102, 68 cm

13-3 (a) 1.88 and 3.26 mm   (b) 2.66 and 3.76 mm

13-4 (a) 0.0346 cm   (b) 833   (c) 20 cm, 6.67 cm, 4 cm

13-6 (a) 0.02 cm   (b) 2500

13-7 (a) 4 ×   (b) very nearly zero   (c) 5; 6

13-8 0.0012%

13-9 (a) 1/100   (b) 50.31 cm

13-10 1.05, 1.48, 1.82 mm

13-12 1.97 mm radius; zero

13-13 14.8 cm

13-14 (a) $0.066I_u$ (b) $1.53I_u$

13-15 (a) $0.0119I_u$ (b) $1.23I_u$

13-16 (a) $0.018I_u$ (b) $0.0538I_u$

13-17 $1.19I_u; 0.86I_u$

13-18 $0.55I_u$

13-19 21%

13-20 (b) 0.145 mm   (c) $0.645I_u$

13-21 19 μm

Chapter 14 **14-2** (a) $\frac{1}{\sqrt{2}}\begin{bmatrix} 1 \\ -1 \end{bmatrix}$: linearly polarized at $-45°$ (b) $\frac{1}{\sqrt{2}}\begin{bmatrix} 1 \\ 1 \end{bmatrix}$: linearly polarized at $+45°$

(c) $\frac{1}{\sqrt{2}}\begin{bmatrix} 1 \\ \frac{1}{\sqrt{2}}(1-i) \end{bmatrix}$: right-elliptically polarized at $+45°$ (d) $\frac{1}{\sqrt{2}}\begin{bmatrix} 1 \\ i \end{bmatrix}$: left-circularly polarized

14-3 (a) linearly polarized along x-direction, traveling in $+z$-direction with amplitude of $2E_0$ (b) linearly polarized at 53.1° relative to the x-axis, traveling in the $+z$-direction with amplitude of $5E_0$ (c) right-circularly polarized, traveling in $-z$-direction, with amplitude of $5E_0$

14-4 75°

14-5 right-circularly polarized light

14-6 (a) $\vec{E} = E_0(\sqrt{3}\hat{y} + \hat{z})e^{i(kx-\omega t)}$ (b) $\vec{E} = E_0(2\hat{z} - i\hat{x})e^{i(ky-\omega t)}$ (c) $\vec{E} = \hat{z}E_0 \exp\{i[(x+y)k/\sqrt{2} - \omega t]\}$

14-7 (a) $C = 0, m\pi$ (b) $B = 0, \left(m + \frac{1}{2}\right)\pi$ (c) $B = 0, A = \pm C, \left(m + \frac{1}{2}\right)\pi$

14-9 (a) linearly polarized, $\alpha = 18.4°, A = \sqrt{10}$ (b) right-circularly polarized, $A = 1$ (c) right-elliptically polarized; semimajor axis = 5 along y-axis, semiminor axis = 4 along x-axis   (d) linearly polarized, horizontal, $A = 5$ (e) left-circularly polarized, $A = 2$ (f) linearly polarized, $\alpha = 56.3°, A = \sqrt{13}$ (g) left-elliptically polarized, $\varepsilon = 53.1°, \alpha = -7°, E_{0x} = 2, E_{0y} = 10$

14-10 right-elliptically polarized, symmetrical with x- and y-axes, $E_{0x}/E_{0y} = \sqrt{3}$

14-13 right-circularly polarized light

14-14 no light emerges

14-15 (a) right-elliptically polarized, major axis along x-axis   (b) vertically linearly polarized

14-16 (a) linearly polarized at $\pm 45°$ (b) elliptically polarized

14-17 $\begin{bmatrix} 1 & -i \\ i & 1 \end{bmatrix}$

14-20 (a) Elliptical polarization with inclination angle $\alpha = -25.097°$: $\frac{1}{\sqrt{13}}\begin{bmatrix} 2 \\ \frac{3}{2} + \frac{i3\sqrt{3}}{2} \end{bmatrix}$

(b) Elliptical polarization with inclination angle $\alpha = 4.903°$, or rotated by 30° $\frac{1}{\sqrt{34.475}}\begin{bmatrix} 2.6519 \\ -0.6651 + i(5.1962) \end{bmatrix}$

14-21 (a) elliptically polarized with semi-axes E_{0x} and E_{0y} aligned with coordinate axes   (b) elliptically polarized with principal axes at 45° to coordinate axes   (c) circularly polarized, centered at origin, radius of E_0 (d) linearly polarized with slope E_{0y}/E_{0x}

14-22 $I = I_0(2 \sin^2 \theta \cos^2 \theta)$

Chapter 15   **15-1** 28.1%

15-2 67.5°; 22.5°

15-3 (a) $I_0\{0.5(\alpha^2 + \beta^2) \cos^2 \theta + \alpha\beta \sin^2 \theta\}$ (b) $0.4525 I_0$ versus $0.5 I_0$; $0.351 I_0$ versus $0.375 I_0$; both $0.25 I_0$; $0.0475 I_0$ versus 0

15-4 0.0633 mm

15-5 (a) single refraction, phase retardation, any polarization possible   (b) single refraction, no phase retardation, unpolarized   (c) same as (a)
(d) double refraction, no phase retardation in each separated beam, each beam linearly polarized   (e) cases (a) and (c)

15-6 (a) difference of 0.121 mm   (b) 0.015 mm

15-7 (b) 0%   (c) 33%

15-8 24

15-9 0.0162 mm

15-10 20°

15-11 0.005

15-12 (a) 53.12°   (b) 11.5°

15-13 (a) mixture of unpolarized and circularly polarized light   (b) elliptically polarized light

15-14 (a) 56.2°   (b) 33.8°

15-15 (a) 0.05 g/ml   (b) about 46°

15-16 (a) 0.200 mm   (b) 50°

15-17 (a) 8.57×10^{-5} cm   (b) green

15-19 (a) 0.0091   (b) 15 μm   (c) 12   (d) 15 μm

15-20 3.15×10^{-4}

15-21 (a) 36.8 μm   (b) 18.4 μm

15-22 14.7°

15-23 (a) 14.8%   (b) 2.03% of I_0 (c) 0.92

15-24 $I_0 \sin^2(2\theta)$

15-25 (a) 42.5°   (b) 0.60°

Chapter 16   **16-2** (b) 0.866

16-3 1000 Gb

16-4 1.82 μm

16-5 63.3 m/s

16-7 1.8×10^{10} bits

16-8 (a) 250 nm   (b) 500 nm   (c) 433 nm

16-9 (b) 365 nm   (c) 38°

16-10 365 nm; blue components shift into ultraviolet and are missing.

16-11 (a) $1.88 \times$ (b) $6330 \times$

Chapter 17   **17-1** 1850 nm

17-2 (a) 0.1   (b) 0.0724 A/W

17-3 0.08 μA

17-4 1.3×10^{-13} W

17-5 (a) 300   (b) No

17-8 40,000

17-9 0.1 mW

Chapter 18   **18-1** $f_1 = -62.05$ cm; $f_2 = 46.66$ cm; $r = 2.91$ cm; $s = -0.98$ cm

18-2 (a) $f_1 = 14.06$ cm $= -f_2$; $r = 1.17$ cm; $s = -0.73$ cm   (b) 8.92 cm left of lens center   (c) 9.78 cm left of lens center, with 9.6% error

18-3 Erect, virtual image at 6.67 cm left of second vertex, 0.556 in. high

18-4 $f_1 = -11.51$ cm; $f_2 = 15.31$ cm; $r = 0.400$ cm; $s = -1.16$ cm; $v = 4.20$ cm; $w = 2.64$ cm; $s_i = 18.9$ cm from H_2; $h_i = -1.18$ cm

18-5 (a) $f_1 = -20.15$ cm $= -f_2$; $r = 10$ cm $= -s$; $v = 10$ cm $= -w$ (b) Image is inverted, real, 61.38 cm from sphere center and $m = -2.05 \times$

18-6 (a) $y = 1$ cm; $\alpha = -5.73°$ (b) $A = 1 - x/10$; $B = 10/3 + 2x/3$; $C = -1/10$; $D = 2/3$ (c) $x = 10$ cm

18-8 $p = -4.17$ cm, $q = +2.17$ cm, $r = -0.83$ cm, $s = -2.17$ cm, $f_1 = -3.33$ cm, $f_2 = 4.33$ cm

18-9 $f_1 = -20$ cm, $f_2 = +20$ cm, $p = -30$ cm, $q = +10$ cm, $r = -10$ cm, $s = -10$ cm

18-10 $f_1 = -16.7$ cm, $f_2 = +23.3$ cm, $q = +18.7$ cm, $p = -18.3$ cm, $r = -1.67$ cm, $s = -4.67$ cm

18-11 (a) $A = -\frac{1}{2}$, $B = 0$, $C = -\frac{1}{10}$, $D = -2$ (b) Input and output planes fall at conjugate object and image positions; A is identical with the linear magnification.

18-12 (a) $p = -2$, $q = +2$, $f_1 = -6$, $f_2 = +6$, $r = 4$, $s = -4$ in.   (b) 2 in. beyond ball

18-13 (a) Elements of system matrix: $A = \frac{16}{15}$, $B = \frac{2}{3}$, $C = -\frac{1}{150}$, $D = \frac{14}{15}$
(b) $p = -140$, $q = 160$, $r = s = 10$, $f_1 = -150$, $f_2 = 150$, all in cm

18-14 (a) $A = 0.9764$, $B = 0.9676$, $C = 0.009182$, $D = 1.033$
(b) $f_1 = 108.9$ cm, $f_2 = -108.9$ cm, $p = 112.5$ cm, $q = -106.3$ cm, $r = 3.62$ cm, $s = 2.57$ cm   (c) -100 cm

18-15 (a) $A = \frac{2}{3} - s'/6$; $B = \frac{25}{3} + 2 - ss'/6 + s'$; $C = -\frac{1}{6}$; $D = -s/6 + 1$
(b) $s' = (4s + 12)/(s - 6)$; $m = \frac{2}{3} - s'/6$ (c) $s' = 6\frac{4}{7}$ cm; $m = -0.429$
(d) $s' = 4$ cm corresponds to second focal plane; $s = 6$ cm   corresponds to first focal plane

18-16 $A = 0.93935$; $B = 22.2212$; $C = -0.009284$; $D = 0.8448$; $r = v = 16.72$ mm; $s = w = -6.53$ mm; $p = -90.99$ mm; $q = 101.18$ mm; $f_1 = -f_2 = -107.71$ mm; the film plane is a distance $q = 101.2$ mm behind the last lens surface.

18-17 $A = \dfrac{n - n_L}{n_L R_1} t + 1$, $B = \dfrac{nt}{n_L}$, $C = \dfrac{n_L - n'}{n' R_2} - \dfrac{n_L - n}{n' R_1} - \dfrac{(n_L - n)(n_L - n')}{n' n_L} \dfrac{t}{R_1 R_2}$, $D = \dfrac{n}{n'} + \left(\dfrac{n_L - n'}{n' R_2}\right)\dfrac{nt}{n_L}$

18-21 For $\alpha = 0°$, $s' = 3.180$ cm and $\alpha' = -23.51°$; for $\alpha = -20°$, $s' = 16.104$ cm and $\alpha' = 6.081°$

18-22 $s' = -49.525$ cm; $\alpha' = 3.371°$; $Q = 2.912$

18-23 For $h = 1$ mm, $s' = 98.20$ mm, $\alpha' = -0.567°$; for $h = 5$ mm, $s' = 102.45$ mm, $\alpha' = -2.723°$

Chapter 19 **19-1** (a) He-Cd appears about $1.3 \times$ brighter   (b) about 2.4 mW

19-2 (a) 900 cd   (b) 85.4 lm/m^2 or lx

19-3 1.055:1

19-4 320 lx

19-5 (a) $1.7 \times 10^9 \text{ cd/m}^2$ (b) $2\pi L$

19-6 0.97 lm

19-7 +41.6 D

19-8 (a) 8.33 mm; +120 D   (b) 42.86 mm; +23.33 D   (c) 43.65 mm, measured from its second principal plane, or 42.38 mm from its second surface; +22.9 D

19-9 (a) 22.34 mm from cornea   (b) 21.60 mm from cornea

19-10 (a) $A = 0.75846$, $B = 5.1050$, $C = -0.05011$, $D = 0.65180$ (b) Focal points are 13.01 mm in front and 22.34 mm behind the cornea; principal points are 1.96 mm behind and 2.38 mm behind the cornea.

19-11 Block-letter sizes are 1.309 in. for 20/300; 0.436 in. for 20/100; 0.262 in. for 20/60; 0.087 in. for 20/20; 0.065 in. for 20/15. Letter details are $\frac{1}{5}$ block letter size in each case.

19-12 (a) 3.45 D   (b) 30.5 cm

19-13 (a) -2.000 D   (b) 21.4 cm   (c) -2.083 D; 19.8 cm

19-14 (a) myopia; astigmatism   (b) myopia   (c) hyperopia   (d) hyperopia; astigmatism

19-15 (a) Right eye: 7.32 cm, 14.3 cm; Left eye: 8.57 cm, 20 cm
(b) Right eye: 11.5 cm, 50.2 cm; Left eye: 21.4 cm, ∞ (strained eye)

19-16 Far vision: -7.41 D; Near vision: -6.10 D, (both for eye to lens distance of 1.5 cm)

Chapter 20 **20-2** (d) $s' = 0, +8, +14.4$ cm

20-3 $a = -0.858$ mm; $b_y = -10.98$ mm; $b_z = -35.15$ mm

20-5 $a = 0.015$ mm; $b_y = 0.49$ mm; $b_z = 3.9$ mm

20-6 (a) 0.0296 mm   (b) 0.021 mm

20-7 (a) 0.60 mm   (b) 1.2 mm

20-8 $b_z = 1.64$ mm; $b_y = 0.164$ mm

20-9 $b_z = 0.974$ mm at $h = 1$ cm, 3.84 mm at $h = 2$ cm, 8.44 mm at $h = 3$ cm, 14.53 mm at $h = 4$ cm, 21.81 mm at $h = 5$ cm

20-10 $b_z = 1.82$ cm; $b_y = 0.970$ mm

20-12 For $\sigma = 0.7$, $r_1 = 17.65$ and $r_2 = -100$ cm; for $\sigma = 3$, $r_1 = 7.50$ and $r_2 = 15.0$ cm

20-13 $r_1 = 18.62$ cm, $r_2 = -33.75$ cm

20-14 optimum $\sigma = +0.867$, closer to $+1$ than to -1

20-15 (a) $+0.714$ (b) $r_1 = 17.5$ cm; $r_2 = -105$ cm   (c) -0.714, reverse the lens

20-16 (a) 0.8   (b) $r_1 = 16.7$ cm; $r_2 = -150$ cm   (c) -0.8, reverse the lens

20-17 $+20$ and -20 cm

20-18 answers the same

20-19 -17.7 cm

20-20 (a) $R = 15.7$ cm   (b) $f_2 = -3.476$ cm

20-21 $r_{11} = 8.5168$ cm, $r_{22} = -434.89$ cm; $f_D = 20.0000$ cm, $f_C = 20.0096$ cm, $f_F = 20.0096$ cm

20-22 (a) $r_{11} = 3.4535$ cm, $r_{22} = -12.6576$ cm   (b) $f_D = 5.0000$ cm, $f_C = 5.0026$ cm, $f_F = 5.0026$ cm
(c) $P_{1D} = 0.3695 \text{ cm}^{-1}$, $P_{2D} = -0.1695 \text{ cm}^{-1}$, $\Delta_{1D} = 0.01802$, $\Delta_{2D} = 0.03928$ (d) yes

20-23 (a) $r_{11} = -5.2415$ cm, $r_{22} = 53.1840$ cm   (b) $f_{1D} = -4.5770$ cm, $f_{2D} = 8.4399$ cm
(c) $f_D = -10.0000$ cm, $f_C = -10.0050$ cm, $f_F = -10.0050$ cm

Chapter 21 **21-1** (a) 0.633, 1.898, 3.164 mm   (b) 0.50 mm   (c) 12.57, 37.70, and 62.83 cycles/mm   (d) $1 : \frac{1}{9} : \frac{1}{25}$

21-2 (a) product

21-3 (c) 32.8%; 0.48

21-4 (a) 5 units of amplitude   (c) $25[1 + \sin(ay)]^2$

21-7 $(\pi A^2/\omega)\cos(\omega\tau)$

21-8 (a) 18.3 kHz   (b) 17.2 Hz

21-9 (a) 0.04 Å   (b) 0.1 Å

21-10 (a) 2.86×10^{-3} cm   (b) 5.59 nm   (c) 224 nm   (d) 0.80 reading/s

21-11 (a) 3.6 cm^{-1} (b) 2450   (c) 0.093 mm/s

Chapter 22 **22-2** (a) 102 nm, 1.22   (b) 0.084%

22-4 (a) 2.81%   (b) 3.17%   (c) 4.26%

22-5 32.3%

22-6 2; 0.25 μm; ZrO_2

22-7 (a) 859 Å of aluminum oxide, 1058 Å of cryolite; 0.0003%   (b) 15.6%

22-8 (a) 227 nm and 370 nm   (b) 10%   (c) 1.2%

22-12 For example, from surface to substrate: MgF_2 ($n = 1.35$), SiO ($n = 1.5$), ZnS ($n = 2.2$)

22-13 (a) 81.1%   (b) 98.4%   (c) 99.99%

22-14 99.96%

22-15 2.24

Chapter 23 **23-2** $61°4'$; $28°56'$

23-3 $\theta_c = 32.9°$, $\theta_p = 61.5°$, $\theta_p' = 28.5°$

23-4 1.272

23-10 (a) 2.55%   (b) 0.233%   (c) 4.26%   (d) 1.26%

23-11 (a) 2.01%, 2.10%, 5.23%, 100%   (b) 2.01%, 1.91%, 0.274%, 100%

23-12 (a) TM: $\theta_p = 67°33'$, no θ_c; TE: no θ_p, no θ_c (b) TM: $\theta_p = 22°27'$, $\theta_c = 24°24'$; TE: no θ_p, $\theta_c = 24°24'$

23-13 (a) $R = 13.85\%$, $T = 86.15\%$ (b) $R = 0.62\%$, $T = 99.38\%$

23-15 (a) $\theta_c = 43.3°$, $\theta_p = 55.6°$, $\theta_p' = 34.4°$ (b) $R = 3.47\%$, $T = 96.53\%$; $R = 8.21\%$, $T = 91.79\%$
(c) $R = 3.47\%$, $T = 96.53\%$; $R = 0.67\%$, $T = 99.33\%$ (d) $0, 0, \pi, 2.43$ rad, 2.65 rad, π

23-16 (a) $59°51'$ (b) 0.457π and 0.539π

23-17 (a) 29.3%, 34.5%, 45.4%, 65.7%, 100%   (b) 29.3%, 24.2%, 14.9%, 5.4%, 100%

23-18 (a) 82.5%, 84.7%, 90.9%   (b) 82.5%, 80.1%, 69.5%

23-19 (a) 0.113 nm^{-1} (b) 41 nm

23-21 (a) 0.164 μm   (b) 5.1×10^{-6}

Chapter 24   **24-1** $\omega, 3\omega$

24-2 $3\omega_1, 3\omega_2, 2\omega_1 + \omega_2, 2\omega_1 - \omega_2, 2\omega_2 + \omega_1, 2\omega_2 - \omega_1, \omega_1, \omega_2$

24-3 $0, 2\omega_1, 2\omega_2, \omega_1 \pm \omega_2, \omega_1 \pm \omega_3, \omega_2 \pm \omega_3$

24-5 (a) 5983 nm   (b) 0.046

24-6 9.84 kV; V_{HW} is independent of the length

24-7 $7.9 \times 10^{-6}; \pi/2$

24-8 (a) $\Phi = 2\pi, 4\pi, \ldots; V = 2V_{HW}, 4V_{HW}, \ldots$ (b) At $V = 0, I = I_{\max}$; at $V = V_{HW}, I = 0$

24-9 $\Phi_{\text{pock}}/\Phi_{\text{kerr}} = (r/K)(n_0^3/\lambda_0 VL)$; 73

24-10 3.47 m; not practical

24-12 The sound wave advances 150 nm, which is $\lambda/3.3$ for $\lambda = 500$ nm. The ultrasonic wave thus appears practically stationary to the light beam.

24-13 (d) 67

24-14 221 MHz

24-16 2.97'

24-17 For a 5-cm length, the current is 31.8 A.

24-18 (a) 0.0647 min/G-cm   (b) 0.0956 min/G-cm

24-19 $14.1°; 0.0712 \, \mu$m^{-1}

Chapter 25   **25-1** (b) $n_I = 0.455\sqrt{K_I}; n_R = 1.099\sqrt{K_I}$

25-4 (a) 4.80×10^{13} s^{-1} (b) 1.38×10^{16} s^{-1} (c) $n_R = 0.0292; n_I = 3.92$

25-6 (a) 0.856 cm   (b) 6.63 μm

25-7 (a) 0.35 mm   (b) 1 m

25-8 1.7 μm

25-9 (a) 0.405 m^{-1} (b) 11.4 m

25-11 $A = (1 + \alpha)^{1/2}$ $B = 2(\pi c/\omega_0)^2(\alpha + 1)^{-1/2}$ $C = 2(\pi c/\omega_0)^4\alpha(3\alpha + 4)(\alpha + 1)^{-3/2}$ where $\alpha = Ne^2/m\varepsilon_0\omega_0^2$

Chapter 26   **26-5** (c) 0   (d) 1100 W/m^2 (e) 1.11×10^5 W/m^2

(f) For $\gamma = 1$/cm: $N_0/N_T = 0.993$; For $\gamma = 0.01$/cm: $N_0/N_T = 0.99993$ reasonable for both   (g) 1100 W/m^2; 1.10×10^5 W/m^2

26-7 (d) 1.6565×10^7 W/m^2 (e) 1.6567×10^7 W/m^2 (f) 1.6788×10^7 W/m^2

26-11 (a) 6.96 W/m^2 (b) 49.7 W/m^2 (c) 221 W/m^2 (d) 1180 W/m^2 (e) 10,200 W/m^2

26-12 (a) 5.96 W/m^2 (b) 39.7 W/m^2 (c) 121 W/m^2 (d) 180 W/m^2 (e) 200 W/m^2

26-13 (a) 2.7 W/m^2, 1.7 W/m^2 (b) 25.2 W/m^2, 15.2 W/m^2 (c) 181 W/m^2, 81 W/m^2 (d) 1170 W/m^2, 170 W/m^2

(e) 10,200 W/m^2, 200 W/m^2

26-16 (a) 0.00513/cm   (b) 103 W/cm^2

26-18 0.0242

26-21 (a) At line center, 9.69×10^{-22} m^2 (b) 3.1×10^{21}/m^3 (c) 1.94×10^6 W/m^2

26-22 1500 MHz

26-27 2.0×10^{-10} s, 6.7×10^{-10} s, 1.5×10^{-14} s

26-31 (a) 100 MHz   (b) 1430 W

26-32 Full-width angular spreads: 90°, 9°

26-33 1.55 eV

26-34 0.45

Chapter 27   **27-4** (a) $-i(1.24 \text{ m})$ (b) for each, approximately, 50 m $-i(1.24 \text{ m})$

27-5 (a) $R_{50} = 50.03$ m; $w_{50} = 20.15 \times 10^{-3}$ m

27-6 (a) center of cavity   (b) 0.51 mm   (c) 0.51 mm   (d) 0.4 mrad   (e) $\geq$64.6 m   (f) 245 μW/cm^2

27-7 (b) $\begin{bmatrix} 1 & 0 \\ -0.53125 & 1 \end{bmatrix}$

27-8 (a) 1.88 m   (b) 1.88 m

27-9 (a) $-i\pi w_1^2/\lambda$ (b) 0.438 mm   (c) $(-0.952 \text{ m})i$

(d) $\dfrac{ACq_1q_1^* + BD + ADq_1^* + BCq_1}{A^2q_1q_1^* + B^2} = \dfrac{1}{R_2(\ell)} + i\dfrac{\lambda}{\pi w_2^2(\ell)}$ (e) $\ell \cong 6$ cm; $w_2(\ell) \cong 0.54$ mm

27-10 (a) $w_{02} = 0.543$ mm; $z_2 = 0.0663$ m

27-11 (a) $z_{FF} \gg 1.46$ m   (b) 0.371 mrad   (c) $w_{\text{in}} = 1.113$ cm; $w_{\text{out}} = 11.13$ cm   (d) $z_2 \cong 0.1$ m; $w_{02} \cong 1.81 \times 10^{-7}$ m

27-16 14.6 m; 58.3 m; 131.2 m; 364.5 m

27-20 (a) $1 - e^{-2a^2/w^2}$ (b) $1 - (1 + 2a^2/w^2)e^{-2a^2/w^2}$ (c) $1 - \left(1 + 2a^2/w^2 + \dfrac{1}{2}(2a^2/w^2)^2\right)e^{-2a^2/w^2}$

(d) $1 - \left(1 + \dfrac{2}{3}(2a^2/w^2) + \dfrac{1}{2}(2a^2/w^2)^2\right)e^{-2a^2/w^2}$

27-23 (a) 0.155 mm   (b) 1.32×10^4 W/cm^2 (c) 3.17 W/cm^2

27-24 (a) 750 MHz   (b) 750 MHz   (c) 154 MHz   (d) 154 MHz   (e) 308 MHz

Chapter 28   **28-5** 2.1×10^9 V/m

28-6 (a) 181 m   (b) 2.42×10^3 V/m

28-7 (b) $R_{NHZ} = 371$ m

28-8 (b) $R_{NHZ} = 29.9$ m

28-10 (a) 3610 nm

28-11 3.33 W, 3.23 W

28-16 (a) 480 m/s   (b) 0.8 m/s

28-17 1.2 MHz

Index